T0252525

SPACE WEATHER
FUNDAMENTALS

SPACE WEATHER
FUNDAMENTALS

EDITED BY

GEORGE V. KHAZANOV

NASA/Goddard Space Flight Center
Greenbelt, MD

CRC Press
Taylor & Francis Group
Boca Raton London New York

CRC Press is an imprint of the
Taylor & Francis Group, an **informa** business

Parts of this book have been authored by an employee of the United States Government.

The United States Government retains a nonexclusive, royalty-free license to publish or reproduce the published form of the Contribution as described herein, or to allow third parties to do so.

CRC Press
Taylor & Francis Group
6000 Broken Sound Parkway NW, Suite 300
Boca Raton, FL 33487-2742

First issued in paperback 2019

© 2016 by United States Government as represented by the Administrator of the National Aeronautics and Space Administration. No copyright is claimed in the United States under Title 17, U.S. Code. All Other Rights Reserved.
CRC Press is an imprint of Taylor & Francis Group, an Informa business

No claim to original U.S. Government works

ISBN-13: 978-1-4987-4907-7 (hbk)
ISBN-13: 978-0-367-87555-8 (pbk)

This book contains information obtained from authentic and highly regarded sources. Reasonable efforts have been made to publish reliable data and information, but the author and publisher cannot assume responsibility for the validity of all materials or the consequences of their use. The authors and publishers have attempted to trace the copyright holders of all material reproduced in this publication and apologize to copyright holders if permission to publish in this form has not been obtained. If any copyright material has not been acknowledged please write and let us know so we may rectify in any future reprint.

Except as permitted under U.S. Copyright Law, no part of this book may be reprinted, reproduced, transmitted, or utilized in any form by any electronic, mechanical, or other means, now known or hereafter invented, including photocopying, microfilming, and recording, or in any information storage or retrieval system, without written permission from the publishers.

For permission to photocopy or use material electronically from this work, please access www.copyright.com (http://www.copyright.com/) or contact the Copyright Clearance Center, Inc. (CCC), 222 Rosewood Drive, Danvers, MA 01923, 978-750-8400. CCC is a not-for-profit organization that provides licenses and registration for a variety of users. For organizations that have been granted a photocopy license by the CCC, a separate system of payment has been arranged.

Trademark Notice: Product or corporate names may be trademarks or registered trademarks, and are used only for identification and explanation without intent to infringe.

Visit the Taylor & Francis Web site at
http://www.taylorandfrancis.com

and the CRC Press Web site at
http://www.crcpress.com

Contents

Editor

George V. Khazanov is a member of the Geospace Physics Laboratory at Goddard Space Flight Center (GSFC) in Maryland, USA. Prior to joining the National Aeronautics and Space Administration in 2001, Khazanov was a full tenured professor of physics at the University of Alaska Fairbanks, Fairbanks, Alaska. The Geospace Physics Laboratory conducts experimental and theoretical research to support the scientific activities of the NASA/GSFC. The principal objective of Dr. Khazanov's research is to develop an understanding of the physical processes that control the geospace plasma environment and its interaction with both natural and man-made bodies in space. He was the dean of the College of Physics, City Barnaul, Russia, and the Theoretical Physics Department chair at Altai State University, Russia, and chief of the Ionospheric Plasma Physics Laboratory at Irkutsk State University, Russia. He supervised and directed more than 30 MS and 15 PhD graduates. Dr. Khazanov is the author or coauthor of six books and more than 300 peer-reviewed publications. His most recent book titled *Kinetic Theory of Inner Magnetospheric Plasma* was published by Springer in 2011.

Contributors

Spiro Antiochos
Heliophysics Science Division
NASA Goddard Space Flight Center
Greenbelt, Maryland

Daniel N. Baker
Laboratory for Atmospheric and Space Physics
University of Colorado
Boulder, Colorado

Joseph E. Borovsky
Center for Space Plasma Physics
Space Science Institute
Boulder, Colorado

Charles S. Carrano
Institute for Scientific Research
Boston College
Chestnut Hill, Massachusetts

William Daughton
Los Alamos National Laboratory
Los Alamos, New Mexico

Eric Donovan
Department of Physics and Astronomy
University of Calgary
Calgary, Alberta, Canada

Yusuke Ebihara
Research Institute for Sustainable
    Humanosphere
Kyoto University
Kyoto, Japan

Henry B. Garrett
Jet Propulsion Laboratory
California Institute of Technology
Pasadena, California

Melvyn L. Goldstein
Heliospheric Physics Laboratory
NASA Goddard Space Flight Center
Greenbelt, Maryland

Tamas I. Gombosi
Department of Climate and Space Sciences and
    Engineering
University of Michigan
Ann Arbor, Michigan

Keith M. Groves
Institute for Scientific Research
Boston College
Chestnut Hill, Massachusetts

Cheryl Y. Huang
Air Force Research Laboratory
Kirtland Air Force Base
Albuquerque, New Mexico

Joseph D. Huba
Plasma Physics Division
Naval Research Laboratory
Washington, DC

Shrikanth G. Kanekal
NASA Goddard Space Flight Center
Greenbelt, Maryland

Homa Karimabadi
Department of Electrical and Computer Engineering
University of California, San Diego
La Jolla, California

Judith Karpen
Space Weather Laboratory
NASA Goddard Space Flight Center
Greenbelt, Maryland

George V. Khazanov
Geospace Physics Laboratory
NASA Goddard Space Flight Center
Greenbelt, Maryland

Jonathan Krall
Beam Physics Branch
Naval Research Laboratory
Washington, DC

Ari Le
Plasma Theory and Applications
Los Alamos National Laboratory
Los Alamos, New Mexico

Nils Olsen
Department of Geomagnetism
Technical University of Denmark
Kongens Lyngby, Denmark

Larry J. Paxton
Space Sector
The Johns Hopkins University Applied Physics
    Laboratory
Laurel, Maryland

Arthur D. Richmond
High Altitude Observatory
The National Center for Atmospheric Research
Boulder, Colorado

Vadim Roytershteyn
Space Science Institute
Boulder, Colorado

Stanislav Sazykin
Department of Physics and Astronomy
Rice University
Houston, Texas

Robert W. Schunk
Center for Atmospheric and Space Sciences
Department of Physics
Utah State University
Logan, Utah

David G. Sibeck
Space Weather Laboratory
NASA Goddard Space Flight Center
Greenbelt, Maryland

Eftyhia Zesta
Geospace Physics Laboratory
Heliophysics Science Division
NASA Goddard Space Flight Center
Greenbelt, Maryland

Yongliang Zhang
Space Sector
The Johns Hopkins University Applied Physics
    Laboratory
Laurel, Maryland

Introduction

George V. Khazanov

"Space weather" refers to conditions on the Sun and in the solar wind, magnetosphere, ionosphere, and thermosphere that can influence the performance and reliability of space-borne and ground-based technological systems and can endanger human life or health. Adverse conditions in the space environment can cause disruption of satellite operations, communications, navigation, and electric power distribution grids, leading to a variety of socioeconomic losses. This book provides a comprehensive overview of our current knowledge and theoretical understanding of space weather formation and covers all major topics of this phenomena starting from the Sun to the Earth's ionosphere and thermosphere.

This book benefits from the material presented by 29 authors. All authors are very well-known researchers in the field of space science, and most of them have a very distinguish level of accomplishment in space weather studies. The predictive nature of space weather research, however, distinguishes it from conventional space physics research, but the conventional space physics research is the basis of space weather fundamentals. Now without further ado, we introduce the team of the authors and provide a short description of each chapter.

Drs. Judith Karpen, chief of the Space Weather Laboratory, and Spiro Antiochos, senior scientist for space weather and an American Geophysical Union (AGU) fellow [both at Goddard Space Flight Center (GSFC), the National Aeronautics and Space Administration (NASA), Greenbelt, Maryland], wrote Chapter 1 titled "The Sun." The Sun is the ultimate source of all space weather in the solar system. Specifically, the twisting of the Sun's magnetic field from the deep interior through the photosphere, its visible surface, transfers the energy

generated by fusion from the core to the plasma and the magnetic field of the solar corona. Both slow and fast drivers of space weather originate in the corona and then propagate and evolve throughout the heliosphere. This chapter discusses the primary phenomena driving space weather on long (day-to-month) and short (second-to-minute) timescales, and the associated mechanisms of radiative and mechanical forcing. Slow drivers include corotating interaction regions in the solar wind, which originate at the interface between closed and open magnetic fluxes on the rotating Sun. On similar timescales, the emergence and evolution of solar active regions, localized regions of strong magnetic flux that wax and wane with the solar cycle, cause significant fluctuations of emissions across the electromagnetic spectrum; enhancements in ultraviolet radiation and X-rays are particularly relevant to space weather because of their profound impact on planetary ionospheres. Explosive energy release in the corona, commonly ascribed to magnetic reconnection, produces massive eruptions of magnetic field and plasma (coronal mass ejections) as well as intense bursts of electromagnetic radiation and energetic particles (flares). This impulsive activity generates the most destructive space weather events in the heliosphere, with complex consequences for planetary magnetospheres, ionospheres, neutral atmospheres, and life and technology in space and on planetary surfaces. Chapter 1 describes the key observed features and fundamental physical processes leading to slow and fast variations in the Sun's radiative, mass, magnetic, and energetic particle outputs, and points out gaps in our understanding that must be addressed by future observations, theory, and modeling.

Dr. Melvyn L. Goldstein, astrophysicist at GSFC, NASA, and American Geophysical Union (AGU) fellow, wrote Chapter 2 titled "The Solar Wind." The solar wind is the medium through which all drivers of space weather propagate. The "wind" is a highly variable, turbulent magnetofluid. Variations can arise from changes at the Sun and the solar corona from which the solar wind arises. The corona is the source of magnetic waves, temperature changes, and so on, all of which can perturb the Earth's magnetosphere, leading to space weather. Even the very energetic drivers of space weather, such as solar flares and accompanying shock waves, and coronal mass injections, are dynamically changed as they propagate outward from the Sun. Without understanding the detailed structure of the solar wind, one cannot predict accurately if, where, and to what effect solar phenomena will modify the near-Earth plasma environment, impacting spacecraft and, perhaps, inducing currents in the crust that can dramatically affect power grids and communications.

Nils Olsen, professor of geophysics and the head of the department of geomagnetism at Technical University of Denmark, Kongens Lyngby, Denmark, wrote Chapter 3 titled "Earth's Magnetic Field." In this chapter, he describes the Earth's internal field caused by processes in the Earth's core and crust, including its change with time (secular variation). He emphasizes on space weather-related aspects of the main field (e.g., regions of weak field intensity such as the South Atlantic Anomaly and how these regions evolve in time). He describes different geomagnetic field models (dipole approximation, International Geomagnetic Reference Field (IGRF) as well as more advanced models) and the magnetic data (satellite and ground-based) and methods used for deriving these models. In this chapter, he also discusses magnetic coordinates and magnetic time, and various approximations.

Dr. Joseph Borovsky, American Geophysical Union (AGU) fellow and research scientist at the Space Science Institute, Boulder, Colorado, wrote Chapter 4 titled "Solar Wind–Magnetosphere Interaction." His current research interests focus on the structure of the solar wind, solar wind–magnetosphere coupling, the dynamics of the magnetosphere, and systems science. This chapter is organized into five sections. The first sections is a discussion of what coupling means: (1) transfer of mass, momentum, and energy and (2) changes in the morphology of magnetosphere and activation of magnetospheric and ionospheric convection. The second section is a review of standard thinking of simple physical processes that act to couple the solar wind to Earth: chiefly (1) dayside reconnection and control of reconnection rate and (2) the viscous interaction. The second section discusses the transformation of the supersonic solar wind into the shocked magnetosheath plasma, which does the actual coupling. The third section is a review about how coupling is studied in the data and how coupling is investigated via computer simulations of the solar wind driving the Earth's magnetosphere. The fourth section discusses newer thinking of coupling: This new thinking concerns the role of solar wind turbulence in the coupling of the solar wind to the magnetosphere, the concept of a complex driving rather than a driving described by a simple formula, and the feedback by magnetospheric plasmas on the rate of reconnection and the amount of coupling of the solar wind to Earth. The fifth section is a discussion of what is known and what is not known, including a list of unsolved problems/outstanding questions. This chapter presents a thorough and modern review of the topic of solar wind coupling to the Earth's magnetosphere that goes beyond previous review articles or textbooks.

Dr. David G. Sibeck, American Geophysical Union (AGU) fellow and NASA's Van Allen Probes mission scientist, wrote Chapter 5 titled "The Magnetosheath and Its Boundaries." This chapter describes the processes that modify the incoming solar wind plasma as it transits from the Earth's bow shock to the Earth's magnetopause. Topics include the steady-state structures at the bow shock and magnetopause, the transmission of solar wind discontinuities and their effects on the magnetosphere, the formation of features by kinetic processes within the foreshock, their transmission through the magnetosheath and its effects on the magnetopause, waves in the magnetosheath, and instabilities at the magnetopause, including the Kelvin–Helmholtz instability and magnetic reconnection. Connections are made to the solar wind (Chapter 2) and magnetosphere chapters (Chapters 6–10) emphasizing our ability to predict conditions within this region and its relevance to space weather studies.

Drs. Homa Karimabadi (SciberQuest, Del Mar, California), Ari Le (Los Alamos National Laboratory, Los Alamos, New Mexico), Vadim Roytershteyn (Space Science Institute), and William Daughton (Los Alamos National Laboratory) wrote Chapter 6 titled "Magnetic Reconnection." He has a wide range of interests, including numerical algorithms, machine learning, cosmology, and space plasmas. Magnetic reconnection is the dominant driver of space weather. It is the primary cause of massive solar storms that release tons of high-energy particles into the solar wind; reconnection is also the mechanism that enables the penetration of these high-energy particles past the protective shield of the Earth's magnetic dipole field. The environment in the Earth's magnetosphere is collisionless and is dominated by kinetic effects. In this chapter, the latest in our understanding of the reconnection process and its implications for space weather are reviewed.

Dr. Stanislav Sazykin wrote Chapter 7 titled "Magnetospheric Electric Fields and Current Systems." He is a senior faculty fellow in the Department of Physics and Astronomy at Rice University, Houston, Texas. This chapter introduces two widely used approaches to understanding the basics of magnetospheric electric fields and currents, ideal magnetohydrodynamics and drift (kinetic) descriptions, as well as the approaches used to obtain quantitative descriptions of magnetospheric electric fields and currents (numerical simulations and empirical data-based models). It discusses the nature and morphology of main current systems. Electric fields are described in terms of their sources, their morphology, and their role in space weather events. These are illustrated with examples from numerical simulations and data analysis, as appropriate. Specifically, the following topics are described: (1) the large-scale convection electric field and its changes during geomagnetic disturbances, (2) subauroral polarization stream structures and their storm time dynamics, (3) the phenomenon of shielding and its nature, and (4) mesoscale impulsive flow channels and their associated electric fields. This chapter explains the space weather consequences of the aforementioned electric field phenomena and their significance, again illustrating with examples. Finally, it provides a list of references to empirical and first-principle models of electric fields and currents that the reader may find useful.

Dr. Eric Donovan, professor of physics at the University of Calgary, Calgary, Alberta, Canada, wrote Chapter 8 titled "Coupling between the Geomagnetic Tail and the Inner Magnetosphere." Plasma waves and mass transport are central features of magnetospheric dynamics and key drivers of space weather. Nightside magnetospheric dynamics involve transport of material through E-cross-B convection, that is, at times steady and at other times dynamic. On the nightside, this transport

is in general earthward. During this transport, particles are energized via adiabatic processes (e.g., conservation of the first adiabatic invariant) and nonadiabatic processes (e.g., resonant interaction with ultralow frequency (ULF) waves and the so-called Lyons–Speiser orbits). This transport becomes especially interesting at the nightside transition region (NTR), which separates the highly stretched magnetotail for the nearly dipolar inner magnetosphere. Because the most important space weather effects result directly from inner magnetospheric processes, understanding how mass and energy are transported from the magnetotail across the NTR to the inner magnetosphere is a critical goal of space weather research. This chapter is a review of mass and energy transport in the magnetotail focusing on elements of this transport through which inner magnetosphere–magnetotail coupling is accomplished. More specifically, the focus is on recent observation, simulation, and theory of the transport across the NTR and the impacts of that transport on the dynamics of the inner magnetosphere.

Dr. Yusuke Ebihara of Kyoto University wrote Chapter 9 titled "Ring Current." Ebihara received Tanakadate Award of the Society of Geomagnetism and Earth, Planetary and Space Sciences in 2012. The ring current is an electric current encircling Earth, causing the long-lasting and worldwide magnetic disturbance that is a manifestation of magnetic storms. From this view, to understand magnetic storms is basically equivalent to understanding the ring current. After presenting an overview of the ring current and brief history of the ring current study, this chapter introduces the structure and dynamics of the ring current. It also describes the source, transport, acceleration, and loss of the charged particles that carry charge in the ring current. The ring current has a large impact on Earth and space. This chapter summarizes the impacts of the ring current on the other regions, such as the ionosphere, the thermosphere, and the radiation belt, as well as space weather, such as the geomagnetically induced current that is a concern for a power grid system.

Drs. Shrikanth G. Kanekal and Daniel N. Baker wrote Chapter 10 titled "Radiation Belts." Kanekal is a research astrophysicist at the NASA, Goddard Space Flight Center, and deputy mission scientist on the Van Allen Probes mission. Baker is a distinguished professor of planetary and space physics, the Moog-BRE Endowed chair of space sciences, director of the Laboratory for Atmospheric and

Space Physics, professor of astrophysical and planetary sciences, and professor in the Department of Physics, University of Colorado (Boulder, CO). Since their discovery in the early 1950s by James Van Allen, Earth's radiation belts have provided a rich environment to the study of charged particle dynamical processes. We now know that these processes occur not only throughout the solar system, including the Sun and other planetary magnetospheres, but also in astrophysical systems such as supernovae. The Earth's radiation belts therefore provide a readily accessible system to study phenomena such as charged particle acceleration and precipitation, wave–particle interactions, cross-scale coupling of physical processes, and connecting, for example, low-energy plasma processes to relativistic energy radiation belts. Since the 1980s, several space missions have explored the radiation belts and related plasma environments and have laid the groundwork revealing a rich phenomenology necessitating the understanding of the details of the underlying physical processes. However, recent advances in spacecraft instrumentation are yielding high-quality measurements, enabling a detailed study of these phenomena with great resolution. The basic morphology, structure, and particle dynamics of the radiation belts are reviewed here, with emphasis on observational aspects. Fundamental properties of charged particle motions in a closed magnetosphere set the framework in which observed phenomena are described. Within the past 2 years, NASA's Van Allen Probes mission has led to discoveries that have challenged long-standing "facts." New opportunities are being provided by sophisticated instrumentation and easy access to space by CubeSats and other small satellite platforms. Terrestrial radiation belts are an important player in space weather affecting key aspects of human activity. As humans rely more and more on space-based technology, the understanding of the physics of radiation belts becomes crucial. Radiation belt energetic particle populations can not only affect optimal functioning but also completely disable satellite systems. One such example is the global positioning system (GPS) satellite systems, with its ubiquitous use in navigation and personal communication. Disruption of the GPS system can have a huge economical and social impact.

Drs. Jonathan Krall and Joseph D. Huba wrote Chapter 11 titled "Plasmasphere." Krall is a research physicist in the Beam Physics Branch at the Naval Research Laboratory, Washington, DC. Huba is American Geophysical Union (AGU) fellow and the head of the Space Plasma Physics Section of the Beam Physics Branch at the Naval Research Laboratory, Washington, DC. Earth's plasmasphere, a region of plasma trapped in the inner magnetosphere by closed geomagnetic field lines, is shaped by the dynamics of the ionosphere, thermosphere, and magnetosphere. The plasmasphere typically erodes during magnetic storms, over a timescale of hours, and refills during quiet times over a timescale of days. Given its responsiveness to the ionosphere/thermosphere/magnetosphere (ITM) system and its affect on electromagnetic waves and energetic particles in the inner magnetosphere, the plasmasphere is both a marker and a component of space weather. The purpose of this chapter is to examine measurements and models of the ITM system during both storms and quiet periods. In so doing, we consider plasmasphere dynamics, poststorm refilling, composition, and the impact of the plasmasphere on magnetospheric physics, such as magnetic reconnection at the bow shock, where the solar wind impacts the magnetosphere. Specific physical processes are illustrated by showing how Earth-based and satellite measurements, which are often sparse, are illuminated by state-of-the-art numerical models, such as the Naval Research Laboratory SAMI3 ionosphere/plasmasphere model and the ring current atmosphere interaction model-cold plasma (RAM-CPL) plasmasphere model. By comparing measurements to models and models to each other, we validate our results and describe the fundamentals of plasmasphere dynamics. By including or excluding specific physical processes in the models, such as winds in the thermosphere or the ring current in the magnetosphere, we illustrate the influences of specific drivers of space weather.

Dr. Robert W. Schunk wrote Chapter 12 titled "Polar Wind. He is American Geophysical Union (AGU) fellow, professor, and the director of Center for Atmospheric and Space Sciences, Utah State University, Logan, Utah. He is also a coauthor of the textbook titled *Ionospheres: Physics, Plasma Physics, and Chemistry* published by Cambridge University Press. The escape of plasma from the Earth's ionosphere in the northern and southern polar caps has a significant effect on space weather. The plasma outflow, which consists of H^+, He^+, and O^+, begins at about 800 km above the Earth's surface. As the escaping plasma flows up and out of the ionosphere along diverging geomagnetic field lines, it becomes supersonic, and this process has been called the "polar wind." In addition, the polar wind plasma drifts horizontally across the high-latitude region because of magnetospheric electric fields, moving into and out of sunlight, the auroral oval, and the polar cap. The polar wind also interacts in a complex way with energization mechanisms that operate at high altitudes over the polar region, including mechanisms associated with escaping photoelectrons, cusp ion beams and conics, hot magnetospheric electrons, electromagnetic wave turbulence, centrifugal acceleration, and anomalous resistivity associated with field-aligned currents. When the outflow is driven by density and temperature variations in the underlying ionosphere, it is usually called the "classical" polar wind, whereas when high-altitude and low-altitude energization processes are included, it is called the "nonclassical" or "generalized" polar wind. When integrated over both polar regions, the polar wind corresponds to an important ionospheric loss process and an important source of plasma for the magnetosphere. This chapter describes the important processes that affect the polar wind and how the polar wind affects space weather.

Dr. Larry J. Paxton is a group supervisor of the Geospace and Earth Science Group, a member of the Principal Professional Staff at the Johns Hopkins University Applied Physics Laboratory, and president-elect of the Space Physics and Aeronomy section of the American Geophysics Union, Washington, DC. Dr. Yongliang Zhang is a senior scientist and section supervisor of the Space Weather section at the Johns Hopkins University Applied Physics Laboratory. Drs. Paxton and Zhang wrote Chapter 13 titled "Far Ultraviolet Imaging of the Aurora." This chapter provides a handy guide to the state of our art in terms of understanding the global behavior of the aurora in terms of "imaging" and what that means for the scientific understanding of the upper atmosphere. Here, we define imaging beyond the sense of imaging the optical emissions—we include a brief description of the advent of a new era understanding that will come about from the use of AMPERE and SuperMAG data to understand the energy inputs as a continually refreshed image of these processes. Optical auroral imaging is reviewed as well as the techniques used by AMPERE and SuperMAG to provide the user with a useful understanding of the techniques. This chapter also describes the current issues in our understanding of the response of the upper atmosphere during disturbed times and how auroral inputs, and the limitations of our understanding of those inputs, affect our ability to nowcast

and forecast the state of the upper atmosphere. Finally, recent advances in assimilative modeling and climatological models of the aurora and auroral energy inputs are described.

Dr. Arthur D. Richmond is American Geophysical Union (AGU) fellow and a senior scientist at the High Altitude Observatory of the National Center for Atmospheric Research, Boulder, Colorado. He wrote Chapter 14 titled "Ionospheric Electrodynamics." This chapter is based on a previously published chapter in a CRC Press book (Ionospheric electrodynamics, in *Handbook of Atmospheric Electrodynamics*, Vol. II, edited by H. Volland, pp. 249–290, CRC Press, Richmond VA, 1995). The topics include ionospheric conductivity, thermospheric winds, the theory of ionospheric electric fields and currents, observations of ionospheric electric fields and currents, and geomagnetic variations. Each of those topics will be updated based on advances in the field since 1995.

Dr. Tamas I. Gombosi is American Geophysical Union (AGU) fellow, Konstantin I. Gringauz distinguished university professor of space science, University of Michigan, Ann Arbor, MI, and Rollin M. Gerstacker professor of engineering. Gombosi wrote Chapter 15 titled "Simulating Space Weather." The idea that conditions in the space environment are a counterpart to meteorology was first suggested in 1959. The domain of space weather extends from the Sun to the terrestrial troposphere, a region that is also called geospace. In the broader sense, space weather can also refer to conditions anywhere in space where technological systems operate, such as at other planets, the interplanetary medium, or even at the edge of the heliosphere (the two Voyager spacecraft are operating beyond the termination shock of the supersonic solar wind). This chapter reviews the evolution of space weather modeling from the early days when the first empirical models were developed to the present when new-generation physics-based models are capable of outperforming empirical models. It also reviews the various physics-based approaches and summarizes their advantages and disadvantages. It summarizes validation studies and the present state of community use. Finally, it describes the first efforts to transition physics-based space weather models to operational space weather forecasting.

Dr. Henry B. Garrett is principal scientist at the Office of Safety and Mission Success, Jet Propulsion Laboratory, Pasadena, California. He has a wide variety of experiences in the space environment and its effects with specific emphasis in the areas of atmospheric physics, low-Earth ionosphere, radiation, micrometeoroids, space plasma environments, and effects on materials and systems in space. He wrote Chapter 16 titled "Space Weather and the Extraterrestrial Planets" and Chapter 17 titled "Spacecraft Charging." In addition to Earth, each of the other planets, comets, and asteroids interacts with the space environment. As our space probes fly by or orbit these bodies, we have accumulated an increasing understanding of how they interact with the solar wind, solar particle events, and galactic cosmic rays. As at Earth, each body's magnetic field (or lack thereof), atmosphere, and, in the case of the larger bodies, rings cause complex effects that need to be considered in evaluating space weather. Like Earth, the planets Jupiter, Saturn, Uranus, Neptune, and even Mercury have measurable magnetic fields that significantly alter the solar space weather. These fields lead to a variety of effects, including the generation of aurora. They have radiation belts that can be affected by interactions with the environment. Mars, Venus, and comets interact differently through their ionospheres and atmospheres. This chapter reviews many of these interesting interactions and describes the role of space weather in causing them.

Spacecraft charging has become a major environmental concern for the spacecraft design community. Charging typically results from the buildup of charge on surfaces (surface charging) or in dielectric surfaces [buried charge or, as used here, internal electrostatic discharge (IESD)]. In the case of Earth, surface charging can lead to surface potentials in excess of 20 kV. Although such potentials in and of themselves are not necessarily dangerous, it is the buildup of differential potentials between surfaces that are. Differential potentials can lead to arc discharges between surfaces that can seriously damage electronic systems and to material damage to solar arrays. Of more concern, however, is IESD as the charge can accumulate on isolated surfaces or in isolated dielectrics inside a spacecraft's protective Faraday cage and arc next to sensitive devices. As a result, spacecraft charging is currently rated as the most serious space weather-related source of damage.

Dr. Eftyhia Zesta is the chief of Geospace Physics Laboratory at NASA, GSFC; principal investigator of Ionospheric Hazards Program at Air Force Research Laboratory (AFRL), Albuquerque, New Mexico. Dr. Cheryl Y. Huang is the senior scientist of AFRL, Albuquerque, New Mexico. They both wrote Chapter 18 titled Satellite "Orbital Drag." A great number of space-based assets needed for Earth monitoring, research, and communications are used by a variety of agencies. Many of these assets orbit at altitudes within the upper thermosphere and ionosphere. At such altitudes, the orbit of satellites can be greatly affected by the neutral atmosphere, whose dynamics are strongly driven by interactions with the ionosphere and ultimately by geospace phenomena such as geomagnetic storms. Maintenance of the space catalog and protection of our space-based assets require a fundamental understanding of this region of space and the coupling between the thermosphere and the ionosphere. We look at all the different physical properties that affect aerodynamic drag on all low-Earth orbit space objects. We explore the fundamental physics of the Earth's thermosphere layer, its dynamic interaction with the Earth's magnetosphere and ionosphere, and all the different types of models that are and can be used for the specification and forecast of the thermosphere and thus aerodynamic drag.

Dr. Keith M. Groves is senior research scientist at the Institute for Scientific Research, Boston College, Chestnut Hill, Massachusetts. As the former program manager and principal investigator for the Air Force program on ionospheric research, Dr. Groves has served to initiate and lead efforts to understand the impact of ionospheric disturbances on ground- and space-based radio systems. Dr. Charles S. Carrano is a senior research physicist at the Institute for Scientific Research, Boston College, Chestnut Hill, Massachusetts, where he investigates ionospheric effects on radio wave propagation. He recently introduced the first course on space weather for undergraduate students at Boston College, Newton, Massachusetts. Previously, he managed the Ionospheric Environments and Impacts group of the Space Weather and Effects Division at Atmospheric and Environmental Research, Inc. They wrote Chapter 19 titled "Space Weather Effects on Communication and Navigation." Numerous communications and navigation services for both commercial and military users are currently provided by space-based platforms. Common examples include automobile navigation systems that use GPS signals for positioning data and credit-card transactions requiring the exchange of information via satellite communications. Such services may be vulnerable to space weather effects as the radio waves used to transmit information pass through the Earth's ionosphere.

The ionosphere is a partially ionized region of the upper atmosphere; the ionization significantly modifies the atmosphere's refractive index, inducing a variety of propagation effects that may distort the phase and/or amplitude of radio signals and therefore degrade the performance of a given technology. Radio waves transiting the ionosphere experience phase advance and group delay, which manifests itself as a ranging error in single-frequency GPS and radar applications. The delay is dispersive with an inverse-square frequency dependence; however, systems that employ two or more appropriately separated frequencies can successfully correct the ionospheric delay. This technique was demonstrated very successfully in dual-frequency GPS receivers and has since been adopted by every other major global navigation satellite system (GLONASS), Galileo, and Beidou. When the ionization exhibits a significant structure or irregularities on scale sizes from tens to hundreds of meters, variable-phase perturbations across the wavefront can result in diffraction characterized by constructive and destructive interference of the wave with itself. This phenomenon, known as scintillation, is characterized by fluctuations in phase and amplitude that affect both single- and dual-frequency systems. This chapter examines the conditions under which scintillations occur in the natural ionosphere and describes the extent to which they can disrupt the performance of space-based communications and navigation systems, as well as both ground- and space-based radar applications. Furthermore, potential approaches are considered to mitigate ionospheric impacts as well as to use the technologies as tools to monitor the state of the ionosphere and detect conditions where system performance may be compromised due to space weather.

I

Space Weather Drivers

CHAPTER 1

The Sun

Judith Karpen and Spiro Antiochos

CONTENTS

1.1 INTRODUCTION

The Sun is the ultimate source of all space weather in the solar system. Specifically, the stressing of the Sun's magnetic field by plasma motions from the deep interior through the photosphere, its visible surface, transfers the energy generated by fusion from the core to the plasma and magnetic field of the solar corona. Both slow and fast drivers of space weather originate in the corona, then propagate and evolve throughout the heliosphere. In this chapter, we will discuss the primary phenomena driving space weather on long (day-to-month) and short (seconds-to-minutes) timescales, and the associated mechanisms of radiative and mechanical forcing. Slow drivers include corotating interaction regions in the solar wind, which originate at the interface between closed and open magnetic flux on the rotating Sun. On similar timescales, the emergence and evolution of solar active regions (ARs), localized regions of

strong magnetic flux that wax and wane with the solar cycle, cause significant fluctuations of emissions across the electromagnetic spectrum; enhancements in UV radiation and X-rays are particularly relevant to space weather because of their profound impact on planetary ionospheres. Explosive energy release in the corona, commonly ascribed to magnetic reconnection, produces massive eruptions of magnetic field and plasma (coronal mass ejections; CMEs) as well as intense bursts of electromagnetic radiation and energetic particles (flares). This impulsive activity generates the most destructive space weather events in the heliosphere, with complex consequences for planetary magnetospheres, ionospheres, neutral atmospheres, and life and technology in space and on planetary surfaces. We will describe the key observed features and fundamental physical processes leading to slow and fast variations in the Sun's radiative, mass, magnetic, and energetic particle outputs, and point out gaps in our understanding that must be addressed by future observations, theory, and modeling.

1.1.1 Solar Variability: The Dynamo

The fundamental origin of all space weather is the variability of the Sun's magnetism. This variability is readily apparent in both the spatial and temporal properties of the solar magnetic field. In terms of spatial properties, the most important feature of the Sun's magnetism that leads to space weather is that the field exhibits a complex, multipolar flux distribution, very different than that of Earth's simple dipole.

Figure 1.1 shows the line-of-sight magnetic flux at the solar surface, the photosphere, during a period of strong solar activity. The structure of the observable flux consists of a myriad of concentrations over a large range of spatial scales. Detailed analysis has shown that the number of flux concentrations exhibits a power-law dependence on spatial scale, $N \sim \Phi^{-1.85}$, over five orders of magnitude or more (Parnell et al. 2009). All these scales may play an important role in the observed solar activity, but for space weather, the most important flux scales are likely to be the global dipolar field of the Sun and the strong concentrations of flux in sunspots with field strengths of 2000 G or more. Sunspots tend to cluster in groups to form so-called *active regions* (ARs),

with spatial scales of order hundreds of Mms. Large complex ARs are the sites of the most energetic form of solar activity, the giant eruptions of magnetic field and plasma known as *coronal mass ejections* (CMEs)/ eruptive flares. For the science of space weather, therefore, understanding the formation of AR magnetic fields and their dynamics in the corona and heliosphere is of paramount importance.

The temporal variability of the field can roughly be considered as having three dominant scales. The shortest scale is given by the characteristic wave travel time in the corona: $\tau \sim L/V_A$, where V_A is the coronal Alfvén speed. Using typical values for length scales of 100 Mm and $V_A \sim 1 \, \mathrm{Mm \, s^{-1}}$ yields a timescale of order 100 s, which is the observed timescale for explosive events such as flares and CMEs. Solar activity on these timescales is discussed in Section 1.1.3. The second temporal scale is that of the period for solar rotation, ~27 days. The rotation drives recurring phenomena such as solar UV variability as ARs rotate from front to back on the disk, and the interaction of fast and slow solar wind streams in the heliosphere. This latter process is highly important for space weather, and will be discussed in detail in Chapter 2. The third timescale, which we will now discuss, is that of the solar cycle, of order a decade or so.

1.1.1.1 The Solar Cycle

It has been known for well over a century (Schwabe 1844) that the appearance of ARs on the Sun is cyclic with a period of roughly 22 years. Figure 1.2 from Hathaway (2015) shows the salient features of the solar magnetic cycle. The daily longitudinally averaged magnetic field as observed by both ground and space-based magnetographs is plotted for the past 30 years. It should be noted that the measurements at the poles are subject to considerable errors due to foreshortening; however, most of the structure in the field, such as ARs, is at lower latitudes. The cycle "begins" with bipolar ARs emerging at intermediate latitudes ~30° with a small latitudinal tilt to the bipoles; for example, in cycle 22 beginning in 1986, the AR bipoles in the north had a "leading" polarity (i.e., closer to the equator) that is negative while the leading polarities in the south were positive. Note that, when the cycle begins, the polar field in each hemisphere has the same sign as the leading AR polarities, while the trailing polarities, which are at slightly higher (~5°) latitude, have a polarity opposite to that of their respective pole. The cycle

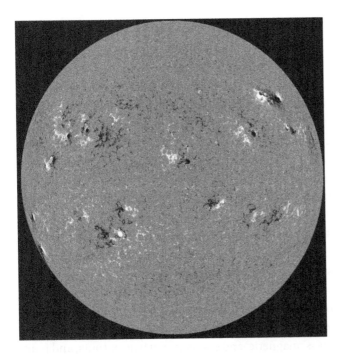

FIGURE 1.1 Image from the HMI magnetograph on SDO showing the positive (white) and negative (black) line-of-sight magnetic flux at the solar photosphere on January 5, 2013.

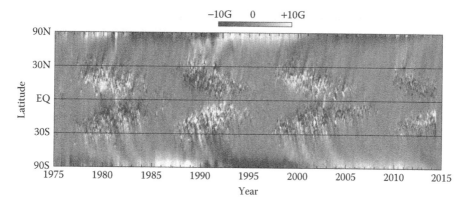

FIGURE 1.2 Longitudinally averaged line-of-sight magnetic field at the solar surface. White corresponds to positive polarity and black to negative. (From Hathaway, D.H., *Living Rev. Solar Phys.*, 12, 4, 2015.)

continues with ARs emerging at successively lower latitudes over the course of a decade or so. After emerging, the trailing polarities undergo a so-called rush to the poles, thereby, canceling the leading polarity of previously emerged ARs and eventually reversing the sign of the polar field. This process can be seen in Figure 1.2 as the broad swaths of color stretching from the AR belt to the poles. AR emergence appears to die out when the emergence reaches the equator and, subsequently, the Sun enters a phase of solar activity minimum lasting a year or two before the new cycle ARs start their emergence again at high latitudes.

Two important features of the solar cycle are apparent in Figure 1.2. Even though only approximately four cycles are shown, it is evident the cycles can vary considerably in strength and in duration. In particular, we note that the present cycle, number 24, is weak (low number of ARs) and was preceded by a long minimum, the so-called deep minimum (e.g., Russell et al. 2010). In fact, the minima sometimes appear to last for decades as inferred from examination of radioisotopes in tree rings and ice cores (Usoskin 2013). For example, Figure 1.3

shows the so-called *Maunder minimum* during which sunspots seemed to disappear for over 50 years.

Predicting strong cycles is clearly important for space weather because of the increased probability of large CMEs and flares, but even predicting the extent of the minima is critical. A deep minimum implies a weak heliospheric magnetic field and, consequently, a larger flux of cosmic rays that can penetrate into the inner heliosphere, posing hazards for deep-space travel. For the science of space weather, therefore, all aspects of the solar cycle seen in Figures 1.2 and 1.3 must be understood. Of course, this is the problem of understanding the solar dynamo, which has long been one of the most difficult physics challenges in all space science (e.g., see review by Charbonneau 2010).

In principle, the solar dynamo can be calculated from first principles using the standard magnetohydrodynamics (MHD) equations for a rotating shell appropriate to the solar convection zone and starting with some seed magnetic field. Given that the cycle sometimes appears to die out, but eventually recovers, as in the Maunder minimum, then the form of the initial field

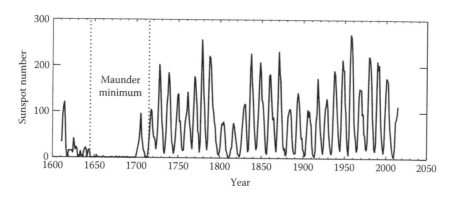

FIGURE 1.3 Sunspot number for the past 400 years. (From Hathaway, D.H., *Living Rev. Solar Phys.*, 12, 4, 2015.)

should be irrelevant. The major difficulty with such a first-principles approach, however, is that the Sun's convection is turbulent. This means that the Sun's internal motions are likely to have important structure over many orders of magnitude, from the global scale down to some eddy dissipation scale. Furthermore, the critical transport coefficients, viscosity, and conductivity are dominated by the turbulence and, hence, very poorly unknown. As a result, even the large-scale motions, such as the form of the meridional circulation in the solar interior, are far from certain. Given this difficulty with a rigorous first-principles dynamo, two general approaches have been pursued in developing a solar dynamo theory: (1) use the observations (e.g., Figure 1.2) as the foundation for building a quasi-empirical theory, the so-called *flux transport models*, or (2) use the basic equations as the foundation for building a statistically averaged or approximate model, the so-called *mean-field models*. We briefly review the salient features of these models below.

1.1.1.2 Flux-Transport Dynamos

For space weather applications, the most successful type of dynamo theory is the flux-transport models that are based on the seminal ideas of Babcock (1961) and Leighton (1969). Babcock (1961) proposed that the solar differential rotation acting on the poloidal flux of an initially dipolar field would generate a toroidal field. It has been known for centuries that the Sun rotates differentially, in that the equatorial rotation rate is approximately 20% faster than that at the poles. Babcock noted that the observed differential rotation profile would produce a growth rate for the toroidal flux that is maximum at roughly 30°, in agreement with the basic solar cycle observations that ARs first emerge at these latitudes. Furthermore, the observed systematic AR tilt, Joy's law, implies that the toroidal flux rotates as it rises through the convection zone, thereby generating a poloidal flux that is opposite to the initial dipole flux. Leighton (1969) greatly advanced these ideas by developing equations for the detailed transport of emerged flux to the poles by surface diffusion, which is essential for reversing the cycle. Key advances to the basic Babcock–Leighton theory have been the addition of meridional flows (Wang et al. 1991) and the extension of the model to include the actual observed radial dependence of the differential rotation (Dikpati and Charbonneau 1999).

The flux transport models are generally 2.5D, assuming a longitudinally averaged structure for the plasma flows and magnetic field, so that the induction equation can be expressed as

$$\frac{\partial \mathbf{B}}{\partial t} = \nabla \times (\mathbf{v} \times \mathbf{B} - \eta \nabla \times \mathbf{B}) + S,$$

and the system is axisymmetric so that **B** can be expressed as

$$\mathbf{B} = B_\varphi(r,\theta,t)\hat{\varphi} + \nabla \times \left(\psi(r,\theta,t)\hat{\varphi}\right),$$

where ψ is the flux function for the poloidal field. Substituting this form into the induction equations results in two coupled equations for the toroidal field B_φ and the poloidal flux ψ. Note that the model is purely kinematic: the velocity **V** consists of the differential rotation and meridional circulation with no feedback from the field on the velocity. The differential rotation is fairly well-constrained by helioseismology (Thompson et al. 2003), but the subsurface meridional flow is poorly known. The meridional circulation is critically important in flux-transport models; assumptions as to its structure with depth and its variation with time play essential roles. The meridional circulation is often likened to a "conveyor belt," carrying poloidal flux poleward above the surface so that it can cancel the polar field and equatorward below the surface so that it can seed the beginning of the new cycle. Furthermore, a dependence of the speed of the meridional circulation on the strength of a cycle may be necessary for understanding how the cycle can recover from a Maunder minimum state (Wang 2004).

Another important feature of the model is the effective resistivity, η, which is due to turbulence and so is weakly constrained. The final major feature of the model is the source term $S(r, \theta, t)$, which represents the generation of poloidal flux by the emergence of tilted toroidal flux. Note that, although expressed as an axisymmetric source, S is actually due to 3D effects and is essential in order to have a dynamo, since by Cowling's (1933) antidynamo theorem, no dynamo is possible for a true axisymmetric system. The source term is the key new feature of the flux transport models; without it the equation above is simply the standard 2D induction equation of resistive MHD. Again this term is poorly constrained and, at present, fairly ad hoc.

The outstanding challenges for the flux transport models are to pin down definitively, either by observation or by theory, the three major unknowns: the meridional circulation, the effective resistivity, and the source term for the poloidal flux. It may also be useful to extend the theory into the fully 3D regime (e.g., Yeates and Munoz-Jaramillo 2013), although it is not clear whether this is truly necessary. In spite of its limitations, however, the theory holds great promise for space weather prediction. Thus far, only flux transport models have been used to calculate the strength of an upcoming solar cycle based on observations of prior cycles (e.g., Dikpati et al. 2006; Choudhuri et al. 2007). Given the uncertainties in these models, it is not surprising that these calculations have yielded results that are far from consistent. On the other hand, it seems likely that the uncertainties will be better constrained by future observation and theory, and that the flux transport model will eventually become a powerful space weather prediction capability.

1.1.1.3 Mean-Field Dynamos

The second type of dynamo model, which is also the oldest, is the mean-field theory pioneered by Parker (1955) in his seminal studies. Although the final mathematical framework for the mean-field models ends up being quite similar to that of the flux transport models, the physical arguments underlying them are quite different. The flux transport models are somewhat heuristic in spirit and are based on the surface observations, whereas the mean-field models attempt to develop a "rigorous" theory for the convection-zone turbulent dynamics and its effects on the subsurface fields.

The starting point for the mean-field models is the separation of the dynamics between the slowly varying, global scale of the differential rotation, for example, and the small-scale, rapidly varying and statistically stationary scales of the convective turbulence. This suggests that the magnetic and velocity fields be broken up into mean fields that are generally assumed to be axisymmetric, and fluctuating fields that are fully 3D but never actually calculated:

$$\mathbf{B} = \langle \mathbf{B} \rangle + b, \quad \mathbf{v} = \langle \mathbf{V} \rangle + v.$$

Note that the fluctuating fields do vanish on average ($\langle b \rangle = \langle v \rangle = 0$), but their magnitudes are of the order

of the mean fields. This is essential because the fluctuating fields are needed to regenerate the mean fields. Substituting these forms into the induction equation above, and dropping the flux transport source term S, yields the induction equation for the mean field:

$$\frac{\partial \langle \mathbf{B} \rangle}{\partial t} = \nabla \times \left(\langle \mathbf{V} \rangle \times \langle \mathbf{B} \rangle - \eta \nabla \times \langle \mathbf{B} \rangle \right) + \nabla \times \langle v \times b \rangle.$$

The key new addition is the last term, where the electric field $\langle v \times b \rangle$ is generally referred to as the *mean electromotive force* ε due to the correlations between the turbulent velocity and magnetic fields. Note that if $\langle B \rangle$ and $\langle V \rangle$ are axisymmetric, and the mean velocity is fixed so the dynamo is kinematic, then the mean-field dynamo equations are formally equivalent to those of the flux transport, except that the source term has a different interpretation. As in the flux-transport models, this source term is also fundamentally due to unresolved 3D effects, which must be present in order to circumvent Cowling's theorem.

As in all statistical physics theories, the primary challenge is to close the equations by determining the large-scale effects of the correlations between the small-scale fluctuating fields; in other words, specifying ε in terms of the mean fields. This clearly depends on the properties of the turbulence, but in the simplest case of quasi-homogeneous, quasi-isotropic turbulence, the mean electromotive force can be expressed as (e.g., Charbonneau 2014):

$$\varepsilon = \alpha \langle \mathbf{B} \rangle - \beta \nabla \times \langle \mathbf{B} \rangle,$$

where α and β are constants. Substituting this expression back into the induction equation and again expressing $\langle B \rangle$ in terms of the mean poloidal (B_φ) and toroidal (ψ) components results in two equations similar to those of the flux transport models. The β term in the mean electromotive force simply represents turbulent diffusion, which again is generally parameterized. The important new addition is the term αB_φ in the equation for the poloidal flux, known as the α effect. This term represents the regeneration of poloidal flux due to systematic twisting of the toroidal flux by convective cells (Parker 1955).

The major advantages of the mean-field model are that it is based on well-studied statistical physics theory and that its simplest form is linear, so it is amenable to a vast range of theoretical, analytic, and numerical studies. As a result, mean-field models have been the subject of an enormous amount of work over the past 50 years or so. The model is especially effective for being tested by numerical experiments and conversely for interpreting them, because it can be applied to a wide range of idealized systems. On the other hand, it is less effective for space weather prediction because it is not clear how to input solar observations directly into the model. Therefore, it appears, that the mean-field and flux transport approaches are highly complementary. The former is enabling advances in understanding the basic physics of turbulent dynamos, in general, whereas the latter is enabling advances in predicting the long-timescale effects of the Sun's dynamo on space weather.

The future of solar dynamo theory is clearly in full 3D MHD simulation. As discussed above, these calculations will be highly challenging, but with the advent of exascale computing in the next decade, and advances in simulating turbulence (e.g., Miesch et al. 2015), it should become possible to perform fully 3D modeling of the solar dynamo. An especially promising approach for long-term space weather prediction would be to use the 3D simulations to pin down the unknowns, such as the source terms, in the flux transport and mean-field models. If the 3D models can provide guidance on how to make the 2D models more rigorous and accurate, this could well lead to a breakthrough in predicting the solar cycle.

1.1.2 Slow Drivers of Space Weather

As discussed in Section 1.1.1, the coronal field is rooted in the solar interior, and is shuffled constantly by photospheric motions. These motions twist and braid the field throughout ARs and quiet Sun alike. In open-field regions such as coronal holes (Section 1.1.2.1), the added energy leaves the Sun in the form of MHD waves. In closed-field regions, however, this twisting complicates the magnetic structure and forms electric current sheets, where magnetic reconnection (see Chapter 6) can occur.

1.1.2.1 Global-Scale Magnetic Structures and Solar Wind

The Sun's large-scale magnetic field is dipolar, becoming multipolar around the peak of the solar cycle. At solar minimum, the dipole is nearly aligned with the Sun's

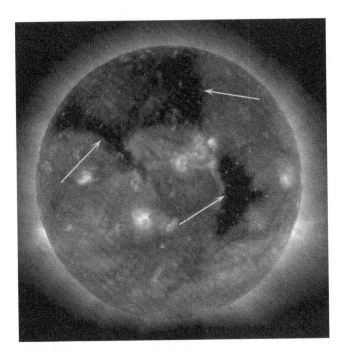

FIGURE 1.4   SDO/AIA 211Å image of the Sun showing emission from ~2 MK plasma that traces the magnetically defined structures. Arrows point to the coronal holes. The brightest emission features are active regions (white) located at mid-latitudes in each hemisphere.

rotational axis, whereas postmaximum it is tilted at an appreciable angle to the rotation axis. Wherever the field is sufficiently weak, for example, in the polar regions around minimum, gas and solar-wind ram pressure force the field to open. These open regions are called *coronal holes* because they appear dark in visible, extreme ultraviolet (EUV), and soft X-ray (SXR) images of the corona (Figure 1.4), but they are merely less dense than the adjacent closed-field regions. The fast solar wind, described in Chapter 2, originates in coronal holes. Because coronal holes link the solar surface to the heliosphere, they are important conduits for waves, flows, and energetic particles generated at or above the solar chromosphere to penetrate interplanetary space. The main sources of energy for the wind must be injected, but not necessarily deposited, at their footpoints (Cranmer 2012). Spectroscopic observations of the lower solar atmosphere in coronal holes suggest that the network—an interconnected, constantly changing web of strong magnetic fields at the perimeters of photospheric granules and supergranules (Figure 1.5; Section 1.1.1)—is the origin of most of the plasma eventually accelerated as the fast wind (Hassler et al. 1999; Tu et al. 2005).

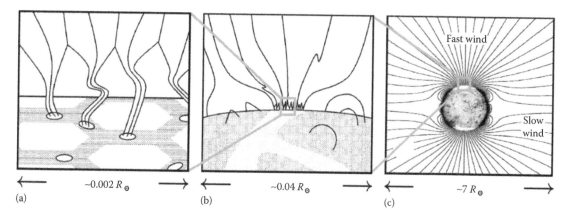

FIGURE 1.5 Magnetic "funnels" in the (a) granular and (b) supergranular network, which provide and channel the plasma of (c) the fast wind. (From Cranmer, S.R., *Living Rev. Solar Phys.*, 6, 3, 2009.)

The boundary between open and closed field evolves constantly, as flux emerges all over the Sun in the form of small bipoles, ARs migrate from low to high latitudes, and photospheric motions twist and braid the field lines in the vicinity of this boundary. Several models propose that this boundary is the source region for the slow solar wind (Chapter 2). The open-closed boundary is a prime location for coronal magnetic reconnection (Chapter 6) to interchange closed and open field lines, thus releasing energy and plasma intermittently from closed to open flux tubes (e.g., Crooker et al. 2004; Wang 2012). Streamer-top models postulate that the slow wind is generated by episodic opening and closing of the unstable cusps of coronal helmet streamers (e.g., Suess et al. 1996; Sheeley et al. 1997; Rappazzo et al. 2005). According to the recent "S-web" model (Antiochos et al. 2011; Linker et al. 2011; Crooker et al. 2012), the slow wind originates in a network of narrow open-field corridors that map into a complex web of separatrices and quasi-separatrix layers in the heliosphere. These narrow corridors link the polar coronal holes with lower-latitude holes of the same polarity, and map to higher-latitude arcs in the heliosphere (Antiochos et al. 2007) attached to the heliospheric current sheet. This is only one example of how the complex magnetic structure at the Sun can be connected to remote locations in interplanetary space, with important effects on the propagation of energetic particles accelerated in the corona and inner heliosphere (Chapter 20).

1.1.2.2 Local-Scale Magnetic Structures (ARs)

As described in Section 1.1.1, ARs appear, evolve, and decay on the solar surface in tune with the solar cycle;

more ARs emerge and persist during solar maximum than during minimum, when it is common to observe no ARs for intervals of a month or more. Since sunspots were first tabulated, the longest spotless (AR-less) period has been 92 days (Janssen, J. http://users.telenet.be/j.janssens/Spotless/Spotless.html). These ARs are composed of one or more sunspot pairs, each consisting of negative and positive polarity concentrations of magnetic flux. When eruptive events (Section 1.1.3) are not occurring, the flux in an AR usually is closed: that is, both footpoints are located on the Sun, although they can be quite far apart. As shown in Figure 1.4, EUV and SXR images of the corona above ARs exhibit bright "coronal loops" that appear to trace narrow magnetic flux tubes rooted in opposing polarities. A diffuse component also is visible in some areas, which is 20%–30% less intense than loop emission at EUV wavelengths (Del Zanna and Mason 2003); the physical properties of the source plasma are not well-known (Zuccarello et al. 2013). Spectroscopic observations of coronal loops show that they are hotter and denser than the corona outside ARs (often referred to as the *quiet Sun*), with typical temperatures of 1–5 MK and electron densities in the range $10^8–10^{11}$ cm^{-3}. Furthermore, the temperature (density) is two orders of magnitude above (below) that of the visible surface, the photosphere, which seems completely counterintuitive for a gaseous sphere with the energy source located at its core.

One of the most vexing unanswered questions in heliophysics is, therefore, why is the corona hot? A thorough review of coronal heating theories is beyond the scope of this book, so here we will discuss only those key points relevant to the Sun's influence on space weather. The reader is referred to the excellent reviews in "Recent advances in coronal heating" (de Moortel and Browning 2015) for an in-depth treatment of this important subject.

It is widely accepted that photospheric motions, driven by subsurface convection, are the ultimate source of the energy that heats the corona. The magnetic field rooted in the photosphere is stressed by these motions, which generate waves and current sheets in the overlying corona. Therefore, the magnetic field transports, stores, and releases the energy needed to maintain the multimillion-degree coronal plasma. Although the AR coronal-heating problem is not fully solved as yet, it is clear from both theory/modeling and observations that the energy-release mechanism must be impulsive and pervasive (Klimchuk 2015). The leading candidates for the coronal heating process are myriad, transient reconnection episodes (e.g., Parker 1988; Cargill 1994; Klimchuk 2006) and a turbulent cascade of Alfvén waves (e.g., Matthaeus et al. 1999; Rappazzo et al. 2008; Dahlburg et al. 2012). Note that these processes are not mutually exclusive, because reconnection generates MHD waves and waves can induce reconnection. Regardless of the heating mechanism, the observed association between ARs and their bright loops indicates that strong magnetic fields yield strong coronal heating.

The radiative output of the entire Sun (the solar irradiance, when measured at 1 AU) varies spectrally and temporally, affecting different components of the Sun–Earth system on timescales ranging from days to decades. (Explosive energy release on shorter timescales is the topic of the next section.) Figure 1.6 shows the total solar irradiance, which is the spectral irradiance summed over wavelength, for the past three solar cycles and the present cycle to date. The background intensity clearly varies over the ~11-year solar cycle (Section 1.1.1) and the solar rotation period of ~27 days. Each visible feature on the solar surface has a unique spectrum, which contributes varying amounts toward the irradiance at each wavelength. Dark features such as sunspots, and the bright webbing of the intergranular network, make the summation of emissions a challenging task at any instant. For example, the appearance of a new AR, with its dark sunspots and bright surrounding areas (called *faculae*), enhances the emissivity at certain wavelengths while reducing the intensity at others. In general, when more ARs face the Earth, the irradiance is higher. During the two weeks that a given feature can be tracked from the east to the west limb of the Sun, an AR can become more complex and magnetically stronger through continued flux emergence, evolve negligibly, or become less intense. At any moment, the amount of ARs on the solar disk

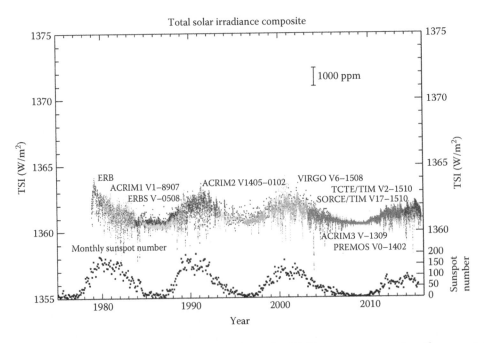

FIGURE 1.6 Observed total solar irradiance over nearly four solar cycles (G. Kopp, pers. comm.), with sunspot number shown at the bottom. The variations in irradiance are due to solar rotation and the appearance/disappearance of active regions and sunspots.

facing the Earth can be zero to several. Some ARs last for many solar rotations, while others disappear in the course of a single rotation. Increases in irradiance over several hours to ~1 day are caused by the formation of new coronal loops associated with flux emergence, with the magnitude of the enhancement proportional to the amount of emerged flux. Figure 1.7 shows the Sun in SXR emission over an entire solar cycle; the changes in area and brightness from minimum (left) to maximum (right) are associated with significant increases in spectral irradiance at short wavelengths.

Except during flares, solar radiation in the extreme and far ultraviolet (EUV and FUV) portions of the spectrum has the greatest effect on Earth and other planets. In particular, this radiation is responsible for ionizing the upper atmosphere, creating the ionosphere (Chapters 15, 21, and 22), and governing its properties and extent. The most important spectral line, in terms of ionospheric impact, is the strong Lyman-α emission line of hydrogen; continuum emission from free-bound recombination and strong emission lines of He, C, N, Fe, and other elements also contribute to the irradiance in this spectral range. Solar UV radiation also affects the ozone layer of the neutral atmosphere directly: higher UV flux generates more ozone in the mid- to upper stratosphere and even influences tropospheric circulation patterns (Hood et al. 2013). The specific

mechanisms by which the nonflaring solar irradiance affects planetary ionospheres are discussed in Chapters 16, 18, 21, and 22. Of course, life on Earth depends crucially on the visible and infrared solar emission for illumination and warmth.

Because the Sun's output varies significantly over a wide range of timescales, operational space weather services need to monitor the variability of the solar EUV/FUV irradiance frequently and consistently, as with the SEE instrument on TIMED (Woods et al. 2005) and the EVE instrument on SDO (Woods et al. 2012). Figure 1.8 shows the degree of variability of the spectral irradiance from 10 to 2000Å as measured by EVE. Numerous spectral and total irradiance monitors have flown since the 1980s (Domingo et al. 2009), but the spectral coverage and lifetime of individual experiments has been limited and cross-calibration among measurements by different instruments has been a major problem. Consequently we also rely on models that use different approaches to predict the irradiance. Modeling the spectral and total solar irradiance is a lively and ever-changing research topic, which has made great strides over the past 20 years. A wide range of proxies and physics-based approaches has been used, with varying degrees of success. One series of studies has reconstructed the daily spectral solar irradiance through a careful multistep modeling process (Fontenla et al. 2007, 2014, 2015; see also Haberreiter 2011): formulating detailed physical models of the atmospheric properties of the key chromospheric and coronal features affecting the irradiance (plage, sunspots, quiet-Sun network); predicting the UV-EUV-SXR spectrum of each type of feature using the CHIANTI database (Dere et al. 1997; Landi et al. 2013) and other atomic data resources; determining the daily spatial coverage of each feature type on the solar disk, at present with SDO/AIA images; and summing the contributions to produce a model spectral irradiance on Earth. The recent results agree well with solar observations from a variety of spaceborne instruments, including HRTS9, SORCE/SOLSTICE, and SDO/EVE (Figure 1.9).

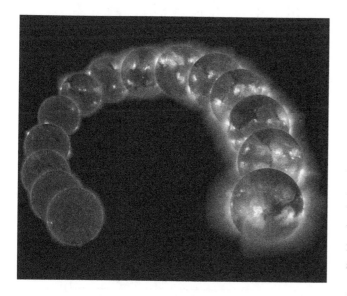

FIGURE 1.7 STEREO A EUVI 284Å images of the Sun's corona throughout solar cycle 23, from minimum in 2007 (left) to maximum in 2014 (right). (B. Thompson, pers. comm.). The sampling is roughly every 6 months, with a larger jump to the final image closest to solar maximum.

1.1.3 Explosive Drivers of Space Weather

Solar eruptions are the most intense energy releases in the heliosphere. These events always originate in very specific locations: the boundaries between positive and negative magnetic flux, denoted polarity inversion lines

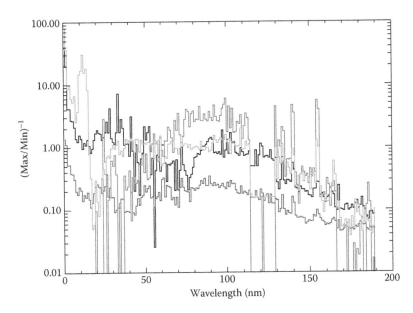

FIGURE 1.8   Comparisons of the irradiance variability due to solar cycle (black), solar rotation (red), and the impulsive (blue) and gradual (green) phases of flares. (P. Chamberlin, pers. comm.; see Chamberlin et al. 2009, Figure 17, for an earlier color version of this figure.)

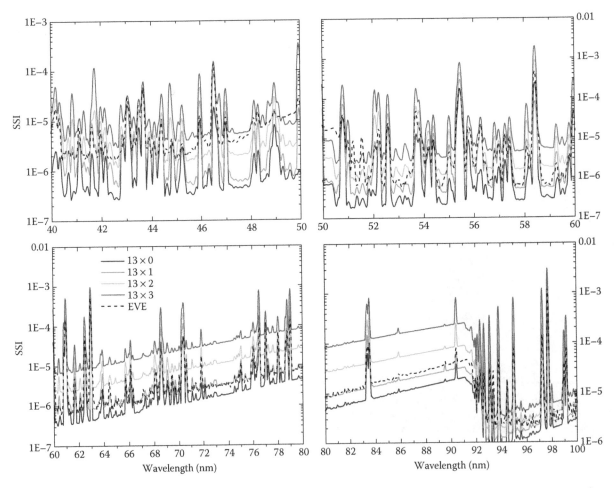

FIGURE 1.9   Spectral irradiance models for different quiet-Sun features, compared with SDO/EVE observations in the range 40–100 nm. The numbered features are dark quiet-Sun internetwork (13 × 0), quiet-Sun internetwork (13 × 1), quiet-Sun network lane (13 × 2), and enhanced network (13 × 3). (From Fontenla, J.M. et al., *Sol. Phys.*, 289, 515, 2014, Figure 13.)

(PILs; Babcock and Babcock 1955). In the magnetogram shown in Figure 1.1, for example, PILs delineate the boundaries between the black and white flux concentrations. We have known for decades that PILs are the primary sites where magnetic free energy (energy above the minimum-energy potential state) builds up in the solar corona, sufficient to power an eruption. A substantial amount of free energy builds up at the PILs, for reasons that are still under debate (Antiochos 2013; Mackay et al. 2014; Knizhnik et al. 2015; Zhao et al. 2015). A well-established observable signature of this concentration of free energy is the presence of a filament channel: highly sheared magnetic field within a narrow zone around the PIL, where the magnetic field lines are nearly aligned with the PIL, rather than perpendicular to it as in a potential field (Foukal 1971; Martin 1998). An example of the distinctive sheared field of a filament channel is shown in Figure 1.10; note how narrow this channel is, and the large contrast between the fibril orientations inside and outside the channel.

Filament channels also host filaments (also known as *prominences* when observed at the solar limb; here, we use the term *filaments* for both)—dense, cool

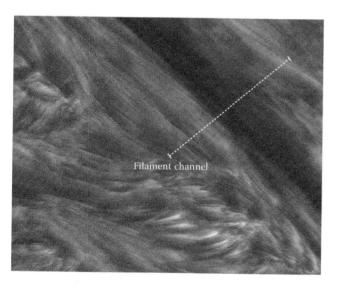

FIGURE 1.10 Part of a chromospheric filament channel observed in the *H*α line center by the NST at Big Bear Solar Observatory on July 25, 2015. The channel is the linear feature extending diagonally across the image, with dark (absorbing) filament material in its center. Note that the thread-like features (fibrils) in the channel are oriented nearly parallel to the axis of the channel, while fine-scale features outside the channel are at various angles to the channel axis. (Courtesy of V. Yurchesyn, BBSO, Big Bear City, CA.)

plasma with chromospheric properties that is suspended in the hot, rarified corona. Not all filament channels contain observable filaments at any given time, but when they do, filaments provide valuable information about the geometry and other properties of the otherwise invisible supporting magnetic field of the channel (e.g., Luna et al. 2014). The distribution of large, stable solar filaments relative to the underlying magnetic field shows definitively that such filaments are much more likely to form in multipolar magnetic configurations (Mackay et al. 2014). In contrast, filaments in bipolar ARs are highly transient, possibly indicating that emerging flux in these regions quickly forms a flux rope that erupts within hours to days. Filaments also are massive (~10^{14-16} g), prompting some speculation that this mass plays a role in the stability of the channel (Low 1999). The coronal extension of the filament-channel field is often denoted as the coronal cavity, due to its dark appearance relative to its surroundings (Gibson 2015).

There are two principal classes of models for the magnetic configuration of filament-channel fields before eruption: twisted flux ropes and sheared arcades (Figure 1.11; see review by Mackay et al. 2010). Most computational studies of flux emergence have shown that subphotospheric flux ropes tend to emerge as sheared arcades, with the flux-rope axis remaining below the surface (Magara 2004; Fang et al. 2010). Partial ionization enhances the reconnection rate in the chromosphere, however, facilitating the formation of a flux rope during emergence (Leake and Linton 2013). We are left with a long-standing dichotomy: to date, numerical simulations of flux emergence yield either large, stable filament channels comprised of sheared arcades, which do not exhibit sufficient shear to drive an eruption, or highly unstable filament channels that form erupting flux ropes while emerging (e.g., Leake et al. 2014), which cannot explain the prevalence of long-lasting, extensive channels.

Regardless of the formation mechanism, filament channels supply the magnetic energy needed to expel mass and magnetic field away from the Sun in the form of CMEs and associated filament eruptions, and to accelerate and heat the particles responsible for flare emissions from the lower solar atmosphere. Because the methods used to detect and characterize CMEs and flares are completely different, the fundamental physical connection between these phenomena was not initially

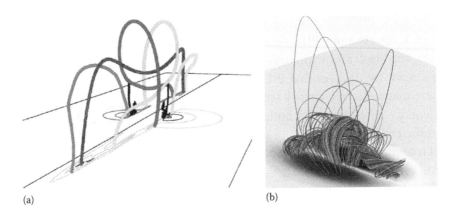

(a)　　　　　　　　　　　　　　　(b)

FIGURE 1.11　Selected magnetic field lines from filament-channel magnetic-structure models. (a) Single sheared-arcade model, formed by shearing footpoint motions followed by relaxation. Greyscale contours on the bottom surface denote lines of constant vertical B. (From DeVore, C.R. and Antiochos, S.K., *ApJ*, 539, 954, 2000.) (b) Strongly twisted flux rope model formed by shearing and converging footpoint motions. (From Amari, T. et al., *ApJ*, 529, L49, 2000.)

recognized. Flares have been observed for centuries, starting with the rare white-light flares visible through the protection of smoked glass or neutral filters. The so-called *Carrington event of 1859*, an extreme eruption that provided the first clues about the Sun's influence on geospace, was seen in white-light continuum emission (Carrington 1859). With the advent of optical spectroscopy (in particular, using the 6563Å emission/absorption line of hydrogen), followed by spectroscopic observations from radio to gamma-ray wavelengths, synoptic coverage of the solar disk allowed flares to be studied and categorized in detail.

A flare is a rapid increase in brightness coming from a small area on and above the visible solar surface; the largest and most common flares originate in filament channels within ARs, but they also occur between ARs. Visible continuum observations also played a key role in the discovery of CMEs by coronagraphs (Tousey et al. 1973), although CMEs are detected through Thompson scattering of the visible photospheric emission by their enhanced densities. Simultaneous observations of flares and CMEs by ground-based optical and radio telescopes, together with spaceborne imaging spectrographs and coronagraphs, have greatly advanced our understanding of these eruptive phenomena over the past 20 years.

After much debate about whether flares generate CMEs or vice versa, heliophysicists now agree that both flares and CMEs are manifestations of a common encompassing phenomenon (e.g., Gosling 1993): explosive release of magnetic energy stored in the corona. The standard picture, shown in Figure 1.12, may not be accurate in all details, but it illustrates the fundamental

link between these primary drivers of space weather. All models of solar eruptions invoke magnetic reconnection (see Chapter 6) as the main energy-release mechanism, and the reconnection site as the separation point between the CME and the flare. In the simplified geometry of these models, the partially disconnected flux above the reconnection site has the configuration of a twisted flux rope, which contains a significant fraction of the magnetic stress originally in the filament channel. Below the reconnection site resides the remaining less-sheared field. Reconnection in this environment liberates energy in several forms: a negligible amount

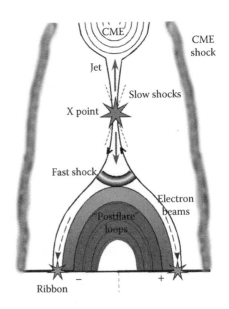

FIGURE 1.12　Authors' illustration of the standard 2D picture of a solar eruption, showing a simplified representation of the relationship between the CME and the flare.

is resistively dissipated, and the DC currents at the reconnection site accelerate some particles, but most of the energy initially goes into strong plasma flows entrained in the retracting reconnected flux. The path of energy transport then diverges for flares and CMEs, leading to markedly different observable signatures and space weather effects, as summarized in Sections 1.1.3.1 and 1.1.3.2.

1.1.3.1 Flares: Impulsive Electromagnetic and Particle Radiation

The sudden brightening of electromagnetic radiation from a constrained volume of the corona, chromosphere, and photosphere surrounding a PIL is the typical definition of a solar flare. Over the course of a flare, this radiation covers a wide swath of the electromagnetic spectrum, but most of the radiative energy is in visible and UV emissions (Woods et al. 2006). Estimates of the radiative losses in flares range from ~10^{29} to 10^{31} erg; in comparison, the total flare energy (in the thermal plasma and nonthermal particles) is at least 10^{32} erg (Canfield et al. 1980; Emslie et al. 2004).

When observed with coarse or no spatial resolution (see Figure 1.13), the so-called *impulsive phase* emission is generally nonthermal in nature, while the longer-lived "gradual" emission tends to be thermal. This rather simplified classification is based on the spectral shape (intensity versus energy or frequency) of the observed radiation, and on certain emission line widths and ratios. Although low-level preflare activity often appears in various wavelengths, flare onset is typically signaled by rapidly rising hard X-rays and microwaves, accompanied by localized lower-energy brightenings in a wide range of continuum and line emissions. As sketched in Figure 1.12, these impulsive emissions come from highly energetic particles accelerated in the flaring corona by processes that are poorly understood at present. Some of these particles appear to be trapped in the vicinity of the reconnection region (Masuda et al. 1994; Krucker et al. 2008, 2010), but most spiral around the magnetic field while traveling down to the footpoints, where the particles plow into the dense chromospheric plasma and lose their energy (Hoyng et al. 1981). The X-rays are bremsstrahlung radiation from ~10 to 300 keV electrons, the microwaves are gyrosynchrotron radiation from electrons at slightly higher energies, and the optical, UV, and EUV footpoint emissions are thermal signatures of the heating of the ambient plasma by the stopped electrons

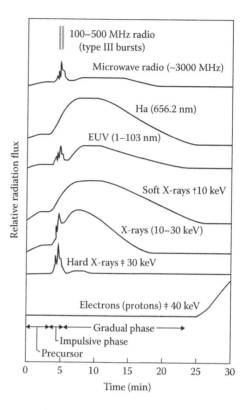

FIGURE 1.13 Electromagnetic radiation intensity versus time in different wavebands from a typical eruptive solar flare. (Data from Lang, K., The violent Sun, in *Sun, Earth and Sky*, Cambridge University Press, Cambridge, UK, 1995. https://ase.tufts.edu/cosmos/print_images.asp?id=27.)

(and possibly protons). The energy thus deposited in the chromosphere drives evaporation upflows at speeds of up to a few hundred kilometers per second, filling the coronal loops with EUV and SXR-emitting plasma at temperatures of order 10–20 MK.

As noted in Section 1.1.2, the UV, EUV, and shorter wavelength emissions have the largest impact on planetary atmospheres. These effects are described in detail in Chapters 11, 15, and 16. Broadly speaking, increased EUV and X-ray emission from flares ionizes more of the neutral atmosphere and ionosphere than during quiet times, while the increased UV penetrates down to the stratosphere, affecting the critical ozone layer; the chemistry and composition of Earth's atmosphere also are altered by flare radiation.

1.1.3.2 CMEs: Impulsive Mass and Magnetic-Field Driving

CMEs consist of plasma and magnetic field expelled from the solar corona at peak speeds ranging from ~200 to 3000 km s^{-1}. Their masses are substantial (~10^{15-16} g), particularly if a preexisting prominence erupts with the

FIGURE 1.14   SOHO/EIT 304Å image of the solar disk surrounded by a SOHO/LASCO C2 coronagraph image of a fast CME ejected from the west (right) limb on April 2, 2001. The bright white patch on the Sun beneath the CME is the associated flare.

CME. Figure 1.14 shows a white-light image of a typical CME, with a dense front (bright white arc) and a trailing prominence core. These events are commonly divided into slow and fast categories, depending on whether they are traveling slower or faster than the ambient solar wind. In general, slow CMEs accelerate gradually until they reach the ambient solar-wind speed, whereas fast CMEs accelerate rapidly in the low corona and decelerate thereafter, again until they are entrained in the solar wind.

Slow CMEs play a negligible role in producing significant space-weather activity for several reasons. Most slow CMEs are incapable of driving shocks, unless the Alfvén speed is exceptionally low; this was the case during the recent solar minimum, when the Sun's magnetic field was unusually weak. Therefore, they are not significant sources of solar energetic particles (SEPs; see Chapter 20). In addition, their sources at the Sun tend to have weak magnetic fields, for example, filament channels around the polar crown or between decaying ARs. The combined effect of weak magnetic fluxes and slow ejection speeds is to generate CMEs with relatively low impact on planetary magnetospheres.

Fast CMEs cause the most destructive space weather, particularly those fast enough to drive strong shocks (e.g., Gopalswamy et al. 2015). Therefore, most of the following discussion pertains to fast CMEs. The 3D magnetic structure of a fast CME is commonly interpreted as a flux rope, as verified by in situ observations of many

(but not all) interplanetary CMEs (e.g., Marubashi et al. 2015). All theoretical models for CMEs agree that the initial ejected field is a flux rope or a more complex twisted configuration, depending on the initial magnetic geometry. Subsequent interactions with the surrounding field as the CME travels into the heliosphere can alter this structure significantly, particularly if a fast CME merges with a slower one (so-called *cannibalism*; Gopalswamy et al. 2001). It is unclear, however, whether the flux rope exists *before* eruption. As a result, different theories assume different initial magnetic conditions in the filament channel, and different processes to destabilize them. In all cases, however, the key challenge is to release suddenly the energy stored in the filament channel.

The eruption onset has been attributed to either a loss of equilibrium/ideal instability (e.g., Forbes and Isenberg 1991; Török and Kliem 2005; Rachmeler et al. 2009) or magnetic reconnection (e.g., Sturrock 1989; Antiochos et al. 1999; Amari et al. 2000; Moore et al. 2001; Roussev et al. 2007; Titov et al. 2008). In general, the ideal models require the presence of a twisted flux rope prior to eruption, whereas the reconnection models can operate with either a twisted flux rope or a sheared arcade (Forbes et al. 2006).

Ideal mechanisms that operate on preexisting flux ropes include kink instability (e.g., Rust and Kumar 1996; Fan and Gibson 2004; Török et al. 2004), torus instability (e.g., Titov and Démoulin 1999; Török and Kliem 2005; Manchester et al. 2008; Démoulin and Aulanier 2010; Fan 2010), and loss of equilibrium (e.g., Forbes and Priest 1995; Roussev et al. 2003; Lin et al. 2010). Flux ropes have the advantage of inherently containing significant magnetic free energy capable of driving an eruption, although coronal EUV observations and nonlinear force-free field extrapolations jointly indicate that only weak twist exists prior to most eruptions (e.g., van Ballegooijen 2004; Bobra et al. 2008). In these models, the flux rope could emerge from below the photosphere or form through cancellation (reconnection) of magnetic flux at the PIL of a coronal arcade (van Ballegooijen and Martens 1989; Amari et al. 2000; Aulanier et al. 2010). Observational evidence has been presented in support of each possibility (e.g., Feynman and Martin 1995; Wang and Sheeley 1999; Lites 2005, 2009; López Ariste et al. 2006; Okamoto et al. 2009). However, it is difficult to determine conclusively from photospheric magnetograms of CME source regions whether a coronal flux rope existed before eruption. The helical kink instability is triggered when the flux rope twist exceeds a critical value of order 1.5 (Fan and Gibson 2004). EUV images of confined eruptions (eruptions that do not produce CMEs that leave the corona) offer compelling evidence of kink-unstable flux ropes occurring in the solar atmosphere (e.g., Chintzoglou et al. 2015). For the torus instability, the system becomes unstable when the external potential field decreases sufficiently rapidly with height (Török and Kliem 2005). Regardless of the origin of the flux rope, a current sheet forms below it during the early stages of the eruption; reconnection in this sheet releases the energy that powers the flare and facilitates, but does not initiate, the CME escape. Reconnection thus plays a secondary role in these ideal models, by enabling the flux rope to grow and rise until the conditions for instability are met.

The primary reconnection-based initiation models invoke the tether cutting and breakout mechanisms. In the tether-cutting scenario, which is based primarily on interpretation of EUV, SXR, and Hα imaging data, reconnection beneath an expanding arcade creates a buoyant flux rope or adds flux to a preexisting flux rope, ultimately triggering eruption (Moore et al. 2001, 2011; Reeves et al. 2015). To the best of our knowledge, though, numerical simulations have yet to demonstrate that this process alone can account for CME onset. As originally defined, the term *tether cutting* only applies to internal reconnection that serves to initiate a CME and the accompanying flare; however, it has been more broadly

defined in the literature to include the secondary reconnection described above for the ideal CME-initiation models (e.g., Fan 2015). The same current sheet beneath the CME is involved in both definitions, but the role in CME initiation is completely different.

The breakout mechanism for CME initiation invokes reconnection in two locations to disrupt the force balance that maintains the highly sheared filament-channel field in the corona (Antiochos et al. 1999; see references in Karpen et al. 2012; Masson et al. 2013). The breakout model begins with the simplest possible multipolar magnetic field: an arcade (Antiochos et al. 1999; DeVore et al. 2005) beneath the ambient dipolar coronal field, as shown in Figure 1.15a. This configuration contains a magnetic null, where current sheets form easily when the magnetic field is stressed, for example, by slow photospheric motions. Shearing footpoint motions parallel to the PIL have been observed in the complex ARs most likely to generate fast CMEs (Deng et al. 2005; Zhang et al. 2007), and appear to be a natural consequence of Lorentz forces during flux emergence through a stratified atmosphere (Manchester et al. 2004; Fang et al. 2010; Leake et al. 2014).

According to the breakout model, the expansion of the sheared inner core pushes the overlying flux systems (Figure 1.15) together, forming the "breakout" current sheet at the null and along the separatrix between these systems (Figure 1.15b). Once this breakout sheet thins down sufficiently, the resistivity "switches on" and reconnection starts to remove overlying flux by transferring it to the side lobes (Figure 1.15b). The downward tension

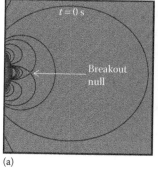 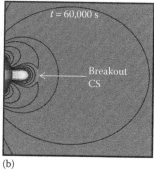 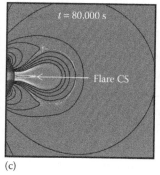 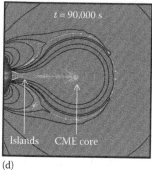

FIGURE 1.15 Illustration of the key stages and features in the breakout model for CME initiation. The images, which show current density perpendicular to the computational plane (dark grey = min., white = max.), are from a high-resolution 2.5D MHD simulation. (a) Initial conditions. (b) Before the onset of breakout reconnection. Note the development of the breakout current sheet (CS). (c) Between the onset of breakout reconnection and the onset of flare reconnection. The extensive breakout CS contains multiple islands, while the elongated flare CS contains no nulls as yet. (d) After the onset of flare reconnection. Note the multiple islands in the breakout and flare CSs, and the complex structure inside the CME. (From Guidoni, S.E. et al., *ApJ*, 820, 60, 2016 [see for a color version of this figure].)

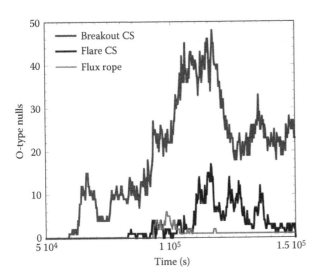

FIGURE 1.16 Number of islands (O-nulls) versus time detected in the Karpen et al. (2012) breakout simulation, separated according to location.

of the overlying field decreases, enabling the sheared field to expand faster and thus creating a feedback loop. When the flux has expanded outward sufficiently, a vertical "flare" current sheet forms in the low corona behind the bulging core (Figure 1.15c). Explosive reconnection at the flare sheet releases energy rapidly, driving the flare and CME, and relaxes the magnetic field back toward its minimum-energy potential state. Flare reconnection also adds to the partially detached CME flux rope, and accelerates the feedback loop between the breakout reconnection and the CME. Simulations of the breakout scenario with ultrahigh spatial and temporal resolution in the current sheets confirmed that the triggering mechanism in this case is resistive (Karpen et al. 2012). These simulations also show that reconnection in both current sheets forms multiple plasmoids/flux ropes (Figure 1.16), in contrast to the Petschek-type reconnection sketched in the standard model (Figure 1.12) but consistent with other high-Lundquist number reconnection studies (Loureiro et al. 2007; Bárta et al. 2008; Bhattacharjee et al. 2009; Guo et al. 2013; Ni et al. 2013). Solar observations are becoming sufficiently well-resolved to reveal dynamic structures resembling plasmoids in flare current sheets (e.g., Savage et al. 2012; Takasao et al. 2012; Kumar and Manoharan 2013; Liu et al. 2013).

1.1.4 Future Prospects

Substantial progress on understanding and predicting the solar origins of space weather will rely on advances in both observations and theory/modeling. Comprehensive, continuous, multipoint observations of the entire Sun, including the poles, would fill in many of the remaining gaps in our knowledge base—for example, the currently unobservable magnetic fields and flows near the poles, and the time-varying spectral and total irradiance throughout the solar cycle. The main observational challenge facing space weather research and prediction is our lack of measurements of the coronal magnetic field. Such measurements are essential for determining the energy buildup process, the preeruption magnetic structure, and the eruption initiation mechanism(s). New techniques (e.g., in spectropolarimetry) currently under development offer great promise for surmounting this long-standing obstacle.

With sufficiently accurate characterization of the initial conditions, existing 3D MHD models of solar eruptions could be transitioned to operational forecasting with much smaller uncertainties than current models allow. One area in which the models still need improvement is in the partitioning of energy among flows, heating, accelerated particles, and radiative losses. These advances await not only more efficient computational methods and speeds, but also greater physical insight into the relevant particle acceleration mechanisms and the kinetic-global coupling inherent in magnetic reconnection. As with dynamo modeling, the anticipated advent of exascale computing will bring breakthroughs in understanding and predicting solar eruptions.

ACKNOWLEDGMENTS

The authors thank P. Chamberlin, S. Cranmer, J. Fontenla, N. Gopalswamy, S. Guidoni, G. Kopp, W. Dean Pesnell, B. Thompson, and V. Yurchesyn for illuminating discussions and figures.

REFERENCES

Amari, T., Luciani, J. F., Mikić, Z., and Linker, J. A. 2000, *ApJ*, 529, L49.
Antiochos, S. K. 2013, *ApJ*, 772, 72.
Antiochos, S. K., DeVore, C. R., Karpen, J. T., and Mikić, Z. 2007, *ApJ*, 671, 936.
Antiochos, S. K., DeVore, C. R., and Klimchuk, J. A. 1999, *ApJ*, 510, 485.
Antiochos, S. K., Mikić, Z., Titov, V. S., Lionello, R., and Linker, J. A. 2011, *ApJ*, 731, 112.
Aulanier, G., Török, T., Démoulin, P., and DeLuca, E. E. 2010, *ApJ*, 708, 314.
Babcock, H. W. 1961, *ApJ*, 133, 572.
Babcock, H. W. and Babcock, H. D. 1955, *ApJ*, 121, 349.

Bárta, M., Karlický, M., and Zĕmlička, R. 2008, *Sol. Phys.*, 253, 173.

Bhattacharjee, A., Huang, Y.-M., Yang, H., and Rogers, B. 2009, *Phys. Plasmas*, 16, 112102, doi:10.1063/1.3264103.

Bobra, M. G., van Ballegooijen, A. A., and DeLuca, E. E. 2008, *ApJ*, 672, 1209.

Canfield, R. C. et al. 1980, in *Solar Flares*, Colorado Associated University Press: Boulder, CO, 451p.

Cargill, P. J. 1994, *ApJ*, 422, 381.

Carrington, R. C. 1859, *MNRAS*, 20, 13.

Chamberlin, P. C., Woods, T. N., Crotser, D. A., Eparvier, F. G., Hock, R.A., and Woodraska, D.L. 2009, *GRL*, 36, L05102. doi:10.1029/2008GL037145.

Charbonneau, P. 2010, *Living Rev. Solar Phys.*, 7, 3.

Charbonneau, P. 2014, *ARA&A*, 52, 251.

Chintzoglou, G., Patsourakos, S., and Vourlidas, A. 2015, *ApJ*, 809, 34.

Choudhuri, A. R., Chatterjee, P., and Jiang, J. 2007, *PRL*, 98, 131103.

Cowling, T. G. 1933, *MNRAS*, 94, 39.

Cranmer, S. R. 2009, *Living Rev. Solar Phys.*, 6, 3.

Cranmer, S. R. 2012, *Space Sci. Rev.*, 172, 145.

Crooker, N. U., Antiochos, S. K., Zhao, X., and Neugebauer, M. 2012, *JGR*, 117, A04104

Crooker, N. U., Huang, C.-L., Lamassa, S. M., Larson, D. E., Kahler, S. W., and Spence, H. E. 2004, *JGR*, 109, A03107.

Dahlburg, R. B., Einaudi, G., Rappazzo, A. F., and Velli, M. 2012, *A&A*, 544, L20.

Del Zanna, G. and Mason, H. E. 2003, *A&A*, 406, 1089.

De Moortel, I. and Browning, P. 2015, *Phil. Trans. Roy. Astr. Soc. A*, 373, 20140269.

Démoulin, P. and Aulanier, G. 2010, *ApJ*, 718, 1388.

Deng, N., Liu, C., Yang, G., Wang, H., and Decker, C. 2005, *ApJ*, 623, 1195.

Dere, K. P., Landi, E., Mason, H. E., Monsignori Fossi, B. C., and Young, P. R. 1997, *A&AS*, 125, 149.

DeVore, C. R. and Antiochos, S. K. 2000, *ApJ*, 539,954.

DeVore, C. R. and Antiochos, S. K. 2005, *ApJ*, 638, 1031.

Dikpati, M. and Charbonneau, P. 1999, *ApJ*, 518, 508.

Dikpati, M., de Toma, G., and Gilman, P. A. 2006, *GeoRL*, 33, L05102.

Domingo, V. et al. 2009, *Space Sci. Rev.*, 145, 337.

Emslie, A. G. et al. 2004, *JGR*, 109, 10104.

Fan, Y. 2010, *ApJ*, 719, 728.

Fan, Y. 2015, in *Solar Prominences*, Astrophysics & Space Science Library, vol. 415, Springer: Cham, Switzerland, p. 297.

Fan, Y. and Gibson, S. E. 2004, *ApJ*, 609, 1123.

Fang, F., Manchester, W. B., IV, Abbett, W. P., and van der Holst, B. 2010, *ApJ*, 714, 1649.

Feynman, J. and Martin, S. F. 1995, *JGR*, 100, 3355.

Fontenla, J. M., Balasubramaniam, K. S., and Harder, J. 2007, *ApJ*, 667, 1243.

Fontenla, J. M., Landi, E., Snow, M., and Woods, T. 2014, *Sol. Phys.*, 289, 515.

Fontenla, J. M., Stancil, P. C., and Landi, E. 2015, *ApJ*, 809, 157.

Forbes, T. G. and Isenberg, P. A. 1991, *ApJ*, 373, 294.

Forbes, T. G. and Priest, E. R. 1995, *ApJ*, 446, 377.

Forbes, T. G. et al. 2006, *Space Sci. Rev.*, 123, 251.

Foukal, P. 1971, *Sol. Phys.*, 19, 59.

Gibson, S. 2015, in *Solar Prominences*, Astrophysics & Space Science Library, vol. 415, Springer: Cham, Switzerland, p. 323.

Gopalswamy, N., Tsurutani, B., and Yan, Y. 2015, *Progr. Earth Planet. Sci.*, 2, 13.

Gopalswamy, N. et al. 2001, *ApJ*, 548, 91.

Gosling, J. T. 1993, *JGR*, 98, 18937.

Guidoni, S. E., DeVore, C. R., Karpen, J. T., and Lynch, B. J. 2016, *ApJ*, 820, 60.

Guo, L.-J., Bhattacharjee, A., and Huang, Y.-M. 2013, *ApJ*, 771, L14.

Haberreiter, M. 2011, *Sol. Phys.*, 274, 473.

Hassler, D. M. et al. 1999, *Science*, 283, 810.

Hathaway, D. H. 2015, *Living Rev. Solar Phys.*, 12, 4.

Hood, L., Schimanke, S., Spangehl, T., Bal, S., and Cubasch, U. 2013, *J. Climate*, 26, 7489.

Hoyng, P. et al. 1981, *ApJ*, 246, 155.

Karpen, J. T. 2015, in *Solar Prominences*, Astrophysics & Space Science Library, vol. 415, Springer: Cham, Switzerland, p. 237.

Karpen, J. T., Antiochos, S. K., and DeVore, C. R. 2012, *ApJ*, 760, 81.

Klimchuk, J. A. 2006, *Sol. Phys.*, 234, 41.

Klimchuk, J.A. 2015, *Phil. Trans. Roy. Astr. Soc. A*, 373, 20140256.

Knizhnik, K. J., Antiochos, S. K., and DeVore, C. R. 2015, *ApJ*, 809, 137.

Krucker, S. et al. 2008, *Astron. Astrophys. Rev.*, 16, 155.

Krucker, S. et al. 2010, *ApJ*, 714, 1108.

Kumar, P. and Manoharan, P. K. 2013, *A&A*, 553, 109.

Landi, E., Young, P. R., Dere, K. P., Del Zanna, G., and Mason, H. E. 2013, *ApJ*, 768, 94.

Lang, K. 1995, The violent Sun, in *Sun, Earth and Sky*, Cambridge University Press, Cambridge, UK.

Leake, J. E. and Linton, M. G. 2013, *ApJ*, 764, 54.

Leake, J. E., Linton, M. G., and Antiochos, S. K. 2014, *ApJ*, 787, 46.

Leighton, R. B. 1969, *ApJ*, 156, 1.

Lin, C.-H., Gallagher, P. T., and Raftery, C. L. 2010, *A&A*, 516, 44.

Linker, J. A., Lionello, R., Mikic, Z., Titov, V. S., and Antiochos, S. K. 2011, *ApJ*, 731, 110.

Lites, B. W. 2005, *ApJ*, 622, 1275.

Lites, B. W. 2009, *Space Sci. Rev.*, 144, 197.

Liu, W., Chen, Q., and Petrosian, V. 2013, *ApJ*, 767, 168.

López Ariste, A., Aulanier, G., Schmieder, B., and Sainz Dalda, A. 2006, *A&A*, 456, 725.

Loureiro, N. F., Schekochihin, A. A., and Cowley, S. C. 2007, *Phys. Plasmas*, 14, 100703.

Low, B. C. 1999, *Sol. Phys.*, 167, 217.

Luna, M. et al. 2014, *ApJ*, 785, 79.

Mackay, D. H., DeVore, C. R., and Antiochos, S. K. 2014, *ApJ*, 784, 164.

Mackay, D. H., Karpen, J. T., Ballester, J. L., Schmieder, B., and Aulanier, G., 2010, *Space Sci. Rev.*, 151, 333.

Magara, T. 2004, *ApJ*, 605, 480.

Manchester, W. B., IV, Gombosi, T., DeZeeuw, D., and Fan, Y. 2004, *ApJ*, 610, 588.

Manchester, W. B., IV et al. 2008, *ApJ*, 684, 1448.

Martin, S. F. 1998, *Sol. Phys.*, 182, 107.

Marubashi, K. et al. 2015, *Sol. Phys.*, 290, 1371.

Masson, S., Antiochos, S. K., and DeVore, C. R. 2013, *ApJ*, 771, 82.

Masuda, S., Kosugi, T., Hara, H., Tsuneta, S., and Ogawara, Y. 1994, *Nature*, 371, 495.

Matthaeus, W. H., Zank, G. P., Oughton, S., Mullan, D. J., and Dmitruk, P. 1999, *ApJ*, 523, L93.

Miesch, M. et al. 2015, *SSRv*, doi:10.1007/s11214-015-0190-7.

Moore, R. L., Sterling, A. C., Hudson, H. S., and Lemen, J. R. 2001, *ApJ*, 552, 833.

Moore, R. L. et al. 2011, *Space Sci. Rev.*, 160, 73.

Ni, L., Lin, J., and Murphy, N. A. 2013, *Phys. Plasmas*, 20, 061206.

Okamoto, T. J. et al. 2009, *ApJ*, 697, 913.

Parker, E. N. 1955, *ApJ*, 122, 293.

Parker, E. N. 1988, *ApJ*, 330, 474.

Parnell, C. E., DeForest, C. E., Hagenaar, H. J., Johnston, B. A., Lamb, D. A., and Welsch, B. T. 2009, *ApJ*, 698, 75.

Rachmeler, L. A., DeForest, C. E., and Kankelborg, C. C. 2009, *ApJ*, 693, 1431.

Rappazzo, A. F., Velli, M., Einaudi, G., and Dahlburg, R. B. 2005, *ApJ*, 633, 474.

Rappazzo, A. F., Velli, M., Einaudi, G., and Dahlburg, R. B. 2008, *ApJ*, 677, 1348.

Reeves, K., McCauley, P. I., and Tian, H. 2015, *ApJ*, 807, 7.

Roussev, I. I. et al. 2003, *ApJ*, 588, L45.

Roussev, I. I., Lugaz, N., and Sokolov, I. 2007, *ApJ*, 668, L87.

Russell, C. T., Luhmann, J. G., and Jian, L. K. 2010, *Rev. Geophys.*, 48, doi:10.1029/2009RG000316.

Rust, D. M. and Kumar, A. 1996, *ApJ*, 464, L199.

Savage, S. L., Holman, G., Reeves, K. K., Seaton, D. B., McKenzie, D. E., and Su, Y. 2012, *ApJ*, 754, 13.

Schwabe, M. 1844, *Astron. Nachr.*, 21(495), 233.

Sheeley, N. R., Jr. et al. 1997, *ApJ*, 484, 472.

Sturrock, P. A. 1989, *Sol. Phys.*, 121, 387.

Suess, S., Wang, A.-H., and Wu, S. T. 1996, *JGR*, 101, 19957.

Takasao, S., Asai, A., Isobe, H., and Shibata, K. 2012, *ApJ*, 745, L6.

Thompson, M. J., Christensen-Dalsgaard, J., Miesch, M. S., and Toomre, J. 2003, *ARA&A*, 41, 599.

Titov, V. S. and Démoulin, P. 1999, *A&A*, 351, 707.

Titov, V. S., Mikić, Z., Linker, J. A., and Lionello, R. 2008, *ApJ*, 675, 1614.

Török, T. and Kliem, B. 2005, *ApJ*, 630, L97.

Török, T., Kliem, B., and Titov, V. S. 2004, *ApJ*, 413, L27.

Tousey, R. et al. 1973, *Sol. Phys.*, 33, 265.

Tsuneta, S. 1997, *ApJ*, 483, 507.

Tu, C.-Y., Zhou, C., Marsch, E., Xia, L.-D., Zhao, L., Wang, J.-X., and Wilhelm, K. 2005, *Science*, 308, 519.

Usoskin, I. G. 2013, *Living Rev. Solar Phys.*, 10, 1.

van Ballegooijen, A. A. 2004, *ApJ*, 612, 519.

van Ballegooijen, A. A. and Martens, P. C. H. 1989, *ApJ*, 343, 971.

Wang, Y.-M. 2004, *Sol. Phys.*, 224, 21.

Wang, Y.-M. 2012, *Space Sci. Rev.*, 172, 123.

Wang, Y.-M. and Sheeley, N. R., Jr. 1999, *ApJ*, 510, 157.

Wang, Y.-M., Sheeley, N. R., Jr., and Nash, A. G. 1991, *ApJ*, 383, 431.

Woods, T. N. et al. 2005, *JGR*, 110, 1312.

Woods, T. N. et al. 2012, *Sol. Phys.*, 275, 115.

Woods, T. N., Kopp, G., and Chamberlin, P. C. 2006, *JGR*, 111, A10S14.

Yeates, A. R. and Muñoz-Jamarillo, A. 2013, *MNRAS*, 436, 3366.

Zhang, J. et al. 2007, *JGR*, 112, A10102.

Zhao, L., DeVore, C. R., Antiochos, S. K., and Zurbuchen, T. H. 2015, *ApJ*, 805, 61.

Zuccarello, F. et al. 2013, *J Space Weather Space Clim*, 3, A18.

CHAPTER 2

Solar Wind

Melvyn L. Goldstein

CONTENTS

2.1 BACKGROUND

The solar wind is the medium through which all drivers of space propagate. The "wind" is a tenuous, but highly variable and turbulent plasma. Variations in the speed and amplitude of the fluctuations in the magnetic field and charged particle velocities can arise from changes at the surface of the Sun, in the corona, where the solar wind is generated, and due to a variety of interactions between the charged particle plasma streams and plasma waves in interplanetary space. Plasma convecting out of the solar corona entrains magnetic fields and waves; the fluctuations in these parameters can be very dynamic. There are many ways that these fluctuations can perturb Earth's magnetosphere and produce dramatic space weather events. In propagating the more than 200 solar radii between the solar surface and one astronomical unit (i.e., to Earth orbit), the energetic drivers of space weather (e.g., solar flares, shock waves, and coronal mass ejections) dynamically change as they propagate outward from the Sun. Without understanding the detailed structure of the solar wind, one cannot predict accurately if, where, and to what effect, solar phenomena will modify the near-Earth

plasma environment, impacting spacecraft and, perhaps, inducing currents on the surface of the Earth that can dramatically affect power. Communication satellites, including the GPS network, can be significantly degraded by scintillation effects that change the index of refraction of the medium through which the waves propagate. The solar wind is a complex phenomenon whose detailed behavior reflects the consequences of fluid behavior, kinetic processes, and solar activity.

2.1.1 Historical Background

There were many hints throughout history that phenomena on the Sun affected Earth. Although interest in that relationship intensified in the West during the Renaissance, there were many hints much earlier. There is evidence from inscriptions on bones that Chinese observers were aware of sunspots more than 2000 years ago (see, e.g., Reference 99). In Europe, observations of sunspots became fairly common only in the seventeenth century (for a short historical description, see Reference 1). Another hint at the relationship between changes on the Sun and Earth came after about

1645 when sunspots essentially disappeared. Activity picked up again around 1710, and the period of time when sunspots were absent has been called the Maunder minimum after Edward Walter Maunder (1851–1928) [29] (see Figure 1.3). In his paper, Eddy claimed that the Maunder minimum occurred during a period when Europe experienced unusually cold weather and he conjectured that the lack of solar activity was to blame. The effect of decreased or increased solar activity on Earth is controversial and, if true, would be independent of warming arising from the addition of greenhouse gases into the atmosphere (see Reference 45 for a summary of the arguments).

By the twentieth century, the effect of solar activity on Earth's magnetic field was fairly well documented and the relationships between that activity and magnetic phenomena were beginning to be understood and documented [73]. In 1905, Maunder published a series of papers that showed clearly the relationship between sunspots and magnetic disturbances on Earth [61]. However, the details were not well-understood. In particular, there was no understanding that there was any continuous magnetic connection between the solar atmosphere and Earth. The relationship between activity on the Sun and the magnetic response of Earth became much clearer following a series of papers by Sydney Chapman, who after finishing his PhD thesis at Cambridge University, became interested in Maunder's research and what were then known as *geomagnetic storms*. His first paper on the subject [19] introduced analysis methods still in use to determine the magnetic currents and how they might be formed in the upper atmosphere. Although recognizing that particles from the Sun were entering the ionosphere, the details of his theory were not correct. In a subsequent paper, however, following a suggestion by F. A. Lindemann [53] that the streams from the Sun would be ionized, the picture of a terrestrial magnetosphere began to take shape. In subsequent papers [21,22] he and Ferraro developed the theory of magnetic storms.

Solar activity also affected the flux of cosmic rays measured on Earth [31,32]. Apparently, the material ejected from the Sun into the interplanetary medium was compressing and distorting Earth's magnetic environment and was also reducing the flux of cosmic rays. Essentially, the material and its embedded magnetic fields was sweeping the cosmic radiation away from Earth. When they occurred, the decreases in cosmic ray intensity could be quite abrupt, but the recovery was typically much slower than the recovery of the magnetic field alone, suggesting that the sweeping effect extended rather far out in heliocentric distance so that it took a finite amount of time for the cosmic ray flux on Earth to recover. In addition to those abrupt "Forbush decreases," a longer cycle of cosmic ray variation was apparent—during the periodic maxima in sunspot number and solar activity, the overall flux of cosmic rays with energies of order 1 GeV was suppressed. The flux recovered during solar minimum. This 11-year (actually 22-year) modulation in cosmic ray flux also indicated that the interstellar material, whatever its nature, extended far out from the Earth's orbit at 1 AU.

The idea that the space between the Sun and Earth might be *continuously* filled with "corpuscular radiation" had not gained any currency, although there was evidence from the scattering and polarization of zodiacal light [12,86] and the behavior of comet tails [14] suggesting that all of interplanetary space might be filled with electrons and/or protons and magnetic fields.

Observations of the solar corona in the 1950s indicated that it was much hotter than the photosphere and had a temperature of $\sim 10^6$ K. One implication of such a hot corona is that, as observed during solar eclipses, it must extend far beyond the photosphere. For an assumed *static* corona [20], the large heat conduction predicted electron densities at 1 AU of 100–1000/cm^3. Although much is wrong in that model (the typical density at 1 AU is 5/cm^3), its significance was that it made clear that the solar atmosphere had to extend into interplanetary space. Thus, whatever the mechanism by which solar magnetic activity influenced Earth's magnetosphere, it was not via some transient introduction of "corpuscular radiation" into interplanetary space.

2.1.2 Solar Wind

The fundamental problem with static models was pointed out by Parker [68–70], who noted that the requirement that the solar atmosphere be static throughout interstellar space implied that its pressure had to be some seven orders of magnitude larger than that estimated from the pressures of the interstellar magnetic field, cosmic rays, and gas. Parker concluded that a hydrostatic solution was not possible. Parker's solution to this problem was to develop a theory of an expanding corona using as a boundary condition the constraint that the pressure and density had to fall to zero at very large distances. Parker also noted that the expansion would entrain the solar magnetic field and the rotation of the Sun would then create an Archimedean spiral magnetic pattern (in the ecliptic plane) (for details of these derivations, see References 47 and 71). It was not long afterwards that in situ observations confirmed many aspects of Parker's theory (see, e.g., References 15 and 37). The instrumentation on the Mariner 2 spacecraft to Venus confirmed that the wind was continuous, consisted of a plasma with an average density of $< n > \sim 5$ cm^{-3} (at 1 AU) and had a supersonic flow speed of $< V > \sim 500$ km/sec near 1 AU. This speed was generally well above both the local sound speed and characteristic wave speed in a magnetofluid, the Alfvén speed, of ~ 40 km/sec, defined as $B / (\mu_o \rho)^{1/2}$ [3].

There was another competing model that was proposed for the solar wind, that is, the exospheric, or evaporative model [17,18]. That model failed to describe the observations (e.g., the predicted plasma flow speed at 1 AU was only ~ 10 km/s in Chamberlain's model, which is about 20 times less than what was observed). This "solar breeze" model clearly failed to account for the observed plasma properties, but, as formulated, the model had a fundamental error in the way it described the initiation of the exospheric solar atmosphere. More recent studies (see, e.g., Reference 74) have shown that it is very feasible to develop a reasonable exospheric model of the solar wind. An exospheric model has the advantage that the paradigm is much closer to the nearly collisionless kinetic state that describes the solar wind plasma above the corona than is a fluid description. Exospheric models can be used to compute heat fluxes, temperature anisotropics, and so on, all of which can now be measured in situ.

2.2 VARIATIONS WITH SOLAR ACTIVITY

The discovery and description of the solar wind in the early 1960s went a long way toward explaining the association between sunspots and solar activity, but many questions remained. Among them was the question of how particles actually entered the magnetosphere. Nor was there any clear understanding of the physical connection between sunspots, solar flares, and geomagnetic storms. Certainly, it was rather obvious that major solar flares produced "corpuscular radiation" that traveled through the interplanetary medium, probably associated with enhanced magnetic fields that affected galactic cosmic rays. The solar ejecta compressed and otherwise deformed the Earth's magnetic field and produced, other features of geomagnetic activity, such as the very common occurrence of active aurora borealis at high latitudes, were related to the arrival of the solar ejecta were unclear. The aurora are often seen rather regularly during times of moderate solar activity even in the absence of large geomagnetic storms.

The key to understanding the connection between interplanetary disturbances and geomagnetic activity became clear following the publication of Jim Dungey's seminal work [28]. In that paper, Dungey expanded on a suggestion by Hoyle [44] that magnetic neutral points and particle acceleration were intimately related. The basic idea is that when the interplanetary magnetic field has a southward component, so that it is directed opposite to the northward polarity of the Earth's magnetic field, the opposing magnetic fields will undergo a topological change during which the field reforms ("reconnects") in such a way that there is then a continuous link between the interplanetary and terrestrial magnetic fields. The reconnected field will be swept tailward as the solar wind flows past the magnetosphere. As the reconnected flux moves into the geomagnetic tail, the northern and southern lobes of the tail will be pushed toward each other by the increasing magnetic pressure. When the opposing fields in the tail are sufficiently compressed, the magnetic field will reconnect again, but this time, because the topological change is in the tail, magnetic tension in the reconnected flux will move the magnetic field and embedded plasma earthward (a "dipolarization" event), depositing energetic plasma into the northern and southern auroral zones and triggering magnetic substorms (see, e.g., Reference 2) that, among other manifestations, generate the auroral displays. The basic geometry is illustrated in Figure 4.4.

This scenario has been confirmed countless times. Spacecraft, such as the Advanced Composition Explorer (ACE) transmit interplanetary magnetic field data to Earth continuously. Once the field turns southward for some time (several minutes or more), it generally takes about an hour for auroral activity to pick up and for auroral forms to become visible in the (nighttime) auroral zone.

As sunspots wax and wane over the 22-year solar cycle, many properties of the solar wind vary in tandem with the sunspots. As shown in Figure 1.7, the variation in extreme ultraviolet flux as observed by the STEREO spacecraft over solar cycle 23 illustrates one aspect of that variation. For reviews of how the solar wind responds to the solar activity cycle, see Marsch [56] and a recent study by Zerbo and Richardson [98].

2.2.1 Solar Minimum

At solar minimum, the wind has well-developed streams and varies from flows that are fast (~400 to more than 700 km/s with a temperature of about 8×10^5 K) to slow (400 km/s with a temperature of order $1.4–1.6 \times 10^6$ K). Fast wind originates in "coronal holes" (for a review, see Reference 26), which are regions of low density plasma on the Sun. As seen during solar eclipses and using coronagraphs, the magnetic fields from coronal holes are open and extend out into the heliosphere. On the solar disk when seen in UV light, coronal holes appear dark. The plasma in these regions can flow along the open magnetic field. This flow is one of the highest speed components of the solar wind and can reach speeds in excess of 700 km/s. When seen on the solar limb, the "holes" have low density.

Coronal holes are most obvious during solar minimum when they often extend from high to low latitudes. The speed of plasma from low latitude coronal holes is fast, while the plasma from other latitudes is slower and denser. The result as observed over a solar rotation (or more) can be a repeating pattern of fast and slow wind. One of the best-studied periods of such flows was during the prime mission of Helios (see, e.g., References 83 and 84). The two Helios spacecraft were launched on December 10, 1974, and January 15, 1976, respectively, and they transmitted data until 1985 (an overview of the plasma environment can be found in Reference 81). Helios explored the heliosphere from 1 AU into 0.3 AU and data from those spacecrafts have provided the most extensive database yet for understanding the evolution of the solar wind plasma in the inner heliosphere. More detailed information will undoubtedly be forthcoming when Solar Orbiter and Solar Probe Plus are launched in about 2018.

2.2.1.1 Fast Wind

High-speed streams during solar minimum can initiate major geomagnetic storms on Earth. Sometimes this happens when the high-speed flow overtakes slow wind, compresses it, and forms (corotating) interaction regions. Such interaction regions may hit and compress the magnetosphere. As the interaction regions evolve in the ecliptic plane [46], the resultant plasma turbulence [25] and particle acceleration can produce significant space weather events. Global models are now incorporating the formation and convection of corotating interaction regions (see, e.g., References 67 and 95) from the corona out into the heliosphere. During solar minimum

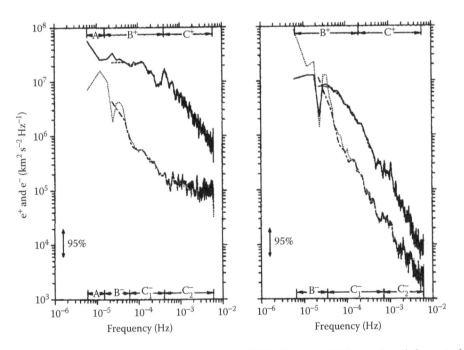

FIGURE 2.1　Power spectra of the Elsässer variables, $z^{\pm}$, (($\delta z^{\pm} \equiv \delta v \pm \delta b$)) for fast wind (left panel) and slow wind (right panel) from data obtained by Helios at 0.3 AU. (Adapted from Tu et al., *Journal of Geophysical Research*, vol. 94, pp. 11739–11759, 1989.)

in fast flows the magnetofluid turbulence in the solar wind is often highly Alfvénic, that is, the fluctuations in velocity and magnetic field are highly correlated in fast wind. That correlation is generally much smaller in slow wind, which, consequently, is less Alfvénic and more amenable to nonlinear turbulent interactions, making slow wind more highly evolved than fast wind. Those variations are described in a series of papers by Marsch and Tu [58,79,91] (also see [57]). During the solar minimum period, data from Helios, Ulysses, and the two Voyager spacecrafts were combined to give a comprehensive picture of the heliosphere and how the embedded turbulence evolves, at least out to 10 AU [34,42,43,62,79].

The high Alfvénicity of fast wind can also give rise to geomagnetic activity simply because those large amplitude Alfvénic fluctuations can drive magnetic reconnection at the nose of the magnetosphere if the magnetic orientations are appropriate [78,90].

One challenge to Parker's fluid theory of the solar wind is that, although it accounts quite well for slow solar wind, it is difficult to find parameters that can provide the fast flows that are observed. Finding solutions of the MHD equations that approximate fast wind requires extra sources of momentum and energy [72], including damping of cyclotron and Alfvén waves [11,55,59,63,65,92]. The momentum and energy carried by an outward propagating flux of Alfvén waves was included in the MHD model developed by Usmanov et al. [94] (Figure 2.2; also see Reference 64). Those solutions do approximate observations, including data from the Ulysses spacecraft when it was out of the ecliptic plane; however, other physical processes beyond heating by the dissipation of waves also must be considered. Related investigations have pointed out [77,88] that there appears to be insufficient energy in the observed waves to have both accelerated and heated the wind to the observed values, which suggests that energy input from, say nanoflares and/or other processes may also be important. (It should be pointed out that although there are observations that appear consistent with the existence of nanoflares [50], there is little evidence that such events either substantially heat the corona [also see Reference 23] or occur in the coronal holes that give rise to fast wind.)

One approach within the fluid context for achieving a solar wind model that has the correct density, speed, and temperature is to incorporate a description of the heat flux [30,33]. In the solar wind, the heat flux is primarily carried by electrons, a fact that makes its effects difficult to include accurately in single-fluid MHD models. Recently, however, heat flux models (see, e.g., References 40 and 41) have been incorporated into multifluid global solar wind models (see, e.g., Reference 95 and Figure 2.3). The figure illustrates the successes and challenges of modeling fast wind. Starting with a source magnetic dipole tilted by 10° (with respect to the solar rotation axis), the simulated (black solid line) profiles are plotted along with the daily averages of plasma and magnetic field parameters from the first fast latitude transit of Ulysses (1994–1995). The radial velocity u_r, B_r, B_{φ} match the data quite well, as does the "T-small" proton temperature. The model estimate of the electron temperature T_E, however, tends to be high.

Additional heating effects arising from dissipation of the ambient turbulence in the solar wind were also included. The relevance of including such effects is to improve the accuracy of the models when trying to forecast the probability that solar events will reach the magnetosphere. At present, since the apparent demise of the second STERO observatory, the reliability of such predictions has decreased measurably.

2.2.1.2 Slow Wind

The slow solar wind is denser and more variable than fast wind and its composition reflects that of the solar corona in contrast with the composition of fast wind which more closely resembles that of the photosphere. Slow wind also has more complex and turbulent flows. The origin of slow wind is not altogether clear, but it is associated with the boundary between fast wind from coronal holes and the Sun's "streamer belt," where closed magnetic loops merge into the plasma that is flowing outward into the heliosphere. This is the region that, in eclipse photographs, shows coronal streamers extending outward from what were closed magnetic loops. The "S-web" model [5,27,54] mentioned in Chapter 1 model, is able to account for the fact that slow solar wind has the composition of the closed-field corona and also is observed to have a large angular width of up to 60°, which indicates that the source of slow wind extends far from the open-closed boundary. A detailed analysis of the properties of slow wind in the vicinity of the heliospheric current sheet and the evolution of that boundary with heliospheric distance was carried out by Roberts et al. [80]. They also discussed the evolution of the heliospheric current sheet with time as the solar maximum of 1980 was approached (see their Figure 7 in [80]).

As mentioned above, the speed of slow wind is about 400 km/s, or less and the temperature is high, of order 1.4–1.6 × 10^6 K. During solar minimum, slow wind

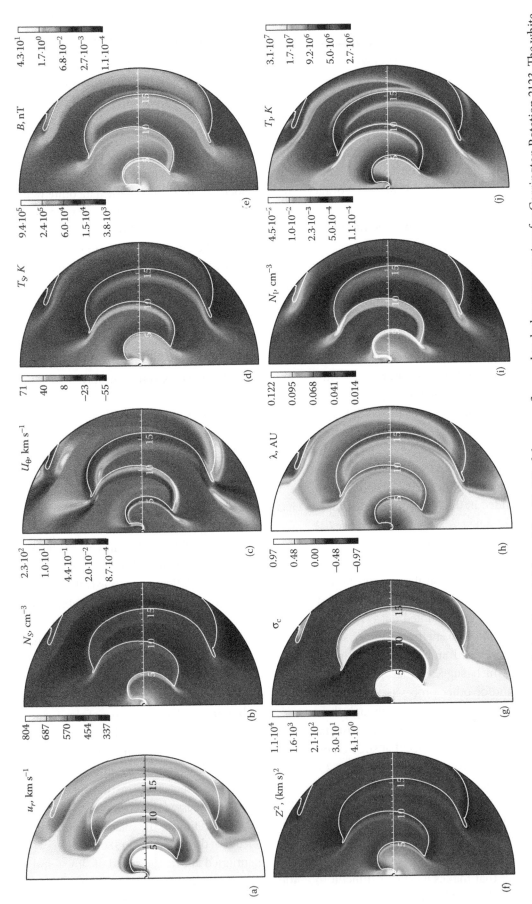

FIGURE 2.2 Contour plots in the meridional plane $\phi = 0.75°$ from 0.3 to 20 AU of the mean flow and turbulence parameters for Carrington Rotation 2123. The white lines depict the heliospheric neutral sheet ($B_r = 0$). The panels (a-e) are: (top) radial velocity, proton number density, meridional velocity, solar wind temperature and magnetic field magnitude. The panels (f-j) are: (bottom) square of the magnetic plus kinetic energy, cross helicity, correlation length, interstellar pickup proton density and pickup proton temperature. (After Usmanov et al., *The Astrophysical Journal*, vol. 788, pp. 43, 2014.)

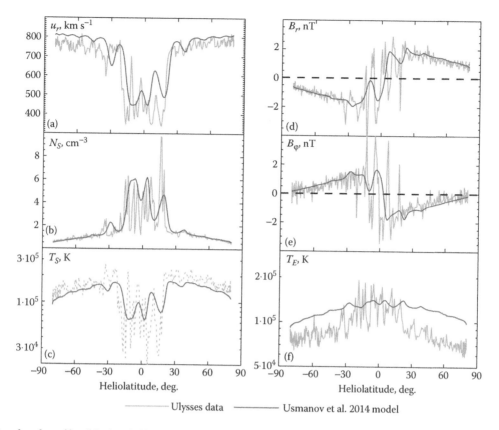

FIGURE 2.3 Simulated profiles (black solid line) from [95] for a source magnetic dipole on the Sun tilted by 10° (with respect to the solar rotation axis) versus Ulysses daily averages of plasma and magnetic field parameters measured during the first fast latitude transit of Ulysses in 1994–1995. The parameters shown are the radial velocity u_r (a), the number density of solar protons N_S (b) and their temperature T_S (c), the radial B_r (d) and azimuthal B_φ (e) magnetic field, and the electron temperature T_E (f). "T—small," of the proton temperature measured by Ulysses are shown by a gray solid lines.

generally comes from latitudes between 30° and 35° around the equator. That equatorial region expands toward the poles as one approaches solar maximum— this has also been described as an increasing tilt of the magnetic dipole axis of the Sun. Data from the HINODE satellite has greatly enhanced our understanding of where and how slow wind emanates from the Sun [7,38,39,60]. It is now known that plasma flowing from active regions on the Sun does have the same composition as that measured in the slow solar wind, at least for some active regions. However, how the complex interactions of active regions and coronal holes and other factors contribute to slow solar wind is not fully understood [39], although recently further progress has been made that relates periodic pressure density structures that are observed in the solar wind to observations using the Sun Earth Connection Coronal and Heliospheric Investigation (SECCHI) suite on the STEREO spacecraft. The periodicities suggest that the density structures might originate from magnetic reconnection at

the top of coronal streamers and, thus, may be tracers of the origin of slow wind [48,96].

As mentioned above, the turbulence that characterizes the solar wind varies considerably between fast flows and slow flows. In contrast to fast wind, which is highly Alfvénic in nature [13], the correlation between velocity fluctuations and magnetic field fluctuations is generally small in slow wind, which makes the nonlinear terms in the MHD equations very important. In fact, the power spectra of the magnetic fluctuations resemble those of fully developed fluid turbulence [51], making slow wind more highly evolved than fast wind. The works cited earlier by Marsch and Tu [58,79,91] discuss observations of slow wind as well as fast wind.

Space weather effects during periods of slow wind are much less predictable than for fast wind, and, in particular, the last period of slow wind during solar minimum was strikingly quiet in terms of space weather events. This last solar minimum, which covered the period from 2006 to 2012 was the deepest and longest ever encountered since the beginning of space observations. There was very little dramatic "space weather" during this period. In investigating the cause of this lack of activity, Kilpua et al. [49] concluded that several factors contributed to the general quiet of that time interval. For one, there was a dearth of long duration interplanetary coronal mass ejections (ICMEs), which might be expected to carry as much southward as northward-directed interplanetary magnetic field to the magnetosphere. But, in fact, there was also a clear north–south asymmetry that reduced geomagnetic activity. In addition, the solar wind density during that period was generally lower than normal and that reduced the effects of interplanetary pressure changes that buffet the magnetosphere.

2.2.2 Solar Maximum

During solar maximum, the pattern is more dynamic and the impact of solar activity on Earth's magnetic environment is more dramatic. The quasiperiodic alternation of fast and slow wind disappears, although even near solar maximum there can well be time intervals when large polar coronal holes still appear and fast wind is detected. The properties of the turbulence also vary considerably, although highly Alfvénic periods may still occur. An example is shown in Figure 2.4 of time series from Helios 1 and the associated power spectrum (Figure 2.5).

The solar magnetic field, which during solar minimum, resembles that of a tilted dipole, evolves. The effective axis of the dipole deviates more and more from being axial and tilts more toward 90°. As solar activity increases, so does the likelihood of explosive eruptive releases of energy. The Sun produces an increasing number of flares, coronal mass ejections (CMEs), magnetic clouds, and other transients (see Chapter 1), all of which can give rise to space weather phenomena on Earth. For an overview of the heliospheric magnetic field, the reader is referred to descriptions of observations by the Ulysses spacecraft as it traversed the polar regions of the heliosphere out to 5 AU [9,10]. The eruptive events originate at the boundaries between positive and negative magnetic flux, which are locations where, via magnetic reconnection, magnetic energy can be converted into energetic particles and heated plasma.

2.2.2.1 Solar Flares, CMEs, and Magnetic Clouds

The two most dramatic manifestations of a solar eruption and release of magnetic energy are solar flares and CMEs. Those phenomena are intimately related, as illustrated in Figure 1.12 [89], which shows one possible configuration of solar magnetic loops that can give rise to both CMEs and flares. Flares are associated with a rapid energy release in the solar corona that occurs when magnetic loops become unstable and the stored magnetic energy is converted into plasma heating and particle acceleration, a reconfiguration of the magnetic fields occurs. Flares also lead to the generation of intense radiation throughout the electromagnetic

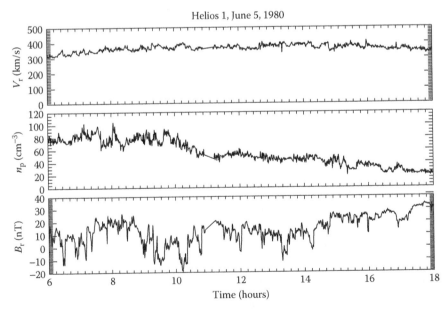

FIGURE 2.4 Time series of velocity, density, and magnetic field for a period near solar maximum in June 1980. Data were by Helios 1 at 0.3 AU.

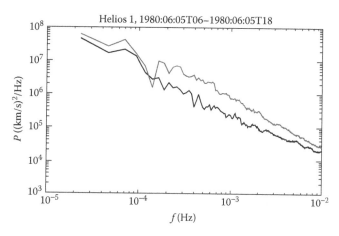

FIGURE 2.5 Power spectra of the Elsässer variables for a period near solar maximum in June 1980. Data were by Helios 1 at 0.3 AU. The power in z^m is gray, while the power in z^p is black.

spectrum including γ-radiation from the largest flares. For a comprehensive discussion of these phenomena, one should see, for example, Reference [6].

The high energy particles and radiation released by solar flares can profoundly affect Earth when they interact with spacecraft electronics. The radiation is also an ever-present danger to humans in space. Thus, the ability to predict the occurrence of flares is a high priority. However, the very high energy particles pass through the solar wind in (the radiation reaches 1 AU in 8 minutes) 30 minutes to an hour as they move along the interplanetary magnetic field—whether or not they impact Earth depends sensitively on our ability to locate the origin of the flare and its local magnetic field topology. Further complicating the task is that it is quite possible that the open field lines along which the particles propagate may not have to be exactly from the same location as the source of the flare. The basic challenge, then, becomes one of trying to predict which active regions will produce flares, when that might occur, and whether or not the interplanetary magnetic fields will connect to Earth's magnetosphere.

In many ways, the behavior of CMEs is more complex and the degree of disruption at Earth more difficult to predict. As the CME moves outward into the corona and solar wind, it is referred to as an ICME. The large-scale coronal structures that give rise to the ICME often exhibit a large-scale rotation of the embedded magnetic field and may be accompanied by smaller-scale twisted magnetic fields, or magnetic flux ropes. The large magnetic field rotations may take a day or more to pass an interplanetary observer and such structures are commonly referred to as *magnetic clouds* [16,87]. ICMEs, however, can have very complicated morphologies [76]:

there are ICMEs that have magnetic clouds embedded in them and in some ICMEs those magnetic clouds encompass the entire ICME. In contrast, other ICMEs show no large-scale magnetic field rotations. Slower ICMEs and/ or magnetic clouds tend not to generate disruptive space weather events; however, fast ones can be very disruptive.

One characteristic of magnetic clouds is that, unlike normal solar wind, the magnetic energy density of magnetic clouds exceeds the energy density of the solar wind flow. It has been noted that the ratio of magnetic energy density to flow energy density, known as the *quasiinvariant*, or QI, can be used as a diagnostic both to identify and to characterize magnetic clouds [97]. QI is defined as $(B^2 / 8\pi)/(\rho \upsilon_{SW}^2 / 2) = M_A^{-2}$ (where M_A is the magnetic Mach number, $\upsilon_{SW} / \upsilon_A$). One reason as to why magnetic clouds can cause significant space weather events is that the large rotation of the total magnetic field of the cloud is often in the meridional plane. Thus, for clouds for which that component is directed southward (cf. Figure 4.4) have the potential of causing significant magnetic activity.

The geoeffectiveness of ICMEs in general, and magnetic clouds in particular, varies considerably from event to event and from solar cycle to solar cycle. For example, cycle 23 produced many more disruptive events than did cycle 24. Since cycle 24 had about 40% fewer sunspots than cycle 23, one might infer that the general ineffectiveness of magnetic clouds during cycle 24 was a direct result of the paucity of sunspots, but, in fact, the number of magnetic clouds in the two cycles was similar. The origins of the striking difference in geomagnetic activity has been analyzed in detail by Gopalswamy et al. [35] who found that among the many differences in characteristics of magnetic clouds observed during cycles 23 and 24, the most significant were the speed of the clouds and the speeds of the accompanying shock waves, which were higher in cycle 23 than in cycle 24, as was the occurrence and strength of southward-directed magnetic fields.

The most energetic and disruptive ICME during the past two solar cycles was the so-called *Bastille Day event* which was initiated by a large and powerful solar flare on July 14, 2000. This complex event included two magnetic clouds as well as two shock waves. At the time of the eruption, there was a very large suite of well-instrumented interplanetary and Earth orbiting space taking data. Lepping et al. [52] have provided an in-depth analysis of those data, taken over a three-day interval by many experiments including those on SOHO, WIND, GEOTAIL, NEAR, ACE, and IMP-8. The energetic particles from the flare arrived on Earth within 15 min. The Bastille Day event had major impacts on the magnetosphere, causing a complex and intense magnetic storm that peaked on July 15. The plasma from the ICME compressed the bowshock and pushed in the magnetopause especially during those time intervals when the interplanetary B_z was southward. One of those compressions pushed the magnetosheath into the orbit of the GOES satellite. One of the southward turnings of B_z lasted for $\approx$ 5 h.

An even larger solar eruption with accompanying ICME occurred on July 23, 2012 [8]. That event is estimated to have been comparable to the very large 1859 Carrington event [8,24,36] that damaged telegraph lines and produced aurora that were visible as far south as 15° latitude. The Carrington event is thought to have been the most powerful magnetic storm known to have hit Earth, but some estimates of the 2012 event indicate that had it hit Earth, the magnitude of the resulting geomagnetic disruptions would have been even greater. Fortunately, the 2012 event missed Earth but hit the STEREO-A spacecraft whose experiments recorded the characteristics of the event. Analysis of data from that event indicates that had Earth been in its path, the disruptions would have been extremely large. For example, as the ICME passed STEREO-A, there was, among other things, a rapid reversal of the B_z component of the interplanetary magnetic field that led to a strongly southward field that lasted for hours [8] that almost certainly would have represented an extremely strong driver of geomagnetic activity.

The magnetic field observations at STEREO-A have been analyzed in detail by Russell et al. [82]. Baker et al. [8] discuss in detail the potential for major disruptions should events such as the July 2012 one hit Earth. It is worth noting, as they do, that the July 2012 event occurred during one of the weakest solar maxima on record—containing relatively few sunspots.

It would appear that large potentially destructive events are not all that rare. Another one that has also been compared to the 1859 Carrington event occurred on August 4, 1972 [4]. That event also resulted in large disruptions of power grids and was accompanied by reports of dramatic auroral displays, electrical surges, radio blackouts, and even effects on the navigation of birds in flight. In addition, the event caused the greatest cosmic ray storm ever observed [75].

Our ability to forecast whether or not a solar eruption will impact Earth depends strongly on the fleet of spacecraft available to monitor the Sun. The STEREO A and B spacecraft provided a unique capability, depending on the orbital configuration of the two spacecrafts, by being able to view the heliosphere from virtually all sides. When one of the STEREOs was able to view the hidden (from Earth) surface of the Sun, one could see active regions form before visible from Earth. With only one spacecraft, it is very difficult to accurately predict if an event will impact Earth [85]—the interplanetary global models are not yet adequate to the task. The effects from events that do impact the magnetosphere are also challenging to predict, although some recent models have made progress in that area [66].

ACKNOWLEDGMENTS

The author acknowledges Lynn Wilson for help with the bibliography, A. V. Usmanov for the global heliospheric simulation figures, and D. A. Roberts for many stimulating conversations and for Figures 2.4 and 2.5.

REFERENCES

1. http://galileo.rice.edu/sci/observations/sunspots.html.
2. S. I. Akasofu, *Polar and Magnetospheric Substorms*, Kluwer Academic Publishers, the Netherlands, 1968.
3. H. Alfvén, Existence of Electromagnetic-Hydrodynamic Waves, *Nature*, vol. 150, pp. 405–406, 1942.
4. C. W. Anderson, L. J. Lanzerotti, and C. G. MacLennan, Outage of the L4 System and the Geomagnetic Disturbances of 4 August 1972, *Bell System Technical Journal*, vol. 53(9), pp. 1817–1837, 1974.

5. S. K. Antiochos, Z. Mikić, V. S. Titov, R. Lionello, and J. A. Linker, A Model for the Sources of the Slow Solar Wind, *The Astrophysical Journal*, vol. 731, 112, 2011.

6. M. J. Aschwanden, *Physics of the Solar Corona. An Introduction with Problems and Solutions (2nd edition)*, Springer, New York, 2005.

7. D. Baker, L. van Driel-Gesztelyi, S. Kamio, J. L. Culhane, L. K. Harra, J. Sun, P. R. Young, and S. A. Matthews, Hinode EIS and XRT observations of hot jets in coronal holes—Does the plasma escape? In *First Results From Hinode, Astronomical Society of the Pacific Conference Series*, vol. 397, edited by S. A. Matthews, J. M. Davis, and L. K. Harra, Astronomical Society of the Pacific, San Francisco, CA, pp. 23, 2008.

8. D. N. Baker, X. Li, A. Pulkkinen, C. M. Ngwira, M. L. Mays, A. B. Galvin, and K. D. C. Simunac, A major solar eruptive event in July 2012: Defining extreme space weather scenarios, *Space Weather*, vol. 11(1), pp. 585–591, 2013.

9. A. Balogh and L. A. Fisk, The heliosphere. In *The Century of Space Science*, Kluwer Academic Publishers, the Netherlands, pp. 1141–1161, 2001.

10. A. Balogh and E. J. Smith, The heliospheric magnetic field at solar maximum: Ulysses observations, *Space Science Reviews*, vol. 97, pp. 147–160, 2001.

11. A. Barnes, P. R. Gazis, and J. L. Phillips, Constraints on solar wind acceleration mechanisms from Ulysses plasma observations: The first polar pass, *Geophysical Research Letters*, vol. 22(23), pp. 3309–3311, 1995.

12. A. Behr and H. Siedentopf, Untersuchungen über Zodiakallicht und Gegenschein nach lichtelektrischen Messungen auf dem Jungfraujoch. Mit 20 Textabbildungen, *Zeitschrift fuer Astrophysik*, vol. 32, pp. 19, 1953.

13. J. W. Belcher, L. Davis, and E. J. Smith, Large-amplitude Alfvén waves in the interplanetary medium: Mariner 5, *Journal of Geophysical Research*, vol. 74, pp. 2302–2308, 1969.

14. L. Biermann, Kometenschweife und solare Korpuskularstrahlung, *Zeitschrift fuer Astrophysik*, vol. 29, pp. 274–286, 1951.

15. A. Bonetti, H. S. Bridge, A. J. Lazarus, B. Rossi, and F. Scherb, Explorer 10 plasma measurements, *Journal of Geophysical Research*, vol. 68(13), pp. 4017–4063, 1963.

16. L. F. Burlaga, *Magnetic Clouds*, vol. 2, Springer-Verlag, Heidelberg, Germany, pp. 1, 1991.

17. J. W. Chamberlain, Interplanetary gas. II. Expansion of a model solar corona, *The Astrophysical Journal*, vol. 131, pp. 47, 1960.

18. J. W. Chamberlain, Interplanetary gas. III. A hydrodynamic model of the corona, *The Astrophysical Journal*, vol. 133, pp. 675–687, 1961.

19. S. Chapman, An outline of a theory of magnetic storms, *Proceedings of the Royal Society of London Series A*, vol. 95, pp. 61–83, 1918.

20. S. Chapman, Notes on the solar corona and the terrestrial ionosphere, *Smithsonian Contributions to Astrophysics*, vol. 2, pp. 1, 1957.

21. S. Chapman and V. C. A. Ferraro, A new theory of magnetic storms, *Terrestrial Magnetism and Atmospheric Electricity (Journal of Geophysical Research)*, vol. 36, pp. 171, 1931.

22. S. Chapman and V. C. A. Ferraro, A new theory of magnetic storms, *Terrestrial Magnetism and Atmospheric Electricity (Journal of Geophysical Research)*, vol. 36, pp. 77, 1931.

23. H. Che and M. L. Goldstein, The origin of non-maxwellian solar wind electron velocity distribution function: Connection to nanoflares in the solar corona, *The Astrophysical Journal Letters*, vol. 795, pp. L38, 2014.

24. E. W. Cliver, The 1859 space weather event: Then and now, *Advances in Space Research*, vol. 38(2), pp. 119–129, 2006.

25. J. T. Coburn, C. W. Smith, B. J. Vasquez, J. E. Stawarz, and M. A. Forman, The turbulent cascade and proton heating in the solar wind during solar minimum, *The Astrophysical Journal*, vol. 754(2), pp. 93, 2012.

26. S. R. Cranmer, Coronal holes, *Living Reviews in Solar Physics*, vol. 6, pp. 3, 2009.

27. N. U. Crooker, S. K. Antiochos, X. Zhao, and M. Neugebauer, Global network of slow solar wind, *Journal of Geophysical Research*, vol. 117(A), pp. A04, 104, 2012.

28. J. W. Dungey, Interplanetary magnetic field and the auroral zones, *Physical Review Letters*, vol. 6(2), pp. 47–48, 1961.

29. J. A. Eddy, The maunder minimum, *Science*, vol. 192(4245), pp. 1189–1202, 1976.

30. W. C. Feldman, J. R. Asbridge, S. J. Bame, M. D. Montgomery, and S. P. Gary, Solar wind electrons, *Journal of Geophysical Research*, vol. 80(31), pp. 4181–4196, 1975.

31. S. E. Forbush, On the effects in cosmic-ray intensity observed during the recent magnetic storm, *Physical Review*, vol. 51(1), pp. 1108–1109, 1937.

32. S. E. Forbush, On cosmic-ray effects associated with magnetic storms, *Terrestrial Magnetism and Atmospheric Electricity*, vol. 43(3), pp. 203, 1938.

33. P. R. Gazis, Observations of plasma bulk parameters and the energy balance of the solar wind between 1 and 10 AU, *Journal of Geophysical Research*, vol. 89(A2), pp. 775–785, 1984.

34. B. E. Goldstein, E. J. Smith, A. Balogh, T. S. Horbury, M. L. Goldstein, and D. A. Roberts, Properties of magnetohydrodynamic turbulence in the solar wind as observed by Ulysses at high heliographic latitudes, *Geophysical Research Letters*, vol. 22(2), pp. 3393–3396, 1995.

35. N. Gopalswamy, S. Yashiro, H. Xie, S. Akiyama, and P. Mäkelä, Properties and geo-effectiveness of magnetic clouds during solar cycles 23 and 24, *Journal of Geophysical Research Space Physics*, vol. 120(11), pp. 9221–9245, 2015.

36. J. L. Green and S. Boardsen, Duration and extent of the great auroral storm of 1859, *Advances in Space Research*, vol. 38(2), pp. 130–135, 2006.

37. K. I. Gringauz, Some results of experiments in interplanetary space by means of charged particle traps on soviet space probes, *Space Research,* vol. 2, pp. 539–553, 1961.

38. L. K. Harra, A new view of the Sun from the hinode space mission, *International Journal of Modern Physics D,* vol. 17, pp. 693–723, 2008.

39. L. K. Harra, The role of coronal hole and active region boundaries in solar wind formation. In *4th Hinode Science Meeting: Unsolved Problems and Recent Insights, Astronomical Society of the Pacific Conference Series,* vol. 455, edited by L. Bellot Rubio, F. Reale, and M. Carlsson, Astronomical Society of the Pacific, San Francisco, CA, pp. 315, 2012.

40. J. V. Hollweg, On electron heat conduction in the solar wind, *Journal of Geophysical Research,* vol. 79(25), pp. 3845–3850, 1974.

41. J. V. Hollweg, Collisionless electron heat conduction in the solar wind, *Journal of Geophysical Research,* vol. 81(10), pp. 1649–1658, 1976.

42. T. S. Horbury and A. Balogh, Evolution of magnetic field fluctuations in high-speed solar wind streams: Ulysses and Helios observations, *Journal of Geophysical Research,* vol. 106(A8), pp. 15929–15940, 2001.

43. T. S. Horbury, A. Balogh, R. J. Forsyth, and E. J. Smith, Ulysses magnetic field observations of fluctuations within polar coronal flows, *Annals of Geophysics,* vol. 13(1), pp. 105–107, 1995.

44. F. Hoyle, *Some Recent Researches in Solar Physics,* Cambridge University Press, Cambridge, 2014.

45. D. V. Hoyt and K. H. Schatten, *The Role of the Sun in Climate Change,* Oxford University Press, New York, 1997.

46. Y. Q. Hu, Evolution of corotating stream structures in the heliospheric equatorial plane, *Journal of Geophysical Research,* vol. 98, pp. 13201–13214, 1993.

47. A. J. Hundhausen, *Solar Wind and Coronal Expansion,* Springer-Verlag, Berlin, Germany, 1972.

48. L. Kepko, N. M. Viall, J. Kasper, and S. Lepri, Using the fingerprints of solar magnetic reconnection to identify the elemental building blocks of the slow solar wind. in *AAS/AGU Triennial Earth-Sun Summit, AAS/AGU Triennial Earth-Sun Summit,* vol. 1, pp. 10802, 2015.

49. E. K. J. Kilpua, J. G. Luhmann, L. K. Jian, C. T. Russell, and Y. Li, Why have geomagnetic storms been so weak during the recent solar minimum and the rising phase of cycle 24? *Journal of Atmospheric and Solar-Terrestrial Physics,* vol. 107, pp. 12–19, 2014.

50. J. A. Klimchuk and S. J. Bradshaw, Are chromospheric nanoflares a primary source of coronal plasma? *The Astrophysical Journal,* vol. 791(1), pp. 60, 2014.

51. A. N. Kolmogorov, The local structure of turbulence in incompressible viscous fluid for very large Reynolds' numbers, *Comptes Rendus (Doklady) de l'Academie des Sciences de l'URSS,* vol. 30(4), pp. 301–305, 1941.

52. R. P. Lepping et al., The Bastille Day magnetic clouds and upstream shocks: Near-Earth interplanetary observations, *Solar Physics,* vol. 204(1–2), pp. 287–305, 524BU Times Cited: 26 Cited References Count:36, 2001.

53. F. A. Lindemann, Note on the theory of magnetic storms, *Philosophical Magazine,* vol. 38, pp. 669–684, 1919.

54. J. A. Linker, R. Lionello, Z. Mikić, V. S. Titov, and S. K. Antiochos, The evolution of open magentic flux driven by photospheric dynamics, *Astrophysical Journal,* vol. 731, 110, 2011.

55. E. Marsch, Theoretical models for the solar wind, *Advances in Space Research,* vol. 14(4), pp. 103–121, 1994.

56. E. Marsch, Solar wind responses to the solar activity cycle, *Advances in Space Research,* vol. 38, pp. 921–930, 2006.

57. E. Marsch, Kinetic physics of the solar corona and solar wind, *Living Reviews in Solar Physics,* vol. 3, 1, 2006.

58. E. Marsch and C. Y. Tu, On the radial evolution of MHD turbulence in the inner heliosphere, *Journal of Geophysical Research,* vol. 95, pp. 8211–8229, 1990.

59. E. Marsch and C. Y. Tu, Heating and acceleration of coronal ions interacting with plasma waves through cyclotron and Landau resonance, *Journal of Geophysical Research,* vol. 106(A1), pp. 227–238, 2001.

60. S. A. Matthews, J. M. Davis, and L. K. Harra (Eds.), *First Results From Hinode, Astronomical Society of the Pacific Conference Series,* Astronomical Society of the Pacific, San Francisco, CA, vol. 397, 2008.

61. E. W. Maunder, The solar origin of terrestrial magnetic disturbances, *The Astronomical Jounal,* vol. 21, pp. 101, 1905.

62. D. J. McComas et al., Solar wind observations over Ulysses' first full polar orbit, *Journal of Geophysical Research,* vol. 105(A5), pp. 10419–10433, 2000.

63. J. F. McKenzie, M. Banaszkiewicz, and W. I. Axford, Acceleration of the high speed solar wind, *Astrophysical Journal,* vol. 303, pp. L45–L48, 1995.

64. J. F. McKenzie, W. I. Axford, and M. Banaszkiewicz, The fast solar wind, *Geophysical Research Letters,* vol. 24(22), pp. 2877–2880, 1997.

65. R. H. Munro and B. V. Jackson, Physical properties of a polar coronal hole from 2 to 5 R_s, *Astrophysical Journal,* vol. 213, pp. 874–886, 1977.

66. C. M. Ngwira, A. Pulkkinen, M. M. Kuznetsova, and A. Glocer, Modeling extreme "Carrington-type" space weather events using three-dimensional global MHD simulations, *Journal of Geophysical Research: Space Physics,* vol. 119(6), pp. 4456–4474, 2014.

67. R. Oran, B. Van Der Holst, E. Landi, M. Jin, I. V. Sokolov, and T. I. Gombosi, A global wave-driven magnetohydrodynamic solar model with a unified treatment of open and closed magnetic field topologies, *The Astrophysical Journal,* vol. 778(2), pp. 176, 2013.

68. E. N. Parker, Dynamics of the interplanetary gas and magnetic fields, *The Astrophysical Journal,* vol. 128, pp. 664–676, 1958.

69. E. N. Parker, Extension of the solar corona into interplanetary space, *Journal of Geophysical Research,* vol. 64(11), pp. 1675–1681, 1959.

70. E. N. Parker, The hydrodynamic theory of solar corpuscular radiation and stellar winds, *The Astrophysical Journal,* vol. 132, pp. 821–866, 1960.

71. E. N. Parker, *Interplanetary Dynamical Processes*, Interscience Publishers, New York–London, 1963.
72. E. N. Parker, Heating solar coronal holes, *The Astrophysical Journal*, vol. 372, pp. 719–727, 1991.
73. E. N. Parker, A history of early work on the heliospheric magnetic field, *Journal of Geophysical Research*, vol. 106(A8), pp. 15797–15801, 2001.
74. V. Pierrard, K. Issautier, N. Meyer-Vernet, and J. Lemaire, Collisionless model of the solar wind in a spiral magnetic field, *Geophysical Research Letters*, vol. 28(2), pp. 223–226, 2001.
75. M. A. Pomerantz and S. P. Duggal, Physical sciences: Record-breaking cosmic ray storm stemming from solar activity in August 1972, *Nature*, vol. 241(5388), pp. 331–333, 1973.
76. I. G. Richardson and H. V. Cane, Near-earth interplanetary coronal mass ejections during solar cycle 23 (1996–2009): Catalog and summary of properties, *Solar Physics*, vol. 264(1), pp. 189–237, 2010.
77. D. A. Roberts, Interplanetary observational constraints on Alfvén wave acceleration of the solar wind, *Journal of Geophysical Research*, vol. 94(A6), pp. 6899–6905, 1989.
78. D. A. Roberts and M. L. Goldstein, Do interplanetary Alfven waves cause auroral activity? *Journal of Geophysical Research*, vol. 95, pp. 4327–4331, 1990.
79. D. A. Roberts, M. L. Goldstein, L. W. Klein, and W. H. Matthaeus, Origin and evolution of fluctuations in the solar wind: Helios observations and Helios-Voyager comparisons, *Journal of Geophysical Research*, vol. 92(A11), pp. 12023–12035, 1987.
80. D. A. Roberts, P. A. Keiter, and M. L. Goldstein, Origin and dynamics of the heliospheric streamer belt and current sheet, *Journal of Geophysical Research*, vol. 110(A6), 2005.
81. H. Rosenbauer et al., A survey on initial results of the Helios plasma experiment, *Journal of Geophysics*, vol. 42, pp. 561–580, 1977.
82. C. T. Russell et al., The very unusual interplanetary coronal mass ejection of 2012 July 23: A blast wave mediated by solar energetic particles, *The Astrophysical Journal*, vol. 770, 38, 2013.
83. R. Schwenn and E. Marsch, *Physics of the Inner Heliosphere I. Large-Scale Phenomena*, Springer-Verlag, New York, 1990.
84. R. Schwenn and E. Marsch, *Physics of the Inner Heliosphere II. Particles, Waves and Turbulence*, Springer-Verlag, New York, 1991.
85. D. G. Sibeck and J. D. Richardson, Toward forecasting space weather in the heliosphere, *Journal of Geophysical Research*, vol. 102(A7), pp. 14721–14733, 1997.
86. H. Siedentopf, A. Behr, and H. Elsässer, Photoelectric observations of the zodiacal light, *Nature*, vol. 171, pp. 1066–1067, 1953.
87. R. M. Skoug, W. C. Feldman, J. T. Gosling, D. J. Mccomas, and C. W. Smith, Solar wind electron characteristics inside and outside coronal mass ejections, *Journal of Geophysical Research*, vol. 105(A), pp. 23069–23084, 2000.
88. E. J. Smith, M. Neugebauer, A. Balogh, S. J. Bame, R. P. Lepping, and B. T. Tsurutani, Ulysses observations of latitude gradients in the heliospheric magnetic field: Radial component and variances, *Space Science Reviews*, vol. 72(1–2), pp. 165–170, 1995.
89. S. Tsuneta, Moving plasmoid and formation of the neutral sheet in a solar flare, *The Astrophysical Journal*, vol. 483(1), pp. 507–514, 1997.
90. B. T. Tsurutani and W. D. Gonzalez, The cause of high-intensity long-duration continuous AE activity (HILDCAAS)—Interplanetary Alfven wave trains, *Planetary and Space Science*, vol. 35, pp. 405, 1987.
91. C. Y. Tu and E. Marsch, MHD structures, waves and turbulence in the solar wind observations and theories, *Space Science Reviews*, vol. 73, pp. 1–200, 1995.
92. C. Y. Tu and E. Marsch, On cyclotron wave heating and acceleration of solar wind ions in the outer corona, *Journal of Geophysical Research*, vol. 106(A5), pp. 8233–8252, 2001.
93. C. Y. Tu, E. Marsch, and K. M. Thieme, Basic properties of solar wind MHD turbulence near 0.3 AU analyzed by means of Elsasser variables, *Journal of Geophysical Research*, vol. 94(A9), pp. 11739–11759, 1989.
94. A. V. Usmanov, M. L. Goldstein, B. P. Besser, and J. M. Fritzer, A global MHD solar wind model with WKB Alfvén waves: Comparison with Ulysses data, *Journal of Geophysical Research*, vol. 105(A6), pp. 12675–12695, 2000.
95. A. V. Usmanov, M. L. Goldstein, and W. H. Matthaeus, Three-fluid, three-dimensional magnetohydrodynamic solar wind model with eddy viscosity and turbulent resistivity, *The Astrophysical Journal*, vol. 788(1), pp. 43, 2014.
96. N. M. Viall and A. Vourlidas, Periodic density structures and the origin of the slow solar wind, *The Astrophysical Journal*, vol. 807, 176, 2015.
97. A. Webb, J. Fainberg, and V. Osherovich, Solar wind quasi-invariant for slow and fast magnetic clouds, *Solar Physics*, vol. 277(2), pp. 375–388, 2012.
98. J. L. Zerbo and J. D. Richardson, The solar wind during current and past solar minima and maxima, *Journal of Geophysical Research: Space Physics*, doi: 10.1002/2015JA021407.
99. X. Zhentao, Solar observations in ancient China and solar variability, *Philosophical Transactions of the Royal Society of London. Series A, Mathematical and Physical Sciences*, vol. 330(1615), pp. 513–515, 1990.

CHAPTER 3

Earth's Magnetic Field

Nils Olsen

CONTENTS

3.1 INTRODUCTION

The Earth has a strong magnetic field that varies both in space and time. Its major part is produced by a self-sustaining dynamo operating in the fluid outer core (at depths greater than 2900 km). But what is measured on ground or in space is the sum of this core field and of fields caused by magnetized rocks in the Earth's crust, by electric currents flowing in the ionosphere, magnetosphere, and oceans, and by currents induced in the Earth by the time-varying external fields. The separation of these various field contributions and the determination of their spatial and temporal structure based on observations of the magnetic field requires advanced modelling techniques (see, e.g., Hulot et al., 2015, for an overview).

"Space weather" concerns the rather rapid variations of the magnetic field due to electric currents in the Earth's ionosphere and magnetosphere (including secondary fields due to electromagnetic induction in the Earth's interior). However, the processes that result in these "external" sources (in the ionosphere and magnetosphere, i.e., external to Earth's surface) depend heavily on the Earth's "internal" field (primarily the sources in the core). The electrical conductivity in the ionosphere is for instance anisotropic: conductivity in the direction of the ambient magnetic field is several orders of magnitude higher than conductivity perpendicular to the field. Knowledge of Earth's own magnetic field and how it varies in space and time is therefore of fundamental importance for a physical understanding of space weather. Providing such information is the aim of this chapter.

3.2 MAGNETIC OBSERVATIONS

The strength of the magnetic induction **B**, in following for simplicity denoted as "magnetic field," varies at Earth's surface between about 25,000 nT near the equator and about 65,000 nT near the poles (1 nT = 10^{-9} T, with 1 T = 1 tesla = 1 $Vs^{-1}m^{-2}$). The magnetic vector **B** is completely described by three independent numbers (also called *magnetic elements*). **B** is typically given in a local *topocentric* (*geodetic*) coordinate system (i.e., relative to a reference ellipsoid as approximation for the geoid), shown in Figure 3.1. The magnetic elements *X*, *Y*, *Z* are the components of the field vector **B** in an orthogonal right-handed coordinate system, the axes

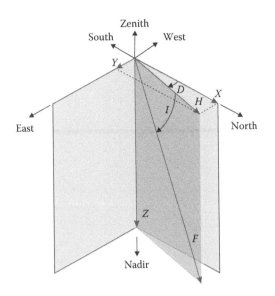

FIGURE 3.1 The magnetic elements in the local topocentric coordinate system, seen from the Northeast.

of which are pointing toward geographic North, geographic East, and vertically down. Derived magnetic elements are the angle between geographic North and the (horizontal) direction in which a compass needle is pointing, denoted as *declination D* = arctan Y/X (for the numerical calculation it is recommended to use $D = \text{atan2}(Y, X)$, to avoid the π ambiguity of the arctan function); the angle between the local horizontal plane and the field vector, denoted as *inclination I* = arctan Z/H; *horizontal intensity* $H = \sqrt{X^2 + Y^2}$; and *total intensity* $F = \sqrt{X^2 + Y^2 + Z^2}$. The latter is simply the strength of the magnetic field, also called *magnetic field intensity*.

Providing the three components of **B** in the geodetic frame, **B** = (*X, Y, Z*) with *X, Y, Z* as defined above, is useful for describing the magnetic field on the Earth's surface. However, when dealing with satellite data, it is often more appropriate to use a spherical coordinate system, also called *North-East-Center (NEC) local Cartesian coordinate frame*, with $\mathbf{B}_{\text{NEC}} = (B_N, B_e, B_C)$, where B_C is the component toward the center of the Earth (as opposed to Z which is approximating the plumb line direction), B_E is the component toward geographic East (and thus is identical to Y), and B_N is the component toward geographic North. Equations for transforming coordinates and magnetic field components between the geodetic and the geocentric frame can be found in Section 5.02.2.1.1 of Hulot et al. (2015). However, the difference between components in the geodetic and the geocentric frame is small (at Earth's surface it is below 160 nT in an

ambient field of up to 60,000 nT) and can be ignored for most space-weather-related purposes, in particular as it manifests as a static offset.

When measuring the magnetic field at ground, it is common to distinguish geomagnetic *observatories*, where the magnetic field is monitored absolutely, and *variometer stations*, where only its temporal variation is measured, which means that the absolute level (the baseline) of the magnetic field measurements is not known, and may even vary with time. Data from variometer stations are therefore mainly used for studying temporal variations of the external field at periods (between seconds and days) shorter than that of the variability of the (unknown) baseline. Studying the slow variation of Earth's magnetic core field requires, however, knowledge of the absolute level of the magnetic field, as monitored by geomagnetic observatories. Presently, the Earth's magnetic field is measured at about 150 geomagnetic observatories, the majority of which is located in the Northern Hemisphere and on continents. The filled circles of Figure 3.2 show their spatial distribution. Observatory data are provided through the INTERMAGNET network (www.intermagnet.org) and through the World Data Center (WDC) system (e.g., www.wdc.bgs.ac.uk). In contrast, the open circles of Figure 3.2 represent the location of variometer stations as given by the SuperMAG repository of magnetic variometer data (see http://supermag.uib.no/). Presently about 450 variometer stations measure fluctuations of the geomagnetic field, mainly at polar latitudes (where ionospheric and magnetospheric processes have largest impact on the Earth's magnetic field).

3.3 SOURCES OF EARTH'S MAGNETIC FIELD

Several sources contribute to the magnetic field at or above Earth's surface. An overview is presented in Figure 3.3. By far the largest part of the geomagnetic field is due to electrical currents generated via induction by fluid motions in the Earth's fluid outer core. This so-called *core field* is responsible for more than 93% of the observed magnetic field at ground.

Magnetized material in the crust (which consists of the uppermost few kilometers of Earth) causes the crustal field; it is relatively weak and accounts on an average only for a few percentage of the observed field at ground (but can exceed several thousands of nT in certain regions, for instance in the area around Kiruna in Northern Sweden where one of the world's largest

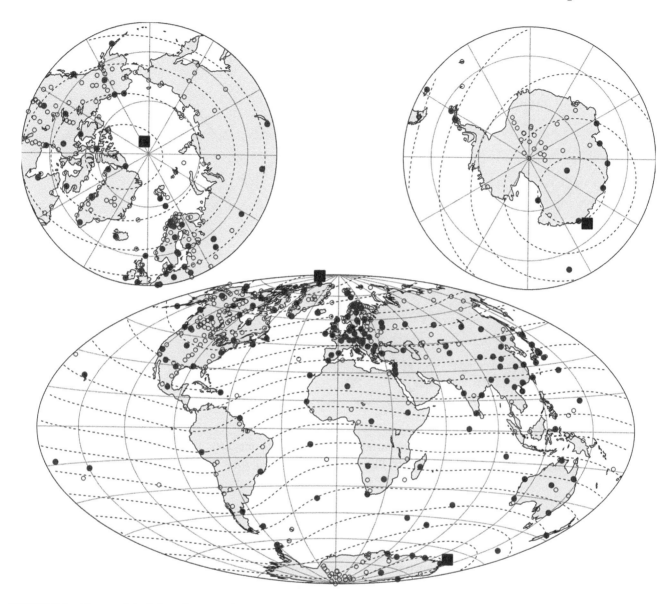

FIGURE 3.2 Geographical distribution of geomagnetic observatories (filled circles) and variometer stations (open circles). Location of the magnetic poles in 2015 is shown by the black squares. Dashed lines show magnetic latitude in steps of 10°.

known iron ore deposits exist). Core and crustal fields together make the internal field (since their sources are internal to the Earth's surface). External magnetic field contributions are caused by electric currents in the ionosphere (at altitudes between 90 and 1000 km) and magnetosphere (at altitudes of several Earth radii). Finally, electric currents induced in the Earth's crust and mantle by the time-varying fields of external origin cause magnetic field contributions that are also of internal origin; however, since they are driven by external sources, typically only the core and crustal fields are understood when one refers to "internal sources." The crustal field is static (at least on the timescales shorter than a few centuries that are interesting for space

weather applications) but the core field changes significantly with time, a process called *secular variation*.

3.4 DESCRIBING EARTH'S MAGNETIC FIELD IN SPACE AND TIME: GEOMAGNETIC FIELD MODELS

A geomagnetic field model is a set of coefficients that describes the Earth's magnetic field vector $\mathbf{B}(t, r, \theta, \phi)$ for a given time t and position (r, θ, ϕ) in spherical coordinates, with r being the distance from the Earth's center, $\theta = 90° - \delta$ the geographic colatitude (δ is geographic latitude) and ϕ the geographic longitude. The magnetic field vector $\mathbf{B} = -\nabla V$ can be derived from a scalar potential $V = V_{\text{int}} + V_{\text{ext}}$ which is the sum of a potential, V_{int},

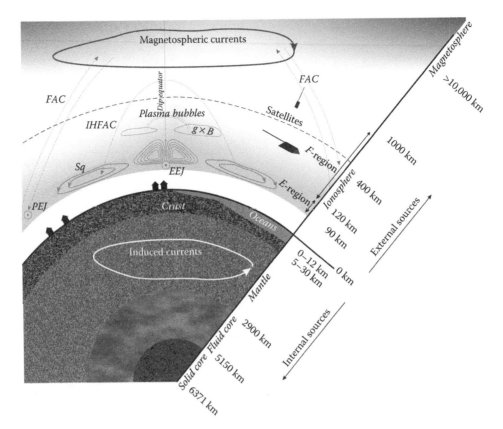

FIGURE 3.3 Sketch of the various sources contributing to the near-Earth magnetic field. FAC: Field Aligned Currents; IHFAC: Inter-hemispheric Field Aligned Currents; PEJ: Polar Electrojet; Sq: solar-driven daily variation; EEJ: Equatorial Electrojet.

describing sources internal to the Earth's surface (e.g., in the Earth's core and crust) and a potential, V_{ext}, describing sources external to the Earth's surface (e.g., due to electric currents in the Earth's ionosphere and magnetosphere). At or above Earth's surface, the magnetic field $\mathbf{B}_{int} = -\nabla V_{int}$ due to internal (core and crust) sources can be derived as the negative gradient of the scalar potential V_{int} which can be expanded in terms of spherical harmonics:

$$V^{int} = a \sum_{n=1}^{N_{int}} \sum_{m=0}^{n} (g_n^m \cos m\phi + h_n^m \sin m\phi) \left(\frac{a}{r}\right)^{n+1} P_n^m(\cos\theta) \tag{3.1}$$

where:

- $a = 6371.2$ km is a reference radius (Earth's mean radius)
- P_n^m are the associated Schmidt seminormalized Legendre functions
- $\{g_n^m, h_n^m\}$ are the Gauss coefficients describing internal sources
- N_{int} is the maximum degree and order of the spherical harmonic expansion

The expansion starts at degree $n = 1$ (representing a dipole); the zero-degree coefficient g_0^0 (which represents a monopole) vanishes because of the nonexistence of magnetic monopoles. Low-degree coefficients of the expansion (terms g_n^m, h_n^m with $n \leq 14$) represent wavelengths of the magnetic field that are dominated by the core field, while higher degrees $n > 14$ are dominated by the crustal field. Due to the core field changes with time (i.e., secular variation), the Gauss coefficients $g_n^m(t), h_n^m(t)$ for $n \leq 14$ depend on time although terms with $n > 14$ are often assumed to be static.

The magnetic field components follow from the expansion Equation 3.1 of the potential as

$$X \approx -B_\theta = +\frac{\partial V}{r\partial\theta} = +\sum_{n,m} (g_n^m \cos m\phi + h_n^m \sin m\phi) \left(\frac{a}{r}\right)^{n+2} \frac{dP_n^m}{d\theta} \tag{3.2}$$

$$Y = +B_\phi = -\frac{1}{r\sin\theta}\frac{\partial V}{\partial\phi}$$
$$= -\sum_{n,m} \left(h_n^m \cos m\phi - g_n^m \sin m\phi\right) \left(\frac{a}{r}\right)^{n+2} \frac{mP_n^m}{\sin\theta} \tag{3.3}$$

$$Z \approx -B_r = +\frac{\partial V}{\partial r}$$

$$= -\sum_{n,m}(n+1)\left(g_n^m \cos m\phi + h_n^m \sin m\phi\right)\left(\frac{a}{r}\right)^{n+2} P_n^m \quad (3.4)$$

The approximations $X \approx -B_\theta$ and $Z \approx -B_r$ become exact if the Earth's surface is assumed to be a sphere rather than an ellipsoid, that is, if the difference between geodetic and geocentric components is ignored. Note the radial dependence $\propto r^{-(n+2)}$ of the magnetic field vector (as opposed to the radial dependence $\propto r^{-(n+1)}$ of the scalar potential V, cf. Equation 3.1). Figure 3.4 shows the strength of the magnetic field, F, at ground (top) and at 6000 km altitude (bottom). In both cases, the field is about twice as strong near the poles compared to the equator, but the field strength at 6000 km altitude (which roughly corresponds to a radius of $r = 2a$) is only about $(1/2)^3 = 1/8$ of its value at Earth's surface ($r = a$). This is the expected decrease with radius for a dipole field (for which $n = 1$, i.e., $(1/2)^{n+2} = (1/2)^3 = 1/8$), thereby indicating the dominance of the dipole terms.

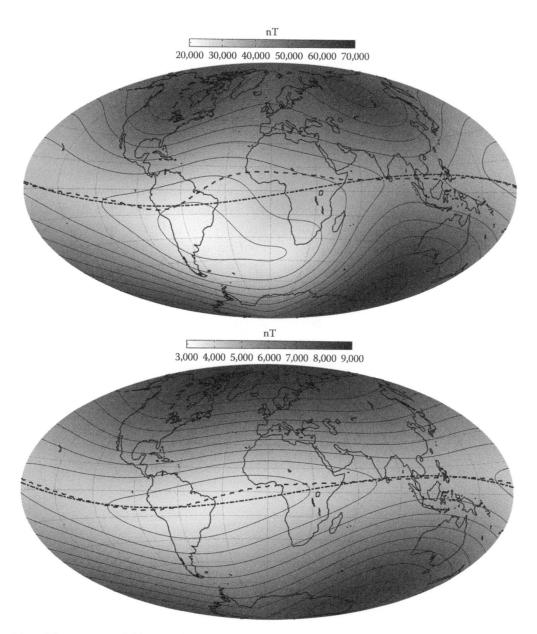

FIGURE 3.4 Map of the magnetic field strength F at ground (top), resp. at 6000 km altitude (bottom). The dashed curve represents the location of the *dip equator* (where the field lines are horizontal) at the respective altitude; the dash-dotted line shows the location of the *dipole equator* (defined as the equatorial plane of the dipole frame).

Terms with higher degrees—for example, coefficients describing a quadrupole ($n = 2$) or octupole ($n = 3$)—attenuate more rapidly with radius. As a consequence, the Earth's magnetic field becomes increasing dipolar with increasing altitude. As an example: 93% of Earth's magnetic field energy at ground is described by the dipole (higher degrees account for the remaining 7%), but at 6000 km altitude the dipole accounts for as much as 98.7% of the total magnetic energy.

It is also of interest to consider the difference between the dipole equator (defined as the equatorial plane of the dipole frame, shown as the dash-dotted line in Figure 3.4) and the dip equator (which is the line where the magnetic field is strictly horizontal, shown by the dashed line). Nondipole contributions are the cause of the difference between the dipole equator and the dip equator, and the fact that they differ more at ground (by up to 15° in latitude at longitudes of about 20° W) compared to 6000 km altitude (where the difference is less than 7° in latitude) indicates the fact that Earth's magnetic field is more dipolar at higher altitudes.

Geomagnetic spectrum. The spatial power spectrum of the geomagnetic field, often called *geomagnetic spectrum* or *Lowes–Mauersberger spectrum* (e.g., Lowes (1966) is a useful way of characterizing the spatial behavior of the Earth's magnetic field. The spectrum of the field of internal origin, R_n, is defined as the *mean square magnetic field* at a sphere of radius r (e.g., at Earth's surface, with $r = a = 6371$ km) due to core and crustal contributions with horizontal wavelength $\lambda_n \sim 2\pi r/n$ corresponding to spherical harmonic degree n. $R_n(r)$ at radius r can be determined from the Gauss coefficients g_n^m, h_n^m by means of

$$R_n(r) = \sum_{m=0}^{n} (n+1)\left[\left(g_n^m\right)^2 + \left(h_n^m\right)^2\right]\left(\frac{a}{r}\right)^{2n+4} \quad (3.5)$$

Figure 3.5 shows $R_n(r)$ at different altitudes. The spectrum on the Earth's surface decreases rapidly with increasing degree n for $n < 14$ but is "flat" (i.e., independent on degree) for higher degrees. The rather sharp "knee" at about degree $n = 14$ marks the transition from dominance of the core field contributions (at spatial scales larger than 3000 km corresponding to degrees $n < 14$) to dominance of the crustal field at smaller spatial scales ($n > 14$).

The steeper the decrease of power with increasing degree n, the more important the low-degree terms. As mentioned above, the Earth's magnetic field becomes more dipolar with altitude, a fact that is confirmed by the altitude dependence of the spectra shown in Figure 3.5.

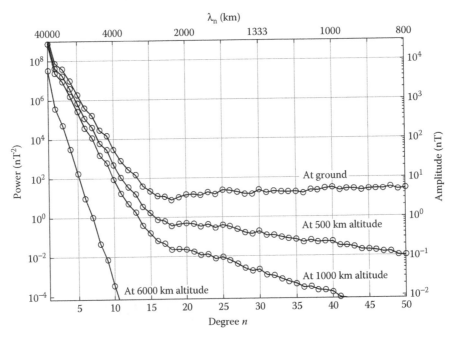

FIGURE 3.5   Power spectrum $R_n(r)$ of Earth's magnetic field in dependence on spherical harmonic degree n (bottom axis), resp. horizontal wavelength λ_n at surface (top axis). Left axis shows power (squared amplitude) while right axis shows amplitude.

3.5 DIPOLE APPROXIMATION OF EARTH'S MAGNETIC FIELD

When analysing data taken at altitudes of several thousand kilometers where nondipole contributions to the geomagnetic field become less important, it is often sufficient and convenient to approximate Earth's internal field by means of a dipole. This approximation is often in particular valid for studying magnetospheric processes.

The first coefficient, g_1^0 of the expansions of Equations 3.1 through 3.4 represents the magnetic field of a dipole at the Earth's centre that is aligned with its rotation axis; such a dipole is called an *axial dipole*. The other degree-1 coefficients g_1^1 and h_1^1 correspond to dipoles at Earth's center located in the equatorial plane and pointing toward the Greenwich meridian (in the case of g_1^1), resp. the 90° E meridian (h_1^1). The superposition of these three terms represent the magnetic field of a centered but tilted dipole. Its dipole axis intersects the Earth's surface at the *dipole poles*, also known as the *geomagnetic poles*; geocentric co-latitude θ_0 and longitude ϕ_0 of the North geomagnetic pole is thus given by

$$\theta_0 = 180° - \arccos\left(\frac{g_1^0}{m_0}\right) \qquad (3.6)$$

$$\phi_0 = -180° + \mathrm{a\,tan\,2}\left(h_1^1, g_1^1\right) \qquad (3.7)$$

where $m_0 = |\tilde{g}_1^0|$ is the dipole strength and

$$\tilde{g}_1^0 = -\sqrt{\left(g_1^0\right)^2 + \left(g_1^1\right)^2 + \left(h_1^1\right)^2} \qquad (3.8)$$

is the Gauss coefficient of an axial dipole in the dipole frame (the negative sign accounts for the fact that the magnetic pole in the Northern Hemisphere is actually a *South pole*, which is often ignored for simplicity). Using the Gauss coefficients $g_1^0 = -29442\,\mathrm{nT}$, $g_1^1 = -1501\,\mathrm{nT}$, $h_1^1 = 4797\,\mathrm{nT}$ of the most recent version of the *International Geomagnetic Reference Field* (IGRF) for epoch 2015 (IGRF-12, see Section 3.6.1 and Thebault et al., 2015b) gives as geocentric coordinates of the North geomagnetic pole: $\theta_0 = 9.63°$, $\phi_0 = -72.63°$ with $\tilde{g}_1^0 = 29868$ nT.

Dipole coordinates and components. A coordinate system with pole at (θ_0, ϕ_0) is referred to as a *dipole or geomagnetic coordinate system* (*cf.* Figure 3.6). *Dipole (or geomagnetic) co-latitude* and *longitude* are defined as

$$\theta_d = \arccos[\cos\theta\cos\theta_0 + \sin\theta\sin\theta_0\cos(\phi - \phi_0)] \quad (3.9)$$

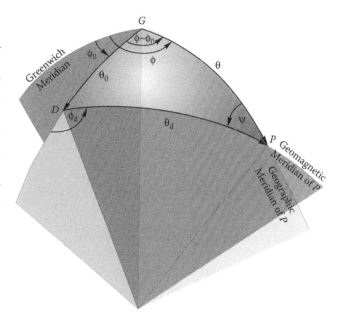

FIGURE 3.6 Relationship between spherical geocentric coordinates θ, φ, dipole (geomagnetic) coordinates θd, φd, and the spherical geocentric coordinates of the North dipole (geomagnetic) pole, θ_0, ϕ_0. G is the North geographic pole, D is the North dipole pole, and P is the location of the point in consideration, given by (θ, φ) in the geographic frame and by (θd, φd) in the geomagnetic (dipole) frame.

$$\phi_d = \mathrm{a\,tan\,2}[\sin\theta\sin(\phi - \phi_0),$$
$$\cos\theta_0\sin\theta\cos(\phi - \phi_0) - \sin\theta_0\cos\theta]. \quad (3.10)$$

The *dipole poles* are then the points for which $\theta_d = 0°$, resp. 180°. *Dipole equator* (also known as *geomagnetic equator*, shown by the dash-dotted curve in Figure 3.4) is defined as the line for which $\theta_d = 90°$.

When working in the dipole approximation it is convenient to rotate, in addition to the *coordinates* (which define the *location* of the point in consideration), also the horizontal vector magnetic field *components* from the geographic to the dipole frame. Rotation of the spherical geographic components B_r, B_θ, B_ϕ, to the dipole frame, B_r', B_θ', B_ϕ' is done by means of

$$\begin{pmatrix} B_r' \\ B_\theta' \\ B_\phi' \end{pmatrix} = \begin{pmatrix} 1 & 0 & 0 \\ 0 & +\cos\psi & +\sin\psi \\ 0 & -\sin\psi & +\cos\psi \end{pmatrix} \begin{pmatrix} B_r \\ B_\theta \\ B_\phi \end{pmatrix} \quad (3.11)$$

where

$$\psi = \mathrm{atan2}(\sin\theta_0\sin(\phi - \phi_0),$$
$$\cos\theta_0\sin\theta - \sin\theta_0\cos\theta\cos(\phi - \phi_0)) \quad (3.12)$$

is the angle between the geographic and the dipole meridian at the location in consideration (cf. Figure 3.6).

In terms of the Gauss coefficients g_1^0, g_1^1, h_1^1 of a centered dipole the magnetic vector components are given by

$$
\begin{pmatrix} X \\ Y \\ Z \end{pmatrix} \approx \begin{pmatrix} -B_\theta \\ +B_\phi \\ -B_r \end{pmatrix}
$$

$$
= \begin{pmatrix} -g_1^0 \sin\theta - \left(g_1^1 \cos\phi + h_1^1 \sin\phi\right)\cos\theta \\ h_1^1 \cos\phi - g_1^1 \sin\phi \\ -2g_1^0 \cos\theta - 2\left(g_1^1 \cos\phi + h_1^1 \sin\phi\right)\sin\theta \end{pmatrix} \left(\frac{a}{r}\right)^3 \quad (3.13)
$$

or, as components in the dipole frame,

$$
\begin{pmatrix} X' \\ Y' \\ Z' \end{pmatrix} \approx \begin{pmatrix} -B'_\theta \\ +B'_\phi \\ -B'_r \end{pmatrix} = \begin{pmatrix} -\tilde{g}_1^0 \sin\theta_d \\ 0 \\ -2\tilde{g}_1^0 \cos\theta_d \end{pmatrix} \left(\frac{a}{r}\right)^3 \quad (3.14)
$$

where, similar to Equations 3.2 through 3.4, the approximation $(X, Y, Z) = (-B_\theta, +B_\phi, -B_r)$ is exact if the ellipticity of the Earth is neglected.

3.6 GEOMAGNETIC FIELD MODELS INCLUDING THE NONDIPOLE FIELD

Describing Earth's magnetic field by an inclined dipole, as in the previous section, is valid at altitudes greater than several Earth radii. It is, however, a rather crude approximation at altitudes below a few thousands of kilometers. For those regions, more advanced descriptions of Earth's magnetic field have to be used.

3.6.1 The International Geomagnetic Reference Field

The *IGRF* is the de facto standard description of the Earth's magnetic field and its secular variation. It describes the large-scale part (spatial scales larger than 3000 km at ground) of internal contributions, a part that is also referred to as *main field*. IGRF does not describe the crustal field (which becomes dominant for spherical harmonic degrees $n > 13$ that are not included in IGRF) or external (e.g., large-scale magnetospheric) sources, nor the Earth-induced magnetic field produced by external sources despite of the fact that they are large scale and of internal origin.

The IGRF model is produced by an international team of scientists under the auspices of the *International Association of Geomagnetism and Aeronomy* (IAGA).

Due to the time changes of the Earth's core field, the IGRF has to be updated regularly, presently every fifth year. The most recent (i.e., 12th) version of the model, called IGRF-12 (Thebault et al., 2015b), is valid until 2020 and consists of definitive field models (DGRFs) for the years 1900 to 2010, a main field model for epoch 2015, and a predictive linear secular variation model for the years 2015 to 2020. The maximum spherical harmonic degree is $N_{int} = 13$ for models after 2000 (and $N_{int} = 10$ for models between 1990 and 2000), with a linear secular variation in time for coefficients g_n^m, h_n^m of degree up to $n = 8$ for 2015 to 2020.

The IGRF is used in a wide variety of studies, including space weather activities, and as a source of orientation information. Thebault et al. (2015a) give an overview about how the IGRF is used for science and applications.

Quasidipole (QD) coordinates. Earth's magnetic field affects the motion of charged particles. As a consequence, processes in the Earth's ionosphere and magnetosphere are naturally organized with respect to the geomagnetic field. Ionospheric conductivity is for instance highly anisotropic, resulting in values that are so high in the direction to the magnetic field that its field lines are practically electric equipotential lines. It is therefore often convenient to work in a coordinate frame that follows the morphology of the Earth's main magnetic field. The use of the dipole-approximation (Equations 3.9 through 3.11) is too crude an approximation at altitudes lower then a few thousand kilometers where the nondipole contributions to the geomagnetic field are important. An example coordinate frame that accounts for nondipole contributions is the *QD system of coordinates* (θ_q, ϕ_q) proposed by Richmond (1995) (see also Emmert et al., 2010). The basic idea is to trace a field line (given by IGRF of a specific epoch) from the point under consideration outward to the highest point of the field line (i.e., its apex). The longitude of that point defines *QD longitude* ϕ_q. *QD colatitude* θ_q is found by following an axial dipole field line from the Apex downward to the ionospheric *E*-layer (typically assumed to be at an altitude of 115 km); the colatitude of the intersection of the dipole field line and that altitude defines QD co-latitude θ_q. The points at the surface where the magnetic field lines are vertical are called *dip poles* or *magnetic poles* and are given by the points for which $\theta_q = 0°$, resp. $\theta_q = 180°$. The *dip equator* (or *magnetic equator*) is the line where the magnetic field lines are horizontal, that is, inclination $I = 0°$ is given by $\theta_q = 90°$. Finally, *magnetic local time* (MLT) is defined as the difference between the magnetic longitudes of the subsolar point and the point under consideration.

In contrast to the dipole frame, which is an orthogonal coordinate system, the QD coordinate system is not orthogonal and therefore mathematical differential operations (like curl, grad, and div) are nontrivial in QD coordinates. Efficient algorithms for calculating QD coordinates and their basis vectors (necessary for rotating components in the QD frame) are provided by Emmert et al. (2010).

Other magnetic coordinate systems that are based on IGRF are Altitude Adjusted Corrected Geomagnetic Coordinates (AACGM) (e.g., Gustafsson et al., 1992), Apex Coordinates (VanZandt et al., 1972), and Modified Apex Coordinates (Richmond, 1995).

3.6.2 High-Resolution Models of the Recent Geomagnetic Field

The IGRF model describes the temporal evolution of the geomagnetic field using a linear time dependence within five years. This is enough to capture slow changes in the geomagnetic field but fails to describe more rapid variations that occur on shorter timescales of less than five years. Sudden changes in the second time derivative of the Earth's magnetic field, so-called *geomagnetic jerks*, are an example of such a short timescale phenomenon that occur in irregular intervals. Prominent examples are the jerks that happened in 1969, 1978, 1991, and 1999 and, more recently, in 2003, 2007, 2011, and in 2014. These events are not captured by IGRF since their description requires a temporal description of the Gauss coefficients $g_n^m(t)$, $h_n^m(t)$ on an annual or even subannual timescales. Geomagnetic jerks presently hamper a more precise prediction of the future evolution of the geomagnetic field.

A continuous monitoring of the Earth's magnetic field from space began in 1999 with the launch of the Danish Ørsted satellite. It was followed by the German CHAMP satellite (2000–2010), the United States/Argentinian/Danish SAC-C satellite (2000–2006) and ESA's 3-satellite constellation mission Swarm (since 2013). In particular, the combination of data from all these satellite missions allows for a determination of the Earth's magnetic field with high resolution in space

and time. Prominent examples of such high-resolution models of the recent geomagnetic field are: the CHAOS series (e.g., Olsen et al., 2006, 2014; Finlay et al., 2015), the GRIMM model series (e.g., Lesur et al., 2008, 2010, 2015), and the POMMME models (e.g., Maus et al., 2005, 2006).

The core field part of the comprehensive model series (e.g., Sabaka et al., 2002, 2004, 2015) is an example of a model that describes the evolution of the Earth's magnetic field over the last 50 years (roughly covering the satellite era).

Models describing the Earth's magnetic field on even longer timescales are based on historical magnetic observations. The *gufm1* model (Jackson et al., 2000), for instance, describes how the Earth's magnetic field has changed over the past four centuries.

3.7 TIME CHANGES OF THE EARTH'S MAIN FIELD

Many space weather phenomena are affected by the geometry and strength of the Earth's magnetic field, and thus time-changes of the geomagnetic field have impact on the space weather. A prominent example is the evolution of the South Atlantic Anomaly (SAA), a region of particularly low magnetic field intensity.

The geomagnetic field is an effective shield against charged particles impinging from outer space onto the Earth. Radiation damage to spacecraft and radiation exposure to humans in space is a matter of increasing concern. On a global scale, the dipole part of the geomagnetic field (*cf.* Equation 3.8) has weakened by 7% during the last century. Weakening of the geomagnetic field is, however, much stronger in certain areas on Earth, for instance in the Southern Atlantic. Since this is the region of lowest field intensity on Earth, any further weakening is of particular concern.

In 1900, the lowest field intensity on Earth was $F = 25,460$ nT close to the East coast of Brazil. The bottom panel of Figure 3.7 shows how this point moved westward by roughly 20 km/year, with field strength decreasing to $F = 24,590$ nT in 1940 and further down to $F = 22,400$ nT in 2015. In addition to this deepening, the spatial size of the SAA also increase: The region where F is below 25,000 nT, shown in Figure 3.7, almost doubled its size every 25 years over the last 60 years, increasing from $5.6 \cdot 10^6$ km^2 in 1960 to $29 \cdot 10^6$ km^2 in 2015.

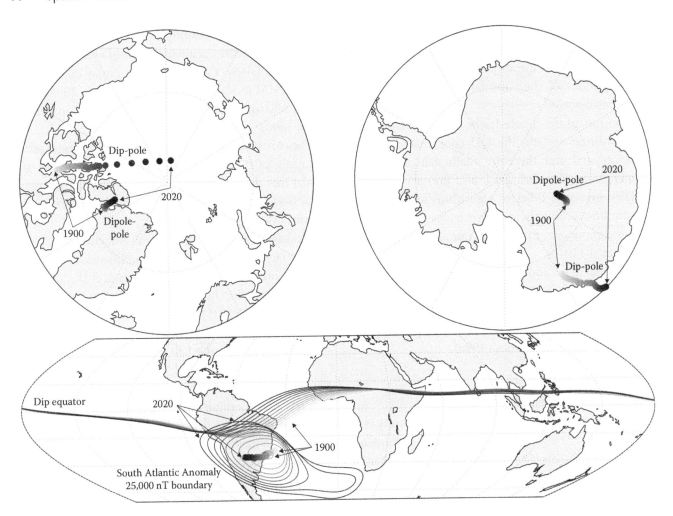

FIGURE 3.7 Top: Movement of the dipole-poles and of the dip-poles between 1900 (light gray symbols) and 2020 (black symbols). Bottom: Location of the dip equator (the line where the magnetic field is strictly horizontal), of the location of weakest field intensity, and of the 25,000 nT contour interval of field intensity. All values are given at ground level.

Another prominent feature of the geomagnetic field is the dip equator, that is, the line where the field is strictly horizontal. This is the location of the *Equatorial Electrojet*, an electric horizontal current in the dayside lower ionosphere. During the last century, the dip equator moved westward by 350 km/decade, with slightly increasing speed.

Looking at polar regions, the dipole poles (defined in Equations 3.9 and 3.10 and shown in the top panel of Figure 3.7) have moved, with a speed of about 1 km/year between 1900 and 1940, since then accelerating to 7 km/year in 2015.

The position of the dip poles (the location where **B** is vertical) moved too, although much faster, with a typical speed between 10 and 20 km/year. Since 1980, the movement of the Northern dip pole accelerated and topped around 2005 with a speed of 55 km/year. It is presently located northward of Canada, is expected to be closest to the geographic pole around 2020 and will thereafter continue its journey toward Siberia.

The magnetic field is of primary importance for the external environment of the Earth. It acts as a shield against high-energy particles from the Sun and from outer space. It controls the radiation belts, and also the trajectories of incoming cosmic ray particles. The movement of regions of low magnetic intensity and hence high radiation, such as the SAA, is a direct consequence of the changing core field. These high radiation environments cause radiation damage to spacecraft and enhanced radiation exposure to humans in space. Recent instrument failures on some low-Earth-orbiting spacecraft confirm that the SAA has shifted to the Northwest. Continuous spacecraft monitoring of the magnetic field at low Earth orbit, and the derivation of field models, plays therefore an important role in predicting radiation hazards within the space environment.

REFERENCES

Emmert, J.T., Richmond, A.D., and Drob, D.P., 2010. A computationally compact representation of magnetic-apex and quasi-dipole coordinates with smooth base vectors, *J. Geophys. Res.*, 115(A8), a08322.

Finlay, C.C., Olsen, N., and Tøffner-Clausen, L., 2015. DTU candidate field models for IGRF-12 and the CHAOS-5 geomagnetic field model, *Earth, Planets Space*, 67, 114.

Gustafsson, G., Papitashvili, N., and Papitashvili, V., 1992. A revised corrected geomagnetic coordinate system for epochs 1985 and 1990, *J. Atmos. Terr. Phys.*, 54, 1609–1631.

Hulot, G., Sabaka, T., Olsen, N., and Fournier, A., 2015. The present and future geomagnetic field, in: G. Schubert, ed., *Treatise on Geophysics,* 2nd Ed., vol. 5. Oxford: Elsevier, Chapter 02, pp. 33–78.

Jackson, A., Jonkers, A.R.T., and Walker, M.R., 2000. Four centuries of geomagnetic secular variation from historical records, *Phil. Trans. R. Soc. Lond.*, A 358, 957–990.

Lesur, V., Rother, M., Wardinski, I., Schachtschneider, R., Hamoudi, M., and Chambodut, A., 2015. Parent magnetic field models for the IGRF-12 GFZ-candidates, *Earth, Planets Space*, 67(1), 87.

Lesur, V., Wardinski, I., Hamoudi, M., and Rother, M., 2010. The second generation of the GFZ reference internal magnetic model: GRIMM-2, *Earth, Planets Space*, 62, 765–773.

Lesur, V., Wardinski, I., Rother, M., and Mandea, M., 2008. GRIMM: The GFZ reference internal magnetic model based on vector satellite and observatory data, *Geophys. J. Int.*, 173, 382–294.

Lowes, F.J., 1966. Mean-square values on sphere of spherical harmonic vector fields, *J. Geophys. Res.*, 71, 2179.

Maus, S., Lühr, H., Balasis, G., Rother, M., and Mandea, M., 2005. Introducing POMME, the POtsdam magnetic model of the Earth, in: C. Reigber, H. Lühr, P. Schwintzer, and J. Wickert, eds., *Earth Observation with CHAMP, Results from Three Years in Orbit.* Berlin, Germany: Springer-Verlag, pp. 293–298.

Maus, S., Rother, M., Stolle, C., Mai, W., Choi, S., Lühr, H., Cooke, D., and Roth, C., 2006. Third generation of the potsdam magnetic model of the Earth (POMME), *Geochem. Geophys. Geosyst.*, 7(7), Q07008.

Olsen, N., Lühr, H., Finlay, C.C., Sabaka, T.J., Michaelis, I., Rauberg, J., and Tøffner-Clausen, L., 2014. The CHAOS-4 geomagnetic field model, *Geophys. J. Int.*, 197, 815–827.

Olsen, N., Lühr, H., Sabaka, T.J., Mandea, M., Rother, M., Tøffner-Clausen, L., and Choi, S., 2006. CHAOS—A model of Earth's magnetic field derived from CHAMP, Ørsted, and SAC-C magnetic satellite data, *Geophys. J. Int.*, 166, 67–75.

Richmond, A.D., 1995. Ionospheric electrodynamics using magnetic Apex coordinates, *J. Geomagn. Geoelectr.*, 47, 191–212.

Sabaka, T.J., Olsen, N., and Langel, R.A., 2002. A comprehensive model of the quiet-time near-Earth magnetic field: Phase 3, *Geophys. J. Int.*, 151, 32–68.

Sabaka, T.J., Olsen, N., and Purucker, M.E., 2004. Extending comprehensive models of the Earth's magnetic field with Ørsted and CHAMP data, *Geophys. J. Int.*, 159, 521–547.

Sabaka, T.J., Olsen, N., Tyler, R.H., and Kuvshinov, A., 2015. CM5, a pre-Swarm comprehensive magnetic field model derived from over 12 years of CHAMP, Ørsted, SAC-C and observatory data, *Geophys. J. Int.*, 200, 1596–1626.

Thebault, E., Finlay, C., and Toh, H., 2015a. International geomagnetic reference field—The twelfth generation (special issue), *Earth, Planets Space*, 67(1), 158.

Thebault, E., Finlay, C.C., Beggan, C.D., Alken, P., Aubert, J., Barrois, O., Bertrand, F. et al., 2015b. International geomagnetic reference field: The 12th generation, *Earth, Planets Space*, 67(1), 79.

VanZandt, T.E., Clark, W.L., and Warnock, J.M., 1972. Magnetic apex coordinates: A magnetic coordinate system for the ionospheric F_2 layer, *J. Geophys. Res.*, 77, 2406.

CHAPTER 4

Solar Wind–Magnetosphere Interaction

Joseph E. Borovsky

CONTENTS

THIS CHAPTER ELABORATES ABOUT the coupling of the solar wind to the Earth's magnetosphere and discusses the multifaceted reaction of the magnetosphere–ionosphere system to that coupling. The magnetosphere's interaction with the solar winds largely controls its morphology and the locations of its plasmas. This chapter further has an appendix entitled "The Geography of the Magnetosphere and Ionosphere."

4.1 WHAT IS SOLAR WIND/ MAGNETOSPHERE COUPLING?

The movement of the solar wind plasma past the Earth activates the Earth's magnetosphere–ionosphere system; solar wind/magnetosphere coupling transfers energy, momentum, and mass from the solar wind into the system. At a simple level the solar wind coupling to the magnetosphere–ionosphere system leads to "geomagnetic activity" that is manifest by increases in the strengths of electrical currents that perturb the Earth's magnetic field. In more detail, the internal convection of the Earth's magnetosphere that is driven by the coupling to the solar wind leads to the evolution and transport of magnetospheric plasmas, to plasma outflow from the ionosphere into the magnetosphere, to the aurora, to the ring current, to the energization of the radiation belts, and to the exhaust of magnetospheric and ionospheric plasma out the dayside magnetosphere into the solar wind. Solar wind/magnetosphere coupling is central to magnetospheric physics and it is a key to understanding and predicting space weather.

Since the 1850s, it has been known that geomagnetic activity reacts to activity on the Sun (cf. Gosling, 1993). Subsequently, it was realized that there is a wind of plasma emanating from the Sun that bathes the Earth; the strongest evidence for the existence of this wind came from observations of the behavior of comet tails (e.g., Alfven, 1957). The study of solar wind/magnetosphere coupling began in earnest in the 1960s with the suggestions by Jim Dungey and by Ian Axford of two mechanisms that could enable the coupling: (1) magnetic field line reconnection between the moving solar wind plasma and the Earth's magnetosphere (Dungey, 1961) and (2) a viscous fluid-like interaction between the moving solar wind plasma and the Earth's magnetosphere (Axford and Hines, 1961). In the early 1960s, the first spacecraft measurements of the solar wind were made and scientific studies connecting the temporally changing properties of the solar wind to temporal changes

in geomagnetic activity began immediately thereafter (Heppner et al., 1963; Snyder et al., 1963).

Research on solar wind/magnetosphere coupling continues today combining theoretical analysis of physical mechanisms, statistical analysis of solar wind, magnetospheric and ground-based measurements, and data analysis of computer simulations. The two suggested mechanisms of Dungey and Axford still provide the core of our thinking about solar wind/magnetosphere coupling.

4.2 THE SUN AND THE ORIGIN OF THE SOLAR WIND

The solar wind, of course, originates from the surface of the Sun. The Earth's magnetosphere–ionosphere system responds to time variations in the properties of the solar wind hitting the Earth. In this section the different types of solar wind coming from the Sun will be overviewed, the nature of the 11-year solar cycle will be discussed, and the properties of the solar wind throughout the heliosphere and at the orbit of Earth will be examined.

Note that the physical mechanisms that accelerate plasma outward from the Sun to create the supersonic solar wind are unknown: active research on the mechanisms creating the solar wind focuses on (1) outward propagating waves and turbulence driven by motions of the solar surface (Cranmer et al., 2007; Hollweg, 2008), (2) magnetic field line reconnection in the magnetic structures above the surface (Fisk and Schwadron, 2001; Shimojo et al., 2007), (3) a hot-electron-driven interplanetary electric field (Lemaire and Scherer, 1971; Pierrard, 2012), and combinations thereof.

4.2.1 Different Types of Solar Wind

Different types of magnetic field regions on the surface of the Sun give rise to different types of solar wind and these magnetic field regions constantly evolve on a rotating Sun (Priest, 2007). The solar wind flows more-or-less radially outward from the Sun so the magnetic field regions near the solar equator are the regions that produce the solar wind that reaches the Earth (Arge et al., 2003). At speeds of 300–1000 km/s, the transit time for the solar wind plasma to travel from the solar surface to the Earth ranges from 140 h (very slow wind) to 40 h (very fast wind).

Two distinct magnetic regions on the solar surface are coronal holes and streamer belts. Coronal holes (Cranmer, 2009) are characterized by having a fraction of their magnetic field lines that are "open," with only one foot point on the Sun. These open magnetic field lines magnetically connect from the surface of the Sun out into the heliosphere and eventually into the interstellar medium. Coronal hole regions of the Sun give rise to a steady flow of high-speed, low-density, quasi-homogeneous solar wind plasma (Zirker, 1977). This is often called the "fast solar wind."

Streamer belts (Eselevich and Eselevich, 2006) are regions where the magnetic field lines are "closed" and have two foot points on the Sun. These magnetic field lines magnetically connect one region of the solar surface to another region on the solar surface. The magnetic field lines in the streamer belts form high arches above the solar surface that magnetically trapped plasma from the solar surface. There is a steady flow of slow, medium-density, inhomogeneous (lumpy) solar wind plasma that is emitted from the edges of streamer belt regions and/or from the edges of coronal holes near the streamer belt regions (Wang and Sheeley, 1990; Feldman et al., 2005). (Exactly where on the Sun this "slow solar wind" originates from is a major unsolved problem of space physics.)

A third source of solar wind is sector-reversal regions, which are boundaries in the streamer belts between regions where the magnetic field points out of the Sun adjacent to regions where the magnetic field points into the Sun. In images of the Sun sector-reversal regions show up as "streamer stalks" and far from the Sun these regions are detected as the vicinity of the heliospheric current sheet. A steady flow of very slow, very dense, very inhomogeneous plasma is emitted from sector-reversal regions (Susino et al., 2008; Suess et al., 2009).

The solar wind plasma that is emitted steadily from regions on the solar surface (i.e., coronal hole plasma, streamer belt plasma, and sector-reversal region plasma) tends to have a "Parker spiral" magnetic field orientation. Plasma that is steadily emitted radially outward from a spot on the rotating Sun will have underlying Archimedes spiral pattern to it. The spiral pattern is quite evident in the magnetic field in the solar wind plasma, since a magnetic field line must thread the stream of plasma that is emitted from the spot on the Sun where that magnetic field line connects. This Archimedes spiral magnetic field pattern is known as the Parker spiral magnetic field. In (r, θ, ϕ) spherical coordinates, the Parker spiral magnetic field **B** at the radius r from the Sun and colatitude θ in the heliosphere is given by (Parker, 1963, 1965)

$$B_r = B_o r_o^2 r^{-2} \tag{4.1}$$

$$B_\theta = 0 \tag{4.2}$$

$$B_\phi = -B_o \omega r_o^2 r^{-1} v_{sw}^{-1} \sin\theta \tag{4.3}$$

where:

B_o is a reference magnetic field strength near the Sun (with $B_o > 0$ for an outward [away] field and $B_o < 0$ for an inward [toward] field)

r_o is the reference radius where B_o is measured

$\omega \approx 2.9 \times 10^{-6}$ radians/s is the angular rotation rate of the solar surface at the solar equator

v_{sw} is the solar wind radial velocity

The colatitute angle θ is $\theta = 0°$ at the North Pole and $\theta = 180°$ at the South Pole. At Earth's orbit ($r = 1$ AU $= 1.5 \times 10^8$ km and $\theta = 90°$), the Parker spiral direction forms an angle of about 45° from radial. Nearer to the Sun, the Parker spiral direction is nearly radial and well beyond the Sun (Jupiter, Saturn); the Parker spiral direction is a ~90° from radial. In reality, the magnetic field direction in the solar wind fluctuates substantially about the Parker spiral direction, but the time-averaged magnetic field direction is very close to the Parker spiral direction.

A fourth source of solar wind is from active regions on the solar surface. The time-evolving magnetic morphology on the Sun can sometimes lead to unstable configurations of magnetic field and plasma, leading to explosive outward eruptions of magnetized plasma. When the eruptions are of large magnitude, they are known as coronal mass ejections (CMEs) (Manchester et al., 2004; Richardson and Cane, 2010). CMEs are emitted at a variety of speeds and the CME plasma can have a variety of densities and can be homogeneous or inhomogeneous. At Earth, CME plasma is characterized by a magnetic field direction that is non-Parker spiral oriented. The emission of a CME can be accompanied by a solar flare.

4.2.2 The 11-Year Solar Cycle

The large-scale magnetic field morphology of the Sun varies over the ~11-year solar cycle: consequently, the types of solar wind plasma hitting the Earth vary over the 11-year solar cycle (Zhao et al., 2009; Xu and Borovsky, 2015).

Each solar cycle is divided into four phases. The phase of the solar cycle is best characterized by a plot of the sunspot number as a function of time. The phase where the sunspot number is maximum is known as "solar maximum." Solar maximum is characterized by a predominance of active regions on the Sun, a lot of CMEs from the Sun, and a lot of CME plasma passing the Earth (Richardson and Cane, 2010). Fast CMEs tend to drive CME geomagnetic storms (see Section 4.6), which are the dominant type of geomagnetic storms occurring during solar maximum. Fast CMEs also tend to launch interplanetary shock waves and solar maximum sees a lot of interplanetary shocks passing the Earth. CMEs also lead to the production of solar energetic particles (SEPs), and solar maximum sees a lot of SEP events at Earth.

The phase of the solar cycle where the sunspot number is declining after solar maximum is known as the "declining phase." During the declining phase of the solar cycle, the magnetic morphology of the Sun is characterized by large-scale polar coronal holes that extend toward the solar equator. These holes produce at Earth long-lived patterns of fast and slow wind that repeat with the 27-day rotation period of the Sun. The intervals of fast wind from the coronal holes at Earth are called "high-speed streams"; these can produce high-speed stream-driven geomagnetic storms (also known as CIR-driven storms, see Section 4.6), which dominate the storms during the declining phase (Richardson et al., 2000). Note that late in solar maximum and early in the declining phase storms can be of a mixed type, driven by a CME followed immediately by a high-speed stream of coronal hole plasma.

The "solar minimum" phase of the solar cycle where the sunspot number is minimum is characterized by very weak (not so fast) high-speed streams and by a lot of slow wind (streamer belt plasma and sector-reversal region plasma) and generally weak geomagnetic activity.

The "ascending phase" of the solar cycle after solar minimum is characterized by the appearance of faster solar winds and of more active regions and CMEs.

4.2.3 Properties of the Solar Wind

In Figure 4.1, measurements of the solar wind velocity, the number density of the solar wind, and the magnetic field strength in the solar wind are plotted as a function of distance r from the Sun. The measurements are from

the Helios 1 and Helios 2 spacecrafts which took measurements between 0.29 and 1 AU and from the Voyager 1 and Voyager 2 spacecraft which took measurements beyond 1 AU: the Voyager 1 and Voyager 2 spacecrafts have continued beyond the realm of the solar wind and into the interstellar medium. Each black point in Figure 4.1 is the average of 1 h of measurements. In the bottom of the bottom panel, the radii of the orbits of the planets are indicated.

In the top panel, the velocity v_{sw} of the solar wind SW is plotted. Note that there is a wide range of velocities,

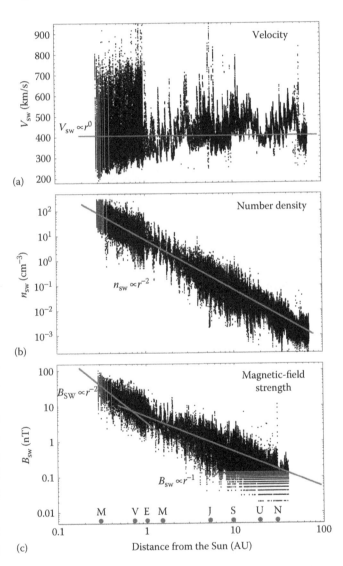

FIGURE 4.1 Using hourly averages of measurements from the spacecraft Helios 1, Helios 2, Voyager 1, and Voyager 2, the speed of the solar wind v_{sw}, the number density of the solar wind n_{sw}, and the magnetic field strength in the solar wind B_{sw} are plotted as a function of distance r from the center of the Sun. The Helios and Voyager data is available at http://cdaweb.sci.gsfc.nasa.gov.

since the properties of the solar wind are a function of time. Traveling repeatedly in and out between 0.29 and 1 AU, the two Helios spacecrafts sampled the solar wind for a long time and the distribution of points in the top panel is quite filled in; on the contrary, the two Voyager spacecrafts traveled rapidly out from 1 AU and so the variation of the velocity versus distance seen in the top panel beyond 1 AU reflects the velocity of the solar wind as a function of time. The velocity of the solar wind is roughly constant with distance from the Sun; this can be seen by the white curve drawn in the top panel, which is a constant velocity. A detailed analysis of the Helios data finds that the slower solar wind is in fact still being accelerated between 0.29 and 1 AU (Arya and Freeman, 1991).

In the second panel of Figure 4.1 the number density n_{sw} of the solar wind is plotted logarithmically. For a radially outward wind that does not change its velocity with distance r from the Sun, it is expected that the density of the wind should fall off as $1/r^2$. This is clearly seen by the white $1/r^2$ curve drawn in the second panel.

In the bottom panel of Figure 4.1 the magnetic field strength B_{sw} in the solar wind plasma is plotted. Owing to the properties of the Parker spiral, it is expected that the magnetic field strength should fall off as $1/r^2$ close to the Sun where the field is nearly radial (cf. Equation 4.1) and should fall off as $1/r$ far from the Sun where the magnetic field direction is nearly perpendicular to radial (cf. Equation 4.3). These two radial dependencies are plotted as the two white curves in the bottom panel.

4.2.4 Solar Wind at Earth

The solar wind is a supersonic plasma flow passing the Earth. It is supersonic in the sense that the solar wind bulk flow speed v_{sw} is greater than the sound speed C_s in the solar wind plasma and greater than the Alfvén speed $v_{Asw} = B/(\mu_o m_p n_{sw})^{1/2}$ in the solar wind plasma, where m_p is the proton mass and n_{sw} is the solar wind number density. The speed of sound is important because it is the velocity at which pressure changes propagate; the Alfvén speed is important because it is the velocity at which electrical current systems propagate. As shown in Figure 4.2, a standing bow shock forms in the supersonic solar wind flow upstream of the Earth's magnetosphere and the shocked (slowed, compressed, and heated) solar wind plasma behind the bow shock is called the magnetosheath (cf. Chapter 5). It is the magnetosheath (shocked

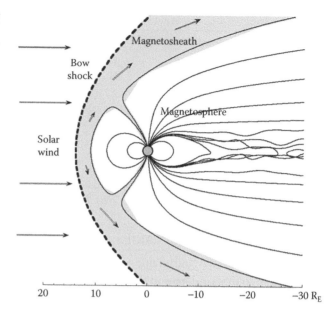

FIGURE 4.2 A sketch of the solar wind flow shocked by the bow shock as it encounters the Earth's magnetosphere. The arrows are flow vectors. The bow shock is drawn as the dashed curve and the shocked solar wind (the magnetosheath) is shaded in gray. The black lines in the magnetosphere are magnetic field lines. The magnetic field is dipolar near the Earth and it is pulled out into a very long magnetotail by the solar wind flow. The scale at the bottom of the sketch is the GSM X– axis (positive X toward the Sun) in units of Earth radii (R_E). Note that the size and shape of the magnetosphere and the position of the bow shock change as the parameters of the solar wind change.

solar wind) plasma that actually interacts directly with the magnetosphere during solar wind/magnetosphere coupling. However, the properties of the upstream solar wind determine the properties of the magnetosheath (Spreiter and Stahara, 1994), so the properties of the upstream solar wind still control the coupling.

The properties of the solar wind at Earth vary with time. Some relevant parameters of the solar wind at Earth are given in Table 4.1, which contains the hourly averages of various solar wind quantities analyzed from 1963 to 2012: the median value, the 5th percentile of the values, and the 95th percentile of the hourly values are given. The data analyzed in Table 4.1 is from the OMNI2 data set of solar wind measurements at Earth (King and Papitashvili, 2005). The OMNI2 data set spans the years 1963 to the present and is easily downloaded from http://omniweb.gsfc.nasa.gov. The data set, assembled from over 20 different spacecraft, has evolved to become a key tool with which to study solar wind/ magnetosphere interactions.

TABLE 4.1   Solar Wind Values at Earth from the 1963–2012 OMNI2 Data Set

Symbol	Median Value	5th Percentile	95th Percentile	Quantity
n_{sw}	5.3 cm⁻³	1.9 cm⁻³	17.1 cm⁻³	Solar wind plasma number density
v_{sw}	414 km/s	309 km/s	643 km/s	Solar wind bulk speed
B_{sw}	5.6 nT	2.7 nT	12.1 nT	Magnetic field strength in plasma
T_{ion}	7.0 eV	1.6 eV	26.5 eV	Plasma ion temperature
θ_{clock}	90.0°	15.6°	164.1°	IMF clock angle (relative to Earth's dipole)
M_A	8.0	4.0	15.9	Alfvén Mach number in the plasma
α/p	3.5%	0.7%	7.9%	He⁺⁺ density/H⁺ density
τ_{age}	100 h	65 h	135 h	Plasma travel time from the Sun to the Earth

A temperature of 1 eV (electron Volt) is equal to a temperature of 11,600 K.

4.3 COUPLING PRELIMINARIES

Before studying solar wind/magnetosphere coupling, it is convenient to first understand the interplanetary magnetic field (IMF) clock angle and the Russell–McPherron effect and to discuss what geomagnetic indices are.

4.3.1 IMF Clock Angle and the Russell–McPherron Effect

For solar wind/magnetosphere coupling the clock angle θ_{clock} of the solar wind magnetic field as seen by the Earth's magnetic dipole is a key parameter. In the geocentric solar magnetic (GSM) coordinate system (Hapgood, 1992; and see Chapter 3) the clock angle is given by

$$\theta_{clock} = \text{invcos}\left(\frac{B_z}{\left(B_y^2 + B_z^2\right)^{1/2}}\right) \quad (4.4)$$

where:

B_z is the dipole north–south component of the solar wind magnetic field vector **B**

B_y is the dipole dawn-dusk component of **B**

The magnetic field in the solar wind plasma is often referred to as the "IMF" and θ_{clock} is often referred to as the "IMF clock angle." The clock angle varies from 0° (purely northward IMF) to 180° (purely southward IMF).

The occurrence distribution of the hourly averaged IMF clock angle θ_{clock} is plotted in Figure 4.3 for the years 1963–2012. The solar wind magnetic field vector usually points in the Parker spiral direction (cf. Equations 4.1 through 4.3), with temporal fluctuations in the direction of the magnetic field vector that are ±45° about the Parker spiral direction. The Parker spiral direction at Earth lies approximately in the ecliptic plane, and in the GSM coordinate system the ecliptic plane wobbles with time in the vicinity of $\theta_{clock} = 90°$, owing to the temporal dawn–dusk rocking of the magnetic dipole as the Earth rotates (θ_{clock} is in the magnetic dipole coordinates, which rock with time). The properties of the solar wind plasma in Table 4.1 usually vary on slow timescales (hours and days); the direction of the solar wind magnetic field vector and hence θ_{clock} vary on much faster timescales (seconds and minutes).

As a general rule, geomagnetic activity is low when the IMF clock angle is northward and geomagnetic activity is high when the IMF clock angle is southward. This is because magnetic field line reconnection between the solar wind and the Earth's magnetosphere is sensitive to the IMF clock angle (Komar et al., 2015): the dayside reconnection rate is strong for southward clock angles ($\theta_{clock} \sim 180°$) and the reconnection rate

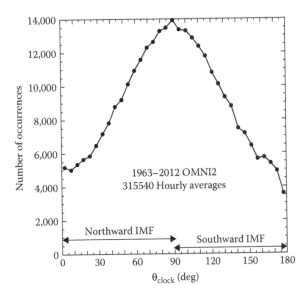

FIGURE 4.3   The IMF clock angle θ_{clock} is binned for 315,540 hourly averages from the years 1963 to 2010.

is very weak for northward clock angles ($\theta_{clock} \sim 0°$). Geomagnetic activity is stronger when the reconnection rate is stronger. The nominal clock angle of the Parker spiral-directed magnetic field in the solar wind is ~90°, however the Russell–McPherron effect (Russell and McPherron, 1973) comes into play to produce southward clock angles if the dawn–dusk tilt of the Earth's magnetic dipole is in the proper direction with respect to the Parker spiral field. The result of the Russell–McPherron effect is that geomagnetic activity tends to be strong (i.e., the solar wind magnetic field tends to be southward as seen by the Earth's dipole) in "toward" magnetic sectors during the Spring of the year and in "away" magnetic sectors during the Fall of the year. Hence, geomagnetic activity tends to be stronger during the Spring and the Fall seasons when the Russell–McPherron effect comes into play and geomagnetic activity tends to be weaker in the Summer and Winter seasons when the Russell–McPherron effect is unimportant.

4.3.2 Geomagnetic Indices

The coupling of the solar wind to the magnetosphere-ionosphere system results in "geomagnetic activity" driven by the solar wind. Geomagnetic activity is measured by a number of geomagnetic indices (Rostoker, 1972), which are measures of the strengths of geomagnetic current systems. Three often-used indices are K_p (planetary K-index), AE (auroral electrojet index), and D_{st} (disturbance storm time index). The K_p index is a gauge of the strength of magnetospheric convection (Thomsen, 2004) and is a good indicator of whether the magnetosphere is quiet (low K_p), having moderate activity (mid-range K_p), or undergoing a geomagnetic storm (high K_p). The AE index is a gauge of high-latitude currents flowing from the magnetosphere into the northern "auroral zone" of the Earth's ionosphere. The AE index is temporally volatile and reacts to surges of auroral activity and is a particularly good indicator of the occurrence, strength, and duration of substorms (Gjerloev et al., 2004), which are ~hour-long activations of the nightside magnetosphere (McPherron et al., 1973) that occur typically a few times per day. The D_{st} index is a gauge of the strength of the Earth's "ring current," a diamagnetic current in the magnetosphere associated with the distortion of the magnetic dipole by plasma pressure (Dessler and Parker, 1959). The D_{st} index is also sensitive to the Chapman–Ferraro current on the dayside magnetopause associated with the ram pressure of the solar wind (cf. Chapter 3); to get a cleaner gauge of the strength of the ring current the Chapman–Ferraro perturbation to D_{st} is removed in a pressure-corrected D_{st}^* index (Siscoe et al., 2005).

In Figure 4.4 the K_p index (lower curve, right axis) is plotted as a function of time for the first 100 days of the year 2005. Also plotted (upper curve, left axis) is the speed of the solar wind v_{sw} for those 100 days. As can be

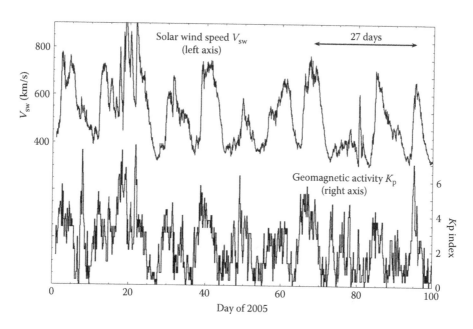

FIGURE 4.4   For the first 100 days of the year 2005, geomagnetic activity as measured by the K_p index is plotted (bottom curve, right axis) and at the same time the speed of the solar wind is measured (upper curve, left axis).

seen, the speed of the solar wind changes on the timescale of a day or so, with multiday intervals of slow wind passing the Earth and multiday intervals of fast wind passing the Earth. Comparing the two curves in Figure 4.4 a relationship can be seen wherein geomagnetic activity (as measured by K_p) tends to be high when v_{sw} is high and the activity tends to be low when v_{sw} is low. But there are clear exceptions to this trend, for example, days 73 and 78. In the upper right portion of Figure 4.4 is an arrow indicator of 27 days; many of the features in the solar wind speed repeat on 27-day (solar rotation period) intervals. Consequently, many aspects of geomagnetic activity (such as calm intervals with very low K_p or storm intervals with high K_p) repeat on a 27-day interval.

In Table 4.2, the Pearson linear correlation coefficients between various solar wind parameters and the geomagnetic indices K_p, AE, $-D_{st}$, and $-D_{st}*$ are displayed. (The linear correlation coefficient r_{corr} between two variables is the covariance divided by the standard deviations, cf. Equation 11.17 of Bevington and Robinson, 1992.) During geomagnetically quiet times the D_{st} index is around 0 nT and D_{st} goes negative as geomagnetic activity increases; hence in Table 4.2 $-D_{st}$ is used so that $-D_{st}$ goes more positive as activity increases, as is the case for K_p and AE. Note in the column labels in Table 4.2 that there are subscripts on the indices: K_{p1}, AE_1, $-D_{st2}$, and $-D_{st2}*$. The geomagnetic indices react to the solar wind, changing when the solar wind changes, but there are time lags in those reactions (e.g., Bargatze et al., 1985). These reaction time lags are associated with the timescales for current system to build up and decay owing to the electrodynamic properties of the magnetosphere–ionosphere system (Goertz et al., 1993). Typical lag times for K_p and AE are about 1 h and typical time lags for D_{st} (and $D_{st}*$) are 2 or 3 h. In cross-correlating the solar wind with the geomagnetic indices in Table 4.2, time lags of AE and K_p of 1 h were used and a time lag of 2 h on $-D_{st}$ and $-D_{st}*$ were used, hence the subscripts indicating the number of hours of time lag for each index.

In Table 4.2 the linear correlation coefficients for v_{sw}, B, B_{perp}, and θ_{clock} are all positive, meaning that geomagnetic activity as measured by the indices tends to increase as each of those four solar wind variables increase. Similarly with F10.7 (the 10.7-cm radio flux from the Sun, an indicator of solar activity). For the solar wind Alfvén Mach number M_A the correlations

TABLE 4.2   For the 1963–2012 OMNI2 Data Set

	r_{corr} with K_{p1}	r_{corr} with AE_1	r_{corr} with D_{st2}	r_{corr} with $D_{st2}*$
v_{sw}	55.4	41.3	44.0	43.9
B	54.3	41.1	39.9	52.2
B_{perp}	41.1	35.0	32.3	42.3
n_{sw}	12.8	5.0	−14.2	8.7
θ_{clock}	32.7	50.2	27.6	25.3
M_A	−21.4	−19.3	−25.4	−18.9
F10.7	15.7	12.1	20.5	15.8

Pearson linear correlation coefficients rcorr (%) between four geomagnetic indices and relevant solar wind parameters are displayed. With more than 280,000 point per correlation, correlation is significant when its magnitude is greater than 0.4%

are negative, meaning geomagnetic activity tends to be weaker when M_A is larger. The trend for the solar wind number density n_{sw} is mixed in Table 4.2: the correlation is negative with $-D_{st}$ but positive with $-D_{st}*$. This is because of the reaction of the Chapman–Ferraro currents to the solar wind density which makes D_{st} positive ($-D_{st}$ negative) when the ram pressure of the solar wind increases, which it does with n_{sw}. $-D_{st}*$ has the Chapman–Ferraro current effects removed, so it does not have the same reaction to n_{sw}.

4.4 STANDARD THINKING ABOUT THE PHYSICS OF THE COUPLING

This section discusses the status of the ideas of Dungey and Axford on the coupling of the solar wind to the magnetosphere via magnetic field line reconnection and via a viscous interaction.

4.4.1 Dayside Reconnection

It is widely accepted that the major mechanism that controls solar wind/magnetosphere coupling is magnetic field line reconnection (also called "magnetic merging") between the magnetospheric plasma and the solar wind plasma across the dayside magnetopause. Whatever controls the rate of dayside reconnection largely controls the amount of coupling and largely controls the driving of the magnetosphere–ionosphere system.

In the "Dungey cycle," a magnetospheric magnetic field line that becomes connected on the dayside to the solar wind plasma is dragged by the moving solar wind from the front of the magnetosphere over the poles and is laid down into the magnetotail. This is depicted in Figure 4.5, where the magnetic field lines of the solar wind and the magnetosphere are sketched in a noon–midnight plane as viewed from the dusk side.

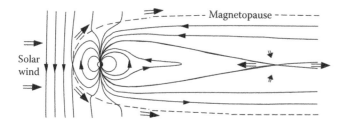

FIGURE 4.5 A sketch (after Dungey) of the solar wind magnetic field lines becoming connected to magnetospheric magnetic field lines at a dayside reconnection site, with those newly connected field lines being dragged into the magnetotail by the motion of the solar wind. A second reconnection site in the magnetotail disconnects the solar wind magnetic field from the magnetospheric magnetic field. The hollow vectors indicate the direction of plasma flow and the vector tips indicate the direction of the magnetic field lines. (Courtesy of Joachim Birn.)

The solar wind magnetic field lines are depicted as purely southward (**B** pointing downward in the figure). As the solar wind magnetic field lines make contact with the northward-pointing magnetic field lines of the magnetosphere at the dayside magnetopause, they "reconnect" (or "merge") and the solar wind becomes magnetically connected to the magnetosphere. In Figure 4.5, the newly connected solar wind magnetic field lines are shown passing the Earth, connected to the polar cap of the Earth in both the northern and southern polar regions of the Earth. With time, the newly connected magnetic field lines are laid down into the magnetotail of the magnetosphere. In Figure 4.5, a second site of magnetic field line reconnection in the magnetotail allows those magnetospheric magnetic field lines to move sunward from the magnetotail into the nightside dipolar region of the magnetosphere. Magnetospheric convection then moves those same field lines around either the dawnside or the duskside of the dipole into the dayside magnetosphere, completing the cycle. Eventually, the field lines will again reconnect with the solar wind and again go through the Dungey cycle. Commensurate with this magnetospheric Dungey cycle of motion, the foot point of the field line in the ionosphere makes a cycle from the high-latitude dayside region, antisunward across the polar cap into the high-latitude nightside region, then either around the dawnside or the duskside at high latitude back into the high-latitude dayside region. This is the high-latitude "two-cell" convection pattern observed in the ionosphere (Cowley and Lockwood, 1992; Weimer, 2005), which is modulated by the properties of the solar wind. Concerning these magnetospheric and ionospheric convection patterns that are driven by the solar wind, two scenarios can be considered: (1) the solar wind drives the magnetosphere which in turn drives the ionosphere or (2) the solar wind drives the ionosphere which in turn drives the magnetosphere. The first scenario is probably more dominant than the second. It is expected that the Dungey convection pattern in the magnetosphere will pull on the ionosphere (to which it is magnetically connected) to move the ionosphere into the two-cell pattern (Gary et al., 1995); however the ionospheric two-cell pattern in the polar cap is also directly driven by the solar wind, to which the polar cap ionosphere is magnetically connected (Siscoe and Siebert, 2006). The inner magnetosphere corotates with the Earth (Mozer, 1973), which is clear evidence that at some times and in some locations the ionosphere can drive magnetospheric convection. At low solar wind Mach numbers the ionosphere can also slow down the solar wind to which it is magnetically connected over the polar caps (Ridley, 2007)

On the dayside, magnetopause magnetic field line reconnection occurs along a line called the "neutral line" or "reconnection line." An equal number of magnetic field lines flow into this reconnection line from the solar wind (magnetosheath) and from the magnetosphere, wherein the field lines are reconnected (or more precisely, they are connected together). The rate of reconnection per unit length of the reconnection line is given by the amount of magnetic flux **B** flowing into the reconnection line from either plasma, which is v_{in}**B** where v_{in} is the plasma inflow velocity. The quantity v_{in}**B** has the units of an electric field and indeed it is the motional electric field $-\mathbf{v} \times \mathbf{B}$ of the inflowing plasma as seen in the reference frame of the reconnection line; hence the reconnection rate v_{in}**B** is often called the "reconnection electric field." The total rate of reconnection between the solar wind and the magnetosphere is given by v_{in}**B** multiplied by the length of the reconnection line, or more precisely it is given by the integral of v_{in}**B** over the length of the reconnection line. This total reconnection rate has the units of electric field times length, which is the units of electrical potential. This total reconnection rate is often called the "reconnection potential."

4.4.2 What Controls Dayside Reconnection?

A very important issue for solar wind/magnetosphere coupling and hence for understanding space weather is: What quantities in the solar wind control the dayside reconnection rate?

Because the dayside reconnection rate has the units of an electric field, in space physics it has been assumed for decades that the solar wind motional electric field $v_{sw}\mathbf{B}$ upstream of the bow shock controls the dayside reconnection rate (Gonzalez and Mozer, 1974). More accurately, it has been assumed that the dawn-to-dusk (east–west) component of the solar wind electric field $-v_{sw}B_z$ controls the dayside reconnection rate, where B_z (GSM coordinates) is the north–south component of the solar wind magnetic field with B_z positive when northward. There is a definite correlation between $-v_{sw}B_z$ in the solar wind and geomagnetic activity, but correlation coefficients are quite low. This is seen in Figure 4.6a where the auroral electrojet index AE is plotted as a function of $-v_{sw}B_z$ of the solar wind. The black points are individual points for 273,050 hourly averages of AE and of $-v_{sw}B_z$ and the light gray points are 300-point running averages of the gray points. The Pearson linear correlation coefficient between $-v_{sw}B_z$ and AE for the 273,050 points is $r_{corr} = +0.575$. There have been several refinements of electric field-type solar wind driver functions for the Earth's magnetosphere, the most famous being the Newell "universal coupling function" Φ_{MP} (Newell et al., 2007) given by

$$\Phi_{MP} = V_{SW}^{4/3} B_{perp}^{2/3} \sin^{8/3}\left(\frac{\theta_{clock}}{2}\right) \qquad (4.5)$$

where B_{perp} is the magnitude of the component of the solar wind magnetic field that is perpendicular to the Sun–Earth line and where θ_{clock} (Equation 4.4) is the clock angle of the B_{perp} vector in the Earth's magnetic dipole reference frame. As can be seen in Figure 4.6b where the hourly averaged values of the Earth's AE index are plotted as a function of the hourly averaged values of the Newell solar wind-driver function F_{MP}, the description of the variations in geomagnetic activity by F_{MP} are much better than for $-v_{sw}B_z$; the correlation coefficient for Φ_{MP} is $r_{corr} = +0.775$ whereas it was only +0.575 for $-vB_z$. The "prediction efficiency" is r_{corr}^2: for the Newell function F_{MP} the prediction efficiency for the AE index is $r_{corr}^2 = 59.7\%$ and whereas r_{corr}^2 is only 33.1% for $-vB_z$.

It has been argued that this long-standing assumption that the solar wind electric field controls the dayside reconnection rate is wrong, since the solar wind electric

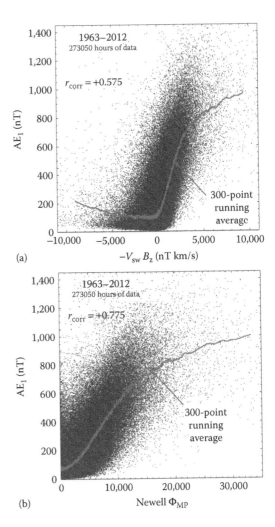

(a)

(b)

FIGURE 4.6 In the top panel (a) hourly values of the auroral electrojet index AE are plotted as functions of the measured value of $-vB_z$ in the solar wind. In the bottom panel (b) hourly values of AE are plotted as functions of the universal coupling function Φ_{MP} of Newell et al. (2007). The gray points are the individual hourly values and the light gray points are 300-point running averages of the black points.

field is greatly modified in the magnetosheath flow pattern before reaching the magnetopause (Borovsky and Birn, 2014). Indeed simulation experiments comparing reconnection rates with the electric fields in the ambient plasmas found no connection between the electric fields in the plasmas and the reconnection rates (Birn and Hesse, 2007). In plasma physics studies of magnetic field line reconnection in collisionless plasmas, the reconnection rate $v_{in}\mathbf{B}$ is given by $0.1v_A\mathbf{B}$ where v_A is the Alfvén speed in the plasma near the reconnection line (Shay et al., 1999). Hence the inflow speed of $0.1v_A$ and the reconnection rate of $0.1v_A\mathbf{B}$ are governed by the local properties v_A and **B** of the plasma near the site of reconnection. Solar wind driver functions for the magnetosphere based on the $0.1v_A$**B** reconnection rate concept have been theoretically derived by asking what solar wind parameters control v_A on both sides of the magnetopause and what solar wind parameters control **B** on both sides of the magnetopause. The starting point of these derivations is the Cassak–Shay equation for the reconnection rate, which is a function of four parameters: the magnetic field strengths B_{mag} and B_{sh} in the magnetosphere and magnetosheath and the mass densities ρ_{mag} and ρ_{sh} in the magnetosphere and magnetosheath (Cassak and Shay, 2007).

$$\text{rate} = (0.2/\mu^{1/2})\sin^2(\theta_{clock}/2)B_{sh}^{3/2}B_{mag}^{3/2}/\{(B_{sh}\rho_{mag}$$
$$+B_{mag}\rho_{sh})^{1/2}(B_{sh}+B_{mag})^{1/2}\} \qquad (4.6)$$

The simplest of these derived reconnection driver functions for the magnetosphere based on the Cassak–Shay equation is (Borovsky and Birn, 2014),

$$R_{quick} = \sin^2(\theta_{clock}/2)\, n_{sw}^{1/2}\, v_{sw}^{2}\, M_A^{-1.35}$$
$$\left[1 + 680 M_A^{-3.30}\right]^{-1/4} \qquad (4.7)$$

where $M_A = v_{sw}/v_{Asw}$ is the Alfvén Mach number of the solar wind upstream of the bow shock, v_{Asw} is the Alfvén speed in the solar wind upstream of the bow shock, and n_{sw} is the number density of the solar wind upstream of the bow shock. Equation 4.7 is valid at the nose of the magnetosphere. In correlation tests, the mathematically derived reconnection-type functions do as well as the best of the electric field-type functions at predicting the variance of geomagnetic indices.

Algebraic analysis of Equation 4.7 in the limits of high and low Mach numbers M_A finds that it has the functional form $R_{quick} \propto n_{sw}^{0.24} v_{sw}^{1.49} B^{0.51}\sin^2(\theta_{clock}/2)$ for very low Mach numbers and the functional form $R_{quick} \propto n_{sw}^{-0.19} v_{sw}^{0.62} B^{1.38}\sin^2(\theta_{clock}/2)$ for very high Mach numbers. These functional forms resemble the functional forms of electric field-based driver functions (cf. Equation 4.5); because of this similarity it has been conjectured that the solar wind electric field does not control the dayside reconnection rate and that it is a matter of coincidence that electric field-type driver functions correlate so well with geomagnetic activity.

4.4.3 The Viscous Interaction

A viscous interaction mechanism (Axford and Hines, 1961) is believed to transport plasma and momentum across the magnetopause from the solar wind into the magnetosphere. Evidence for the viscous interaction becomes clearest when the solar wind magnetic field has very northward clock angles ($\theta_{clock} \sim 0°$) when dayside reconnection should be very weak or nonexistent (cf. Equation 4.7). Under such northward conditions a residual geomagnetic activity is still seen (cf. Figure 4.6a) which is attributable to the viscous interaction of the solar wind with the magnetosphere, but which also may be partially driven by magnetic field line reconnection with the solar wind along the edges of the nightside magnetosphere (cf. Lockwood and Moen, 1999). There are ionospheric convection features (Drake et al., 2009) and magnetospheric convection features (Lundin et al., 1995) that can be attributed to a viscous interaction with the solar wind and that can be used to quantify the strength of the interaction. Measurements indicate that the viscous contribution to geomagnetic activity is small and that the viscous interaction only comes to dominate when the IMF clock angles are northward and geomagnetic activity is weak. Even though the viscous interaction may not be so important for driving geomagnetic activity, the viscous interaction may be very important for the transport of mass from the solar wind into the magnetosphere. The magnetospheric region where solar wind plasma enters is known as the low-latitude boundary layer (LLBL). The two main viscous-type mechanisms considered for the transport of plasma and momentum in from the solar wind are (1) plasma diffusion across the magnetopause mediated by plasma waves (Eviatar and Wolf, 1968) and (2) large-scale Kelvin–Helmholtz waves on the magnetopause that exchange plasma parcels and that may lead to reconnection in the LLBL (Nykyri and Otto, 2001).

Solar wind driver functions attempting to describe the effect of the viscous interaction on geomagnetic activity have been mathematically derived based on simplified diffusion models for the viscous interaction (e.g., Vasyliunas et al., 1982). When added to reconnection driver functions these viscous driver functions improve the correlations with geomagnetic indices slightly. The main impediment to obtaining realistic viscous driver functions is a lack of understanding of the physics underlying the viscous interaction.

4.4.4 Mass Coupling of the Solar Wind to the Magnetosphere

There are two sources for the plasma in the Earth's magnetosphere (Welling et al., 2015): the solar wind (i.e., the magnetosheath) and the ionosphere. Magnetospheric plasma from the solar wind has the same ionic composition H^+ and He^{++} as does the solar wind; ionospheric outflows into the magnetosphere have the ionic composition H^+, He^+, and O^+. The hot plasmas of the magnetosphere (the ion plasma sheet and the LLBL) are dominantly of solar wind origin, particularly when geomagnetic activity is low. The cool plasmas of the magnetosphere (the plasmasphere and the warm plasma cloak) are of ionospheric origin.

There are two major pathways for solar wind ions to enter the magnetosphere: (1) via the LLBL on the dawn and dusk flanks of the magnetosphere and (2) via the cusp mantle over the polar regions. Both pathways deliver solar wind ions into the magnetotail where magnetospheric convection (aided by substorm-enhanced impulsive convection) delivers some of these ions from the magnetotail into the nightside dipolar region of the magnetosphere. The entry mechanisms for solar wind plasma into the LLBL are probably related to the physical mechanisms of the viscous interaction; entry into the mantle is associated with dayside reconnection wherein magnetospheric magnetic field lines become connected into the solar wind (magnetosheath) plasma.

With time lags of a few hours, the number density of the hot plasma in the magnetosphere n_{mag} is related to the number density of the solar wind n_{sw} and the temperature of the hot magnetospheric plasma T_{mag} is related to the speed of the solar wind v_{sw} (Borovsky et al., 1998). This can be written as

$$n_{mag} \propto n_{sw} \tag{4.8}$$

$$T_{mag} \propto v_{sw} \tag{4.9}$$

Both of these relationships are associated with the properties of the magnetosheath. First the density relationship. The magnetosheath plasma is the solar wind plasma shocked by the bow shock. At high Mach numbers the density compression ratio of the bow shock is ~4, meaning that the plasma density downstream of the shock (magnetosheath) is about four times higher than the density of the plasma upstream of the shock

(solar wind). Hence, $n_{sh} \sim 4\, n_{sw}$, so as the solar wind density increases, the magnetosheath density increases, and as that magnetosheath plasma leaks into the magnetosphere, the magnetospheric plasma density increases. Second, the temperature relationship. In the reference frame of the bow shock, the shock transforms a supersonic flow upstream into a subsonic downstream flow. The ram kinetic energy of the upstream flow $m_p v_{sw}^2$ gets converted into thermal energy $k_B T$ in the downstream. Hence the downstream (magnetosheath) plasma temperature is $k_B T \sim m_p v_{sw}^2$ and so the magnetosheath temperature increases as the solar wind speed increases. That magnetosheath plasma leaking into the magnetosphere produces a magnetospheric temperature that increases with increasing solar wind speed.

The transport timescales for solar wind plasma to enter the magnetosphere and travel via magnetospheric convection to various locations has been statistically measured (Denton and Borovsky, 2009). In Figure 4.7 a sketch of the equatorial magnetosphere displays the approximate times for solar wind plasma to reach the magnetotail (~2 h), the nightside dipolar region (~4 h), the dawn and dusk regions of the dipole (~7 h), and the dayside magnetosphere (~12 h). Note that the transport timescales follows the Dungey cycle convection pattern: sunward from the magnetotail to the nightside dipolar

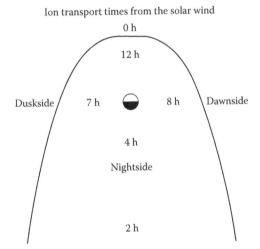

Ion transport times from the solar wind

FIGURE 4.7 A sketch of the equatorial plane of the Earth's magnetosphere viewed from high over the North Pole of the Earth. The black curve is the magnetopause bounding the magnetosphere and the Sun is off the top of the figure. The time values are the transport times from the solar wind to the various locations in the equatorial magnetosphere. (After Denton, M.H. and Borovsky, J.E., *J. Atmos. Solar-Terr. Phys.*, 71, 1045, 2009. With Permission.)

region, then sunward around the dawn and dusk regions of the dipole into the dayside magnetosphere.

For the ionospheric source of magnetospheric plasma the rate of global plasma outflow from the ionosphere into the magnetosphere typically increases as geomagnetic activity increases (Yau et al., 2011), so in a sense solar wind/magnetospheric coupling leads to enhanced outflow of ionospheric plasma into the magnetosphere (Welling et al., 2015). The ion plasma sheet is typically dominantly of solar wind origin with a composition of H^+ and He^{++}, but during geomagnetic storms O^+ ions from the ionosphere can dominate the ion number density of the magnetosphere. These O^+ ions in the magnetosphere are heavy and will have consequences for dayside reconnection and for solar wind/magnetosphere coupling (see Section 4.8.3).

4.5 REACTION OF THE MAGNETOSPHERE–IONOSPHERE SYSTEM TO SOLAR WIND DRIVING

When the coupling of the solar wind to the Earth's magnetosphere becomes strong, the magnetosphere–ionosphere system reacts in a number of ways. Intervals of strong coupling are most often associated with intervals wherein the IMF has a southward clock angle and the rate of dayside reconnection is strong. The IMF clock angle varies at all timescales in the solar wind at Earth. Such southward intervals can last a few minutes, in which case not many of the reaction phenomena discussed below will be seen. Southward IMF intervals can last an hour or so, in which case most of the phenomena below will occur, including the occurrence of a substorm. Southward IMF intervals can last for a day or so, in which case a geomagnetic storm may occur (see Section 4.6). The major features of the reaction of the magnetosphere–ionosphere system to strong driving by the solar wind are discussed in items (a) to (l) below. The order in which the various phenomena are discussed does not represent a temporal ordering in which they occur and does not represent a ranking of their importance.

When the rate of dayside reconnection increases the magnetosphere–ionosphere system has the following reactions.

a. *The total amount of magnetic flux in the magnetotail increases*: This increase is owed to newly reconnected magnetic field lines from the dayside magnetosphere being pulled over the poles of the Earth by the solar wind and laid down into the magnetotail (cf. Figure 4.5). This added flux causes an increase in the diameter of the magnetotail and a squeezing of the tail by the aerodynamic flow of the solar wind past a fattened (and flaring) magnetotail.

b. *There is an enlargement of the polar cap*: The polar cap is the region of magnetic field lines on the Earth that connect to the solar wind. This is the so-called "open flux" of the magnetosphere. The enlargement of the polar cap is observable as an equatorward migration of the auroral oval, a ring around each magnetic pole where the aurora light up the upper atmosphere. The northern edge of that ring is generally taken as the boundary between open flux (field lines connecting to the solar wind) and closed flux (field lines connecting back to the Earth in the opposite Northern–Southern Hemisphere).

c. *Magnetospheric convection increases*: This is the convection associated with the Dungey cycle (see Figure 4.5). The convection pattern carries plasma from the magnetotail into the nightside dipolar region of the magnetosphere, then sunward around the dawn and dusk dipolar region, then sunward into the dayside magnetopause where it flows into the dayside reconnection region and is exhausted into the solar wind.

d. *Ionospheric convection increases*: This is the "two-cell pattern" in the ionosphere that is magnetically connected to the Dungey convection pattern in the magnetosphere. That pattern is antisunward across the polar cap from the dayside ionosphere into the nightside ionosphere and then sunward at lower latitudes on the dawn side and on the dusk side. This convection pattern is best seen with a global radar network that observes the convective motions of small-scale plasma structure in the ionosphere.

e. *There is a stretching of the nightside dipolar magnetic field into a more tail-like morphology*: This change in the magnetic field morphology is associated with the formation of a cross-tail current system flowing in the dawn-to-dusk direction and with an increase in the magnetic pressure in the magnetotail.

f. *The ion and electron plasma sheets move deeper into the dipolar field regions*: These are hot tenuous plasmas originating in the magnetotail that are carried by the magnetospheric convection pattern from the magnetotail, around the dipole, into the dayside magnetopause. The stronger the convection, the deeper these plasmas penetrate into the dipole; as they are convected into the stronger magnetic field of the dipole these plasma are squeezed adiabatically, which heats them and increases their density.

g. *The ring current increases in strength*: Historically, the ring current was considered to be a ring of energetic ions orbiting the Earth in the dipolar region that produced an electric current; now it is realized that orbiting energetic ions is only part of the story, that currents associated with pressure gradients of plasmas passing the dipole moving from the nightside to the dayside are also important. The intensification of the ring current is related to the convection of the ion plasma sheet into the dipole (item [f] above) and its strength is best measured by the geomagnetic index D_{st}^{*}.

h. *The aurora intensifies and moves equatorward*: The aurora is an airglow in the upper atmosphere produced by the impacts of magnetospheric electrons; the auroral oval is magnetically connected to the electron plasma sheet in the magnetosphere. The intensification of the aurora is owed to magnetosphere–ionosphere currents associated with enhanced convection and to a heating of the electron plasma sheet. The equatorward migration of the aurora is owed both to the expansion of the polar cap (item [b] above) and to the deeper penetration of the electron plasma sheet into the dipole (item [f] above).

i. *A substorm may occur if the stronger driving occurs for a fraction of an hour or more*: Substorms will be discussed in the next paragraph.

j. *A population of substorm-injected energetic electrons and ions appears in the dipolar magnetosphere if a substorm occurs*: This is discussed further in the next paragraph.

k. *There is an enhancement of plasma waves in the dipolar magnetosphere*: Some of these waves are generated by plasma anisotropies associated with plasma convection in the magnetosphere and some of these waves are associated with the substorm-injected energetic electrons.

l. *The electron radiation belt will intensify if the strong driving is sustained for a day or more*: This point will be discussed further in Section 4.6.

An important reaction of the magnetosphere-ionosphere system to driving by the solar wind is the occurrence of a substorm (variously called a magnetospheric substorm or an auroral substorm). A substorm is a rapid morphological relaxation of the magnetotail from a stretched configuration with excess magnetic flux to a more dipolar magnetic field morphology; this morphology change is enabled by the formation of a "near-Earth neutral line" (McPherron et al., 1973), which is a new site of magnetic field line reconnection in the near-Earth portion of the magnetotail. A large downtail ejection of a "plasmoid" of magnetic flux and plasma occurs during the substorm (Hones et al., 1986), along with a large fast injection of plasma from the magnetotail sunward into the nightside dipolar magnetosphere. There is an extreme intensification of the nightside aurora during a substorm, including brightening, dynamic motions of auroral forms, and an expansion of the geographical extent of the aurora. Along with sudden plasma transport into the dipolar regions, energetic ions, and electrons from the magnetotail are injected into the dipolar portions of the magnetosphere (Moore et al., 1981); the substorm-injected ions drift westward around the Earth and the substorm-injected electrons drift eastward around the Earth. The onset time of a substorm can variously be detected as a rapid dipolarization (rotation into a more dipolar orientation) of the nightside magnetic field, a brightening of the aurora, the sudden appearance of injected energetic ions and electrons, or the rapid strengthening of specific magnetosphere–ionosphere current systems that have ground-based magnetic signatures. The occurrence of a substorm is associated with the occurrence of an interval of enhanced solar wind driving of the magnetosphere (i.e., enhanced dayside reconnection) (Newell and Liou, 2011). Substorms occur on average a few times per day; if strong driving persists the magnetosphere–ionosphere system can exhibit the occurrence of periodic substorms with about a 3-h recurrence period and if solar wind driving is weak a substorm may not occur for a day or two.

4.6 SOLAR WIND DRIVING OF GEOMAGNETIC STORMS

Geomagnetic storms are intervals of very strong geomagnetic activity caused by intervals of very strong solar wind driving of the magnetosphere–ionosphere system. The strong driving involves elevated velocities v_{sw}, high densities n_{sw}, and strong magnetic fields $\mathbf{B}$ (cf. Table 4.2). Southward IMF clock angles θ_{clock} are critical to whether or not a storm occurs. Geomagnetic storms are generally associated with two types of large-scale solar wind features: corotating interaction regions (CIRs) and CMEs (Borovsky and Denton, 2006).

A CIR (cf. Chapters 1 and 3) is a solar rotation effect wherein a region on the Sun that emits slow solar wind is adjacent to a region that emits fast wind and the slow wind moving radially outward from the Sun is followed by fast wind moving radially outward from the Sun. The fast wind pushes into the back of the slow wind causing a compression region wherein the plasma density and magnetic field strength are often a few times higher than normal. The large-scale compression region takes about a day to pass the Earth, and then several days of fast solar wind will often follow. (Because of the intervals of fast wind, CIR-driven storms are also called high-speed stream-driven storms.) The compression region leads to high densities in the magnetosphere (cf. Equation 4.8), which in turn can lead to a strong ring current if the solar wind also drives strong magnetospheric convection. During the several days of fast wind following the CIR the electron radiation belt can become energized if the solar wind is able to drive magnetospheric convection. During the CIR compression and the high-speed wind that follows, the magnetic field in the solar wind plasma is Parker spiral oriented with a nominal clock angle of ~90°; however, the Russell–McPherron effect can come into play to produce southward clock angles. Hence, CIR-driven geomagnetic storms tend to occur during the Spring and during the Fall and tend to be less prevalent in the Summer and Winter seasons. Also, since they are a solar-rotation effect, CIR storms tend to repeat with the 27-day solar rotation period. The boundary between fast solar wind and slow solar wind in a CIR is often located near a "magnetic sector reversal" of the IMF (Gosling et al., 1978). Across the

sector reversal, there is a reversal of the favorability of the Russell–McPherron effect for geomagnetic activity; hence, if there is a geomagnetic storm in the fast wind by the Russell–McPherron effect, there will tend to be a geomagnetic calm in the slow wind beforehand. Hence, CIR storms tend to be preceded by a "calm before the storm" that lasts several days; the calm before the storm can precondition the magnetosphere for the ensuing storm (Borovsky and Steinberg, 2006).

CME (cf. Chapters 1 and 2) is an impulsive emission of plasma from the Sun's corona, often associated with a solar flare. If the velocity of the ejected plasma is high, the ejected plasma will compress the slower solar wind plasma ahead of it and often produce an outward-traveling interplanetary shock wave leading that compression. The compressed solar wind plasma between the shock wave and the ejecta is denoted as the "CME sheath." Because the sheath plasma is compressed, it has a high density and a strong magnetic field. Whether the compressed (and sped up) sheath plasma with its high density and strong magnetic field drives a geomagnetic storm depends to some degree on the Russell–McPherron effect since the field is Parker spiral oriented in the sheath plasma. The magnetic field in the CME plasma is not Parker spiral oriented, and whether or not the fast ejecta drives a geomagnetic storm depends on the orientation of the magnetic field in the ejecta plasma, without the Russell–McPherron effect coming into play. The onset of the CME storm often occurs when the interplanetary shock passes the Earth; because the D_{st} index is sensitive to the ram pressure of the solar wind, there is often a sudden positive jump in the D_{st} index when this shock passes. This positive jump of D_{st} is known as the "storm sudden commencement" (SSC).

The two types of storms (CME-driven storms and CIR-driven storms) tend to have different properties. CME storms typically last a day or so; CIR storms typically last several days. CME storms are most prevalent during the maximum of the solar cycle; CIR storms are most prevalent in the declining phase of the solar cycle and tend to repeat with a 27-day period. CME storms often begin with an SSC; CIR storms tend not to have SSCs. CME-driven storms are associated with very strong ring currents (large $-D_{st}^*$), with the formation of new radiation belts, with "great aurora," with geomagnetically induced currents, and with polar cap potential saturation. CIR-driven storms are associated with high levels of spacecraft charging and with high energization levels of the outer electron radiation belt. CIR-driven storms tend to be preceded by a "calm before the storm"; CME-driven storms tend not to be.

Note that especially at the end of the solar maximum into the early declining phase of the solar cycle, storms are often of a mixed type, with a CME followed immediately by a long interval of high-speed solar wind.

4.7 WHEN SOLAR WIND DRIVING BECOMES WEAK

When the solar wind coupling to the magnetosphere is weak for an extended period of time, typically during intervals of low solar wind speed v_{sw} and northward IMF clock angles θ_{clock}, the magnetospheric–ionosphere system becomes geomagnetically "quiet." The weak coupling is associated with the absence (or great reduction) of dayside reconnection.

A number of phenomena follow when the driving by the solar wind is reduced (Borovsky and Steinberg, 2006). (1) Geomagnetic activity subsides. The timescale for activity to fall off when the IMF goes strongly northward is about 30 min (Rostoker et al., 1991). (2) Substorms cease to occur, or only occur infrequently and with small amplitude. (3) The areas of the polar caps shrink: the amount of open flux reduces since dayside reconnection is no longer creating open flux. (4) The auroral oval contracts poleward. This is associated with the shrinkage of the polar cap, but also associated with a weakening of magnetospheric convection and the electron plasma sheet (home of the aurora) not being convected into the dipolar magnetosphere. (5) The length of the magnetotail shrinks (since magnetic flux is not being added to the magnetotail when dayside reconnection is not operating). (6) The cool dense plasma sheet appears in the outer portions of the magnetosphere; this is probably the geographical spreading of the LLBL deeper into the magnetosphere under northward IMF, resulting in plasma that is denser and cooler than the normal ion plasma sheet filling the outer magnetosphere (Wing et al., 2006). (7) If the absence of coupling persists for a day or more, a "refilling" of the outer plasmasphere from the ionosphere will be observed (cf. Chapter 5). This refilling increases the radius of the plasmasphere and moves the plasmapause outward. (8) If the absence of coupling persists long enough for the outer plasmasphere to refill, the outer electron radiation belt will decay as its electrons scatter into the atmosphere (Meredith et al., 2006).

4.8 NEW THINKING ABOUT SOLAR WIND/MAGNETOSPHERE COUPLING

This section discusses new findings about the coupling of the solar wind to the magnetosphere.

4.8.1 "Turbulence Effect" in the Solar Wind/Magnetosphere Coupling

The time series of solar wind magnetic field measurements and velocity measurements is filled with fluctuations at all timescales (see Chapter 2); prominent in the temporal fluctuations are current sheets and abrupt velocity shears. It has been observed that geomagnetic activity increases when the fluctuation amplitudes δB of the solar wind magnetic field upstream of the bow shock increase (Borovsky and Funsten, 2003); in determining these geomagnetic activity correlations care was taken to ensure that δB is not acting as a proxy for another solar wind variable that might be important for solar wind/magnetosphere coupling. Other studies (D'Amicis et al., 2007) found that high-latitude geomagnetic activity was correlated with solar wind that had highly Alfvénic fluctuations (which is typical of the fast solar wind of coronal hole origin). It has been argued (Tsurutani and Gonzalez, 1987) that on-and-off dayside reconnection associated with southward–northward temporal fluctuations of the magnetic field clock angle θ_{clock} is the mechanism for the coupling associated with Alfvénic fluctuations in the solar wind. However, the turbulence effect persists even when the fluctuations are confined to northward clock angles. A suggested physical mechanism for the coupling is the "freestream turbulence effect" of aerodynamics (Volino, 1998), wherein turbulence in the magnetosheath plasma more effectively transports momentum from the magnetosheath flow to the Earth's magnetopause, boosting the effectiveness of any viscous interaction acting there. The physical mechanisms underlying the turbulence effect remains an open question.

4.8.2 Mach-Number Effects on the Magnetosheath and the Magnetosphere

The Alfvén Mach number $M_A = v_{sw}/v_{Asw}$ of the solar wind varies with time. Typically, the Mach number is high (cf. Table 4.1), but it can go as low as ~1. Often M_A is low when coronal-mass ejection plasma is passing the Earth. The flow pattern of the shocked solar wind plasma (the magnetosheath) around the Earth varies with the solar wind Mach number, and the properties of the magnetosheath plasma that bathes the magnetosphere vary with the solar wind Mach number. At the typical high Mach numbers, the bow shock is close to the Earth and the magnetosheath is a narrow flow pattern; when the Mach number becomes low the bow shock's position moves sunward and the magnetosheath becomes a broad flow pattern (Chapman et al., 2004; Lavraud and Borovsky, 2008). The bow shock operates differently for high versus low Mach number (Kennel, 1994). At high Mach numbers (above 6) the shocked magnetosheath plasma has its pressure dominated by particle pressure and the magnetic field lines in the magnetosheath are pliable and easily bent by the flow; at low Mach numbers (below 6) the magnetic pressure in the magnetosheath plasma is comparable with the particle pressure and the magnetic field lines in the magnetosheath flow are stiff and not easily bent. Further, at low Mach numbers the pressure in the magnetosheath plasma is not isotropic: the stiff magnetic field lines act to squeeze the magnetosphere into a flattened shape as viewed from the Sun (Lavraud et al., 2013). The dominance of the particle pressure in the magnetosheath at high Mach numbers also means a weaker magnetic field in the magnetosheath at higher Mach numbers; in the $0.1 v_A \mathbf{B} = 0.1 \mathbf{B}^2/(\mu_o m_p n)^{1/2}$ reconnection rate, the weakening of $\mathbf{B}$ at high Mach numbers weakens the dayside reconnection rate. This is the origin of the $M_A^{-1.35} (1 + 680 M_A^{-3.30})^{-1/4}$ factor in Equation 4.7 that lowers the reconnection rate as the Mach number increases. Global magnetohydrodynamics (MHD) simulations of the magnetosphere have demonstrated a low-Mach-number "density effect" wherein a change in the solar wind density can change the solar wind Mach number, which changes the compression ratio of the bow shock, which results in a change in the magnetic field strength of the magnetosheath, which in turn changes the dayside reconnection rate (Lopez et al., 2004).

Owing to the nonpliable magnetic field lines in the magnetosheath at low Mach numbers, high-speed flow jets in the magnetosheath are produced by a "magnetic slingshot" effect (Lavraud et al., 2007) and these flow jets change the nature of the Kelvin–Helmholtz waves on the magnetopause, potentially changing the nature of the viscous interaction and mass exchange between the solar wind and the magnetosphere.

Additionally, low-Mach-number driving by the solar wind is also associated with polar cap potential saturation, with a weakening of the magnetic field strength in the dayside magnetosphere, and with the occurrence of global sawtooth oscillations of the magnetosphere (Lavraud and Borovsky, 2008).

4.8.3 Feedback by the Magnetosphere on the Coupling

In the Cassak–Shay equation for the dayside reconnection rate (Equation 4.6), the variable ρ_{mag} enters the reconnection rate via the term $(B_{sh} \rho_{mag} + B_{mag} \rho_{sh})$, where ρ_{mag} and B_{mag} are the mass density and magnetic field strength of the magnetospheric plasma near the reconnection site and ρ_{sh} and B_{sh} are the mass density and magnetic field strength of the magnetosheath plasma near the reconnection site. Under typical conditions $B_{sh} < B_{mag}$ and $\rho_{mag} < \rho_{sh}$ so the term $\rho_{mag} B_{sh}$ can be ignored in $(B_{sh} \rho_{mag} + B_{mag} \rho_{sh})$. Hence, under typical conditions ρ_{mag} does not play a role in the dayside reconnection rate. (For this reason Equation 4.7 does not contain the ρ_{mag} term.) However, during the times when $\rho_{mag} \geq \rho_{sh} B_{mag}/B_{sh}$ the mass density of the magnetosphere has an important influence on the dayside reconnection rate, and thus the mass density of the dayside magnetosphere influences the amount of solar wind/magnetosphere coupling. The case $\rho_{mag} \geq \rho_{sh} B_{mag}/B_{sh}$ often occurs during geomagnetic storms. One region along the reconnection line where this condition occurs is in the storm time "plasmaspheric drainage plume." The enhanced magnetospheric convection during a geomagnetic storm will transport dense plasma from the plasmasphere in the dipolar region of the magnetosphere sunward into the dayside reconnection line in a plume a few R_E wide. (One Earth radius R_E is 6370 km.) Whereas ordinarily the number density of dayside magnetospheric plasma is $n_{mag} \sim 1$ cm^{-3}, the plasmaspheric plasma has a number density $n_{mag} \sim 100$ cm^{-3} where it reaches the reconnection line. A reduced rate of dayside reconnection caused by mass loading of the reconnection line by plasmaspheric drainage plumes has been observed by spacecraft in the plumes (Walsh et al., 2014a). Additionally, during geomagnetic storms heavy oxygen ions O$^+$ flow out of the ionosphere into the magnetosphere; energetic (keV) O$^+$ ions join the "ion plasma sheet" and lower-energy (10's of eV) O$^+$ ions form the "warm plasma cloak" (Chappell et al., 2008). These two plasmas are transported into the dayside reconnection line by magnetospheric convection where they too can act to mass load the dayside reconnection rate during storms (Borovsky et al., 2013). Feedback in this situation occurs as follows: when the solar wind drives the magnetosphere hard enough, heavy plasmas of the magnetosphere flow into the day-side reconnection site to mass load the reconnection rate and reduce solar wind/magnetosphere coupling.

4.8.4 Solar Wind Transients

There are a variety of transients associated with structures in the solar wind plasma and with the behavior of the Earth's bow shock.

As the solar wind passes the Earth, sudden changes in the solar wind density produce sudden changes in the ram pressure of the solar wind which produce impulsive reactions of the magnetosphere. Interplanetary shocks passing the Earth produce similar (often stronger) impulsive reactions.

Rotational discontinuities (current sheets) in the solar wind plasma produce sudden changes in the direction of the solar wind magnetic field vector as they pass the Earth. If the geometry of the current sheet if favorable, a "hot flow anomaly" can form as the discontinuity contacts the Earth's bow shock, wherein solar wind ions reflecting from the bow shock are guided upstream along the current sheet where their thermal energy causes the plasma at the current sheet to pressure expand, producing a magnetic cavity and density rarefaction at the current sheet and an over-dense region on either side of the sheet. This density rarefaction produces a ram-pressure rarefaction at Earth, causing a transient reaction of the magnetosphere–ionosphere system. A similar current sheet phenomena are foreshock bubbles.

The current sheets in the solar wind plasma are often co-located with abrupt wind shears, which are sudden large changes in the solar wind velocity vector. These wind shears can severely distort the morphology of the Earth's magnetosphere.

Temporal variations of the Earth's bow shock produce sudden large amplitude magnetic structures (SLAMS) in the vicinity of the shock, which advect with the plasma to impact the Earth's magnetosphere.

These many transient features and their impacts on the magnetosphere–ionosphere system are discussed in Chapter 5.

4.9 WHAT IS KNOWN AND WHAT IS NOT KNOWN

We know that the Sun and the solar wind affect geomagnetic activity and we have some ideas about the solar wind variables that affect the coupling. However, cause and effect between the set of coupled solar wind variables and the reaction of the Earth is difficult to sort out.

We know that reconnection on the dayside magnetopause between the solar wind magnetic field lines and the magnetospheric magnetic field lines is a dominant process controlling solar wind/magnetosphere coupling. However, we do not have a good understanding of the post-reconnection physical processes that couple the moving solar wind plasma to the magnetosphere-ionosphere system once the magnetic field lines are interconnected. An example of our lack of understanding of post-reconnection coupling is our disagreement about how polar-cap potential saturation works. Additionally, we do not have an understanding of the physical mechanisms underlying the viscous interaction and do not have good quantifications of the strengths of the viscous interaction.

The reaction of the magnetosphere–ionosphere system to driving is complicated, with multiple plasmas interacting with each other with multiple independent time lags, feedback loops, and hysteresis effects. Most of our analysis of coupling is algebraic (a state of the solar wind corresponding to a state of the magnetosphere), but it is clear that some aspects of the system are driven in an integrodifferential fashion. An example of this is the cumulative energization of the electron radiation belt that depends on the duration and strength of solar wind driving. Another example is the thermospheric flywheel effect wherein geomagnetic current systems persist after solar wind driving is turned off.

New ideas about solar wind/magnetosphere coupling are being explored: the upstream turbulence effect, Mach number effects, feedback by magnetospheric mass density, and the effects of solar wind transients. For the turbulence effect, we lack a physics understanding of why solar wind turbulence affects geomagnetic activity. For the feedback of magnetospheric density on the dayside reconnection rate, we have not built a knowledge base about the properties and global evolution of the warm plasma cloak and we don't have good surveys of the mass density of the dayside magnetosphere.

Most of the data analysis for studying solar wind/magnetosphere coupling is statistical, looking for trends in the data and using data to test our understanding. Fortunately for statistics, there is usually a lot of data. Unfortunately, the reaction of the magnetosphere–ionosphere system is complicated. Correlative studies between the solar wind and geomagnetic activity have been a core method for studying solar wind/magnetosphere coupling. Solar wind driver functions have steadily improved in their predictive accuracy. In the solar wind all variables are correlated or anticorrelated with each other; because of these intercorrelations, cause and effect between the solar wind and the magnetosphere is very difficult to sort out and the physical correctness of driver functions is difficult to judge. In the magnetosphere–ionosphere system, all measurements we have are imperfect. Typically, we gauge solar wind coupling by the reaction of various geomagnetic indices (measures of ground-based magnetic perturbations), but those indices are not direct measures of coupling. We would like to directly measure the dayside reconnection rate, directly measure system-wide energy-input rates, directly measure global plasma entry, and directly measure the strength of global current systems. Likewise, the solar wind data used in correlation studies is imperfect. The solar wind measurements we use are generally taken by solar wind monitors (spacecraft) that are ~235 R_E upstream of the Earth and not directly on the Sun–Earth line. The solar wind plasma and magnetic field have a lot of small-scale structure and the structure that hits the solar wind monitor is not always the structure that hits the Earth.

Global MHD computer simulations have been helpful for understanding the internal complexities of the magnetosphere–ionosphere system as driven by the solar wind. However, MHD codes lack the correct physics to simulate the viscous interaction and mass entry through the magnetopause. Further, ordinary MHD does not in general get reconnection rates correct (Birn et al., 2001) and so these simulations have drawbacks for quantifying the total amount of coupling and the total energy input into the magnetosphere–ionosphere system.

From the 1961 ideas of Jim Dungey and Ian Axford a lot of progress has been made in understanding solar wind/magnetosphere coupling. New ideas are emerging. There is a need for physicists, mathematicians, and numerical simulators to answer outstanding major questions such as the following:

- What physically controls the dayside reconnection rate?

- How strong is the viscous interaction and how does it work?

- Why does geomagnetic activity increase with increasing levels of solar wind turbulence?

- What are the entry mechanisms of solar wind plasma into the magnetosphere?

- What solar wind parameters govern the amount of plasma entry?

4.10 FOR FURTHER STUDY

Review articles on solar wind/magnetosphere coupling emphasizing various aspects can be found in McPherron (1995), Russell (2000, 2007), Farrugia et al. (2001), Cowley et al. (2003), Otto (2005), Baker (2010), Koskinen (2011), and Walsh et al. (2014b). A large compendium of articles that shows the state of research in the 1980s appears in the book edited by Kamide and Slavin (1986). Several articles exploring the physics of solar wind/magnetosphere coupling through computer simulations are Lyon (2000), Siscoe et al. (2004), Borovsky et al. (2009), Toffoletto and Siscoe (2009), Ridley et al. (2010), Siscoe (2011), Lopez et al. (2011), Chi et al. (2013), and Pulkkinen et al. (2015). Some statistical studies of the magnetosphere's reaction to the solar wind can be found in Hsu and McPherron (2003), Fung and Shao (2008), Balikhin et al. (2011), Milan et al. (2012), Tenfjord and Østgaard (2013), Luo et al. (2013), Borovsky (2014), Owens et al. (2014), and McPherron et al. (2015).

A.1 APPENDIX: GEOGRAPHY OF THE MAGNETOSPHERE AND THE IONOSPHERE

A.1.1 Magnetic Morphology of the Magnetosphere

Compressed dayside: Because of the ram pressure of the solar wind, the dayside magnetosphere is compressed to a magnetic field strength that is slightly larger than the dipole magnetic field value. Under unusual conditions (low-Mach-number solar wind) the field strength in the dayside magnetosphere can be weaker than the dipole magnetic field value.

Magnetotail: Owing to the interaction of the magnetosphere with the moving solar wind plasma, the nightside of the magnetosphere is pulled out into a long magnetotail. The length of the tail varies with the conditions of the solar wind.

Dipolar region or inner magnetosphere: The magnetic field of the magnetosphere near the Earth is almost undisturbed from the dipole field. Hence, the inner magnetosphere (inwards of ~6 R_E) can be described as the dipolar region.

Neutral sheet. (See the discussion of the ion and electron plasma sheet below.)

Cusps or clefts: The northern and southern cusps (or clefts) of the magnetosphere are locations that separate regions where magnetic field lines from the Earth are directed sunward from regions where the magnetic field lines from the Earth are directed tailward. The northern cusp of the magnetosphere is magnetically connected to the surface of the Earth near the geographic North Magnetic Pole and the southern cusp of the magnetosphere is magnetically connected to the surface of the Earth near the geographic South Magnetic Pole. Magnetosheath plasma can penetrate through the magnetopause into the cusps to reach the upper atmosphere. The cusps are sites of intensified ionospheric ion outflows known as the cleft ion fountain.

Magnetopause: The magnetopause is the outer boundary of the magnetosphere. At lower latitudes the magnetopause is the boundary between magnetic field lines that connect to the Earth on both ends and magnetic field lines that are not connected to the Earth. At high latitudes where magnetic field lines have one foot point on the Earth and connect into the distant solar wind, the magnetopause may be difficult to identify.

Dayside reconnection line: The curve or curves on the dayside magnetopause where magnetic field line reconnection between the magnetosheath plasma and the magnetospheric plasma is occurring. Magnetospheric magnetic field lines that flow into this site of reconnection becoming connected into the solar wind.

Distant neutral line: The distant neutral line is a reconnection site typically located 100 R_E or more behind the Earth. Here magnetospheric magnetic field lines that were recently connected into the solar wind plasma at the dayside reconnection line are carried antisunward over the poles by the solar wind flow and those field lines joining the magnetotail via reconnection in the distant neutral line.

Near-Earth neutral line (NENL): During a substorm, a second neutral line forms in the magnetotail closer to the Earth (10–20 R_E) than the distant neutral line. The reconnection at the NENL changes the magnetic morphology of the magnetotail and magnetically disconnects part of the tail from the Earth. The disconnected portion of the magnetotail is in the form of a magnetic plasmoid, which moves downtail and is lost from the Earth.

A.1.2 Regions External to the Magnetosphere

Bow shock: Upstream of the Earth's magnetosphere a curved bow shock converts the supersonic flow of the solar wind into a subsonic flow. Since collisions between the particles of the plasma are unimportant, the bow shock is categorized as a "collisionless shock." The region of the bow shock where the solar wind magnetic field is nearly parallel to the shock normal direction is known as the quasi-parallel shock and the region where the solar wind magnetic field is nearly parallel to the surface of the shock is known as the quasi-perpendicular shock.

Magnetosheath: The shocked solar wind plasma downstream of the bow shock is called the magnetosheath. It is comprised of H^+ and He^{++} ions. The magnetosheath plasma is hotter (~1 keV), denser (~4 n_{sw}), and slower flowing than the unshocked solar wind plasma. The magnetosheath plasma bathes the magnetosphere. The magnetosheath flow is slow near the nose (sunward end) of the magnetosphere and the magnetosheath flow is accelerated around the flanks of the magnetosphere to become a supersonic flow again. The magnetosheath is characterized by large-amplitude magnetic fluctuations and velocity fluctuations.

Ion foreshock and electron foreshock: As the solar wind plasma flows through the bow shock, most of the solar wind ions pass from the upstream solar wind through the bow shock into the downstream magnetosheath. However, a small fraction of the solar wind ions are reflected off of the bow shock and travel back upstream (sunward) along the solar wind magnetic field lines. These back-streaming ions can excite intense plasma waves in the solar wind upstream of the bow shock. The region upstream of the bow shock where back-streaming ions are present is known as the ion foreshock. The plasma (magnetosheath) downstream of the bow shock is warm (~1 keV). Some of the downstream electrons can escape through the bow shock and travel upstream along the magnetic field into the unshocked solar wind. These sunward-streaming electrons can also excite plasma waves in the solar wind. The region upstream of the bow shock where sunward-streaming electrons are present is known as the electron foreshock.

First Lagrangian point (*L1*): A point 235 R_E (1.5′10⁶ km) upstream of the Earth in the solar wind where the gravitational pull of the Sun balances that of the Earth. Solar wind-monitoring spacecraft typically reside near the L1 point. It is the warning point for oncoming space weather.

A.1.3 Plasmas and Particle Populations of the Magnetosphere

Plasmasphere, plasmapause, and plasmaspheric drainage plume: The plasmasphere is a dense, low-temperature (~1 eV) plasma that resides in the dipolar regions of the inner magnetosphere. The outer boundary of the plasmasphere is known as the "plasmapause" and the size of the plasmasphere (location of the plasmapause) varies with the time history of geomagnetic activity. The plasmasphere tends to corotate with the Earth. The plasmasphere's shape is distorted so that it extends further from the Earth on the duskside than it does on the dayside, dawnside, and nightside: this is the duskside bulge. The origin of the plasmaspheric plasma is the ionosphere and it is comprises H^+, He^+, and O^+ ions. When geomagnetic activity becomes high, the outer portions of the plasmasphere cease to corotate and plasma of the outer plasmasphere is carried by sunward magnetospheric convection to the dayside magnetopause in the form of a narrow "plasmaspheric drainage plume" in the dayside magnetosphere.

Ion plasma sheet and electron plasma sheet: The plasma sheet is a hot (multi-keV), low-density (~1 cm⁻³) plasma that forms a flat horizontal sheet in the Earth's magnetotail. The plasma sheet is ~6 R_E thick in the north–south direction and the

plasma sheet extends across the magnetotail (20–30 R_E) in the dawn–dusk direction. In the middle of the flat plasma sheet is a "neutral sheet" that bifurcates the plasma sheet into a northern portion and a southern portion. Across the neutral sheet the direction of the magnetic field in the plasma sheet reverses: the magnetic field points sunward (Earthward) north of the neutral sheet and the field points antisunward (downtail) south of the neutral sheet. The neutral sheet is the location of a dawn-to-dusk "cross-tail current sheet." The origin of the plasma-sheet plasma is chiefly from the solar wind (magnetosheath) H^+ and He^{++}, but outflows of energized H^+ and O^+ ions from the ionosphere can make significant contributions to the total ion density when geomagnetic activity is high. The plasma sheet extends from the magnetotail into the dipolar magnetosphere: because of differences in temperatures the ions (hotter) are on different orbits in the dipolar magnetic field than are the electrons (less hot). Because of the orbital differences, the near-Earth boundaries of the ion plasma sheet and the electron plasma sheet differ, with the ion plasma sheet reaching deeper into the dipolar region than the electron plasma sheet. The inner edge of the electron plasma sheet is approximately colocated with the plasmapause of the plasmasphere; inward of this common boundary is cool plasmaspheric plasma, outward of this boundary is electron–plasma–sheet plasma.

Lobes: In the magnetotail above (northward of) and below (southward of) the flat plasma sheet there are regions devoid of hot plasma. These are the northern and southern lobes. The sunward-pointing magnetic field lines of the northern lobe connect to the Earth's northern polar cap and the antisunward-pointing magnetic field lines of the southern lobe connect to the Earth's southern polar cap. The mantle resides in the lobe but not everywhere in the lobe: one could separate the geography of the lobe into the "empty lobe" versus the "mantle."

Plasma sheet boundary layer (*PSBL*): The northern and southern boundaries of the flat plasma sheet in the magnetotail are characterized by magneticfield-aligned beams of electrons and ions. These PSBLs separate the plasma sheet from the lobes.

Ring current and partial ring current: As the ion plasma sheet convects into the dipolar magnetosphere it is adiabatically compressed and heated and the resulting plasma pressure in the dipole produces a diamagnetic electrical current that distorts the Earth's magnetic field. The strength of this current (which can be monitored by the geomagnetic index D_{st}) is a common measure of the strength of a geomagnetic storm. As plasma sheet ions move from the nightside to the dipole, some of the ions are captured into orbits where they circle the Earth in the dipole magnetic field: this is the "ring current proper." The partial ring current is the ion plasma sheet moving from the nightside to the dayside through the outer regions of the dipole without being captured.

Substorm-injected electrons and ions: The strong convection of plasma from the magnetotail into the nightside dipolar regions during a substorm expansion adiabatically energizes the tail of the particle distribution functions of electrons and ions to several 10s of keV. The energetic populations suddenly appear in the dipolar regions at the start of substorms, where the energetic electrons and ions are trapped and orbit around the Earth. These particle populations drive important plasma waves and the substorm electrons are the likely seed population for the electron radiation belt.

Trough or electron trough: As the electron plasma sheet is carried by magnetospheric convection from the nightside magnetosphere to the dayside magnetosphere, most of the electrons precipitate along the magnetic field lines into the atmosphere where they are lost (and where they make diffuse aurora). As a result of this loss of electrons, the electron plasma sheet has a much lower density in the dayside magnetosphere than it does in the nightside magnetosphere. Because of the low density of hot electrons, the dayside magnetosphere is often referred to as the electron trough. A more-accurate name would be the "dayside electron plasma sheet."

LLBL: Inside the magnetopause on the dawn and dusk sides of the magnetosphere is a layer of plasma of solar wind origin (H^+ and He^{++}) that has properties similar to the magnetosheath plasma, but less dense and hotter. The plasma of the LLBL flows antisunward. This plasma is undoubtedly magnetosheath plasma that has recently been captured into the magnetosphere.

Cool dense plasma sheet: Plasma that is cooler (~1 keV) and denser (few cm^{-3}) than the typical ion plasma sheet fills the outer magnetosphere during geomagnetically quiet times. This plasma is probably an expansion of LLBL plasma deep into the magnetosphere.

Mantle: Magnetosheath plasma can enter the magnetosphere in the magnetospheric cusps. The cusps are also sites of intensified ion outflow from the ionosphere. This magnetosheath plasma and these ionospheric outflows are carried tailward by the convection of magnetic field lines in the Dungey cycle. In the lobes of the magnetotail cold antisunward-flowing ions can be seen from these cusp sources. This cold, sometimes dense, antisunward flowing plasma in the lobes is called the mantle.

Warm plasma cloak: The warm plasma cloak is a population of cool (~10 eV) ions, primarily magnetic field aligned, seen throughout the nightside and dayside regions of the dipolar magnetosphere outside of the plasmasphere. The warm plasma cloak is of ionospheric origin (H^+ and O^+), with the ionospheric ions probably emanating from the nightside polar cap or from the nightside auroral zone, wherein the ions are carried into the dayside regions by the sunward magnetospheric convection. The full geographical extent and the spatial-temporal evolution of the warm plasma cloak are unknown. Early papers referred to this population as the "oxygen torus."

Electron radiation belts: Hot electrons are energized (to MeV) in the Earth's magnetosphere and can exist on trapped orbits in the dipolar regions circulating around the Earth. A "slot region" of low fluxes at ~3 R_E separates the high-flux inner electron radiation belt from the high-flux outer electron radiation belt. The inner electron radiation belt is very stable and is only perturbed during very large geomagnetic storms. The electron outer radiation belt is very dynamic and changes over timescales of hours and days. The outer electron radiation belt geographically overlaps the regions of the cold plasmasphere and the hot plasma sheet.

Proton radiation belt: The proton radiation belt is less intense (in terms of particle fluxes) then the electron radiation belts and the proton radiation belt is concentrated closer to the Earth then the electron belts are. Like the outer electron radiation belt, the outer portions of the proton radiation belt are temporally dynamic.

A.1.4 Regions of the Ionosphere

D-Region: The ionosphere below 90-km altitude is denoted as the D-region or D-layer.

E-Region: The altitude range from 90 to 150 km is termed the E-region or E-layer.

F-Region: The altitude region from 150 to 500 km is termed the F-region. The altitude where the ionization density is maximum is denoted as the F-peak.

Topside ionosphere: The ionosphere above the F-region peak.

Polar caps: The regions of magnetic field lines that go out of the ionosphere in one hemisphere and do not return to the Earth. These field lines connect into the solar wind. The polar cap is approximately bounded by the high-latitude edge of the auroral oval.

Auroral zone or Auroral oval: The auroral zone is a ring around the northern (and the southern) polar region of the Earth where magnetospheric charged particles precipitate along the magnetic field lines into the upper atmosphere and produce visible airglow. Most aurora is produced by electron precipitation, so for most part the auroral zone in the atmosphere is magnetically connected to the electron plasma sheet in the magnetosphere. However, proton aurora, which can occur equatorward of the electron aurora, are magnetically connected to the ion plasma sheet in the magnetosphere. Because of ionization by the particle precipitation, the auroral zone ionosphere can have enhanced densities compared to regions devoid of particle precipitation.

Auroral electrojets: Global current systems flowing in the auroral zone ionosphere. On the nightside the electrojet current flows westward and on the dayside the electrojet current flows eastward: both flow in the dawn-to-dusk sense.

Equatorial electrojet: A narrow band of eastward-flowing current in the dayside ionosphere at the magnetic dipole equator where the Earth's magnetic field lines are approximately horizontal.

Subauroral polarization stream (SAPS) and subauroral ion drift (SAID): A band of enhanced westward plasma drift in the evening sector near the equator toward the edge of the auroral oval during geomagnetically active times. The SAPS is a long-lived broad band of convection and the SAID is a shorter-lived narrow more-intense band of convection within the SAPS.

Storm-enhanced density (SED) and tongue of ionization (TOI): A poleward-flowing plume of enhanced ionization (enhanced TEC) in the dayside ionosphere during storms. The plume of enhanced ionospheric plasma density flows poleward from mid-latitudes into the polar cap. This plume is typically called the SED at lower latitudes and the TOI at higher latitudes.

Cleft ion fountain: An upward outflow of ionospheric ions associated with the interaction of magnetosheath plasma with the ionosphere in the cusps. The cleft ion fountain feeds the mantle.

Equatorial plasma fountain: At the magnetic dipole equator where the magnetic field lines are horizontal electric field effects that convect ionospheric plasma vertically upward whereupon it is gravitationally pulled down along the dipole magnetic field lines to land back in the ionosphere at $10°$–$20°$ of latitude north and south of the equator. The resulting bands of high-density ionospheric plasma at $\pm10°$–$20°$ north and south of the equator are known as the Appleton anomalies.

A.1.5 Neutral Gas

Thermosphere: The region of the upper atmosphere (above about 85 km) where the temperature increases with altitude.

Hydrogen geocorona: Neutral (not ionized) hydrogen escape upward from the upper atmosphere, forming a low-density cloud of neutral hydrogen around the Earth. Owing to "charge exchange collisions," the hydrogen geocorona can be important for the evolution of the ring current/ion plasma sheet.

REFERENCES

Alfven, H., On the theory of comet tails, *Tellus, 9*, 92, 1957.

Arge, C. N., D. Odstrcil, V. J. Pizzo, and L. R. Mayer, Improved method for specifying solar wind speed near the Sun, *AIP Conf. Proc., 679*, 190, 2003.

Arya, S. and J. W. Freeman, Estimates of solar wind velocity gradients between 0.3 and 1 AU based on velocity probability distributions from Helios 1 at perihelion and aphelion, *J. Geophys. Res., 96*, 14183, 1991.

Axford, W. I. and C. O. Hines, A unifying theory of high-latitude geophysical phenomena and geomagnetic storms, *Canad. J. Phys., 39*, 1433, 1961.

Baker, D. N., Perspectives on geospace plasma coupling, *AIP Conf. Proc., 1320*, 10, 2010.

Balikhin, M. A., R. J. Boynton, S. N. Walker, J. E. Borovsky, S. A. Billings, and H. L. Wei, Using the NARMAX approach to model the evolution of energetic electrons fluxes at geostationary orbit, *Geophys. Res. Lett., 38*, L18105, 2011.

Bargatze, L. L., D. N. Baker, R. L. McPherron, and E. W. Hones, Magnetospheric impulse response for many levels of geomagnetic activity, *J. Geophys. Res., 90*, 6387, 1985.

Bevington, P. R. and D. K. Robinson, *Data Reduction and Error Analysis for the Physical Sciences*, Second Edition, McGraw-Hill, New York, 1992.

Birn, J., J. F. Drake, M. A. Shay, B. N. Rogers, R. E. Denton, M. Hesse, M. Kuznetsova et al., Geospace environment modeling (GEM) magnetic reconnection challenge, *J. Geophys. Res., 106*, 3715, 2001.

Birn, J. and M. Hesse, Reconnection rates in driven magnetic reconnection, *Phys. Plasmas, 14*, 082306, 2007.

Borovsky, J. E., Canonical correlation analysis of the combined solar-wind and geomagnetic-index data sets, *J. Geophys. Res., 119*, 5364, 2014.

Borovsky, J. E. and J. Birn, The solar-wind electric field does not control the dayside reconnection rate, *J. Geophys. Res., 119*, 751, 2014.

Borovsky, J. E. and M. H. Denton, The differences between CME-driven storms and CIR-driven storms, *J. Geophys. Res., 111*, A07S08, 2006.

Borovsky, J. E., M. H. Denton, R. E. Denton, V. K. Jordanova, and J. Krall, Estimating the effects of ionospheric plasma on solar-wind/magnetosphere coupling via mass loading of dayside reconnection: Ion-plasma-sheet oxygen, plasmaspheric drainage plumes, and the plasma cloak, *J. Geophys. Res., 118*, 5695, 2013.

Borovsky, J. E. and H. O. Funsten, Role of solar wind turbulence in the coupling of the solar wind to the Earth's magnetosphere, *J. Geophys. Res., 108*, 1246, 2003.

Borovsky, J. E., B. Lavraud, and M. M. Kuznetsova, Polar cap potential saturation, dayside reconnection, and changes to the magnetosphere, *J. Geophys. Res., 114*, A03224, 2009.

Borovsky, J. E. and J. T. Steinberg, The "calm before the storm" in CIR/magnetosphere interactions: Occurrence statistics, solar-wind statistics, and magnetospheric preconditioning, *J. Geophys. Res., 111*, A07S10, 2006.

Borovsky, J. E., M. F. Thomsen, and R. C. Elphic, The driving of the plasma sheet by the solar wind, *J. Geophys. Res., 103*, 17617, 1998.

Cassak, P. A. and M. A. Shay, Scaling of asymmetric magnetic reconnection: General theory and collisional simulations, *Phys. Plasmas, 14*, 102114, 2007.

Chapman, J. F., I. H. Cairns, J. G. Lyon, and C. R. Boshuizen, MHD simulations of Earth's bow shock: Interplanetary magnetic field orientation effects on shape and position, *J. Geophys. Res., 109*, A04215, 2004.

Chappell, C. R., M. M. Huddleston, T. E. Moore, B. L. Giles, and D. C. Delcourt, Observations of the warm plasma cloak and an explanation of its formation in the magnetosphere, *J. Geophys. Res., 113*, A09206, 2008.

Chi, W., G. XiaoCheng, P. Zhong, T. BinBin, S. TianRan, L. WenYa, and H. YuQiu, Magnetohydrodynamics (MHD) numerical simulations on the interaction of the solar wind with the magnetosphere: A review, *Sci. China: Earth Sci., 56*, 1141, 2013.

Cowley, S. W. H., J. A. Davies, A. Grocott, H. Khan, M. Lester, K. A. McWilliams, S. E. Milan et al., Solar-wind-magnetosphere-ionosphere interactions in the Earth's plasma environment, *Phil. Trans. R. Soc. Lond., 361*, 113, 2003.

Cowley, S. W. H. and M. Lockwood, Excitation and decay of solar wind-driven flows in the magnetosphere-ionosphere system, *Ann. Geophys., 10*, 103, 1992.

Cranmer, S. R., Coronal holes, *Living Rev. Solar Phys., 6*, 3, 2009.

Cranmer, S. R., A. A. van Ballegooijen, and R. J. Edgar, Self-consistent coronal heating and solar wind acceleration from anisotropic magnetohydrodynamic turbulence, *Astrophys. J. Suppl. Series, 171*, 520, 2007.

D'Amicis, R., R. Bruno, and B. Bavassano, Is geomagnetic activity driven by solar wind turbulence? *Geophys. Res. Lett., 34*, L05108, 2007.

Denton, M. H. and J. E. Borovsky, The superdense plasma sheet in the magnetosphere during high-speed-steam-driven storms: Plasma transport timescales, *J. Atmos. Solar-Terr. Phys., 71*, 1045, 2009.

Dessler, A. J. and E. N. Parker, Hydromagnetic theory of geomagnetic storms, *J. Geophys. Res., 64*, 2239, 1959.

Drake, K. A., R. A. Heelis, M. R. Hairston, and P. C. Anderson, Electrostatic potential drop across the ionospheric signature of the low-latitude boundary layer, *J. Geophys. Res., 114*, A04215, 2009.

Dungey, J. W., Interplanetary magnetic field and the auroral zones, *Phys. Rev. Lett., 6*, 47, 1961.

Eselevich, M. V. and V. G. Eselevich, Some features of the streamer belt in the solar corona and at the Earth's orbit, *Astron. Reports., 50*, 748, 2006.

Eviatar, A. and R. A. Wolf, Transfer processes in the magnetopause, *J. Geophys. Res., 73*, 5561, 1968.

Farrugia, C. J., F. T. Gratton, and R. B. Torbert, Viscous-type processes in the solar wind-magmnetosphere interaction, *Space Sci. Rev., 95*, 443, 2001.

Feldman, U., E. Landi, and N. A. Schwadron, On the sources of fast and slow solar wind, *J. Geophys. Res., 110*, A07109, 2005.

Fisk, L. A. and N. A. Schwadron, Origin of the solar wind: Theory, *Space Sci. Rev., 97,* 21, 2001.

Fung, S. F. and X. Shao, Specification of multiple geomagnetic responses to variable solar wind and IMF input, *Ann. Geophys., 26,* 639, 2008.

Gary, J. B., R. A. Heelis, and J. P. Thayer, Summary of field-aligned Poynting flux observations from DE 2, *Geophys. Res. Lett., 22,* 1861, 1995.

Gjerloev, J. W., R. A. Hoffman, M. M. Friel, L. A. Frank, and J. B. Sigwarth, Substorm behavior of the auroral electrojet indices, *Ann. Geophys., 22,* 2135, 2004.

Goertz, C. K., L.-H. Shan, and R. A. Smith, Prediction of geomagnetic activity, *J. Geophys. Res., 98,* 7673, 1993.

Gonzalez, W. D. and F. S. Mozer, A quantitative model for the potential resulting from reconnection with an arbitrary interplanetary magnetic field, *J. Geophys. Res., 79,* 4186, 1974.

Gosling, J. T., The solar flare myth, *J. Geophys. Res., 98,* 18937, 1993.

Gosling, J. T., J. R. Asbridge, S. J. Bame, and W. C. Feldman, Solar wind stream interfaces, *J. Geophys. Res., 83,* 1401, 1978.

Hapgood, M. A., Space physics coordinate transformations: A user guide, *Planet. Space Sci., 40,* 711, 1992.

Heppner, J. P., N. F. Ness, C. S. Scearce, and T. L. Skillman, Explorer 10 magnetic field measurements, *J. Geophys. Res., 68,* 1, 1963.

Hollweg, J. V., The solar wind: Our current understanding and how we hot here, *J. Astrophys. Astron., 29,* 217, 2008.

Hones, E. W., T. A. Fritz, J. Birn, J. Cooney, and S. J. Bame, Detailed observations of the plasma sheet during a substorm on April 24, 1979, *J. Geophys. Res., 91,* 6845, 1986.

Hsu, T.-S. and R. L. McPherron, Occurrence frequencies of IMF triggered and nontriggered substorms, *J. Geophys. Res., 108,* 1207, 2003.

Kamide, Y. and J. A. Slavin (eds.), *Solar Wind-Magnetosphere Coupling,* Terra Scientific, Tokyo, Japan, 1986.

Kennel, C. F., The magnetohydrodynamic Rankine-Hugoniot relations, *AIP Conf. Proc., 314,* 180, 1994.

King, J. H. and N. E. Papitashvili, Solar wind spatial scales in and comparisons of hourly wind and ACE plasma and magnetic field data, *J. Geophys. Res., 110,* 2104, 2005.

Komar, C. M., R. L. Fermo, and P. A. Cassak, Comparative analysis of dayside magnetic reconnection models in global magnetosphere simulations, *J. Geophys. Res., 120,* 276, 2015.

Koskinen, H. E. J., *Physics of Space Storms,* Section 1.3, Springer Praxis Books, Chichester, UK, 2011.

Lavraud, B. and J. E. Borovsky, Altered solar wind-magnetosphere interaction at low Mach numbers: Coronal mass ejections, *J. Geophys. Res., 113,* A00B08, 2008.

Lavraud, B., J. E. Borovsky, A. J. Ridley, E. W. Pogue, M. F. Thomsen, H. Reme, A. N. Fazakerley, and E. A. Lucek, Strong bulk plasma acceleration in Earth's magnetosheath: A magnetic slingshot effect? *Geophys. Res. Lett., 34,* L14102, 2007.

Lavraud, B., E. Larroque, E. Budnik, V. Genot, J. E. Borovsky, M. W. Dunlop, C. Foullon et al., Asymmetry of magnetosheath flows and magnetopause shape during low Alfven Mach number solar wind, *J. Geophys. Res., 118,* 1089, 2013.

Lemaire, J. and M. Scherer, Kinetic models of the solar wind, *J. Geophys. Res., 76,* 7479, 1971.

Lockwood, M. and J. Moen, Reconfiguration and closure of lobe flux by reconnection during northward IMF: Possible evidence for signatures in cusp/cleft auroral emissions, *Ann. Geophys., 17,* 996, 1999.

Lopez, R. E., V. G. Merkin, and J. G. Lyon, The role of the bow shock in solar wind-magnetosphere coupling, *Ann. Geophys., 29,* 1129, 2011.

Lopez, R. E., M. Wiltberger, S. Hernandez, and J. G. Lyon, Solar wind density control of energy transfer to the magnetopause, *Geophys. Res. Lett., 31,* L08804, 2004.

Lundin, R., M. Yamauchi, J. Woch, and G. Marklund, Boundary layer polarization and voltage in the 14 MLT region, *J. Geophys. Res., 100,* 7587, 1995.

Luo, B., X. Li, M. Temerin, and S. Liu, Prediction of the AU, AL, and AE indices using solar wind parameters, *J. Geophys. Res., 118,* A019188, 2013.

Lyon, J. G., The solar wind-magnetosphere-ionosphere system, *Science, 288,* 1987, 2000.

Manchester, W. B., T. I. Gombosi, I. Roussev, A. Ridley, D. L. De Zeeuw, I. V. Sokolov, and K. G. Powell, Modeling a space weather event from the Sun to the Earth: CME generation and interplanetary propagation, *J. Geophys. Res., 109,* A02107, 2004.

McPherron, R. L., Magnetospheric dynamics, in *Introduction to Space Physics,* M. G. Kivelson and C. T. Russell (eds.), Cambridge University Press, New York, 1995, p. 400.

McPherron, R. L., T.-S. Hsu, and X. Chu, An optimum solar wind coupling function for the AL index, *J. Geophys. Res., 120,* 2494, 2015.

McPherron, R. L., C. T. Russell, and M. P. Aubry, Satellite studies of magnetospheric substorms on August 15, 1968: 9. Phenomenological model for substorms, *J. Geophys. Res., 78,* 3131, 1973.

Meredith, N. P., R. B. Horne, S. A. Glaurt, R. M. Thorne, D. Summers, J. M. Albert, and R. R. Anderson, Energetic outer zone electron loss timescales during low geomagnetic activity, *J. Geophys. Res., 111,* A05212, 2006.

Milan, S. E., J. S. Gosling, and B. Hubert, Relationship between interplanetary parameters and the magnetopause reconnection rate quantified from observations of the expanding polar cap, *J. Geophys. Res., 117,* A03226, 2012.

Moore, T. E., R. L. Arnoldy, J. Feynman, and D. A. Hardy, Propagating substorm injection fronts, *J. Geophys. Res., 86,* 6713, 1981.

Mozer, F. S., Electric fields and plasma convection in the plasmasphere, *Rev. Geophys. Space Phys., 11,* 755, 1973.

Newell, P. T. and K. Liou, Solar wind driving and substorm triggering, *J. Geophys. Res., 116,* A03229, 2011.

Newell, P. T., T. Sotirelis, K. Liou, C.-I. Meng, and F. J. Rich, A nearly universal solar wind-magnetosphere coupling function inferred from 10 magnetospheric state variables, *J. Geophys. Res., 112*, A01206, 2007.

Nykyri, K. and A. Otto, Plasma transport at the magnetospheric boundary due to reconnection in Kelvin-Helmholtz vortices, *Geophys. Res. Lett., 28*, 3565, 2001.

Otto, A., The magnetosphere, *Lect. Notes Phys., 656*, 133, 2005.

Owens, M. J., T. S. Horbury, R. T. Wicks, S. L. McGrgor, N. P. Savani, and M. Xiong, Ensemble downscaling in coupled solar wind-magnetosphere modeling for space weather forecasting, *Space Weather, 12*, 395, 2014.

Parker, E. N., *Interplanetary Dynamical Processes*, Wiley Interscience, New York, 1963, p. 138.

Parker, E. N., Dynamical theory of the solar wind, *Space Sci. Rev., 4*, 666, 1965.

Pierrard, V., Solar wind electron transport: Interplanetary electric field and heat conduction, *Space Sci. Rev., 172*, 315, 2012.

Priest, E. R., Solar atmosphere, in *Handbook of the Solar-Terrestrial Environment*, Y. Kamide and A. Chian (eds.), Springer-Verlag, Berlin, Germany, 2007, pp. 55–93.

Pulkkinen, T. I., A. P. Dimmock, A. Osmane, and K. Nykyri, Solar wind energy input to the magnetosheath and at the magnetopause, *Geophys. Res. Lett., 42*, 4723, 2015.

Richardson, I. G. and H. V. Cane, Near-Earth interplanetary coronal mass ejections during Solar Cycle 23 (1996–2009) Catalog and summary of properties, *Solar Phys., 264*, 189, 2010.

Richardson, I. G., E. W. Cliver, and H. V. Cane, Sources of geomagnetic activity over the solar cycle: Relative importance of coronal mass ejections, high-speed streams, and slow solar wind, *J. Geophys. Res., 105*, 18203, 2000.

Ridley, A. J., Alfven wings at Earth's magnetosphere under strong interplanetary magnetic fields, *Ann. Geophys., 25*, 533, 2007.

Ridley, A. J., T. I. Gombosi, I. V. Sokolov, G. Toth, and D. T. Welling, Numerical considerations in simulating the global magnetosphere, *Ann. Geophys., 28*, 1589, 2010.

Rostoker, G., Geomagnetic indices, *Rev. Geophys. Space Phys., 10*, 935, 1972.

Rostoker, G., T. D. Phan, and F. Pascal, Inference of magnetospheric and ionospheric electrical properties from the decay of geomagnetic activity, *Can. J. Phys., 69*, 921, 1991.

Russell, C. T., The solar wind interaction with the Earth's magnetosphere: A tutorial, *IEEE Trans. Plasma Sci., 28*, 1818, 2000.

Russell, C. T., The coupling of the solar wind to the Earth's magnetosphere, in *Space Weather—Physics and Effects*, V. Bothmer and I. A. Daglis (eds.), Springer Praxis Books, Chichester, UK, 2007, p. 103.

Russell, C. T. and R. L. McPherron, Semiannual variation of geomagnetic activity, *J. Geophys. Res., 78*, 92, 1973.

Shay, M. A., J. F. Drake, B. N. Rogers, and R. E. Denton, The scaling of collisionless, magnetic reconnection for large systems, *Geophys. Res. Lett., 26*, 2163, 1999.

Shimojo, M., N. Narukage, R. Kano, T. Sakao, S. Tsuneta, K. Shibasaki, J. W. Cirtain, L. L. Lundquist, K. K. Reeves, and A. Savcheva, Fine structures of solar X-ray jets observed with the X-ray telescope aboard Hinode, *Publ. Astron. Soc. Japan, 59*, S745, 2007.

Siscoe, G. L., Aspects of global coherence of magnetospheric behavior, *J. Atmos. Solar-Terr. Phys., 73*, 402, 2011.

Siscoe, G. L., R. L. McPherron, and V. K. Jordanova, Diminished contribution of ram pressure to Dst during magnetic storms, *J. Geophys. Res., 110*, A12227, 2005.

Siscoe, G., J. Rader, and A. J. Ridley, Transpolar potential saturation models compared, *J. Geophys. Res., 109*, A09203, 2004.

Siscoe, G. L. and K. D. Siebert, Bimodal nature of solar wind-magnetosphere-ionosphere-thermosphere coupling, *J. Atmos. Solar-Terr. Phys., 68*, 911, 2006.

Snyder, C. W., M. Neugebauer, and U. R. Rao, The solar wind velocity and its correlation with cosmic-ray variations and with solar and geomagnetic activity, *J. Geophys. Res., 68*, 6361, 1963.

Spreiter, J. R. and S. S. Stahara, Gas dynamic and magnetohydrodynamic modeling of the magnetosheath: A tutorial, *Adv. Space Sci., 14*(7), 5, 1994.

Suess, S. T., Y.-K. Ko, R. Von Steiger, and R. L. Moore, Quiescent current sheets in the solar wind and origins of slow wind, *J. Geophys. Res., 114*, A04103, 2009.

Susino, R., R. Ventura, D. Spadaro, A. Vourlidas, and E. Landi, Physical parameters along the boundaries of a mid-latitude streamer and its adjacent regions, *Astron. Astrophys., 488*, 303, 2008.

Tenfjord, P. and N. Østgaard, Energy transfer and flow in the solar wind-magnetosphere-ionosphere system: A new coupling function, *J. Geophys. Res., 118*, 5659, 2013.

Thomsen, M. F., Why Kp is such a good measure of magnetospheric convection, *Space Weather, 2*, S11044, 2004.

Toffoletto, F. R. and G. L. Siscoe, Solar-wind-magnetosphere coupling: An MHD perspective, in *Heliophysics: Plasma Physics of the Local Cosmos*, C. J. Schrijver and G. L. Siscoe (eds.), Cambridge University Press, New York, 2009, p. 295.

Tsurutani, B. T. and W. D. Gonzalez, The cause of high-intensity long-duration continuous AE activity (HILDCAAS): Interplanetary Alfven wave trains, *Planet. Space Sci., 35*, 405, 1987.

Vasyliunas, V. M., J. R. Kan, G. L. Siscoe, and S.-I. Akasofu, Scaling relations governing magnetospheric energy transfer, *Planet. Space Sci., 30*, 359, 1982.

Volino, R. J., A new model for free-stream turbulence effects on boundary layers, *J. Turbomach., 120*, 613, 1998.

Walsh, B. M., J. C. Foster, P. J. Erickson, and D. G. Sibeck, Simultaneous ground- and space-based observations of the plasmaspheric plume and reconnection, *Science, 343*, 1122, 2014a.

Walsh, A. P., S. Haaland, C. Forsyth, A. M. Keesee, J. Kissinger, K. Li, A. Runov et al., Dawn-dusk asymmetries in the coupled solar wind-magnetosphere-ionosphere system: A review, *Ann. Geophys., 32*, 705, 2014b.

Wang, Y.-M., and N. R. Sheeley, Solar wind speed and coronal flux-tube expansion, *Astrophys. J., 355,* 726, 1990.

Weimer, D. R., Improved ionospheric electrodynamic models and application to calculating Joule heating rates, *J. Geophys. Res., 110,* A05306, 2005.

Welling, D. T., M. Andre, I. Dandouras, D. Delcourt, A. Fazakerley, D. Fontaine, J. Foster et al., The Earth: Plasma sources, losses, and transport processes, *Space Sci. Rev., 192,* 145, 2015.

Wing, S., J. R. Johnson, and M. Fujimoto, Timescale for the formation of the cold-dense plasma sheet: A case study, *Geophys. Res. Lett., 33,* L23106, 2006.

Xu, F. and J. E. Borovsky, A new 4-plasma categorization scheme for the solar wind, *J. Geophys. Res., 120,* 70, 2015.

Yau, A. W., W. K. Peterson, and T. Abe, Influences of the ionosphere, thermosphere and magnetosphere on ion outflows, in *The Dynamic Magnetosphere*, W. Liu and M. Fujimoto (eds.), Springer, Berlin, Germany, 2011.

Zhao, L., T. H. Zurbuchen, and L. A. Fisk, Global distribution of the solar wind during solar cycle 23: ACE observations, *Geophys. Res. Lett., 36,* L14104, 2009.

Zirker, J. B., Coronal holes and high-speed wind streams, *Rev. Geophys. Space Phys., 15,* 257, 1977.

CHAPTER 5

The Magnetosheath and Its Boundaries

David G. Sibeck

CONTENTS

5.1 INTRODUCTION

Quantifying the flow of mass, energy, and momentum from the solar wind into the magnetosphere and determining the causes(s) of space weather are the core objectives of the heliophysics discipline. As described in Chapter 4, several processes that occur at the magnetopause, including magnetic reconnection, the Kelvin–Helmholtz instability, and pressure pulse-driven magnetopause boundary waves regulate this flow. Theory provides us with predictions concerning the magnetosheath conditions under which each of these process occurs. In the absence of *in situ* magnetosheath measurements, these predictions are often tested using more readily available solar wind parameters under an implicit assumption that magnetosheath conditions faithfully reflect those seen in the pristine solar wind. This chapter describes processes in the solar wind, foreshock, and magnetosheath that modify incoming solar wind plasma prior to its interaction with the magnetosphere. We begin by identifying which magnetosheath parameters need to be understood to predict the nature of each of the above interaction mechanisms.

A host of indirect measurements demonstrate that reconnection is the most important process governing the solar wind–magnetosphere interaction (see Chapter 6). As predicted by reconnection models (e.g., Dungey, 1961), the orientation of the interplanetary magnetic field (IMF) plays an essential role in determining the location and nature of magnetic reconnection on the magnetopause. Southward IMF turnings enhance the occurrence of flows accelerated by reconnection (Scurry et al., 1994b) and flux transfer events (FTEs) generated by bursty reconnection (Berchem and Russell, 1984) on the dayside equatorial magnetopause, open closed dayside magnetospheric magnetic field lines, erode the dayside magnetopause earthward (Aubry et al., 1970) and the cusps equatorward (Newell et al., 1989), increase the radius of the auroral oval (Holzworth and Meng, 1975) and therefore the quantity of open magnetic flux in the polar caps, strengthen field-aligned currents into and out of the ionosphere (Weimer, 2001), enhance cross-polar potential drops and therefore high-latitude ionospheric convection (Reiff et al., 1981), and ultimately enhance geomagnetic activity (Fairfield and Cahill, 1966; Arnoldy, 1971). Northward IMF turnings may cause reconnection to move to latitudes poleward of the cusps (e.g., Crooker, 1979).

These observations make it clear that a firm understanding of the magnetosheath magnetic field orientation is essential in determining when and where magnetopause reconnection occurs. However, factors other than the IMF orientation also determine the characteristics of reconnection on the magnetopause. Low, high, and variable β favor subsolar, tearing, and patchy magnetopause reconnection, respectively (Crooker, 1990). High plasma beta (Phan et al., 2013) and low IMF cone angles (Scurry et al., 1994a) may inhibit reconnection altogether. The Alfvén velocity may determine the speed at which the reconnection line grows (Shepherd and Cassak, 2012). Steady reconnection lines become impossible in supersonic/super-Alfvénic magnetosheath flows (Fuselier et al., 2000). Hybrid code simulation results (N. Omidi, pers. comm., 2015) indicate that turbulent magnetosheath conditions favor bursty reconnection and the formation of FTEs (Russell and Elphic, 1978).

Waves generated by the Kelvin–Helmholtz instability on the magnetopause can transfer magnetosheath energy and momentum to the magnetosphere (Axford and Hines, 1964; Pu and Kivelson, 1983). Nonlinear waves can trigger magnetic reconnection and thereby abet the transfer of magnetosheath mass into the magnetosphere (Ma et al., 2014a, b). Large flow shears (i.e., high magnetosheath velocities) perpendicular to ambient magnetic fields and small Alfvén velocities favor the occurrence of the linear instability (Southwood, 1968).

Pressure variations within the magnetosheath drive waves on the magnetopause that also transfer energy and momentum to the magnetosphere. The pressure variations may originate in the solar wind (e.g., Sibeck, 1990), due to processes within the foreshock (Fairfield et al., 1990), at the bow shock (Hietala et al., 2009), or in the magnetosheath itself. Intrinsic density and dynamic pressure variations are most common at stream–stream interactions within the solar wind. Alfvénic fluctuations, which may be converted to pressure variations within the magnetosheath (Lin et al., 1996a, b), are most common in the high speed solar wind. The kinetic processes that modify solar wind parameters prior to their interaction with the bow shock occur within (Blanco-Cano et al., 2009) or at the edges of the foreshock (Omidi et al., 2009b), which is located upstream from the subsolar magnetopause during periods of nearly radial IMF orientation. Temperature anisotropies result in ion cyclotron waves when the plasma β is low, but mirror mode waves when it is high (Gary et al., 1993; Schwartz et al., 1996).

Summarizing, to understand the nature of the processes that govern the solar wind-magnetosphere interaction at the magnetopause, we need to be able to predict steady-state and transient magnetosheath magnetic field strengths and orientations, magnetosheath plasma and Alfvén velocities, magnetosheath pressures and plasma β as a function of time, solar wind conditions, and location on the magnetopause. The remainder of this paper is structured to determine which of these predictions can be successfully accomplished with the aid of gas dynamic, magnetohydrodynamic (MHD), and hybrid kinetic models. For other aspects of the magnetosheath, please see excellent reviews of waves by Denton (2000) and of large-structures by Siscoe et al. (2002) and Song and Russell (2002).

5.2 THE STRUCTURE OF THE STEADY-STATE MAGNETOSHEATH AND ITS BOUNDARIES

Thanks to the frozen-in condition, the Earth's magnetic field carves out the magnetospheric cavity in the oncoming solar wind. Because the solar wind flow is supersonic and super-Alfvénic, a bow shock is required to slow and divert the oncoming plasma around the magnetosphere and through the magnetosheath. This section compares the predictions of numerical simulations with observations of the magnetosheath for steady solar wind conditions. It progresses from results from gas dynamic through MHD to hybrid code simulations.

5.2.1 Gas Dynamic Models

No paper has had, or continues to have, as great an impact on magnetosheath studies than that of Spreiter et al. (1966), who presented the results of a gas dynamic simulation for magnetosheath plasma parameters. Begin by considering the location of the gas dynamic magnetopause and bow shock. At rest, the dayside magnetopause lies along the locus of points where the sum of the thermal and magnetic pressures within the magnetosheath and magnetosphere balance. The very successful Newtonian approximation indicates that the pressure applied by the magnetosheath to any point on the magnetopause is proportional to $\rho_{SW} \cos^2\theta$, where ρ_{SW} is the solar wind dynamic pressure and θ is the angle between the normal to the magnetopause and the flow direction of the (unshocked) solar wind. By contrast, the magnetic pressure predominates throughout most of the outer magnetosphere. With reasonable assumption that the Earth's dipole and Chapman–Ferraro magnetopause currents make predominant contributions to the magnetospheric magnetic field, and noting that the contributions of both current systems to the magnetic field strength just inside the magnetopause vary as the inverse cube of distance from Earth, the location of the magnetopause can be determined by equating magnetosheath and magnetospheric pressures (Martyn, 1951). This leads to the expectation that the magnetopause moves earthward when solar wind dynamic pressures increase. Furthermore, the shape of the magnetopause should remain nearly self-similar to variations in the solar wind Mach number and dynamic pressure, a point confirmed by Sibeck et al. (1991).

Turning to the bow shock, this boundary lies at rest along the locus of points where the magnetosheath sound speeds balance the component of the solar wind velocity normal to the bow shock within the gas dynamic framework. The bow shock moves earthward, and the width of the magnetosheath decreases, as the solar wind Mach number increases and the solar wind dynamic pressure remains constant. Magnetosheath densities, velocities, and temperatures applied to the magnetopause all increase with increasing solar wind Mach number.

The Spreiter et al. model employs a rigid axisymmetric tangential discontinuity magnetopause. The magnetic field has no effect upon the gas dynamic flow. Figure 5.1 shows their results for a solar wind sonic Mach number of 8 and an adiabatic constant of 5/3. Densities and temperatures peak in the subsolar magnetosheath, where velocities vanish. Flows increase, but temperatures decrease, antisunward toward the flanks, where there is a strong positive outward radial density gradient. Figure 5.2 shows how spiral IMF lines drape around the magnetopause within the equatorial plane. The sense of the magnetic field component along the Sun–Earth line reverses across local noon, whereas the sense of the component transverse to this line remains the same.

The gas dynamic model explains the basic features of the magnetosheath very well, in particular the draping of magnetosheath magnetic field lines around the magnetopause (Fairfield, 1967) and the diversion of flows around the magnetosphere (Howe and Binsack, 1972). Nevertheless, it is not perfect. Following a statistical study that employed simultaneous solar wind and magnetosheath measurements, Coleman (2005) concluded "it is not safe to rely on the orientation of the magnetosheath magnetic field at any given patch within 2 R_E of the magnetopause to be similar to that observed in the upstream IMF or predicted by any simple gas dynamic or analytical model." Fortunately, the chance of predicting the north/south orientation of the magnetosheath magnetic field increases with increasing strength of this IMF component (Safrankova et al., 2009).

More recently, Dimmock and Nykyri (2013) have reported results from a statistical study of THEMIS observations. They confirm the predicted steady increase in magnetosheath velocities from minimal values just outside the subsolar magnetopause to values near those in the solar wind on the magnetotail flanks and note the expected decrease in magnetosheath densities and magnetic field strengths from the subsolar magnetopause toward the flanks. To complete the survey, Dimmock et al. (2015) reported the anticipated decrease in magnetosheath temperatures from the dayside to the nightside magnetosheath.

Despite the availability of the more modern, and more self-consistent, global MHD and hybrid code simulations

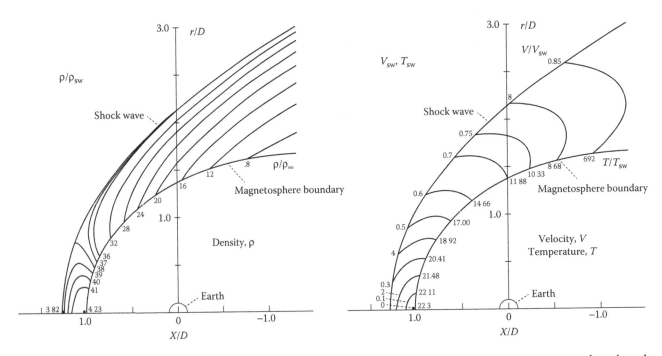

FIGURE 5.1 Predictions of a gas dynamic model for magnetosheath densities, velocities, and temperatures when the solar wind Mach number is 8 and the adiabatic constant is 5/3. (From Spreiter, J.R. et al., *Planet. Space Sci.*, 14, 223–253, 1966. With Permission.)

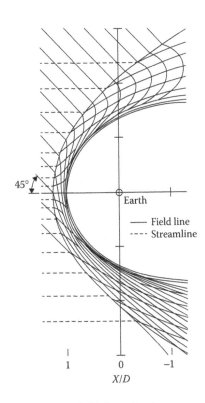

FIGURE 5.2 Magnetic field line draping around the magnetosphere for a spiral interplanetary magnetic field in a gas dynamic model. (From Spreiter, J.R. et al., *Planet. Space Sci.*, 14, 223–253, 1966. With Permission.)

that will be discussed momentarily, researchers continue to use the Spreiter et al. (1966) model to predict conditions just outside the magnetopause (e.g., Fear et al., 2007; Sibeck and Lin, 2011) and particle precipitation in the cusps (Onsager et al., 1993; Wing et al., 2001). Its continued use is primarily due to the work of Cooling et al. (2001), who generated and popularized easy-to-use equations for curves fit to the predictions of the Spreiter et al. model for plasma parameters just outside the magnetopause.

For corresponding steady-state magnetic field strengths and directions just outside the magnetopause, Cooling et al. (2001) adopted the algorithms for a current-free magnetosheath originally reported by Kobel and Flückiger (1994). The gas dynamic and analytical models for the magnetosheath magnetic field find utility in predicting paths for energetic particle entry and exit from the magnetosheath (Luhmann et al., 1984b; Trattner et al., 2003) and possible reconnection locations on the magnetopause (e.g., Luhmann et al., 1984a; Trattner et al., 2007a, b). However, they fail to predict the degree to which the magnetosheath magnetic field is compressed (Petrinec, 2013). And the magnetic nulls predicted by the models for locations outside the pre-noon magnetopause during periods of spiral IMF orientation (e.g., Crooker et al., 1985) have yet to be identified in observations.

5.2.2 MHD Models

Despite the successes of the gas dynamic plasma and analytical magnetic field models, there are many features of the magnetosheath that they cannot predict, because they do not include the effects of the interplanetary, magnetosheath, and magnetospheric magnetic fields self-consistently. Here, we consider several effects seen in global MHD models, and one which is not. The first is the effect of the north/south component of the IMF upon the locations of the bow shock and magnetopause. The second is the presence of a depletion layer with depressed densities outside the dayside equatorial magnetopause. The third is a closely related effect, namely enhanced antisunward acceleration of magnetosheath plasma on the flanks of the magnetosphere. The fourth addresses the effects of a radial IMF upon magnetosheath plasma parameters and the locations of the bow shock and magnetopause. The fifth concerns the effects of magnetopause reconnection on the magnetosheath magnetic field and plasma configuration. MHD models that invoke anisotropic temperatures (e.g., Samsonov et al., 2001) can also be used to predict the magnetosheath regions where ion cyclotron and mirror mode instabilities occur, a topic discussed later in Section 5.2.3. Finally, we consider the evidence for standing slow mode waves in the magnetosheath, a phenomenon predicted on the basis of observations (Southwood and Kivelson, 1992, 1995), but not observed in simulations.

5.2.2.1 Effects of the North/South IMF Orientation on the Locations of the Bow Shock and Magnetopause

When the IMF turns southward, the chance of reconnection occurring on the equatorial dayside magnetopause increases. Reconnection diverts a portion of the magnetopause current to flow into and out of the ionosphere in the Region 1 sense, namely downward prior to local noon and upward after local noon. The effect of these currents is to diminish magnetic field strengths within the outer dayside magnetosphere. Decreases in magnetospheric magnetic field strengths upset the balance between magnetosheath and magnetospheric

pressures at the magnetopause. The dayside magnetopause moves earthward, or "erodes," until it reaches locations where pressure balance can be restored (Hill and Rassbach, 1975; Maltsev and Lyatsky, 1975).

Erosion is seen in both the output of MHD simulations and observations. MHD simulations indicate that the subsolar magnetopause erodes inward and the magnetotail magnetopause flares outward when the IMF turns southward IMF during the growth phase of geomagnetic substorms (e.g., Elsen and Winglee, 1997; Siscoe et al., 2002; Wang et al., 2004), just as observed (Aubry et al., 1970; Maezawa, 1975). Simulations also predict that the subsolar bow shock moves earthward during intervals of southward IMF orientation (e.g., Siscoe et al., 2002). The density profiles shown in Figure 5.3 indicate bow shock and magnetopause locations as a function of IMF clock angle.

5.2.2.2 Effects of the IMF Cone Angle on the Bow Shock, Magnetopause, and Magnetosheath

The bow shock, magnetopause, and magnetosheath also respond to variations in the IMF cone angle, the angle between the IMF and the Sun–Earth line. These responses result from the different ways by which the bow shock processes solar wind plasma with magnetic field orientation nearly parallel and perpendicular to the shock normal, that is, quasi-parallel and quasi-perpendicular bow shocks.

Within the MHD framework, the bow shock lies at rest along the locus of points where the components of the fast mode velocities within the magnetosheath and the solar wind velocities normal to the bow shock balance. Since fast mode velocities perpendicular to magnetic field lines exceed those parallel to magnetic field lines, the distance that the quasi-perpendicular bow shock stands off from the magnetopause should exceed that for the quasi-parallel bow shock. Thus, as shown in Figure 5.4, the post-noon bow shock should lie further from Earth than the pre-noon bow shock for spiral IMF orientations (Chapman et al., 2004), an effect yet to be reported at Earth.

Furthermore, one might expect the subsolar bow shock to move outward as the IMF orientation varies from quasi-parallel to quasi-perpendicular, as seen in the simulation results reported by Chapman et al. (2004). Verigin et al. (2001) confirmed an interesting prediction that the bow

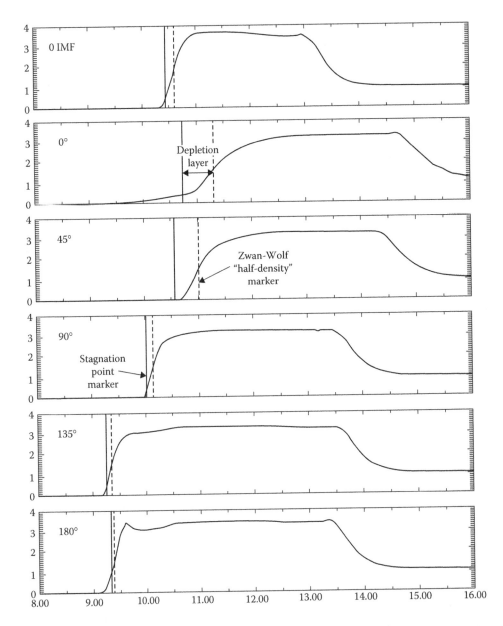

FIGURE 5.3   MHD model magnetosheath density profiles for no IMF (top panel) and clock angles of 0°, 45°, 90°, 135°, and 180°. The IMF clock angle is the angle between the IMF and the northward direction in the plane perpendicular to the Sun–Earth line. (From Siscoe, G.L. et al., *Planet. Space Sci.*, 50, 461–471, 2002. With Permission.)

shock moves Earthward with increasing Mach number for IMF orientations transverse to the Sun–Earth line, but outward for IMF orientations parallel to the Sun–Earth (Cairns and Lyon, 1996).

Now consider magnetosheath plasma parameters. Walters (1964) argued that the presence of a Parker spiral IMF embedded in the flowing solar wind plasma results in greater dawnside (quasi-parallel bow shock) than duskside (quasi-perpendicular bow shock) magnetosheath densities, particularly during intervals of low solar wind Mach number. Global MHD models support

this prediction (Chapman et al., 2004). Observationally, Sibeck (1995) and Sibeck and Gosling (1996) presented case studies demonstrating greater magnetosheath densities behind the quasi-perpendicular than quasi-parallel bow shock. While Walsh et al. (2012) reported observations indicating greater dawnside than duskside densities for spiral IMF orientations, Dimmock and Nykyri (2013) found little evidence for any such density asymmetry. Both sets of authors reported greater magnetic field strengths behind the quasi-perpendicular than quasi-parallel bow shock. Dimmock and Nykyri

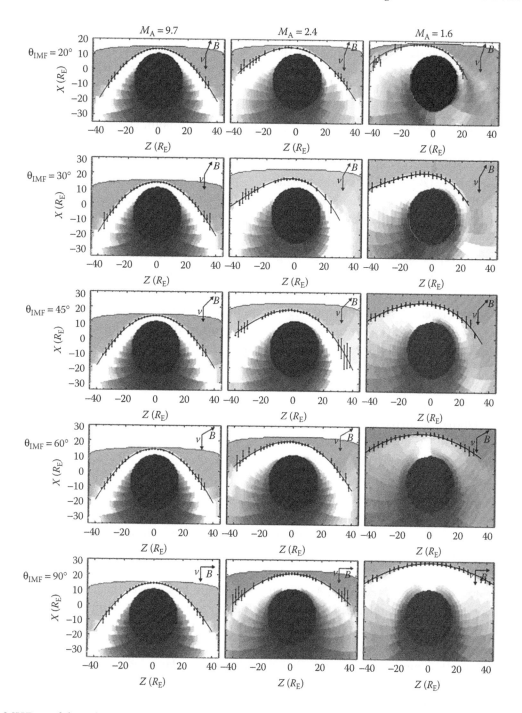

FIGURE 5.4 MHD model predictions for magnetosheath densities (white high, gray low) and the location of the bow shock as a function of solar wind Mach number and the angle between the IMF and the Sun–Earth line. The dark inner ellipse is the magnetopause obstacle. (From Chapman, J.F. et al., *J. Geophys. Res.*, 109, doi:10.1029/2003JA010235, 2004. With Permission.)

(2013) also reported greater velocities behind the quasi-perpendicular than quasi-parallel bow shock.

We might therefore also expect the IMF orientation to control the pressure applied to the dayside magnetopause. Samsonov et al. (2012) first inspected results from global MHD simulations employing isotropic pressures to demonstrate that sum of the thermal and

magnetic pressures applied to the subsolar magnetopause diminishes from 97% to 86% of the solar wind dynamic pressure as the IMF rotates from northward to radial orientations. He then demonstrated that the pressure applied to the subsolar magnetopause diminishes from 99% to 78% of the solar wind dynamic pressure for anisotropic temperatures. The decrease in pressure for radial IMF orientations might explain previously reported case and statistical studies indicating that the dayside magnetopause moves outward under these circumstances (Dusik et al., 2010; Suvorova et al., 2010).

5.2.2.3 Plasma Depletion Layer

In one of magnetospheric physics most striking but least heralded successes, Zwan and Wolf (1976) predicted the formation of a narrow (700–1300 km wide) plasma depletion layer (PDL) with factor of ~2 depleted densities and enhanced magnetic field strengths just outside the dayside magnetosphere when plasma is squeezed out from IMF fields draped against the dayside magnetosphere during intervals when no reconnection is occurring locally. Case studies immediately verified the existence of such a layer (Paschmann et al., 1978; Crooker et al., 1979). Phan et al. (1994) then reported results from a superposed epoch study indicating the absence of the layer for southward IMF orientations and its presence with a typical thickness of 0.3 R_E for northward IMF orientations.

Pudovkin et al. (2001) concurred that a narrow PDL forms during intervals of northward IMF orientation, but went on to predict a broad gradual magnetic barrier transition from low magnetic field strengths and high densities in the mid-magnetosheath to high magnetic field strengths and low densities just outside the magnetopause during periods of southward IMF orientation when reconnection is occurring on the subsolar magnetopause. Results from global magnetosheath hydrodynamic models do not confirm this prediction. As shown in Figure 5.3, Siscoe et al. (2002) found no such magnetic barrier for southward IMF orientations in global MHD simulations. Wang et al. (2004) reported results from an extensive study of MHD model results for depletion layer thicknesses and density depressions as a function of solar wind conditions, showing that layer thicknesses decrease as the IMF rotates away from northward. The thickness of the layer also increases for low solar wind Alfvénic Mach number (Farrugia et al., 1997).

However, the layer can sometimes be observed even during periods with southward IMF orientations. Although less pronounced than those which occurred for northward magnetosheath magnetic field orientations, some of the PDL cases reported by Fuselier et al. (1991) and Anderson and Fuselier (1993) occurred for southward magnetosheath fields. Since these events were observed by the AMPTE/CCE spacecraft, with an apogee of only 8.8 R_E, they also occurred for enhanced solar wind dynamic pressures and magnetosheath plasma beta, the latter of which presumably suppresses reconnection enabling the drainage that forms the depletion layer (Anderson et al., 1997). Moretto et al. (2005) reported multipoint cluster observations indicating the presence of a prominent PDL at high latitudes even during periods when the IMF has a southward component.

The significance of the layer lies in the fact that it enables stable reconnection to occur at locations further away from the subsolar point on the magnetopause than might otherwise be expected by enhancing Alfvén velocities in the inner magnetosheath and thereby depressing Alfvénic Mach numbers. The presence of the layer has been invoked to explain remote observations indicating steady reconnection on the magnetopause poleward of the cusps during northward IMF orientations (Fuselier et al., 2000; Petrinec et al., 2003). Both case (Avanov et al., 2001) and statistical studies (Panov et al., 2008) of *in situ* observations confirm that the presence of the layer enables quasi-steady reconnection on the high latitude magnetopause.

5.2.2.4 Curvature Forces, Accelerated Flows, and Magnetotail Flattening

The ends of the magnetic field lines that lie draped over the dayside magnetopause continue to move antisunward as plasma drains from the PDL. Consequently, these field lines are subjected to both strong antisunward magnetic curvature and pressure gradient forces. Once they begin to slip over the flanks of the magnetosphere, they accelerate rapidly. Numerical simulations (Lavraud et al., 2007) indicate that within a narrow layer just outside the flank magnetopause can even exceed those within the solar wind itself (Chen et al., 1993; Lavraud et al., 2007).

Theory indicates that curvature forces associated with the draped magnetosheath magnetic fields should flatten the magnetotail cross section in the direction perpendicular of the component of the IMF in the plane transverse to the Sun–Earth line and elongate the magnetotail cross section in the direction parallel to this component (Michel and Dessler, 1970). This prediction was ultimately confirmed by both global MHD simulations (Berchem et al., 1998) and observations (Sibeck et al., 1986). However, the cross section of the distant bow shock does not exhibit the same pronounced flattening. Greater fast mode speeds perpendicular than

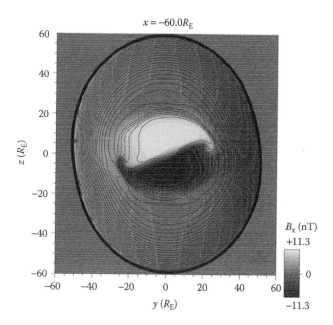

FIGURE 5.5 MHD model predictions for the magnetotail cross section at lunar distance when there is a strong (7 nT) duskward IMF orientation. Colors indicate the strength of the magnetic field component parallel to the Sun–Earth line, while contours indicate electric current strengths. (From From Sibeck, D.G. and Lin, R.-Q., *J. Geophys. Res.*, 119, 1028–1043, 2014. With Permission.)

parallel to the IMF lines draped over the distant magnetotail cause the bow shock to lie further from the magnetotail in the direction transverse to the IMF component in the plane transverse to the Sun–Earth line than in the direction parallel to this component. As shown in Figure 5.5, this can result in marked latitudinal differences in the magnetosheath thickness (Sibeck and Lin, 2014).

5.2.2.5 Effects of Reconnection on the Magnetosheath Magnetic Field Configuration

Some of the magnetosheath magnetic field lines that drape against the dayside magnetopause reconnect with magnetospheric magnetic field lines. Whether reconnection occurs along a line passing through the subsolar point whose tilt depends upon the IMF orientation (Gonzalez and Mozer, 1974; Sonnerup, 1974), in regions where magnetopause current strengths peak (Semenov and Pudovkin, 1985; Alexeev et al., 1998), along lines tracing peaks in magnetic shear across the magnetopause (Trattner et al., 2007a, b), or along the portions of these curves where magnetosheath and magnetospheric magnetic fields lie nearly antiparallel

(Crooker, 1979), it can have a profound effect upon the structure of the magnetosheath. For example, White et al. (1998) presented results from global MHD models indicating the presence of thin "sashes" of weak magnetic field strengths over regions of the magnetopause where magnetosheath and magnetospheric magnetic fields lie nearly antiparallel, that is, the locus of points where antiparallel reconnection is expected. Eriksson et al. (2004) reported evidence for both accelerated and deccelarated flows with the directions and velocities predicted for antiparallel reconnection in locations where the sash is expected.

More generally, and more broadly, reconnection launches a set of MHD waves that propagate into both the magnetosheath and magnetosphere, as shown in Figure 5.6 (Coroniti and Kennel, 1979). These waves serve to mediate the transition between high density flowing magnetosheath plasmas with magnetic field strengths and orientations that depend upon those in the solar wind and lower density, slower moving magnetospheric plasmas with magnetic field strengths and orientations determined by magnetospheric current systems. The fast mode compressional (FTCP) waves propagate most rapidly away from the magnetopause, but are generally not needed to mitigate the transition and have therefore not been reported. By contrast, an intermediate mode wave is needed to effect a rotation in the magnetic field orientation from that in the magnetosheath to the magnetosphere. A slow mode rarefaction (SLRF) wave fan is needed to produce a gradual transition from magnetosheath to magnetospheric densities, velocities, and magnetic field strengths.

The effects of the intermediate mode wave can already be seen at distances of 30–40 Earth radii (R_E) downstream from Earth. Here, Kaymaz et al. (1992) have shown that the pattern of magnetosheath magnetic field line draping about the magnetotail when viewed from the Earth rotates clockwise for dawnward IMF orientations and counterclockwise for duskward IMF orientations. The sense of rotation is consistent with that for the component reconnection line on the dayside magnetopause (Gonzalez and Mozer, 1974) and magnetotail twisting (Cowley, 1981) for the given IMF conditions, that is, it facilitates the transition between magnetotail and magnetosheath magnetic field orientations. At similar downstream distances, the structure of the magnetotail magnetopause is consistent with predictions

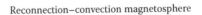

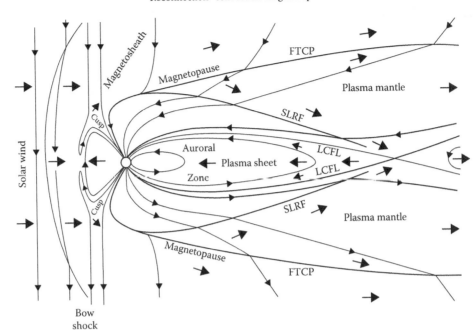

FIGURE 5.6   A reconnection model for the magnetosphere showing both regions and wave fronts. FTCP stands for fast compression wave, SLRF for slow rarefaction wave, LCFL for last closed magnetic field line. (From Coroniti, F.V. and Kennel, C.F., *Magnetospheric reconnection, substorms, and energetic particle acceleration, AIP Conference Proceedings*, American Institute of Physics, New York, 56, pp. 169–178, 1979. With Permission.)

for a transition mediated by slow and intermediate mode waves (Sanchez and Siscoe, 1990). The slow mode transition fan broadens with downstream distance. Magnetosheath and magnetospheric magnetic field lines connect through the apices of the flattened magnetotail via a transition region many Earth radii wide (Berchem et al., 1998). The magnetotail flattens, twists, and flares in response to the imposed interplanetary and magnetosheath magnetic field orientations (Sibeck et al., 1986). Defining the location of the magnetopause becomes difficult (Sibeck and Lin, 2014).

5.2.2.6 Standing Slow Mode Wave Fronts
As a final MHD effect, consider the standing compressional slow mode wave fronts proposed by Southwood and Kivelson (1992, 1995). By analogy to the standing bow shock (a standing fast mode wave front), and the slow rarefaction fan invoked to explain the magnetopause plasma boundary layer, these authors suggested that a compressional slow mode wave might stand upstream from the magnetopause in the magnetosheath flow. Such a wave might exhibit a spatially localized density increase and magnetic field strength decrease; Song et al. (1990, 1992b) had reported features

such as these. However, no evidence for such isolated density and magnetic field strength variations could be found in global MHD simulations (Wang et al., 2004) and the localized density increases and magnetic field strength decreases could readily be explained in terms of solar wind features transmitted through the bow shock into the magnetosheath (Hubert and Samsonov, 2004, 2005; Samsonov and Hubert, 2004; Song et al. 2005).

5.2.3 Hybrid Code Models
Within recent years, results from 2.5- and occasionally 3-dimensional global hybrid code models have become available. Because they can account for wave–particle (ion) interactions, these models afford an opportunity to incorporate the effects of the foreshock upon the solar wind–magnetosphere interaction and the generation of both ion cyclotron and mirror mode waves via temperature/pressure anisotropies within the Earth's magnetosphere.

5.2.3.1 Waves in the Foreshock
Bale et al. (2005), Burgess et al. (2005), Eastwood et al. (2005), and Wilson (2016) provide excellent extensive

reviews of foreshock waves and particle distributions. Although the quasi-perpendicular foreshock is restricted to a narrow region near the shock foot upstream from regions of the bow shock where IMF lines make angles greater than 45° with shock normals, the quasi-parallel foreshock can extend many Earth radii upstream from the bow shock along IMF lines that make angles less than 45° with shock normals. Different regions of the quasi-parallel foreshock are populated by reflected, intermediate, and diffuse suprathermal ion populations reflected from the bow shock and energized/thermalized by repeated interactions with a wealth of waves and structures, including 30s, 10s, 3s, and 1s waves, shocklets, and short large-amplitude magnetic structures (Blanco-Cano et al., 2009; Kempf et al., 2015).

Solar wind and magnetosheath flows sweep the particles distributions, waves, and structures across the magnetosheath and up to the magnetopause. The suprathermal particles appear at locations in the inner or outer magnetosheath that can be predicted on the basis of the IMF orientation, but only on magnetosheath magnetic field lines that lead to the quasi-parallel bow shock (Crooker et al., 1981). Simultaneous foreshock and magnetospheric observations demonstrate that foreshock waves with periods of 10–100 s succeed in entering the magnetosphere, perhaps via the cusps (Engebretson et al., 1987).

The microphysical processes that occur within the foreshock can have a significant effect upon the overall solar wind–magnetosphere interaction. As shown in Figure 5.7 and noted by Thomas and Brecht (1988), counterstreaming solar wind and reflected ion populations are susceptible to beam–beam interactions that generate waves which in turn thermalize both solar wind and reflected ion populations. The enhanced temperatures and pressures associated with the thermalized ion populations excavate foreshock cavities with depressed densities and magnetic field strengths bounded by regions of enhanced density and magnetic field strengths. Features with these characteristics were noted by Sibeck et al. (1989) and demonstrated

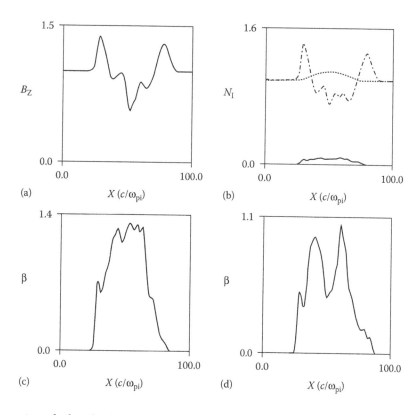

FIGURE 5.7 The cross section of a foreshock cavity as seen in a hybrid code simulation. (a) Magnetic field strength, B_z, (b) ion density, (c) ion beta for the beam of ions reflected from the bow shock, and (d) ion beta for the background solar wind plasma. Background (beam) ions are given by dashed-dot (solid) lines. The dotted line shows initial densities. The beam of reflected ions has a width of $50c/\omega_{pi}$ where c is the speed of light and ω_{pi} is the ion plasma frequency. (From Thomas, V.A. and Brecht, S.H., *J. Geophys. Res.*, 93, 11341–11353, 1988. With Permission.)

to result from foreshock phenomena by Fairfield et al. (1990). Sibeck et al. (2001) reported a statistical study of the density depressions within the cavities, while Rojas-Castillo et al. (2013) reported a survey of the enhanced densities bounding them.

If transmitted through the magnetosheath, the density decreases that occur within the foreshock should diminish the solar wind dynamic pressure applied to the magnetosphere and allow the magnetopause to move outward. Both Sibeck et al. (1989) and Fairfield et al. (1990) reported observations indicating a direct correspondence between the foreshock density and dayside magnetospheric magnetic field strengths. Since the foreshock lies immediately upstream from the subsolar bow shock during intervals of radial IMF, one might expect the magnetopause to move outward during periods of radial IMF, as indeed was reported by Dusik et al. (2010) and Suvorova et al. (2010). Hence both the anisotropic pressures discussed earlier within the context of MHD model predictions and the foreshock density cavities generated by kinetic processes contribute to reducing pressures upon the dayside magnetosphere during the period of radial IMF (Samsonov et al., 2015).

5.2.3.2 Waves in the Magnetosheath

The solar wind and magnetosheath flow sweeps all of the waves generated in the upstream region across the bow shock and into the magnetosheath. In addition to these waves generated in the upstream region, there are waves generated by temperature/pressure anisotropies within the magnetosheath. Consequently, the magnetosheath is replete with waves over a wide range of frequencies (e.g., Song et al., 1992a, 1993a, b).

Anderson and Fuselier (1993) summarize the conditions governing the generation of mirror mode and electromagnetic ion cyclotron (EMIC) waves in the dayside magnetosheath. Parcels of solar wind plasma are compressed when they cross the quasi-perpendicular bow shock, resulting in enhanced temperatures perpendicular to the magnetic field. The newly created temperature anisotropies are susceptible to the mirror mode instability, which creates bundles of magnetic field lines with anticorrelated thermal and magnetic pressures. Plasma drains from magnetosheath magnetic field lines draped over the magnetopause in the PDL, depressing the plasma β and further enhancing

temperature anisotropies to the point where the conditions required to generate EMIC waves are satisfied. Hubert et al. (1998) reported a case study in which ion cyclotron waves were also observed in a narrow layer just inside the bow shock. Finally, mirror mode and EMIC waves are absent for quasi-parallel IMF orientations, leaving only broad-band noise in the dayside magnetosheath.

Omidi et al. (2014) recently reported the results of 2.5-D hybrid code simulations demonstrating the serial formation of mirror mode-like structures in the magnetosheath. Figure 5.8 shows some examples. Although the structures form at the quasi-parallel bow shock, they can extend deep into the magnetosheath behind the quasi-perpendicular bow shock. They exhibit anticorrelated densities and magnetic field strengths. Their existence was even more recently confirmed by Gutynska et al. (2015), who reported the frequent occurrence of such structures with dimensions on the order of $0.4 \, R_E$ behind the quasi-parallel bow shock. By contrast, it is expected that EMIC waves will generally only be observed in the spatially limited depletion layer region just outside the dayside magnetopause. Consequently, it is not surprising that a magnetosheath survey by Denton et al. (1995) found only

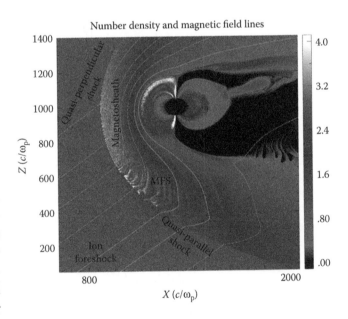

FIGURE 5.8 Field-aligned density structures in the dayside magnetosheath seen in results from a global hybrid code model. The ion skin depth, c/ω_{pi}, is defined as in Figure 5.7. (From Omidi, N. et al., *J. Geophys. Res.*, 119, 2593–2604, 2014. With Permission.)

mirror mode (and no ion cyclotron) waves in the frequency range from 0.01 to 0.04 Hz.

Whatever their origin is, the significance of the mirror mode and mirror mode-like density structures within the magnetosheath lies in their ability to trigger transient, patchy reconnection at the dayside magnetopause. As noted earlier (Phan et al., 2013), high plasma β suppresses reconnection. The arrival of a series of plasma filaments with alternating plasma β might therefore be expected to trigger a series of reconnection bursts (Crooker, 1990). Suspecting such a scenario, N. Omidi (pers. comm., 2015) spatially smoothed magnetosheath parameters in global hybrid code simulations, finding that smoothing reduced the occurrence of bursty reconnection and the production of FTEs, as can be seen in Figure 5.9. Whether mirror modes actually trigger bursty reconnection at the magnetopause remains to be determined, as Song et al. (1993a) have noted that strong slow mode structures with frequencies in the 10–100 s range decay rapidly upon approaching the magnetopause boundary layer. Consistent with this, Genot et al. (2009) report that mirror mode waves are most prominent in the mid-magnetosheath at low (subsolar) zenith angles. However, they go on to note that the waves only reach the magnetopause at larger zenith angles.

5.3 TRANSMISSION OF SOLAR WIND DISCONTINUITIES AND FORESHOCK-GENERATED STRUCTURES INTO THE MAGNETOSHEATH

Since the onset of a wide range of magnetospheric phenomena, including bursty reconnection at the dayside magnetopause (Lockwood and Wild, 1993), substorms (e.g., Liou, 2007; Wild et al., 2009), and geomagnetic storms (e.g., Gosling et al., 1967), have been attributed to the arrival of solar wind discontinuities such as rotations in the IMF orientation and abrupt variations in the solar wind dynamic pressure, it is important to determine how processes at the bow shock and within the magnetosheath modify incoming solar wind discontinuities and foreshock-generated structures. Even setting aside the need to determine whether individual solar wind features actually arrive (Crooker et al., 1982) and then the times at which they arrive (Weimer et al., 2002), the literature on this subject is immense. Here, we simply provide pointers to representative articles which may serve as starting points for more extensive literature searches. We consider interplanetary shocks, rotational discontinuities, and tangential discontinuities, sequentially.

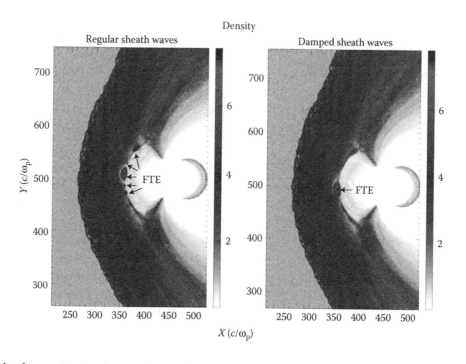

FIGURE 5.9 Results from a 2.5D hybrid code simulation indicating the formation of fewer flux transfer events on the magnetopause when spatial smoothing is applied to waves in the magnetosheath. The ion skin depth, c/ω_{pi}, is defined as in Figure 5.7. (From Omidi, N., pers. comm., 2015.)

5.3.1 Interplanetary Shocks

Within gas dynamic models, shocks move at the sum of the sound speed and the background plasma flow velocity. Since this quantity is approximately conserved, shock fronts remain nearly planer when entering the magnetosheath in gas dynamic models (Spreiter and Stahara, 1994). Results from MHD simulations and observations indicate that front speeds lag in the inner subsolar magnetosheath (Koval et al., 2005, 2006). The transmitted shock drives bow shock and magnetopause motion (Zhang et al., 2009).

5.3.2 Rotational Discontinuities

Within MHD, the interaction of any two structures, such as a rotational discontinuity and the Earth's bow shock, has the potential to launch a set of forward and backward propagating fast, slow, and intermediate mode waves (Völk and Auer, 1974). Lin et al. (1996a, b) employed both MHD and hybrid code simulations to study the transmission of rotational discontinuities through the bow shock. Within MHD, the interaction launches pairs of intermediate and slow mode shocks that transmit a strong density pulse through the magnetosheath toward the magnetopause. Within the hybrid code simulation, the pulse is bounded by rotational discontinuities.

Observationally, Hubert and Harvey (2000) tracked the effects of interplanetary discontinuities from the solar wind into the magnetosphere through the magnetosheath, reporting that their interaction with the bow shock results in a density enhancement and magnetic field strength decrease that propagates toward and strikes the magnetopause. Korotova et al. (2012) reported simultaneous observations of the forward fast mode wave propagating inward through the magnetosphere and the outward-propagating reverse fast mode wave forming the new bow shock.

More recent work, employing hybrid code simulations, indicates that the arrival of certain rotational discontinuities can result in the formation of foreshock bubbles, vast regions of solar wind upstream from the bow shock with greatly enhanced temperatures and depressed densities and magnetic field strengths (Omidi et al., 2010; Turner et al., 2013). By contrast to hot flow anomalies (HFAs), foreshock bubbles have a global impact upon the magnetosphere (Archer et al., 2015).

5.3.3 Tangential Discontinuities

Within MHD simulations, the transmission of tangential discontinuities across the bow shock might be considered a relatively simple matter. As in the case of rotational discontinuities, a set of fast, slow, and intermediate mode wave may be generated (e.g., Wu et al., 1993). The forward fast mode wave propagates across magnetosheath magnetic field lines to strike the magnetopause. The transmitted tangential discontinuity drapes about the magnetosphere (Keika et al., 2009).

Furthermore, the compression of the tangential discontinuities that occurs when they transit the bow shock and become draped against the magnetopause can trigger magnetic reconnection (Omidi et al., 2009a; Phan et al., 2011).

However, the most surprising feature associated with the interaction of certain tangential discontinuities with the bow shock is the formation of HFA sin the region immediately upstream from the Earth's bow shock. The features can be identified on the basis of greatly heated ion and electron plasmas exhibiting flows deflected far from the Sun–Earth line centered on sharp rotations in the magnetic field direction (Schwartz et al., 1985; Thomsen et al., 1986). The events exhibit depressed densities and magnetic field strengths, but are often bounded by regions of greatly enhanced densities and magnetic field strengths.

HFAs were soon seen in the output of numerical simulations. As shown in Figure 5.10, Thomas et al. (1991) employed a hybrid code simulation with a planar shock geometry to obtain HFAs with the requisite characteristics, while Lin (1997) and Omidi and Sibeck (2007) employed global hybrid code simulations to demonstrate the formation of HFAs with similar properties at curved bow shocks.

It has been known for some time that HFAs can have a significant effect upon both the magnetosphere and the ionosphere (Sitar et al., 1998; Sibeck et al., 1999). Although HFAs had been observed within the magnetosheath (Paschmann et al., 1988), the method by which their pressure variations were transmitted to the magnetopause long remained a mystery. That situation abruptly changed with the remarkable suggestion that transmission occurred via jets behind the oblique portions of rippled bow shocks (Hietala et al., 2009, 2012). As writing for this review came to a close, a number of researchers were devoting considerable attention to the characteristics of these fascinating jets

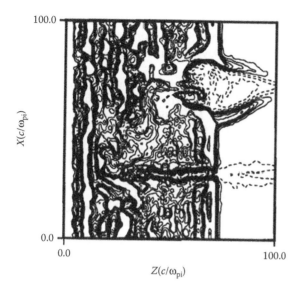

FIGURE 5.10 Predictions of a hybrid code simulation for the formation of a HFA at a planar bow shock. The figure shows density contours in the $X–Z$ plane with a bow shock in the $Z = 70$ c/ω_{pi} plane. The magnetosheath lies inside the bow shock at $Z < 70$ c/ω_{pi}, and the solar wind lies beyond. Two tangential discontinuities intersect the bow shock at $X = 33$ and 66 c/ω_{pi}. A HFA bounded by enhanced densities forms at the intersection of the 66 c/ω_{pi} tangential discontinuity because it has inward pointing electric fields. No HFA forms at the intersection of the 33 c/ω_{pi} tangential discontinuity because it has outward pointing electric fields. The ion skin depth, c/ω_{pi}, is defined as in Figure 5.7. (From Thomas, V.A. et al., *J. Geophys. Res.*, 96, 11625–11632, 1991. With Permission.)

and their effects on the magnetosphere (Archer et al., 2012, 2013; Hietala and Plaschke, 2013; Plaschke et al., 2013; Gunell et al., 2014; Dmitriev and Suvorova, 2015).

5.4 CONCLUSION

Much has been learned in the 50 years since Spreiter et al. (1966) employed a gas dynamic model to predict the plasma and magnetic field characteristics of the magnetosheath. A succession of ever more sophisticated models have appeared in the literature. Observations have both verified most of these model predictions and motivated further developments. It cannot be long until the first accurate global, high-resolution, simulations incorporating electron microphysics appear.

Nevertheless, we still cannot accurately predict some very basic things, namely the locations of the bow shock and magnetopause, or the precise plasma and magnetic field parameters applied to the magnetosphere, for any given solar wind condition. In fact, as the recent discoveries

of jets in the magnetosheath and foreshock bubbles in the foreshock demonstrate, our catalogue of magnetosheath phenomena remains incomplete. We are still discovering significant new phenomena that may play very important roles in the solar wind—magnetosphere interaction.

Finally, our views of the magnetosheath have been derived from multipoint *in situ* measurements. The prospects of obtaining global views have never been brighter. Recently reported energetic-neutral atom images of the magnetosheath (Fuselier et al., 2010; Ogasawara et al., 2013), planned observations of the magnetosheath in soft X-rays (Branduardi-Raymont et al., 2012), and possible observations in solar visible light Thomson scattered by geospace electrons (Meier et al., 2009) could lead to a revolutionary revision not only of our understanding of the magnetosheath, but also of the solar wind's interaction with the magnetosphere.

REFERENCES

Alexeev, I.I., D.G. Sibeck, and S.Y. Bobrovnikov. 1998. Concerning the Location of Magnetopause Merging as a Function of Magnetopause Current Strength. *J. Geophys. Res.* 103: 6675–6684.

Anderson, B.J. and S.A. Fuselier. 1993. Magnetic Pulsations from 0.1 to 4.0 Hz and Associated Plasma Properties in the Earth's Subsolar Magnetosheath and Plasma Depletion Layer. *J. Geophys. Res.* 98: 1461–1479.

Anderson, B.J., T.D. Phan, and S.A. Fuselier. 1997. Relationship between Plasma Depletion and Subsolar Reconnection. *J. Geophys. Res.* 102: 9531–9542.

Archer, M.O., T.S. Horbury, and J.P. Eastwood. 2012. Magnetosheath Pressure Pulses: Generation Downstream of the Bow Shock From Solar Wind Discontinuities. *J. Geophys. Res.* 117: A05228, doi:10.1029/2011JA017468.

Archer, M.O., T.S. Horbury, J.P. Eastwood. J.M. Weygand, and T.K. Yeoman. 2013. Magnetospheric Response to Magnetosheath Pressure Pulses: A Low-Pass Filter Effect. *J. Geophys. Res.* 118: 5454–5466.

Archer, M.O., D.L. Turner, J.P. Eastwood, S.J. Schwartz, and T.S. Horbury. 2015. Global Impacts of a Foreshock Bubble: Magnetosheath, Magnetopause and Ground-Based Observations. *Planet Space Sci.* 106: 56–66.

Arnoldy, R.L. 1971. Signature in the Interplanetary Medium for Substorms. *J. Geophys. Res.* 76: 5189–5201.

Aubry, M.P., C.T. Russell, and M.G. Kivelson. 1970. Inward Motion of the Magnetopause before a Substorm. *J. Geophys. Res.* 75: 7018–7031.

Avanov, L.A., S.A. Fuselier, and O.L. Vaisberg. 2001. High-Latitude Magnetic Reconnection in Sub-Alfvénic Flow: Interball Tail Observations on May 29, 1996. *J. Geophys. Res.* 106: 29491–29502.

Axford, W.I. and C.O. Hines. 1964. Comments on "A Hydro-magnetic Theory of Geomagnetic Storms" by J. H. Piddington. *Planet. Space Sci.* 12: 660–661.

Bale, S.D., M.A. Balikhin, T.S. Horbury et al. 2005. The Quasi-Perpendicular Shock Structure and Processes. *Space Sci. Rev.* 118: 161–203.

Berchem, J., J. Raeder, M. Ashour-Abdalla et al. 1998. The Distant Tail at 200 R_E: Comparison between Geotail Observations and the Results from a Global Magnetohydrodynamic Simulation. *J. Geophys. Res.* 103: 9121–9142.

Berchem, J. and C.T. Russell. 1984. Flux Transfer Events on the Magnetopause—Spatial Distribution and Controlling Factors. *J. Geophys. Res.* 89: 6689–6703.

Blanco-Cano, X., N. Omidi, and C.T. Russell. 2009. Global Hybrid Simulations: Foreshock Waves and Cavitons under Radial Interplanetary Magnetic Field Geometry. *J. Geophys. Res.* 114: doi:10.1029/2008JA013406.

Branduardi-Raymont, G., S.F. Sembay, J.P. Eastwood et al. 2012. AXIOM: Advanced X-Ray Imaging of the Magnetosphere. *Experiment. Astron.* 33: 403–443.

Burgess, D., E.A. Lucek, M. Scholer et al. 2005. Quasi-parallel Shock Structure and Processes. *Space Sci. Rev.* 118: 205–22.

Cairns, I.H. and J.G. Lyon. 1996. Magnetic Field Orientation Effects on the Standoff Distance of Earth's Bow Shock. *Geophys. Res. Lett.* 23: 2883–2886.

Chapman, J.F., I.H. Cairns, J.G. Lyon, and C.R. Boshuizen. 2004. MHD Simulations of Earth's Bow Shock: Interplanetary Magnetic Field Orientation Effects on Shape and Position. *J. Geophys. Res.* 109: doi:10.1029/2003JA010235.

Chen, S.-H., M.G. Kivelson, J.T. Gosling, R.J. Walker, and A.J. Lazarus. 1993. Anomalous Aspects of Magnetosheath Flow and of the Shape and Oscillations of the Magnetopause during an Interval of Strongly Northward Interplanetary Magnetic Field. *J. Geophys. Res.* 98: 5727–5742.

Coleman, I.J. 2005. A Multi-Spacecraft Survey of Magnetic Field Line Draping in the Dayside Magnetosheath. *Ann. Geophys.* 23: 885–900.

Cooling, B.M.A., C.J. Owen, and S.J. Schwartz. 2001. Role of the Magnetosheath Flow in Determining the Motion of Open Flux Tubes. *J. Geophys. Res.* 106: 18763–18776.

Coroniti, F.V. and C.F. Kennel. 1979. Magnetospheric Reconnection, Substorms, and Energetic Particle Acceleration. *AIP Conference Proceedings*, American Institute of Physics, New York, 56, pp. 169–178.

Cowley, S.W.H. 1981. Magnetospheric Asymmetries Associated with the Y-Component of the IMF. *Planet. Space Sci.* 29: 79–96.

Crooker, N.U. 1979. Dayside Merging and Cusp Geometry. *J. Geophys. Res.* 84: 951–959.

Crooker, N.U. 1990. Morphology of Magnetic Merging at the Magnetopause. *J. Atmo. Terr. Phys.* 52: 1123–1134.

Crooker, N.U., T.E. Eastman, and G.S. Stiles. 1979. Observations of Plasma Depletion in the Magnetosheath at the Dayside Magnetopause. *J. Geophys. Res.* 84: 869–874.

Crooker, N.U., J.G. Luhmann, C.T. Russell, E.J. Smith, J.R. Spreiter, and S.S. Stahara. 1985. Magnetic Field Draping against the Dayside Magnetopause. *J. Geophys. Res.* 90: 3505–3510.

Crooker, N.U., C.T. Russell, T.E. Eastman, L.A. Frank, and E.J. Smith. 1981. Energetic Magnetosheath Ions and the Interplanetary Magnetic Field Orientation. *J Geophys. Res.* 86: 4455–4460.

Crooker, N.U., G.L. Siscoe, C.T. Russell, and E.J. Smith. 1982. Factors Controlling Degree of Correlation between ISEE 1 and ISEE 3 Interplanetary Magnetic Field Measurements. *J. Geophys. Res.* 87: 2224–2230.

Denton, R.E. 2000. ULF Waves in the Magnetosheath. *Int. J. Geomagn. Aeronomy* 2: 45–55.

Denton, R.E., S.P. Gary, X. Li, B.J. Anderson, J.W. Labelle, and M. Lessard. 1995. Low-Frequency Fluctuations in the Magnetosheath near the Magnetopause. *J. Geophys. Res.* 100: 5665–5679.

Dimmock, A.P. and K. Nykyri. 2013. The Statistical Mapping of Magnetosheath Plasma Properties Based on THEMIS Measurements in the Magnetosheath Interplanetary Medium Reference Frame. *J. Geophys. Res.* 118: 4963–4976.

Dimmock, A.P., K. Nykyri, H. Karimabadi, A. Osmane, and T.I. Pulkkinen. 2015. A Statistical Study into the Spatial Distribution and Dawn-Dusk Asymmetry of Dayside Magnetosheath Ion Temperatures as a Function of Upstream Solar Wind Conditions. *J. Geophys. Res.* 120: 2767–2782.

Dmitriev, A.V. and A.V. Suvorova. 2015. Large-Scale Jets in the Magnetosheath and Plasma Penetration across the Magnetopause: THEMIS Observations. *J. Geophys. Res.* 120: 4423–4437.

Dungey, J.W. 1961. Interplanetary Magnetic Field and the Auroral Zones. *Phys. Rev. Lett.* 6: 47–48.

Dusik, S., G. Granko, J. Safrankova, Z. Nemecek, and K. Jelinek. 2010. IMF Cone Angle Control of the Magnetopause Location: Statistical Study. *Geophys. Res. Lett.* 37: doi:10.1029/2010GL044965.

Eastwood, J.P., E.A. Lucek, C. Mazelle, K. Meziane, Y. Narita, J. Pickett, and R.A. Treumann. 2005. The Foreshock. *Space Sci. Rev.* 118: 41–94.

Elsen, R.K. and R.M. Winglee. 1997. The Average Shape of the Magnetopause: A Comparison of Three-Dimensional Global MHD and Empirical Models. *J. Geophys. Res.* 102: 4799–4820.

Engebretson, M.J., L.J. Zanetti, T.A. Potemra, W. Baumjohann, H. Lühr, and M.H. Acuña. 1987. Simultaneous Observation of Pc 3–4 Pulsations in the Solar Wind and in the Earth's Magnetosphere. *J. Geophys. Res.* 92: 10053–10062.

Eriksson, S., S.R. Elkington, T.D. Phan et al. 2004. Global Control of Merging by the Interplanetary Magnetic Field: Cluster Observations of Dawnside Flank Magnetopause Reconnection. *J. Geophys. Res.* 109: doi:10.1029/2003JA010346.

Fairfield, D.H. 1967. The Ordered Magnetic Field of the Magnetosheath. *J. Geophys. Res.* 72: 5865–5877.

Fairfield, D.H., W. Baumjohann, G. Paschmann, H. Lühr, and D.G. Sibeck. 1990. Up-stream pressure variations associated with the bow shock and their effects on the magnetosphere. *J. Geophys. Res.* 95: 3773–3786.

Fairfield, D.H. and L.J. Cahill, Jr. 1966. Transition Region Magnetic Field and Polar Magnetic Disturbances. *J. Geophys. Res.* 71: 155–169.

Farrugia, C.J., N.V. Erkaev, H.K. Bienat, G.R. Lawrence, and R.C. Elphic. 1997. Plasma Depletion Layer Model for Low Alfvén Mach Number: Comparison with ISEE Observations. *J. Geophys. Res.* 102: 11315–11324.

Fear, R.C., S.E. Milan, A.N. Fazakerley et al. 2007. Motion of Flux Transfer Events: A Test of the Cooling Model. *Ann. Geophys.* 25: 1669–1690.

Fuselier, S.A., H.O. Funsten, D. Heirtzler et al. 2010. Energetic Neutral Atoms from the Earth's Subsolar Magnetopause. *Geophys. Res. Lett.* 37: doi:10.1029/2010GL044140.

Fuselier, S.A., D.M. Klumpar, E.G. Shelley, B.J. Anderson, and A.J. Coates. 1991. He^{2+} and H^+ Dynamics in the Subsolar Magnetosheath and Plasma Depletion Layer. *J. Geophys. Res.* 96: 21095–21104.

Fuselier, S.A., S.M. Petrinec, and K.J. Trattner. 2000. Stability of the High-Latitude Reconnection Site for Steady Northward IMF. *Geophys. Res. Lett.* 27: 473–476.

Genot, V., E. Budnik, C. Jacquey, I. Dandouras, and E. Lucek. 2009. Mirror Modes Observed with Cluster in the Earth's Magnetosheath: Statistical Study and IMF/Solar Wind Dependence. In *Advances in Geosciences*, vol. 14, M. Duldig (ed.), World Scientific, Singapore, pp. 263–283.

Gonzalez, W.D. and F.S. Mozer. 1974. A Quantitative Model for the Potential Resulting from Reconnection with an Arbitrary Interplanetary Magnetic Field. *J. Geophys. Res.* 79: doi:10.1029/JA079i028p04186.

Gosling, J.T., J.R. Asbridge, S.J. Bame, A.J. Hundhausen, and I.B. Strong. 1967. Measurements of the Interplanetary Solar Wind during the Large Geomagnetic Storm of April 17–18, 1965. *J. Geophys. Res.* 72: 1813–1821.

Gunell, H., G. Stenberg Wieser, M. Mella et al. 2014. Waves in High-Speed Plasmoids in the Magnetosheath and at the Magnetopause. *Ann. Geophys.* 32: 991–1009.

Gutynska, O., D.G. Sibeck, and N. Omidi. 2015. Magnetosheath Plasma Structures and Their Relation to Foreshock Processes. *J. Geophys. Res.* 120: doi:10.1002/2014JA020880.

Hietala, H., T.V. Laitenen, K. Andréeová et al. 2009. Supermagnetosonic Jets behind a Collisionless Quasiparallel Shock. *Phys. Rev. Lett.*, 103: doi:10.1103/PhysRevLett.103.245001.

Hietala, H., N. Partamies, T.V. Laitinan et al. 2012. Supermagnetosonic Subsolar Magnetosheath Jets and Their Effects: From the Solar Wind to the Ionospheric Convection. *Ann. Geophys.* 30: 33–48.

Hietala, H. and F. Plaschke. 2013. On the Generation of Magnetosheath High-Speed Jets by Bow Shock Ripples. *J. Geophys. Res.* 118: 7237–7245.

Hill, T.W. and M.E. Rassbach. 1975. Interplanetary Magnetic Field Direction and the Configuration of the Dayside Magnetosphere. *J. Geophys. Res.* 80: 1–6.

Holzworth, R.H. and C.-I. Meng. 1975. Mathematical Representation of the Auroral Oval. *Geophys. Res. Lett.* 2: 377–380.

Howe, H.C. Jr. and J.H. Binsack. 1972. Explorer 33 and 35 Plasma Observations of Magnetosheath Flow. *J. Geophys. Res.* 77: 3334–3344.

Hubert, D. and C.C. Harvey. 2000. Interplanetary Rotational Discontinuities: From the Solar Wind to the Magnetosphere through the Magnetosheath. *Geophys. Res. Lett.* 27: 3149–3152.

Hubert, D., C. Lacombe, C.C. Harvey, M. Moncuquet, C.T. Russell, and M.F. Thomsen. 1998. Nature, Properties, and Origin of Low-Frequency Waves from an Oblique Shock to the Inner Magnetosheath. *J. Geophys. Res.* 103: 26783–26798.

Hubert, D. and A. Samsonov. 2004. Steady State Slow Shock Inside the Earth's Magnetosheath: To Be or Not to Be? 1. The Original Observations Revisited. *J. Geophys. Res.* 109: doi:10.1029/2003JA010008.

Hubert, D. and A. Samsonov. 2005. Reply to the Comment by P. Song et al. on "Steady State Slow Shock Inside the Earth's Magnetosheath: To Be or Not to Be? 1. The Original Observations Revisited" by D. Hubert and A. Samsonov. *J. Geophys. Res.* 110: doi:10.1029/2005JA011224.

Kaymaz, Z., G. Siscoe, and J.G. Luhmann. 1992. IMF Draping around the Geotail: IMP 8 Observations. *Geophys. Res. Lett.* 19: 829–832.

Kempf, Y., D. Pokhotelov, O. Gutynska et al. 2015. Ion Distributions in the Earth's Foreshock: Hybrid-Vlasov Simulation and THEMIS Observations. *J. Geophys. Res.* 120: 3684–3701.

Keika, K., R. Nakamura, W. Baumjohann et al. 2009. Deformation and Evolution of Solar Wind Discontinuities through their Interactions with the Earth's Bow Shock. *J. Geophys. Res.* 114: doi:10.1029/2008JA013481.

Kobel, E. and E.O. Flückiger. 1994. A Model of the Steady State Magnetic Field in the Magnetosheath, *J. Geophys. Res.* 99: 23617–23622.

Korotova, G.I., D.G. Sibeck, N. Omidi, and V. Angelopoulos. 2012. THEMIS Observations of Unusual Bow Shock Motion Attending a Transient Magnetospheric Event. *J. Geophys. Res.* 117: doi:10.1029/2012JA017510.

Koval, A., J. Safrankova, Z. Nemecek, L. Prech, A.A. Samsonov, and J.D. Richardson. 2005. Deformation of Interplanetary Shock Fronts in the Magnetosheath. *Geophys. Res. Lett.* 32: doi:10.1029/2005GL023009.

Koval, A., J. Safrankova, Z. Nemecek, A.A. Samsonov. L. Prech, J.D. Richardson, and M. Hayosh. 2006. Interplanetary Shocks in the Magnetosheath: Comparison of Experimental Data with MHD Modeling. *Geophys. Res. Lett.* 33: doi:10.1029/2006GL025707.

Lavraud, B., J.E. Borovsky, A.J. Ridley et al. 2007. Strong Bulk Plasma Acceleration in Earth's Magnetosheath: A Magnetic Slingshot Effect? *J. Geophys. Res.* 34: doi:10.1029/2007GL030024.

Lin, Y. 1997. Generation of Anomalous Flows near the Bow Shock by Its Interaction with Interplanetary Discontinuities. *J. Geophys. Res.* 102: 24265–24282.

Lin, Y., L.C. Lee, and M. Yan. 1996a. Generation of Dynamic Pressure Pulses Downstream of the Bow Shock by Variations in the Interplanetary Magnetic Field Orientation. *J. Geophys. Res.* 101: 479–494.

Lin, Y., D.W. Swift, and L.C. Lee. 1996b. Simulation of Pressure Pulses in the Bow Shock and Magnetosheath Driven by Variations in Interplanetary Magnetic Field Direction. *J. Geophys. Res.* 101: 27251–27270.

Liou, K. 2007. Large, Abrupt Pressure Decreases as a Substorm Onset Trigger. *Geophys. Res. Lett.* 34: doi:10.1029/2007GL029909.

Lockwood, M. and M.N. Wild. 1993. On the Quasi-Periodic Nature of Magnetopause Flux Transfer Events. *J. Geophys. Res.* 98: 5935–5940.

Luhmann, J.G., R.J. Walker, C.T. Russell, N.U. Crooker, J.R. Spreiter, and S.S. Stahara. 1984a. Patterns of Potential Magnetic Field Merging Sites on the Dayside Magnetopause. *J. Geophys. Res.* 89: 1741–1744.

Luhmann, J.G., R.J. Walker, C.T. Russell, J.R. Spreiter, S.S.Stahara, and D.J. Williams. 1984b. Mapping the Magnetosheath Field between the Magnetopause and the Bow Shock-Implications for Magnetospheric Particle Leakage. *J. Geophys. Res.* 89: 6829–6834.

Ma, X., A. Otto, and P.A. Delamere. 2014a. Interaction of Magnetic Reconnection and Kelvin-Helmholtz Modes for Large Magnetic Shear: 1. Kelvin-Helmholtz Trigger. *J. Geophys. Res.* 119: 781–797.

Ma, X., A. Otto, and P.A. Delamere. 2014b. Interaction of Magnetic Reconnection and Kelvin-Helmholtz Modes for Large Magnetic Shear: 2. Reconnection Trigger. *J. Geophys. Res.* 119: 808–820, 3543–3548.

Maezawa, K. 1975. Magnetotail Boundary Motion Associated with Geomagnetic Substorms. *J. Geophys. Res.* 80: 3543–3548.

Maltsev, Y.P. and W.B. Lyatsky. 1975. Field Aligned Currents and Erosion of the Dayside Magnetosphere. *Planet. Space Sci.* 23: 1257–1260.

Martyn, D.F. 1951. The Theory of Magnetic Storms and Auroras. *Nature* 167: 92–94.

Meier, R.R., C. Englert, D. Chua et al. 2009. Geospace Imaging Using Thomson Scattering. *J. Atmo. Solar-Terr. Phys.* 71: 132–142.

Michel, F.C. and A.J. Dessler. 1970. Diffusive Entry of Solar-Flare Particles into the Geomagnetic Tail. *J. Geophys. Res.* 75: 6061.

Moretto, T., D.G. Sibeck, B. Lavraud, K.J. Trattner, H. Reme, and A. Balogh. 2005. Flux Pile-Up and Plasma Depletion at the High Latitude Dayside Magnetopause during Southward Interplanetary Magnetic Field: A Cluster Event Study. *Ann. Geophys.* 23: 2259–2264.

Newell, P.T., C.-I. Meng, D.G. Sibeck, and R. Lepping. 1989. Some Low-Altitude Cusp Dependencies on the Interplanetary Magnetic Field. *J. Geophys. Res.* 94: 8921–8927.

Ogasawara, K., V. Angelopoulos, M.A. Dayeh, S.A. Fuselier, G. Livadiotis, D.J. McComas, and J.P. McFadden. 2013. Characterizing the Dayside Magnetosheath Using Energetic Neutral Atoms: IBEX and THEMIS Observations. *J. Geophys. Res.* 118: 3126–3137.

Omidi, N., J.P. Eastwood, and D.G. Sibeck. 2010. Foreshock Bubbles and Their Global Magnetospheric Impacts. *J. Geophys. Res.* 115. doi:10.1029/2009JA014828.

Omidi, N., T. Phan, and D.G. Sibeck. 2009a. Hybrid Simulations of Magnetic Reconnection Initiated in the Magnetosheath. *J. Geophys. Res.* 114: doi:10.1029/2008JA013647.

Omidi, N. and D.G. Sibeck. 2007. Formation of Hot Flow Anomalies and Solitary Shocks. *J. Geophys. Res.* 112: doi:10.1029/2006JA011663.

Omidi, N., D.G. Sibeck, and X. Blanco-Cano. 2009b. Foreshock Compressional Boundary. *J. Geophys. Res.* 114: doi:10.1029/2008JA013950.

Omidi, N., D. Sibeck, O. Gutynska, and K.J. Trattner. 2014. Magnetosheath Filamentary Structures Formed by Ion Acceleration at the Quasi-Parallel Bow Shock. *J. Geophys. Res.* 119: 2593–2604.

Onsager, T.G., C.A. Kletzing, J.B. Austin, and H. Mackiernan. 1993. Model of Magnetosheath Plasma in the Magnetosphere-Cusp and Mantle. *Geophys. Res. Lett.* 20: 479–482.

Panov, E.V., J.Büchner, M.Fränzel et al. 2008. High-Latitude Earth's Magnetopause Outside the Cusp: Cluster Observations. *J. Geophys. Res.* 113: doi:10.1029/2006JA012123.

Paschmann, G., G. Haerendel, N. Sckopke, E. Möbius, H. Lühr, and C.W. Carlson. 1988. Three-Dimensional Plasma Structures with Anomalous Flow Directions near the Earth's Bow Shock. *J. Geophys. Res.* 93: 11279–11294.

Paschmann, G., N. Sckopke, G. Haerendel et al. 1978. ISEE Plasma Observations near the Subsolar Magnetopause. *Space Sci. Rev.* 22: 717–727.

Petrinec, S.M. 2013. On the Magnetic Field Configuration of the Magnetosheath. *Terr. Atmos. Ocean. Sci.* 24: 265–272.

Petrinec, S.M., K.J. Trattner, and S.A. Fuselier. 2003. Steady Reconnection during Intervals of Northward IMF: Implications for Magnetosheath Properties. *J. Geophys. Res.* 108: doi:10.1029/2003JA009979.

Phan, T.-D., T.E. Love, J.T. Gosling et al. 2011. Triggering of Magnetic Reconnection in a Magnetosheath Current Sheet due to Compression against the Magnetopause. *Geophys. Res. Lett.* 38: doi:10.1029/2011GL048586.

Phan, T.-D., G. Paschmann, W. Baumjohann, N. Sckopke, and H. Lühr. 1994. The Magnetosheath Region Adjacent to the Dayside Magnetopause: AMPTE/IRM Observations. *J. Geophys. Res.* 99: 121–141.

Phan, T.D., G. Paschmann, J.T. Gosling et al. 2013. The Dependence of Magnetic Reconnection on Plasma Beta and Magnetic Shear: Evidence from Magnetopause Observations. *Geophys. Res. Lett.* 40: 11–16.

Plaschke, E., H. Hietala, and V. Angelopoulos. 2013. Anti-Sunward High-Speed Jets in the Subsolar Magnetosheath. *Ann. Geophys.* 31: 1877–1889.

Pu, Z.-Y. and M.G. Kivelson. 1983. Kelvin-Helmholtz Instability at the Magnetopause: Energy Flux into the Magnetosphere, *J. Geophys. Res.* 88: 853–862.

Pudovkin, M.I., B.P. Besser, S.A. Zaitseva, V.V. Lebedeva, and C.-V. Meister. 2001. Magnetic Barrier in Case of a Southward Interplanetary Magnetic Field. *J. Atmo. Solar-Terr. Phys.* 63: 1075–1083.

Reiff, P.H., R.W. Spiro, and T.W. Hill. 1981. Dependence of Polar Cap Potential Drop on Interplanetary Parameters. *J. Geophys. Res.* 86: 7639–7648.

Rojas-Castillo, D., X. Blanco-Cano, P. Kajdic, and N. Omidi. 2013. Foreshock Compressional Boundaries Observed by Cluster. *J. Geophys. Res.* 118: 698–715.

Russell, C.T. and R.C. Elphic. 1978. Initial ISEE Magnetometer Results—Magnetopause Observations. *Space Sci. Rev.* 22: 681–715.

Safrankova, J., M. Hayosh, O. Gutynska, Z. Nemecek, and L. Prech. 2009. Reliability of Prediction of the Magnetosheath Bz Component from Interplanetary Magnetic Field Observations. *J. Geophys. Res.* 114: doi:10.1029/2009JA014552.

Samsonov, A. and D. Hubert. 2004. Steady State Slow Shock Inside the Earth's Magnetosheath: To Be or Not to Be? 2. Numerical Three-Dimensional MHD Modeling. *J. Geophys. Res.* 109: doi:10.1029/2003JA010006.

Samsonov, A.A., Z. Nemecek, J. Safrankova, and K. Jelinek. 2012. Why Does the Subsolar Magnetopause Move Sunward for Radial Interplanetary Magnetic Field? *J. Geophys. Res.* 117: doi:10.1029/2011JA017429.

Samsonov, A.A., M.I. Pudovkin, S.P. Gary, and D. Hubert. 2001. Anisotropic MHD Model of the Dayside Magnetosheath Downstream of the Oblique Bow Shock. *J. Geophys. Res.* 106: 21689–21700.

Samsonov, A.A., D.G. Sibeck, Z. Nemecek, and J. Safrankova. 2015. Magnetopause Position for a Radial IMF: Using MHD Simulations to Reproduce the 16 July 2007 Event. *Adv. Space Res* (in press).

Sanchez, E.R. and G.L. Siscoe. 1990. IMP 8 Magnetotail Boundary Crossings—A Test of the MHD Models for an Open Magnetosphere. *J. Geophys. Res.* 95: 20771–20779.

Schwartz, S.J., D. Burgess, and J.J. Moses. 1996. Low-Frequency Waves in the Earth's Magnetosheath: Present Status. *Ann. Geophys.* 14: 1134–1150.

Schwartz, S.J., C.P. Chaloner, D.S. Hall, P.J. Christiansen, and A.D. Johnstone. 1985. An Active Current Sheet in the Solar Wind. *Nature.* 318: 269–271.

Scurry, L., C.T. Russell, and J.T. Gosling. 1994a. Geomagnetic Activity and the Beta Dependence of the Dayside Reconnection Rate. *J. Geophys. Res.* 99: 14811–14814.

Scurry, L., C.T. Russell, and J.T. Gosling. 1994b, A Statistical Study of Accelerated Flow Events at the Dayside Magnetopause. *J. Geophys. Res.* 99: 14815–14829.

Semenov, V.S. and M.I. Pudovkin. 1985. Localization and Features of the Development of Reconnection Processes at the Magnetopause. *Geomag. Aeron.* 25: 592–597.

Shepherd, L.S. and P.A. Cassak. 2012. Guide Field Dependence of 3-D X-Line Spreading during Collisionless Magnetic Reconnection. *J. Geophys. Res.* 117: doi:10.1029/2012JA017867.

Sibeck, D.G. 1990. A Model for the Transient Magnetospheric Response to Sudden Solar Wind Dynamic Pressure Variations. *J. Geophys. Res.* 95: 3755–3771.

Sibeck, D.G. 1995. The Magnetospheric Response to Foreshock Pressure Pulses. In *Physics of the Magnetopause*, P. Song, B.U.O. Sonnerup, and M. Thomsen (eds.), AGU, Washington, DC, pp. 293–302.

Sibeck, D.G., W. Baumjohann, R.C. Elphic, D.H. Fairfield, and J.F. Fennell. 1989. The Magnetospheric Response to 8-Minute Period Strong-Amplitude Upstream Pressure Variations. *J. Geophys. Res.* 94: 2505–2519.

Sibeck, D.G., N.L. Borodkova, S.J. Schwartz et al. 1999. Comprehensive Study of the Magnetospheric Response to a Hot Flow Anomaly. *J. Geophys. Res.* 104: 4577–4593.

Sibeck, D.G., R.B. Decker, D.G. Mitchell, A.J. Lazarus, R.P. Lepping, and A. Szabo. 2001. Solar Wind Preconditioning in the Flank Foreshock: IMP 8 Observations. *J. Geophys. Res.* 106: 21675–21688.

Sibeck, D.G. and J.T. Gosling. 1996. Magnetosheath Density Fluctuations and Magnetopause Motion. *J. Geophys. Res.* 101: 31–40.

Sibeck, D.G. and R.-Q. Lin. 2011. Concerning the Motion and Orientation of Flux Transfer Events Produced by Component and Antiparallel Reconnection. *J. Geophys. Res.* 116: doi:10.1029/2011JA016560.

Sibeck, D.G. and R.-Q. Lin. 2014. Size and Shape of the Distant Magnetotail. *J. Geophys. Res.* 119: 1028–1043.

Sibeck, D.G., R.E. Lopez, and E.C. Roelof. 1991. Solar Wind Control of the Magnetopause Shape, Location, and Motion. *J. Geophys. Res.* 96: 5489–5495.

Sibeck, D.G., G.L. Siscoe, J.A. Slavin, and R.P. Lepping. 1986. Major Flattening of the Distant Geomagnetic Tail. *J. Geophys. Res.* 91: 4223–4237.

Siscoe, G.L., N.U. Crooker, G.M. Erickson et al. 2002. MHD Properties of Magnetosheath Flow. *Planet. Space Sci.* 50: 461–471.

Sitar, R.J., J.B. Baker, C.R. Clauer et al. 1998. Multi-Instrument Analysis of the Ionospheric Signatures of a Hot Flow Anomaly Occurring on July 24, 1996. *J. Geophys. Res.* 103: 23357–23372.

Song, P. and C.T. Russell. 2002. Flow in the Magnetosheath: The Legacy of John Spreiter. *Planet. Space Sci.* 50: 447–460.

Song, P., C.T. Russell, J.T. Gosling, M. Thomsen, and R.C. Elphic. 1990. Observations of the Density Profile in the Magnetosheath near the Stagnation Streamline. *Geophys. Res. Lett.* 17: 2035–2038.

Song, P., C.T. Russell, J.T. Gosling, M.F. Thomsen, and R.C. Elphic. 2005. Comment on "Steady State Slow Shock Inside the Earth's Magnetosheath: To Be or Not to Be? 1. The Original Observations Revisited" by D. Hubert and A. Samsonov. *J. Geophys. Res.* 110: doi:10.1029/2005JA011161.

Song, P., C.T. Russell, and C.Y. Huang. 1993b. Wave Properties near the Subsolar Magnetopause-Pc1 Waves in the Sheath Transition Layer. *J. Geophys. Res.* 98. 5907–5923.

Song, P., C.T. Russell, R.J. Strangeway, J.R. Wygant, C.A. Cattell, R.J. Fitzenreiter, and R.R. Anderson. 1993a. Wave Properties near the Subsolar Magnetopause- Pc 3–4 Energy Coupling for Northward Interplanetary Magnetic Field. *J. Geophys. Res.* 98: 187–196.

Song, P., C.T. Russell, and M.F. Thomsen. 1992a. Waves in the Inner Magnetosheath—A Case Study. *Geophys. Res. Lett.* 19: 2191–2194.

Song, P., C.T. Russell, and M.F. Thomson. 1992b. Slow Mode Transition in the Frontside Magnetosheath. *J. Geophys. Res.* 97: 8295–8305.

Sonnerup, B.U.Ö. 1974. Magnetopause Reconnection Rate. *J. Geophys. Res.* 79: 1546–1549.

Southwood, D.J. 1968. The Hydromagnetic Stability of the Magnetospheric Boundary. *Planet. Space Sci.* 16: 587–605.

Southwood, D.J. and M.G. Kivelson. 1992. On the Form of the Flow in the Magnetosheath. *J. Geophys. Res.* 97: 2873–2879.

Southwood, D.J. and M.G. Kivelson. 1995. Magnetosheath Flow near the Subsolar Magnetopause: Zwan-Wolf and Southwood-Kivelson Theories Reconciled. *Geophys. Res. Lett.* 22: 3275–3278.

Spreiter, J.R. and S.S. Stahara. 1994. Gasdynamic and Magnetohydrodynamic Modeling of the Magnetosheath: A Tutorial. *Adv. Space Res.* 14: 5–19.

Spreiter, J.R., A.L. Summers, and A.Y. Alksne. 1966. Hydromagnetic Flow around the Magnetosphere. *Planet. Space Sci.* 14: 223–253.

Suvorova, A.V., J.-H. Shue, A.V. Dmitriev et al. 2010. Magnetopause Expansions for Quasi-Radial Interplanetary Magnetic Field: THEMIS and Geotail Observations. *J. Geophys. Res.* 115: doi:10.1029/2010JA015404.

Thomas, V.A. and S.H. Brecht. 1988. Evolution of Diamagnetic Cavities in the Solar Wind. *J. Geophys. Res.* 93: 11341–11353.

Thomas, V.A., D. Winske, M.F. Thomsen, and T.G. Onsager. 1991. Hybrid Simulation of the Formation of a Hot Flow Anomaly. *J. Geophys. Res.* 96: 11625–11632.

Thomsen, M.F., J.T. Gosling, S.A. Fuselier, S.J. Bame, and C.T. Russell. 1986. Hot, Diamagnetic Cavities Upstream from the Earth's Bow Shock. *J. Geophys. Res.* 91: 2961–2973.

Trattner, K.J., S.A. Fuselier, W.K. Peterson, S.-W. Chang, R. Friedel, and M.R. Aellig. 2003. Reply to comment on "Origins of Energetic Ions in the Cusp" by R. Sheldon, J. Chen, and T.A. Fritz. *J. Geophys. Res.* 108: doi:10.1029/2002JA009781.

Trattner, K.J., J.S. Mulcock, S.M. Petrinec, and S.A. Fuselier. 2007a. Location of the Reconnection Line at the Magnetopause during Southward IMF Conditions. *Geophys. Res. Lett.* 34: doi:10.1029/2006GL028397.

Trattner, K.J., J.S. Mulcock, S.M. Petrinec, and S.A. Fuselier. 2007b. Probing the Boundary between Antiparallel and Component Reconnection during Southward Interplanetary Magnetic Field Conditions. *J. Geophys. Res.* 112: doi:10.1029/2007JA012270.

Turner, D.L., N. Omidi, D.G. Sibeck, and V. Angelopoulos. 2013. First Observation of Foreshock Bubbles Upstream of Earth's Bow Shock: Characteristics and Comparison to HFAs. *J. Geophys. Res.* 118: 1552–1570.

Verigin, M., G. Kotova, A. Szabo et al. 2001. Wind Observations of the Terrestrial Bow Shock: 3-D Shape and Motion. *Earth, Planets, Space.* 53: 1001–1009.

Völk, H.J. and R.-D. Auer. 1974. Motions of the Bow Shock Induced by Interplanetary Disturbances. *J. Geophys. Res.* 79: 40–48.

Walsh, B.M., D.G.Sibeck, Y. Wang, and D.H. Fairfield. 2012. Dawn-Dusk Asymmetries in the Earth's Magnetosheath. *J. Geophys. Res.* 117: doi:10.1029/2012JA018240.

Walters, G.K. 1964. Effect of Oblique Interplanetary Magnetic Field on Shape and Behavior of the Magnetosphere. *J. Geophys. Res.* 69: 1769–1783.

Wang, Y., J. Raeder, and C. Russell. 2004. Plasma Depletion Layer: Its Dependence on Solar Wind Conditions and the Earth Dipole Tilt. *Ann. Geophys.* 22: 4273–4290.

Weimer, D.R. 2001. Map of Ionospheric Field-Aligned Currents as a Function of the Interplanetary Magnetic Field Derived from Dynamics Explorer 2 Data. *J. Geophys. Res.* 106: 12889–12902.

Weimer, D.R., D.M. Ober, N.C. Maynard et al. 2002. Variable Time Delays in the Propagation of the Interplanetary Magnetic Field. *J. Geophys. Res.* 1078: doi:10.1029/s001JA009102.

White, W.W., G.L. Siscoe, G.M. Erickson et al. 1998. The Magnetospheric Sash and the Cross-Tail S. *Geophys. Res. Lett.* 25: 1605–1608.

Wild, J.A., E.E. Woodfield, and S.K. Morley. 2009. On the Triggering of Auroral Substorms by Northward Turnings of the Interplanetary Magnetic Field. *J. Geophys. Res.* 27: 3559–3570.

Wilson, L.B. III. 2016. Low Frequency Waves at and Upstream of Collisionless Shocks, in *Low Frequency Waves in Space Plasmas*, A. Keiling, D.-H. Lee, and V. Nakariakov (eds.), AGU, Washington, DC, pp. 269–291.

Wing, S., P.T. Newell, and J.M. Ruohoniemi. 2001. Double Cusp: Model Predictions and Observational Verification. *J. Geophys. Res.* 106: 25571–25594.

Wu, B.H., M.E. Mandt, L.C. Lee, and J.K. Chao. 1993. Magnetospheric Response to Solar Wind Dynamic Pressure Variations: Interaction of Interplanetary Tangential Discontinuities with the Bow Shock. *J. Geophys. Res.* 98: 21297–21311.

Zhang, H., Q.-G. Zong, D.G. Sibeck, T.A. Fritz, J.P. McFadden, K.-H. Glassmeier, and D. Larson. 2009. *J. Geophys. Res.* 114: doi:10.1029/2008JA013488.

Zwan, B.J. and R.A. Wolf. 1976. Depletion of Solar Wind Plasma near a Planetary Boundary. *J. Geophys. Res.* 81: 1636–1648.

CHAPTER 6

Magnetic Reconnection

Homa Karimabadi, Ari Le, Vadim Roytershteyn, and William Daughton

CONTENTS

6.1 INTRODUCTION

Magnetic reconnection is a basic relaxation process in magnetized plasmas that enables the release of magnetic energy accumulated in a system, often in explosive form (Priest and Forbes 2000, Yamada et al. 2010). The energy is released through energetic particles, heat, and coherent structures such as jets and magnetic flux ropes. The energy build up and its subsequent

relaxation due to reconnection are one of the main drivers of space weather. Reconnection is also closely tied to turbulence in magnetized plasmas and facilitates the cascade of energy from large scales to smaller kinetic scales.

Many of the space weather events are caused by reconnection-driven processes or electromagnetic storms on the Sun and near the Earth. On the Sun, the dynamical plasma energy is stored in the magnetic fields, which then gets released in so-called solar flares and coronal mass ejections (CMEs) (Forbes 2000). The plasma is ejected into space, and, upon reaching the Earth, may wreak havoc on technological systems. The energetic particles released from a large solar storm can carry over 10^{25} J into the solar wind. Over \$4 billion in satellite losses have been traced to space weather damage. It is estimated that a solar storm of the magnitude of the 1859 Solar Superstorm (the Carrington Event) would cause over \$2 trillion in damage today.

In the face of this threat, there is an urgent need to develop accurate space weather forecasting capabilities. It takes 1–5 days for the energetic particles released during a solar storm to reach the Earth. This provides sufficient time to take pre-emptive measures (e.g., shutting down the power grid). However, there are currently no reliable forecasting models that can predict the severity of the effects and/or the location of the impact on Earth for a given solar storm. For example, none of the models can explain the dramatic ground magnetic field variations that caused the March 1989 blackout in the Hydro-Quebec power grid. The goal is to improve space weather forecasts to levels similar to those in the field of atmospheric weather forecasting, where the strength of a storm (e.g., hurricane category) and the location of impact can be predicted so that evasive action can be taken.

One key factor that limits the performance of global magnetosphere models is the lack of accurate models of reconnection. Local fully kinetic simulations that simulate a small region of the magnetosphere have been instrumental in developing a better understanding of the reconnection process. The challenge is to incorporate this physics into global models. The computational challenge results from the extreme multiscale nature of reconnection. The onset of reconnection occurs on electron scales, but reconnection leads to a global rearrangement of magnetic field lines. To provide perspective, the Earth's magnetosphere extends to about $200\,R_E \sim 1.28 \times 10^6$ km, whereas electron kinetic scales are on the order of a few kilometers. Resolving electron scales in global models of the Earth would require coverage of a factor of 10^6 separation in spatial scales. Similar range is also needed in temporal scales. Such simulations would be out of reach of even exascale computers.

Some important questions regarding reconnection and its relation to space weather are

1. What are the conditions for fast reconnection as a function of collisionality?

   Here we describe the various regimes of reconnection and discuss the effects of turbulence on reconnection.

2. How are energetic particles generated in the magnetotail and on the Sun?

   Here we describe our perspective on these issues, including some of our previously unpublished work.

3. How are we to model reconnection in global magnetosphere simulations?

   Here we discuss the various attempts at this important problem and highlight some promising venues to correctly model the onset, rate, and temporal behavior of magnetic reconnection.

We will give our outlook on these questions in the course of a review of the basic phenomenology of magnetic reconnection, a discussion of the nonideal physical processes that may be included in closure models, and a survey of some recent results on incorporating reconnection physics into global space weather models.

6.2 WHAT IS MAGNETIC RECONNECTION?

The term "reconnection" derives from a topological rearrangement of the magnetic field, as sketched in Figure 6.1 in its most basic form. Initially, plasma on the left and right sides of Figure 6.1a are nowhere linked magnetically. The two sides are separated by a current layer with a sheared magnetic field, with the shear measured by the angle between the magnetic fields on either side. In anti-parallel reconnection, the magnetic field reverses direction across the current sheet with a shear angle of 180°. In general, the shear angle may be less than 180°, which implies that there is an additional component of the magnetic field out of the plane of field reversal, called the guide magnetic field. The transport of particles, momentum, or energy between the two regions in Figure 6.1a must proceed across the magnetic field. In well-magnetized weakly collisional plasmas,

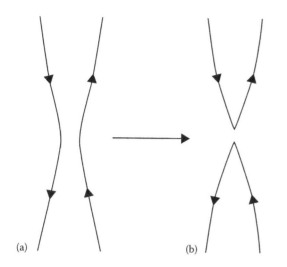

FIGURE 6.1 Sketch of the magnetic reconnection process.

cross-field transport is typically a relatively slow process. In contrast, parallel transport along the field lines can be orders of magnitude faster. If magnetic reconnection takes place, a localized change in the magnetic field (near the center of Figure 6.1a) occurs with the development of an X-line, where oppositely directed magnetic field lines converge and the in-plane magnetic field has a null point. Plasma from the left and right is now directly connected by magnetic field lines, which allows rapid parallel plasma transport between the two regions.

By allowing the global topology of the magnetic field to change, magnetic reconnection also provides a pathway to release the energy stored in the magnetic field. In ideal magnetohydrodynamic (MHD), where the plasma is considered as perfectly conducting, the frozen-in condition holds:

$$\mathbf{E} + \mathbf{u} \times \mathbf{B} = \mathbf{0},$$

where $\mathbf{u}$ is the plasma flow velocity. The magnetic field evolves as though it was convected with the plasma flow, and as a result plasma fluid elements that connected along a field line remain connected as the system evolves.

An immediate consequence of the frozen-in condition is that there is zero parallel electric field, $E_\parallel = \mathbf{E} \cdot \mathbf{B}/B$. In general, for magnetic reconnection to occur, a nonzero $E_\parallel$ must exist in a finite region of plasma. However, the reverse it not necessarily true, as finite $E_\parallel$ may appear without magnetic reconnection. More careful analysis shows that it is the integral of $E_\parallel$ along magnetic field lines

that plays a special role. For example, in a configuration with no magnetic nulls (points where $\mathbf{B} = \mathbf{0}$), the reconnection pseudopotential $\Phi_\parallel$ is defined as

$$\Phi_\parallel = \int\limits_{X1}^{X2} \mathbf{E} \cdot d\mathbf{l},$$

where the integral is taken along magnetic field line connecting two points in the ambient ideal region with $E_\parallel = 0$ and passing through the nonideal region with finite $E_\parallel$ (Schindler et al. 1988). The reconnection rate is then given by the maximum of $\Phi_\parallel$. The pseudopotential is useful for diagnosing analytical and numerical models of reconnection. In space observations, however, it is generally not possible to measure $\Phi_\parallel$ directly. Consequently, a substantial body of work has been dedicated to finding local proxies for identification of reconnection events (see, e.g., Scudder et al. 2015 for a recent discussion).

The localized region of nonideal behavior near the X-line is referred to as the "diffusion region." In the classic picture, the diffusion region is further broken down into an ion diffusion region where ion kinetics play a role and a smaller electron diffusion region where electron kinetic effects become important. This configuration is sketched in Figure 6.2. The ion diffusion region has characteristic ion kinetic scales, typically on the order of a few ion inertial lengths $d_i = c/\omega_{pi}$ for anti-parallel reconnection or ion sound Larmor radii $\rho_s = \sqrt{T_e/m_i}/\Omega_{ci}$ when there is a strong guide magnetic field. The electron diffusion region nested within the ion diffusion region typically has length scales of several electron inertial lengths $d_e = c/\omega_{pe}$ or electron meandering orbit widths.

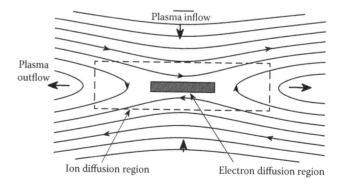

FIGURE 6.2 The nonideal behavior that allows reconnection to proceed may be localized to small diffusion regions.

In two-dimensional configurations, the rate of reconnection is measured by the strength of the reconnection electric field E_{rec}, the component out of the reconnection plane. Plasma from the external ideal MHD region convects toward the X-line at the velocity $v_{in} \sim E_{rec}/B_0$, where B_0 is a typical value of the reconnecting component of the magnetic field. Under a very wide range of conditions, reconnection in plasma regimes typical of space weather has been found to proceed at the relatively fast rate of $E_{rec} \sim 0.1 v_A B_0$ or $v_{in} \sim 0.1 v_A$, where v_A is the Alfvén speed based on B_0.

6.3 ROLE OF MAGNETIC RECONNECTION IN SPACE WEATHER

6.3.1 Reconnection on the Sun

Reconnection was first proposed as part of the dynamics of the solar corona, where the magnetic fields that emerge from Sun spots and other coronal features interact and drive currents in the solar plasma. Typically, the thermal energy density of the coronal plasma is much less than the magnetic field energy density, with $\beta <$ 0.01 (where β is the plasma pressure normalized to the magnetic field pressure). Thus, it is believed that solar flares and other disruptive events in the corona derive their energy, which can be upward of 10^{25} J for a single large solar flare, from the magnetic field. Although there is still no complete model for how solar reconnection occurs, there is ample evidence for reconnection and concomitant electron and ion energization during flares and CMEs. CMEs are of particular importance for space weather because they entail an eruption of a large flux of solar plasma (upward of 10^{16} g) into the solar wind that may be directed toward the Earth.

A basic model for disruptive magnetic reconnection on the Sun is that magnetic field lines emerge from the photosphere, and their convective foot point motion drives electrical currents in the coronal plasma. This stores magnetic energy over the course of several hours or days. During reconnection, the energy may then be released in minutes, as evidenced by soft X-ray imaging of the evolving magnetic topology. The reconnection produces accelerated plasma jets, consistent, for example, with X-ray emission produced by a large flare outflow striking the underlying coronal looptop (Masuda et al. 1994). Meanwhile, observations near the foot points of reconnecting loops suggest that energetic particles (ions up to 1 GeV and electrons up to 100 MeV) produced by

reconnection stream along the magnetic field and strike the photosphere. These nonthermal particles may contain half the released energy. There has also been evidence for magnetic reconnection within the solar wind (Gosling et al. 2005). In particular, plasma jetting consistent with reconnection outflows was observed within the flux tubes ejected during CMEs as they passed near 1 AU.

6.3.2 Reconnection near the Earth

The Earth's global dipole field forms a shield, the magnetosphere, that prevents most of the charged particles in the solar wind from entering the Earth's atmosphere. Magnetic reconnection controls the evolution of the magnetosphere by altering the global topology of the magnetic field, allowing the direct transport of solar wind plasma into the magnetosphere. The main features of the magnetosphere are sketched in Figure 6.3. The interaction of the solar wind with the Earth's dipole field begins at the bow shock, which forms because the solar wind flows at supersonic and super-Alfvénic speeds. The shocked plasma forms a turbulent boundary layer, the magnetosheath, between the Earth's magnetosphere and the solar wind. The magnetosphere dominated by particles trapped in the Earth's dipole field begins at the magnetopause. The nightside magnetosphere is stretched into an elongated magnetic tail that contains two lobes separated by a current-carrying plasma sheet of higher density.

One site that is susceptible to magnetic reconnection is the magnetopause, particularly during periods when the incoming solar wind carries a southward interplanetary magnetic field (IMF). Reconnection at the magnetopause can either be quasi-steady or occur in patchy bursts. Transient bursts of reconnection are called flux

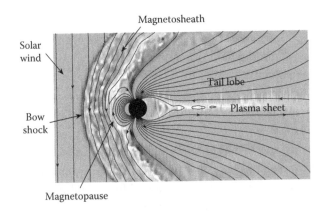

FIGURE 6.3 Global structure of a planetary magnetosphere.

transfer events (FTEs), and they have been found to occur every few minutes near the Earth. FTEs generate a magnetic flux tube carrying solar wind plasma that can then be transported downwind over the Earth's poles. One property of the magnetopause boundary that affects the reconnection dynamics is that there is usually a large asymmetry in plasma parameters across the magnetopause (Cassak and Shay 2007), with the magnetic field strength varying by a factor of 2–3 and the density varying by a factor >10.

The nightside magnetosphere contains the extended magnetic tail, elongated by the pressure of the impinging solar wind. This stretching induces an electrical current in the dawn to dusk direction in the central plasma sheet of the tail, which is another site susceptible to reconnection. Reconnection breaks up the tail current channel into filaments and joins field lines from the northern lobe to the southern lobe. Examples of these filaments, also called magnetic islands or plasmoids, are apparent in Figure 6.3. Energetic electrons produced during tail reconnection may stream along the Earth's field lines and strike the upper atmosphere in the polar regions, generating the auroral phenomena. Reconnected tail magnetic field may convect back toward the Earth as part of dipolarization fronts, when the magnetic field relaxes to conform to the dipole field (Runov et al. 2009).

6.3.3 Particle Energization

Besides resulting in a global rearrangement of the magnetic fields, reconnection is also a source of kinetic energy in the forms of bulk flows, heating, and energetic nonthermal particles. A variety of models have been proposed to explain how magnetic reconnection accelerates both electrons and ions, including effective heating of the bulk populations and the generation of nonthermal power law tails in the particle energy distributions (Ono et al. 1996, Zenitani and Hoshino 2001, Jaroschek et al. 2004, Guo et al. 2015). Because a parallel electric field $E_{\parallel}$ is always present at the X-line during reconnection, particles may become energized by direct acceleration by $E_{\parallel}$. Direct acceleration by $E_{\parallel}$, however, is limited to a small region near the X-line, and a main challenge is to describe how a large population of particles gains energy during reconnection.

Except for the case of anti-parallel reconnection, the electrons usually follow well-magnetized orbits, and it can be useful to analyze the electron motion

in terms of two adiabatic invariants. The first is the magnetic moment

$$\mu = \frac{m v_{\perp}^2}{2B}$$

which is well-conserved when the magnetic field varies slowly compared to the electron Larmor frequency Ω_{ce} and the Larmor radius and field line radius of curvature satisfy $\rho/R_C < 1$. When μ is conserved, the perpendicular electron kinetic energy $K_{\perp}$ is proportional to the magnetic field strength, $K_{\perp} \propto B$. This underlies the betatron heating effect: when magnetized electrons move to regions of stronger magnetic field, such as the flux pile-up region of a reconnection exhaust, their perpendicular energy increases (Ashour-Abdalla et al. 2011, Fu et al. 2011).

In many cases, there is an additional adiabatic invariant associated with the electron motion along the magnetic field lines. Because the electron thermal speed v_{te} typically satisfies $v_{te} > v_A$, the electrons rapidly transit the length of a reconnection layer much faster than the time is takes for the magnetic field to evolve. If some process keeps the electrons on closed or bouncing orbits, then the second adiabatic invariant

$$J = \oint m v_{\parallel} dl$$

is also conserved. In this case, the average parallel kinetic energy $K_{\parallel} \propto l^{-2}$, where l is a typical length scale of the bouncing motion. Drake et al. (2006) proposed that electrons may be energized during reconnection because they encircle contracting magnetic islands. As the island contracts with J conserved, the electron parallel energy $K_{\parallel}$ increases by the Fermi mechanism. If a large number of volume-filling islands develop near a reconnection site, this mechanism would efficiently energize a large population of electrons. This type of magnetic field geometry is posited to be generated in turbulent regions or when a current sheet breaks up into a large number of filaments. Such a volume-filling configuration, however, has not yet been observed directly or reproduced by global space weather models.

The two adiabatic invariants may be used to analyze the motion of electrons in rather general geometries. A new physical process is introduced in the model for reconnection, the trapping of electrons by the pseudopotential $\Phi_\parallel(X) = \int_\infty^X \mathbf{E} \cdot d\mathbf{l}$. During collisionless reconnection, an ambipolar electric field with a parallel component develops near the X-line to maintain quasi-neutrality (equal average charge densities of electrons and ions). Without trapping by the associated $\Phi_\parallel$, a large portion of the electrons would rapidly exit the regions of weak magnetic field because of the magnetic mirror force. When the electron motion is adiabatic, a solution of the electron kinetic equation is

$$f = \begin{cases} f_0(K - e\Phi_\parallel), & K_{\parallel 0} > 0 \\ f_0(\mu B_0), & K_{\parallel 0} \leq 0 \end{cases} \qquad (6.1)$$

where $f_0(K)$ is the electron distribution (assumed here for simplicity to be isotropic and to depend only on kinetic energy K), $K_{\parallel,\perp} = 1/2 m v_{\parallel,\perp}^2$, $K_{\parallel 0} = K - \mu B_0 - e\Phi_\parallel$, $K_{\parallel 0} = 0$ defines the trapped/passing boundary, and electrons with $K_{\parallel 0} > 0$ are passing while electrons with $K_{\parallel 0} \leq 0$ are trapped. Contours of the distribution are plotted in Figure 6.4 for the typical case with both electric and magnetic trapping, and the distribution becomes highly anisotropic when $e\Phi_\parallel > T_e$. The model distribution agrees well with observations taken by the Wind and Cluster spacecraft and with a large number of fully kinetic simulations (Egedal et al. 2005, 2010).

While the adiabatic assumption breaks down in the anti-parallel reconnection limit, the model distribution still usually holds in the reconnection inflow. The electrons ejected into the exhaust near the X-line have been pre-energized by $\Phi_\parallel$. The electron orbits are chaotic in the center of the exhaust when the total magnetic field is very weak, and this leads to a mixing in velocity space of electrons with different velocity pitch angles. When the distribution of the type in Equation 6.1 undergoes pitch angle mixing, it generates flat-top distributions. Flat-top distributions (Asano et al. 2008), which are observed in reconnection exhausts in the magnetotail, are isotropic and flat in energy up to a knee (an energy at which they fall off), and the peak values of $e\Phi_\parallel$ give the energies of the knees (Egedal et al. 2008).

In kinetic simulations, the adiabatic model also breaks down in very low β plasmas. When the electron β is lower than ~0.02, $e\Phi_\parallel$ becomes tens times larger than the ambient electron temperature. In extreme cases, double layers develop along the boundaries of the reconnection exhaust (Egedal et al. 2015). Double layers contain strong parallel electric fields that extend up to 10s of Debye lengths along the magnetic field, and they have been observed in the magnetotail (Ergun et al. 2009). These layers accelerate essentially all of the upstream electrons into the exhaust

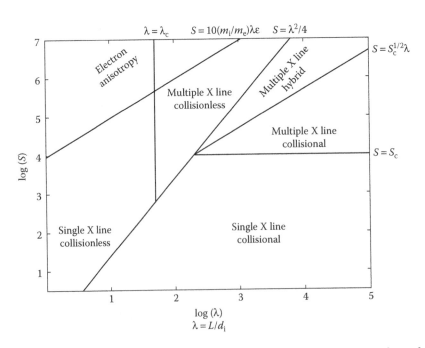

FIGURE 6.4 Phase diagram for near-anti-parallel reconnection. The dynamics of reconnection depend on both the Lundquist number S and the system size normalized to the ion inertial length $\lambda = L/d_i$. The critical Lundquist number is taken as $S_c \sim 10^4$, and the critical system size is $\lambda \sim 50$.

as a beam, which then disperses in velocity space as the electrons undergo chaotic motion in the center of the exhaust. The parallel electric field bordering the exhaust also confines the heated electrons, allowing further energization by other mechanisms. For example, the curvature drift of electrons in the exhaust in the presence of the reconnection electric field leads to substantial additional electron energization (Dahlin et al. 2014, Guo et al. 2014, Egedal et al. 2015).

Phan et al. (2013, 2014) recently surveyed THEMIS observations of a large number of reconnection events at the Earth's magnetopause. One focus was to determine how the bulk heating resulting from reconnection depends on the plasma conditions. A fit to the increment in the electron and ion temperatures from the inflow into the exhaust suggested power laws of the form

$$\Delta T_e \cong 0.017 m_i v_A^2 \quad \text{and} \quad \Delta T_i \cong 0.13 m_i v_A^2$$

where the Alfvén speed v_A is an average based on upstream conditions on either side of the reconnection layer. This scaling applied to a broad range of reconnection events that varied significantly in plasma asymmetries and magnetic shear angles, and a similar scaling law was found in a series of fully kinetic simulations (Shay et al. 2014). This dependence on v_A^2 suggests that a nearly fixed portion of the available magnetic energy goes into heating each species, with the ions receiving a larger share.

6.4 CHALLENGES IN INCORPORATION OF RECONNECTION IN FORECASTING MODELS

6.4.1 Overview of the Problem

Because of the wide separation of scales relevant to magnetic reconnection, it is typically unfeasible to directly resolve all of the relevant physics in global simulation models. Thus, development of subgrid or closure models capable of accurately describing magnetic reconnection has been a grand challenge problem in space sciences and theoretical plasma physics.

At a minimum, a predictive global magnetosphere model should include a mechanism that allows the onset of reconnection and correctly models the reconnection rate. Because reconnection is strongly tied to the electron dynamics, a substantial effort has been undertaken to study various closure models for the electrons.

In particular, the electron momentum balance equation can be recast as a generalized Ohm's law:

$$\mathbf{E} + \mathbf{u} \times \mathbf{B} = \eta \mathbf{J} + \frac{1}{ne} \mathbf{J} \times \mathbf{B} - \frac{1}{ne} \nabla \cdot \overleftrightarrow{P}_e + \frac{m_e}{ne^2} \, d\mathbf{J}/dt,$$

where nonzero terms on the right-hand side may break the frozen-in law. These terms represent physical processes that become important on smaller kinetic scales that are not resolved by the ideal MHD model. Below, we will review how these nonideal effects have been found to alter reconnection dynamics.

6.4.2 Including Resistivity

The first term added to the Ohm's law includes the resistivity η, which may either be generated by ion–electron collisions or be an "anomalous" resistivity produced by microscopic instabilities. A common measure of the resistivity is the Lundquist number

$$S = \frac{\mu_0 L v_A}{\eta},$$

where:

L is a characteristic length scale of the plasma
v_A is the Alfvén speed
S is the ratio of an Alfvén wave transit time across the system to the typical time it would take for the magnetic field to diffuse due to resistivity

In space plasmas, typical values are $S > 10^{10}$ if only Coulomb collisions are taken into account. The corresponding resistive diffusion times are typically many orders of magnitude smaller than the observed reconnection rates.

Resistive MHD, which includes only the resistive term in the Ohm's law, was the foundation for early models of magnetic reconnection such as the influential Sweet–Parker model. In the Sweet–Parker model, resistivity is important only in a narrow boundary layer width of $\delta_{SP} = L_{SP}/\sqrt{S}$, where L_{SP} is a macroscopic length associated with the global system size. The model predicts that plasma flows into the layer at a reconnection rate of $v_{inc} = v_A/\sqrt{S}$. While this is significantly faster than purely resistive diffusion ($\propto 1/S$), the model grossly underestimates the timescales for the observed reconnection events. Petschek introduced a modified model in an attempt to account for the discrepancy between the Sweet–Parker model and observations. In this model, the acceleration of the plasma to an Alfvénic outflow speed occurs not in a thin current sheet, but rather along a set of slow mode shocks that border the reconnection exhaust. While this model predicts a faster reconnection rate with a weak scaling of $\propto 1/\ln S$, there is little evidence in space observations or numerical simulations (Biskamp 1986, Uzdensky and Kulsrud 2000) that the shock geometry of the Petschek model develops in reconnecting current sheets.

Computer simulations typically demonstrate that at modest values of S reconnection proceeds according to Sweet–Parker model, unless resistivity is spatially localized. As the collisionality is reduced and the Lundquist number S increases, the behavior of the reconnection layers crucially depends on the ratio between the Sweet–Parker width of the reconnection current layer $\delta_{SP} = L_{SP}/\sqrt{S}$ and a characteristic ion kinetic scale (d_i or $\rho_s = \sqrt{T_e/m_i}/\Omega_{ci}$ for the cases with strong magnetic field) (Daughton et al. 2009). In some systems, δ_{SP} can become comparable to ion kinetic scales while the Lundquist number S remains below critical value S_{crit} for the development of plasmoid instability (see below). In such cases, there is a transition in the structure of the reconnecting current sheet and an increase in the rate of reconnection. The transition occurs when the electron and ion fluids decouple, and the reconnection rate rises to the typical rate of $E_{rec} \sim 0.1 v_A B_0$.

The reconnection dynamics of collisional systems depends also on the system size. Indeed, the aspect ratio of Sweet–Parker current sheet L/δ_{SP} increases with increasing S. Consequently, in large systems the reconnection layer can become unstable to secondary tearing instabilities that generate multiple magnetic islands or plasmoids (Loureiro et al. 2007). Commonly referred to as the plasmoid instability, this powerful instability limits the extent of Sweet–Parker current sheet. Crucially, its nonlinear development allows the reconnection rate to become fast, that is, independent on the resistivity of system size, even within the framework of MHD (e.g., Huang and Bhattacharjee 2013). Under most conditions however, the development of the instability will lead to a rapid transition to a kinetic regime characterized by values of δ_{SP} below ion kinetic scales (e.g., Ji and Daughton 2011). Development of multiple X-line, patchy reconnection appears to be a common feature of many reconnection regimes. For weakly collisional systems, multiple X-line formation was found to occur when the system size is $>50 \, d_i$ or ρ_s. Another transition is related to electron trapping generating pressure anisotropy, which

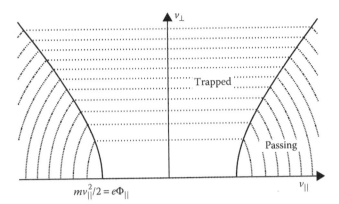

FIGURE 6.5 Contours of the distribution including trapped particles of Equation 6.1, drawn for the typical case of electric trapping $\Phi_{\parallel} > 0$ and a magnetic field weaker than the upstream field $B/B_0 < 1$.

requires the electron collision time to be smaller than the full transit time of an electron fluid element through the reconnection exhaust (Le at al. 2015). A summary of the results is plotted in the "phase diagram" in Figure 6.5, which categorizes the regimes of reconnection based on Lundquist number and a normalized system size (Ji and Daughton 2011). Under usual conditions, the current sheets found in the magnetopause, magnetotail, and solar wind, characterized by $S \sim 10^{12}-10^{18}$ and nominal system sizes of $10^2-10^6 \, d_i$, fall into the collisionless regime with multiple X-lines expected. The collisional regime becomes important in the solar photosphere, and also in laboratory astrophysics experiments devoted to studying reconnection.

6.4.3 The Hall Effect

The first multifluid effect that becomes evident in weakly collisional plasmas is the Hall effect, which stems from the second term in the generalized Ohm's law, $(1/ne)\mathbf{J}\times\mathbf{B}$. The Hall term is related to the separation of scales between the electrons and ions, and it allows the propagation of dispersive whistler waves with frequencies between the ion and electron cyclotron frequencies. The Hall effect by itself does not permit magnetic reconnection to occur because the generalized Ohm's law including only the Hall term is equivalent to $\mathbf{E}+\mathbf{u}_e\times\mathbf{B}=0$, where $\mathbf{u}_e$ is the electron flow velocity. The magnetic field is therefore frozen into the electron flow. An additional dissipation mechanism, such as resistivity, is therefore necessary to allow the onset of reconnection in Hall models. Nevertheless, the Hall effect alters the structure of the reconnection region.

A large effort was made for the Geospace environmental modeling (GEM) challenge to determine what models reproduce the rate of reconnection determined from kinetic modeling (Birn et al. 2001). Based on a large comparison of fully kinetic, hybrid kinetic/fluid, multispecies fluid, and single fluid frameworks, a minimal model sufficient to allow the onset of reconnection and to give a fast reconnection rate was found to be the resistive Hall model. Although Hall models result in fast reconnection, there is evidence that it is not the Hall effect *per se* that makes reconnection fast. In particular, fully kinetic simulations have been carried out to study reconnection in pair plasmas composed of electrons and positrons, where there is thus no separation of kinetic scales between the two species (Daughton and Karimabadi 2007). Nevertheless, reconnection in pair plasmas is fast. Other evidence against the crucial role of the Hall effect and dispersive waves as the cause of fast reconnection has come from Hall-less hybrid simulations (Karimabadi et al. 2004) and, more recently, simulations with large guide fields (Liu et al. 2014, TenBarge et al. 2014, Stanier et al. 2015b) where reconnection remains fast even in the absence of dispersive waves.

6.4.4 The Electron Pressure Tensor

Fluid formulations that use the collisional limit as a starting point typically include an isotropic pressure characterized by a single scalar pressure p. In collisionless space plasmas, however, the pressure can become highly anisotropic. In most regions of a magnetized plasma, the electron pressure tensor is nevertheless "gyrotropic" (isotropic about the magnetic field direction), with just two independent pressure components $p_{\parallel}$ and $p_{\perp}$. The main anisotropy during reconnection is typically an anisotropy $p_{\parallel}/p_{\perp}$ different from one. Note, however, that kinetic simulations have demonstrated that gradients in the electron pressure tensor are often dominant in supporting the reconnection electric field at the X-line (Kuznetsova et al. 2001). A gyrotropic pressure tensor in 2D cannot produce magnetic reconnection. If the electron pressure tensor is responsible for "breaking" magnetic field lines, then it must have a nongyrotropic part, which means there are two different perpendicular pressure components $p_{\perp 1}$ and $p_{\perp 2}$. The agyrotropy measure

$$A = 2\frac{(p_{\perp 1} - p_{\perp 2})}{(p_{\perp 1} + p_{\perp 2})}$$

has been suggested as a signature of the electron diffusion region in spacecraft observations. A well-resolved electron diffusion region measured by the POLAR spacecraft at the Earth's magnetopause exhibited both anisotropy $p_{\parallel}/p_{\perp} \sim 7$ and agyrotropy $A \sim 1$ (Scudder et al. 2012).

Based on the electron energization model presented in Section 6.3.3, equations of state for the electron pressure tensor may be derived when the electrons follow adiabatic orbits (Le et al. 2009). This generally requires a guide magnetic field. While no closed analytical form exists for these equations of state, the following approximate forms are good to a few percent for a range of densities and magnetic fields typical of reconnection in space:

$$\tilde{p}_{\parallel} = \tilde{n}\frac{2}{\alpha+2} + \frac{\pi\tilde{n}^3}{6B^2}\frac{2\alpha}{2\alpha+1}$$

$$\tilde{p}_{\perp} = \tilde{n}\frac{1}{\alpha+1} + \tilde{n}\tilde{B}\frac{\alpha}{\alpha+1}$$

where $\alpha = \tilde{n}^3/\tilde{B}^2$ and, for any quantity Q, $\tilde{Q} = Q/Q_0$ is normalized to its value in the ambient ideal plasma at the end of a given flux tube. For $\alpha \ll 1$, there are practically no trapped electrons, and the equations of state reduce to an isothermal equation of state with $p_{\parallel} \sim p_{\perp} \sim nT_0$. In the opposite limit, $\alpha \gg 1$, the electron population consists mainly of trapped particles, and the CGL equations of state $p_{\parallel} \propto n^3/B^2$ and $p_{\perp} \propto nB$ are recovered (Figure 6.6).

Although the anisotropy predicted by the equations of state is not sufficient in itself to break the frozen-in condition, the anisotropy largely governs the currents and magnetic structure of the electron diffusion region. In particular, pressure anisotropy near the firehose instability threshold, $p_{\parallel} - p_{\perp} = B^2/\mu_0$, allows the formation of a quasi-1D current sheet with a normal component of magnetic field. For anti-parallel reconnection, the pressure anisotropy drives electron jets that extend from the X-line at nearly the electron thermal speed (Karimabadi et al. 2007, Shay et al. 2007, Hesse et al. 2008, Le et al. 2010). These jets are typically limited to $\sim 100\ d_e$ in length, and they are deflected by weak guide fields (Goldman et al. 2011). Another regime exists with intermediate values of the guide fields where the electron pressure anisotropy extends into the reconnection exhaust and an elongated current sheet forms, which can stretch 10s of d_i, and is not confined to usual electron kinetic scales (Le et al. 2013).

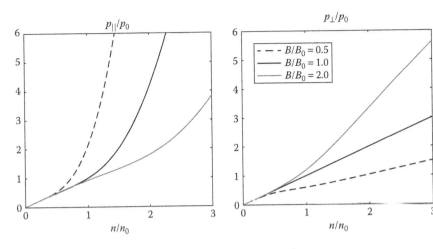

FIGURE 6.6   Equations of state for the electron pressure components $p_\parallel$ and $p_\perp$.

6.4.5 Electron Inertia

The final term in the generalized Ohm's law takes into account the inertia of the electrons. The first proposal for reconnection in Earth's magnetic tail suggested it could occur through an instability including the effects of electron inertia (Coppi et al. 1966). Typically, this requires gradients on the electron inertial length $d_e = c/\omega_{pe}$, which is on the order of ~1–10 km in the magnetosphere. While these scales were eventually rejected as being too small to account for an appreciable acceleration of large populations of electrons, electron inertia could contribute to microscopic instabilities that generate the anomalous transport coefficients needed for other subgrid models.

6.4.6 Ion Kinetic Effects

While electron kinetic effects are necessary to break the frozen-in condition of ideal MHD and to allow reconnection to occur, the reconnection dynamics depends heavily on ion kinetic effects. In a large PIC simulation of Harris sheet reconnection run to late time, the ion profiles of the reconnection exhaust were found to have a multiscale structure (Le et al. 2014). Near the X-line, meandering ions may form two distinct counterstreaming populations composed of ions in opposite phases of their meandering motion, as seen in Figure 6.7. The ion dynamics may be modeled as a pick-up mechanism whereby ions drift from the inflow into the exhaust across a narrow boundary at the magnetic separator and are eventually accelerated to the outflow $\mathbf{E} \times \mathbf{B}$ velocity (Drake et al. 2009). The counterstreaming populations lead to an anisotropic pressure, and the anisotropy in Figure 6.7 with $p_\parallel/p_\perp \sim 3$ is strong enough to approach

the ion firehose threshold $p_\parallel - p_\perp = B^2/\mu_0$ and support an elongated current sheet in the exhaust center. Further downstream, the ion orbits become chaotic, and the distinct populations in velocity space are mixed, resulting in a nearly isotropic distribution with an increased effective temperature.

The ion pressure anisotropy plays into overall momentum balance of the reconnection exhaust. The anisotropy cancels a portion of the "tension" force of the exhaust magnetic field, which is the main drive accelerating the flow of exhaust plasma. In the kinetic simulation, the ion anisotropy is as important as the ion inertia in setting the outflow speed of the plasma. As a result, the ion anisotropy alters the Walen condition, based on MHD jump conditions, which predicts the outflow speed to be

$$\mathbf{v}_W = \pm\sqrt{\frac{n_0 m_i}{\mu_0}}\left[\frac{\mathbf{B}}{nm_i} - \frac{\mathbf{B}_0}{n_0 m_i}\right],$$

where quantities with a 0 subscript are upstream conditions. In the simulation, the Walen condition overestimates the ion flow by a factor of ~2. It turns out, however, that the Walen condition holds for the electrons (Scudder et al. 1999) in this region except for narrow (on the order of the electron meandering orbit width) layers at the center of the exhaust and near the magnetic separators where the electron pressure is agyrotropic.

More recently, using the island coalescence problem, Stanier et al. (2015a) compared the reconnection rate and the evolution of the system in fully kinetic, hybrid, and Hall MHD simulations. They found that the Hall MHD model fails to reproduce the kinetic results and

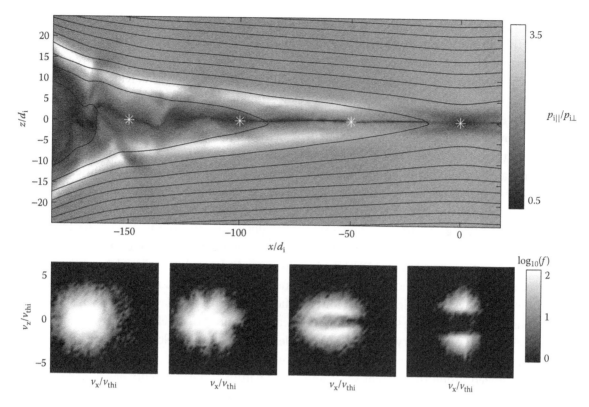

FIGURE 6.7 Ion kinetic effects in a fully kinetic PIC simulation of anti-parallel reconnection. (Top) The ion pressure may become highly anisotropic. (Bottom) Ion velocity distributions (each sampled at a "*" in the top figure).

yielded an incorrect rate of reconnection. In contrast, the hybrid simulations yielded a much closer match with the fully kinetic results. The authors concluded that the minimum physics required for proper modeling of collisionless systems is ion kinetic effects.

6.4.7 Instabilities

A broad spectrum of instabilities can develop in the vicinity of reconnection sites. Lower-hybrid drift instability (LHDI) has been extensively investigated as a possible source of anomalous resistivity and viscosity (e.g., Papadopoulos 1977). The Harris current sheet represents a convenient exact kinetic equilibrium that allows accurate linear analysis. In such a configuration, the instability occupies a relatively broad range of wavenumbers $\rho^{-1}_e \le k \le (\rho_e\rho_i)^{-1/2}$ with a nearly perpendicular orientation of the wavevector with respect to the magnetic field. The short-wavelength modes tend to be electrostatic and are stabilized at relatively large values of plasma $\beta \sim 1$. For that reason, such modes are localized on the edges of the current sheet unless the guide field is strong. The longer wavelength instabilities with characteristic wavenumber $k \sim (\rho_e\rho_i)^{-1/2}$ are electromagnetic and can penetrate the central region of the current sheet even in the

antiparallel configuration (Daughton 2003). Numerical simulations suggest that lower-hybrid modes can play a significant role in the onset of magnetic reconnection. For the example, even though the short-wavelengths LHDI tends to be localized on the edges of the current sheet, its nonlinear development leads to significant modifications in the structure of the current sheet that may result in a rapid onset of reconnection (e.g., Daughton et al. 2004). Similarly, nonlinear development of the long-wavelength mode leads to large-amplitude perturbations in the normal component of magnetic field, which may modify the timing of the reconnection onset. Configurations with curved magnetic field, as is typical of the Earth's magnetotail, can also become unstable against a ballooning-interchange mode that can be thought of as a lower-frequency extension of the lower-hybrid drift mode (Pritchett and Coroniti 2010). In 3D kinetic simulations, nonlinear development of this instability leads to onset of localized reconnection and violent disruption of the current sheet.

The influence of lower-hybrid modes on the properties of fully developed reconnection regions (past onset) remains an area of active research. In part, this is because of the complexity and high computational cost of three-dimensional simulations required to simultaneously capture the reconnection dynamics together with a full spectrum of current-aligned instabilities. Existing numerical simulations suggest that lower-hybrid modes do not dramatically affect properties of a fully developed reconnection region (e.g., Pritchett and Mozer 2011; Roytershteyn et al. 2012). In particular, the overall structure of the reconnection region and the reconnection rate remain quite similar between 3D simulations where the modes reach relatively large amplitudes and 2D simulations where the instabilities are entirely suppressed by the geometry. This conclusion is consistent with laboratory measurements performed in the magnetic reconnection experiment (MRX) device (Roytershteyn et al. 2013). In a typical reconnection current sheet, the short-wavelength electrostatic lower-hybrid fluctuations are localized along some or all of the separatrices (depending on asymmetry and magnitude of the guide field), while the longer-wavelength fluctuations tend to be localized inside the current sheet and may lead to its kinking (Figure 6.8).

A variety of other instabilities have also been discussed in the literature. In quasi-2D collisionless reconnection, the separatrices are characterized by the presence of intense non-Maxwellian features in both electron and ion distribution functions as well as by a significant relative drift (current) between the two populations. Consequently, the separatrices are unstable against a variety of modes, especially in configurations with significant guide field. Kinetic simulations reported development of whistler modes, streaming, and electron flow instabilities (Divin et al. 2012). Nonlinear development of these modes may lead to the formation of electron holes propagating in the direction parallel to the magnetic field. While such instabilities play a significant role in relaxation of the local unstable distributions, their influence on the overall energy conversion rate or the global structure of the current sheet remains unclear. Under some conditions, for example, when reconnection is strongly driven by external drivers, the primary reconnection layer may become very thin. If significant guide field is present, such layers have been shown by numerical simulations to develop a filamentation instability that leads to a rapid development of small-scale turbulence and subsequent broadening of the layer (Che et al. 2011). The nature of the instability has been somewhat controversial. The initial ideas suggested the gradients in the electron current density as the primary driver, while a subsequent kinetic investigation found no such modes and suggested that the layers become unstable only when the relative streaming between electrons and ions exceeds the threshold for streaming instability (Liu et al. 2013). In a force-free configuration considered by Liu et al., the corresponding half-thickness of the current layer is given by $\lambda < d_e / \sqrt{2\beta_e}$, where β_e is defined with the reconnecting component of the magnetic field. Furthermore, in large-scale fully kinetic simulations the development of instability and the broadening of the layer have been shown to be transient effects, with tearing instabilities of secondary layers dominating over longer timescales.

In contrast to the instabilities discussed above, tearing instability of self-consistently generated secondary reconnection layers may have a profound influence on the overall structure of the reconnection region. Large-scale 3D kinetic simulations have demonstrated that in the most common case of reconnection with a guide field, development of the tearing instabilities often results in the formation of interacting flux ropes (Figure 6.9) and onset of global fully developed turbulence within a large region (Daughton et al. 2011). The turbulence leads to stochastization of the magnetic field, which in turn enables rapid mixing of plasma populations from the two sides of the reconnection region via fast parallel streaming. Despite the dramatic effect of

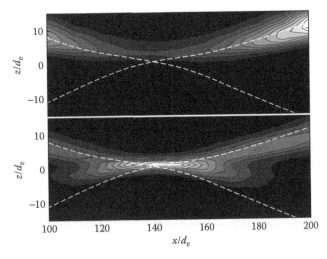

FIGURE 6.8 Typical distribution of lower-hybrid electromagnetic fluctuation power in the vicinity of a reconnection region. Fluctuations of the electric field are predominantly localized around separatrices away from the X-line (Top). In contrast, fluctuations of the magnetic field are localized in the vicinity of the X-line (Bottom). Results from a fully kinetic 3D simulation. The dashed line in both panels shows the position of separatrix averaged over the out-of-plane direction.

FIGURE 6.9 Formation of flux ropes and onset of turbulence in a 3D simulation of magnetic reconnection in asymmetric configuration with a guide field. The figure shows vorticity of the ion flow.

turbulence on the electron-scale structure of the reconnection region, the overall reconnection rate in such simulations remained fairly close to the corresponding 2D simulations.

All of the instabilities discussed above are of kinetic origin, or operate in the regimes where a kinetic treatment is necessary. For example, the thickness of secondary current sheets is typically comparable to the local electron gyroradius, so that electron kinetic physics must be considered. This presents formidable challenges to faithfully incorporating the effects of the instabilities discussed above into large-scale models.

6.4.8 3D Reconnection

Most early reconnection models focused on 2D geometries. When fully 3D systems are considered, however, a much wider array of scenarios becomes possible, and even the definition of magnetic reconnection becomes problematic. Magnetic fields without 2D symmetry will generically exhibit chaotic field lines that exponentially diverge from each other (Boozer 2012). While 2D reconnection is characterized by the locations of X-lines and separator flux surfaces, these topological features generally do not exist in 3D systems. Three-dimensional magnetic field topologies may sometimes be characterized by the locations of the nulls and special spine field lines and separator surfaces that connect them to each other or to sources of flux in the system. These topological "skeletons" have been used to analyze the braided magnetic field configurations of the solar corona. For bounded systems with a guide magnetic field, it has been proposed that quasi-separatrix layers (QSLs) (Démoulin et al. 1996) may be used to characterize reconnection.

QSLs are not topological in nature, but rather they are determined by the geometric property that field lines on either side very rapidly diverge. Because it is so far not possible to perform global fully kinetic simulations of large 3D systems, it is still not known how kinetic effects play into the complex effects described above.

6.4.9 External Turbulence

There has been much discussion about the effect of turbulence on reconnection, that is, reconnection in current sheets embedded in turbulent medium. Various models have been devised which obtain rates as a function of underlying turbulence. In best case scenarios, they obtain rates that are fast and independent of resistivity. The reader is referred to a recent review article (Karimabadi and Lazarian 2013) on the assessment of these models. There is currently no evidence for these models in the solar wind which is known to have embedded turbulence. And since the large-scale Alfvénic turbulence assumed in such models is absent both at the dayside magnetopause and in the magnetotail, we will not discuss these models further here. We should point out, however, that within the collisionless regime, the self-generated turbulence due to the reconnection process seems to have minimal impact on the macroscopic details of reconnection such as energization and reconnection rate.

6.5 ASSESSMENT OF APPROACHES TO CLOSURE MODELS

6.5.1 Hall MHD

As mentioned previously, a main result of the GEM challenge was that a minimal model that reproduces a correct rate of reconnection in single current sheets is the Hall MHD model. Because the Hall MHD equations resolve smaller time and space scales, particularly those associated with the whistler wave, global Hall MHD modeling requires substantially more computational effort than traditional MHD models. Global Hall MHD simulations have recently been carried out of the interaction of the rotating Jovian magnetosphere with Jupiter's moon Ganymede, which possesses its own dipole magnetic field (Dorelli et al. 2015). The Ganymede magnetosphere provides a useful test case because, under typical conditions, the radius of Ganymede is $R_G \sim 2600\,\text{km} \sim 6d_i$, which is accessible to high-resolution 3D simulation (Jia et al. 2008). The Jovian plasma flow at Ganymede is sub-Alfvénic with an Alfvén Mach number $M_A \sim 0.3 - 0.6$, and Ganymede's magnetosphere therefore has a different structure from the Earth's. There is no bow shock, but rather a pair of "Alfvén wings" that stretch from Ganymede's surface at an angle determined by M_A. In any case, reconnection still occurs at Ganymede's magnetopause and in the tail. It was found that the Hall fields generate $\mathbf{J} \times \mathbf{B}$ forces that drive convection of Jovian plasma through the equatorial region of Ganymede's magnetosphere. These convection patterns are asymmetric in the dawn/dusk direction and are not seen in standard MHD models. There is evidence for these Hall structures in magnetometer data collected during flybys of the Galileo spacecraft.

6.5.2 Hybrid Modeling

Hybrid modeling refers to treating one species (usually the ions) kinetically while using a fluid model for another species (usually the electrons). In this way, the hybrid method retains ion kinetic effects without being constrained to resolve the smallest electron kinetic time and space scales. While fully kinetic simulations continue to be limited to modeling small regions such as boundary layers at the magnetopause or in the magnetic tail, global modeling is now within reach of hybrid simulations.

One major issue with the global modeling of the magnetosphere using Hall MHD or any fluid model is the fact that the Earth's magnetosphere is dominated by ion kinetic effects. For example, the ion foreshock which leads to magnetosheath turbulence and associated space weather effects, is absent from any fluid treatment. An example of ion kinetic effects with significant space weather implications was demonstrated recently by Karimabadi et al. (2014). Figure 6.10 shows the formation and penetration of large velocity plasma jets that enter the magnetosheath and their interaction with the magnetopause. Only a small segment of the global simulation is shown.

6.5.3 Anisotropic Pressure Closure for the Electrons

The anisotropic equations of state for the electrons presented in Section 6.4.4 have been incorporated into both a two-fluid and a hybrid code. Initial studies focused on reconnection in an initially force-free current sheet. In both the hybrid and the two-fluid simulations, the most obvious difference when electron pressure anisotropy is included is the development of extended electron current sheets in the reconnection exhaust, mentioned in Section 6.4.4, that are upward of 10 d_i long. An example is in the top panel of Figure 6.11 from a fully kinetic simulation and in the middle panel from a hybrid H3D simulation with the anisotropic electron closure (Le et al. 2016). The bottom panel shows a hybrid run that used a scalar electron pressure, and the exhaust electron current sheet did not develop. A parameter scan was performed using a two-fluid model implemented with the HiFi framework (Ohia et al. 2012), and comparisons of simulations with different electron closures (anisotropic or scalar pressure) showed that the electron anisotropy feeds back on the larger-scale structure of the reconnection exhaust.

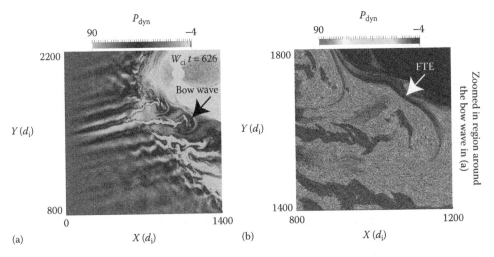

FIGURE 6.10 Plot of dynamical pressure in an area of the simulation zoomed in around the quasi-parallel bowshock. Some of the jets reach the magnetopause while others terminate closer to the bow shock. (b) LIC of magnetic field colored by dynamical pressure for an area zoomed in around a strong jet marked by the bow wave in panel (a) that has triggered a flux transfer event at the magnetopause. (From Karimabadi, H. et al., *Phys. Plasmas,* 21, 062308, 2014. With Permission.)

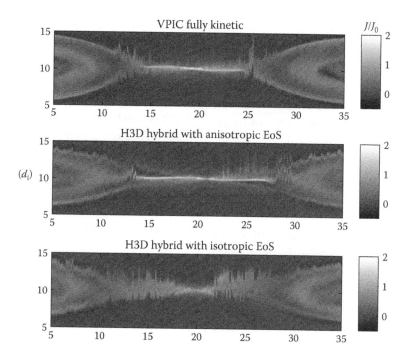

FIGURE 6.11 The out-of-plane current density from (top) a fully kinetic simulation, (middle) a hybrid simulation with kinetic ions and an anisotropic pressure closure for the electrons, and (bottom) a hybrid simulation with isotropic electron pressure. The anisotropic closure allows the formation of an elongated current sheet supported by electron pressure anisotropy.

6.5.4 High-Moment Fluid Closures

A general approach to obtain a fluid closure is to take velocity moments of the particle distributions, with the lowest moments resulting in the hydrodynamic conservation equations for particle number, momentum, and energy. Each moment equation depends on the next higher-order moments, and the process must be truncated at some point. For modeling reconnection, models retaining 5, 10, and 20 moments for all particle species have recently been implemented by Wang et al. (2015). The 5-moment model includes only a scalar pressure, while the 10- and 20-moment models include the full anisotropic pressure tensor. The truncated equations call for an approximation of the thermal heat flux, which is not correctly described by any model in collisionless plasmas. For preliminary tests, therefore, the higher-moment models were closed using a single global collision frequency based on typical collisionless damping scales with a Krook-type operator to relax pressure anisotropies. Although this initial closure is somewhat crude, the higher-order fluid models with a tuned collision frequency were able to reproduce properties of the kinetic simulations, such as peak currents and reconnection layer widths.

6.5.5 Gyrokinetics

When the magnetic field is nowhere too weak, a gyrokinetic approximation of the kinetic equations can be obtained (e.g., Brizard and Hahm 2007). The gyrokinetic approximation assumes that the frequency ω of typical dynamical processes of interest is smaller than the particle cyclotron frequencies, $\Omega_{ci,ce}$. A formal expansion of the kinetic equations along with taking "gyroaverages" of the electromagnetic fields results in the gyrokinetic equations, where fast temporal scales are removed and the phase space dimensionality is reduced by one. Gyrokinetics has been applied to studies of reconnection with a strong guide field (Rogers et al. 2007, Püschel et al. 2011, TenBarge et al. 2014, Munoz et al. 2015). Direct comparison with fully kinetic simulations demonstrated that the gyrokinetic model successfully reproduces fully kinetic results when the guide field is large enough (TenBarge et al. 2014). This also implies the existence of asymptotic solution in the limit of very large guide fields. Gyrokinetics may thus supplement fully kinetic simulations in local studies of reconnection with very strong guide fields, or be applied to global hybrid modeling with the ions treated gyrokinetically and with fluid electrons.

6.6 PROPOSED ROADMAP AND SUMMARY

Magnetic reconnection is a robust mechanism that occurs in a large variety of parameter regimes and conditions. This robustness is perhaps not too surprising since in the absence of magnetic reconnection the plasma dynamics would be quite restricted. And as recent studies indicate, there is a close link between collisionless shocks, turbulence, and reconnection. For example, reconnection can generate turbulence and vice versa, and fast magnetosonic shocks can generate microreconnection (Retino et al. 2007, Karimabadi et al. 2014).

There are as yet no first principle theories of magnetic reconnection, even for simple geometries. In the absence of such a theory, much of the research has centered on simulations and analysis of in situ spacecraft observations. And despite the advent of petascale computing, studies of 3D reconnection under realistic conditions remain quite challenging. One ray of hope is that macroscopic details of reconnection such as its rate are range bound and are not sensitive to many details such as the dimensionality of the simulations or the plasma parameters.

With regard to modeling reconnection in global magnetosphere models, a number of facts point to a 3D global hybrid simulation as the best target for the next generation global models beyond MHD. First, the magnetosphere is dominated by ion kinetic effects, and processes such as ion foreshock, turbulence in the magnetosheath, among others are forever beyond fluid models. Figure 6.10 showed an example of plasma jets inside the magnetosheath due to ion scale turbulence associated with the quasi-parallel bow shock. Such plasma jets have been reported in spacecraft data and can have significant space weather effects. Second, as first pointed out by Karimabadi et al. (2004), ion kinetic effects are the minimal physics required for the proper description of reconnection. Strong evidence for this has come recently from a detailed study that compared simulations of island coalescence using fully kinetic, hybrid and Hall MHD simulations. Third, the inclusion of any electron scale effects in global fluid codes would be prohibitively expensive and even then, such simulations would be missing all the critical ion kinetic effects.

Three-dimensional global hybrid simulations are now possible using petascale simulations (e.g., Karimabadi et al. 2014). And use of discrete event methodology, which is ideally suited for multiscale problems, offers a promising path to the future of global simulations. In discrete event methodology, each computational cell gets updated based on the level of "activity" in the cell. For example, the segment of the simulation in the magnetosheath gets updated much more frequently than the lobe region in the magnetosphere where the ion inertial length is much larger and the ion cyclotron frequency is smaller. And the pristine solar wind gets updated much less frequently. Discrete event-based hybrid simulations have now been successfully performed for studies of shocks, global magnetosphere, and laboratory plasmas. Such temporal adaptivity on computational cell level can lead to speedups of a factor of 10 or more.

REFERENCES

Asano, Y., Nakamura, R., Shinohara, I., Fujimoto, M., Takada, T., Baumjohann, W., Owen, C. J. et al. 2008. Electron flat-top distributions around the magnetic reconnection region. *J. Geophys. Res.* 113(A1): A01207.

Ashour-Abdalla, M., El-Alaoui, M., Goldstein, M. L., Zhou, M., Schriver, D., Richard, R., Walker, R., Kivelson, M. G., and Hwang, K.-J. 2011. Observations and simulations of non-local acceleration of electrons in magnetotail magnetic reconnection events. *Nat. Phys.* 7(4): 360–365.

Birn, J., Drake, J. F., Shay, M. A., Rogers, B. N., Denton, R. E., Hesse, M., Kuznetsova, M. et al. 2001. Geospace environmental modeling (GEM) magnetic reconnection challenge. *J. Geophys. Res.* 106: 3715–3720.

Biskamp, D. 1986. Magnetic reconnection via current sheets. *Phys. Fluids* 29(5): 1520.

Boozer, A. H. 2012. Separation of magnetic field lines. *Phys. Plasmas* 19(11): 112901.

Brizard, A. J. and Hahm, T. S. 2007. Foundations of nonlinear gyrokinetic theory. *Rev. Mod. Phys.* 79: 421.

Cassak, P. A. and Shay, M. A. 2007. Scaling of asymmetric magnetic reconnection: General theory and collisional simulations. *Phys. Plasmas* 14(10): 102114.

Che, H., Drake, J. F., and Swisdak, M. 2011. A current filamentation mechanism for breaking magnetic field lines during reconnection. *Nature* 474: 184–187.

Coppi, B., Laval, G., and Pellat, R. 1966. Dynamics of the geomagnetic tail. *Phys. Rev. Lett.* 16(26): 1207–1210.

Dahlin, J. T., Drake, J. F., and Swisdak, M. 2014. The mechanisms of electron heating and acceleration during magnetic reconnection. *Phys. Plasmas* 21(9): 092304.

Daughton, W. 2003. Electromagnetic properties of the lower-hybrid drift instability in a thin current sheet. *Phys. Plasmas* 10: 3103–3119.

Daughton, W., Giovanni, L., and Paolo, R. 2004. Nonlinear evolution of the lower-hybrid drift instability in a current sheet. *Phys. Rev. Lett.* 93: 105004.

Daughton, W. and Karimabadi, H. 2007. Collisionless magnetic reconnection in large-scale electron-positron plasmas. *Phys. Plasmas* 14(7): 072303.

Daughton, W., Roytershteyn, V., Albright, B. J., Karimabadi, H., Yin, L., and Bowers, K. J. 2009. Transition from collisional to kinetic regimes in large-scale reconnection layers. *Phys. Rev. Lett.* 103(6): 065004.

Daughton, W., Roytershteyn, V., Karimabadi, H., Yin, L., Albright, B. J., Bergen, B., and Bowers, K. J. 2011. Role of electron physics in the development of turbulent magnetic reconnection in collisionless plasmas. *Nat. Phys.* 7: 539.

Démoulin, P., Hénoux, J. C., Priest, E. R., and Mandrini, C. H. 1996. Quasi-Separatrix layers in solar flares. I. Method. *Astron. Astrophys.* 308: 643–655.

Divin, A., Lapenta, G., Markidis, S., Newman, D. L., and Goldman, M. V. 2012. Numerical simulations of separatrix instabilities in collisionless magnetic reconnection. *Phys. Plasmas* 19: 042110.

Dorelli, J. C., Glocer, A., Collinson, G., and Tóth, G. 2015. The role of the Hall effect in the global structure and dynamics of planetary magnetospheres: Ganymede as a case study. *J. Geophys. Res. Space Phys.* 120: 5377–5392.

Drake, J. F., Swisdak, M., Che, H., and Shay, M. A. 2006. Electron acceleration from contracting magnetic islands during reconnection. *Nature* 443(7111): 553–556.

Drake, J. F., Swisdak, M., Phan, T. D., Cassak, P. A., Shay, M. A., Lepri, S. T., Lin, R. P., Quataert, E., and Zurbuchen, T. H. 2009. Ion heating resulting from pickup in magnetic reconnection exhausts. *J. Geophys. Res. Space Phys.* 114: 5111.

Egedal, J., Daughton, W., Le, A., and Borg, A. L. 2015. Double layer electric fields aiding the production of energetic flat-top distributions and superthermal electrons within magnetic reconnection exhausts. *Phys. Plasmas* 22(10): 101208.

Egedal, J., Fox, W., Katz, N., Porkolab, M., Oieroset, M., Lin, R. P., Daughton, W., and Drake J. F. 2008. Evidence and theory for trapped electrons in guide field magnetotail reconnection. *J. Geophys. Res.* 113(6): A12207.

Egedal, J., Fox, W., Porkolab, M., and Fasoli, A. 2005. Eigenmode response to driven magnetic reconnection in a collisionless plasma. *Phys. Plasmas* 12(5): 052107.

Egedal, J., Le, A., Katz, N., Chen, L. J., Lefebvre, B., Daughton, W., and Fazakerley, A. 2010. Cluster observations of bidirectional beams caused by electron trapping during antiparallel reconnection. *J. Geophys. Res.* 115: A03214.

Ergun, R. E., Andersson, L., Tao, J., Angelopoulos, V., Bonnell, J., McFadden, J. P., Larson, D. E. et al. 2009. Observations of double layers in Earth's plasma sheet. *Phys. Rev. Lett.* 102(15): 155002.

Forbes, T. 2000. A review on the genesis of coronal mass ejections. *JGR* 105(A10): 23153.

Fu, H. S., Khotyaintsev, Y. V., André, M., and Vaivads, A. 2011. Fermi and betatron acceleration of suprathermal electrons behind dipolarization fronts. *Geophys. Res. Lett.* 38(16): 16104.

Goldman, M. V., Lapenta, G., Newman, D. L., Markidis, S., and Che, H. 2011. Jet deflection by very weak guide fields during magnetic reconnection. *Phys. Rev. Lett.* 107:135001.

Gosling, J. T., Skoug, R. M., McComas, D. J., and Smith, C. W. 2005. Direct evidence for magnetic reconnection in the solar wind near 1 AU. *J. Geophys. Res. Space Phys. (1978–2012)* 110(A1): A01107.

Guo, F., Hui, L., Daughton, W., and Liu, Y.-H. 2014. Formation of hard power laws in the energetic particle spectra resulting from relativistic magnetic reconnection. *Phys. Rev. Lett.* 113: 155005.

Guo, F., Liu, Y.-H., Daughton, W., and Li, H. 2015. Particle acceleration and plasma dynamics during magnetic reconnection in the magnetically-dominated regime. *Astrophys. J.* 806: 167.

Hesse, M., Zenitani, S., and Klimas, A. 2008. The structure of the electron outflow jet in collisionless magnetic reconnection. *Phys. Plasmas* 15(11): 112102.

Huang, Y.-M. and Bhattacharjee, A. 2013. Plasmjoid Instability in high-lundquist-numbermagnetic reconnection. *Phys. Plasmas* 20: 055702.

Jaroschek, C. H., Treumann, R. A., Lesch, H., and Scholer, M. 2004. Fast reconnection in relativistic pair plasmas: Analysis of particle acceleration in self-consistent full particle simulations. *Phys. Plasmas* 11(3): 1151–1163.

Ji, H. and Daughton, W. 2011. Phase diagram for magnetic reconnection in heliophysical, astrophysical, and laboratory plasmas. *Phys. Plasmas* 18(11): 111207.

Jia, X., Walker, R. J., Kivelson, M. G., Khurana, K. K., and Linker, J. A. 2008. Three-dimensional MHD simulations of Ganymede's magnetosphere. *J. Geophys. Res. Space Phys. (1978–2012)* 113(A6): 6212.

Karimabadi, H., Daughton, W., and Scudder, J. 2007. Multiscale structure of the electron diffusion region. *Geophys. Res. Lett.* 34: L13104.

Karimabadi, H., Krauss-Varban, D., Huba, J. D., and Vu, H. X. 2004. On magnetic reconnection regimes and associated three-dimensional asymmetries: Hybrid, Hall-less hybrid, and Hall-MHD simulations. *J. Geophys. Res. Space Phys. (1978–2012)* 109: A09205.

Karimabadi, H. and Lazarian, A. 2013. Magnetic reconnection in the presence of externally driven and self-generated turbulence. *Phys. Plasmas* 20(11): 112102.

Karimabadi, H., Roytershteyn, V., Vu, H. X., Omelchenko, Y. A., Scudder, J., Daughton, W., Dimmock, A. et al. 2014. The link between shocks, turbulence, and magnetic reconnection in collisionless plasmas. *Phys. Plasmas* 21(6): 062308.

Kuznetsova, M. M., Hesse, M., and Winske, D. 2001. Collisionless reconnection supported by nongyrotropic pressure effects in hybrid and particle simulations. *J. Geophys. Res. Space Phys. (1978–2012)* 106(A3): 3799–3810.

Le, A., Daughton, W., Karimabadi, H., and Egedal, J. 2016. Hybrid simulations of magnetic reconnection with kinetic ions and fluid electrons pressure anisotropy. *Phys. Plasmas* 23(3): 032114.

Le, A., Egedal, J., Daughton, W., Drake, J. F., Fox, W., and Katz, N. 2010. Magnitude of the Hall fields during magnetic reconnection. *Geophys. Res. Lett.* 37: L03106.

Le, A., Egedal, J., Daughton, W., Fox, W., and Katz, N. 2009. Equations of state for collisionless guide-field reconnection. *Phys. Rev. Lett.* 102(8): 085001.

Le, A., Egedal, J., Daughton, W., Roytershteyn, V., Karimabadi, H., and Forest, C. 2015. Transition in electron physics of magnetic reconnection in weakly collisional plasma. *J. Plasma Phys.* 81(01): 305810108.

Le, A., Egedal, J., Ng, J., Karimabadi, H., Scudder, J., Roytershteyn, V., Daughton, W., and Liu, Y.-H. 2014. Current sheets and pressure anisotropy in the reconnection exhaust. *Phys. Plasmas* 21(1): 012103.

Le, A., Egedal, J., Ohia, O., Daughton, W., Karimabadi, H., and Lukin, V. S. 2013. Regimes of the electron diffusion region in magnetic reconnection. *Phys. Rev. Lett.* 110: 135004.

Liu, Y.-H., Daughton, W., Karimabadi, H., Li, H., and Gary, S. P. 2014. Do dispersive waves play a role in collisionless magnetic reconnection? *Phys. Plasmas* 21(2): 022113.

Liu, Y.-H., Daughton, W., Karimabadi, H., Li, H., and Roytershteyn, V. 2013. Bifurcated structure of the electron diffusion region in three-dimensional magnetic reconnection. *Phys. Rev. Lett.* 110: 265004

Loureiro, N. F., Schekochihin, A. A., and Cowley, S. C. 2007. Instability of current sheets and formation of plasmoid chains. *Phys. Plasmas* 14(10): 100703.

Masuda, S., Kosugi, T., Hara, H., and Ogawaray, Y. 1994. A loop-top hard X-ray source in a compact solar-flare as evidence for magnetic reconnection. *Nature* 371(6497): 495–497.

Munoz, P. A., Told, D., Kilian, P., Büchner, J., and Jenko, F. 2015. Gyrokinetic and kinetic particle-in-cell simulations of guide-field reconnection. I. Macroscopic effects of the electron flows. *Phys. Plasmas* 22(8): 082110.

Ohia, O., Egedal, J., Lukin, V. S., Daughton, W., and Le, A. 2012. Demonstration of anisotropic fluid closure capturing the kinetic structure of magnetic reconnection. *Phys. Rev. Lett.* 109: 115004.

Ono, Y., Yamada, M., Akao, T., Tajima, T., and Matsumoto, R. 1996. Ion acceleration and direct ion heating in three-component magnetic reconnection. *Phys. Rev. Lett.* 76(18): 3328.

Papadopoulos, K. 1977. A review of anomalous resistivity for the ionosphere. *Rev. Geophys.* 15(1): 113–127.

Phan, T. D., Drake, J. F., Shay, M. A., Gosling, J. T., Paschmann, G., Eastwood, J. P., Oieroset, M., Fujimoto, M., and Angelopoulos, V. 2014. Ion bulk heating in magnetic reconnection exhausts at Earth's magnetopause: Dependence on the inflow Alfvén speed and magnetic shear angle. *Geophys. Res. Lett.* 41(20): 7002–7010.

Phan, T. D., Shay, M. A., Gosling, J. T., Fujimoto, M., Drake, J. F., Paschmann, G., Oieroset, M., Eastwood, J. P., and Angelopoulos, V. 2013. Electron bulk heating in magnetic reconnection at Earth's magnetopause: Dependence on the inflow Alfvén speed and magnetic shear. *Geophys. Res. Lett.* 40(17): 4475–4480.

Priest, E. and Forbes, T. 2000. *Magnetic Reconnection.* Cambridge University Press, New York.

Pritchett, P. L. and Coroniti, F. V. 2010. A kinetic ballooning/interchange instability in the magnetotail. *J. Geophys. Res.* 115: A06301.

Pritchett, P. L. and Mozer, F. S. 2011. Rippling mode in the subsolar magnetopause current layer and its influence on three-dimensional magnetic reconnection. *J. Geophys. Res.* 116: A04215.

Püschel, M. J., Jenko, F., Told, D., and Büchner, J. 2011. Gyrokinetic simulations of magnetic reconnection. *Phys. Plasmas* 18(11): 112102.

Retino, A., Sundkvist, D., Vaivads, A., Mozer, F., André, M., and Owen, C. J. 2007. In situ evidence of magnetic reconnection in turbulent plasma. *Nat. Phys.* 3(4): 236–238.

Rogers, B. N., Kobayashi, S., Ricci, P., Dorland, W., Drake, J., and Tatsuno, T. 2007. Gyrokinetic simulations of collisionless magnetic reconnection. *Phys. Plasmas* 14(9): 092110.

Roytershteyn, V., Daughton, W., Karimabadi, H., and Mozer, F. S. 2012. Influence of the lower-hybrid drift instability on magnetic reconnection in asymmetric configurations. *Phys. Rev. Lett.* 108: 185001.

Roytershteyn, V., Dorfman, S., Daughton, W., Ji, H., Yamada, M., and Karimabadi, H. 2013. Electromagnetic instability of thin reconnectionlayers: Comparison of 3D simulations with MRX observations. *Phys. Plasmas* 20: 061212.

Runov, A., Angelopoulos, V., Sitnov, M. I., Sergeev, V. A., Bonnell, J., McFadden, J. P., Larson, D., Glassmeier, K.-H., and Auster, U. 2009. THEMIS observations of an earthward-propagating dipolarization front. *Geophys. Res. Lett.* 36(14): L14106.

Schindler, K., Hesse, M., and Birn, J. 1988. General magnetic reconnection, parallel electric fields, and helicity. *JGR* 93: 5547–5557.

Scudder, J. D., Holdaway, R. D., Daughton, W. S., Karimabadi, H., Roytershteyn, V., Russell, C. T., and Lopez, J. Y. 2012. First resolved observations of the demagnetized electron-diffusion region of an astrophysical magnetic-reconnection site. *Phys. Rev. Lett.* 108(22): 225005.

Scudder, J. D., Karimabadi, H., Daughton, W., and Roytershteyn, V. 2015. Frozen flux violation, electron demagnetization and magnetic reconnection. *Phys. Plasmas* 22(10): 101204.

Scudder, J. D., Puhl-Quinn, P., Mozer, F. S., Ogilvie, K. W., and Russell, C. T. 1999. Generalized Walén tests through Alfvén waves and rotational discontinuities using electron flow velocities. *J. Geophys. Res.* 104: 19.

Shay, M. A., Drake, J. F., and Swisdak, M. 2007. Two-scale structure of the electron dissipation region during collisionless magnetic reconnection. *Phys. Rev. Lett.* 99(15): 155002.

Shay, M. A., Haggerty, C. C., Phan, T. D., Drake, J. F., Cassak, P. A., Wu, P., Oieroset, M., Swisdak, M., and Malakit, K. 2014. Electron heating during magnetic reconnection: A simulation scaling study. *Phys. Plasmas* 21(12): 122902.

Stanier, A., Daughton, W., Chacón, L., Karimabadi, H., Ng, J., Huang, Y.-M., Hakim, A., and Bhattacharjee, A. 2015a. Role of ion kinetic physics in the interaction of magnetic flux ropes. *Phys. Rev. Lett.* 115(17): 175004.

Stanier, A., Simakov, A. N., Chacón, L., and Daughton, W. 2015b. Fast magnetic reconnection with large guide fields. *Phys. Plasmas* 22(1): 010701.

TenBarge, J. M., Daughton, W., Karimabadi, H., Howes, G. G., and Dorland, W. 2014. Collisionless reconnection in the large guide field regime: Gyrokinetic versus particle-in-cell simulations. *Phys. Plasmas* 21(2): 020708.

Uzdensky, D. A. and Kulsrud, R. M. 2000. Two-dimensional numerical simulation of the resistive reconnection layer. *Phys. Plasmas* 7(10): 4018–4030.

Wang, L., Hakim, A. H., Bhattacharjee, A., and Germaschewski, K. 2015. Comparison of multi-fluid moment models with particle-in-cell simulations of collisionless magnetic reconnection. *Phys. Plasmas* 22(1): 012108.

Yamada, M., Kulsrud, R., and Ji, H. 2010. Magnetic reconnection. *Rev. Mod. Phys.* 82(1): 603–664.

Zenitani, S. and Hoshino, M. 2001. The generation of nonthermal particles in the relativistic magnetic reconnection of pair plasmas. *Astrophys. J. Lett.* 562(1): L63.

CHAPTER 7

Magnetospheric Electric Fields and Current Systems

Stanislav Sazykin

CONTENTS

Terrestrial magnetosphere is a cavity-shaped region bounded by streaming solar wind-magnetized plasma from above, and by collisional partially ionized particle population of the ionospheric layers. As in many astrophysical phenomena, interaction of plasmas having different origins and energies is accompanied by formation of boundaries, cellular structures, and associated electrical currents in the form of sheets. Earth's magnetosphere is a driven system: reconnection electric field sets up a large-scale plasma convection. However, existence of current systems in Earth's magnetosphere implies the presence of large-scale electric fields that modify the reconnection-driven electric field. The dynamics of electric fields and currents occurs on several different timescales determined by both solar wind "driving" and plasma properties and configuration in the magnetosphere. The primary goal of this chapter is to explain those aspects of magnetospheric large-scale electrodynamics that are known to have potentially significant space weather consequences. To accomplish this, we will first introduce two widely used approaches to mathematically describe the basics of magnetospheric

electric fields and currents: numerical modeling based on the equations of ideal magnetohydrodynamics (MHD) complimented by kinetic guiding-center drift models. We will use these approaches to obtain quantitative descriptions of magnetospheric electric fields and currents to illustrate the nature and morphology of main current systems. Electric fields will be described in terms of their sources, morphology, and their role in space weather events by using examples from numerical simulations. Specifically, we will describe (1) the large-scale convection electric field and its changes during geomagnetic disturbances, (2) subauroral polarization stream (SAPS) structures and their storm time dynamics, (3) the phenomenon of shielding and its nature, and (4) mesoscale impulsive flow channels and associated electric fields.

Further, we explain the space weather consequences of the above-mentioned electric field phenomena and their significance, again illustrating with examples. Finally, we provide a list of references to empirical and first-principle models of electric fields and currents that the reader may find useful.

7.1 APPROACHES TO DESCRIPTION OF FIELDS AND CURRENTS

Excellent textbook descriptions of large-scale magnetospheric electric fields and currents exist in the literature (e.g., Gombosi, 1998) that can be consulted for comprehensive and thorough introduction to the topic. Here, we limit ourselves to presenting the minimum background required to understand the methods used to obtain solutions and examples used later in the chapter. The simplest mathematically closed set of equations that yield realistic solutions of global magnetospheric configurations varying with the solar wind conditions is based on the equations of ideal MHD. In this approximation, magnetosphere is a single-fluid plasma with mass density ρ, momentum density $\rho\mathbf{v}$, and pressure p that relates to the internal energy density via the ideal gas law $u = p/(\gamma-1)$ with the adiabatic exponent taken to be 5/3. In the gas dynamic (semi-conservative) form, the three conservation laws

$$\frac{\partial\rho}{\partial t} + \nabla\cdot(\rho\mathbf{v}) = 0 \qquad (7.1)$$

$$\frac{\partial(\rho\mathbf{v})}{\partial t} + \nabla\cdot(\rho\mathbf{vv} + p\mathbf{I}) = \frac{1}{\mu_0}\mathbf{j}\times\mathbf{B} \qquad (7.2)$$

$$\frac{\partial}{\partial t}\left(\frac{\rho v^2}{2} + \frac{p}{\gamma-1}\right) + \nabla\cdot\left(\mathbf{v}\left(\frac{\rho v^2}{2} + \frac{p}{\gamma-1} + p\right)\right) = \frac{1}{\mu_0}\mathbf{j}\cdot\mathbf{E} \quad (7.3)$$

are combined with the Maxwell's equations for electromagnetic fields $\mathbf{E}$ and $\mathbf{B}$ that relate the fields to the current density $\mathbf{j}$ (neglecting the displacement current):

$$\frac{\partial\mathbf{B}}{\partial t} = -\nabla\times\mathbf{E} \qquad (7.4)$$

$$\nabla\cdot\mathbf{B} = 0 \qquad (7.5)$$

$$\mathbf{j} = \frac{1}{\mu_0}\nabla\times\mathbf{B} \qquad (7.6)$$

The electric field is assumed to obey the ideal Ohm's law:

$$\mathbf{E} = -\mathbf{v}\times\mathbf{B} \qquad (7.7)$$

The "outer" boundary conditions outside of the magnetosphere are representative of the inflowing supersonic solar wind with embedded interplanetary magnetic field (IMF) and are specified in the usual way (e.g., Gombosi et al., 2003; Raeder, 2003). The "inner" boundary conditions, however, have to take into account the fact that the ionosphere conduct currents and any field-aligned currents $J_\parallel$ flowing in or out of the ionosphere must close horizontally

via Pedersen and Hall currents. The boundary condition is an elliptic equation

$$\nabla\cdot\left(-\Sigma\cdot\nabla\Phi\right) = J_\parallel\sin I \qquad (7.8)$$

solved on the ionospheric shell ("inner boundary") for the electrostatic potential Φ related to the velocity $\mathbf{v}$ via electric field $\mathbf{E} = -\nabla\Phi$ by Equation 7.7. I is the inclination angle of the magnetic field at ionospheric height, and Σ is the conductance tensor.

Numerical codes solving the set of Equations 7.1 through 7.8 are referred to as resistive MHD. In this chapter, we will use examples obtained with one such code, BATS-R-US (Gombosi et al., 2003). Global MHD solutions begin to deviate from observations significantly in the inner magnetosphere, due to lack of spatial resolution and, more importantly, because they lack the drift physics that becomes increasingly important. The inner magnetosphere is where kinetic approach is needed.

In the inner magnetosphere, where magnetic field lines are closed, a good fraction of charged particles are trapped on closed drift orbits. Guiding-center description based on various forms of the collisionless Vlasov equation is usually the approach of choice, although the specific assumptions about the particle distribution function and various loss processes result in a number of various numerical codes used in this field. Here we will use one such approach, where particle distribution function is assumed to be isotropic in pitch angle, and the drift velocities are assumed to be smaller than typical characteristic speed (usually, the fast mode speed). There is an elegant mathematical theory of drift physics that was derived for quasi-steady slow-flow approximation (Wolf, 1983). It was implemented in the Rice Convection Model (RCM) (e.g., Toffoletto et al., 2003), which is briefly described next and will be used in Sections 7.3 and 7.4 in this chapter.

On closed magnetic field lines, the volume of a flux tube of unit magnetic flux is $V = \int ds/B$ (the integral taken from one ionospheric footprint of the tube to the other, conjugate one), and the particle number density n and pressure P are constant along field lines. The two main results of the adiabatic drift theory for isotropic pitch angle distribution are

1. Particles of species s (protons, electrons, or O$^+$) and charge q_s drift with average velocity

$$\mathbf{v_s} = \frac{\lambda_s}{q_s B^2}\mathbf{B}\times\nabla V^{-2/3} + \frac{\mathbf{E}\times\mathbf{B}}{B^2} \qquad (7.9)$$

where the energy invariant λ_s is related to the particle kinetic energy W_s as $W_s = \lambda_s V^{-2/3}$ and is constant along drift paths. Having such an invariant allows it to be used as an independent variable instead of energy. This and further equations in the RCM are assumed to be written in the polar coordinates on the ionospheric spherical shell.

2. The specific form of the Vlasov equation is

$$\left(\frac{\partial}{\partial t} + \mathbf{v_s} \cdot \nabla\right)\delta\eta_s = S(\eta_s) - L(\eta_s) \qquad (7.10)$$

for the time evolution of the flux tube content (number of particles per unit magnetic flux) $\delta\eta_s$. The flux tube content $\delta\eta_s$ is closely related to the particle distribution function $f_s(\lambda_s)$ by

$$\delta\eta_s = \frac{4\pi\sqrt{2}}{m_s^{3/2}}\sqrt{\lambda_s} f(\lambda_s)\delta\lambda \qquad (7.11)$$

and plasma moments can be obtained by integration over all "invariant energies," that is, pressure is

$$P_s = \frac{2}{3}\int \delta\eta_s \lambda_s \, d\lambda_s \qquad (7.12)$$

If the time-dependent electromagnetic fields are known, Equation 7.10 (with Equation 7.9) can be advanced in time for a range of appropriate values of λ_s spanning the energy range of typical plasma populations in the inner magnetosphere. The right-hand side of Equation 7.10 includes parameterized loss processes due to precipitation into the atmosphere (for electrons), charge exchange with geocorona (for positive ions), and, when suitably prescribed, source outflows.

The magnetic field used in Equation 7.9 is typically prescribed as a time sequence of static configurations varying (slowly) with solar wind and geomagnetic conditions (e.g., Hilmer and Voigt, 1995). Although this is not entirely consistent (the currents implied by the chosen magnetic field will not in general be the same as the ones computed from the solutions of Equation 7.10), the magnetic field in the inner magnetosphere is dominated by the dipole component of the Earth's intrinsic field, and the discrepancy can be considered as a first-order correction. The same cannot be said, however, about the electric field, which to a very significant extent is determined by the spatial distribution of plasma pressure. Therefore, the electric field, assumed to be potential at the ionosphere heights (but not at higher altitudes), is solved for by considering the current conservation Equation 7.8, with the exception that the field-aligned current density here is related to plasma pressure gradients and magnetic field via a well-known equation (e.g., Wolf, 1983):

$$J_{\parallel} = \hat{b} \cdot \nabla V \times \nabla P \qquad (7.13)$$

where b is the unit vector along the magnetic field at the ionospheric altitude and the current is a sum for both hemispheres. Thus, the system of Equations 7.9 and 7.8 with 7.13 are integrated in time inside the region enclosing inner magnetosphere, with the driving "parameter" being the potential distribution along the high-latitude boundary of this region taken to represent the polar cap potential.

The two approaches (global MHD and RCM) are in fact somewhat complimentary, and De Zeeuw et al. (2004) took advantage of this fact to develop a coupled MHD–RCM model that is based on global MHD solutions that are "corrected" in the inner magnetosphere to account for gradient and curvature drift effects, with the two codes using a common solution of Equation 7.8. This model was further incorporated into the more advanced space weather modeling framework (Toth et al., 2005). In this chapter, we will use results obtained with space weather modeling framework and RCM.

7.2 MORPHOLOGY OF LARGE-SCALE MAGNETOSPHERIC CURRENTS

Figure 7.1 is shows a typical magnetospheric configuration obtained from a steady-state solution of the BATS-R-US-RCM equations for steady nominal solar wind ($n = 5$ cm^{-3} protons and $V_{sw} = 400$ km/s) and IMF ($B_x = B_y = 0$, $B_z = -5$ nT) conditions. The simulation builds up a solution that is typical of the magnetosphere driven by the southward (negative) IMF B_z component.

Several major current systems were obtained by calculating and plotting current density in the magnetosphere. These are

1. Chapman–Ferraro currents, flowing and fully closed on the surface of the dayside magnetopause. These currents are confined to the magnetopause.

2. Tail currents, flowing around the magnetotail part of the magnetopause and closing across the neutral sheet.

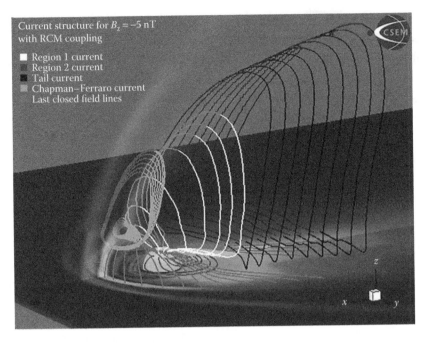

FIGURE 7.1 Typical morphology of the four major magnetospheric current systems (Chapman–Ferraro currents, tail current, region 1, and region 2 field-aligned currents). The currents were obtained by visualization of current density in a steady-state solution obtained with the BATS-R-US-RCM model (see text for details).

3. Region 1 field-aligned (Birkeland) currents flow along magnetic field lines connecting the magnetospheric outer boundary layers and part of the plasma sheet to the ionosphere. These currents are on the average upward on the dusk side and downward on the dawn side.

4. Region 2 field-aligned currents flow along magnetic field lines connecting the inner magnetosphere to the ionosphere. They tend to form a pair of sheets of currents of polarity opposing that of region 1 currents (i.e., region 2 currents are downward on the dusk side and are upward on the dawn side).

The overall plasma flow (convection) is anti-sunward in the outermost magnetosphere and is sunward in the interior. The formation of region 1 Birkeland currents is primarily due to the interaction of the solar wind with the magnetosphere, and they can be thought of as "driving currents" that set up the canonical two-cell convection pattern associated with the dawn-to-dusk directed large-scale electric field across the magnetosphere. Region 2 currents, on the other hand, are strongly affected by the internal magnetospheric dynamics.

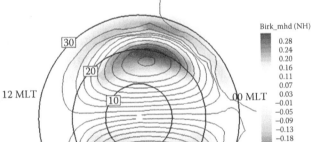

FIGURE 7.2 Field-aligned current density into the ionosphere (greyscale) superimposed on the contours of constant ionospheric potential (lines), for simialr time and conditions as in Figure 7.1.

Figure 7.2 is a plot of the field-aligned currents mapped to the ionosphere, together with the pattern of electrostatic ionospheric potential (in the frame of reference rotating with the Earth).

7.3 NATURE OF MAGNETOSPHERIC ELECTRIC FIELDS

7.3.1 Large-Scale Convection and Shielding

With the basic understanding of the structure of magnetospheric currents, we now turn to convection electric field. Prior to discussing the specifics, we note that the z-component (or, more generally, $B_T = \sqrt{B_x^2 + B_y^2}$ if $B_y \neq 0$) of the IMF is the "geoeffective" quantity that controls many electrodynamic large-scale phenomena. The cross polar cap potential drop can be related to the solar wind dynamic pressure P_{SW} (in nPa) and IMF values in nT (here the IMF "clock angle" $\theta = \arctan\left(\left|B_y\right| / B_z\right)$ taken at the "nose" of the magnetopause as (e.g., Siscoe et al., 2002):

$$\Phi_{PCP} = \frac{57.6 P_{sw}^{1/3} V_{sw} \left(B_y^2 + B_z^2\right)^{1/2} \sin^3(\theta/2)}{P_{sw}^{1/2} + 0.43 \cdot V_{sw} \left(B_y^2 + B_z^2\right)^{1/2} \sin^3(\theta/2)} \quad (7.14)$$

$$+ \min\left(16, 0.00011 \cdot V_{sw}^2\right)$$

The second term (with V_{sw} in km/s) is an estimated contribution to the polar cap potential drop during times of northward (positive) IMF B_z and is taken from empirical estimates of Boyle et al. (1997). In the RCM, it is the value of Φ_{PCP} (estimated via Equation 7.14 by using upstream values of the solar wind and IMF) that "drives" the solution of the current-conservation Equation 7.8.

Under steady magnetospheric conditions, region 2 Birkeland currents shield the inner magnetosphere from the convection E-field set up by the region 1 currents.

This effect can be easily deduced from Figure 7.2: the convection pattern (equipotentials) is mostly confined to the region 2 poleward of currents. This result can be understood in terms of the drift paths of plasma sheet particles and Equation 7.8 (e.g., Wolf et al., 2007 and references therein): region 2 currents are formed at the inner edge of the plasma sheet where large gradients of the plasma sheet and ring current pressure exist.

To understand the nature of time-varying convection electric fields in the magnetosphere, we now consider what happens to the electric fields when the IMF B_z component undergoes sudden step-wise changes. Response to a step-wise idealized change in B_z can be considered as elemental systems response. In the simulation, during the first 8 h B_z at the upstream boundary (in the solar wind, at $X_{GSM} = 32 R_E$) is maintained constant at −5 nT, then at $T = 8$ h it is changed instantaneously to +5 nT, and again to −5 nT at $T = 12$ h.

Response of the convection pattern mapped to the northern ionosphere is shown in Figure 7.3 for three times (8:09, 8:20, and 8:50). The initial high-latitude convection is almost entirely confined to the region whose equatorward most extent is the edge of region 2 (shielding) currents (although Figure 7.3 does not have field-aligned currents, this can be seen clearly in Figure 7.2). In this situation, region 2 currents set up a dusk-to-dawn electric field that (at lower latitudes) opposes the region 1-associated dawn-to-dusk convection field. It is said that convection electric field is shielded at lower latitudes (outside the auroral zone) (e.g., Wolf et al., 2005).

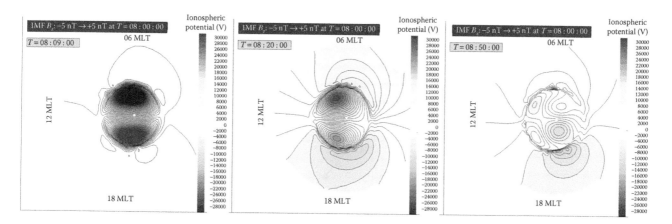

FIGURE 7.3 Response of ionospheric convection pattern to a sudden northward IMF turning shown at three times (ionospheric projects viewed from above the North Pole). Both lines and greyscale indicate the values of ionospheric convection potential. The transient enhanced convection at lower latitudes implying dusk-to-dawn electric field is termed "overshielding."

A sudden northward IMF turning renders region 2 currents too strong, which generates a short-lived (~1 h) electric field at low-latitude ionosphere that is "dusk–dawn" and backward convection in inner magnetosphere. This phenomenon is termed "overshielding"; it peaks at approximately 20 min following the change in the upstream of IMF B_z, and the system relaxes to its new equilibrium solution with a much weaker convection (consistent with Equation 7.14) in ~30 min.

Figure 7.4 indicates in more detail the time-dependent nature of the magnetospheric electric field to the IMF B_z change. The total potential difference is computed along the low-latitude boundary of RCM (mapping to 9.84° magnetic latitude in the ionosphere) and plotted as a function of time. Superimposed is the time history of IMF B_z. The ~25 min delay in the response is due to the propagation time of the change through the solar wind and the magnetosheath, as well as an MLT-dependent reconfiguration within the magnetosphere. The decay time is ~30 min and is determined by the properties of the plasma sheet generating region 2 shielding currents.

For reference, we note that in the simulation, the value of the cross polar cap potential is ~72 kV for IMF $B_z = -5$ nT and ~20 kV for IMF $B_z = +5$ nT. Therefore, during the idealized overshielding event, the peak of the penetration potential is ~20% of the "source" high-latitude potential difference.

Next, we consider a sudden southward IMF turning (change of B_z to −5 nT) at $T = 12$ in the simulation. This situation is perhaps of more interest from the space weather point of view, as it is an idealized (although not as large in magnitude) picture of what happens during magnetic storms driven by coronal mass ejections (CMEs). A southward IMF turning causes the opposite effect (undershielding) because the initial region 2 currents are too weak, which then adjusts. This transition is displayed in Figure 7.5 in the same format as Figure 7.3.

In this case, the response is similar to the case of the overshielding, but with two differences: it is in the opposite sense, and the new steady-state solution indicates a stronger residual convection electric field even after the shielding is re-established.

The change in the shielding and its re-establishing can also be seen in Figure 7.6 by looking at the potential difference along the equator. The delay in the response is somewhat longer (~30 min) but the characteristic nature of the response is the same as for the case of overshielding. The ratio of the potential differences along the equator and along the high-latitude boundary is ~20%.

The transient nature of the shielding phenomenon is because it is governed by region 2 Birkeland currents, which depend on the inner edge of the plasma sheet. As the dynamics of kilovolt plasma sheet particles changes on the timescales of typical drift periods (tens of minutes or longer), the response of region 2 currents occurs on these timescales, explaining the gradual dropoff in the curves in Figures 7.4 and 7.6.

Both overshielding and undershielding result in an appearance of magnetospherically generated electric fields in the inner magnetosphere and ionosphere

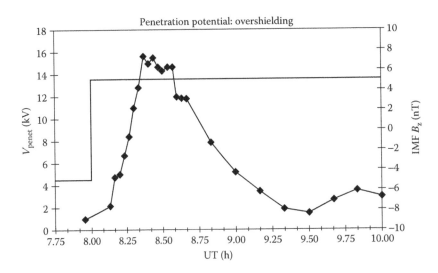

FIGURE 7.4 Time response of the magnetospheric convection potential along the equator to the northward IMF turning (curve with symbols). IMF B_z is shown for reference (curve without symbols).

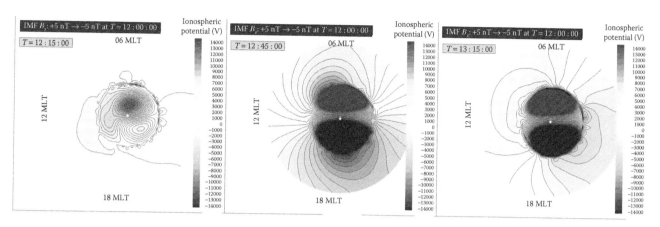

FIGURE 7.5   Response of ionospheric convection pattern to a sudden southward IMF turning; same format as Figure 7.3. The existence of a transient dawn-to-dusk electric field at lower latitudes is an example of "undershielding."

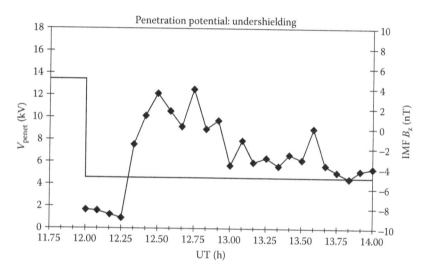

FIGURE 7.6   Time response of the magnetospheric convection potential along the equator to the southward IMF turning (curve with symbols). IMF B_z is shown for reference (curve without symbols).

called prompt penetration electric fields. In the ionosphere, prompt-penetration electric fields can cause substantial re-distributions of ionization (Huba and Sazykin, 2014), upset the thermal balance, and affect generation and evolution of a variety of plasma instabilities (e.g., equatorial spread-F). Although changes in the IMF B_z orientation are not the only solar wind cause of prompt-penetration electric fields, they may be among the most efficient one.

7.3.2 Storm Effects

During geomagnetic storms, periods of strong and sustained negative IMF B_z result (see Equation 7.14) in periods of enhanced magnetospheric convection. Southward turnings cause undershielding due to initial region 2 currents being too weak, which then adjusts. However,

the sustained nature of enhanced high-latitude convection has another consequence not seen in the solutions presented in the previous section, namely, storm-time penetration electric fields can be long lasting without any signs of shielding. For strong storms, massive undershielding has been observed, indicating that there is an entirely different regime.

Figure 7.7 shows statistical patterns of the E_y component (approximately dawn-to-dusk) electric field measured by the CRRES satellite (Rowland and Wygant, 1998). The measurements were transformed into the frame that rotates with the Earth, and binned by the radial distance and the K_p index of magnetic activity. These data were taken near the magnetospheric equatorial plane and indicate several consistent features of the electric fields:

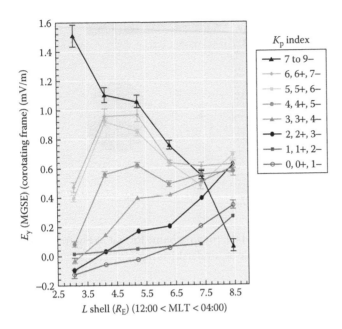

FIGURE 7.7 Dependence of the dawn-to-dusk component of the electric field in the magnetosphere on the radial distance and geomagnetic activity. The data are from CRRES spacecraft and are displayed in the frame rotating with Earth. (From Rowland, D.E. and Wygant, J.R., *J. Geophys. Res.*, 103, 14959–14964, 1998. With Permission.)

1. At larger radial distances, the dawn-to-dusk component of the electric field is on average 0.2–0.4 mV/m and varies by a factor of ~3 for storm time compared to quiet-time values. This is qualitatively consistent with the overall picture of enhanced convection during active times, and is consistent with the fact that the inner edge of the plasma sheet and related region 2 shielding currents are earthward of $L = 8$.

2. At smaller radial distances, the values of the electric field vary from <0.1 mV/m during quiet times to ~0.4 mV/m during active times (excluding the highest K_p bin likely subject to low statistics). This is consistent with the picture of undershielding (dawn-to-dusk field penetrating to lower L-values) as the shielding layer under most conditions is located beyond $L = 3$.

3. There is also a local peak in the region $3.5 < L < 5.5$ for higher K_p values indicating that the strongest electric fields occur inside the inner magnetosphere. This phenomenon will be further discussed below in Section 7.3.4.

To discuss time dependence of storm-time electric fields, we need to turn to ionospheric measurements, which provide better spatial coverage and time span. Figure 7.8 displays ion drift measurements from the DMSP F13 polar orbiting spacecraft at 840 km altitude during one of the largest magnetic storms (the D_{st} index reached almost −400 nT). The orbit of the F13 spacecraft crosses at approximately 1800–0600 magnetic local time (MLT). In this case, the east–west component of the measured ion drift can be used as a proxy for the electric field assuming that ions in the topside ionosphere undergo E-cross-B (ExB) drift. Each vertical "streak" is one polar orbit, and the degree of shading indicates the absolute value of the measured drift velocities. The outer "envelope" of the pattern is the anti-sunward convection in the polar cap, while the "inner" (lower-latitude) pattern immediately equatorward of it is the auroral sunward return flow (compare with Figure 7.2). The fuzziness of the equatorward region is a direct indication of the penetration electric field.

The most striking feature of this plot is expansion of the high-latitude convection reaching low latitudes (all the way to the equator) during the main phase of the storm (as indicated by both the D_{st} index and the IMF B_z plotted in the upper panel). This particular storm had two periods of very large and sustained southward IMF B_z, each one resulting in a main phase (hence the double-peaked time history of the D_{st} index). One can see a clear correlation with periods of southward B_z and penetration electric fields. The sign of the values (sunward) indicates undershielding, but there is no indication of shielding re-establishing itself during periods lasting many hours.

Are these long-duration penetration events (see also Huang et al., 2007, and references therein) consistent with the theory developed in Sections 7.1 and 7.3.1? An additional factor at work during magnetic storms is that continuous strong convection results in injection of the storm-time ring current (Chapter 9). Magnetic perturbations from the newly injected plasma in the ring current (as well as in the plasma sheet at larger distances on the nightside) cause the magnetospheric magnetic field on the nightside to stretch. This continuous process of stretching magnetic field during main phases results in an inductive electric field in the magnetosphere that opposes the convection dawn-to-dusk field implied by Equation 7.14. As a result, the inner edge

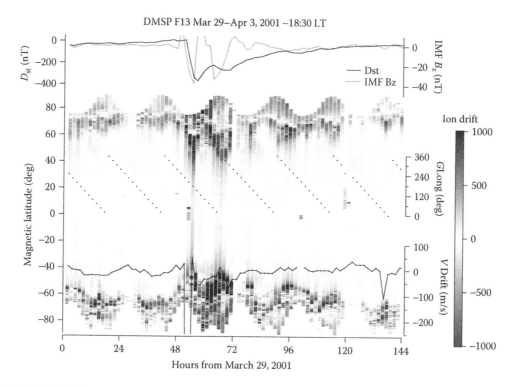

FIGURE 7.8   (Top) The IMF B_z component and the D_{st} index during a period of several days that included a very large magnetic storm on March 31, 2001. (bottom) The zonal ion drift measured by the polar-orbiting DMSP F13 spacecraft along each orbit. See text for further explanation. (Courtesy of R. Heelis, University of Texas at Dallas, Richardson, TX.)

of the plasma sheet does not come to an equilibrium position, and region 2 currents remain ineffective for establishing shielding. RCM simulations (e.g., Huang et al., 2007) largely confirm this picture, although exact agreement between model results and observations also depends on other factors such as thermospheric neutral wind dynamo.

7.3.3 Mesoscale Electric Fields and Effects of Substorms

The previous section addressed the role of enhanced large-scale convection during geomagnetic storms. Models such as RCM and BATS-R-US-RCM reproduce the picture of the overall smoothly varying magnetospheric convection that is enhanced during active times, and many modeling predictions agree quite well with observations. Yet this picture is not completely accurate. It was discovered in the early 1990s (Baumjohann et al., 1990; Angelopoulos et al., 1992) that, though the average earthward flow velocity of convecting plasma was only a few km/s (consistent with a fraction of a mV/m electric field in the plasma sheet), a large fraction of the earthward flow came in brief flow bursts with velocities of

up to 100–400 km/s ("bursty bulk flows" [BBFs]). These flow bursts last only 1–2 min, and tend to have enhanced B_z and decreased B_x meaning that magnetic field lines are more dipolar. Earthward of about 20 R_E, BBFs are almost entirely earthward. Measurements by the Cluster spacecraft suggested that flow bursts have dawn–dusk dimension of 1–2 R_E confirming the idea that these are narrow "channels" (e.g., Nakamura et al., 2004). Chen and Wolf (1993) proposed that these regions of fast localized flows were physically "bubbles" (underpopulated flux tubes created in the magnetotail via either sporadic reconnection or other nonadiabatic processes), and explained that such bubbles would move earthward relative to background via interchange motions. Charge builds up on the sides of a bubble, creating an enhanced dawn–dusk electric field in the bubble. To maintain quasi-neutrality, field-aligned currents flow down to the ionosphere on the dawn side of the bubble, up from the ionosphere on the dusk side.

The overall convection then appears to be bi-modal: most of the time it is relatively slow and smooth sunward flow, but superimposed on it are sporadic in time (and also in location) bursts of locally enhanced flows. It is still

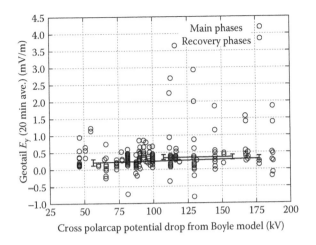

FIGURE 7.9 Variation of the dusk-to-dawn component of the magnetospheric electric field measured by Geotail spacecraft in the plasma sheet with geomagnetic activity as measured by the cross polar cap potential drop. With increasing levels of activity, spread in the measured electric field values increases. (Adapted from Hori, T. et al., *J. Geophys. Res.*, 110, A04213, 2005. With Permission.)

not understood how these mesoscale regions contribute to the overall average convection electric field. Figure 7.9 (Hori et al., 2005) shows one way to illustrate the problem: electric field measurements made by instruments on board the Geotail spacecraft in the nightside plasma sheet are plotted against estimated values for the cross polar cap potential drop estimated using an empirical formula derived by Boyle et al. (1997) (this produces results similar to expression [Equation 7.14]) for the conditions at the time of measurements. The data during active times used to produce Figure 7.9 shows that the measured convection electric field is not directly proportional to the estimated polar cap potential drop;

instead, during more active times there are larger values measured but the spread in values also increases, indicating non-negligible role of convection transients.

There is increasing evidence that plasma sheet bubbles and fast flow channels play a role in leading to conditions favorable to substorm triggering (e.g., Sergeev el al., 2014, and references therein); it is also possible that the largest of these events cause substorms directly. Here, we limit ourselves to a brief discussion of the salient features of the entropy bubble and a depleted channel.

The equations of the RCM (Section 7.1) can be adapted to simulating a time evolution of a localized channel, by imposing the appropriate boundary conditions on the distribution function in Equation 7.10 and on the electrostatic potential (Equation 7.8). At the same time, the magnetic field also needs to be adjusted in order to account for a collapse of the field lines within the channel. Figure 7.10 shows examples from one such RCM simulation (the setup was similar to one used in Zhang et al., 2009). Shown are electrostatic potential (lines) and field-aligned current density (shading) both mapped to the equatorial plane (the Sun is on the left, view is from above). The potential is in the co-rotating frame. A depleted fast flow channel is imposed at $T = 04:25$ simulation time, and the three snapshots are for $+1, +4,$ and $+9$ min (the system relaxes back to its steady-state solution after about 20–30 min).

The presence of a depleted channel centered at midnight causes a region of inflow, with a pair of field-aligned currents in the region 1 sense on the edges of the channel. This is an area of active research, and there is no clear understanding of how transient fast flows contribute to the overall convection. It is clear, though, that the mesoscale BBF phenomenon, occurring on the

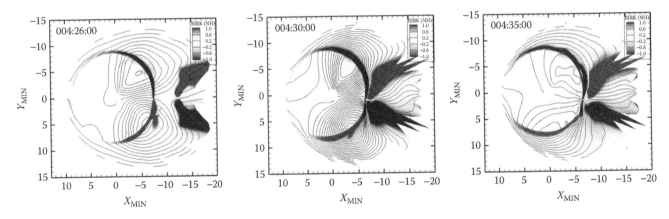

FIGURE 7.10 Contours of electrostatic potential (lines) and field-aligned current density into the Northern hemisphere (shaded) mapped out and displayed in the equatorial plane during an RCM simulation of a fast flow channel. Potential is in the co-rotating frame. The channel is imposed at $T = 04:25$ simulation time, centered at midnight.

timescale of minutes and on spatial scales of a few R_E in width doesn't fit comfortably within the picture of quasi-steady earthward convection and cannot be ignored.

7.3.4 Subauroral Polarization Streams

In Section 7.3.2, it was briefly mentioned that there is a peak in the y-component of the magnetospheric electric field at $L = 3.5$–5.5. This magnetospheric feature is due to the specific nature of the magnetosphere–ionosphere electrodynamic coupling, and is called SAPS (see Foster and Burke, 2002 for a review). SAPS appear equatorward of the auroral oval electron precipitation boundary during geomagnetically active times and drive plasma drifts that are noticeably detached from the high latitude convection system. These structures are observed between 1800 and 0300 MLT (dusk-to-dawn sector) equatorward of the auroral oval; they are extended in

MLT but narrow in latitude (0.5°–3°). As geomagnetic activity increases, SAPS are observed at progressively lower latitudes. Some of the alternative names used for SAPS are polarization jets and subauroral ion drifts. Figure 7.11 is an example of the SAPS electric field structure observed with ground-based incoherent-scatter radar at the Millstone Hill Observatory during a large magnetic storm. The top panel shows westward ExB drift velocity and electron density measured in the ionospheric F-region by the radar: there is a clear secondary peak coinciding with the main ionospheric trough. The trough is a dusk-to-midnight feature that is outside the auroral zone. The MLT is approximately 1800.

Coincident satellite measurements of the drift velocity above the F-region (840 km) in the bottom panel show the same spatial structure, while electron and ion precipitation data measured by the DMSP satellite

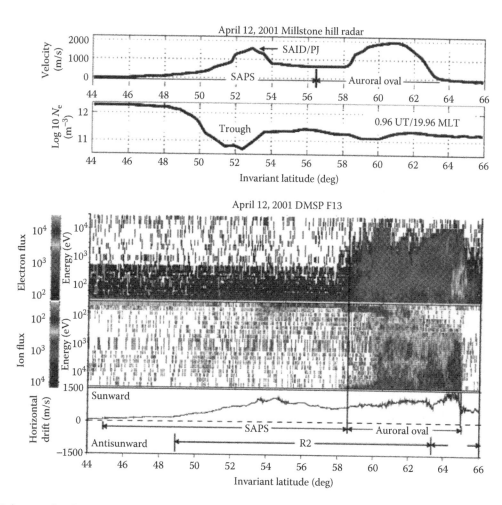

FIGURE 7.11 Subauroral polarization stream (SAPS) are narrow channels of fast westward flow in the ionosphere just equatorward of the auroral zone as seen in simultaneous ground-based incoherent scatter radar (top two panels) and topside polar orbiting DMSP F13 spacecraft (bottom three panels). MLT is 1800. (From Foster, J.C. and Burke, W.J., *Eos Trans. AGU*, 83(36), 393, 2002. With Permission.)

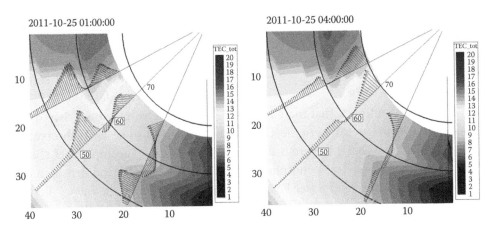

FIGURE 7.12   Results from RCM simulations of the October 25, 2011, geomagnetic storm. Shown are ionospheric views (from above the North Pole) of the dusk-to-midnight sector (the Sun is at the top, midnight at the bottom, and 18 MLT is on the left). Greyscale plots are contours of constant total electron content (TEC). Vectors show two-dimensional ionospheric ExB convection velocities; vectors are displayed along three lines of constant MLT for clarity (the RCM computes the full two-dimensional pattern). Left panel shows a well-developed SAPS, which is the more equatorward peak in the sunward flow; right panel shows the solution 3 h later with almost no sign of SAPS.

confirm that this secondary peak is outside of the main convection region (auroral oval).

RCM simulations often predict the existence of these structures. An example is shown in Figure 7.12, which displays the ionospheric projection of the dusk-to-midnight sector of the Northern Hemisphere (the simulation domain includes all local times). ExB drift velocities calculated from the model-predicted electric field are shown only for three selected MLT slices, with the rest of the vector field suppressed for plotting purposes. The shaded quantity is contours of total electron content (TEC); variations of it outside the auroral zone are prescribed, but the auroral contribution to ionization is calculated in the model.

The main phase of this storm started around 20:00 UT on October 24, 2010, and ended shortly after 01:00 UT on October 25, 2010, and was driven by a period of strong southward IMF B_z. The left panel of Figure 7.12 is a plot of the ionospheric convection in the dusk sector at the peak of the storm main phase, and it shows a well-pronounced secondary peak in the westward convection that is equatorward of the auroral zone. The signature of the auroral zone in these plot is a ridge in TEC, caused by electron precipitation and enhanced ionization. Convection velocities in the SAPS region at earlier MLT are larger than those in the auroral zone. Shortly after 01:00 UT, IMF B_z turned sharply northward and remained near zero for many hours, triggering the recovery phase of the storm. The right panel of Figure 7.12 shows the RCM solution at 04:00 UT, indicating a weak remnant SAPS.

The reason for that can be understood once we explain the mechanism for SAPS formation.

The diagram of the inner magnetospheric configuration shown in the left panel of Figure 7.13 displays the ionospheric projections of several magnetospheric boundaries viewed from above the North Pole. The shaded oval is the auroral oval, with its equatorward edge mapping to the inner edge of the electron plasma sheet is the magnetosphere. The ion inner edge is displayed as a line with crosses. Because of the differences in drift paths of electrons and ions in the magnetosphere, the two inner edges are slightly displaced in latitude relative to each other, with the ion inner edge being closer to the Earth on the dusk side and farther away on the dawn side. The electron plasma sheet contributes to the auroral precipitation, enhancing the conductance in the auroral zone. The ion plasma sheet inner edge contributes dominantly to the region 2 field-aligned currents (Equation 7.13). As a result, there is a gap between the field-aligned currents and the region of enhanced auroral conductance, and on the dusk side the gap has low conductance. Therefore, the solution to the current conservation Equation 7.8 turns out to be a strong poleward-directed electric field across the gap. In other words, some region 2 current flows into low-conductance, subauroral ionospheric region in the pre-midnight sector, causing strong electric fields in SAPS.

This mechanism was first proposed by Southwood and Wolf (1978) with an idealized analytic solution. The SAPS occurs in RCM solutions exactly by this mechanism.

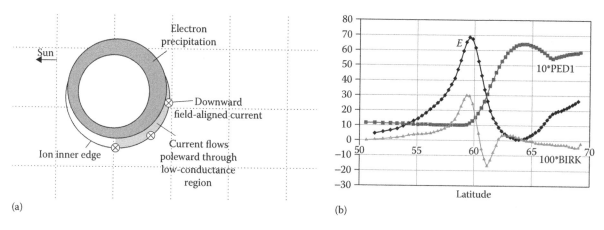

(a)   (b)

FIGURE 7.13   (a) Diagram showing the mechanism of SAPS formation. Ionospheric view is from above the North Pole, with the Sun on the left and dusk at the bottom. SAPS is the crescent-shaped gap region between 18 and 00 MLT where the inner edge of the ion plasma sheet (indicated with crosses) extends equatorward of the electron inner edge (auroral oval). (Courtesy of R. Wolf, Rice University, Houston, TX.) (b) A latitudinal profile through an idealized RCM simulation with SAPS (MLT = 20). Electric-field peak sits on the equatorward edge of the diffuse aurora and in a region of mostly positive (downward) current. (From Wolf, R. A., et al., *J. Atmos. Sol.-Terr. Phys.*, 69(3), 288, 2007.)

The right panel of Figure 7.13 demonstrates that with an idealized RCM simulation. The simulation used the inputs corresponding to a stepwise southward IMF B_z turning typical of a large storm. A latitudinal profile (at 20 MLT) of ionospheric poleward electric field component, field-aligned Birkeland current density, and Pedersen conductance are overlapped to show that the morphology of the model-predicted SAPS is consistent with the theoretical picture (left panel of Figure 7.13) and observations (compare with Figure 7.11). The peak of the electric field sits on the equatorward edge of the diffuse aurora and in a region of mostly positive (downward) current.

The separation between inner edges of plasma sheet ions and electrons is greater for major storms than for shorter bursts of strong convection. SAPS are known to be highly dynamic and seem to occur in response to southward IMF B_z turnings (Clausen et al., 2012). However, a northward B_z turning is not expected to cause the SAPS to disappear immediately since the timescales for the electron and ion inner edges are different and are at least tens of minutes.

7.4 SPACE WEATHER CONSEQUENCES

Penetration electric fields (Sections 7.3.1 and 7.3.2) are an important factor for space weather. For example, Basu et al. (2001) proposed that during the main phase of large magnetic storms, eastward penetration electric fields result in massive uplifting (via ExB drift) of the equatorial ionospheric F-layer above 840 km altitude. The interpretation is that the F-layer is lifted above

this altitude, resulting in elimination of the pre-existing F-layer at the equator and enhanced electron densities at mid-latitudes due to the fountain effect. Such uplifts are often accompanied by strong scintillations, because eastward electric field causes a downward drag that acts like increased gravity and encourages the Rayleigh–Taylor instability that causes spread F.

The sustained driving of enhanced convection during magnetic storms, discussed in Section 7.3.2, not only leads to penetration of electric fields to the innermost region of the magnetosphere, but also causes intensification of auroral electrojets and other ionospheric horizontal currents. Ground-induced currents associated with enhanced electrojets, and especially during abrupt changes in the Hall ionospheric currents, can cause significant induced currents in power distribution systems and are a major space weather concern of societal importance (e.g., Pulkkinen et al., 2012; Weimer, 2013).

Strong localized electric fields associated with fast flows discussed in Section 7.3.3, and related substorm phenomena, cause energetic particle injections to the region within geosynchronous orbit (Yang et al., 2012). Substorm-related injections occur for electrons of energies of 1–500 keV; electrons of these energies are capable of causing differential surface charging of spacecraft in the geosynchronous orbit, potentially leading to discharges damaging to electronics (e.g., Thomsen et al., 2013).

The role of SAPS in space weather is at least twofold. First, within SAPS flow channels there is often fine-scale plasma and electric field structuring accompanied by waves and instabilities (e.g., Makarevich and Bristow, 2014, and references therein). Local ionospheric irregularities caused by SAPS can be of potential concern for radio wave propagation. Second, SAPS electric fields contribute to ionospheric electron density redistributions during storms, combining with overall penetration electric fields to produce highly structured plumes of ionization known as storm-enhanced densities (SEDs) (Foster and Coster, 2007). Gradients in the ionospheric TEC around SED plumes are a major factor in the loss of accuracy of navigational systems using global positioning system (GPS) satellite signals (Coster and Komjathy, 2008; Astafyeva et al., 2014).

7.5 RESOURCES FOR MAGNETOSPHERIC MODELING

In many situations of interest to space weather effects, it may be desirable to have realistic descriptions of magnetospheric electric fields and currents. Here, we list some of the resources available for this purpose. Rather than provide a comprehensive compendium, a task made impossible by space limitations, we focus on the most recent and accessible relevant magnetospheric models, both empirical and first principles, and we hope that this will serve as a useful starting point.

There exists a wide variety of empirical (i.e., database-based) models describing the average configuration and dynamical variations of the magnetospheric magnetic field (since currents are not measured directly but rather via measurements of magnetic fields induced by currents, it is magnetic field models that are typically constructed.) The model of Tsyganenko and Andreeva (2015) is the latest in a long-standing effort to represent magnetospheric global magnetic field with contributions from the ring current, region 2 and region 1 field-aligned currents, and the tail current. The model is constructed by using a very large database of magnetometer data from a number of recent spacecraft missions, and is driven by interplanetary parameters. The Fortran source code is freely available from the website indicated in the paper.

It is difficult to construct an empirical model of the magnetospheric electric field pattern, as electric fields measured close to the equatorial plane often include significant inductive component. Therefore, ionospheric electric field (or convection) patterns are commonly used. The empirical model of Weimer (2005) is probably the most widely use parameterized representation of the ionospheric convection electrostatic potential. The model is derived from the large database of electric field and magnetic field measurements on board the topside polar orbiting Dynamics Explorer 2 satellite, and varies as a function of the solar wind parameters. In addition to the convection potential patterns, the model of Weimer (2005) provides statistical patterns of field-aligned currents.

When the model of Weimer (2005) is combined with that of Tsyganenko and Andreeva (2015) and if one assumes that magnetic field lines are largely equipotential, electric fields in the magnetosphere can be obtained by mapping the ionospheric values along the magnetic field. At this time, this is probably the best available empirical description of the large-scale electric fields and currents.

There are two aspects that empirical models do not represent well. First, local structuring in the electric fields that are MLT-dependent, such as SAPS flow channels, cannot be adequately included in the empirical models. Second, due to the statistical nature of empirical models, they do not contain realistic representation of time-varying phenomena, such as reestablishing of the shielding by the inner edge of the plasma sheet, or responses of the convection due to large sudden changes in the solar wind or IMF values, due to the transient nature of these phenomena. This is where first-principles models have a clear advantage. Significant parts of the discussion in this chapter was based on two such models: the RCM and the coupled global MHD-inner magnetosphere code BATS-R-US/RCM. Both of these codes (as well as the two empirical models mentioned earlier in this section) are available for "runs-on-request" at the Community Coordinated Modeling Center (CCMC) by using the electronic portal http://ccmc.gsfs.nasa.gov, where users can request simulation runs for specific event intervals or for idealized conditions. The CCMC facilities include long-term archiving of simulations output and very extensive set of visualization tools that require only a browser. CCMC provides access to a large variety of other models, including global MHD codes and inner magnetospheric ("ring current") models, and can be used as an excellent resource for obtaining the latest information on available models and as a tool for understanding space weather phenomena.

REFERENCES

Angelopoulos, V., W. Baumjohann, C. F. Kennel et al. (1992), Bursty bulk flows in the inner central plasma sheet, *J. Geophys. Res.*, 97(A4), 4027–4039.

Astafyeva, E., Y. Yasyukevich, A. Maksikov, and I. Zhivetiev (2014), Geomagnetic storms, super-storms, and their impacts on GPS-based navigation systems, *Space Weather*, 12(7), 2014SW001072, doi:10.1002/2014SW001072.

Basu, S., K. M. Groves, H. C. Yeh et al. (2001), Response of the equatorial ionosphere in the South Atlantic region to the great magnetic storm of July 15, 2000, *Geophys. Res. Lett.*, 28(18), 3577–3580.

Baumjohann, W., G. Paschmann, and H. Luhr (1990), Characteristics of high-speed ion flows in the plasma sheet, *J. Geophys. Res.*, 95, 3801–3809.

Boyle, C. B., P. H. Reiff, and M. R. Hairston (1997), Empirical polar cap potentials, *J. Geophys. Res.*, 102(A1), 111–125, doi:10.1029/96ja01742.

Chen, C. X. and R. A. Wolf (1993), Interpretation of high speed flows in the plasma sheet, *J. Geophys. Res.*, 98, 21409–21419, doi:10.1029/93ja02080.

Clausen, L. B. N., J. B. H. Baker, J. M. Ruohoniemi et al. (2012), Large-scale observations of a subauroral polarization stream by midlatitude SuperDARN radars: Instantaneous longitudinal velocity variations, *J. Geophys. Res.*, 117, doi:10.1029/2011JA017232.

Coster, A. and A. Komjathy (2008), Space weather and the global positioning system, *Space Weather*, 6(6), S06D04, doi:10.1029/2008SW000400.

De Zeeuw, D. L., S. Sazykin, R. A. Wolf, T. I. Gombosi, A. J. Ridley, and G. Toth (2004), Coupling of a global MHD code and an inner magnetospheric model: Initial results, *J. Geophys. Res.*, 109(A12), A12219, doi:10.1029/2003JA010366.

Foster, J. C. and W. J. Burke (2002), SAPS: A new categorization for sub-auroral electric fields, *Eos Trans. AGU*, 83(36), 393.

Foster, J. C. and A. J. Coster (2007), Conjugate localized enhancement of total electron content at low latitudes in the American sector, *J. Atmos. Sol.-Terr. Phys.*, 69(10–11), 1241–1252, doi:10.1016/j.jastp.2006.09.012.

Gombosi, T. (1998), *Physics of the Space Environment*, Cambridge University Press, New York.

Gombosi, T. I., D. L. De Zeeuw, K. G. Powell, A. J. Ridley, I. V. Sokolov, Q. F. Stout, and G. Toth (2003), Adaptive mesh refinement for global magnetohydrodynamic simulation, in *Space Plasma Simulation*, edited by J. Buchner, C. T. Dum and M. Scholer, pp. 247–274, Springer, Berlin, Germany.

Hilmer, R. V. and G. H. Voigt (1995), A magnetospheric magnetic field model with flexible current systems driven by independent physical parameters, *J. Geophys. Res.*, 100(A4), 5613–5626.

Hori, T., A. T. Y. Lui, S. Ohtani et al. (2005), Storm-time convection electric field in the near-Earth plasma sheet, *J. Geophys. Res.*, 110(A4), A04213, doi:10.1029/2004ja010449.

Huang, C. S., S. Sazykin, J. L. Chau, N. Maruyama, and M. C. Kelley (2007), Penetration electric fields: Efficiency and characteristic time scale, *J. Atmos. Sol.-Terr. Phys.*, 69(10–11), 1135–1146, doi:10.1016/j.jastp.2006.08.016.

Huba, J. D. and S. Sazykin (2014), Storm time ionosphere and plasmasphere structuring: SAMI3-RCM simulation of the 31 March 2001 geomagnetic storm, *Geophys. Res. Lett.*, 41(23), 8208–8214, doi:10.1002/2014gl062110.

Makarevich, R. A. and W. A. Bristow (2014), Fine structure of subauroral electric field and electron content, *J. Geophys. Res.*, 119(5), 3789–3802, doi:10.1002/2014JA019821.

Nakamura, R., W. Baumjohann, C. Mouikis et al. (2004), Spatial scale of high-speed flows in the plasma sheet observed by Cluster, *Geophys. Res. Lett.*, 31(9), L09804, doi:10.1029/2004gl019558.

Pulkkinen, A., E. Bernabeu, J. Eichner, C. Beggan, and A. W. P. Thomson (2012), Generation of 100-year geomagnetically induced current scenarios, *Space Weather*, 10(4), S04003, doi:10.1029/2011SW000750.

Raeder, J. (2003), Global magnetohydrodynamics–A tutorial, in *Space Plasma Simulation*, edited by J. Buchner, C. T. Dum and M. Scholer, pp. 212–246, Springer, Berlin, Germany.

Rowland, D. E. and J. R. Wygant (1998), Dependence of the large-scale, inner magnetospheric electric field on geomagnetic activity, *J. Geophys. Res.*, 103(A7), 14959–14964.

Sergeev, V. A., I. A. Chernyaev, V. Angelopoulos, A. V. Runov, and R. Nakamura (2014), Stopping flow bursts and their role in the generation of the substorm current wedge, *Geophys. Res. Lett.*, 41(4), 1106–1112, doi:10.1002/2014gl059309.

Siscoe, G. L., N. U. Crooker, and K. D. Siebert (2002), Transpolar potential saturation: Roles of region 1 current system and solar wind ram pressure, *J. Geophys. Res.*, 107(A10), doi:10.1029/2001ja009176.

Southwood, D. J. and R. A. Wolf (1978), An assessment of the role of precipitation in magnetospheric convection, *J. Geophys. Res.*, 83, 5227–5232.

Thomsen, M. F., M. G. Henderson, and V. K. Jordanova (2013), Statistical properties of the surface-charging environment at geosynchronous orbit, *Space Weather*, 11(5), 237–244, doi:10.1002/swe.20049.

Toffoletto, F., S. Sazykin, R. Spiro, and R. Wolf (2003), Inner magnetospheric modeling with the Rice Convection Model, *Space Sci. Rev.*, 107(1–2), 175–196, doi:10.1023/A:1025532008047.

Toth, G., I. V. Sokolov, T. I. Gombosi et al. (2005), Space weather modeling framework: A new tool for the space science community, *J. Geophys. Res.*, 110(A12), A12226, doi:10.1029/2005JA011126.

Tsyganenko, N. A. and V. A. Andreeva (2015), A forecasting model of the magnetosphere driven by an optimal solar wind coupling function, *J. Geophys. Res.*, 120, doi:10.1002/2015JA021641.

Weimer, D. R. (2005), Improved ionospheric electrodynamic models and application to calculating Joule heating rates, *J. Geophys. Res.*, 110(A5), A05306, doi:10.1029/2004ja010884.

Weimer, D. R. (2013), An empirical model of ground-level geomagnetic perturbations, *Space Weather*, 11(3), 107–120, doi:10.1002/swe.20030.

Wolf, R. A. (1983), The quasi-static (slow-flow) region of the magnetosphere, in *Solar Terrestrial Physics*, edited by R. L. Carovillano and J. M. Forbes, pp. 303–368, D. Reidel, Hingham, MA.

Wolf, R. A., S. Sazykin, X. Xing et al. (2005), Direct effects of the IMF on the Inner Magnetosphere, in *Inner Magnetosphere Interactions: New Perspectives from Imaging*, edited by J. Burch, M. Schulz, and H. Spence, pp. 127–139, American Geophysical Union, Washington, DC.

Wolf, R. A., R. W. Spiro, S. Sazykin, and F. R. Toffoletto (2007), How the Earth's inner magnetosphere works: An evolving picture, *J. Atmos. Sol.-Terr. Phys.*, 69(3), 288–302, doi:10.1016/j.jastp.2006.07.026.

Yang, J., F. R. Toffoletto, R. A. Wolf, S. Sazykin, P. A. Ontiveros, and J. M. Weygand (2012), Large-scale current systems and ground magnetic disturbance during deep substorm injections, *J. Geophys. Res.*, 117, doi:10.1029/2011JA017415.

Zhang, J. C., R. A. Wolf, G. M. Erickson et al. (2009), Rice convection model simulation of the substorm-associated injection of an observed bubble into the inner magnetosphere: 1. Magnetic field and other inputs, *J. Geophys. Res.*, 114, A08218, doi:10.1029/2009ja014130.

CHAPTER 8

Coupling between the Geomagnetic Tail and the Inner Magnetosphere

Eric Donovan

CONTENTS

8.1 INTRODUCTION

The magnetosphere arises as a direct result of the interaction between the solar wind and Earth's magnetic field. On the dayside, its shape is determined primarily by the balance of forces, wherein magnetic pressure inside the magnetopause roughly matches the dynamic pressure of the solar wind plasma, a situation that is complicated by the fact that the magnetosheath plasma has been thermalized upon passage across the bow shock. Progressing along the magnetopause antisunward of the Earth, the situation changes, such that while pressure balance is still important, so too is the stretching of the magnetic field as a consequence of reconnection (on the dayside) and the frozen-in condition applying to the magnetosheath plasma. Ultimately, the overall effect is the shaping of the magnetosphere in the grossest sense, with the result being a structure that is compressed (from the otherwise roughly dipolar topology) on the dayside and stretched to great distances on the nightside (Behannon and Ness, 1968). For obvious reasons, we call this region behind the Earth the magnetotail.

Imagine the cross section of the magnetosphere as it appears to the oncoming solar wind. The column of solar wind impinging upon that cross section carries with it an amount of momentum that is removed from the solar wind. While the plasma in that column is ultimately accelerated back to the solar wind speed, that restored momentum is extracted from the adjacent solar wind. The lost momentum is imparted to the magnetosphere, and represents a net anti-sunward directed force upon it. Siscoe (1966) pointed out that this momentum transfer amounts to a force that is trying to "rip" the magnetotail away from the Earth. This does not happen, so that means there must be a balancing force on the magnetotail toward Earth. This force is accomplished via the integrated magnetic force on the cross-tail current. The magnetic field within the magnetotail closes from the southern to Northern Hemisphere, so this cross-tail current must, predominantly, flow from dawn to dusk. Keeping in mind that in the magnetotail region, the dominant effects of the combined magnetic fields due to the Earth's internal dynamo and the ring and Chapman–Ferraro currents is to make the field on the nightside less, not more, tail-like, it is not surprising that the cross-tail current produces a magnetic field perturbation that sustains the tail-like topology.

Looked at in the broadest sense, mass transport within Earth's magnetosphere can be understood as being accomplished as part of what we now know as the Dungey convection cycle (Dungey, 1961). This cycle is enabled by magnetic reconnection at the magnetopause, and powered by the large-scale electric field that permeates the magnetosphere and drives the global transport, and which is to first approximate ExB convection. The electric field corresponds to the motion of the solar wind with embedded magnetic field that can be traced back to the magnetosphere along open magnetic field lines. Looking, for example, at the Northern Hemisphere, the field (on open field lines that thread the solar wind) is directed toward Earth (e.g., "southward"), and the motion is anti-sunward. This corresponds to a direction of electric field (consistent with the motion being largely ExB drift) that points from dawn to dusk. As discussed below, this global transport is consistent with the large-scale distribution of magnetospheric electric field and the topology of the magnetospheric magnetic field.

While the large-scale geospace transport is mostly consistent with the ExB convection of cold plasma, other effects complicate the picture. And these are not just small complications or perturbations; rather, these are a collection of effects that at any one instant and location usually represent a small deviation from true (or an "average" picture of) convection, but the sum total of which produce many of the most important effects that must be accounted for is we are working toward being able to predict and mitigate serious space weather. Energy-dependent adiabatic charged particle motions such as the gradient and curvature drifts lead to differential motions that in turn produce some of the large-scale currents including the ring current, and are responsible for energizing magnetospheric plasma (e.g., through conservation of the first adiabatic invariant). Transient adiabatic and nonadiabatic effects including spatially limited bursts of fast convection, breaking of adiabatic invariance due to strong spatial gradients, changes of the magnetic and/or electric fields, and turbulence and wave–particle interactions all lead to coupling between different plasma regimes (e.g., across energy, from open to closed trajectories or vice versa, radial transport, and loss via precipitation or to the solar wind).

The magnetic field topology in the magnetosphere can be understood as comprising different regions that are in turn identified by some combination of properties of the magnetic field and plasma therein. In this way, the lobes are threaded by open magnetic field lines and populated by relatively low-density plasma. On the nightside, the plasma sheet separates the north and south lobes, and is populated by plasma of both solar wind (see, e.g., Hill, 1974, and references therein) and ionospheric origin (see, e.g., Chappell et al., 1987, and references therein). Also on the nightside, at large distances from the Earth, the magnetic field is directed more or less sunward in the north lobe and tailward in the south lobe, and the plasma sheet is thin and sustains the above-mentioned cross-tail current. This is the magnetotail. Much closer to the Earth, in the inner magnetosphere, the magnetic field is roughly dipolar. Separating the highly stretched magnetotail and the largely dipolar inner magnetosphere is the nightside transition region (NTR). Figure 8.1, reproduced here from Chappell et al. (1987), illustrates the potential sources of magnetotail plasma as well as the basic topology of the nightside magnetosphere, at least in terms of a projection on the meridional plane.

From the perspective of coupling between the magnetotail and the inner magnetosphere, the NTR is an important region. Physical processes within and adjacent to the NTR determines the destiny of virtually all magnetospheric plasma. In quiet times, it is a "magnetic wall," deflecting plasma around the inner magnetosphere, leaving it on convection paths that return to the solar wind. In active times, a plethora of physical processes within the NTR energize and transport plasma from the stretched magnetotail into the inner magnetosphere. This injected plasma can be trapped in the ring current and radiation belts, creating the largest space weather disturbances and setting the stage for the greatest space weather hazards. Precipitation mechanisms that lead to the loss of magnetospheric charged particles are different outside of, within, and inside of the NTR. In the magnetotail, this precipitation is dominated by soft diffuse precipitation of plasma sheet electrons and discrete aurora caused by field-aligned acceleration (e.g., "arcs"). Inside of the NTR, wave–particle interactions are responsible for precipitation of much higher energy electrons' ring current and radiation belt electrons and positive ions, with subsequent significant effects on the upper atmosphere. The NTR itself gives rise to the most

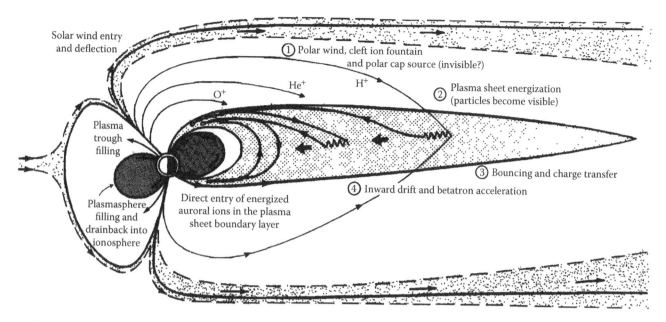

FIGURE 8.1 Sources of plasma in the magnetotail. Magnetotail plasma originates in the terrestrial ionosphere (ion outflow) and from the magnetosheath (solar wind plasma). (Reproduced from Chappell, C.R. et al., *J. Geophys. Res.*, 92, 5896–5910, 1987. With Permission.)

intense proton precipitation, and corresponds to the location of the bright proton aurora. To the best of our knowledge discrete aurora never occurs on magnetic field lines that are inside of (earthward of) the NTR (Yahnin et al., 1997).

Where, in radial distance, is the NTR? Earth's magnetosphere is a highly dynamic system, so there is not just one answer to this question. That said, it is clear from *in situ* magnetic field observations that, on the nightside, the magnetic field in the neutral sheet, beyond roughly 10 R_E, is almost always much smaller in magnitude than that in the high latitude plasma sheet, indicating tightly curved magnetic field lines which in turn indicate a thin current sheet (see, e.g., bottom panel of Figure 2 of Shen et al., 2003). This is thus the region of the magnetotail. Although there are notable exceptions during times of extreme activity, it is generally the case that inside of geosynchronous orbit, also on the night-side, the magnetic field topology is qualitatively (e.g., curvature, inclination) to that of an appropriate Earth-centered dipole. Further, the inner edge of the plasma sheet is usually around or slightly earthward of geosyn-chronous distance (see, e.g., Figure 8.3 of Jiang et al., 2011). Thus, it is reasonable to consider the NTR to be usually (and roughly) between geosynchronous distance and ~10 R_E downtail.

Figure 8.2 is meant to illustrate the concept of the NTR and the transition from magnetotail to inner magnetosphere. The bottom panel of Figure 8.2 shows the magnetic topology in the noon–midnight/dipole plane, according to the "T89" empirical model (Tsyganenko, 1989) combined with the magnetic field of an appropri-ate Earth-centered dipole that is oriented perpendicular to the Sun–Earth line (so the model is symmetric about the equatorial and the noon–midnight/dipole planes). Along any given field line the curvature is tightest as the field line crosses the neutral sheet, which in this case is the equatorial plane. One can estimate the local radius of curvature at any point on a field line by the radius of an inscribed circle. The middle panel shows such inscribed circles for three field lines as they cross the neutral sheet at 8, 13, and 18 R_E downtail. These circles have radii of 1.2, 0.3, and 0.1 R_E, respectively.

A key issue related to the transition between the mag-netotail and inner magnetosphere is how the field line curvature in the neutral sheet compares to the gyrora-dii of electrons and protons with various energies that are typical for the plasma sheet. Tsyganenko (1982) and Sergeev et al. (1983) used test particle simulations to show that if the radius of curvature of the field line (in the vicinity of the neutral sheet) was smaller than about a factor of 9 times the gyroradius of the particle (again, in the vicinity of the neutral sheet), then there was a slight breaking of the first adiabatic invariant as the particle crossed the neutral sheet. This nonadia-baticity, they argued, resulted in a stochastic "kick"

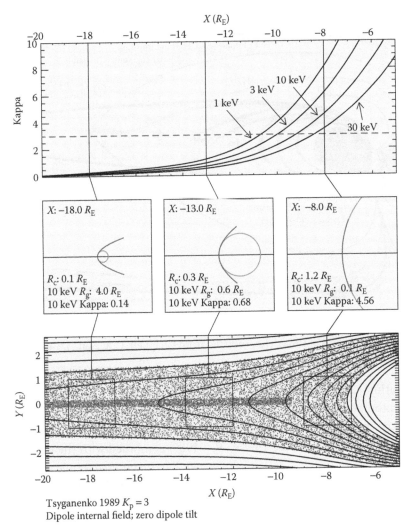

FIGURE 8.2 As described in the text, the "kappa" parameter can be used as one way of differentiating the magnetotail and inner magnetosphere.

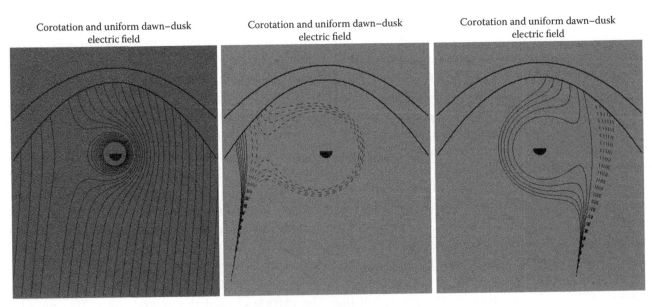

FIGURE 8.3 (Left) Contours of constant electric potential in the magnetospheric equatorial plane (dawn is to the right).

to the pitch angle such that the loss cone is constantly filled. Büchner and Zelenyi (1987) developed the concept of the "kappa parameter" (κ) which is the square root of the ratio between the radius of curvature and the local gyroradius of a particle. They also showed that if $\kappa < 3$, then the particles are in strong pitch angle diffusion, consistent with the earlier results of Tsyganenko and Sergeev. A key point is that since the gyroradius depends on energy, so too does κ.

Looking back to Figure 8.2, the top panel shows κ (in the neutral sheet) for protons with four energies typical of the plasma sheet ion population. The results shown here are typical of those run for all empirical models, namely that for distances beyond about 11 R_E we can expect the bulk of the plasma sheet keV and above ion population to be in strong pitch angle diffusion. For these energies, and at these distances, the loss cone is filled on a timescale comparable to or faster than typical bounce times, and so this is the region of proton auroral precipitation (see, e.g., Donovan et al., 2012, and references therein). Inside of about 6 R_E, on the other hand, essentially all protons (ions) with typical plasma sheet energies will be stably bounce-trapped. The NTR is the region of the ion-trapping boundary, which in turn is energy dependent. The trapping boundary is also referred to as the ion iosotropy boundary (the "ion IB"), and depends on energy such that the IB is successively earthward (equatorward in the ionosphere) for higher energies, a phenomenon that has been observed in topside *in situ* ion measurements. For example, Sergeev and Gvozdevsky (1995) showed that the ion IB appears at lower latitudes for increasing energy under most circumstances.

An interesting question is what is the radial extent of the NTR? In other words, how "sharp" is the transition between the stretched magnetotail and the inner magnetosphere. The observations necessary to address this have yet to be made. It is known, however, that in the late growth phase of the substorm, this transition can become radially confined such that it is very sharp. Donovan et al. (2012) found an event where the ion IB passed over (moving outward) several THEMIS satellites during a substorm expansion phase, and the transition was sudden in time, very likely corresponding to a dramatic, radially narrow, transition. Sergeev et al. (2011) also argued that in the late growth phase, this radial transition became in effect a "magnetic wall," a sharp transition between nearly dipolar and very highly stretched field lines.

What constitutes coupling between the magnetotail and inner magnetosphere? This can be thought of in a number of ways and within different frameworks. One can discuss forces that act between regions on global scales, and consider field-aligned currents that couple regions. An alternate approach is to consider transport between regions and coupling between plasma regimes. In this chapter, the latter is the focus.

8.2 TEST PARTICLE SIMULATIONS OF RELEVANT TRANSPORT PROCESSES

As stated above, the transport of plasma is accomplished via ExB convection and other adiabatic drifts (e.g., the gradient and curvature drifts). Consider first a highly simplified picture that is nonetheless quite instructive regarding transport in the night-time magnetosphere as it pertains to magnetotail inner magnetosphere coupling. In this picture, the Earth's magnetic dipole axis is perpendicular to the Sun–Earth line, and the global magnetic and electric field topologies are static and correspond to average states inferred from *in situ* measurements of the magnetic field and of convection, and based on convection observed in the ionosphere via top-side *in situ* measurements or radar measurements from the ground. Such a treatment has been used for decades to explore the basics of transport in the terrestrial magnetosphere. For example, this basic approach was used by Chen (1970) to explore forbidden regions for protons in the nightside magnetosphere, and Ejiri (1978) to explain aspects of the plasmapause.

For the magnetic field, we can look to the family of models developed by Tsyganenko and coworkers, and to keep things simple we can restrict our attention to, for example, and easy to work with early generation model like the "T87" model (Tsyganenko, 1987). For the large-scale electric field, we can consider simple empirical models like that put forward by both Stern and Volland (the electric field consistent with the so-called Stern–Volland potentials) (Stern, 1973; Volland, 1973), or something even simpler, such as uniform cross-tail potential added to a pure corotation potential. Both of these are roughly consistent with surveys of *in situ* measurements, and of ionospheric convection, the latter being relevant if one considers the implications of ionospheric potential patterns mapped along magnetic field lines to the equatorial plane, assuming field lines to be equipotentials (see e.g., Foster, 1984; Weimer, 1996).

The left-hand panel of Figure 8.3 shows the equipotentials in the equatorial plane for such a simplified model. The uniform cross-tail electric field is 1 keV/R_E, and the magnetic field is given by the sum of that due to an Earth-centered dipole and the T87 $K_p = 0$ magnetic field model. The equipotentials in the panel are separated by 2.5 keV. Charged particles in the equatorial plane with zero kinetic energy move only via the ExB drift. Their motion is thus perpendicular to the magnetic field, so they remain in the equatorial plane; since they also move perpendicular to the electric field, the equipotentials correspond to drift paths of these very low energy particles.

The pattern of equipotentials in the left-hand panel of Figure 8.3 shows two qualitatively different types of drift paths. Close to the Earth, the equipotentials, and hence the drift paths of very low energy charged particles, are closed. This is the region known as the plasmasphere, with the ExB motion dominated by the effects of the corotation electric field, and wherein cold plasma largely of ionospheric origin is effectively trapped (see, e.g., Carpenter, 1963; Gingrauz, 1963). Further from Earth, the drift paths are open. Charged particles from deep in the tail move earthward, past the Earth, and out to the magnetosheath where they are lost from the magnetosphere. The teardrop shaped contour, known as the magnetic separatrix, separates closed and open drift paths for very low energy charged particles, and corresponds to the plasmapause if the large-scale potential pattern has remained relatively unchanged for tens of hours.

If we consider somewhat more energetic charged particles, but with their motion still restricted to the equatorial plane (meaning they have 90° pitch angles and are referred to as "equatorially mirroring"), then the gradient drift also plays a role in their motion. The middle and right-hand panels of Figure 8.3 show the trajectories of such equatorially mirroring particles Trajectories of equatorially mirroring (e.g., 90° pitch angle) electrons and protons determined using this model. The teardrop shape on both panels corresponds to the idealized plasmapause for these model electric and magnetic fields. The middle panel shows trajectories beginning at an initial position of $x = -18\ R_E$ and $y = 14\ R_E$ (GSM and SM coordinate system given 0° dipole tilt). The thick solid trajectory corresponds to a zero energy electron or proton, and thus also to an equipotential. The solid black curves indicate the trajectories of protons with initial energies of between 0.05 and 0.3 keV. The black dashed curves indicate the trajectories of electrons with those same starting energies. Trajectories that deviate further from the zero energy trajectory correspond to progressively higher starting energies (and thus relatively greater effects due to the gradient drift). The right-hand panel shows the same, but for trajectories starting from $x = -18\ R_E$ and $y = -5\ R_E$.

As a general rule, electrons and ions are swept dawnward and duskward, respectively, by the effects of the gradient drift. This drift moves particles across equipotentials and is thus responsible for energization as the particles move in the sunward direction. Another way to look at this is that while (ExB) convection is driving the particles into a region of larger magnetic field strength, the convection velocity is always perpendicular to the electric field and so can accomplish no energization. The gradient drift (and the curvature drift if the particles have other than 90° pitch angles) serves to affect the changes in kinetic energy required for conservation of the first adiabatic invariant (see Cully and Donovan, 1998).

Figures 8.4 and 8.5 show the results of a large number of test particle traces, in this simple magnetic and electric field configuration, for protons and electrons, respectively. Looking at Figure 8.4, each panel shows the positions of 500 protons that have been traced from random starting positions between 15 and 19 R_E downtail. The shading indicates the kinetic energy of the proton at each position, with black representing zero kinetic energy, and the lightest shaded points corresponding to about 30 keV. The symbols indicate the locations of the different protons at equally spaced times along their trajectories. Looking at the panel for a first adiabatic invariant of zero (meaning zero kinetic energy), the shading indicates these particles remain at zero kinetic energy as they move through the system. As well, the protons are excluded from a region corresponding to the idealized plasmapause (shown as above with the teardrop shaped contour). Looked at in this way, one can consider the plasmapause to be the inner edge of the "thermal" (very low energy) plasma sheet. Such an inner boundary, defining an excluded region for charged particles of a given adiabatic invariant, has historically been called an Alfvén layer (Korth et al., 1999; Zhang et al., 2015). As in the specific case of the plasmapause which is the Aflvén layer for zero energy protons (and electrons), an Alfvén layer separates closed and open drift paths for particles of a given species (e.g., protons and electrons), and a specific range of first adiabatic invariant.

For different values of the first adiabatic invariant, the Alfvén layer is of different size and shape. Furthermore,

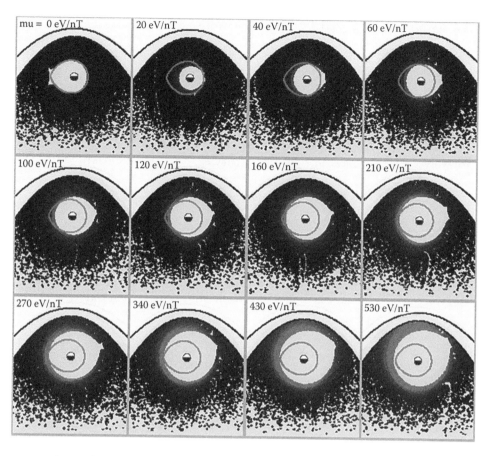

FIGURE 8.4   To create each panel, protons with a given value of the first adiabatic invariant (the particular value is indicated on each panel) are traced from starting positions in the T87 model magnetic field and an electric field given by the Stern–Volland electric potential distribution (see, e.g., Volland, 1973). For each panel, 500 equatorially mirroring protons were traced starting from random locations between $x = -15\ R_E$ and $x = -19\ R_E$. The location of the protons are shown at equally spaced time intervals along the trajectories, and the symbols are color coded according to their instantaneous kinetic energies. The shading is described in the text.

for larger values of the first adiabatic invariant, it is clear that as particles approach the Alfvén layer they gain energy, consistent with conservation of the first adiabatic invariant and the magnitude of the magnetic field which increases rapidly with decreasing distance from Earth in the NTR and inner magnetosphere. As well, it is generally the case that for larger values of the first adiabatic invariant, the Alfvén layer is further from Earth. This too is consistent with conservation of the first adiabatic invariant and the fact that all of the change in kinetic energy of the particles (in this static model) comes from work done by the electric field. If one considers, for example, a 1 keV proton (90° pitch angle) which is at a location where the field is 5 nT. As this particle approaches the inner magnetosphere, it might have access to a change in electric potential of 30 keV. This means that it cannot access any location where the magnetic field is larger than around 150 nT; however, a particle starting from

the same location (e.g., 5 nT) but with a kinetic energy of 3 keV cannot access regions where the magnetic field magnitude is larger than around 50 nT. Hence, the Alfvén layer is expected to be further out for higher values of the first adiabatic invariant. Figure 8.5 shows the results of the same test particle simulation, but in this case for electrons.

Considering Figures 8.4 and 8.5 (and consistent with Figure 8.3), it is clear that the effects of convection, the gradient drift (and the curvature drift which does not change this picture qualitatively), and conservation of the first adiabatic invariant leads to protons and electrons being preferentially swept duskward and dawnward, respectively. As well, one expects the inner edge of the ion and electron plasma sheets to be populated by the most energetic plasma in the late evening and morning sectors, respectively. In this time stationary model, all of this plasma moves through the system and out to

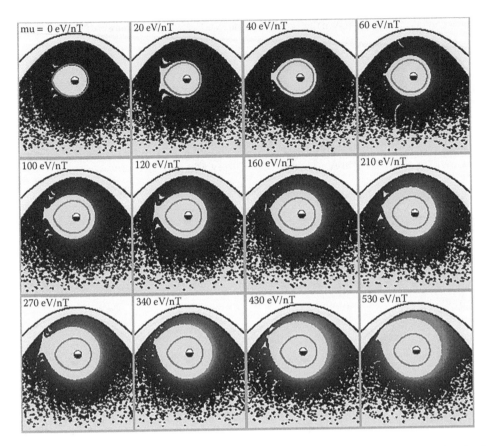

FIGURE 8.5 Same as Figure 8.4, but for equatorially mirroring electrons, and with energy indicated by shading of the points, darker and lighter corresponding to lower and higher energies, respectively.

the dayside magnetopause where it is subsequently lost from the magnetosphere. However, it should be clear that this nightside steady-state transport delivers moderately energetic (several keV to several tens of keV) electrons, protons, and other ions, to the NTR. As discussed below, this in effect stages plasma for transport via other processes inward across the NTR to the inner magnetosphere where it can become trapped on closed drift paths, contributing to the ring current and radiation belts.

Up to this point in the discussion, only the effects of a time stationary electric and magnetic field configuration have been considered. The left-hand panel of Figure 8.6 shows the trajectory of one electron through this time stationary system. The electron starts downtail with an initial energy of 0.3 keV. Four hours later it is near or in the Alfvén layer for electrons with this value of first adiabatic invariant, and it has the highest kinetic energy it will have on this trajectory (~24 keV), and 5 h after that it is lost to the magnetopause. The trajectories of all charged particles that are outside of their respective Alfvén layers are open in this sense.

The middle panel of Figure 8.6 follows that same electron, however 3 h into the simulation, its kinetic energy is arbitrarily increased by 10 keV. The subsequent motion of the electron shows that it is now on a closed drift path. In this case, the increase in kinetic energy has led to an increase in the gradient drift relative to the ExB drift, such that the gradient drift now dominates the motion, and the drift path is closed. In this case, the electron all of a sudden has a different value of its first adiabatic invariant, and is inside its new Alfvén layer. It is important to note that this is fundamentally different than the case where a drift path of a very low energy electron or proton that is inside the plasmasphere is closed because the corotation electric field dominates the motion.

The right-hand panel of Figure 8.6 shows a variation on this theme. The electron again starts out as it did in the two other panels, but in this case, 3 h into the simulation, the cross-tail electric field is arbitrarily increased by a factor of 3. It stays at that elevated level for 30 min, after which it returns to its original value. The effect of this transient increase in the cross-tail field is qualitatively similar to what is illustrated in the middle panel of

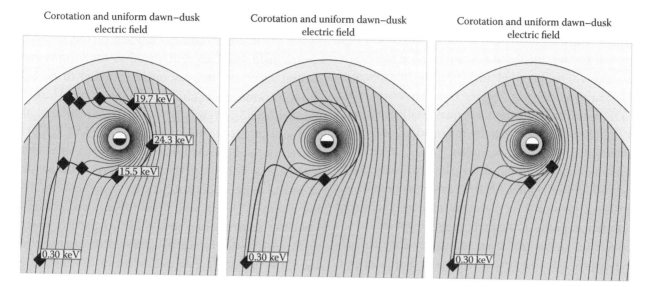

FIGURE 8.6 (Left) Trajectory of an electron with an initial kinetic energy of 0.3 keV in the time stationary electric and magnetic field models as discussed above. The diamonds are separated by 1 h along the trajectory. (Middle) This again shows the trajectory of an electron moving in the same electric and magnetic fields, and with the same starting kinetic energy. However, 3 h into the simulation, indicated by the second diamond, its kinetic energy is arbitrarily increased by 10 keV. (Right) This situation is the same as that which produced the middle panel, up until 3 h into the simulation (second diamond). Starting at that time, and for the following 30 min (e.g., the time corresponding to the third diamond along the trajectory), the cross-tail electric field is increased by a factor of 3. After that 30 min the electric field returns to its initial value.

Figure 8.6. While the electric field is larger, the effect of the relatively larger ExB drift is to transport the electron closer to Earth. This moves the electron into a larger magnetic field region that it would otherwise have accessed, but does not lead to a change in the first adiabatic invariant. Looked at another way, the electron has access to a larger change in electric potential than it otherwise would have because of the larger cross-tail electric field. When the electric field decreases after the 30 min, the electron then follows a closed trajectory where again the gradient drift dominates its motion. In this case, the first adiabatic invariant has not changed, but the electron has been transported inward across its Alfvén layer.

8.3 TRANSPORT ACROSS THE NTR

From the previous section, on the nightside, and as a baseline, plasma transport can be understood to be largely earthward, a process by which plasma from deep in the magnetotail is staged to the inner edge of the plasma sheet. From there, it can be deflected around the inner magnetosphere and is subsequently lost to the solar wind on the dayside. In this situation, the NTR serves as a magnetic wall, excluding plasma of magnetotail origin from accessing the inner magnetosphere. On the other hand, there are a host of processes that can transport these particles of

magnetotail origin across their (ultimate) respective Alfvén boundaries, leaving them on closed drift paths in the inner magnetosphere. It is these particles that cross the NTR that have the greatest space weather effects, and the NTR and physical processes therein serve as regulators of the space weather impacts of large-scale geospace dynamics.

Geospace as a system is decidedly complex. Things that matter at the system level are often accomplished in a number of ways. In many instances we do not know how different fundamental processes accomplish a particular aspect of the larger-scale dynamics. For example, in general we know that plasma comes to the plasma sheet via a number of pathways (dayside merging, direct entry through the cusp, the cusp ion fountain, the polar wind, and transverse heating in the ionosphere), but in general we cannot say how much any one of these mechanisms, each of which are fundamental processes, is contributing relative to the rest either on average or in specific cases. The same is true for coupling via transport between the magnetotail and inner magnetosphere.

The problem of plasma sheet access to the inner magnetosphere is analogous to the formation of the plasma sheet. It would be an uninteresting system if the electric and magnetic fields were truly time stationary, and if every particle motion conserved adiabatic invariants.

However, even a cursory consideration of test particle motion in realistic magnetic and electric fields (see, e.g., Figure 1 of Speiser, 1965) and a cursory look at real *in situ* observations of fields and plasmas near the inner edge of the plasma sheet (see, e.g., Figure 1 of Henderson et al., 2006) tells us that while there are quasi-stationary periods of relative quiescence in the magnetotail and inner magnetosphere (see, e.g., Rostoker, 2000), nonadiabaticity is important, and the electric and magnetic fields are often highly dynamic. A key consequence of these effects is that plasma sheet plasma is frequently captured on closed drift paths in the inner magnetosphere.

Three such deviations from the steady state considered above are discussed in the remainder of this section. These are as follows:

- The convection electric field is often highly structured in time and space, and it has been argued that most earthward transport on the nightside is accomplished via "transients" or so-called bursty bulk flows (BBFs; see Baumjohann et al., 1989; Angelopoulus et al., 1992).

- Not unrelated to the previous point (though the interrelationship between these two phenomena is far from understood), there are transient events near the inner edge of the plasma sheet wherein the local fluxes of ions and electrons, across a wide range of energies (several to a >100 keV) increase dramatically and suddenly. These sudden increases observed *in situ* are called injections, and occur (most frequently) around the time of substorm onset (DeForest and McIlwain, 1971; Kivelson, 1979).

- The Alfvén layer for a given first adiabatic invariant contracts and expands with increasing and decreasing magnetic activity (see, e.g., Figure 5 of Burke et al., 1995). These changes are really a manifestation of the increasing and decreasing importance of the cross-tail electric field relative to that of the corotation field, such that the more important the cross-tail field is, the closer to Earth particles of a given first adiabatic invariant can access. If the cross-tail field is changing on a timescale of tens of minutes to hours, then this can lead to plasma sheet particles being supplied to the inner magnetosphere (or the inner magnetosphere being drained of particles with these energies) (Korth et al., 1999).

8.3.1 Bursty Bulk Flows

Based on standard average models of the convection electric field, the ExB drift speed in the vicinity of the neutral sheet can be reasonably expected to be below about 100 km/s under almost all circumstances (note this number would correspond to a magnetic field strength of 5 nT and an electric field magnitude of 2 kV/ R_E). *In situ* observations of plasma flows in the plasma sheet and its high latitude boundary (the plasma sheet boundary layer or PSBL) were understood to be consistent with this. For example, Huang and Frank (1986) used convection velocities inferred from *in situ* ISEE 1 particle observations to survey the typical plasma velocities in the plasma sheet (at downtail distances between 10 and 25 R_E). They argued that the velocities were essentially always small (<50 km/s) provided the spacecraft is definitely within the plasma sheet. They attributed large velocities (>150 km/s) to the spacecraft being in the PSBL, meaning that such large speeds were likely field-aligned streaming.

Baumjohann et al. (1988) carried out a careful examination of AMPTE/IRM data from the plasma sheet, beyond 9 R_E (outside of the NTR), and earthward of the 19 R_E satellite apogee. They had the advantage of more comprehensive (than available for previous studies) measurements from which one could determine whether or not the spacecraft was actually in the plasma sheet proper, and from which the direction of the flows could be inferred relative to the local magnetic field direction. They found that flows in excess of 400 km/s occurred 2% or more of the time, and that this was true in the plasma sheet, and in the PSBL. Further, they found that in the plasma sheet the fast flows were almost exclusively nearly orthogonal to the magnetic field, and directed earthward.

Following on the Baumjohann et al. study, Angelopoulos et al. (1992) also used AMPTE/IRM data to explore earthward transport in the plasma sheet. Their Figure 2 is an excellent illustration of a fast earthward flow in the plasma sheet. Note that while the instances of the fastest flows (>500 km/s) are of very short duration (Baumjohann et al., 1990, found that the fast flows typically lasted less than 10 s), there is a period of sustained >400 km/s flow that lasts around 10 min. Angelopoulos et al. (1992) used the term BBF to describe the prolonged (~10 min) period of enhanced flow. Figure 8.7, a reproduction of Figure 1 of Runov et al. (2009), shows THEMIS constellation *in situ* magnetic field observations of an earthward propagating dipolarization front, which is also a BBF.

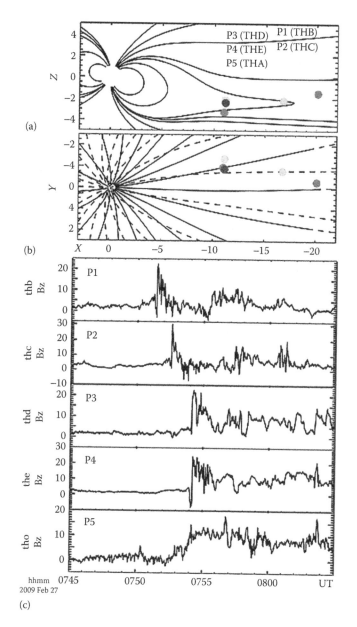

FIGURE 8.7   The five THEMIS satellites observed a dipolarization front, ostensibly a BBF, on February 27, 2009. The satellite locations at the time are indicated in the top two panels, while the bottom five panels show the magnetic field measured *in situ* by the five satellites. (Reproduced from figure 1 of Runov, A. et al., *Geophys. Res. Lett.*, 36, L14106, 2009. With Permission.)

Angelopoulos et al. (1992) go further, suggesting that it is not unreasonable to infer that BBFs are as important (or more important perhaps) in achieving earthward transport in the nightside magnetosphere as the background, much slower, steady convection. They further argued that although it might be compelling to think that these BBFs were being launched by episodes of reconnection in the plasmasheet, the BBFs occur under such a wide range of geomagnetic conditions that this cannot be their only source, even if it might sometimes be. In subsequent studies, multisatellite and other complimenting observations have been used to demonstrate that BBFs are only 1 or 2 R_E in dawn–dusk extent (Angelopolous et al., 1997; Nakamura et al., 2004).

Baker et al. (1996) raised the question of how BBFs brake as they approach the inner magnetosphere. Shiokawa et al. (1997) suggested that, in effect, the BBFs will encounter the larger magnetic field in the transition between stretched and dipolar (e.g., the NTR), and slow down and eventually stop, accordingly. The momentum change would be affected by the magnetic force due to a transient dawnward (inertia) current that would form on the leading edge of the BBF as it encounters the large magnetic field region. Given the BBF is of limited dawn–dusk or azimuthal extent, so too would be the dawnward current. The possibility was also raised that this dawnward current might form part of the substorm current wedge (Haerendel, 1992).

Henderson et al. (1998) identified a north–south structure in expansion phase aurora imaged by the UV instrument on the Viking satellite. He argued that this would be an expected proxy signature of a fast flow in the plasma sheet, and that the auroral structure was probably discrete aurora corresponding to the upward field-aligned current that would be expected on the dusk flank of such a fast flow. Later, Zesta et al. (2000) demonstrated a one–one correspondence between such north–south structures, also known as auroral streamers, and fast flows in the plasma sheet observed *in situ* by Geotail.

It has become common, and perhaps too much so, to look at every transient north–south-aligned auroral streamer and associate it one-to-one with a fast plasma sheet flow. The advantage of using the aurora, of course, to track the 2D spatiotemporal evolution of BBFs is that they provide a true two-dimensional (global in fact, though projected into 2D along magnetic field lines) picture. The disadvantage is that it is a proxy observation, and the one-to-one connection proposed by Zesta et al. (2000) has not been shown to be true. Nevertheless, new ground-based auroral imaging data, especially from the THEMIS-ASI network (Donovan et al., 2006), have provided an exciting new opportunity for looking at the large-scale evolution of auroral streamers, within the context of an increasingly well-understood mapping between the auroral distribution and the magnetotail. Certainly, most noteworthy of the work along these lines is that of Nishimura and colleagues, who have used THEMIS-ASI mosaics to argue that auroral streamers, and hence BBFs, precede virtually every substorm onset (Nishimura et al., 2010).

At first blush, Nishimura's idea appears to be consistent with the idea that BBFs are launched from a reconnection site in the mid-tail region, and brake as they enter the NTR. This braking causes flux pileup (dipolarization), a substorm current wedge, and traditional onset, as laid out in the flux pileup picture of Shiokawa (1997) and others (e.g., Kepko and Kivelson, 1999). Indeed, a later study by Kepko et al. (2009) showed an auroral streamer in multiwavelength auroral observations that directly preceded an onset. Furthermore, in that event, the streamer approached the onset arc at the longitude at which it

brightened, as would be expected in the flux pileup picture. However, there is a significant issue that needs to be resolved. In the Nishimura picture, the streamer emerges in the auroral zone from near the ionospheric footprint of the open–closed (polar cap) boundary, and so, presumably is launched from well tailward of where mid-tail reconnection is known to occur. Observations from ARTEMIS, at lunar distances deep in the magnetotail, have shown fast flows roughly 30 min prior to onsets and associated with such streamers (Nishimura et al., 2013). Perhaps, more perplexing is that there are precursors to these flows that are now being observed via radar and other observations deep in the polar cap (Zou et al., 2015). Thus, although Nishimura's streamers are undoubtedly associated with both BBFs and onsets, they do not fit the flux pileup picture argued for by others.

As final note, in keeping with the topic of magnetotail inner magnetosphere coupling, the BBFs are increasingly seen as flux tubes that move rapidly earthward due to an interchange process, powered by the plasma properties on that flux tube being very different from the surrounding ones (e.g., the "bubble model" of Wolf et al., 2009). Dubyagin et al. (2010), and subsequently Sergev et al. (2014), showed that BBFs move inward until the properties of the surrounding plasma, represented by the property of flux tube entropy ($pV^{5/3}$), is similar to that characterizing the BBF flux tube. The BBFs modify the plasma pressure and entropy in the region around where they come to rest. The changes in plasma properties that result from a series of such BBF arrivals sustain the larger scale substorm phenomena including the current wedge (see also Lyons et al., 2013).

8.3.2 Injection

Near the inner edge of the plasma sheet, and near the onset of a substorm, sudden and dramatic increases in the fluxes of electrons and ions with energies from several to 100 or more keV are frequently observed. These distinctive plasma signatures were called injections (DeForest and McIlwain, 1971; McIlwain, 1974; Moore, 1981). When the flux increase occurs more or less simultaneously across all energies, the injection is termed dispersionless. When the flux increase occurs first at the highest energies, and then successively later at lower and lower energies, the injection is termed dispersed. Figure 8.8, reproduced from Figure 2 of Spanswick et al. (2007), shows three examples of dispersionless electron injections and one of a dispersed electron injection.

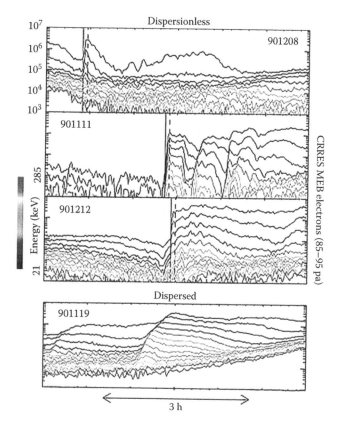

FIGURE 8.8 Dispersionless and dispersed injections as seen in CRRES MEB *in situ* electron observations. (Reproduced from figure 2 of Spanswick, E. et al., *Geophys. Res. Lett.*, 34, L03101, 2007. With Permission.)

The dispersionless injection gave rise to the idea of a source or injection region, wherein the energization process is occurring and simultaneously affecting all relevant energies at once. The increased fluxes across a wide range of energies indicate that this is a local increase of energies of all the particles in that energy range, and the increase in fluxes is consistent with a distribution function that is decreasing with increasing energy. The dispersed injection was understood to be a consequence of a satellite being outside of the injection region when the process began. Subsequently, charged particles within the injection region would drift azimuthally away from it (eastward and westward for electrons and ions, respectively). The satellite then "sees" higher energy particles before lower energy particles (of either species).

The earliest studies of, and in fact to date most of the studies of, injections used data collected from satellites in geosynchronous orbits (or geosynchronous transfer orbits) like ATS-5, CRRES, and the LANL spacecraft. Injections were thus investigated at radial distances that were typically near to the inner edge of the plasma sheet or inside it. It was clear, early on, that this process was

acting on plasma sheet particles (DeForest and McIlwain, 1971). This led to the concept of the injection boundary, an earthward edge of the injection region (with no real concept of what was happening further downtail), from the region where the injection is observed as dispersed (McIlwain, 1974).

By looking at dispersed injections and particle tracing back to (meaning following their energy-dependent trajectories back in time) the injection region, numerous researchers have built up a picture of the injection region as extending several R_E in the dawn–dusk direction (Friedel et al., 1996; Thomsen et al., 2001). Given that it is believed that injections accompany every substorm onset, but not all substorms produce injections at geosynchronous orbit (e.g., during the early years of the "THEMIS era," many of the substorms were small, and produced no injection signature at geosynchronous orbit); it is clear that the injection boundary is sometimes outside of geosynchronous orbit. Further, the injection boundary can be well inside of geosynchronous, as far in as inside of $4\,R_E$ (Friedel et al., 1996; Turner et al., 2015).

The injection has proven to be an elusive process to explain and characterize. The process is by definition dynamic, and the dawn–dusk extent of the injection region (the region of dispersionless injection) is limited. Consequently, *in situ* measurements which are sparse and localized cannot identify the beginning of injection, and can only provide a kind of average picture of the phenomenon. Reeves et al. (1996) used close radial conjunctions of CRRES and a LANL satellite to show that in most cases the injection is moving earthward at geosynchronous orbit. Spanswick et al. (2010) used radial alignments between Polar or Geotail (meaning one or the other but never both) and a LANL satellite, and found that beyond about 9 R_E, the injection appeared to be evolving tailward. Spanswick et al. (2007) used a proxy signature for dispersionless injection in riometer time series to show that the dispersionless injection appears to begin (typically) outside of geosynchronous and to evolve both earthward and tailward from that starting point. This picture is consistent with the current disruption model of substorms (Lui, 1996), a paradigm wherein the substorm instability evolves near the inner edge of the plasma sheet, and precedes reconnection in the mid-tail.

Moore et al. (1981) proposed that an injection boundary forms at mid-tail distances, and moves radially inward. This has proven a compelling idea, and is consistent with the idea of BBFs emanating from the mid-tail (near-Earth reconnection) region, and is consistent with the near-Earth neutral line substorm model (Baker et al., 1996). In this scenario, the injection is the consequence of a convection surge. There is considerable evidence to support this view. For example, the effects of convection surges on the high-energy particle population in simulations are consistent with the evolution of the distribution function during injections as observed *in situ* (Li et al., 1998; Gabrielse et al., 2012). More work needs to be done to resolve the apparent inconsistency between the work of Spanswick et al. (2007, 2010) and Li et al. (1998) and Gabrielse et al. (2012).

There is also uncertainty as to how the electrons and ions are energized during an injection. On the one hand, injections are closely related to dipolarization, and hence potentially significant induction electric fields (Aggson and Heppner, 1977). Kabin et al. (2011) modeled the collapse of a thin current sheet and showed that a current disruption-type scenario can provide changes to the distribution function similar to what is observed during an injection. On the other hand, as discussed at length in Gabrielse et al. (2012), the potential drop across a convection surge will be large enough, and the gradient drift fast enough, such that electrons and ions can gain enough energy to explain what is seen during injections by simply moving across the convection surge during the time of inward transport.

8.3.3 Changes in Large-Scale Convection (Storms)

Magnetic storms are large and frequent space weather disturbances. The storm is characterized by a dramatically enhanced ring current carried by 10s to >100 keV ions on closed drift paths, which are thus inside of their respective Alfvén layers. A significant fraction of the ring current plasma originates in the magnetotail, and so a key element of understanding storms is the understanding of magnetotail inner magnetosphere coupling. The plasma density in the inner magnetosphere is controlled by a convolution of the density in the plasma sheet (Thomsen et al., 1998), and the time evolving and spatially structured convection electric field (see, e.g., Liemohn et al., 1998). Storms arise during periods of strong solar wind driving of the magnetosphere, conditions which lead to enhanced plasma sheet densities and the required variations of the electric field.

Enhanced solar wind driving produces stronger cross-tail electric fields which in turn move plasma inward more quickly and closer to the Earth. Temporal variations and spatial structuring of the electric field accomplish the "stranding" of plasma of plasma sheet origin inside the relevant Alfvén layers (see, e.g., Ganushkina et al., 2006).

8.4 DISCUSSION

This chapter is meant to provide a basic overview of magnetotail inner magnetosphere coupling that is in turn accomplished via plasma transport. The focus has been on some basic concepts that are fundamental to this transport and its impact in terms of space weather.

The greatest space weather effects due to magnetospheric processes come from activity in the inner magnetosphere. One can define what one means by "inner" a number of ways including the region inside of the NTR, the region where particles are stably bounce trapped (meaning inside of the $\kappa = 3$ boundary which is more or less equivalent to the NTR), and inside of the Alfvén boundary. The NTR (or $\kappa = 3$ boundary) matters because it represents the physical barrier to inward transport, and somehow provides the source of energy for plasma as it transports inward. The Alfvén boundary matters because the most impactful particles, in terms of space weather, are in most cases on closed drift paths. These are the particles that support the ring current and constitute the radiation belts.

A time stationary convection electric field does not allow magnetotail particles to cross the NTR and relevant Alfvén boundaries. Dynamic processes that accomplish this inward transport are the above-mentioned BBFs, injections, and variations in convection associated with storms. In all cases, the trapping of particles in the inner magnetosphere is achieved via a mechanism that must be qualitatively similar to one or the other of those illustrated in the middle and right-hand panels of Figure 8.6. On the one hand, a particle can be "suddenly" energized (middle panel), as might be expected to occur in the injection region (though it is not clear this is how injected particles are energized). On the other hand, the particle could be subject to a changing electric field as one would expect during a convection surge (the other possibility for injection) or as a response to changes in the solar wind driver (as is the case in magnetic storms).

These processes serve to stage plasma to be trapped in the inner magnetosphere, where it serves as a source population for the ring current and radiation belts. For the ring current, the source plasma arrives with more or less the necessary energy. For the radiation belts, the particles enter a complex interaction between chorus waves (which is a part of the population driver and another part is accelerated by) and ULF waves, with the result being radial diffusion and acceleration to very high energies (see, e.g., Rostoker et al., 1998; Ukhorskiy et al., 2009; Jaynes et al., 2015).

There is still much that must be learned. At present, it is not known exactly how BBFs and injections are related. It is compelling to think they are the same process viewed from two different ways, however BBFs are common phenomena under most conditions, and injections are restricted to the minutes around substorm expansion phase onset. Further, the origin of BBFs, especially those that are related to substorm onset, is not known. Determining this is of fundamental importance to understanding how magnetotail inner magnetosphere coupling is controlled by external factors. As well, knowing how and where the dispersionless injection begins would be a significant step toward understanding the substorm onset mechanism.

Looking forward, the fleet of satellites probing geospace, including RBSP, THEMIS, MMS, Geotail, GOES, Cluster, and LANL, is providing point measurements in regions where key processes that accomplish magnetotail inner

magnetosphere coupling. Ground-based observatories including THEMIS-ASI are complimenting the space fleet providing observations of geospace "at the system level" albeit projected along magnetic field lines into a 2D projection. However, observations of certain parameters are not adequate to address pressing questions including those listed above. A significant problem is the lack of high-quality "maps" of convection that will be needed to explore the physics of the injection, and the details of BBFs and their interaction with the inner magnetosphere. This is almost certainly not possible *in situ*, and so must be thought of as an objective for the ground-based observing community.

REFERENCES

Aggson, T. L. and J. P. Heppner (1977), Observations of large transient magnetospheric electric fields, *J. Geophys. Res.*, 82, 5155–5164.

Angelopoulos, V., W. Baumjohann, C. F. Kennel, F. V. Coroniti, M. G. Kivelson, R. Pellat, R. J. Walker, H. Lühr, and G. Paschmann (1992), Bursty bulk flows in the inner central plasma sheet, *J. Geophys. Res.*, 97(A4), 4027–4039.

Angelopoulos, V. et al. (1997), Magnetotail flow bursts: Association to global magnetospheric circulation, relationship to ionospheric activity, and direct evidence for localization, *Geophys. Res. Lett.*, 24, 2271.

Baker, D. N., T. I. Pulkkinen, V. Angelopoulos, W. Baumjohann, and R. L. McPherron (1996), Neutral line model of substorms: Past results and present view, *J. Geophys. Res.*, 101, 12975–13010.

Baumjohann, W., G. Paschmann, and C. A. Cattell (1989), Average plasma properties in the central plasma sheet, *J. Geophys. Res.*, 94(A6), 6597–6606, doi:10.1029/JA094iA06p06597.

Baumjohann, W., G. Paschmann, and H. Lühr (1990), Characteristics of high-speed ion flows in the plasma sheet, *J. Geophys. Res.*, 95(A4), 3801–3809, doi:10.1029/JA095iA04p03801.

Baumjohann, W., G. Paschmann, N. Sckopke, C. A. Cattell, and C. W. Carlson (1988), Average ion moments in the plasma sheet boundary layer, *J. Geophys. Res.*, 93(A10), 11507–11520, doi:10.1029/JA093iA10p11507.

Behannon, K. W. and N. F. Ness (1968), Satellite studies of the Earth's magnetic tail, in *Physics of the Magnetosphere*, edited by R. L. Carovillano, J. F. McClay, and H. R. Radoski, pp. 409–434, D. Reidel, Dordrecht, the Netherlands.

Büchner, J., and L. M. Zelenyi (1987), Chaotization of the electron motion as the cause of an internal magnetotail instability and substorm onset, *J. Geophys. Res.*, 92(A12), 13456–13466, doi:10.1029/JA092iA12p13456.

Burke, W. J., A. G. Rubin, D. A. Hardy, and E. G. Holeman (1995), Banded electron structures in the plasmasphere, *J. Geophys. Res.*, 100(A5), 7759–7769, doi:10.1029/94JA03148.

Carpenter, D. L. (1963), Whistler evidence of a 'knee' in the magnetospheric ionization density profile, *J. Geophys. Res.*, 686, 1675–1682.

Chappell, C. R., T. E. Moore, and J. H. Waite Jr. (1987), The ionosphere as a fully adequate source of plasma for the Earth's magnetosphere, *J. Geophys. Res.*, 92(A6), 5896–5910.

Chen, A. J. (1970), Penetration of low-energy protons deep into the magnetosphere, *J. Geophys. Res.*, 75(13), 2458–2467, doi:10.1029/JA075i013p02458.

Cully, C. M., and E. F. Donovan (1998), A derivation of the gradient drift based on energy conservation, *Am. J. Phys.*, 67, 909–911, doi:10.1119/1.19147.

DeForest, S. E. and C. E. McIlwain (1971), Plasma clouds in the magnetosphere, *J. Geophys. Res.*, 76(16), 3587–3611, doi:10.1029/JA076i016p03587.

Donovan, E. et al. (2006), The THEMIS All-sky imager array—System design and initial results from the prototype imager, *J. Atmosph. Solar Terr. Phys.*, 68, 1472–1487.

Donovan, E., E. Spanswick, J. Liang, J. Grant, B. Jackel, and M. Greffen (2012), Magnetospheric dynamics and proton aurora, in *Auroral Phenomenology and Magnetospheric Processes: Earth and Other Planets*, 443pp., AGU Monograph, WA.

Dubyagin, S., V. Sergeev, S. Apatenkov, V. Angelopoulos, R. Nakamura, J. McFadden, D. Larson, and J. Bonnell (2010), Pressure and entropy changes in the flow-braking region during magnetic field dipolarization, *J. Geophys. Res.*, 115, A10225, doi:10.1029/2010JA015625.

Dungey, J. W. (1961), Interplanetary magnetic field and the auroral zones, *Phys. Rev. Lett.*, 6, 47–48.

Ejiri, M. (1978), Trajectory traces of charged particles in the magnetosphere, *J. Geophys. Res.*, 83(A10), 4798–4810, doi:10.1029/JA083iA10p04798.

Foster, J. C. (1984), Ionospheric signatures of magnetospheric convection, *J. Geophys. Res.*, 89(A2), 855–865, doi:10.1029/JA089iA02p00855.

Friedel, R. H. W., A. Korth, and G. Kremser (1996), Substorm onsets observed by CRRES: Determination of energetic particle source regions, *J. Geophys. Res.*, 101(A6), 13137–13154, doi:10.1029/96JA00399.

Gabrielse, C., V. Angelopoulos, A. Runov, and D. L. Turner (2012), The effects of transient, localized electric fields on equatorial electron acceleration and transport toward the inner magnetosphere, *J. Geophys. Res.*, 117, A10213, doi:10.1029/2012JA017873.

Ganushkina, N. Y., T. I. Pulkkinen, A. Milillo, and M. W. Liemohn (2006), Evolution of the proton ring current energy distribution during 21–25 April 2001 storm, *J. Geophys. Res.*, 111, A11S08, doi:10.1029/2006JA011609.

Gringauz, K. I. (1963), The structure of the ionized gas envelope of Earth from direct measurements in the U.S.S.R. of local charged particle concentrations, *Planet. Space Sci.*, 11, 281–296.

Haerendel, G. (1992), Disruption, ballooning, or auroral avalanche – On the cause of substorms, in Substorms 1, *Proceedings of the First International Conference on Substorms (ICS-1)*, European Space Agency Spec. Publ. SP-335, 417.

Henderson, M. G., G. D. Reeves, and J. S. Murphree (1998), Are north-south aligned auroral structures an ionospheric manifestation of bursty bulk flows?, *Geophys. Res. Lett.*, 25, 3737–3740, doi:10.1029/98GL02692.

Henderson, M. G. et al. (2006), Substorms during the 10–11 August 2000 sawtooth event, *J. Geophys. Res.*, 111, A06206, doi:10.1029/2005JA011366.

Hill, T. W. (1974), Origin of the plasma sheet, *Rev. Geophys.*, 12, 379–388.

Huang, C. Y. and L. A. Frank (1986), A statistical study of the central plasma sheet: Implications for substorm models, *Geophys. Res. Lett.*, 13, 652–655.

Jaynes, A. N. et al. (2015), Source and seed populations for relativistic electrons: Their roles in radiation belt changes, *J. Geophys. Res. Space Physics*, 120, 7240–7254, doi:10.1002/2015JA021234.

Jiang, F., M. G. Kivelson, R. J. Walker, K. K. Khurana, V. Angelopoulos, and T. Hsu (2011), A statistical study of the inner edge of the electron plasma sheet and the net convection potential as a function of geomagnetic activity, *J. Geophys. Res.*, 116, A06215, doi:10.1029/2010JA016179.

Kabin, K., E. Spanswick, R. Rankin, E. Donovan, and J. C. Samson (2011), Modeling the relationship between substorm dipolarization and dispersionless injection, *J. Geophys. Res.*, 116, A04201, doi:10.1029/2010JA015736.

Kepko, L. and M. Kivelson (1999), Generation of Pi2 pulsations by bursty bulk flows, *J. Geophys. Res.*, 104, 25021–25034.

Kepko, L., E. Spanswick, V. Angelopoulos, E. Donovan, J. McFadden, K.-H. Glassmeier, J. Raeder, and H. J. Singer (2009), Equatorward moving auroral signatures of a flow burst observed prior to auroral onset, *Geophys. Res. Lett.*, 36, L24104, doi:10.1029/2009GL041476.

Khazanov, G. V., M. W. Liemohn, T. S. Newman, M.-C. Fok, and A. J. Ridley (2004), *Annal. Geophys.*, 22, 497–510.

Kivelson, M. G. (1979), The physics of plasma injection events, in *Dynamics of the Magnetosphere, edited by S.-I. Akasofu*, pp. 385–405, D. Reidel Publishing, Dordrecht, the Netherlands.

Korth, H., M. F. Thomsen, J. E. Borovsky, and D. J. McComas (1999), Plasma sheet access to geosynchronous orbit, *J. Geophys. Res.*, 104(A11), 25047–25061, doi:10.1029/1999JA900292.

Li, X., D. N. Baker, M. Temerin, D. Reeves, and R. D. Belian (1998), Simulation of dispersionless injections and drift echoes of energetic electrons associated with substorms, *Geophys. Res. Lett.*, 25(20), 3763–3766, doi:10.1029/1998GL900001.

Liemohn, M. W., G. V. Khazanov, and J. U. Kozyra (1998), Banded electron structure formation in the inner magnetosphere, *Geophys. Res. Lett.*, 25, 877–880.

Lui, A. T. Y. (1996), Current disruption in the Earth's magnetosphere: Observations and models, *J. Geophys. Res.*, 101(A6), 13067–13088, doi:10.1029/96JA00079.

Lyons, L. R., Y. Nishimura, E. Donovan, and V. Angelopoulos (2013), Distinction between auroral substorm onset and traditional ground magnetic onset signatures, *J. Geophys. Res. Space Phys*, 118, 4080–4092, doi:10.1002/jgra.50384.

McIlwain, C. E. (1974), Substorm injection boundaries, in *Magnetospheric Physics*, edited by B. M. McCormac, p. 143, D. Reidel, Norwell, MA.

Moore, T. E., R. L. Arnoldy, J. Feynman, and D. A. Hardy (1981), Propagating substorm injection fronts, *J. Geophys. Res.*, 86(A8), 6713–6726, doi:10.1029/JA086iA08p06713.

Nakamura, R. et al. (2004), Spatial scale of high-speed flows in the plasma sheet oberved by Cluster, *Geophys. Res. Lett.*, 31, L09804, doi:10.1029/2004GL019558.

Nishimura, Y., L. Lyons, S. Zou, V. Angelopoulos, and S. Mende (2010), Substorm triggering by new plasma intrusion: THEMIS all-sky imager observations, *J. Geophys. Res.*, 115, A07222, doi:10.1029/2009JA015166.

Nishimura, Y., L. R. Lyons, X. Xing, V. Angelopoulos, E. F. Donovan, S. B. Mende, J. W. Bonnell, and U. Auster (2013), Tail reconnection region versus auroral activity inferred from conjugate ARTEMIS plasma sheet flow and auroral observations, *J. Geophys. Res. Space Phys*, 118, 5758–5766, doi:10.1002/jgra.50549.

Reeves, G. D., M. G. Henderson, P. S. McLachlan, R. D. Belian, R. H. W. Friedel, and A. Korth (1996), Radial propagation of substorm injections, in *Proceedings of the 3rd International Conference on Substorms*, Versailles, France, European Space Agency, ESA SP-389, p. 579.

Rostoker, G. (2000), Ground magnetic signatures of ULF and substorm activity during an interval of abnormally weak solar wind on May 11, 1999, *Geophys. Res. Lett.*, 27, 3789–3792.

Rostoker, G., S. Skone, and D. N. Baker (1998), On the origin of relativistic electrons in the magnetosphere associated with some geomagnetic storms, *Geophys. Res. Lett.*, 25, 3701–3704.

Runov, A., V. Angelopoulos, M. I. Sitnov, V. A. Sergeev, J. Bonnell, J. P. McFadden, D. Larson, K.-H. Glassmeier, and U. Auster (2009), THEMIS observations of an earthward-propagating dipolarization front, *Geophys. Res. Lett.*, 36, L14106, doi:10.1029/2009GL038980.

Sergeev, V., V. Angelopoulos, M. Kubyshkina, E. Donovan, X.-Z. Zhou, A. Runov, H. Singer, J. McFadden, and R. Nakamura (2011), Substorm growth and expansion onset as observed with ideal ground-spacecraft THEMIS coverage, *J. Geophys. Res.*, 116, A00I26, doi:10.1029/2010JA015689.

Sergeev, V. A., and B. B. Gvozdevsky (1995), MT-index: A possible new index to characterize the magnetic configuration of the magnetotail, *Ann. Geophys.*, 13, 1093–1103, doi:10.1007/s00585-995-1093-9.

Sergeev, V. A., E. M. Sazhina, N. A. Tsyganenko, J. Å. Lundblad, F. Søraas (1983), Pitch-angle scattering of energetic protons in the magnetotail current sheet as the dominant source of their isotropic precipitation into the nightside ionosphere, *Planet. Space Sci.*, 31, 1147–1155.

Sergeev, V. A., I. A. Chernyaev, V. Angelopoulos, A. V. Runov, and R. Nakamura (2014), Stopping flow bursts and their role in the generation of the substorm current wedge, *Geophys. Res. Lett.*, 41, 4, 1106.

Shen, C., X. Li, M. Dunlop, Z. X. Liu, A. Balogh, D. N. Baker, M. Hapgood, and X. Wang (2003), Analyses on the geometrical structure of magnetic field in the current sheet based on cluster measurements, *J. Geophys. Res.*, 108, 1168, doi:10.1029/2002JA009612.

Shiokawa, K., W. Baumjohann, and G. Haerendel. (1997), Braking of high-speed flows in the near-Earth tail, *Geophys. Res. Lett.*, 24, 1179–1182.

Siscoe, G. L. (1966), A unified treatment of magnetospheric dynamics with applications to magnetic storms, *Planet. Space Sci.*, 14, 947–967.

Spanswick, E., E. Donovan, R. Friedel, and A. Korth (2007), Ground based identification of dispersionless electron injections, *Geophys. Res. Lett.*, 34, L03101, doi:10.1029/2006GL028329.

Spanswick, E., G. D. Reeves, E. Donovan, and R. H. W. Friedel (2010), Injection region propagation outside of geosynchronous orbit, *J. Geophys. Res.*, 115, A11214, doi:10.1029/2009JA015066.

Speiser, T. W. (1965), Particle trajectories in model current sheets, 1, Analytical solutions, *J. Geophys. Res.*, 70, 4219–4226.

Stern, D. P. (1975), Motion of a proton in the equatorial magnetosphere, *J. Geophys. Res.*, 80, 595–599.

Thomsen, M. F., J. Birn, J. E. Borovsky, K. Morzinski, D. J. McComas, and G. D. Reeves (2001), Two-satellite observations of substorm injections at geosynchronous orbit, *J. Geophys. Res.*, 106(A5), 8405–8416, doi:10.1029/2000JA000080.

Thomsen, M. F., J. E. Borovsky, D. J. McComas, and M. R. Collier (1998), Variability of the ring current source population, *Geophys. Res. Lett.*, 25, 3481–3484, doi:10.1029/98GL02633.

Tsyganenko, N. A. (1982), Pitch-angle scattering of energetic particles in the current sheet of the magnetospheric tail and stationary distribution functions, *Planet. Space Sci.*, 30, 433–437.

Tsyganenko, N. A. (1987), Global quantitative models of geomagnetic field in the cislunar magnetosphere for different disturbance levels, *Planet. Space Sci.*, 35, 1347.

Tsyganenko, N. A. (1989), A magnetospheric magnetic field model with a warped tail current sheet, *Planet. Space Sci.*, 37, 5–20.

Turner, D. L., S. G. Claudepierre, J. F. Fennell, T. P. O'Brien, J. B. Blake, C. Lemon, M. Gkioulidou, K. Takahashi, G. D. Reeves, S. Thaller, A. Breneman, J. R. Wygant, W. Li, A. Runov, and V. Angelopoulos (2015), Energetic electron injections deep into the inner magnetosphere associated with substorm activity, *Geophys. Res. Lett.*, 42, 2079–2087, doi:10.1002/2015GL063225.

Ukhorskiy, A. Y., M. I. Sitnov, K. Takahashi, and B. J. Anderson (2009), Radial transport of radiation belt electrons due to storm time Pc5 waves *Ann. Geophys.*, 27, 2173–2181, doi:10.5194/angeo-27-2173-2009.

Volland, H. (1973), A semiempirical model of large-scale magnetospheric electric fields, *J. Geophys. Res.*, 78, 171–180.

Weimer, D. R. (1996), A flexible, IMF dependent model of high-latitude electric potentials having "Space Weather" applications, *Geophys. Res. Lett.*, 23, 2549-2552, doi:10.1029/96GL02255.

Wolf, R. A., Y. Wan, X. Xing, J.-C. Zhang, and S. Sazykin (2009), Entropy and plasma sheet transport, *J. Geophys. Res.*, 114, A00D05, doi:10.1029/2009JA014044.

Yahnin, A. G., V. A. Sergeev, B. V. Gvozddevsky, and S. Vennerstrøm (1997), Magnetospheric source region of discrete auroras inferred from their relationship with isotropy boundaries of energetic particles, *Annal. Geophys.*, *15*, 943–958.

Zesta, E., L. R. Lyons, and E. Donovan (2000), The auroral signature of earthward flow bursts observed in the magnetotail, *Geophys. Res. Lett.*, 27, 3241–3244, doi: 10.1029/2000GL000027.

Zhang, D., J. B. Cao, X. H. Wei, and L. Y. Li (2015), New technique to calculate electron Alfvén layer and its application in interpreting geosynchronous access of PS energetic electrons, *J. Geophys. Res.: Space Phys.*, 120, 1675–1683, doi: 10.1002/2014JA020670.

Zou, Y., Y. Nishimura, L. R. Lyons, E. F. Donovan, K. Shiokawa, J. M. Ruohoniemi, K. A. McWilliams, and N. Nishitani (2015), Polar cap precursor of nightside auroral oval intensifications using polar cap arcs, *J. Geophys. Res. Space Phys.*, 120, doi:10.1002/2015JA021816.

CHAPTER 9

Ring Current

Yusuke Ebihara

CONTENTS

9.1   INTRODUCTION AND GENERAL OVERVIEW

The term "ring current" is used to describe the electric current encircling the Earth in the near-Earth space environment called magnetosphere. In the early twentieth century, Carl Størmer predicted the existence of the ring current to explain the equatorward movement of the auroral zone during large magnetic storms. The auroral zone is the region where auroral phenomena frequently occur, and is surrounding the north and south geomagnetic poles. In 1917, Adolf Schmidt applied the concept of the ring current to explain a worldwide decrease in the magnetic field on the ground. The long-lasting, worldwide decrease in the magnetic field is nowadays known as a magnetic storm. The magnetic disturbance on the ground was first recorded by George Graham in London. Soon after, the magnetic disturbance was found to occur simultaneously at two different locations by Anders Celsius in the eighteenth century. This was the discovery of magnetic storms (Egeland and Burke 2012).

In many cases, a magnetic storm starts with a sudden increase in the horizontal component of magnetic disturbance ΔH at low- and middle latitudes. The sudden increase in ΔH is called a sudden commencement (SC), or a sudden impulse (SI) (Chapman 1917), and it remains higher than usual for a few hours. This is called an initial phase of the magnetic storm. After a while, ΔH starts to decrease. ΔH keeps decreasing for a few hours to a half day. This period is called a main phase. For large storms, the horizontal component of the magnetic field is decreased by ~1% of the intrinsic magnetic field of the Earth (a few 100s nT). The large decrease in ΔH is primarily caused by the intensification of the westward flowing ring current. Then, ΔH starts to increase and recovers to the nondisturbed condition. It takes a few to several days to completely recover. This is called a recovery phase. During the magnetic storms, large-amplitude, short-duration magnetic disturbances are often observed in the polar region. The magnetic disturbance usually lasts 1–2 h, and the amplitude is much larger than that at low- and middle latitudes. At the beginning of the sudden decrease in ΔH, luminosity of aurora is suddenly intensified, called auroral breakup. The disturbance pronounced in the polar region is called a substorm. The term substorm comes from the idea that a storm consists of a succession of substorms (Akasofu 1968). According to this idea, when substorms take place frequently, hot particles are frequently injected into the ring current, which could intensify the ring current. This idea is still controversial because an opposite tendency is observed. For example, when a substorm occurs in the polar region, ΔH increases at low latitude during magnetic storms (Iyemori and Rao 1996). Another example is that ΔH keeps decreasing at low latitude without substorms (McPherron 1997). These observations may imply that a substorm is not a necessary condition to develop the ring current, and hence to cause a storm. As mentioned below, other processes may intensify the ring current more gradually, and globally.

The magnitude of the decrease in ΔH depends on magnetic local time (MLT) (Akasofu and Chapman 1964). This implies that the ring current is not axisymmetric. Occasionally, a decrease in ΔH is observed most significantly on the nightside whereas it increases on the dayside at the beginning of the magnetic storms (Hashimoto et al. 2002). This may mean that the ring current is primarily intensified on the nightside, whereas is weakened on the dayside. The same tendency is also found in the global imaging data from the IMAGE satellite (Brandt et al. 2002). In order to obtain a unified measure of the

strength of the ring current, a D_{st} index (Sugiura 1964) is suggested, and is widely used in the science community. The D_{st} index is a weighted average of ΔH observed at four different longitudes at low latitudes (Honolulu, San Juan, Hermanus, and Kakioka). ΔH at magnetic equator is not used for this purpose because the influence of the equatorial electrojet flowing in the ionosphere is large at the magnetic equator. The D_{st} index is officially calculated by World Data Center for Geomagnetism, Kyoto, Japan, and is available at http://wdc.kugi.kyoto-u.ac.jp/. The D_{st} index is provided in a real-time manner, but the real-time D_{st} index is subject to change. Three classes of the D_{st} index are provided, including real-time D_{st}, provisional D_{st}, and final D_{st}. The smallest value of the D_{st} ever observed since 1957 is −589 nT, which was recorded on March 14, 1989. We have to note that the contribution from the other current systems may also change ΔH, such as the magnetopause current, the tail current, the field-aligned current (FAC), and the ionospheric current. Thus, the D_{st} index is not a perfect measure of the ring current strength. The SYM index is also used, which are almost a 1-min version of the D_{st} index. The D_{st} index is calculated based on the magnetic field disturbance at four longitudes at low latitudes, whereas the SYM index is based on six longitudes at mid-latitudes. Based on a historical magnetic record acquired in Colaba (Bombay), a minimum D_{st} value of approximately −1760 nT is estimated for an extreme large storm that occurred on September 1, 1859 (Tsurutani et al. 2003). This storm is followed by a large solar flare reported with a delay of ~17 h and 40 min (Carrington 1859).

The main phase of the magnetic storm is caused by the ring current flowing westward at 2–4 R_E. The existence of the ring current had been predicted since the early twentieth century, but the carrier of the ring current had not been confirmed until 1967. A hypothesis was proposed in the mid-1950s that the carrier of the ring current is keV-range protons that originate from the Sun (Singer 1957). When the radiation belt was discovered in 1958, the newly found relativistic particles were thought to be the prime carrier of the ring current. However, the total kinetic energy of the relativistic particles is too small to decrease the ground magnetic field (Akasofu and Chapman 1961). Later on, a particle detector on board the OGO 3 satellite carried out measurements of the spectrograms of protons and electrons with energy from ~200 eV to 50 keV. The intensity of the protons with energy 31–49 keV was significantly increased by factors of >30 during magnetic storms (Frank 1967). This was the first identification of the prime carrier of the storm-time ring current. The OGO 3 satellite measurements were limited to the energy range from ~200 eV to 50 keV. The Explorer 45 satellite acquired the ions and electrons over a broad energy range from 0.8 to 3800 keV (Smith and Hoffman 1973). It was found that the ions with energy around 200 keV dominate the energy density during magnetically quiet time, whereas the ions with energy less than 100 keV become dominant during storm time. Hereinafter, we call the ions with energy range from ~1 to ~100 keV "ring current ions" for the sake of simplicity. The increase in the ring current ions depends on energy and position. In the evening sector, the ions with a few 10s keV are observed to increase first, followed by higher and lower energy ions (Smith and Hoffman 1974). This is called a nose dispersion because it looks like a nose when the intensify of the ion flux is displayed in an energy versus time spectrogram. A significant loss was thought to remove the ions with higher and lower energies to remain in a constant phase. Based on a computer simulation, the nose dispersion was reasonably explained in terms of energy-dependent drift trajectories of the ions (Ejiri et al. 1980). The nose dispersion is commonly observed in the inner edge of the plasma sheet (or storm-time ring current) in the dusk–midnight sector during the storm main phase.

In the late 1970s, particle detectors were advanced to resolve mass. Large amount of O^+ ions were found in the energy below 16 keV/q in the ring current by the GEOS and Prognoz 7 satellites (Balsiger et al. 1980; Lundin et al. 1980). Sometimes, the O^+ ions were found to dominate that of the H^+ ions during the storm time. The AMPTE/CCE satellite that was launched on August 16, 1984, carried out ion measurements with broader energy range from a few eV to 5 MeV with resolving mass. It was found that the H^+ ions are the dominant species in the quiet time ring current, but the energy density of the O^+ ions can dominate that of the H^+ ions during large storms. The ring current is well recognized as being primarily consist of the H^+ and O^+ ions with energy range from ~1 keV to ~a few hundred keV, followed by minor ionic species including He^{++} and He^+ (Hamilton et al. 1988; Daglis 2006 for reviews). Since the O^+ ions are rarely observed in the solar wind, the ring current is populated by the ions originating from the Sun and the Earth. This implies that large amount of ions can be supplied from the Earth. The same tendency

is also seen in the magnetospheres at Jupiter and Saturn. At Jupiter, the ring current (as seen by the plasma pressure) is dominated by sulfur (Mauk et al. 2004), which is probably supplied from Io. At Saturn, the plasma pressure is dominated by water group ions (OH^+, H_2O^+, and H_3O^+) in the inner magnetosphere, which is probably supplied from Enceladus.

The ultimate cause of the magnetic storms is the Sun. A magnetic storm tends to occur when the southward component of the interplanetary magnetic field (IMF) is present (Kokubun 1972). The prolonged southward IMF is brought to 1 AU (astronomical unit) by the following two drivers: a coronal mass ejection (CME) and a corotating interaction region (CIR). The CME contains a magnetic cloud inside. Ahead of the magnetic cloud, a sheath compress the solar wind and has a large-amplitude magnetic field. An interplanetary shock is located in front of the sheath. Thus, the interplanetary shock comes first, followed by the sheath and the magnetic cloud in time. The interplanetary shock results in SC, or SI. When the southward component of IMF is embedded in the sheath and/or the magnetic cloud, the ring current is intensified. The CIR is followed by high-speed solar wind streams coming from the coronal hole. The IMF tends to fluctuate in the high-speed stream (Tsurutani and Gonzalez 1987). The southward component of the fluctuated field can also intensify the ring current. CME-driven storms take place frequently around solar maxima with some exceptions. Statistical studies have shown that CIR-driven storms tend to take place in the declining phase of the solar activity. When the coronal hole structures are kept for a long time, recurrent storms occur owing to the rotation of the Sun. Both the CME-driver and CIR-driver cause magnetic storms, but on average, the minimum D_{st} index is smaller in the CME-driven storms than in the CIR-driven storms (Borovsky and Denton 2006; Denton et al. 2006). Detailed explanation on the solar wind drivers for the magnetospheric activities is described in Chapters 2 and 4.

The intensity of the ring current decreases gradually during the recovery phase of a magnetic storm. The decay of the ring current is attributed to loss of the ring current ions. One such loss process is charge exchange between the ions with neutral hydrogen (Dessler and Parker 1959). The number density of the neutral hydrogen is fairly large in the inner magnetosphere. The density is ~1000 cm^{-3} at 2 R_E, and ~200 cm^{-3} at 4 R_E (Rairden et al. 1986). The charge exchange cross section depends on energy and mass, which may be a primary cause of the population of the ring current ions during quiet times.

9.2 STRUCTURE AND IONIC COMPOSITION

Satellite observations of the magnetic field show that, on average, the ring current consists of a westward current in the outer part of the ring current and an eastward current in the inner part of it (Le et al. 2004). The net westward current dominates the net eastward current, so that the magnetic field is decreased on the ground (Ebihara and Ejiri 2000). Satellite observations also show that the ring current is not always axisymmetric about the center of the Earth. There is a tendency that the ring current is fairly symmetric during the quiet time, whereas it is asymmetric during the storm time. The term "partial ring current" is often used to describe the asymmetric ring current, and to distinguish from the "ring current." Hereinafter, we call both of them a ring current regardless of its symmetry because of the following reason. First, a completely symmetric ring current is unlikely in the real magnetosphere from a theoretical view point. The strength of the magnetic field is axisymmetric. The dayside magnetosphere is compressed by the solar wind, and the nightside magnetosphere is stretched by the tail current. Under this circumstance, the radial distance of the drift trajectories depend on equatorial pitch angles. Second, the large-scale electric field is present in the inner magnetosphere, which gives rise to asymmetric trajectory of the particles.

What is the entity of the ring current? Of course, there is no conductor surrounding the Earth in the magnetosphere. The charge in the ring current is carried by charged particles including light ions such as H^+, O^+, He^+, and He^{++} as well as electrons. The particles undergo cyclotron motion in which they encircle a magnetic field, and bounce motion in which they travel along a magnetic field line between two mirror points in the Northern and the Southern Hemispheres. The particles also drift in the direction perpendicular to the magnetic field. Positively charged particles drift westward, whereas negatively charged particles drift eastward in the dipolar magnetic field. By definition, an electric current is the number of charges passing through a surface per unit time. The motion of the charged particles is in general complicated, so that calculation of the current is not straightforward. A convenient way to calculate the current was proposed by Parker (1957) who divided the motion of the charged particles into cyclotron motion

and drift motion. Drift motion can be further divided into the $E \times B$ drift, the curvature drift, and the ∇B drift. The $E \times B$ drift velocity is independent of charge, while the direction of the curvature drift and the ∇B drift of positive charge is opposite to that of negative charge. Thus, a net current can be generated by the ∇B drift and the curvature drift. The drift velocities of the ∇B drift $\mathbf{J}_B$ and the curvature drift $\mathbf{J}_C$ can be written by

$$\mathbf{J}_B = \frac{P_\perp}{B^3} \mathbf{B} \times \nabla B \tag{9.1}$$

and

$$\mathbf{J}_C = \frac{P_\parallel}{B^4} \mathbf{B} \times (\mathbf{B} \cdot \nabla) \mathbf{B} \tag{9.2}$$

where $P_\perp$, $P_\parallel$, and $\mathbf{B}$ are plasma pressure perpendicular to the magnetic field, plasma pressure parallel to the magnetic field, and the magnetic field, respectively. It is important to take into account the current associated with cyclotron motion. The current density associated with cyclotron motion $\mathbf{J}_M$ is given by

$$\mathbf{J}_M = -\nabla \times \left(P_\perp \frac{\mathbf{B}}{B^2} \right)$$
$$= \frac{\mathbf{B}}{B^2} \times \nabla P_\perp - \frac{P_\perp}{B^3} \mathbf{B} \times \nabla B - \frac{P_\perp}{B^4} \mathbf{B} \times (\mathbf{B} \cdot \nabla) \mathbf{B} \tag{9.3}$$

The second term of the right hand of Equation 9.3 is obviously canceled by Equation 9.1. The third term of Equation 9.3 is canceled by Equation 9.2 when the plasma pressure is isotropic, that is, $P_\perp = P_\parallel$. The sum of these three currents, $\mathbf{J}_B$, $\mathbf{J}_C$, and $\mathbf{J}_M$, is then given by

$$\mathbf{J}_\perp = \mathbf{J}_M + \mathbf{J}_B + \mathbf{J}_c$$
$$= \frac{\mathbf{B}}{B^2} \times \left[\nabla P_\perp + \left(P_\parallel - P_\perp \right) \frac{(\mathbf{B} \cdot \nabla) \mathbf{B}}{B^2} \right] \tag{9.4}$$

This equation implies that the current flows when the gradient of the plasma pressure is nonzero, or the plasma pressure is anisotropic ($P_\perp = P_\parallel$) (with the presence of the curved magnetic field line). In the ring current, the first term on the right-hand side of Equation 9.4 dominates the second term, so that the pressure gradient is the major contributor to the ring current. The first term of Equation 9.4 is also called a diamagnetic current, which can be understood as follows. When the density of particles is uniform, the current generated by a charged particle is fully canceled by neighbor ones. When the

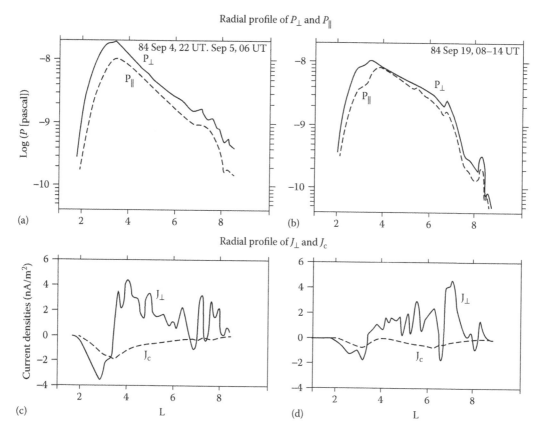

FIGURE 9.1   Radial profiles of the plasma pressure ($P_\perp = P_\parallel$) and the azimuth component of the ring current observed in a large magnetic storm that took place in September 1984. In the bottom panel, the solid line indicates the perpendicular current, and the dashed line indicates the second term of the right-hand side of Equation 9.4. (Adapted from Lui, A.T.Y. et al., *J. Geophys. Res.*, 92, 7459, 1987. With Permission.)

density is not uniform in the direction perpendicular to a magnetic field line, the current is not fully canceled by the others, so that a net current is generated in the direction perpendicular to the gradient of the density and the magnetic field. Figure 9.1 shows an example of the plasma pressure and the current density as a function of the L-value. There are four points to be noted: (1) the pressure peaks at $L \approx 3$, (2) $P_\perp > P_\parallel$, (3) the westward current flows in the outer part of the ring current and the eastward current flows in the inner part of it. (4) The first term of the right-hand side of Equation 9.4 is significant. The second term of it is not negligible and contributes to the eastward current.

The maximum plasma pressure is of the order of 10^{-8} Pa in Earth's ring current, which is about 10 orders of magnitude lower than the atmospheric pressure on the ground. The plasma pressure is too low to be measured directly like by a barometer. Instead, we can derive the plasma pressure by integrating the distribution function of charged particles in the velocity space as

$$P_\perp = \frac{1}{2} \sum_i \int F_i(\mathbf{v}) m_i v^2 \cos^2 \alpha \, dv d\alpha \qquad (9.5)$$

$$P_\parallel = \sum_i \int F_i(\mathbf{v}) m_i v^2 \sin^2 \alpha \, dv d\alpha \qquad (9.6)$$

where F, i, m, v, and α are the distribution function of particles, particle species, mass, speed, and pitch angle, respectively. The distribution function of charged particles is observed by an electrostatic analyzer, or a solid-state telescope, depending on energy to cover. In order to obtain the plasma pressure precisely, particle measurements with broad energy, ranging between eV to MeV, are required. Since the current density is associated with the plasma pressure, the structure of the ring current can be described in terms of pressure. Usually, the plasma pressure has a single peak at $L = 2.5$–3. During storm time, the plasma pressure has multiple peaks in some cases. Multiple peaks of the pressure imply multiple layer of the ring current (Lui et al. 1987).

By integrating the plasma pressure over the entire volume of the ring current, one may obtain the total kinetic energy of particles. The Dessler–Parker–Sckopke relationship describes that the magnetic disturbance at the center of the Earth is proportional to the total kinetic energy of particles (Dessler and Parker 1959; Sckopke 1966). On average, the D_{st} index can be used as a proxy for the magnetic field perturbation at the center of the Earth (Greenspan and Hamilton 2000). Further investigations show that the tail current may also contribute to the storm-time variation of the D_{st} index (Turner et al. 2001).

To understand the ring current is equivalent to understanding the ions that constitute the plasma pressure. The H^+ ions usually make the greatest contribution to the plasma pressure, though the O^+ ions occasionally become dominant during magnetic storms (Lundin et al. 1980, Lennartsson et al. 1981, Krimigis et al. 1985, Hamilton et al. 1988, Daglis et al. 1999). It should be emphasized that the energy density of the O^+ ions do not always dominate that of the H^+ ions during the magnetic storms. There is a tendency that the energy density of the O^+ ions increases with the solar radiation flux ($F_{10.7}$) and the magnetic activities (Young et al. 1982). Other ionic species such as He^+, He^{++}, O^{++} (Krimigis et al. 1985), N^+ (Liu et al. 2005b), $O^{\geq+3}$ (Ebihara et al. 2009), Fe, Mg, Si (Grande et al. 1996), and molecular ions (Klecker et al. 1986) have also been observed in the ring current region. The composition ratio is highly variable. The N^+/O^+ ratio is ~0.314 during quiet times, and it decreases with $F_{10.7}$ (Liu et al. 2005a). Single-charged ions, He^+ and O^+ ions, are thought to originate in the Earth, while He^{++}, $O^{\geq+3}$, Fe, Mg, and Si ions are thought to originate from the Sun. This implies that the composition of the ring current is basically a mixture of the ions originating in the solar wind and in the ionosphere. The contribution of electrons to the plasma pressure is uncertain when compared to ions. During quiet times, electrons with energy ~1–50 keV contribute to 1% of the ring current (Liu et al. 2005a). During active times, these electrons contribute to ~25% (Frank 1967), and 8%–19% (Liu et al. 2005a) of the ring current. The inequality of the contribution may come from the difference in the temperature of ions and electrons. In the magnetosphere, the ion temperature is larger than the electron temperature by a factor of 5–7 (Baumjohann et al. 1989). Because the ion density and electron density are almost equal due to quasi-neutrality of charge, the ion pressure is larger than the electron pressure.

Observations show that the plasma pressure is fairly symmetric about the dipole axis of the Earth during geomagnetically quiet times, whereas it (or the energy density) becomes asymmetric during high AE (De Michelis et al. 1999), low D_{st} (Ebihara 2002), and high K_p (Lui 2003) periods. The degree of the asymmetry depends on the storm phase (Ebihara 2002). During the storm main phase, the energy density of the H^+ ions increases on the nightside and decreases on the dayside. During the storm recovery phase, the energy density decreases on the nightside, and increases on the dayside. The same tendency has been observed by in situ particle measurements (Stüdemann et al. 1987; Korth and Thomsen 2001) and by measurements of energetic neutral atoms that are emitted from the ring current (Brandt 2002). There is a tendency that the anisotropy of the plasma pressure ($A = P_\perp/P_\parallel - 1$) is large on the dayside (or inner region), and small on the nightside (or outer region) (Lui and Hamilton 1992; De Michelis et al. 1999). At midnight, the anisotropy index A is ~2, ~1, and ~0.5 at $L = 3$, 4, and 6, respectively, in quiet times. Thus, the second term of Equation 9.4 may be non-negligible.

An electric current must be closed in the magnetosphere. If we consider the timescale of the order of second or longer, we can neglect the displacement current. When the plasma pressure distribution is axisymmetric, like a storm main phase, the large amount of current cannot be closed in the magnetosphere. To satisfy the continuity of the current, the divergence of the ring current must be connected to a FAC. The FAC tends to flow into the ionosphere on the duskside and out on the dawnside when the plasma pressure is maximized near midnight. This sense of the FAC is called region 2 current. An example of the FAC associated with the ring current is shown in Figure 9.2.

The plasma pressure of the ring current consists of ions and electrons. The source, trajectory, acceleration, and loss processes depend on charge as well as energy and pitch angle of particles. To understand the evolution of the plasma pressure, we have to know motion of the particles. In Section 9.4, we discuss fundamentals of motion of the particles that are composed of the plasma pressure (ring current) in the inner magnetosphere.

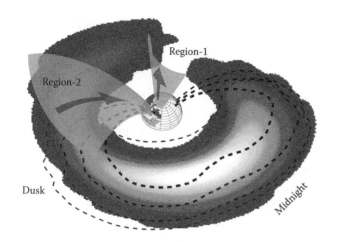

FIGURE 9.2 Perspective view of the plasma pressure of the ring current and a current system that connects the ring current and the ionosphere. (Adapted from Brandt, P. C. et al., *Geophysical Monograph*, vol. 181, pp. 135–143, American Geophysical Union, Washington, D C, 2008. With Permission.)

9.3 ORIGIN OF RING CURRENT PARTICLES

9.3.1 Solar Wind Origin

The abundance of H is the highest in the Solar System, followed by He, O, C, Ne, N, Mg, Si, Fe, S, Ar, and other species (Anders and Grevesse 1989). These elements that exist in the interior of the Sun are ionized by heating in the vicinity of the solar surface, and released into the interplanetary space. After traveling in the interplanetary space, they are heated when they pass through the bow shock in front of Earth's magnetosphere. After being heated, they enter the magnetosphere by some processes, including magnetic reconnection near the subsolar point under a southward IMF (Dungey 1961), double-lobe reconnection under a northward IMF (Song and Russell 1992), diffusive entry (Terasawa et al. 1997), Kelvin–Helmholz instability (Hasegawa et al. 2004), and cusp entry (Fritz et al. 2003). Once they enter the magnetosphere, some of them are transported to the near-Earth plasma sheet by the large-scale convection electric field. The ions arriving at the near-Earth plasma can be a direct source of the ring current because they can move earthward further into the inner magnetosphere by the convection electric field. Thus, the plasma sheet density is thought to be of importance for the ring current development.

At geosynchronous altitude, the plasma sheet density is usually 0.4–2 cm^{-3}, though it sometimes exceeds 2 cm^{-3} (Borovsky et al. 1997). The plasma sheet density at geosynchronous orbit is high in strong magnetic storms, and the peak of the plasma sheet density takes place about 9 h before the storm peak in super storms (Liemohn et al. 2008). Statistical study shows that the correlation coefficient between the solar wind electric field and the minimum D_{st} value is 0.70. The coefficient increases to 0.86 when the plasma sheet density is taken into account (Thomsen et al. 1998). The statistical study suggests that the plasma sheet density may be one of the important factors that may contribute to the development of the ring current during magnetic storms. It may be natural to think that the plasma sheet density is related to the solar wind density. Indeed, the plasma sheet density is known to be correlated to the solar wind density (Terasawa et al. 1997; Ebihara and Ejiri 2000). Terasawa et al. (1997) found that the plasma sheet becomes dense and cold when the IMF is northward. The best correlations between the solar wind density and the plasma sheet density is obtained when the solar wind density is averaged over 5–12 h prior to the plasma sheet observations. Thus, it is speculated that the development of the ring current would be correlated to the solar wind density. A statistical study shows that the D_{st} index is well correlated with the solar wind electric field with a lag of <1 h and the solar wind density with a lag of 5 h (Smith et al. 1999). However, O'Brien and McPherron (2000) pointed out that the correlation between the solar wind density and the D_{st} index is not so high except for the particular period from November 1994 and September 1995. This particular period includes the period on which Smith et al. (1999) focused. The effectiveness of the solar wind density on the ring current development may depend on period, or some other unknown factors.

9.3.2 Ionosphere Origin

The existence of O$^+$ ions in the ring current means that the ionosphere can be a source of the ring current because the O$^+$ ions are rare in the solar wind. The energy of the O$^+$ ions exceeds 100 keV in the heart of the ring current, whereas the energy of the O$^+$ ions is of the order of a few eV in the ionosphere. The process that accelerates the O$^+$ ions from eV range to 100 keV range is a subject of debate. The temporal variation of the O$^+$ ions in the ionosphere may have three distinct timescales: ~10s of min, ~days, and ~years.

The O$^+$ ions are rapidly increased in the ring current in association with a substorm expansion on a timescale of the order of ~10s of min. The rapid increase in the

O^+ ions during a substorm expansion is attributed to a rapid feeding of ions from the ionosphere to the near-Earth plasma sheet (Daglis and Axford 1996). During the substorm expansion, auroral electrojets are strongly intensified in the polar ionosphere, which give rise to an increase in Joule heating and ionospheric ion outflow. This process may be unlikely to explain the rapid increase in the O^+ ions in the ring current because the transport time of cold O^+ ions from the ionosphere to the equatorial plane is ~1–2 h. It is suggested that the ions preexisting in the near-Earth plasma sheet are rapidly accelerated by strong electric field during the substorm expansion (Mitchell et al. 2003). In this view, it is reasonably explained that the ions are simultaneously increased just after the substorm expansion. The magnetic field lines are stretched in the near-Earth plasma sheet near the substorm expansion onset. Most of the ions with energy greater than keV would undergo non-adiabatic motion when the gyroradius of the ions is comparable to the curvature radius of the magnetic field line. Under this circumstance, the ions are efficiently accelerated by the strong dawn-to-dusk electric field in the plasma sheet.

Observations have shown that O^+ ions are also increased during magnetic storms on a timescale of the order of a day (Krimigis et al. 1985; Hamilton et al. 1988). Near the D_{st} minima of large magnetic storms, the energy density of the O^+ ions occasionally exceeds that of the H^+ ions. This implies that the majority of the ring current consists of the ions originating from the Earth. There are a few paths to feed ionospheric ions to the ring current. First, the ionospheric ions are fed into the ring current by way of the plasma sheet on the nightside. The O^+ concentration is also increased in the plasma sheet, which can be attributed to an enhancement of the auroral and polar outflow of O^+ (Yau et al. 1985b). Second, the ionospheric ions are fed into the ring current directly. O^+ ions with beam-like pitch angle distribution may be an indirect evidence for direct feeding (Kaye et al. 1981; Sheldon et al. 1998).

The O^+ ions are known to show solar cycle dependence. Outflowing O^+ from the topside ionosphere is increased with the $F_{10.7}$ index (Yau et al. 1985a). O^+ concentration is also increased with increasing $F_{10.7}$ in the plasma sheet (Young et al. 1982; Lennartsson 1989). A possible reason is that the increase in the heating of the neutral atmosphere may increase the outflow of the O^+ ions from the ionosphere.

9.4 SUPPLY OF RING CURRENT PARTICLES

9.4.1 Convective Transport

A large-scale electric field is necessary to convey the majority of particles from the plasma sheet to the inner magnetosphere. One of the most important electric fields for the formation of the ring current is the convection electric field, which is a manifestation of the global circulation of plasma in the magnetosphere. Figure 9.3 shows an example of the electric potential of the convection electric field derived from the observation in the inner magnetosphere. The interplanetary electric field ($IEF = V(B_y^2 + B_z^2)^{1/2} \sin^2(\theta/2)$) is used to sort the potential pattern, where V is the solar wind speed, B_y is the Y-component of IMF, B_z is the Z-component of IMF, and θ is the clock angle of the IMF $\left(= \tan^{-1}\left(B_y/B_z\right)\right)$. The electric potential tends to be highest on the dawnside, and lowest on the duskside. Thus, the electric field points duskward in the inner magnetosphere. Statistical studies demonstrate that the potential drop between the highest and lowest potential increases with IEF.

If one assumes that a magnetic field line is equipotential, the ionospheric electric potential will be a projection of the magnetospheric electric potential. Figure 9.4 shows an example of the ionospheric electric potential derived from an empirical model (Weimer 2001). Twin vorticities are found. One is located on the dawnside, and the other one is on the duskside. The difference between the maximum and the minimum of the potential is called a cross polar cap potential drop, or a polar cap potential (PCP) drop. For this example, the PCP is 132 kV. The PCP is used to be a measure for the strength of the convection electric field. The electric potential is 20–30 kV in quiet times, where it is observed to increase to 200 kV in large magnetic storms. Large PCP gives rise to strong ionospheric and magnetospheric convection electric field, and then the ring current. However, the PCP is known to saturate for large amplitude of the solar wind electric field (Shepherd 2007). Readers may refer to Chapter 7 for the detailed explanation on the magnetospheric electric field.

The driving mechanism of the convection was first described by Dungey (1961). He introduced two neutral points of the magnetic field. One is on the dayside and the other one is on the nightside. Plasma is accelerated away from the neutral point by the Maxwell stress. On the dayside, the plasma is accelerated away from the equatorial plane toward north and south directions.

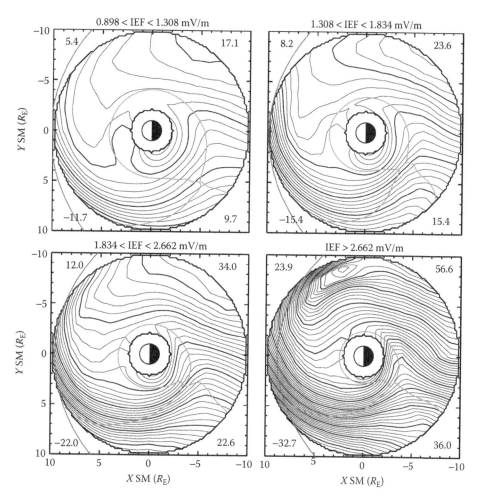

FIGURE 9.3 Electric potential of the large-scale convection electric field in the equatorial plane for different interplanetary electric field (IEF). The Sun is to the left. (From Matsui, H. et al., *J. Geophys. Res.: Space Phys.*, 118, 4119–4134, 2013. With Permission.)

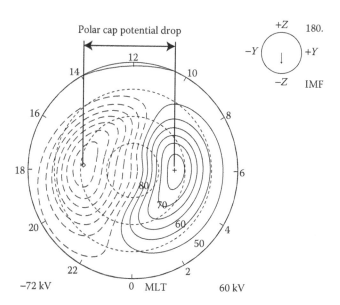

FIGURE 9.4 Example of ionospheric electric potential for southward IMF and fast solar wind. The Sun is to the top. The outer circle corresponds to the magnetic latitude of 60°.

The plasma then moves anti-sunward through high latitudes. On the nightside, the plasma is pulled into the neutral point, and is discharged away from it in the sunward and anti-sunward directions. The plasma that is ejected sunward pass through the inner magnetosphere, and reaches the dayside magnetopause. Then, a cycle of the convection is completed. The motion of the plasma is associated with the electric field. The sunward motion of the plasma corresponds to the duskward electric field. The Dungey model provides essentials to explain the global circulation of plasma in the magnetosphere under the presence of the southward component of IMF, although the model does not take into account the force balance acting on plasma.

Ground-based observations show that the response of the ionospheric convection to the southward turning of IMF is much shorter than expected from the traditional Dungey model. Based on the result of a global magnetohydrodynamics (MHD) simulation, Tanaka

(2007) pointed out the importance of the FAC to explain the rapid response of the magnetospheric convection. In his model, the role of the tangential Maxwell stress at the dayside neutral point is to accelerate plasma anti-sunward, and to form high-pressure region at high latitudes. The high-pressure region at high latitudes is known as a cusp. The simulation result shows that the tangential Maxwell stress is too weak to drive the overall convection in the entire magnetosphere. The tailward part of the cusp or the mantle transmits electromagnetic energy, that is, a dynamo. The current originating from the cusp/mantle flows into the ionosphere on the dawnside, and it flows out of the ionosphere on the duskside. In order to cancel space charge deposited by the FAC, the electric field is redistributed in the ionosphere. The redistributed electric field is transferred to the magnetosphere along a magnetic field line. The redistributed electric field conveys the magnetospheric plasma. In this view, the ionospheric convection may respond to the change in the solar wind immediately from pole to pole. The horizontal propagation of the electric field in the ionosphere may be explained by a transmission line model (Kikuchi 2014), or compressional wave model (Nishitani et al. 2002). The transmission line model describes that the electromagnetic energy that is transmitted by the FAC propagates horizontally in the atmosphere sandwiched by two conducting layers: the ionosphere and the ground. The propagation speed is equal to the speed of light, which is consistent with the observational facts that the ionospheric current (and the ionospheric electric field) responds almost simultaneously at polar and equatorial regions.

Another important electric field for the development of the ring current is a corotation electric field. The corotation electric field is generated at an altitude above ~60 km where neutral atoms are ionized by the solar EUV radiation. In the lower ionosphere, the ionized elements may corotate with the Earth's rotation because of viscous drag with neutral atmosphere. The corotation electric field $\mathbf{E}_2$ is given by

$$\mathbf{E}_2 = -(\Omega \times \mathbf{R}) \times \mathbf{B} \qquad (9.7)$$

where Ω and $\mathbf{R}$ are the angular velocity vector of Earth's rotation, and the position vector from the center of the Earth. The corotation electric field is absent in a frame of reference corotating with the Earth. The electric potential of the corotation electric field is the lowest at $L = 1$, which is about -90 kV. This means that the Earth is negatively charged in the quasi-inertial frame of reference.

Let us consider a low energy particle placed in the plasma sheet on the nightside. When the curvature drift speed and the ∇B drift speed are negligible, the particle undergoes the $\mathbf{E} \times \mathbf{B}$ drift with its drift velocity determined by $\mathbf{E}_c \times \mathbf{B}/B^2$, where $\mathbf{E}_c = \mathbf{E}_1 + \mathbf{E}_2$. The low energy particle moves toward the Sun primarily by convection electric field. As the particle approaches the Earth, it tends to drift eastward by corotation electric field, and may reach the dayside magnetopause. When the low energy particle is initially placed closed to the Earth, the particle would drift around the Earth primarily by corotation electric field. The separatrix between the open and closed drift trajectories is called a last-closed equipotential line. Inside the separatrix, cold plasma will be continuously supplied from the ionosphere, and the density will increase. When the electric field is steady, the location of the spearatrix is also steady, so that a sharp density gradient is expected at the separatrix. This may correspond to the plasmapause at which the number density of cold plasma has a sharp gradient. The separatrix is also considered as the inner boundary of the particles coming from the nightside magnetosphere. Observations show that the inner edge of the newly injected protons coincided with the plasmapause (Cornwall et al. 1970). This coincidence may be reasonably explained in terms of the separatrix.

It should be noted that the separatrix depends on energy and charge. The total energy of the charged particles is given by

$$q\Phi + \mu B + \frac{mv_{\parallel}^2}{2} = \text{const.} \qquad (9.8)$$

where:

μ is the magnetic moment of a particle ($= mv_{\perp}^2/2B$)
v is the speed of the particle

For the equatorially mirroring particles ($v_{\parallel} = 0$), they follow the isocontour of $q\phi + \mu B = \text{const.}$ Figure 9.5 shows an example of the equipotential lines for a singly charged ion with magnetic moment of 0.266 keV/nT. This ion has kinetic energy of 125 keV at $L = 4$, and 37 keV at $L = 6$, so that this ion is regarded as a typical ion constituting the storm-time ring current. An empirical convection electric field model is used to draw the equipotential

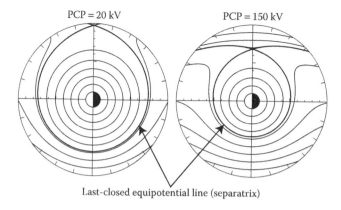

PCP = 20 kV   PCP = 150 kV

Last-closed equipotential line (separatrix)

FIGURE 9.5   Example of equipotential lines for an ion with magnetic moment of 0.266 keV/nT for large PCP (left) and small PCP (right).

lines (Volland 1973; Stern 1975). The left panel is for the condition that PCP = 150 kV, and the right one is for the condition that PCP = 20 kV. The separatrix shrinks with increasing PCP, meaning that the ions can penetrate deep into the inner magnetosphere from the plasma sheet for high PCP. In other words, high PCP is basically required to convey the plasma sheet ions deep into the magnetosphere to energize the ions, and to develop the strong ring current.

When the convection electric field is steady, the rate of increase of kinetic energy is given by $q\mathbf{V}\cdot\mathbf{E}$, where q is charge and $\mathbf{V}$ is the drift velocity. Let us consider the steady condition and the ion coming from the nightside plasma sheet. As the ion proceeds earthward, it gains kinetic energy as it drifts westward. In the case shown in Figure 9.5, the kinetic energy reaches its maximum at dusk. This can also be explained in terms of the conservation of the magnetic moment μ. When μ is conserved, the kinetic energy of the ion becomes large when the magnitude of the magnetic field is large. The equipotential lines shown in Figure 9.5 indicate that the separatrix is closest to the Earth at 6.8 R_E for low PCP condition. It is located at 4.6 R_E for high PCP case. Since the magnitude of the magnetic field increases with decreasing radial distance, the ion can gain higher energy for high PCP case than for low PCP case. The simple convection electric field model predicts that the kinetic energy is highest at dusk. This may explain the reason why the plasma pressure is the highest on the duskside. The upper limit of the kinetic energy to be accelerated is determined by the PCP. The PCP increases as high as 200–300 kV for large magnetic storm. This implies that the particles can gain kinetic energy as high as 200–300 kV due to the enhanced convection electric field. In order to gain kinetic energy greater than 200–300 kV, the electric field, other than the convection electric field, would be needed.

9.4.2 Substorm-Time Transport

A substorm is one of the most drastic disturbances taking place in the magnetosphere and the ionosphere. The visible manifestation of the substorm is a sudden brightening of aurora in the polar region. Sometimes, this moment is referred to as onset of the substorm expansion. After the onset, the bright auroral region expands poleward, westward, and eastward, accompanied with a surge. The magnetic field is largely decreased by the ionospheric current associated with a substorm. The amplitude of the magnetic disturbance can exceed several tens thousands of nT. Drastic changes take place in the magnetosphere. The stretched magnetic field line is suddenly collapsed (or relaxed), and becomes dipole-like. This transition is called dipolarization. In the course of dipolarization, a strong induction electric field is induced. The induction electric field can have an amplitude of 30 mV/m and larger. Low frequency (with a period of ~10 s) and high frequency (with a period less than 1 s) are simultaneously observed (Aggson et al. 1983). Simultaneously, a large number of particles rapidly increase near the substorm onset; frequently, these increase is observed in the magnetosphere, and is called an injection of hot plasma (Konradi 1967; DeForest and McIlwain 1971). The increase in the particles is observed at all MLTs. The drift speed depends on the kinetic energy, so that, usually, high-energy particles arrive at an observer first, followed by low-energy particles. The time-dependent increase in the particles, which is called a dispersion, has rich information about travel time between the newly arrived particles and a source location of them. With the aid of computation of trajectory traces, a common source on the nightside was discovered for the newly arrived particles. The source is called an injection boundary. The location of the injection boundary depends on MLT and K_p, and has a spiral-shaped boundary (Mauk and McIlwain 1974). At the injection boundary, the increase in the particles takes place independently of energy, namely, dispersionless injection. The dispersionless injection is found at $4 \leq L \leq 8$ (Friedel et al. 1996). This may imply that the substorm injection affects the ring current directly.

The dispersionless injection is thought to result from either local acceleration or rapid transport. The local acceleration is divided into three: adiabatic acceleration, nonadiabatic acceleration, and parallel acceleration. The adiabatic acceleration is expressed by

$$\frac{\partial K}{\partial t} = q\mathbf{E} \bullet \mathbf{V} + \mu \frac{\partial B}{\partial t} \tag{9.9}$$

where K is the kinetic energy. The first term on the right-hand side (called drift-betatron acceleration) represents that the kinetic energy increases when the drift velocity is parallel (anti-parallel) to the electric field for a positively (negatively) charged particle. The second term (called gyrobetatron acceleration) implies that the kinetic energy increases when the local magnetic field increases. During the substorm expansion, both terms may contribute to the acceleration (Ebihara and Tanaka 2013). The nonadiabatic acceleration may take place when the gyroradius of a particle is comparable to the curvature radius of a magnetic field line, or the gyroperiod is comparable to the timescale of changes in the magnetic field. Under such conditions, the magnetic moment μ is no longer conserved. The particles may also be accelerated by electric field parallel to the magnetic field line. The parallel electric field is thought to be present in the substorm expansion, and accelerate the particles in the direction parallel to the magnetic field. A cloud of particles is generated by the parallel acceleration, which undergoes bounce motion between the mirror points in the Northern and Southern Hemispheres. The cloud of particles, also called bouncing particle cluster, are observed by satellites near geosynchronous orbit on the nightside. The particles can also move in the course of the acceleration. Test particle simulations show that H+ ions with energy >20 keV are effectively accelerated by the nonadiabatic acceleration, whereas H+ ions with energy <20 keV are not because the low energy H+ ions rapidly escape from the region where the electric field is strong due to the $E \times B$ drift (Birn et al. 1997b). Thus, the combined effect, local acceleration, and transport should be taken into consideration.

At geosynchronous orbit, ions and electrons are not simultaneously increased during substorm (Birn et al. 1997a). The ion injection region is slightly displaced westward in comparison with the electron injection region. The temperature of the ions and electrons increases, whereas the number density remains almost constant during the substorm. This may imply that the local acceleration is dominant, rather than transport from the tail region. Overall increase in the H+ and O+ ions is observed by global imaging of energetic neutral atom emitted from the ions in the ring current (Mitchell et al. 2003). When substorm onset takes place, the H+ and the O+ ions are observed to increase. During some substorms, the H+ ions and the O+ ions are not simultaneously increased. Mass-dependent acceleration processes are successfully explained by simulation (Fok et al. 2006). On the basis of the ENA observation, the energy density of the H+ ions is shown to be enhanced by a factor of ~2 by a substorm. The structure of the high-energy density region expands westward and inward, and the energy density then drops back to the pre-injection levels within 1–2 h (Grimes et al. 2010). A substorm is also known to result in a decrease in the number of particles. The decrease in the electron flux rapidly starts near midnight. The decrease in the electron flux propagates eastward. This is called a drifting electron hole (Sergeev et al. 1992). The drifting electron hole is energy dependent; high energy electrons are observed first, followed by low energy ones. It is absent below certain energy. Energy-dependent processes probably take place, but the mechanism for the drifting electron hole is unknown.

The dipolarization is a transition from the stretched, tail-like configuration of the magnetic field line to the collapsed, dipole-like configuration of it. Multiple intensification of the electric field and particle flux are observed on the nightside (Sergeev et al. 1998). Each pulse of the electric field is 1–2 min long, and is correlated with the increase in the electron flux. This may imply that the transition does not take place smoothly. Recently, the multiple intensification of the electric field is successfully reproduced by a global MHD simulation, and is explained in terms of force balance between the Lorentz force and the pressure gradient force (Ebihara and Tanaka 2013). The simulated electron flux shows a gradual increase in the electron flux together with short-lived multiple increase in association with the multiple pulse of the electric field.

9.4.3 Impact of Interplanetary Shock

When the solar wind dynamic pressure suddenly increases, the solar wind compresses the magnetosphere. The compression of the magnetosphere is accompanied with the intensification of the magnetopause current, known as the Chapman–Ferraro current.

The intensification of the magnetopause current is easily noticed by looking at horizontal component of ground magnetic fields (ΔH), known as SI, or SC. ΔH is of the order of 10 nT for usual storms. ΔH exceeding 100 nT is rarely observed. A large ΔH of 202 nT was observed in the Kakioka Magnetic Observatory in Japan at 0341 UT on March 24, 1991. The largest ΔH is found to occur on March 24, 1940, in Kakioka with an amplitude of 273 nT (Araki 2014). The rapid compression of the magnetopause transmits a compressional wave toward the magnetotail. The electric field shows a bipolar variation. There is a tendency that the first excursion is westward and the second one is eastward on the dayside magnetosphere. The peak-to-peak amplitude of the electric field variation is 0.2–40 mV/m in or in the vicinity of the plasmapause (Shinbori et al. 2004). A large amplitude of the electric field (~80 mV/m) was detected at 0341 UT on March 24, 1991, deep in the inner magnetosphere (Wygant et al. 1994). Electrons with energy ~15 MeV were observed to increase, suggesting that the large-amplitude electric field may accelerate electrons to ~15 MeV. The initial pulse of the electric field points westward, which transport particles inward. Thus, the SI/SC may have a large impact on the ring current. The variation of the ring current can be monitored by the ground magnetic field, but will be masked by the variation of the magnetopause current. The intensification of the ring current causes a negative variation of ΔH, but the intensification of the magnetopause current causes a positive variation of ΔH. Global imaging of the ENA provides a vital clue for estimating the impact of SI/SC on the ring current. The ENA observations show that total emission rate of ENA with energy between keV and 100s keV are increased by ~25%–40% due to adiabatic acceleration (Lee et al. 2007).

9.5 LOSS OF RING CURRENT PARTICLES

9.5.1 Charge Exchange

Before the discovery of the carrier of the storm-time ring current, charge exchange with neutral hydrogen was suggested to remove the ions trapped in the magnetic field line (Dessler and Parker 1959; Stuart 1959). When the charge exchange takes place, the H^+ ion and the O^+ ion become fast H and O, respectively, which are free of any control of a magnetic field. The fast neutral atoms can be observed, and are used to monitor the global distribution of the ions. The characteristic lifetime of the ions for the charge exchange is given by

$$\tau = \frac{1}{n_H v \sigma_{ch}} \qquad (9.10)$$

where:

n_H is the density of neutral H
v is the velocity of the ion
σ_{ch} is the charge exchange cross section

The charge exchange is short where the neutral density is high. The Earth is surrounded by a cloud of neutral H, called the geocorona (Rairden et al. 1986). The density of neutral H decreases with altitude, so that the H^+ ions can be efficiently lost by the charge exchange at low L-value, or at low altitude. For bouncing ions, the H^+ ions undergo the charge exchange effectively near mirror points where the neutral density is higher than near the equatorial plane. Thus, the ions with equatorial pitch angles α_0 close to 0° or 180° decreases significantly. As a consequence, the pitch angle distribution becomes round-top, in which particle concentration is the highest near the equatorial pitch angle α_0 of 90°. The following equation is useful to evaluate the characteristic lifetime for equatorially mirroring particles with arbitrary α_0

$$\langle \tau \rangle = \frac{\cos^{3.5 \pm 0.2} \lambda_m}{n_H v \sigma_{ch}} \qquad (9.11)$$

where λ_m is the magnetic latitude of the mirror point (Smith and Bewtra 1976). λ_m increases with decreasing α_0. For the dipole magnetic field, λ_m and α_0 have the following relationship

$$\sin^2 \alpha_0 = \frac{\cos^6 \lambda_m}{\sqrt{4 - 3\cos^2 \lambda_m}} \qquad (9.12)$$

In the topside ionosphere, the concentration of neutral O is large, so that the fast H^+ can exchange charge with O at low altitude with short lifetime. Because of this, strong emission of fast neutral atom is observed near the surface of the Earth (Valek et al. 2010).

The charge exchange lifetime for equatorially mirroring H^+, He^+, and O^+ ions is summarized in Figure 9.6. The charge exchange lifetime of the H^+ ions is shorter than that of the O^+ ions for energy less than ~50 keV. The AMPTE/CCE satellite observed a large fraction of the O^+ ions in the energy density during the large magnetic storm that took place in February 1986

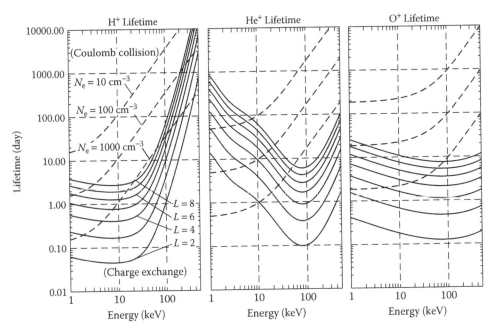

FIGURE 9.6 Charge exchange lifetime (solid line) and Coulomb collision lifetime (dashed line) for equatorially mirroring H^+, He^+, and O^+ ions. The neutral hydrogen density corresponding to different L-values are assumed. The cold plasma density of 10, 100, and 1000 cm^{-3} is assumed to calculate the Coulomb collision lifetime. Detailed in the calculation is found in Ebihara and Ejiri (2003), and is based on Fok et al. (1991).

(Hamilton et al. 1988). During this storm, the D_{st} index showed a rapid recovery with e-folding timescale of ~9.3 h, followed by a slow recovery. The two-step recovery has been attributed to the charge exchange lifetime depending on mass. For energy greater than ~50 keV, the charge exchange lifetime for the O^+ ions is shorter than that for the H^+ ions. Indeed, the rapid loss of the O^+ energy density was observed by the AMPTE CCE during the rapid recovery. However, numerical simulations have shown that the charge exchange of the O^+ ions hardly explains the rapid decay of the ring current. Precipitation loss may play an important role in the rapid loss of the ring current (Fok et al. 1995; Kozyra et al. 1998a). The rapid recovery of the D_{st} index is still in a controversy.

9.5.2 Coulomb Collision

The ions are decelerated and scattered by Coulomb collisions with thermal plasma. The former process is called Coulomb drag, and the latter one is called Coulomb scattering. According to simulation, the Coulomb drag is effective for low-energy ions. An example of the Coulomb collision lifetime is shown in Figure 9.6. The decay lifetime for 1 keV (10 keV) H^+ is a few hours (~a day) when the thermal electron density is 1000 cm^{-3} (Fok et al. 1991). If the energy density

of the ring current largely consists of >10 keV H^+, the Coulomb drag alone may be insufficient to explain the decay of the ring current as inferred from the D_{st} index. Coulomb drag redistributes the ions in the velocity space. One of the significant processes to enhance low-energy ions inside the plasmasphere (Jordanova et al. 1996). The kinetic energy of the colliding ions is transferred the thermal electrons. The heating of the plasmaspheric electrons may be a source of stable auroral red (SAR) arcs that are emitted at 630.0 nm (Cole 1965). The decay rates for the Coulomb scattering are much smaller than those for the Coulomb drag by two orders of magnitude (Jordanova et al. 1996).

9.5.3 Wave–Particle Interaction

The temperature anisotropy (perpendicular temperature being larger than parallel temperature) of the ions can provide free energy to excite the electromagnetic ion cyclotron (EMIC) waves (Kennel and Petschek 1966). The EMIC waves are favorably excited near the equatorial plane where a magnetic field line has a minimum of the magnetic field strength. The wave frequency is below the gyrofrequency of the H^+ ions in the equatorial plane, so that the EMIC waves are in the frequency range from 0.1 to 5.0 Hz in the inner magnetosphere. The ions can also be scattered by resonant interactions with the EMIC waves (Kennel and Petschek 1966; Cornwall et al. 1970). When the scattered ions enter the loss cone, they will precipitate into the atmosphere. A localized, detached precipitation of ions with energy >30 keV is frequently observed equatorward of the main auroral oval associated with the EMIC waves (Yahnina et al. 2003), which is consistent with simulation (Jordanova et al. 2001). The occurrence probability of the EMIC waves is highest in the afternoon sector (Anderson et al. 1992). The temperature anisotropy that is necessary for the growth of the EMIC may be provided by fresh ions coming from the nightside. The convective motion of the trapped ions as well as charge exchange loss may generate the temperature anisotropy of the hot ions. A sudden compression of the magnetosphere, associated with SI/SC, may also result in the growth of the EMIC waves (Anderson and Hamilton 1993). EMIC rising tone emissions are identified by the Cluster satellite (Pickett et al. 2010). Non-linear wave–particle interactions are suggested for the explanation of the EMIC rising tone emissions, in which the EMIC rising tone emission is excited when a coherent wave amplitude exceeds a threshold (Omura et al. 2010). The presence of the EMIC-triggered emission may result in the redistribution and scattering of the ions rapidly (Shoji and Omura 2012).

9.5.4 Field Line Curvature Scattering

When a magnetic field line is strongly bended and the curvature radius of the field line is comparable to the gyroradius of the particle, the first adiabatic invariant of them is violated. When the first adiabatic invariant is violated, the particles are scattered to the loss cone. They precipitate into the atmosphere. This is called field-line curvature (FLC) scattering, and is thought to efficiently fill the loss cone (Sergeev et al. 1983; Young et al. 2002). The FLC scattering may take place in the outer ring current where the magnetic field line is strongly bended and the intensity of the magnetic field is small. Simulation results show that the ring current decays rapidly, and that the timescale of the FLC scattering is found to be comparable to that of the charge exchange lifetime (Ebihara et al. 2011). The FLC scattering is unlikely to occur in the deep inner magnetosphere where the strength of the magnetic field is strong. It is expected that the FLC scattering may not work efficiently for large magnetic storms because the majority of the ring current shifts deep inside the inner magnetosphere.

9.5.5 Outflow through Magnetopause

Energetic O^+ ions are observed in the solar wind and the magnetosheath (Möbius et al. 1986; Zong and Wilken 1999). Since O^+ ions are rare in the solar wind, they most likely originate in the magnetosphere. The equipotential lines shown in Figure 9.5 show that the drift trajectories are open in the outer magnetopause and some of them may intersect the magnetopause. The fate of the ions intersecting the magnetopause is unknown, but the observation shows an evidence that some of them can escape from the magnetosphere.

The ions originating in the nightside magnetosphere pass through the inner magnetosphere and contribute to the enhancement of the ring current. Then, they will move toward the dayside magnetopause when the drift path is open. Simulation results show that such ions on the drift path have a major contribution to the storm-time ring current (Liemohn et al. 2001). If there are no source and sink in the inner magnetosphere, the difference between the number of incoming particles and the number of outgoing particles will be equal to the rate of the change in the number of particles (change in the ring current). When the number of incoming particles becomes equal to the number of outgoing particles, the ring current is saturated. The excess of the outgoing particles may occur when the number density of the incoming particle decreases. The sudden decrease in the number density would result in rapid decay of the ring current (Kozyra et al. 1998b; Ebihara and Ejiri 2000). The excess of the outgoing particles may also occur when the standoff distance of the dayside magnetopause decreases. Statistical studies show that the outgoing O^+ ions is correlated with the solar wind dynamic pressure that compresses the magnetosphere (Keika et al. 2005).

9.6 IMPACT OF THE RING CURRENT

9.6.1 Impact on the Radiation Belt

The flux of relativistic particles in the radiation belt tends to be correlated with the D_{st} index. This is called a ring current effect (McIlwain 1966).The ring current has a sufficient energy to inflate the magnetic field in the inner magnetosphere (Akasofu and Chapman 1961; Hoffman and Cahill 1968; Berko et al. 1975). The decrease in the equatorial magnetic field has some consequences on the trapped particles. During the storm main phase, the particles move outward, and the energy of the particles decreases due to the induction electric field. This process can be explained in terms of the conservation of the first and third adiabatic invariants, and be used to explain the decrease in the relativistic particle flux during the storm time. To reduce the influence of the ring current effect, the corrected L-value (L^*) is used to specify the distribution of relativistic particles, which is given by

$$L^* = -\frac{2\pi k_0}{\Phi a} \quad (9.13)$$

where:

k_0 is the dipole moment of the Earth
a is the radius of the Earth
ϕ is the magnetic flux inside the loop as

$$\Phi = \oint \mathbf{B} \cdot d\mathbf{S} \quad (9.14)$$

The temperature anisotropy of the ions provide a free energy to excite and grow the EMIC waves. The EMIC waves can be resonant with relativistic electrons, scattering them into the loss cone (Thorne and Kennel 1971; Summers 2003). When the EMIC rising tone emission is present, the pitch angle of relativistic electrons is changed rapidly from a high value to a low value, falling into the loss cone (Omura and Zhao 2013).

9.6.2 Impact on the Ring Current

The ring current has also an impact on the ions constituting the ring current itself. The decrease in the equatorial magnetic field is known to reduce the flux of the ions with energy >200 keV near the equatorial pitch angle α_0 ~ 90° (Lyons 1977). This is understood to adiabatic deceleration. The pitch angle distribution of the ions become butterfly-like, in which the flux has a minimum near α_0 ~ 90°. The ions with small α_0 may be accelerated by the conservation of the first two adiabatic invariants. The Polar satellite observation shows that the ion flux with α_0 ~ 90° decreases, whereas the ion flux with α_0 ~ 0° and ~180° increases (Ebihara et al. 2008). There is a tendency that the ions with energy >100 keV with equatorial pitch angle of 90° decrease during the storm main phase and increase during the storm recovery phase. The increase in the H$^+$ flux with energy >100 keV occurs on the dayside, and the flux frequently exceeds the pre-storm level (Temporin and Ebihara 2011). Thus, the process may not be fully adiabatic, and some other nonadiabatic processes, for example, wave–particle interaction may participate in the transport and acceleration processes of the ions with energy >100 keV.

The ring current is the current flowing perpendicular to the magnetic field line. To satisfy the current continuity, the ring current is suggested to generate the FACs. Some of the FACs can be connected to the ionosphere, redistributing the electric potential in the ionosphere. The electric field is then transferred to the magnetosphere. Simulation results show that the redistributed electric field transports magnetospheric ions toward dawn, which is different from the expectation based on the traditional pattern of equipotential (Figure 9.5). This result is consistent with the global ENA images showing a peak flux of keV ions on the dawnside (Fok et al. 2003; Ebihara and Fok 2004).

9.6.3 Impact on the Plasmasphere

The redistributed electric field is also suggested to influence the plasmasphere and form an indentation of the plasmapause (Goldstein 2003; Fok et al. 2005). Interacting with the cold plasma in the plasmasphere, the kinetic energy of the ions constituting the ring current are transferred to the cold plasma, and propagating along a field line toward the ionosphere. SAR arcs then glows (Cole 1965).

9.6.4 Impact on the Ionosphere

The ionospheric electric potential can be redistributed by the region 2 FAC that is probably connected to the ring current. On average, the FAC flows into the ionosphere on the duskside, and out on the dawnside in the region 2 current system (Zmuda and Armstrong 1974; Iijima and Potemra 1976). The current that flows in the opposite direction is called region 1 current system, and is probably connected to the high-latitude mantle region (Siscoe et al. 2000; Tanaka 2000; Tanaka et al. 2010). On the duskside, the region 1 current flows into the ionosphere at high latitude, and the region 2 current flows out of the ionosphere at low latitude. When the FACs flow in a sheet elongated toward the east and west directions, a poleward electric field is generated to flow the current in the slab sandwiched by the FACs. The current flows in the direction of the electric field, and is called the Pedersen current. The upward region 1 current coincides with a part of the auroral oval where electrons are precipitating into the atmosphere, so that the ionospheric conductivity is high. However, the ionospheric conductivity is in general small under the downward region 2 current. Thus, there appears the poleward gradient of the ionospheric conductivity. The poleward electric field is large in the equatorward part of it. The poleward electric field with large intensity gives rise to fast ionospheric plasma flow pointing westward. The fast, westward flow may correspond to subauroral ion drift (SAID) (Fejer et al. 1979; Spiro et al. 1988; Anderson et al. 1993), or subauroral polarization stream (SAPS) (Foster and Vo 2002). The speed of the ionospheric plasma exceeds 1 km/s. The strong electric field is suggested to give rise to decrease in the ionospheric density and to form the trough, and to have a large impact on the mid-latitude ionosphere (Rodger et al. 1992). The magnetic latitude of the SAPS channel decreases with decreasing D_{st} index (Huang and Foster 2007). This may suggest that the SAPS is related to the region current because the inner edge of the newly injected ions moves earthward for large magnetic storms. The SAPS is intensified after a substorm (Anderson et al. 2001), which may be an indicative that the ions are injected into the inner magnetosphere, and the region 2 FAC is temporally enhanced. A radar observation shows multiple intensifications of the SAPS, which may be attributed to complicated structure of the ring current (Ebihara et al. 2009).

9.6.5 Impact on Aurora

When the trapped protons enter the loss cone, they fall into the atmosphere along a field line. As they proceed toward the thermosphere, the density of neutral atoms and molecules increases. The proton may experience electron capture, neutral excitation, and electron loss processes. Finally, they become an H atom. Since the H atom is in an excited state, it radiates photon, known as Lyman, Balmer, or other series. These emissions are called proton aurora, and are observable. In the main auroral oval, the H+ ions are known to precipitate into the atmosphere together with electrons, in particular, on the duskside. Usually, the energy input of precipitating ions into the atmosphere is about one order of magnitude smaller than that of precipitating electrons (Hardy et al. 1989). The energy input of ions becomes significant near the equatorward edge of the auroral oval (Senior et al. 1987; Galand and Richmond 2001). That is because the equatorward boundary of the proton precipitation is a few degrees displaced equatorward in comparison with that of the electron precipitation. The displacement between the proton and electron precipitation boundary may be attributed to the different drift trajectories of magnetospheric ions and electrons.

An isolated spot of the proton aurora equatorward of the main auroral oval is thought to be caused by precipitating protons (Ono et al. 1987; Sakaguchi et al. 2007; Yahnin et al. 2007). Some of them are observed with a burst of P_c 1 magnetic pulsations. The good coincidence between the proton auroral spot and the burst of P_c 1 pulsation may suggest that the H+ ions are scattered by the EMIC waves that excited near equatorial plane.

The ENA emitted from the ring current may strike the atmosphere and cause aurora. Anomalous auroral emissions are observed in the nightside thermosphere at low latitudes during a large magnetic storm (Zhang et al. 2006). The intensity of the aurora is anti-correlated with the D_{st} index, so that the aurora is probably caused by precipitating neutral hydrogen emitted from the ring current.

9.7 IMPACT ON SPACE WEATHER

The ring current may have large influences in view of space weather. One of the striking impacts of the ring current would appear in geomagnetically induced current (GIC) on power grid system. Quasi-DC currents with frequency <1 Hz are known to cause saturation of a magnetic core in a transformer. The saturation of the magnetic core results in amplified AC current, and the damage transformers connected to the grid. Large amplitude of GIC is a concern for power grid at high latitude where auroral electrojet flows in the ionosphere. Recently, large amplitude of GIC was observed at magnetically low latitude during magnetic storms. The GIC observed in central Japan increases with decreasing D_{st} index (Kappenman 2004). This may mean that the ring current generates large GIC at magnetically low latitude during large magnetic storms.

The electric potential of a spacecraft is floating against the ambient potential of plasma. When the spacecraft potential is different from the ambient potential, this situation is called spacecraft charging. The spacecraft charging damages the electronic devices onboard the satellite. At geosynchronous orbit, many communication satellites are in operation, and they are exposed to hot electrons that constitute the ring current and/or the plasma sheet. Anomalies of a satellite are frequently recorded in the post-midnight sector (Rosen 1976), indicating that newly injected electrons cause the charging as they drift eastward. During magnetic storms, the plasmasphere is shrunk, and many geosynchronous satellites are exposed to hot electrons. At geosynchronous orbit, the spacecraft potential becomes significantly large during CIR-driven storms than during CME-driven storms (Denton et al. 2006). Detailed explanation on the spacecraft charging is found in Chapter 19.

During large magnetic storms, large scintillations are observed in the signal from global positioning system in association with a mid-latitude trough and SAPS (Basu et al. 2001; Ledvina et al. 2002). Probably, the scintillations are caused by small-scale irregularities that are excited in the trough and/or SAPS (Keskinen et al. 2004). Although the causality between the mid-latitude trough and SAPS is not well identified, there may be a possibility that the ring current may intensify the region 2 FAC intensifying the SAPS. SAPS may result in the reduction of the ionospheric density, causing a trough. If this is the case, the ring current may also affect the navigation system at mid-latitude.

REFERENCES

Aggson, T. L., J. P. Heppner, and N. C. Maynard. 1983. Observations of Large Magnetospheric Electric Fields during the Onset Phase of a Substorm. *Journal of Geophysical Research* 88(A5): 3981. doi:10.1029/JA088iA05p03981.

Akasofu, S. I. 1968. *Polar and Magnetospheric Substorms*. Dordrecht, the Netherlands: D. Reidel Publishing.

Akasofu, S. I. and S. Chapman. 1961. The Ring Current, Geomagnetic Disturbance, and the Van Allen Radiation Belts. *Journal of Geophysical Research* 66(5): 1321–1350. doi:10.1029/JZ066i005p01321.

Akasofu, S. I. and S. Chapman. 1964. On the Asymmetric Development of Magnetic Storm Fields in Low and Middle Latitudes. *Planetary and Space Science* 12(6): 607–626. doi:10.1016/0032-0633(64)90008-X.

Anders, E. and N. Grevesse. 1989. Abundances of the Elements: Meteoritic and Solar. *Geochimica et Cosmochimica Acta* 53(1): 197–214. doi:10.1016/0016-7037(89)90286-X.

Anderson, B. J., R. E. Erlandson, and L. J. Zanetti. 1992. A Statistical Study of Pc 1–2 Magnetic Pulsations in the Equatorial Magnetosphere: 1. Equatorial Occurrence Distributions. *Journal of Geophysical Research* 97(A3): 3075. doi:10.1029/91JA02706.

Anderson, B. J. and D. C. Hamilton. 1993. Electromagnetic Ion Cyclotron Waves Stimulated by Modest Magnetospheric Compressions. *Journal of Geophysical Research* 98(A7): 11369. doi:10.1029/93JA00605.

Anderson, P. C., D. L. Carpenter, K. Tsuruda, T. Mukai, and F. J. Rich. 2001. Multisatellite Observations of Rapid Subauroral Ion Drifts (SAID). *Journal of Geophysical Research* 106(A12): 29585. doi:10.1029/2001JA000128.

Anderson, P. C., W. B. Hanson, R. A. Heelis, J. D. Craven, D. N. Baker, and L. A. Frank. 1993. A Proposed Production Model of Rapid Subauroral Ion Drifts and Their Relationship to Substorm Evolution. *Journal of Geophysical Research* 98(A4): 6069. doi:10.1029/92JA01975.

Araki, T. 2014. Historically Largest Geomagnetic Sudden Commencement (SC) since 1868. *Earth, Planets and Space* 66(1): 164. doi:10.1186/s40623-014-0164-0.

Balsiger, H., P. Eberhardt, J. Geiss, and D. T. Young. 1980. Magnetic Storm Injection of 0.9- to 16-keV/e Solar and Terrestrial Ions into the High-Altitude Magnetosphere. *Journal of Geophysical Research* 85(A4): 1645. doi:10.1029/JA085iA04p01645.

Basu, S., S. Basu, C. E. Valladares, H.-C. Yeh, S.-Y. Su, E. MacKenzie, P. J. Sultan et al. 2001. Ionospheric Effects of Major Magnetic Storms during the International Space Weather Period of September and October 1999: GPS Observations, VHF/UHF Scintillations, and in Situ Density Structures at Middle and Equatorial Latitudes. *Journal of Geophysical Research* 106(A12): 30389. doi:10.1029/2001JA001116.

Baumjohann, W., G. Paschmann, and C. A. Cattell. 1989. Average Plasma Properties in the Central Plasma Sheet. *Journal of Geophysical Research* 94: 6597. doi:10.1029/JA094iA06p06597.

Berko, F. W., L. J. Cahill, and T. A. Fritz. 1975. Protons as the Prime Contributors to Storm Time Ring Current. *Journal of Geophysical Research* 80(25): 3549–3552. doi:10.1029/JA080i025p03549.

Birn, J., M. F. Thomsen, J. E. Borovsky, G. D. Reeves, D. J. McComas, and R. D. Belian. 1997a. Characteristic Plasma Properties during Dispersionless Substorm Injections at Geosynchronous Orbit. *Journal of Geophysical Research* 102(A2): 2309. doi:10.1029/96JA02870.

Birn, J., M. F. Thomsen, J. E. Borovsky, G. D. Reeves, D. J. McComas, R. D. Belian, and M. Hesse. 1997b. Substorm Ion Injections: Geosynchronous Observations and Test Particle Orbits in Three-Dimensional Dynamic MHD Fields. *Journal of Geophysical Research* 102(A2): 2325. doi:10.1029/96JA03032.

Borovsky, J. E. and M. H. Denton. 2006. Differences between CME-Driven Storms and CIR-Driven Storms. *Journal of Geophysical Research: Space Physics* 111(7): 1–17. doi:10.1029/2005JA011447.

Borovsky, J. E., M. F. Thomsen, and D. J. McComas. 1997. The Superdense Plasma Sheet: Plasmaspheric Origin, Solar Wind Origin, or Ionospheric Origin? *Journal of Geophysical Research* 102(A10): 22089. doi:10.1029/96JA02469.

Brandt, P. C. 2002. Global IMAGE/HENA Observations of the Ring Current: Examples of Rapid Response to IMF and Ring Current-Plasmasphere Interaction. *Journal of Geophysical Research* 107(A11): 1359. doi:10.1029/2001JA000084.

Brandt, P. C., D. G. Mitchell, Y. Ebihara, B. R. Sandel, E. C. Roelof, J. L. Burch, and R. Demajistre. 2002. Global IMAGE/HENA Observations of the Ring Current-Examples of Rapid Response to IMF and Ring Current-Plasmasphere Interaction. *Journal of Geophysical Research: Space Physics* 107(A11). doi:10.1029/2001JA000084.

Brandt, P. C., Y. Zheng, T. S. Sotirelis, K. Oksavik, F. J. Rich. 2008. The linkage between the ring current and the ionosphere system. In *Geophysical Monograph*, edited by P. M. Kintner, A. J. Coster, T. Fuller-Rowell, A. J. Mannucci, M. Mendillo, and R. Heelis, vol. 181, pp. 135–143. Washington, D C: American Geophysical Union.

Carrington, R. C. 1859. Description of a Singular Appearance Seen in the Sun on September 1, 1859. *Monthly Notices of the Royal Astronomical Society* 20: 13–15.

Chapman, S. 1917. On the Times of Sudden Commencement of Magnetic Storms. *Proceedings of the Physical Society of London* 30(1):205–214.doi:10.1088/1478-7814/30/1/317.

Cole, K. D. 1965. Stable Auroral Red Arcs, Sinks for Energy of D St Main Phase. *Journal of Geophysical Research* 70(7): 1689–1706. doi:10.1029/JZ070i007p01689.

Cornwall, J. M., F. V. Coroniti, and R. M. Thorne. 1970. Turbulent Loss of Ring Current Protons. *Journal of Geophysical Research* 75(25): 4699–4709. doi:10.1029/JA075i025p04699.

Daglis, I. A. 2006. Ring Current Dynamics. *Space Science Reviews* 124: 183–202. doi:10.1007/s11214-006-9104-z.

Daglis, I. A. and W. I. Axford. 1996. Fast Ionospheric Response to Enhanced Activity in Geospace: Ion Feeding of the Inner Magnetotail. *Journal of Geophysical Research* 101: 5047–5065. doi:10.1029/95JA02592.

Daglis, I. A., G. Kasotakis, E. T. Sarris, Y. Kamide, S. Livi, B. Wilken. 1999. Variations of the ion composition during an intense magnetic storm and their consequences. *Phys. Chem. Earth* 24: 229–232. doi:10.1016/S1464-1917(98)00033-6.

De Michelis, P., I. A. Daglis, and G. Consolini. 1999. An Average Image of Proton Plasma Pressure and of Current Systems in the Equatorial Plane Derived from AMPTE/CCE-CHEM Measurements. *Journal of Geophysical Research* 104: 28615. doi:10.1029/1999JA900310.

DeForest, S. E. and C. E. McIlwain. 1971. Plasma Clouds in the Magnetosphere. *Journal of Geophysical Research* 76(16): 3587–3611. doi:10.1029/JA076i016p03587.

Denton, M. H., J. E. Borovsky, R. M. Skoug, M. F. Thomsen, B. Lavraud, M. G. Henderson, R. L. McPherron, J. C. Zhang, and M. W. Liemohn. 2006. Geomagnetic Storms Driven by ICME- and CIR-Dominated Solar Wind. *Journal of Geophysical Research: Space Physics* 111(7): 1–12. doi:10.1029/2005JA011436.

Dessler, A. J. and E. N. Parker. 1959. Hydromagnetic Theory of Geomagnetic Storms. *Journal of Geophysical Research* 64(12): 2239–2252. doi:10.1029/JZ064i012p02239.

Dungey, J. W. 1961. Interplanetary Magnetic Field and the Auroral Zones. *Physical Review Letters* 6(2): 47–48. doi:10.1103/PhysRevLett.6.47.

Ebihara, Y. 2002. Statistical Distribution of the Storm-Time Proton Ring Current: POLAR Measurements. *Geophysical Research Letters* 29(20). doi:10.1029/2002GL015430.

Ebihara, Y. and M. Ejiri. 2000. Simulation Study on Fundamental Properties of the Storm-Time Ring Current. *Journal of Geophysical Research: Space Physics* 105: 15843–15859.

Ebihara, Y., M. Ejiri. 2003. Numerical simulation of the ring current: Review. *Space Sci. Rev.* 105(1–2): 377–452. doi:10.1023/A:1023905607888.

Ebihara, Y. and M. C. Fok. 2004. Postmidnight Storm-Time Enhancement of Tens-of-keV Proton Flux. *Journal of Geophysical Research: Space Physics* 109. doi:10.1029/2004ja010523.

Ebihara, Y., M.-C. C. Fok, J. B. Blake, and J. F. Fennell. 2008. Magnetic Coupling of the Ring Current and the Radiation Belt. *Journal of Geophysical Research: Space Physics* 113: 1–10. doi:10.1029/2008ja013267.

Ebihara, Y., M.-C. Fok, T. J. Immel, P. C. Brandt. 2011. Rapid decay of storm time ring current due to pitch angle scattering in curved field line. *J. Geophys. Res.* 116: A03218. doi:10.1029/2010JA016000.

Ebihara, Y., S. Kasahara, K. Seki, Y. Miyoshi, T. A. Fritz, J. Chen, M. Grande, and T. H. Zurbuchen. 2009. Simultaneous Entry of Oxygen Ions Originating from the Sun and Earth into the Inner Magnetosphere during Magnetic Storms. *Journal of Geophysical Research: Space Physics* 114: 1–10. doi:10.1029/2009JA014120.

Ebihara, Y. and T. Tanaka. 2013. Fundamental Properties of Substorm Time Energetic Electrons in the Inner Magnetosphere. *Journal of Geophysical Research: Space Physics* 118(4): 1589–1603. doi:10.1002/jgra.50115.

Egeland, A. and W. J. Burke. 2012. The Ring Current: A Short Biography. *History of Geo- and Space Sciences* 3: 131–42. doi:10.5194/hgss-3-131-2012.

Ejiri, M., R. A. Hoffman, and Paul H. Smith. 1980. Energetic Particle Penetrations into the Inner Magnetosphere. *Journal of Geophysical Research* 85(A2): 653. doi:10.1029/JA085iA02p00653.

Fejer, B. G., C. A. Gonzales, D. T. Farley, M. C. Kelley, and R. F. Woodman. 1979. Equatorial Electric Fields during Magnetically Disturbed Conditions 1. The Effect of the Interplanetary Magnetic Field. *Journal of Geophysical Research* 84(A10): 5797. doi:10.1029/JA084iA10p05797.

Fok, M. C., Y. Ebihara, T. E. Moore, D. M. Ober, and K. A. Keller. 2005. Geospace Storm Processes Coupling the Ring Current, Radiation Belt and Plasmasphere. In *Geophysical Monograph*, edited by J. Burch, vol. 159, pp. 207–220. Washington, DC: American Geophysical Union.

Fok, M. C., J. U. Kozyra, A. F. Nagy, and T. E. Cravens. 1991. Lifetime of Ring Current Particles due to Coulomb Collisions in the Plasmasphere. *Journal of Geophysical Research* 96(A5): 7861. doi:10.1029/90JA02620.

Fok, M. C., T. E. Moore, P. C. Brandt, D. C. Delcourt, S. P. Slinker, and J. A. Fedder. 2006. Impulsive Enhancements of Oxygen Ions during Substorms. *Journal of Geophysical Research: Space Physics* 111: A10222. doi:10.1029/2006JA011839.

Fok, M.-C., T. E. Moore, J. U. Kozyra, G. C. Ho, and D. C. Hamilton. 1995. Three-Dimensional Ring Current Decay Model. *Journal of Geophysical Research* 100(A6): 9619. doi:10.1029/94JA03029.

Fok, M. C., T. E. Moore, G. R. Wilson, J. D. Perez, X. X. Zhang, P. C. Son Brandt, D. G. Mitchell et al. 2003. Global Ena Image Simulations. *Space Science Reviews* 109(1–4): 77–103. doi:10.1023/B:SPAC.0000007514.56380.fd.

Foster, J. C. and H. B. Vo. 2002. Average Characteristics and Activity Dependence of the Subauroral Polarization Stream. *Journal of Geophysical Research: Space Physics* 107: 1–10. doi:10.1029/2002JA009409.

Frank, L. A. 1967. On the Extraterrestrial Ring Current during Geomagnetic Storms. *Journal of Geophysical Research* 72(15): 3753–3767. doi:10.1029/JZ072i015p03753.

Friedel, R. H. W., A. Korth, and G. Kremser. 1996. Substorm Onsets Observed by CRRES: Determination of Energetic Particle Source Regions. *Journal of Geophysical Research* 101(A6): 13137. doi:10.1029/96JA00399.

Fritz, T. A., J. Chen, G. Siscoe. 2003. Energetic ions, large diamagnetic cavities, and Chapman-Ferraro cusp. *J. Geophys. Res.* 108(A1): 2156–2202. doi:10.1029/2002JA009476.

Galand, M. and A. D. Richmond. 2001. Ionospheric Electrical Conductances Produced by Auroral Proton Precipitation. *Journal of Geophysical Research* 106(A1): 117. doi:10.1029/1999JA002001.

Goldstein, J. 2003. Overshielding Event of 28–29 July 2000. *Geophysical Research Letters* 30(8): 1421. doi:10.1029/2002GL016644.

Grande, M., C. H. Perry, J. B. Blake, M. W. Chen, J. F. Fennell, and B. Wilken. 1996. Observations of Iron, Silicon, and Other Heavy Ions in the Geostationary Altitude Region during Late March 1991. *Journal of Geophysical Research* 101: 24707–24718. doi:10.1029/96JA00044.

Greenspan, M. E. and D. C. Hamilton. 2000. A Test of the Dessler-Parker-Sckopke Relation during Magnetic Storms. *Journal of Geophysical Research* 105(A3): 5419. doi:10.1029/1999JA000284.

Grimes, E. W., J. D. Perez, J. Goldstein, D. J. McComas, and P. Valek. 2010. Global Observations of Ring Current Dynamics during Corotating Interaction Region–Driven Geomagnetic Storms in 2008. *Journal of Geophysical Research* 115(A11): A11207. doi:10.1029/2010JA015409.

Hamilton, D. C., G. Gloeckler, F. M. Ipavich, W. Stüdemann, B. Wilken, and G. Kremser. 1988. Ring Current Development during the Great Geomagnetic Storm of February 1986. *Journal of Geophysical Research* 93(A12): 14343. doi:10.1029/JA093iA12p14343.

Hardy, D. A., M. S. Gussenhoven, and D. Brautigam. 1989. A Statistical Model of Auroral Ion Precipitation. *Journal of Geophysical Research* 94(A1): 370. doi:10.1029/JA094iA01p00370.

Hasegawa, H., M. Fujimoto, T.-D. Phan, H. Rème, A. Balogh, M. W. Dunlop, C. Hashimoto, and R. Tandokoro. 2004. Transport of Solar Wind into Earth's Magnetosphere through Rolled-Up Kelvin-Helmholtz Vortices. *Nature* 430(7001): 755–758. doi:10.1038/nature02799.

Hashimoto, K. K., T. Kikuchi, and Y. Ebihara. 2002. Response of the Magnetospheric Convection to Sudden Interplanetary Magnetic Field Changes as Deduced from the Evolution of Partial Ring Currents. *Journal of Geophysical Research: Space Physics* 107: 1–14. doi:10.1029/2001JA009228.

Hoffman, R. A. and L. J. Cahill. 1968. Ring Current Particle Distributions Derived from Ring Current Magnetic Field Measurements. *Journal of Geophysical Research* 73(21): 6711–6722. doi:10.1029/JA073i021p06711.

Huang, C.-S. and J. C. Foster. 2007. Correlation of the Subauroral Polarization Streams (SAPS) with the Dst Index during Severe Magnetic Storms. *Journal of Geophysical Research* 112(A11): A11302. doi:10.1029/2007JA012584.

Iijima, T. and T. A. Potemra. 1976. The Amplitude Distribution of Field-Aligned Currents at Northern High Latitudes Observed by Triad. *Journal of Geophysical Research* 81(13): 2165. doi:10.1029/JA081i013p02165.

Iyemori, T. and D. R. K. Rao. 1996. Decay of the Dst Field of Geomagnetic Disturbance after Substorm Onset and Its Implication to Storm-Substorm Relation. *Annales Geophysicae* 14: 608. doi:10.1007/s005850050325.

Jordanova, V. K., C. J. Farrugia, R. M. Thorne, G. V. Khazanov, G. D. Reeves, and M. F. Thomsen. 2001. Modeling Ring Current Proton Precipitation by Electromagnetic Ion Cyclotron Waves during the May 14–16, 1997, Storm. *Journal of Geophysical Research* 106(A1): 7. doi:10.1029/2000JA002008.

Jordanova, V. K., L. M. Kistler, J. U. Kozyra, G. V. Khazanov, and A. F. Nagy. 1996. Collisional Losses of Ring Current Ions. *Journal of Geophysical Research* 101(A1): 111. doi:10.1029/95JA02000.

Kappenman, J. G. 2004. Space Weather and the Vulnerability of Electric Power Grids. In *Effects of Space Weather on Technology Infrastructure*, edited by I. A. Daglis, pp. 257–99. Dordrecht, the Netherlands: Kluwer.

Kaye, S. M., E. G. Shelley, R. D. Sharp, and R. G. Johnson. 1981. Ion Composition of Zipper Events. *Journal of Geophysical Research* 86(A5): 3383. doi:10.1029/JA086iA05p03383.

Keika, K., M. Nosé, S. Ohtani, K. Takahashi, S. P. Christon, and R. W. McEntire. 2005. Outflow of Energetic Ions from the Magnetosphere and Its Contribution to the Decay of the Storm Time Ring Current. *Journal of Geophysical Research: Space Physics* 110(A9): 1–13. doi:10.1029/2004JA010970.

Kennel, C. F. and H. E. Petschek. 1966. Limit on Stably Trapped Particle Fluxes. *Journal of Geophysical Research* 71(1): 1–28. doi:10.1029/JZ071i001p00001.

Keskinen, M. J., S. Basu, and S. Basu. 2004. Midlatitude Sub-Auroral Ionospheric Small Scale Structure during a Magnetic Storm. *Geophysical Research Letters* 31(9). doi:10.1029/2003GL019368.

Kikuchi, T. 2014. Transmission Line Model for the Near-Instantaneous Transmission of the Ionospheric Electric Field and Currents to the Equator. *Journal of Geophysical Research: Space Physics* 119(2): 1131–1156. doi:10.1002/2013JA019515.

Klecker, B., E. Möbius, D. Hovestadt, M. Scholer, G. Gloeckler, and F. M. Ipavich. 1986. Discovery of Energetic Molecular Ions (NO^+ and $O2^+$) in the Storm Time Ring Current. *Geophysical Research Letters*. doi:10.1029/GL013i007p00632.

Kokubun, S. 1972. Relationship of Interplanetary Magnetic Field Structure with Development of Substorm and Storm Main Phase. *Planetary and Space Science* 20(7): 1033–1049. doi:10.1016/0032-0633(72)90214-0.

Konradi, A. 1967. Proton Events in the Magnetosphere Associated with Magnetic Bays. *Journal of Geophysical Research* 72(15): 3829–3841. doi:10.1029/JZ072i015p03829.

Korth, H. and M. F. Thomsen. 2001. Plasma Sheet Access to Geosynchronous Orbit: Generalization to Numerical Global Field Models. *Journal of Geophysical Research* 106: 29655. doi:10.1029/2000JA000373.

Kozyra, J. U., M.-C. Fok, E. R. Sanchez, D. S. Evans, D. C. Hamilton, and A. F. Nagy. 1998a. The Role of Precipitation Losses in Producing the Rapid Early Recovery Phase of the Great Magnetic Storm of February 1986. *Journal of Geophysical Research* 103(A4): 6801. doi:10.1029/97JA03330.

Kozyra, J. U., V. K. Jordanova, J. E. Borovsky, M. F. Thomsen, D. J. Knipp, D. S. Evans, D. J. McComas, and T. E. Cayton. 1998b. Effects of a High-Density Plasma Sheet on Ring Current Development during the November 2–6, 1993, Magnetic Storm. *Journal of Geophysical Research* 103: 26285. doi:10.1029/98JA01964.

Krimigis, S. M., G. Gloeckler, R. W. McEntire, T. A. Potemra, F. L. Scarf, and E. G. Shelley. 1985. Magnetic Storm of September 4, 1984: A Synthesis of Ring Current Spectra and Energy Densities Measured with AMPTE/CCE. *Geophysical Research Letters* 12(5): 329–332. doi:10.1029/GL012i005p00329.

Le, G., C. T. Russell, and K. Takahashi. 2004. Morphology of the Ring Current Derived from Magnetic Field Observations. *Annales Geophysicae* 22(4): 1267–1295. doi:10.5194/angeo-22-1267-2004.

Ledvina, B. M., J. J. Makela, and P. M. Kintner. 2002. First Observations of Intense GPS L1 Amplitude Scintillations at Midlatitude. *Geophysical Research Letters* 29(14): 29–32. doi:10.1029/2002GL014770.

Lee, D.-Y., S. Ohtani, P. C. Brandt, and L. R. Lyons. 2007. Energetic Neutral Atom Response to Solar Wind Dynamic Pressure Enhancements. *Journal of Geophysical Research* 112(A9): A09210. doi:10.1029/2007JA012399.

Lennartsson, W. 1989. Energetic (0.1- to 16-keV/ E) Magnetospheric Ion Composition at Different Levels of Solar F 10.7. *Journal of Geophysical Research* 94(A4): 3600. doi:10.1029/JA094iA04p03600.

Lennartsson, W., R. D. Sharp, E. G. Shelley, R. G. Johnson, H. Balsiger. 1981. Ion composition and energy distribution during 10 magnetic storms. *J. Geophys. Res.* 86(A6): 4628–4638. doi:10.1029/JA086iA06p04628.

Liemohn, M. W., J. U. Kozyra, M. F. Thomsen, J. L. Roeder, G. Lu, J. E. Borovsky, and T. E. Cayton. 2001. Dominant Role of the Asymmetric Ring Current in Producing the Stormtime Dst*. *Journal of Geo* 106(A6): 10883. doi:10.1029/2000JA000326.

Liemohn, M. W., J. C. Zhang, M. F. Thomsen, J. E. Borovsky, J. U. Kozyra, and R. Ilie. 2008. Plasma Properties of

Superstorms at Geosynchronous Orbit: How Different Are They? *Geophysical Research Letters* 35(6): 2–6. doi:10.1029/2007GL031717.

Liu, S., M. W. Chen, J. L. Roeder, L. R. Lyons, and M. Schulz. 2005a. Relative Contribution of Electrons to the Stormtime Total Ring Current Energy Content. *Geophysical Research Letters* 32(3): 1–5. doi:10.1029/2004GL021672.

Liu, W. L., S. Y. Fu, Q. G. Zong, Z. Y. Pu, J. Yang, and P. Ruan. 2005b. Variations of N$^+$/O$^+$ in the Ring Current during Magnetic Storms. *Geophysical Research Letters* 32(15): 2–5. doi:10.1029/2005GL023038.

Lui, A. T. Y. 2003. Inner Magnetospheric Plasma Pressure Distribution and Its Local Time Asymmetry. *Geophysical Research Letters* 30(16): 7–10. doi:10.1029/2003GL017596.

Lui, A. T. Y. and D. C. Hamilton. 1992. Radial Profiles of Quiet Time Magnetospheric Parameters. *Journal of Geophysical Research* 97(A12): 19325. doi:10.1029/92JA01539.

Lui, A. T. Y., R. W. McEntire, and S. M. Krimigis. 1987. Evolution of the Ring Current during Two Geomagnetic Storms. *Journal of Geophysical Research* 92(A7): 7459. doi:10.1029/JA092iA07p07459.

Lundin, R., L. R. Lyons, and N. Pissarenko. 1980. Observations of the Ring Current Composition at L Less than 4. *Geophysical Research Letters* 7(6): 425–428. doi:10.1029/GL007i006p00425.

Lyons, L. R. 1977. Adiabatic Evolution of Trapped Particle Pitch Angle Distributions during a Storm Main Phase. *Journal of Geophysical Research* 82(16): 2428–2432. doi:10.1029/JA082i016p02428.

Matsui, H., R. B. Torbert, H. E. Spence, Yu, V. Khotyaintsev, P.-A. Lindqvist. 2013. Revision of empirical electric field modeling in the inner magnetosphere using Cluster data. *J. Geophys. Res.: Space Phys.*, 118: 4119–4134. doi:10.1002/jgra.50373.

Mauk, B. H. and C. E. McIlwain. 1974. Correlation of Kp with the Substorm-Injected Plasma Boundary. *Journal of Geophysical Research* 79(22): 3193–3196. doi:10.1029/JA079i022p03193.

Mauk, B. H., D. G. Mitchell, R. W. McEntire, R. W. Paranicas, E. C. Roelof, D. J. Williams, and S. M. Krimigis. 2004. Energetic Ion Characteristics and Neutral Gas Interactions in Jupiter's Magnetosphere. *Journal of Geophysical Research* 109(A9): A09S12. doi:10.1029/2003JA010270.

McIlwain, C. E. 1966. Ring Current Effects on Trapped Particles. *Journal of Geophysical Research* 71(15): 3623–3628. doi:10.1029/JZ071i015p03623.

McPherron, R. L. 1997. The Role of Substorms in the Generation of Magnetic Storms. In *Magnetic Storms*, edited by B. T. Tsurutani, D. W. Gonzalez, Y. Kamide, and J. K. Arballo, pp. 131–147. Washington, DC: American Geophysical Union.

Mitchell, D. G., P. C:son Brandt, E. C. Roelof, D. C. Hamilton, K. C. Retterer, and S. Mende. 2003. Global Imaging of O + from IMAGE/HENA. *Space Science Reviews* 109(1–4): 63–75. doi:10.1023/B:SPAC.0000007513.55076.00.

Möbius, E., D. Hovestadt, B. Klecker, M. Scholer, F. M. Ipavich, C. W. Carlson, and R. P. Lin. 1986. A Burst of Energetic O + Ions during an Upstream Particle Event. *Geophysical Research Letters* 13(13): 1372–1375. doi:10.1029/GL013i013p01372.

Nishitani, N., T. Ogawa, N. Sato, H. Yamagishi, M. Pinnock, J. P. Villain, G. Sofko, and O. Troshichev. 2002. A Study of the Dusk Convection Cell's Response to an IMF Southward Turning. *Journal of Geophysical Research: Space Physics* 107(A3). doi:10.1029/2001JA900095.

O'Brien, T. P. and R. L. McPherron. 2000. Evidence against an Independent Solar Wind Density Driver of the Terrestrial Ring Current. *Geophysical Research Letters* 27(23): 3797–3799. doi:10.1029/2000GL012125.

Omura, Y., J. Pickett, B. Grison, O. Santolik, I. Dandouras, M. Engebretson, P. M. E. Décéreau, and A. Masson. 2010. Theory and Observation of Electromagnetic Ion Cyclotron Triggered Emissions in the Magnetosphere. *Journal of Geophysical Research: Space Physics* 115(7): 1–13. doi:10.1029/2010JA015300.

Omura, Y. and Q. Zhao. 2013. Relativistic Electron Microbursts due to Nonlinear Pitch Angle Scattering by EMIC Triggered Emissions. *Journal of Geophysical Research: Space Physics* 118(8): 5008–5020. doi:10.1002/jgra.50477.

Ono, T., T. Hirasawa, and C. I. Meng. 1987. Proton Auroras Observed at the Equatorward Edge of the Duskside Auroral Oval. *Geophysical Research Letters* 14(6): 660–663. doi:10.1029/GL014i006p00660.

Parker, E. N. 1957. Newtonian Development of the Dynamical Properties of Ionized Gases of Low Density. *Physical Review* 107(4): 924–933. doi:10.1103/PhysRev.107.924.

Pickett, J. S., B. Grison, Y. Omura, M. J. Engebretson, I. Dandouras, A. Masson, M. L. Adrian et al. 2010. Cluster Observations of EMIC Triggered Emissions in Association with Pc1 Waves near Earth's Plasmapause. *Geophysical Research Letters* 37(9): 1–5. doi:10.1029/2010GL042648.

Rairden, R. L., L. A. Frank, and J. D. Craven. 1986. Geocoronal Imaging with Dynamics Explorer. *Journal of Geophysical Research* 91(A12): 13613. doi:10.1029/JA091iA12p13613.

Rodger, A. S., R. J. Moffett, and S. Quegan. 1992. The Role of Ion Drift in the Formation of Ionisation Troughs in the Mid- and High-Latitude Ionosphere—A Review. *Journal of Atmospheric and Terrestrial Physics* 54(1): 1–30. doi:10.1016/0021-9169(92)90082-V.

Rosen, A. 1976. Spacecraft Charging by Magnetospheric Plasmas. *IEEE Transactions on Nuclear Science* 23(6): 1762–1768. doi:10.1109/TNS.1976.4328575.

Sakaguchi, K., K. Shiokawa, A. Ieda, Y. Miyoshi, Y. Otsuka, T. Ogawa, M. Connors, E. F. Donovan, and F. J. Rich. 2007. Simultaneous Ground and Satellite Observations of an Isolated Proton Arc at Subauroral Latitudes. *Journal of Geophysical Research* 112(A4): A04202. doi:10.1029/2006JA012135.

Sckopke, N. 1966. A General Relation between the Energy of Trapped Particles and the Disturbance Field near the Earth. *Journal of Geophysical Research* 71(13): 3125–3130. doi:10.1029/JZ071i013p03125.

Senior, C., J. R. Sharber, O. de la Beaujardière, R. A. Heelis, D. S. Evans, J. D. Winningham, M. Sugiura, and W. R. Hoegy. 1987. E and F Region Study of the Evening Sector Auroral Oval: A Chatanika/Dynamics Explorer 2/ NOAA 6 Comparison. *Journal of Geophysical Research* 92(A3): 2477. doi:10.1029/JA092iA03p02477.

Sergeev, V. A., T. Bösinger, R. D. Belian, G. D. Reeves, and T. E. Cayton. 1992. Drifting Holes in the Energetic Electron Flux at Geosynchronous Orbit Following Substorm Onset. *Journal of Geophysical Research* 97(A5): 6541. doi:10.1029/92JA00182.

Sergeev, V. A., E. M. Sazhina, N. A. Tsyganenko, J. Å. Lundblad, and F. Søraas. 1983. Pitch-Angle Scattering of Energetic Protons in the Magnetotail Current Sheet as the Dominant Source of Their Isotropic Precipitation into the Nightside Ionosphere. *Planetary and Space Science* 31(10): 1147–1155. doi:10.1016/0032-0633(83)90103-4.

Sergeev, V. A., M. A. Shukhtina, R. Rasinkangas, A. Korth, G. D. Reeves, H. J. Singer, M. F. Thomsen, and L. I. Vagina. 1998. Event Study of Deep Energetic Particle Injections during Substorm. *Journal of Geophysical Research* 103(A5): 9217. doi:10.1029/97JA03686.

Sheldon, R. B., H. E. Spence, and J. F. Fennell. 1998. Observation of the 40 keV Field-aligned Ion Beams. *Geophysical Research Letters* 25(10): 1617. doi:10.1029/98GL01054.

Shepherd, S. G. 2007. Polar Cap Potential Saturation: Observations, Theory, and Modeling. *Journal of Atmospheric and Solar-Terrestrial Physics* 69(3): 234–248. doi:10.1016/j.jastp.2006.07.022.

Shinbori, A., T. Ono, M. Iizima, and A. Kumamoto. 2004. SC Related Electric and Magnetic Field Phenomena Observed by the Akebono Satellite inside the Plasmasphere. *Earth, Planets and Space* 56(2): 269–282. doi:10.1186/BF03353409.

Shoji, M. and Y. Omura. 2012. Precipitation of Highly Energetic Protons by Helium Branch Electromagnetic Ion Cyclotron Triggered Emissions. *Journal of Geophysical Research: Space Physics* 117(12): 1–8. doi:10.1029/2012JA017933.

Singer, S. F. 1957. A New Model of Magnetic Storms and Aurorae. *Transactions, American Geophysical Union* 38(2): 175. doi:10.1029/TR038i002p00175.

Siscoe, G. L., N. U. Crooker, G. M. Erickson, B. U. Sonnerup, K. D. Siebert, D. R. Weimer, W. W. White, and N. C. Maynard. 2000. Global Geometry of Magnetospheric Currents Inferred from MHD Simulations. *Magnetospheric Current Systems* 118: 41–52. doi:10.1029/GM118p0041.

Smith, J. P., M. F. Thomsen, J. E. Borovsky, and M. Collier. 1999. Solar Wind Density as a Driver for the Ring Current in Mild Storms. *Geophysical Research Letters* 26(13): 1797–1800. doi:10.1029/1999GL900341.

Smith, P. H. and N. K. Bewtra. 1976. Dependence of the Charge Exchange Lifetimes on Mirror Latitude. *Geophysical Research Letters* 3(11): 689–692. doi:10.1029/GL003i011p00689.

Smith, P. H. and R. A. Hoffman. 1973. Ring Current Particle Distributions during the Magnetic Storms of December 16–18, 1971. *Journal of Geophysical Research* 78(22): 4731–4737. doi:10.1029/JA078i022p04731.

Smith, P. H. and R. A. Hoffman. 1974. Direct Observations in the Dusk Hours of the Characteristics of the Storm Time Ring Current Particles during the Beginning of Magnetic Storms. *Journal of Geophysical Research* 79(7): 966–971. doi:10.1029/JA079i007p00966.

Song, P. and C. T. Russell. 1992. Model of the Formation of the Low-Latitude Boundary Layer for Strongly Northward Interplanetary Magnetic Field. *Journal of Geophysical Research* 97(A2): 1411. doi:10.1029/91JA02377.

Spiro, R. W., R. A. Wolf, and B. G. Fejer. 1988. Penetration of High-Latitude-Electric-Field Effects to Low Latitudes during SUNDIAL 1984. *Annales Geophysicae*, 6, 39–49.

Stern, D. P. 1975. The Motion of a Proton in the Equatorial Magnetosphere. *Journal of Geophysical Research* 80(4). doi:10.1029/JA080i004p00595.

Stuart, G. W. 1959. Satellite-Measured Radiation. *Physical Review Letters* 2(10): 417–418. doi:10.1103/PhysRevLett.2.417.

Stüdemann, W., B. Wilken, G. Kremser, A. Korth, J. F. Fennell, B. Blake, R. Koga et al. 1987. The May 2–3, 1986 Magnetic Storm: First Energetic Ion Composition Observations with the Mics Instrument on Viking. *Geophysical Research Letters* 14(4): 455–458. doi:10.1029/GL014i004p00455.

Sugiura, M. 1964. Hourly values of equatorial Dst for IGY. *Ann. Int. Geophys. Year*, 35: 945–948, Oxford: Pergamon Press.

Summers, D. 2003. Relativistic Electron Pitch-Angle Scattering by Electromagnetic Ion Cyclotron Waves during Geomagnetic Storms. *Journal of Geophysical Research* 108(A4): 1143. doi:10.1029/2002JA009489.

Tanaka, T. 2000. Field-Aligned-Current Systems in the Numerically Simulated Magnetosphere. *Magnetospheric Current Systems, Geophysical Monograph Series 118*: 53–59.

Tanaka, T. 2007. *Magnetosphere-Ionosphere Convection as a Compound System. Space Science Reviews.* 133. doi:10.1007/s11214-007-9168-4.

Tanaka, T., A. Nakamizo, A. Yoshikawa, S. Fujita, H. Shinagawa, H. Shimazu, T. Kikuchi, and K. K. Hashimoto. 2010. Substorm Convection and Current System Deduced from the Global Simulation. *Journal of Geophysical Research* 115: 1–26. doi:10.1029/2009JA014676.

Temporin, A. and Y. Ebihara. 2011. Energy-Dependent Evolution of Ring Current Protons during Magnetic Storms. *Journal of Geophysical Research* 116(A10): A10201. doi:10.1029/2011JA016692.

Terasawa, T., M. Fujimoto, T. Mukai, I. Shinohara, Y. Saito, T. Yamamoto, S. Machida et al. 1997. Solar Wind Control of Density and Temperature in the near-Earth Plasma Sheet: WIND/GEOTAIL Collaboration. *Geophysical Research Letters* 24(8): 935–938. doi:10.1029/96GL04018.

Thomsen, M., J. Borovsky, D. J. McComas, and M. R. Collier. 1998. Variability of the Ring Current Source Population. *Geophysical Research Letters* 25(18): 3481. http://adsabs.harvard.edu/cgi-bin/nph-data_query?bibcode=1998GeoRL..25.3481T&link_type=ABSTRACT\npapers://7cc7c90c-7160-4bff-82fe-6e5baf637e14/Paper/p1263.

Thorne, R. M. and C. F. Kennel. 1971. Relativistic Electron Precipitation during Magnetic Storm Main Phase. *Journal of Geophysical Research* 76(19): 4446–4453. doi:10.1029/JA076i019p04446.

Tsurutani, B. T. and W. D. Gonzalez. 1987. The Cause of High-Intensity Long-Duration Continuous AE Activity (HILDCAAs): Interplanetary Alfvén Wave Trains. *Planetary and Space Science* 35(4): 405–412. doi:10.1016/0032-0633(87)90097-3.

Tsurutani, B. T., W. D. Gonzalez, G. S. Lakhina, and S. Alex. 2003. The Extreme Magnetic Storm of 1–2 September 1859. *Journal of Geophysical Research* 108(A7): 1268. doi:10.1029/2002JA009504.

Turner, N. E., D. N. Baker, T. I. Pulkkinen, J. L. Roeder, J. F. Fennell, and V. K. Jordanova. 2001. Energy Content in the Storm Time Ring Current. *Journal of Geophysical Research.* doi:10.1029/2000JA003025.

Valek, P., P. C. Brandt, N. Buzulukova, M.-C. Fok, J. Goldstein, D. J. McComas, J. D. Perez, E. Roelof, and R. Skoug. 2010. Evolution of Low-Altitude and Ring Current ENA Emissions from a Moderate Magnetospheric Storm: Continuous and Simultaneous TWINS Observations. *Journal of Geophysical Research* 115(A11): A11209. doi:10.1029/2010JA015429.

Volland, H. 1973. A Semiempirical Model of Large-Scale Magnetospheric Electric Fields. *Journal of Geophysical Research* 78(1): 171. doi:10.1029/JA078i001p00171.

Weimer, D. R. 2001. An Improved Model of Ionospheric Electric Potentials Including Substorm Perturbations and Application to the Geospace Environment Modeling November 24, 1996, Event. *Journal of Geophysical Research* 106(2000): 407. doi:10.1029/2000JA000604.

Wygant, J., F. Mozer, M. Temerin, J. Blake, N. Maynard, H. Singer, and M. Smiddy. 1994. Large Amplitude Electric and Magnetic Field Signatures in the Inner Magnetosphere during Injection of 15 MeV Electron Drift Echoes. *Geophysical Research Letters* 21(16): 1739–1742. doi:10.1029/94GL00375.

Yahnin, A. G., T. A. Yahnina, and H. U. Frey. 2007. Subauroral Proton Spots Visualize the Pc1 Source. *Journal of Geophysical Research* 112(A10): A10223. doi:10.1029/2007JA012501.

Yahnina, T. A., A. G. Yahnin, J. Kangas, J. Manninen, D. S. Evans, A. G. Demekhov, V. Y. Trakhtengerts, M. F. Thomsen, G. D. Reeves, and B. B. Gvozdevsky. 2003. Evnergetic Particle Counterparts for Geomagnetic Pulsations of Pc1 and IPDP Types. *Annales Geophysicae* 21(12): 2281–2292. doi:10.5194/angeo-21-2281-2003.

Yau, A. W., P. H. Beckwith, W. K. Peterson, and E. G. Shelley. 1985a. Long-Term (Solar Cycle) and Seasonal Variations of Upflowing Ionospheric Ion Events at DE 1 Altitudes. *Journal of Geophysical Research* 90(A7): 6395. doi:10.1029/JA090iA07p06395.

Yau, A. W., E. G. Shelley, W. K. Peterson, and L. Lenchyshyn. 1985b. Energetic Auroral and Polar Ion Outflow at DE 1 Altitudes: Magnitude, Composition, Magnetic Activity Dependence, and Long-Term Variations. *Journal of Geophysical Research* 90(A9): 8417. doi:10.1029/JA090iA09p08417.

Young, D. T., H. Balsiger, and J. Geiss. 1982. Correlations of Magnetospheric Ion Composition with Geomagnetic and Solar Activity. *Journal of Geophysical Research* 87(A11): 9077. doi:10.1029/JA087iA11p09077.

Young, S. L., R. E. Denton, B. J. Anderson, and M. K. Hudson. 2002. Empirical Model for μ Scattering Caused by Field Line Curvature in a Realistic Magnetosphere. *Journal of Geophysical Research: Space Physics* 107(A6). doi:10.1029/2000JA000294.

Zhang, Y., L. J. Paxton, J. U. Kozyra, H. Kil, and P. C. Brandt. 2006. Nightside Thermospheric FUV Emissions due to Energetic Neutral Atom Precipitation during Magnetic Superstorms. *Journal of Geophysical Research* 111(A9): A09307. doi:10.1029/2005JA011152.

Zmuda, A. J. and J. C. Armstrong. 1974. The Diurnal Flow Pattern of Field-Aligned Currents. *Journal of Geophysical Research* 79(31): 4611. doi:10.1029/JA079i031p04611.

Zong, Q. G. and B. Wilken. 1999. Bursty Energetic Oxygen Events in the Dayside Magnetosheath: GEOTAIL Observations. *Geophysical Research Letters* 26(22): 3349–3352. doi:10.1029/1999GL003634.

CHAPTER 10

Radiation Belts

Shrikanth G. Kanekal and Daniel N. Baker

CONTENTS

10.1 INTRODUCTION: DISCOVERY AND HISTORICAL PERSPECTIVE

In 1912, Victor Hess discovered cosmic rays by measuring the response of an electroscope at different altitudes. Since his measurements showed increased rate of ionization with altitude, he correctly inferred that this radiation was coming from space. The discovery was followed by research into their nature, for example, composition and energy spectra. In 1952, under James Van Allen's leadership a group of researchers started investigating the lower energy end of cosmic radiation and detected auroral soft electrons using Geiger tubes flown on rockets (Van Allen 1957). Subsequently in 1958, again using Geiger counters on Explorer series of rockets, Van Allen and his group established the existence of trapped natural radiation above the Earth; as Van Allen notes (Van Allen 1983), "...I further concluded that the causative particles were present in trapped orbits in the geomagnetic field, moving in spiral paths back and forth between the Northern and Southern Hemispheres and drifting slowly around the Earth. The intensity of such trapped particles would be diminished at low altitudes by the cumulative effect of atmospheric absorption and scattering." The Geiger counters onboard the spacecraft saturated resulting in null counts, which were correctly interpreted as due to the presence of large number of trapped particles (Van Allen et al. 1958). Since the discovery, the study of radiation belts has developed greatly with observations of many complex physical processes leading to the formation and decay of the radiation or Van Allen belts. The field continues to produce exciting and new discoveries with many dedicated satellite missions including the most recent twin-spacecraft mission launched in 2012 by NASA and named in honor of the discoverer of radiation belts, the Van Allen Probes.

10.2 STRUCTURE AND MORPHOLOGY

The terrestrial radiation belts surround the Earth as two doughnut shaped rings separated by a region with few energetic charged particles. The outer ring comprises mainly electrons ranging in energies from a few hundred keV to multi MeV, while the inner ring comprises mainly protons of energies ranging from a few MeV to several hundreds of MeV. The region between the two rings is called the "slot" and contains very few energetic particles, although the regions still contain substantial numbers of lower energy electrons. The reason for the paucity of energetic electrons is due to the fact they are precipitated into the atmosphere by interaction with plasma whistler mode waves (Lyons et al. 1972). The Van Allen belts are of course within the magnetosphere, which is formed due to interaction of the solar wind with the Earth's magnetic field and is the region downstream of the bow shock where the supersonic solar wind slows down. The terrestrial magnetosphere is a long tear-shaped structure, that is, with a magnetotail, and contains, apart from the radiation belts, other morphological features such as the plasmasphere, various boundary layers, and the plasma sheet. These contain lower energy particles and plasmas which are involved in the dynamics of the radiation belts (see Section 10.3). While the two belt structure is the "usual and normal" condition, recent discoveries by the Van Allen Probes mission have shown a more complex structure, which will be discussed in the later sections.

Figure 10.1 shows a schematic of the magnetosphere and the radiation belts together with other morphological features such as the bow shock, magnetopause, plasmasphere, and the plasma sheet.

10.3 RADIATION BELT PARTICLE POPULATIONS

As mentioned before, the inner belt comprises mostly protons and the outer belt contains chiefly electrons.

The source of the inner belt protons are both solar and cosmic in origin; the so-called cosmic ray albedo neutron decay (CRAND) and trapped solar energetic particles (SEPs). The CRAND population results from the decay of neutrons produced by the interaction of primary cosmic ray protons with the Earth's atmosphere. The neutrons being unstable decay into protons, electrons, and neutrinos which escape the magnetosphere. Energetic protons from the neutron decay however, can get trapped and form a stable population that varies only slowly on timescales of years. Solar events such as coronal mass ejections (CMEs) and strong interplanetary (IP) shocks can accelerate protons which enter the magnetosphere via the open field lines over the polar regions and can get trapped.

The outer radiation belt is much more dynamic and electron fluxes often wax and wane on many timescales ranging from minutes to years. The energetic electron population results from the multiplicity of energization processes within the magnetosphere acting on a "seed" population injected into the radiation belt region, that is, inner magnetosphere from the

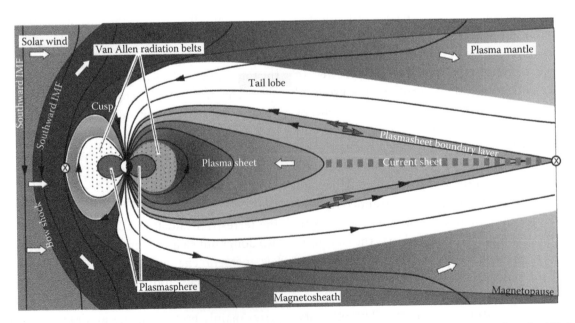

FIGURE 10.1 A schematic illustrating some of the prominent morphological features of the magnetosphere highlighting the radiation belts, bow shock, magnetopause, plasmasphere, and the plasma sheet.

distant magnetotail. The low energy electron population is of ionospheric and solar origin.

In addition to trapped electrons and protons, there are a much lesser populations of various ions either of terrestrial ionospheric or solar origin which may be transient or trapped for longer durations. SEPs, of which protons are the most abundant, can enter the magnetosphere on open field lines and may be trapped for short durations, although occasionally a large solar disturbance (e.g., strong earthward directed CME) may cause the SEP to become trapped. Other particles sources include the anomalous cosmic rays which have been observed by low Earth orbiting spacecraft, such as the Solar Anomalous Magnetospheric Explorer (SAMPEX) (Baker et al. 1993) to become trapped and form a belt; however, these particle fluxes are much lower than either the electrons in the outer belt or the protons in the inner belt and do not contribute significantly to the particle dynamics.

10.4 CHARGED PARTICLE MOTION AND INVARIANTS

The Earth's magnetic field is mostly dipolar within a few Earth radii (R_E), although a multipole representation has to be used for accurate description of the geomagnetic field. To keep matters simple, we can use the dipole approximation and describe the motion of radiation belt particles with a view to gain a physical understanding into their dynamics. Within this approximately dipolar field, charged particle motion can be conveniently described as a superposition of gyromotion, bounce motion, and drift around the Earth. Charged particles gyrate around the field lines and bounce along the meridional direction and drift azimuthally around the Earth. The associated timescales of these motions are approximately milliseconds, seconds, and tens of minutes, respectively. Associated with each motion are adiabatic invariants, which are only truly invariant, if the magnetic field changes on timescales comparable to those of the particle motion itself. Full details of the adiabatic invariants and the motion of charged particles in the radiation belts can be found in Shulz and Lanzerotti (1974) and Roderer (1970). Here, we merely define the three invariants of the motion and note that the dynamics of the particle populations, namely their energization, and loss results from one of the invariants to be broken. For example, breaking the first invariant leads to diffusion in pitch angle scattering, whereas radial

transport conserving the first invariant leads to diffusion in energy, that is, particle energization if the transport is inward from a region of weak geomagnetic field to a region of strong field and vice versa.

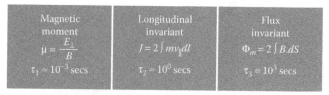

Magnetic moment	Longitudinal invariant	Flux invariant
$\mu = \dfrac{E_\perp}{B}$	$J = 2 \int m v_\parallel dl$	$\Phi_m = 2 \int B.dS$
$\tau_1 \approx 10^{-3}$ secs	$\tau_2 \approx 10^{0}$ secs	$\tau_3 \approx 10^{3}$ secs

(1 MeV electron 6 Re 60° pitch angle)

10.5 PHYSICAL PROCESSES AND DYNAMICS OF RADIATION BELT PARTICLES

While the inner belt comprising mainly of energetic protons is relatively stable, the outer belt comprising chiefly of energetic electrons is highly dynamic and shows variability over multiple timescales. Energetic electron fluxes can change over several orders of magnitude in as short a timescale as minutes; they also are variable over other characteristic timescales ranging from hours, days, seasons, and years (for a review see Kanekal 2006, Millan and Baker 2012). This variability suggests that a rich and varied physics underlies electron dynamics in the outer belt. Figure 10.2 shows long-term measurements made by SAMPEX and Polar satellites, which illustrate the changing electron populations on timescales of days. The figure shows color-coded electron fluxes as a function of the McIlwain L parameter (cf. McIlwain 1961) or in a roughly equivalent sense the radial distance from the Earth (the distance is measured in units of Earth radii, $R_E \sim 6372$ km).

From Figure 10.2 several important characteristics of electron energization can be inferred. We note that during the years 1992–1994, which was the declining phase of the solar cycle, recurrent energization events occurred as a result of interaction with a high-speed solar wind stream (HSS), as they are prevalent during such times; later during the ascending part of the solar cycle, energization events are more sporadic as they are driven in this period by CMEs (Kanekal et al. 2003). Note also that after 1995 when Polar was launched and collecting data from high altitude regions, a comparison of SAMPEX and Polar data show that these fluxes tend to track one another indicating a coherent behavior of electron dynamics (Kanekal et al. 2001).

Current understanding suggests that electron energization processes are of two broad classes, namely, radial

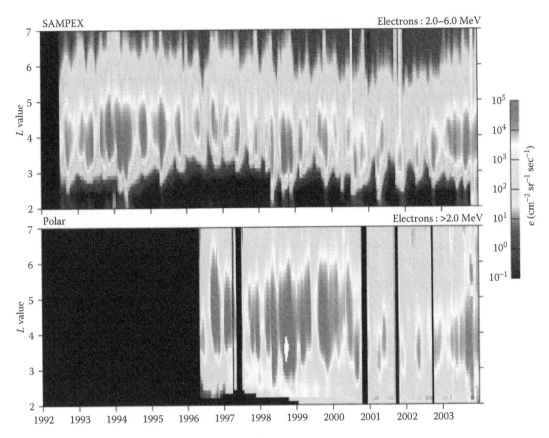

FIGURE 10.2   SAMPEX and Polar measurements of energetic electrons. The top panel shows electron flux measured by the PET sensor onboard SAMPEX and the bottom panel shows electron flux measured by the HIST instrument onboard Polar spacecraft, which is at a much higher altitude than SAMPEX, which is in a low Earth orbit at about 600 km.

transport and local energization via wave–particle inter-action (for reviews, see Friedel et al. 2002; Kanekal 2006; Hudson et al. 2008; Reeves 2015). These processes act on a "seed" population of ~100s of keV electrons that is injected by substorms from the distant tail to the inner magneto-sphere (Baker et al. 1998). A variety of electromagnetic waves occur in the magnetosphere which either by reso-nant or stochastic interactions can energize electrons either by affecting radial transport or *in situ*. For example, ULF waves can lead to strengthened radial diffusion (Elkington 2006), while chorus waves can locally energize electrons (Summers et al. 1998, Horne et al. 2005, Reeves et al. 2013).

The radiation belts respond to IP shocks, which also often energize electrons. This response to IP shocks is in a very different manner from the above two processes which typically have timescales of a few to several days for the energized fluxes to reach their maximum or peak values. IP shocks, on the other hand, result in a more dramatic and very rapid energization of electrons (Blake et al. 1992, Kanekal et al. 2016), driven by electric fields resulting from the rapid shock compression of the

magnetosphere (Li et al. 1993). Discriminating the dom-inant acceleration process in any given event is difficult and remains an open question. For example, Kanekal et al. (2015) have recently shown that both radial diffu-sion and local energization can occur concurrently in the same energization episode.

While the energization processes increase the number of energetic electrons, loss processes will decrease their number. Thus, the net flux in the outer zone is a result of the balance between these two processes. The importance of taking loss processes into account is shown in Figure 10.3, which shows the evolution of energetic electron fluxes during three geomagnetic storms of similar strengths, as measured by the minimum value of the D_{st} index.

The pre- and post-storm electron flux levels in these three events increase, decrease, and show little change, respectively (Reeves et al. 2003). This implies clearly that during the first event, energization processes were dominant, during the second event loss processes were dominant, and in the third event, the two processes were of comparable strength. These loss processes are

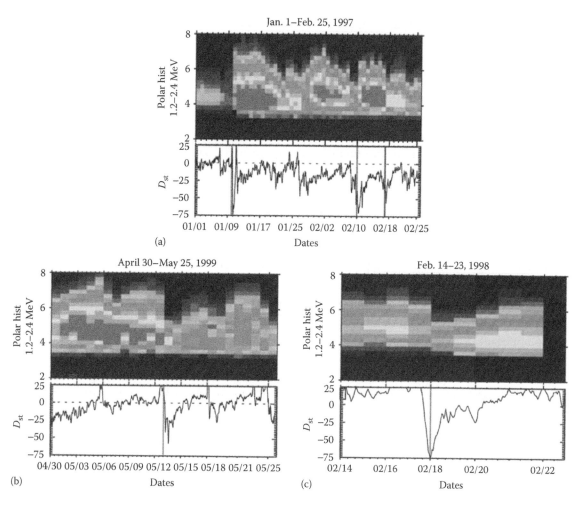

FIGURE 10.3 Outer zone electron response during geomagnetic storms showing the three types of responses. (a) The January 1997 geomagnetic storm with increased fluxes of relativistic electron fluxes. (b) The May 1999 storm displaying a clear loss of electrons throughout the outer belt. (c) The February 1998 storm of comparable strength to (a) and (b) with unchanged pre- and post-storm electron fluxes. (From Reeves, G.D. et al., *Geophys. Res. Lett.*, 30, 1529, 2003.)

due either to *in situ* pitch angle scattering into the loss cone (Thorne 2010) or escape to IP space through the magnetopause (Turner et al. 2014). The former may be of a steady persistent kind or a more rapid short-lived phenomenon termed microbursts.

10.6 PARTICLE MEASUREMENTS AND INSTRUMENTATION

Charged particle instruments essentially measure particle count rates, their energies, and angular direction of approach. These measurements are converted into physical quantities, that is, fluxes and angular distributions with rest to the local magnetic field. Particle spectra and pitch angle distributions thus obtained can then be used to analyze and study the radiation belt responses to IP drivers. In this section, we limit ourselves to briefly describing two types of energetic

charged particle instruments which measure electrons, protons, and ions in the few hundred keV to tens of MeV range, namely, solid-state detector (SSD)-based particle telescopes and particle spectrometers that employ magnetic fields to bend particles. Specifically, we will describe the REPT (relativistic electron and proton telescope) (Baker et al. 2012) and the MagEIS (magnetic electrons and ion spectrometer) (Blake et al. 2013). We note that these types of instrumentation have a long history in space physics, not only in magnetospheric but also in heliospheric physics as well.

The REPT instrument comprises a stack of SSDs surrounded by thick double-layered shielding, an inner tungsten layer to stop bremsstrahlung photons and an outer aluminum layer to stop energetic electrons and protons. A thin beryllium window in the front of the stack stops the lower energy electrons (<~1 MeV). Fast front-end electronics combined with coincidence requirements ensure high-fidelity operation during high-flux events. The instrument design was optimized using the Geant4 particle interaction tool (Agostinelli et al. 2003). During the mission lifetime of Van Allen Probes, the REPT instrument has performed admirably resulting in several important discoveries and increased understanding of electron dynamics of the radiation belts; a full description of the REPT instrument may be found in Baker et al. (2012).

The MagEIS instrument is a state-of-the-art particle spectrometer and uses the knowledge that charged particles bend within a magnetic field with a radius determined by the particle energy and the field strength. The MagEIS instrument also uses a SSD stack at the focal plane, which enable a full pulse-height measurement and thus provides a double measurement of the energy of the incident electron. This method greatly reduces background contamination. There are four MagEIS units which cover an extended energy range; a low-energy unit which measures electrons of 20–240 keV, two medium-energy units covering energies from 80 to 1200 keV, and a high-energy unit which covers 800–4800 keV electron. In addition, the high unit also has a proton telescope, which measures protons from 55 keV to 20 MeV. Front-end electronics include a custom multichip module which amplifies and digitizes the charge pulse from the SSD stacks. The overlapping energy ranges of MagEIS and the REPT instruments enables the measurement of electron spectra from 10s of keV to multi MeV in a well-calibrated manner. Thus, a complete understanding of the electron dynamics can be obtained, since the seed as well as the energized population spectra are measured. Furthermore, both instruments measure particle pitch angle distributions as well, which together with the knowledge of the global geomagnetic field enables the determination of the electron phase space density (PSD). MagEIS has also performed extremely well and made significant contributions to electron dynamics, such as the first direct observation of wave–particle interactions (Fennell et al. 2014). The MagEIS instrument is described in detail by Blake et al. (2013).

10.7 OVERVIEW OF RECENT OBSERVATIONAL ASPECTS

In the recent past, several spacecraft missions dedicated to the understanding of radiation belt particle dynamics have been launched, notable among them the Van Allen Probes mission launched by NASA. Since the most recent advances in our understanding has come from the observations made by the Van Allen Probes, this section will highlight results from that mission.

Several other missions, however, must be mentioned in regard to radiation belts, in particular the SAMPEX (Baker et al. 1993), which was the first of NASA's small explorer missions. With a prime mission of 1 year and designed to last at most 3 years, SAMPEX from a high inclination low-Earth orbit (LEO) provided valuable measurements for nearly two decades from its launch in July 1992. Prior to SAMPEX, NASA's CRRES, and overlapping with SAMPEX, and launched in 1995, the Polar satellites also made significant contributions to the study of radiation belts, particularly to the dynamics of energetic electrons in the outer belt. CRRES observation of rapid energization of electrons to ultrarelativistic energies and penetration deep into the magnetosphere during the famous March 1991 event (Blake et al. 1992) was very significant though serendipitous. The demonstration using both SAMPEX and Polar observations of the "remarkable global coherence" of electron energization (Kanekal et al. 2001, 2005) established an important characteristic of electron energization, namely, that energization and flux isotropization by pitch angle scattering were most likely to occur simultaneously. The long duration high-quality measurements of energetic electrons enabled investigation into the dependence of electron energization on different solar wind drivers such as high-speed streams which are prevalent during the descending part of the solar cycle and CMEs which are more common during the ascending part of the solar cycle (Kanekal 2006). The seasonal dependence of electron fluxes due perhaps to the so-called Russell–McPherron effect (Russell and McPherron 1973) was measured using data that covered multiple years (Kanekal et al. 2010). The clear dependence of electron energization upon the southward turning of the IP magnetic field (IMF) was also demonstrated (Blake et al. 1997). These and many other studies while establishing many characteristics of electron dynamics also revealed the incomplete nature of our understanding of the many physical processes occurring in the radiation belts. The need for a dedicated mission, preferably a multipoint one, was clear and lead to the concept and eventual realization of the Van Allen Probes mission.

The Van Allen Probes (formerly called Radiation Belt Storm Probes, or RBSP) was launched on October 30, 2012, and is a twin-spacecraft mission with comprehensive and state-of–the-art instrumentation to study the Earth's radiation belts (Mauk et al. 2012). The spacecraft carried identical suites of instrumentation that measure electric and magnetic fields and waves, plasma populations, and energetic charged particles. In particular, the ECT suite (Spence et al. 2013) measures electrons from a few eV to tens of MeV, and we highlight here results from two instruments, the REPT (Baker et al. 2012) and the MagEIS (Blake et al. 2013).

One of the most remarkable results, very early on during the mission, was the discovery by the REPT team of a long-lived storage ring or the third belt of energetic electrons (Baker et al. 2013). Measurements of high-energy (>2 MeV) electrons revealed that a stable ring of electrons formed in space at about ~3.5 R_E on September 2, 2012, and persisted for more than a month before being wiped out by a strong shock. It must be emphasized that models of radiation belts did not anticipate the formation and persistence of such a morphological feature. Subsequent modeling has come up with several possible explanations (Thorne et al. 2013) ranging from radial transport to slow flux decay.

Figure 10.4 shows REPT measurements of electrons in the three differential energy channels, in the energy ranges of 2.0–2.5, 4.0–5.0, and 6.2–7.5 MeV as the top, middle, and bottom panels, respectively. The electron fluxes are shown as a function of time L-shell from September 1, 2012, to October 4, 2012, and are color coded as shown in the color bar to the right. A third belt or the "storage ring" is clearly visible during the period from ~2 September and lasted to the end of October 1, 2012.

A second unexpected discovery, again made possible by measurements of ultrarelativistic electrons by the REPT instrument, was that of a "barrier" for these electrons. It was found that very energetic electrons do not come closer than about 2.8 R_E, even during strong geomagnetic disturbances (Baker et al. 2014). This barrier is energy-dependent, in that lower energy electrons of up to about 800 keV or so do enter the slot region quite often in response to solar drivers; however, the most energetic electrons seem unable to radially diffuse inward of 2.8 R_E. It must be noted that the Van Allen Probes era has been one of relatively quiet geomagnetic conditions; however, even during the large geomagnetic storm of March 2015 the ultrarelativistic electrons were observed not to cross this "barrier." Baker et al. (2014) conjectured that such as barrier might exist due to a balance between inner radial diffusion and electron precipitation, although other interesting possibilities such as anthropogenic plasma waves may account for the observations as well.

Figure 10.5 shows a yearlong observations of ultrarelativistic electrons, $E = 7$ MeV, in a similar format to that of Figure 10.4. It is evident from the figure, during this period, these high-energy electrons were not observed within 2.8 R_E; even a strong IP shock that occurred on October 1, 2013, resulted in merely a step-like displacement with the electron population still unable to reach distances less than 2.8 Earth radii.

One of the major open questions in radiation belt physics is the identification and quantification the relative contributions of radial transport and *in situ*

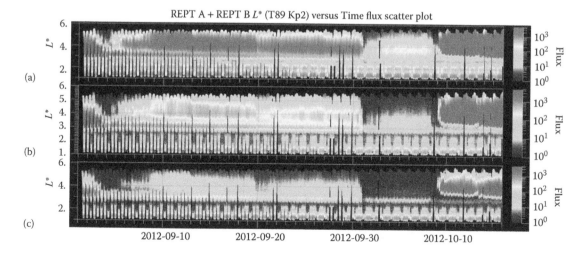

REPT A + REPT B L^* (T89 Kp2) versus Time flux scatter plot

FIGURE 10.4 Relativistic electron fluxes as measured by the REPT instrument from September 1, 2012 to October 4, 2012. Logarithm of electron fluxes (electrons/cm²-s-sr-MeV) color-coded according to the color bar at right are shown as a function of L* and time. (a through c) Show electrons in the energy range 2.0–2.5, 4.0–5.0, and 6.2–7.5 MeV, respectively. A third belt of electrons can be clearly seen in all the three energy-ranges at L* ~ 3.5. The third belt formed on ~2 September and lasted until the end of October 1, 2012. (Reproduced from Baker, D.N. et al., *Science*, 340(6), 186–190, 2013. With Permission.)

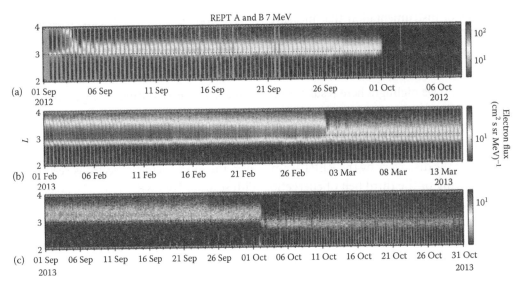

FIGURE 10.5 REPT measurements of ultra-relativistic electron fluxes showing the "impenetrable barrier." Color-coded electron fluxes are shown as a function of L-shell and time in three separate panels covering the time duration from September 1, 2012 to October 31, 2013. An IP shock on October 1, 2013 pushed the electrons slightly inward, although not past 2.8 Earth radii. (Reproduced from Baker, D.N. et al., *Nature*, 515, 7528, 2014. With Permission.)

wave–particle interactions to electron energization. Detailed analysis of a geomagnetic storm event that occurred in October 2012 (Reeves et al. *Science*) established clearly that electron energization during this event was due to wave–particle interactions. Simultaneous measurements by REPT and MagEIS instruments on both the Van Allen probes spacecraft showed a clear localized peak in the radial profiles of electron PSD.

Figure 10.6 shows simultaneous measurements of PSD by the two REPT instruments (adapted from

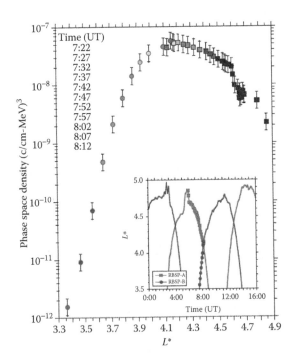

FIGURE 10.6 Measurements of radial profiles of phase space density by Van Allen Probes A and B. Probe A measurements are shown as squares and Probe B as circles, with the times of measurements and the corresponding data points being shown in the same color. The times range from 07:22 to 08:12 UT on October 9, 2012. Shown in the Inset are the spacecraft orbits of probes A(B) inbound(outbound) from apogee(perigee). (Reproduced from Reeves, G. D. et al. *Science*, 341, 991–994, 2013. With Permission.)

Reeves et al.) as they approached the energization region from opposite directions, that is, one of the probes (A) was inbound from the apogee while the second (B) was outbound from the perigee. Thus, the acceleration process was unambiguously identified as a localized wave–particle interaction process.

As mentioned in Section 10.5, one of the two broad classes of energization of electrons is the *in situ* wave–particle interactions. Recent measurements by the MagEIS and EMFISIS (Kletzing et al. 2013) instruments enabled the direct observation of wave–particle interactions in the outer zone. Bursts of low energy electrons in the 17–26 keV energy range of a quasi-periodic nature were observed to occur in conjunction with burst of chorus wave activity. The particle and wave properties were independently measured and model calculations of the energy of electrons that would resonate with the observed waves agreed very well with the energies of the observed electron bursts (Fennell et al. 2014).

Figure 10.7 shows electron bursts and chorus wave observation on January 13, 2013. The high-resolution pitch angle wave spectrograms for the duration of about 1 h clearly show increased wave activity and their simultaneous increased burst of enhanced fluxes of electrons. The wave data are from the EMFISIS suite on Van Allen Probes. Both MagEIS instruments on each of the Van Allen Probes spacecraft observed the electron bursts.

In this section, we have described a select few of the many significant observations made by the Van Allen Probes mission, the most recent comprehensively instrumented mission dedicated to the study of radiation belt phenomena. These observations have spurred theoretical and modeling work aimed at a more complete understanding of the underlying physics of the radiation belts.

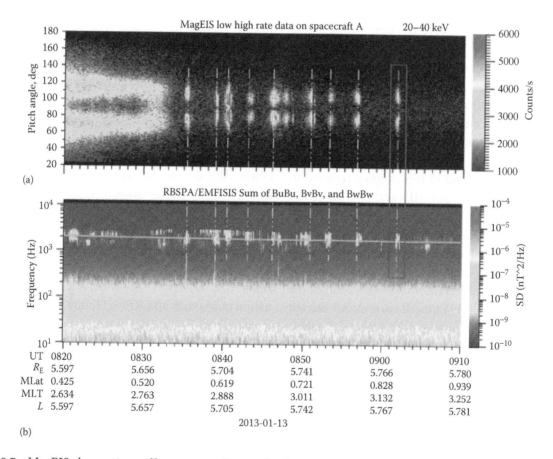

FIGURE 10.7 MagEIS observations of low-energy, electron flux bursts and EMFISIS observations of chorus waves during the 0820–0910 UT on January 13, 2013. The top panel shows the 20–40 keV, electron pitch angle spectrogram with electron bursts indicated by vertical dash-dotted lines. The lower panel is a frequency-time spectrogram obtained from the EMFISIS instrument showing bursts of chorus emissions, which occur simultaneously as the electron bursts. One pair of electron and wave bursts is enclosed in a gray rectangle. The gray horizontal line in the bottom panel shows the location of half the electron cyclotron frequency. (Reproduced from Fennell, J.F. et al., *Geophys. Res. Lett.*, doi:10.1002/2013gl059165, 2014. With Permission.)

It is evident that the study of radiation belts is not only a vibrant field with much to be understood but also one of exciting new discoveries, a full 50 years after their discovery.

10.8 RADIATION BELTS AND SPACE WEATHER

As the previous sections have delineated, radiation belts contain trapped high-energy particles whose fluxes often increase by several orders of magnitude due to interaction with solar events. Increasingly, modern society relies on space-based technology for not only day-to-day activities, for example, GPS but also to warn of potential catastrophic weather conditions such as hurricanes. The effects of events in space on the Earth and human technology come under the rubric of space weather, which is defined by the National Space Weather Program Council (2010): "The term 'space weather' refers to variable conditions on the Sun, throughout space, and in the Earth's magnetic field and upper atmosphere that can influence the performance and reliability of space-borne and ground-based technological systems and endanger human life and health. Adverse conditions in the space environment can cause disruption of satellite operations, communications, navigation, and electric power distribution grids, leading to a variety of socioeconomic losses and impacts on our security." The radiation belts, as discussed in this chapter, respond to various solar drivers, such as CMEs and HSS as well as IP shocks; the relativistic electron fluxes in outer belt particularly can reach very high levels (of the order of 10^6 [cm^2-sr-sec-MeV]$^{-1}$) and remain there for a duration of several weeks. Such high fluxes can cause anomalies and even failures of spacecraft. The compression of the geomagnetic field results in induced current that disrupt long conductors such as power grids, communication cables, and pipelines. These so-called telluric currents affect different types of "long conductors" in different manners. For example, in electrical power grids, space weather-induced currents may cause transformers to fail, and power switches to trip cutting off power supplies to large fractions of human populations. It is well known now that the existence of an ionized layer above the atmosphere, namely, the ionosphere makes radio communication over long distances possible. When the ionospheric state is disturbed, for example, by increased ionization due to strong SEP penetration, radio communication can be disrupted. Radiation belt particles can also precipitate in increased numbers driven by solar events and lead to a similarly disturbed ionosphere. A comprehensive survey of space weather effects in general can be found in Baker and Lanzerotti (2015).

10.9 FUTURE OF RADIATION BELT RESEARCH

Recent innovations in charged particle instrumentation and multipoint measurements, such as the ones available from recent radiation belt missions such as the Van Allen Probes, have re-invigorated the study of radiation belt dynamics. The realization of the extent of human dependence on space weather has also provided a substantial impetus to the study of radiation belts. The discoveries and detailed understanding of radiation belt dynamics have revealed the incompleteness in our current understanding, though much advanced compared to just a decade earlier.

The future of radiation belt physics will make use of new and innovative detection techniques, advances in electronics and computing, and increased data rates to download data. New space platforms such as CubeSats and SmallSats enable multipoint measurements and most importantly provide an inexpensive access to space. The instrumentation on these platforms can be quite sophisticated and provide detailed information. Several CubeSats such as CSSWE (Li et al. 2012), FIREBIRD (http://firebird.unh.edu/), and the AeroCube (Blake and O'Brien 2015) have already been launched and returned measurements of radiation belt electron energization and loss. Currently planned CubeSats include ELFIN (http://elfin.igpp.ucla.edu/) and the Compact Radiation Belt Explorer (CeREs), which will carry an innovative SSD-based instrument that is capable measuring electrons from 10s of keV to ~10 MeV, similar to the combination of MagEIS and REPT (Kanekal et al. 2014).

10.10 SUMMARY

In this chapter, we have provided a basic description of the radiation belts, with emphasis on the outer belt, which is highly variable and dynamic driven by the multifarious solar drivers such as CME, HSS, and IP shocks. These drivers elicit a variety of responses in the outer Van Allen belt via multiple physical processes, which fall into two broad classes, namely, radial transport and *in situ* wave–particle interactions. Physical understanding of electron dynamics is facilitated by regarding charged particle motion as a combination of gyration, bounce, and azimuthal drift around the Earth; each with its own associated adiabatic invariant. Breaking these invariants leads to particle energization and loss, the balance between which determines the net flux levels of energetic electrons in the outer belt. We have described two types of particle instrumentation on the most recent comprehensive instruments mission to study the Earth's radiation belts, namely, the Van Allen Probes. Select measurements and discoveries regarding electron dynamics are presented with a view to bring to one's attention the vibrant and evolving nature of understanding of the variety of phenomena associated with particle energization and loss. Practical implications of radiation belt dynamics, that is, space weather aspects are overviewed and shown to be significant both in a terrestrial and a geo-space context. We also share our views of the possible future of the study of radiation belt science and conclude that it is bright with much to be learned and its clear importance in the context of space science.

REFERENCES

Agostinelli, S. et al. (2003), Nuclear instruments and methods in physics research section A: Accelerators, spectrometers, detectors and associated equipment, *Nucl. Instrum. Meth. A*, 506, 250–303, doi:10.1016/S0168-9002(03)01368-8.

Baker, D. N., G. M. Mason, O. Figueroa, G. Colon, J. G. Watzin, and R. M. Aleman (1993), An overview of the solar, anomalous, and magnetospheric particle explorer (SAMPEX) mission, *IEEE Trans. Geosci. Remote Sens.*, 31(3), 531–541.

Baker, D. N., T. Pulkinnen, X. Li, S. G. Kanekal, J. B. Blake, R. S. Selesnick, M. G. Henderson, G. D. Reeves, H. E. Spence, and G. Rostoker (1998), Coronal mass ejections, magnetic clouds, and relativistic magnetospheric electron events: ISTP, *J. Geophys. Res.*, 103(17), 279–291.

Baker, D. N. and Lanzerotti, L. J. (2013), Resource letter SW1: Space weather, *Am. J. Phys.*, 84, 166(2016); doi: 10.1119/1.4938403.

Baker, D. N. et al. (2012), The relativistic electron-proton telescope (REPT) instrument on board the radiation belt storm probes (RBSP) spacecraft: Characterization of Earth's radiation belt high-energy particle populations, *Space Sci. Rev.*, 179, 337–381, doi:10.1007/s11214-012-9950-9.

Baker, D. N. et al. (2013), A long-lived relativistic electron storage ring embedded in Earth's outer Van Allen belt, *Science*, 340(6), 186–190, doi:10.1126/science.1233518.

Baker, D. N. et al. (2014), An impenetrable barrier to ultra-relativistic electrons in the Van Allen radiation belts, *Nature*, 515, 7528, doi:10.1038/nature13956.

Blake, J. B. and T. P. O'Brien (2016), Observations of small-scale latitudinal structure in energetic electron precipitation, *J. Geophys. Res.: Space Phys.*, 121, doi:10.1002/2015JA021815.

Blake, J. B., D. N. Baker, N. Turner, K. W. Ogilvie, and R. P. Lepping (*April 15*, 1997), Correlation of changes in the outer-zone relativistic-electron population with upstream solar wind and magnetic field measurements, *Geophys. Res. Lett.*, 24(8), 927–929.

Blake, J. B., W. A. Kolasinski, R. W. Fillius, and E. G. Mullen (1992), Injection of electrons and protons with energies of tens of MeV into L < 3 on 24 March 1991, *Geophys. Res. Lett.*, 19, 821.

Blake, J. B. et al. (2013), The magnetic electron ion spectrometer (MagEIS) instruments aboard the radiation belt storm probes (RBSP) spacecraft, *Space Sci. Rev.*, 179, 383–421, doi:10.1007/s11214-013-9991-8.

Elkington, S. (2006), A review of ULF interactions with radiation belt electrons, *Solar Eruptions and Energetic Particles Geophysical Monograph Series 165*, 169, 177, American Geophysical Union, Washington DC.

Fennell, J. F. et al. (2014), Van Allen probes observations of possible direct wave particle interactions, *Geophys. Res. Lett.*, doi:10.1002/2013gl059165.

Friedel, R. H. W., G. D. Reeves, and T. Obara (2002), Relativistic electron dynamics in the inner Magnetosphere—A review, *J. Atmos. Solar Terr. Phys.*, 64(2), 265.

Horne, R. B., R. M. Thorne, Y. Y. Shprits, N. P. Meredith, S. A. Glauert, A. J. Smith, S. G. Kanekal, D. N. Baker, M. J. Engebretson, J. L. Posch, M. Spasojevic, U. S. Inan, J. S. Pickett, P. M. E. Decreau (2005), Wave acceleration of electrons in the Van Allen radiation belts, *Nature*, 437, 227–230, doi:10.1038/nature03939.

Hudson, M. K., B. T. Kress, H. R. Mueller, J. A. Zastrow, and J. Bernard Blake (2008), Relationship of the Van Allen radiation belts to solar wind drivers, *J. Atmos. Solar Terr. Phys.*, 70, 708–729, doi:10.1016/j.jastp.2007.11.003.

Kanekal, S. et al. (2014), CeREs: A compact radiation bElt explorer, in *Proceedings of the AIAA/USU Conference on Small Satellites*, SSC14-P2-7. http://digitalcommons.usu.edu/smallsat/ 2014/Poster/16/.

Kanekal, S. G., et al. (2015), Relativistic electron response to the combined magnetospheric impact of a coronal mass ejection overlapping with a high-speed stream: Van Allen probes observations, *J. Geophys. Res.: Space Phys.*, 120, 7629–7641, doi:10.1002/2015JA021395.

Kanekal, S. G. (2006), A review of recent observations of relativistic electron energization in the Earth's outer Van Allen radiation belt, in *Proceedings of the ILWS Workshop*, Goa, India, edited by N. Gopalswamy and A. Bhattacharyya, p. 274, Quest Publications, Athens, Greece.

Kanekal, S. G., D. N. Baker, and J. B. Blake (2001), Multisatellite measurements of relativistic electron: Global coherence, *J. Geophys. Res.*, 106(29), 29721.

Kanekal, S. G., D. N. Baker, and J. B. Blake (2003), A comparison of relativistic electron in response in the Earth's radiation belts to high speed solar wind streams and coronal mass ejections, in *ISEC 2003, Radiation Belt Science*, Toulouse, France.

S. G. Kanekal, D. N. Baker, J. F. Fennell, A. Jones, Q. Schiller et al. (2016), Near-instantaneous acceleration of magnetospheric electrons to ultra-relativistic energies by the powerful interplanetary shock during 17 March 2015, *J. Geophys. Res.: Space Phys.*, in press.

Kanekal, S. G., R. H. W. Friedel, G. D. Reeves, D. N. Baker, and J. B. Blake (2005), Relativistic electron events in 2002: Studies of pitch angle isotropization, *J. Geophys. Res. (Space Phys.)*, 110(A9), 12224, doi:10.1029/2004JA010974.

Kletzing, C. A. et al. (2013), The electric and magnetic field instrument suite and integrated science (EMFISIS) on RBSP, *Space Sci. Rev.*, 179, 127–181, doi:10.1007/s11214-013-9993-6.

Li, X., S. Palo, R. Kohnert, D. Gerhardt, L. Blum, Q. Schiller, D. Turner, W. Tu, N. Sheiko, and C. S. Cooper (2012), Colorado student space weather experiment: Differential flux measurements of energetic particles in a highly inclined low Earth orbit, in *Dynamics of the Earth's Radiation Belts and Inner Magnetosphere*, Geophysical Monograph Series, vol. 199, edited by D. Summers et al., pp. 385–404, AGU, Washington, DC.

Li, X., I. Roth, M. Temerin, J. R. Wygant, M. K. Hudson, and J. B. Blake (1993), Simulation of the prompt energization and transport of radiation belt particles during the March 24, 1991 SSC, *Geophys. Res. Lett.*, 20(22), 24232426, doi:10.1029/93GL02701.

Lyons, L. R., R. M. Thorne, and C. F. Kennel (1972), Pitch-angle diffusion of radiation belt electrons within the plasmasphere, *J. Geophys. Res.*, 77(19), 3455–3474.

Mauk, B. H., N. J. Fox, S. G. Kanekal, R. L. Kessel, D. G. Sibeck, and A. Ukhorskiy (2012), Science objectives and rationale for the radiation belt storm probes mission, *Space Sci. Rev.*, 179, 3–27, doi:10.1007/s11214-012-9908-y.

McIlwain, C. E. (1961), Coordinates for mapping the distribution of magnetically trapped particles, *J. Geophys. Res.*, 66, 3681–3691.

Millan, R. M. and D. N. Baker (2012), Acceleration of particles to high energies in Earth's radiation belts. *Space Sci. Rev.*, 173, 103–131, doi:10.1007/s11214-012-9941-x.

National Space Weather Program Council (2010), National Space Weather Program, Strategic Plan, National Space Weather Program Council, FCM-P30-2010, Washington, DC, June 2010, http://www.ofcm.gov/nswp-sp/fcm-p30.htm.

Reeves, G. D. (2015), *Radiation Belt Electron Acceleration and Role of Magnetotail, Keiling/Magnetotails in the Solar System*, John Wiley & Sons, Hoboken, NJ.

Reeves, G. D., K. L. McAdams, R. H. W. Friedel, and T. P. O'Brien (2003), Acceleration and loss of relativistic electrons during geomagnetic storms, *Geophys. Res. Lett.*, 30, 1529.

Reeves, G. D. et al. (2013), Electron acceleration in the heart of the Van Allen radiation belts, *Science*, 341, 991–994, doi:10.1126/science.1237743.

Roederer, J. G. (1970), *Dynamics of Geomagnetically Trapped Radiation*, Springer, New York.

Russell, C. T. and McPherron, R. L. (1973), Semiannual variation of geomagnetic acivity, *J. Geophys. Res.*, 78, 92.

Spence, H. E. et al. (2013), Science goals and overview of the radiation belt storm probes (RBSP) energetic particle, composition, and thermal plasma (ECT) suite on NASA's Van Allen probes mission, *Space Sci. Rev.*, 179, 311–336, doi:10.1007/s11214-013-0007-5.

Schulz, M. and L. J. Lanzerotti (1974), *Particle Diffusion in the Radiation Belts*, 215pp., Springer-Verlag, New York.

Summers, D., R. M. Thorne, and F. Xiao (1998), Relativistic theory of wave-particle resonant diffusion with application to electron acceleration in the magnetosphere, *J. Geophys. Res.*, 103, 20487–20500.

Thorne, R. M. (2010), Radiation belt dynamics: The importance of wave-particle interactions, *Geophys. Res. Lett.*, 37, L22107, doi:10.1029/2010GL044990.

Turner, D. L. et al. (2014), Competing source and loss mechanisms due to wave-particle interactions in Earth's outer radiation belt during the 30 September to 3 October 2012 geomagnetic storm, *J. Geophys .Res. (Space Phys.)*, 119(3), 1960–1979, doi:10.1002/2014JA019770.

Van Allen, J. A. (1957), Direct detection of auroral radiation with rocket equipment, *Proc. Natl. Acad. Sci. USA*, 43(1), 57–62.

Van Allen, J. A. (1983), *Origins of Magnetospheric Physics*, Smithsonian Institution Press, Washington, DC.

Van Allen, J. A., G. H. Ludwig, E. C. Ray, and C. E. McIlwain (1958), Observation of high intensity radiation by satellites 1958 alpha and gamma, *J. Jet Propul.*, 28, 588–592.

CHAPTER 11

Plasmasphere

Jonathan Krall and Joseph D. Huba

CONTENTS

11.1 INTRODUCTION

The Earth's plasmasphere is a region of cold, dense plasma trapped on closed geomagnetic field lines in the inner magnetosphere. The outer boundary of the plasmasphere is called the plasmapause and is typically in the range $3 < L < 7$. Although the plasmasphere is a relatively benign region of space, it does play an important role in a number of space weather processes.

The plasmasphere is a low β plasma, meaning that magnetic forces dominate thermal forces, and the plasma is effectively "tied" to the geomagnetic field. Even as the geomagnetic field interacts with the solar wind, field lines closest to the Earth remain "closed." It is these field lines that contain the plasmasphere. High-latitude field lines that extend far enough into space to reconnect with the solar wind magnetic field can become "open." Light or energetic ions on these open field lines can escape into space. The opening and closing of the magnetic field, along with the emptying and refilling of individual "field lines," shapes the boundary, called the plasmapause, that separates the relatively high (>10–100 cm^{-3}) plasmasphere and

the lower density (<1–10 cm^{-3}) background plasma within the magnetosphere.

The plasmasphere is shaped by the dynamics of the magnetosphere (Carpenter, 1966; Nishida, 1966), ionosphere (Galvan et al., 2008), and thermosphere (Krall et al., 2014). It typically erodes during geomagnetic storms, with a timescale of hours (Goldstein et al., 2003), and refills during quiet times with a timescale of many days (Singh and Horwitz, 1992; Krall et al., 2008).

The plasmasphere, as well as the plasmapause, play a role in the development of strong ionospheric density enhancements and gradients that form in the low-to-mid latitude ionosphere during magnetic storms. A manifestation of large magnetic storms in the ionosphere are storm-enhanced densities (SEDs) (Foster, 2013) and intensified subauroral polarization streams (SAPs) (Huang and Foster, 2007). For example, during the very strong Halloween 2003 storm, the total electron content (TEC) in the mid-latitude ionosphere increased dramatically within 3 h (Mannucci et al., 2005), from nominal levels of 80 TECU to over 300 TECU in the Southern hemisphere. TEC represents a vertical line integration of the electron density, a quantity having units of electrons per unit area. SAPs are an ionospheric feature positioned near the foot points of the outermost closed field lines of the plasmapause, and are regions of high flow (>1 km s^{-1}) in the ionosphere.

Associated with these ionospheric manifestations of magnetic storms, there is a corresponding density structure in the magnetosphere referred to as "drainage plume" (Foster, 2002) that represents the sunward extension of the plasmasphere during a storm. The plume can be seen in Figure 11.1, which shows an extreme ultraviolet (EUV) image of the plasmasphere taken by the IMAGE spacecraft (Sandel at al., 2001).

Both the position of the plasmapause and the plume itself can affect space weather. The plume can extend sunward as far as the magnetopause, injecting plasma into the reconnection region, where the magnetosphere reconnects with the solar wind magnetic field, injecting energy into near-Earth space. The plume tends to reduce the reconnection rate (Borovsky et al., 2013), lessening the impact of a storm. Additionally, during magnetic storms the plasmasphere undergoes an erosion process, that is, the plasmapause moves earthward. (Goldstein et al., 2002, 2005). This is very important because erosion of the plasmasphere modifies plasma waves that scatter

radiation belt particles, allowing the radiation belts to move earthward (Bortnik and Thorne, 2007; Millan and Thorne, 2007).

11.2 PLASMASPHERE DYNAMICS

Bulk motions within the plasmasphere are driven by corotation and convection $\mathbf{E} \times \mathbf{B}$ drifts, and field-aligned plasma flows. As the Earth rotates, the geomagnetic field, and the plasma trapped on that field, corotates with it. The interaction of the geomagnetic and solar wind magnetic fields involves reconnection of solar and geomagnetic field lines, even during quiet times. Newly opened field lines convect from the bow to the magnetotail, where these elongated field lines can reconnect, again becoming closed. Newly closed magnetic fields in the Earth's magnetotail tend to move earthward. These solar wind-driven convective motions of the plasma are aligned with the Earth–Sun line.

The erosion of the plasmasphere during a storm is shown in Fig. 11.2 using the SAMI3 ionosphere/plasmasphere code (Huba and Krall, 2013), illustrates the erosion of the plasmasphere during a storm. Figure 11.2 shows the log of the electron density during (left panel) and following (right) the storm. Contour lines indicate the electrostatic potential, including the "corotation potential," mapped into the equatorial plane. Under the influence of the Earth's rotation and solar wind-driven convection, magnetic "field lines" effectively move along these contours. Thus, open contours indicate open field lines; plasma on these open field lines will be swept out of the model system. Following the storm, closed contour lines with little or no electron density indicate regions of closed field lines that await refilling.

The combination of corotation and convection leads to a steady-state plasmasphere configuration with a bulge on the dusk side (Nishida, 1966). This bulge becomes a sunward-directed plume upon storm onset (Grebowsky, 1970; Sandel et al., 2001). Observations of the combined corotation and convection effects are summarized by Darrouzet et al. (2009) and by Singh et al. (2011), who describe such features as shoulders, fingers, notches, crenulation, and channels.

Additional bulk motions occur as pressure forces along a given field line cause plasma to move parallel to the field, such as when a field line opens. Finally, wind-driven electric fields in the ionosphere can cause $\mathbf{E} \times \mathbf{B}$ drifts, moving plasma across magnetic field lines.

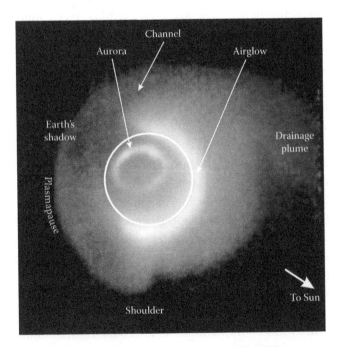

FIGURE 11.1 Plasmasphere image from the IMAGE spacecraft. (From Sandel, B.R. et al., *Space Sci. Rev.*, 109, 1–4, 2003. With permission.)

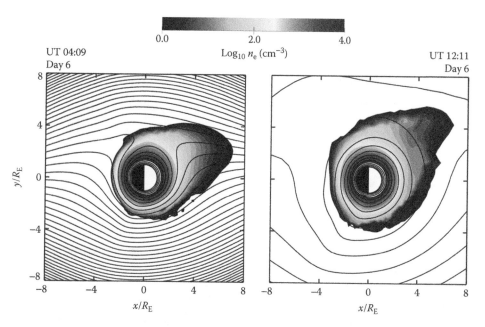

FIGURE 11.2 Log of the electron density (cm⁻³, shaded area) during and following a model storm. Contour lines indicate the electrostatic potential, including the "corotation potential," mapped into the equatorial plane. Following the storm, closed contour lines with no electron density indicate regions of closed field lines that await refilling. (From Krall, J. and Huba, J.D., *Geophys. Res. Lett.*, 40, 2484–2488, 2013. With permission.)

These drifts are large enough to affect the shape of plasmasphere but not large enough to significantly distort the shape of the geomagnetic field (Krall et al., 2014).

11.3 PLUMES, SEDs, AND SAPs

During magnetic storms, the plasmasphere undergoes a complex reconfiguration (Goldstein et al., 2002, 2005). One of the more dramatic features of the plasmasphere is the development of an extended plume in the late afternoon/dusk sector shown in Figure 11.1. This plume can extend to the magnetopause and can modify the reconnection process at the nose of the magnetopause (Borovsky and Denton, 2006). Furthermore, there are indications that this drainage plume is connected to features in the ionosphere (Foster et al., 2002).

Other manifestations of large magnetic storms in the ionosphere are SEDs (Foster, 1993) and intensified SAPs (Huang and Foster, 2007). For example, Coster and Skone (2009) show an example of a large plume of plasma in the ionosphere extending from the Great Lakes into Canada. This demonstrates the electrodynamic connection between the low- and high-latitude ionosphere during large storms. This connection is mediated by the SAPs region. The detailed quantitative relationship between SEDs and SAPs is an outstanding issue to be studied.

Recently, the latest versions of the SAMI3 (ionosphere/plasmasphere) and the ring current models (RCMs) have been coupled and a simulation of the March 29, 2001, storm has been carried out. This version of the coupled model includes the self-consistent wind-driven electric field. The results are shown in Figure 11.3. The top of each frame shows shaded contours of the log of the electron density in the equatorial plane and contour lines of the electrostatic potential. The white contour line denotes $n_e = 30\,\text{cm}^{-3}$. The bottom of each frame shows shaded contours of the TEC; the dark black contour line is for TECU = 75. The triangles in the top frames show where this TEC contour level maps along the geomagnetic field into the plasmasphere.

Figure 11.3a is at the beginning of the storm and Figure 11.3b is roughly 19 h later after the storm has subsided. In the lower panel of Figure 11.3a we find a large increase in the TEC in the mid-latitude ionosphere that is extending toward the polar cap region, reminiscent of an SED. This enhancement is similar to that shown in Coster and Skone (2009) and is caused by the penetration electric field computed by RCM. Following the storm this enhancement dissipates. Additionally, we find that the location of the enhanced TEC in Figure 11.3a maps to a plume-like structure in the plasmasphere. This is consistent with the observations and data presented in Foster et al. (2002).

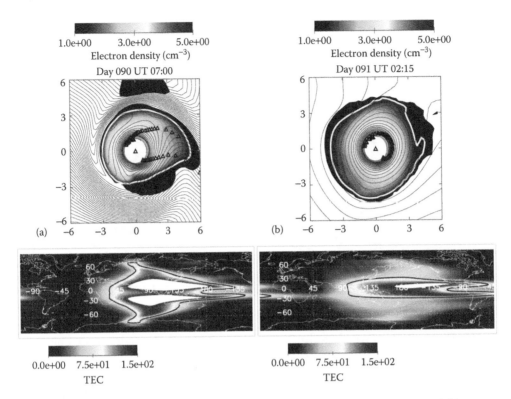

FIGURE 11.3 Simulation of the March 29, 2001, storm using SAMI3/RCM. (a) is at storm onset and (b) is post-storm (details of the figure are given in the text). (From Huba, J.D. and Sazykin, S., *Geophys. Res. Lett.*, 41, 8208–8214, 2014. With permission.)

11.4 PLUMES AND GEOMAGNETIC STORM STRENGTH

A primary driver of space weather is reconnection of southward-directed solar wind magnetic field, if present, with the northward geomagnetic field at the magnetopause. When the two reconnecting collisionless plasmas are asymmetric (differing magnetic field strengths and differing mass densities), the antiparallel reconnection rate is given by the Cassak–Shay equation (Cassak and Shay, 2007; Birn et al., 2008, 2010, 2012),

$$R = \frac{0.1 B_m^{3/2} B_s^{3/2}}{\pi^{1/2} \left(B_m \rho_s + B_s \rho_m \right)^{1/2} \left(B_m + B_s \right)^{1/2}} \quad (11.1)$$

where B and ρ are the magnetic field strength and mass density of the two plasmas and where the subscript m denotes the magnetospheric plasma and the subscript s denotes the magnetosheath plasma. Four plasma parameters determine the dayside reconnection rate: B_m, B_s, ρ_m, and ρ_s.

Ionospheric outflows can increase magnetospheric ion densities near magnetopause from typical values of order 1 cm^{-1} to values 10 to 100 times greater. One can see from expression (11.1) that large values of ρ_m lead to reductions in the dayside reconnection rate. This effect is a mass loading of the reconnection site: lowering Alfvén speeds, lowering reconnection outflow velocities, and lowering reconnection inflow rates. Here, we will focus on the effect of the plasmaspheric plume, but outflows of mildly energetic O$^+$ ions from the polar cap can play a similar role.

Normally, the magnetospheric density is much lower than the magnetosheath density, such that the reconnection rate can be estimated by assuming $\rho_m = 0$. When the plume enters the reconnection region, however, ρ_m can be significant. The correction factor to the $\rho_m = 0$ limit of the Cassak–Shay reconnection rate, to account for non-zero magnetospheric mass density, can be expressed as

$$M = \frac{\left(B_m \rho_s \right)^{1/2}}{\left(B_m \rho_s + B_s \rho_m \right)^{1/2}} \quad (11.2)$$

Accounting for compression of the solar wind at the magnetosphere bow shock and assuming B_{SW} perpendicular to the Earth–Sun line, Borovsky et al. (2013) compute the reduction of the reconnection rate as a function of the solar wind Alfvén Mach number M_A. This is plotted in Figure 11.4 for four different values of ρ_m/ρ_0 where ρ_0

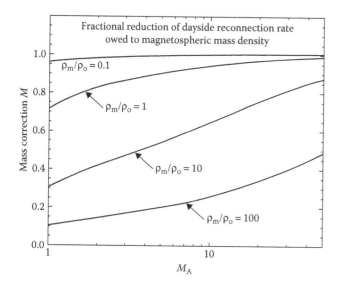

FIGURE 11.4 Reconnection mass correction factor versus solar wind Alfvén Mach number M_A for four values of the magnetosphere to solar wind mass density ratio: 0.1, 1, 10, and 100. (From Borovsky, J.E. et al., *J. Geophys. Res. Space Physics*, 118, 5695–5719, 2013. With permission.)

is the mass density of the solar wind plasma. As can be seen in Figure 11.4, the local reconnection rate is further reduced for lower Mach numbers. This is because a higher Mach number increases plasma β in the magnetosheath and an increase in β decreases the degree to which the solar wind field is compressed to make up the sheath field. As a result, higher M_A is associated with lower B_S and higher reconnection rates. Thus, during geomagnetic storms plasmaspheric plumes may extend to the magnetopause and affect the reconnection process, which in turn affects the dynamics of the magnetosphere. This results in a highly complex nonlinear interaction of the solar wind–magnetosphere–plasmasphere–ionosphere system.

11.5 IMPACT ON THE RADIATION BELTS AND RING CURRENT

The onset, propagation, and saturation of plasma waves in the plasmasphere is known to have a significant impact on the radiation belts and the energetic ring current ions. Specifically, electromagnetic ion cyclotron (EMIC) waves, plasmaspheric hiss, and ultralow frequency (ULF) waves can play a critical role in the evolution of radiation belt particles. These waves enhance acceleration, energization, and diffusion of radiation belt and ring current ions. For example, erosion of the plasmasphere during a storm allows the radiation belts to move earthward, modifying plasma waves that scatter radiation belt particles (Bortnik and Thorne, 2007; Millan and Thorne, 2007; Masson et al.,

2009). The plasmasphere provides the mediating environment for these plasma waves. Although it is not responsible for wave generation *per se*, it plays a significant role in wave dispersion and propagation.

EMIC waves are generated by the temperature anisotropy associated with ring current ions (Kennel and Petschek, 1966). These waves scatter ring current ions into the loss cone and cause the ring current to decay (Cornwall et al., 1970). Decay times of <1 h are possible (Gonzalez et al., 1989) during the main phase of a geomagnetic storm. Satellite measurements show that the admixture of thermal He⁺ plays a significant role in the generation and propagation of EMIC waves (Young et al., 1981; Roux et al., 1982). The complex interaction between the multi-ion plasmasphere and EMIC waves has been studied using theoretical and modeling methods (e.g., Khazanov et al., 2006; Khazanov, 2011).

Plasmaspheric hiss is an electromagnetic plasma wave that propagates in the whistler mode regime (Thorne et al., 1973). The wave frequencies are in the range $f_{cH} < f < f_{ce}$, where f_{cH} and f_{ce} are the local proton and electron gyrofrequencies, respectively. These waves are generally confined within the plasmasphere ($L \sim 1.6$ to the plasmapause). The importance of plasmaspheric hiss is that it plays a key role in controlling the dynamics of the radiation belts and creating the slot region that separates the inner and outer belts (Bortnik et al., 2011).

ULF waves are low-frequency (1–100 mHz) MHD waves that propagate in the inner magnetosphere. They can be generated by energetic particles or by solar wind processes (Chen and Hasegawa, 1974; Southwood, 1974; Allan et al., 1986; Mathie and Mann, 2000). ULF waves are important because they can accelerate radiation belt particles (Hudson et al., 1999, 2000).

11.6 REFILLING

The rate of post-storm refilling of the plasmasphere depends on ions (i.e., densities) in the topside ionosphere, on the flow of these ions into the plasmasphere, and on the capture of these ions via collisions. If left undisturbed following a storm, H⁺ ions can continue to refill the plasmasphere for days and even weeks (Rasmussen et al., 1993), eventually reaching saturation densities that are larger than typical observed densities (Krall et al., 2008). However, the plasmasphere is rarely left in an undisturbed state for this length of time (Corcuff et al., 1972; Lemaire and Schunk, 1992) and there is a competition between erosion and refilling.

In general, H+ refilling rates decrease with increasing solar activity. This effect has been attributed to reduced neutral H in the H+ source region, where H+ is produced via a charge-exchange reaction with O+ (Richards and Torr, 1985). Noting that the topside ionosphere O+ density increases with solar activity, Krall et al. (2008) speculated that, because O+ acts as a diffusive barrier to H+ upflow (Lemaire and Gringauz, 1998), the increase in O+ with sunspot number might explain a corresponding reduction in H+ refilling rates.

Additionally, photoionization is the primary source of He+. As a result, He+ refilling generally increases with solar activity while generally remaining a minor ion in the plasmasphere.

11.6.1 Forces Shaping the Refilling Plasmasphere

To lowest order, corotation, magnetospheric convection and refilling control the size and shape of the plasmasphere. Numerical modeling suggests that thermosphere winds also play a role.

Krall et al. (2014) show that winds affect the plasmasphere in two fundamental ways. First, winds shape the ionosphere through vertical $\mathbf{E} \times \mathbf{B}$ drifts, affecting the source of the plasmasphere. Over the course of a day, the wind pattern changes so that the mid-to-high latitude ionosphere, the source of refilling for $L > 3$, varies in strength. This effect is shown in Figure 11.5, where the longitudinally averaged equatorial plasmasphere density is plotted versus time for a SAMI3 simulation of a quiet period following the February 1, 2001, storm. With no winds in the simulation (top panel), refilling is optimum and is faster than the measured rates, indicated by the dashed lines. When winds are included from any of the three different wind models (HWM07, HWM93, TIMEGCM; next three panels), refilling rates are lower.

The hypothesis that winds affect refilling is supported by Chi et al. (2013), who find that the plasmaspheric density versus local time behaves differently than one might expect merely from solar influences. For example, density increases often continue past sunset, an effect that they attribute to thermospheric influences.

Second, winds affect zonal $\mathbf{E} \times \mathbf{B}$ streamlines as in Figure 11.6, which shows the effect of winds on the combined convection and corotation potentials. Plotted is the electrostatic potential, mapped to the equatorial plane at three times, 0100 UT on day 32, 0800 UT on day 32, and 1200 UT on day 36 with no winds (top row) and when using the HWM93 wind model (bottom row). The leftmost panels

are on day 32, near the end of the storm, while the latter two panels are during the quiet period plotted in Figure 11.5.

In Figure 11.6, three contributions to the potential are in evidence. During the storm, the convection potential is strong (left-hand panels). When the convection field is weak, the corotation potential can dominate, creating closed potential contours out to about $L = 5$ (right-hand panels). Without winds, the potential contours near the Earth are round, indicating that the plasmasphere is simply corotating with the Earth (upper right). With winds, these contours are no longer round and the system is more dynamic. The wind-driven dynamo potential distorts potential contours out to about $L = 4$, even during this moderate storm (lower left). During quiet times, the effect of the winds extends out as far as $L = 6$ (lower right).

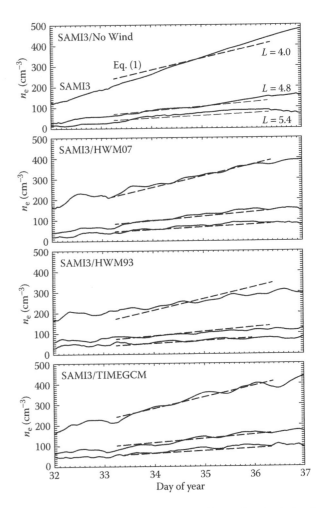

FIGURE 11.5 Electron density averaged over longitude in the equatorial plane plotted versus time for $L = 4.0$, 4.8, and 5.4 (solid curves) for SAMI3/No Wind, SAMI3/HWM07, SAMI3/HWM93, and SAMI3/TIMEGCM. Dashed lines in each plot indicate rates determined from *in situ* measurements. (From Krall, J. et al., *J. Geophys. Res.*, 119, 5032–5048, 2014. With permission.)

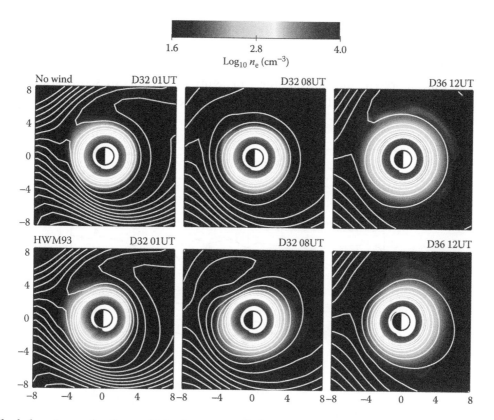

FIGURE 11.6 Shaded contours of n_e (log scale) in the equatorial plane at three different times for the no wind case (upper row) and the HWM93 case (lower row). Contour lines, separated by 1.5 kV, indicate the potential. (From Krall, J. et al., *J. Geophys. Res.*, 119, 5032–5048, 2014. With permission.)

11.6.2 Diurnal Oscillations in the Plasmasphere Density

Simulations suggest that measurements of plasmasphere density at fixed longitude or fixed local time will exhibit diurnal oscillations. To date, however, only the oscillation at fixed longitude has been clearly observed.

Variations in the plasmasphere density versus time at fixed longitude are directly tied to corresponding variations in the ionosphere, which decays by several orders of magnitude at night. Plasmaspheric densities can be inferred from images taken from well above the North Pole by the IMAGE spacecraft (Sandel et al., 2001). Such analysis indicates a diurnal "breathing" in which the density increases during the day and decreases at night (Galvan et al., 2008). Like the dusk-side bulge, the breathing of the plasmasphere can be described as a global standing wave that is fixed relative to the Earth–Sun line.

Modeling using the SAMI2 code (Huba et al., 2000), which simulates a single magnetic longitude versus time, shows this effect. Figure 11.7 shows density versus time at the apex of the $L = 3$ and $L = 4$ field lines during a very long quiet period following a model storm

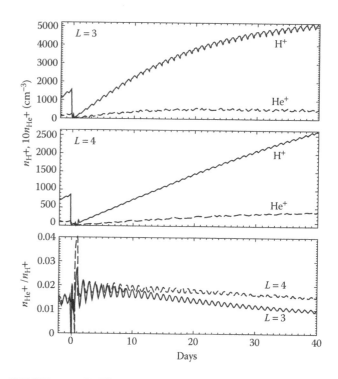

FIGURE 11.7 Refilling on two selected field lines after model storm. (From Krall, J. et al., *Ann. Geophys.*, 26, 1507–1516, 2008. With permission.)

(Krall et al., 2008). Diurnal oscillations are evident in both the H⁺ and He⁺ densities.

Diurnal oscillations in the plasmasphere density versus time at fixed local time, if present, indicate that the plasmasphere has a density structure that is rotating with the Earth. Figure 11.8 shows n_e from SAMI3 plotted versus time at fixed magnetic latitude and magnetic local time (MLT) at $L = 4.0$ (dashed) and 5.4 (solid). As in Figure 11.5, results are shown with no winds and with each of three different wind models (HWM07, HWM93,

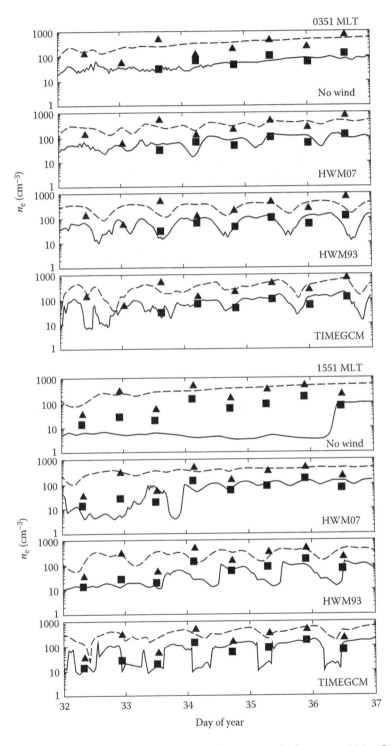

FIGURE 11.8   Electron density versus time during and following the storm of February 1, 2001. Curves show SAMI3 model results for each of three different wind models and the "no wind" case. Points show corresponding *in situ* measurements. (From Krall, J. et al., *J. Geophys. Res.*, 119, 5032–5048, 2014. With permission.)

and TIMEGCM). These latitude and local time positions were chosen to correspond to *in situ* measurements taken by the IMAGE spacecraft. During this time, IMAGE passed through the plasmasphere once every 14 h, sweeping through the post-midnight sector (MLT 0350) and the afternoon sector (MLT 1550). Measurements at $L = 4.0$ (triangles) and 5.4 (squares) are shown in each panel.

Except in the no wind case, SAMI3 curves in Figure 11.8 show diurnal oscillations corresponding to the corotation of the plasmasphere, with one cycle per day. The measured density varies to a degree consistent with the model result, but measurements do not occur often enough to resolve the diurnal oscillations, if present. The measured density variations are not clearly in phase with any of SAMI3 curves, possibly as a result of sub-corotation of the plasmasphere versus nearly exact corotation of the model plasmasphere. Galvan et al. (2010) have used cross-correlations of IMAGE EUV images to find time-varying corotation rates that are 88%–95% of exact.

Modeling suggests that wind-driven processes support a density profile that corotates with the Earth. The winds affect ions both directly, by collisions as ions move along field lines, and indirectly, through the wind-driven dynamo **E** fields and consequent **E** × **B** drifts. Further simulations, selectively including these direct and indirect wind effects, suggests that **E** × **B** drifts are responsible for these variations in the density versus time at fixed local time.

11.7 CONTRIBUTION TO TEC

A common diagnostic of space weather in the ionosphere is TEC, a vertical line integration of the electron density. TEC is routinely measured via ground detection of GPS satellite signals. This gives TEC between the Earth and about 20,000 km ($L = 4.1$).

One method of computing the plasmasphere contribution to TEC is to separately measure TEC from GPS satellites and from a low-Earth orbit satellite, such as JASON or TOPEX, orbiting at about 1340 km. Computing the difference between these two results gives a measure of the plasmasphere contribution to TEC, sometimes called pTEC (Lee et al., 2013). In studies of space weather, the globally averaged TEC is sometimes used as a measure of the impact of a geomagnetic storm, with typical values being $10 < \langle \text{TEC} \rangle < 60$ TECU (Lean et al., 2011), where one TECU $= 10^{16}$ electrons/m². Modeling suggests that $\langle \text{pTEC} \rangle$ can account for up to 30% of $\langle \text{TEC} \rangle$.

An important distinction must be made between TEC, the number of electrons in 1 m² vertical column, and the number of ions or electrons above 1 m² of Earth. The latter number reflects a volume that expands with r^2, where r is the distance to the center of the Earth. Because the plasmasphere extends from the topside ionosphere to altitudes as large as $6R_E$, the contribution of plasmaspheric ions to TEC strongly depends on their altitude.

The plasmaspheric ion population that most strongly contributes to TEC may not be the ion population that is removed and replaced during and after a geomagnetic storm. To illustrate, we present a SAMI3 simulation of a model storm, monitoring the ion population in a plasmaspheric volume. The volume of interest is illustrated in Figure 11.9. It is bounded by an outer surface defined by $L = 6$ dipole field lines in the simple model geomagnetic field used in this case (solid curve). An inner boundary is at $L = 2$ (dotted curve). Northern (dashed curve) and southern (dashed curve) topside ionosphere boundaries are shown at the 1340 km height sometimes used for pTEC calculations. As the model storm proceeds, **E** × **B** drifts are monitored at the inner and outer boundaries to determine ion fluxes into and out of this volume. Similarly, field-aligned flows are monitored at the northern and southern topside ionosphere boundaries.

Results are shown in Figure 11.10 (upper left), where the number of ions inside the volume is plotted for H⁺

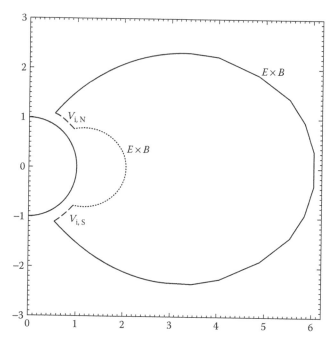

FIGURE 11.9 Plasmasphere boundaries for analysis of ion fluxes during and following a model storm.

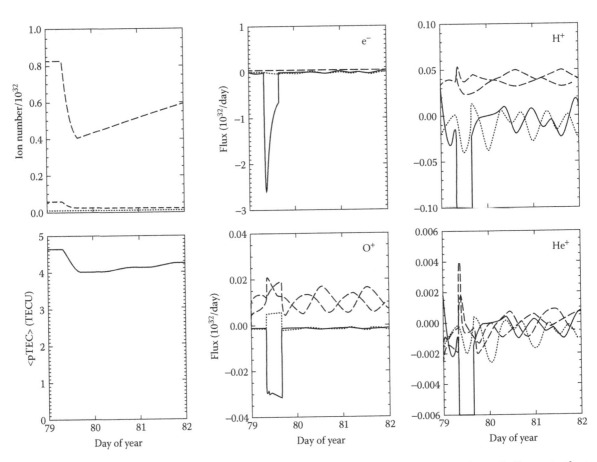

FIGURE 11.10 The plasmasphere during and following a model storm on day 79 (equinox conditions). Shown in the top row are total ion numbers within the volume specified in Figure 11.9 for hydrogen (long-dashed curve), helium (dashed), and oxygen (dotted) ions; electron fluxes through the outer (solid curve), inner (dotted), northern (dashed), and southern (dashed) boundaries; and the hydrogen ion fluxes. Shown in the bottom row is the globally averaged plasmasphere contribution to the total electron content, the oxygen ion fluxes, and the helium ion fluxes.

(long-dashed line), He$^+$ (dashed line), and O$^+$ (dotted line). In this low-solar-activity example, with daily and 80-day average solar EUV indices each set to 80, He$^+$ and O$^+$ have very low densities relative to H$^+$. In the panels to the right, fluxes through the four boundaries are shown for H$^+$, He$^+$, O$^+$, and the electrons (the sum of the ions). Shown are fluxes at the $L = 6$ (solid curve), $L = 2$ (dotted curve), northern (dashed curve), and southern (dashed curve) boundaries, with a positive flux indicating an increase in the corresponding ion number. For comparison, $\langle$pTEC$\rangle$ is plotted in the lower left.

In Figure 11.10, the model storm is included in the simulation via the Volland–Stern–Manyard–Chen model potential (Volland, 1973; Maynard and Chen, 1975; Stern, 1975), which represents the convection potential field as a function of geomagnetic index K_p.

Specifically, $K_p = 9$ was specified for 6 h on day 79 and was otherwise set to a geomagnetically quiet $K_p = 2$. The storm on day 79 is followed by 2 days of calmness, during which time the plasmasphere begins to refill.

As discussed above, the storm has a much larger affect on the ion numbers, reduced by 50%, than on pTEC, reduced by 15%. The ion number is a useful metric because changes to its value can be accounted for by examining ion fluxes into and out of the plasmasphere. TEC, on the other hand, varies with the distribution of plasma within the plasmasphere. For example, consider a fixed number of electrons concentrated near the lower boundary of the plasmasphere within an altitude range of thickness L and with a density n_0. If that same number of electrons were located at a higher altitude, still within the plasmasphere and still within an altitude range of thickness L, its density would be $n_1 < n_0$ and TEC would be lowered.

Other aspects of a geomagnetic storm are evident in Figure 11.10, particularly the sudden loss of plasma density at the storm onset (top center panel) and the diminishing rate of the plasma loss rate versus time as the storm continues. As expected, refilling following the model storm is dominated by field-aligned H+ fluxes from the topside ionosphere (top right panel).

11.8 COMPOSITION

The plasmaspheric mass density, a function of the ion composition, is important because of its impact on space weather. In particular, radiation belt energetic particle populations are very sensitive to the core plasmasphere distribution and specifically to the position of the plasmapause (Pierrard and Voiculescu, 2011).

It is generally thought that the slot region between the two radiation belts is formed because pitch angle scattering by plasma waves removes energetic electrons. These waves, mainly whistler and plasmaspheric hiss (Lyons at al., 1972), are commonly observed in the plasmasphere (Meredith et al., 2004; Golden et al., 2012). However, Darrouzet et al. (2013) show that the plasmapause position is more variable than the radiation belt boundary positions, especially during small geomagnetic activity enhancements.

Because ULF waves are affected by the presence of heavy ions, the plasmasphere mass density can be inferred from wave measurements. Field-line resonance eigenfrequencies, for example, have been used to estimate magnetospheric plasma mass densities (e.g., Denton et al., 2004; Menk et al., 2004; Takahashi et al., 2006; Nosè et al., 2011). Similarly, measurements of EMIC waves have been used to infer the He+ density (Kim et al, 2015), under the assumption of a plasma containing only H+ and He+ ions.

Plasmasphere composition has also been measured *in situ*. Mass spectrometer measurements from the GEOS satellite during the 1977–1978 solar minimum period found typical values of 2%–6% for the He+ fraction (Farrugia et al., 1989). Measurements of He+ via IMAGE/EUV images of the He+ component of the plasmasphere, when compared to measurements of the local electron density via IMAGE/RPI, similarly provide a determination of the He+ fraction. In these 2001 (solar maximum) measurements, the typical He+ fraction was 9% (Sandel, 2011).

Plasmasphere refilling and composition have been simulated using SUPIM (Sheffield University plasmasphere ionosphere model) (Denton et al., 2002), SAMI2 (Krall et al., 2008), and SAMI3 (Krall et al., 2014). During H+ and He+ refilling following a model storm, SAMI2 He+ fractions were typically 2% for moderate solar activity (see Figure 11.7) and 10% for high solar activity, similar to measured values. At high solar activity, He+ refilling exhibits a strong diurnal variation (about a factor of 2), presumably due to photoionization of helium. As a result, the He+/H+ density ratio is increased at dusk and reduced at dawn.

11.9 SUMMARY

The role of the plasmasphere in space weather is multifaceted. It provides a connection between the ionosphere and the inner magnetosphere (i.e., radiation belts, ring current). As such it is not an independent entity but an integral part of a larger, complex system. It has a direct impact on the dynamics and morphology of the radiation belts and ring current, especially during geomagnetic storms when the plasmasphere undergoes a significant reconfiguration.

During a storm, the plasmasphere contracts and the plasmapause moves earthward, allowing the radiation belts to move earthward, and modify plasma waves that scatter radiation belt and ring current particles. Additionally, a pronounced dusk sector plume can develop. This can have consequences both in the ionosphere, such as storm time enhanced densities and subauroral plasma streams, and at the magnetopause, where reconnection rate can be reduced via mass loading.

REFERENCES

Allan, W., S. P. White, and E. M. Poulter (1986), Hydromagnetic wave coupling in the magnetosphere—Plasmapause effects on impulse-excited resonances, *Planet. Space Sci.*, 34, 1189–1200.

Birn, J., J. E. Borovsky, and M. Hesse (2008), Properties of asymmetric magnetic reconnection, *Phys. Plasmas*, 15, 032101.

Birn, J., J. E. Borovsky, and M. Hesse (2012), The role of compressibility in energy release by magnetic reconnection, *Phys. Plasmas*, 19, 082109.

Birn, J., J. E. Borovsky, M. Hesse, and K. Schindler (2010), Scaling of asymmetric reconnection in compressible plasmas, *Phys. Plasmas*, 17, 052108.

Borovsky, J. E. and M. H. Denton (2006), Effect of plasmaspheric drainage plumes on solar-wind/magnetosphere coupling, *Geophys. Res. Lett.*, 33, L20101, doi:10.1029/GL026519.

Borovsky, J. E., M. H. Denton, R. E. Denton, V. K. Jordanova, and J. Krall (2013), Estimating the effects of ionospheric plasma on solar wind/magnetosphere coupling via mass loading of dayside reconnection: Ion-plasma-sheet oxygen, plasmaspheric drainage plumes, and the plasma cloak, *J. Geophys. Res. Space Physics*, 118, 5695–5719, doi:10.1002/jgra.50527.

Bortnik, J., L. Chen, W. Li, R. M. Thorne, and R. B. Horne (2011), Modeling the evolution of chorus waves into plasmaspheric hiss, *J. Geophys. Res.*, 116, A08221, doi:10.1029/2011JA016499.

Bortnik, J. and R. M. Thorne (2007), The dual role of ELF/VLF chorus waves in the acceleration and precipitation of radiation belt electrons, *J. Atmos. Solar Terr. Phys.*, 69, 378.

Carpenter, D. L. (1966), Whistler studies of the plasmapause in the magnetosphere, 1, temporal variations in the position of the knee and some evidence on plasma motions near the knee, *J. Geophys. Res.*, 71, 693–709.

Cassak, P. A. and M. A. Shay (2007), Scaling of asymmetric magnetic reconnection: General theory and collisional simulations, *Phys. Plasmas*, 14, 102114.

Chen, L. and A. Hasegawa (1974), A theory of long-period magnetic pulsations: (2) Impulse excitation of surface eigenmode, *J. Geophys. Res.*, 79, 1033–1037.

Chi, P. J. et al. (2013), Sounding of the plasmasphere by mid-continent magnetoseismic chain (mcmac) magnetometers, *J. Geophy. Res. Space Phys.*, 118(6), 3077–3086, doi:10.1002/jgra.50274.

Corcuff, P., Y. Corcuff, D. L. Carpenter, C. R. Chappell, J. Vigneron, and N. Kleimenova (1972), La plasmasphere en periode de recouvrement magnetique. Etude combinee des donnees des satellites OGO 4, OGO 5, et des sifflements recus au sol, *Ann. Geophys.*, 28, 679–696.

Cornwall, J. M., F. V. Coroniti, and R. M. Thorne (1970), Turbulent loss of ring current protons, *J. Geophys. Res.*, 75, 4699.

Coster, A. and S. Skone (2009), Monitoring storm-enhanced density using IGS reference station data, *J. Geod.*, 83, 345–351, doi:10.1007/s00190-008-0272-3.

Darrouzet, F. et al. (2009), Plasmaspheric density structures and dynamics: Properties observed by the CLUSTER and IMAGE missions, *Space Sci. Rev.*, 145, 55–106, doi:10.1007/s11214-008-9438-9.

Darrouzet, F., V. Pierrard, S. Benck, G. Lointier, J. Cabrera, K. Borremans, N. Yu Ganushkina, and J. De Keyser (2013), Links between the plasmapause and the radiation belt boundaries as observed by the instruments CIS, RAPID and WHISPER onboard Cluster, *J. Geophys. Res. Space Phys.*, 118, 4176–4188, doi:10.1002/jgra.50239.

Denton, M. H., G. J. Bailey, C. R. Wilford, A. S. Rodger, and S. Venkatraman (2002), He+ dominance in the plasmasphere during geomagnetically disturbed periods: 1. Observational results, *Annal. Geophys.*, 20, 461–470.

Denton, R. E., J. D. Menietti, J. Goldstein, S. L. Young, and R. R. Anderson (2004), Electron density in the magnetosphere, *J. Geophys. Res.*, 109(09), 215.

Farrugia, C. J., D. T. Young, J. Geiss, and H. Balsiger (1989), The composition, temperature, and density structure of cold ions in the quiet terrestrial plasmasphere: GEOS 1 results, *J. Geophys. Res.*, 94(A9), 11865–11891, doi:10.1029/JA094iA09p11865.

Foster, J.C. (1993), Storm-time plasma transport at middle and high latitudes, *J. Geophys. Res.*, 98, 1675–1689.

Foster, J. C. (2013), Ionospheric-magnetospheric-heliospheric coupling: Storm-time thermal plasma redistribution, in *Midlatitude Ionospheric Dynamics and Disturbances*, P. M. Kintner, A. J. Coster, T. Fuller-Rowell, A. J. Mannucci, M. Mendillo, and R. Heelis (eds), pp. 121–134, American Geophysical Union, Washington, DC.

Foster, J. C., P. J. Erickson, A. J. Coster, and F. J. Rich (2002), Ionospheric signatures of plasmaspheric tails, *Geophys. Res. Lett.*, 29 (13), 1623, 10.1029/2002GL015067.

Galvan, D. A., M. B. Moldwin, and B. R. Sandel (2008), Diurnal variations in plasmaspheric He+ inferred from extreme ultraviolet images, *J. Geophys. Res.*, 113, A09216, doi:10.1029/2007JA013013.

Galvan, D. A., M. B. Moldwin, B. R. Sandel, and G. Crowley (2010), On the causes of plasmaspheric rotation variability: IMAGE EUV observations, *J. Geophys. Res.*, 115, A01214, doi:10.1029/2009JA014321.

Golden, D. I., M. Spasojevic, W. Li, and Y. Nishimura (2012), Statistical modeling of in situ hiss amplitudes using ground measurements, *J. Geophys. Res.*, 117, A05218, doi:10.1029/2011JA017376.

Goldstein, J. et al. (2002), IMF-driven overshielding electric field and the origin of the plasmaspheric shoulder of May 24, 2000, *Geophys. Res. Lett.*, 29, 1819, doi:10.1029/2001GL014534.

Goldstein, J., S. G. Kanekal, D. N. Baker, and B. R. Sandel (2005), Dynamic relationship between the outer radiation belt and the plasmapause during March–May 2001, *Geophys. Res. Lett.*, 32, L15104, doi:10.1029/2005GL023431.

Goldstein, J., B. R. Sandel, W. T. Forrester, and P. H. Reiff (2003), IMF-driven plasmasphere erosion of 10 July 2000, *Geophys. Res. Lett.*, 30, 1146, doi:10.1029/2002GL016478.

Gonzalez, W. D., B. T. Tsurutani, A. L. C. Gonzalez, E. J. Smith, F. Tang, and S.-I. Akasofu (1989), Solar wind-magnetosphere coupling during intense magnetic storms (1978–1979), *J. Geophys. Res.*, 94, 8835.

Grebowsky, J. M. (1970), Model study of plasmapause motion, *J. Geophys. Res.*, 75, 4329–4333, doi:10.1029/JA075i022p04329.

Huang, C.-S. and J. C. Foster (2007), Correlation of the subauroral polarization streams (SAPS) with the Dst index during severe magnetic storms, *J. Geophys. Res.*, 112, A11302, doi:10.1029/2007JA012584.

Huba, J. D., G. Joyce, and J. A. Fedder (2000), SAMI2 (Sami2 is another model of the ionosphere): A new low-latitude ionosphere model, *J. Geophys. Res.*, 105(A10), 23035–23053, doi:10.1029/2000JA000035.

Huba, J. D. and J. Krall (2013), Modeling the plasmasphere with SAMI3, *Geophys. Res. Lett.*, 40, 6–10, doi:10.1029/2012GL054300.

Huba, J. D. and S. Sazykin (2014), Storm time ionosphere and plasmasphere structuring: SAMI3-RCM simulation of the 31 March 2001 geomagnetic storm, *Geophys. Res. Lett.*, 41, 8208–8214, doi:10.1002/2014GL062110.

Hudson, M. K., S. R. Elkington, J. G. Lyon, and C. C. Goodrich (2000), Increase in relativistic electron flux in the inner magnetosphere: ULF wave mode structure, *Adv. Space Res.*, 25, 2327, doi:10.1016/S0273-1177(99)00518-9.

Hudson, M. K., S. R. Elkington, J. G. Lyon, C. C. Goodrich, and T. J. Rosenberg (1999), Simulation of radiation belt dynamics driven by solar wind variations, in *Geophysical Monograph*, p. 171–182, American Geophysical Union, Washington, DC.

Kennel, C. F. and H. E. Petschek (1966), Limit on stably trapped particle fluxes, *J. Geophys. Res., 71*, 1.

Khazanov, G.V. (2011), *Kinetic Theory of the Inner Magnetospheric Plasma*, Astrophysics and Space Science Library, Springer, New York.

Khazanov, G. V., K. V. Gamayunov, D. L. Gallagher, and J. U. Kozyra (2006), Self-consistent model of magnetospheric ring current and propagating electromagnetic ion cyclotron waves: Waves in multi-ion magnetosphere, *J. Geophys. Res.*, 111, A10202, doi:10.1029/2006JA011833.

Kim, E.-H., J. R. Johnson, H. Kim, and D.-H. Lee (2015), Inferring magnetospheric heavy ion density using EMIC waves, *J. Geophys. Res. Space Phys.*, 120, 6464–6473, doi:10.1002/2015JA021092.

Krall, J. and J. D. Huba (2013), SAMI3 simulation of plasmasphere refilling, *Geophys. Res. Lett.*, 40, 2484–2488, doi:10.1002/GRL.50458.

Krall, J., J. D. Huba, R. E. Denton, G. Crowley, and T.-W. Wu (2014), The effect of the thermosphere on quiet time plasmasphere morphology, *J. Geophys. Res.*, 119, 5032–5048, doi:10.1002/2014JA019850.

Krall, J., J. D. Huba, and J. A. Fedder (2008), Simulation of field-aligned H^+ and He^+ dynamics during late-stage plasmasphere refilling, *Ann. Geophys.*, 26, 1507–1516, doi:10.5194/angeo-26-1507-2008.

Lean, J. L., J. T. Emmert, J. M. Picone, and R. R. Meier (2011), Global and regional trends in ionospheric total electron content, *J. Geophys. Res.*, 116, A00H04, doi:10.1029/2010JA016378.

Lee, H.-B., G. Jee, Y. H. Kim, and J. S. Shim (2013), Characteristics of global plasmaspheric TEC in comparison with the ionosphere simultaneously observed by Jason-1 satellite, *J. Geophys. Res. Space Phys.*, 118, 935–946, doi:10.1002/jgra.50130.

Lemaire, J. and Gringauz, K. I. (1998), *The Earth's Plasmasphere*, Cambridge University Press, New York, p 167.

Lemaire, J. and Schunk, R. W. (1992), Plasmaspheric wind, *J. Atmos. Solar-Terr. Phys.*, 54, 467–477.

Lyons, L. R., R. M. Thorne, and C. F. Kennel (1972), Pitch-angle diffusion of radiation belt electrons within the plasmasphere, *J. Geophys. Res.*, 77(19), 3455–3474, doi:10.1029/JA077i019p03455.

Mannucci, A J., B. T. Tsurutani, B. A. Iijima, A. Komjathy, A. Saito, W. D. Gonzalez, F. L. Guarnieri, J. U. Kozyra, and R. Skoug (2005), Dayside global ionospheric response to the major interplanetary events of October 29–30, 2003 "Halloween Storms", *Geophys. Res. Lett.*, 32, L12S02, doi10.1029/2004GL021467.

Masson, A. et al. (2009), Advances in plasmaspheric wave research with CLUSTER and IMAGE observations, *Space Sci. Rev.*, 145, 137–191, doi:10.1007/s11214-009-9508-7.

Mathie, R. A. and I. R. Mann (2000), Observations of Pc5 field line resonance azimuthal phase speeds: A diagnostic of their excitation mechanism, *J. Geophys. Res.*, 105, 10713.

Maynard, N. C. and A. J. Chen (1975), Isolated cold plasma regions: Observations and their relation to possible production mechanisms, *J. Geophys. Res.*, 80(7), 1009–1013, doi:10.1029/JA080i007p01009.

Menk, F. W., I. R. Mann, A. J. Smith, C. L. Waters, M. A. Clilverd, and D. K. Milling (2004), Monitoring the plasmapause using geomagnetic field line resonances, *J. Geophys. Res.*, 109, A04216, doi:10.1029/2003JA010097.

Meredith, N. P., R. B. Horne, R. M. Thorne, D. Summers, and R. R. Anderson (2004), Substorm dependence of plasmaspheric hiss, *J. Geophys. Res.*, 109, A06209, doi:10.1029/2004JA010387.

Millan, R. M. and R. M. Thorne (2007), Review of radiation belt relativistic electron losses, *J. Atmos. Solar Terr. Phys.*, 69, 362.

Nishida, A. (1966), Formation of plasmapause, or magnetospheric plasma knee, by combined action of magnetospheric convections and plasma escape from the tail, *J. Geophys. Res.*, 71, 5669–5679, doi:10.1029/JZ071i023p05669.

Nosè, M., K. Takahashi, R. R. Anderson, and H. J. Singer (2011), Oxygen torus in the deep inner magnetosphere and its contribution to recurrent process of O^+-rich ring current formation, *J. Geophys. Res.*, 116, A10224, doi:10.1029/2011JA016651.

Pierrard, V. and M. Voiculescu (2011), The 3D model of the plasmasphere coupled to the ionosphere, *Geophys. Res. Lett.*, 38, L12104, doi:10.1029/2011GL047767.

Rasmussen, C. E., S. M. Guiter, and S. G. Thomas (1993), A two-dimensional model of the plasmasphere: Refilling time constants, *Planet. Space Sci.*, 41, 35–43.

Richards, P. G. and D. G. Torr (1985), Seasonal, diurnal, and solar cyclical variations of the limiting H^+ flux in the Earth's topside ionosphere, *J. Geophys. Res.*, 90(A6), 5261–5268, doi:10.1029/JA090iA06p05261.

Roux, A., S. Perraut, J. L. Rouch, C. de Villedary, G. Kremser, A. Korth, and D. T. Young (1982), Wave-particle interactions near Ω_{He^+} observed on board GEOS 1 and 2, 2, Generation of ion cyclotron waves and heating of He^+ ions, *J. Geophys. Res.*, 87, 8174.

Sandel, B. R. (2011), Composition of the plasmasphere and implications for refilling, *Geophys. Res. Lett.*, 38, L14104, doi:10.1029/2011GL048022.

Sandel, B. R., J. Goldstein, D. L. Gallagher, and M. Spasojevic (2003), Extreme ultraviolet imager observations of the structure and dynamics of the plasmasphere, *Space Sci. Rev.*, 109, 1–4, doi:10.1023/B:SPAC.0000007511.47727.5b.

Sandel, B. R., R. A. King, W. T. Forrester, D. L. Gallagher, A. L. Broadfoot, and C. C. Curtis (2001), Initial results from the IMAGE Extreme Ultraviolet Imager, *Geophys. Res. Lett.*, 28, 1439–1442, doi:10.1029/2001GL012885.

Singh, A. K., R. P. Singh, and D. Siingh (2011), State studies of Earth's plasmasphere: A review, *Planet. Space Sci.*, 59, 810–834, doi:10.1016/j.pss.2011.03.013.

Singh, N. and J. L. Horwitz (1992), Plasmasphere refilling: Recent observations and modeling, *J. Geophys. Res. Space Phys.*, 97(A2), 1049–1079, doi:10.1029/91JA02602.

Southwood, D. J. (1974), Some features of field line resonances in the magnetosphere, *Planet. Space Sci.*, 22, 483–491.

Stern, D. P. (1975), The motion of a proton in the equatorial magnetosphere, *J. Geophys. Res.*, 80(4), 595–599, doi:10.1029/JA080i004p00595.

Thorne, R.M., E.J. Smith, R.K. Burton, and R.E. Holzer (1973), Plasmaspheric hiss, *J. Geophys. Res.*, 78, 2156–2202.

Takahashi, K., R.E. Denton, R.R. Anderson, and J.W. Hughes (2006), Mass density inferred from toroidal wave frequencies and its comparison to electron density, *J. Geophys. Res.*, 111, 2156–2202, doi:10.1029/2005JA011286.

Volland, H. (1973), A semiempirical model of large-scale magnetospheric electric fields, *J. Geophys. Res.*, 78(1), 171–180, doi:10.1029/JA078i001p00171.

Young, D. T., S. Perraut, A. Roux, C. de Villedary, R. Gendrin, A. Korth, G. Kremser, and D. Jones (1981), Wave-particle interactions near Ω_{He^+} observed on GEOS 1 and 2, 1, Propagations of ion cyclotron waves in He$^+$-rich plasma, *J. Geophys. Res.*, 86, 6755.

CHAPTER 12

Polar Wind

Robert W. Schunk

CONTENTS

12.1 INTRODUCTION

In the early 1960s, Axford and Hines (1961) and Dungey (1961) proposed that the interaction of the solar wind plasma with Earth's strong intrinsic magnetic field should lead to the formation of a magnetosphere (Figure 12.1). As the solar wind flows past Earth, currents are induced that cause the near-Earth magnetic field to extend to great distances in the anti-sunward direction (past the orbit of the moon). In the polar cap, the magnetic field lines are basically "open," and it was suggested that light thermal ions (H^+, He^+) should be able to escape the topside ionosphere along these open **B**-field lines because the pressure in the ionosphere is higher than that in the overlying magnetosphere. The original models used to describe this process were based on thermal evaporation (Bauer, 1966; Dessler and Michel, 1966). However, shortly thereafter, it was noted that the light ion outflow should be supersonic like the solar wind, and it was called the "polar wind" (Axford, 1968). Subsequently, a hydrodynamic formulation was used to model the supersonic character of the plasma outflow, and the basic features of the "classical" polar wind were then described (Banks and Holzer, 1968, 1969). In these supersonic outflow simulations, it was assumed that the polar wind was driven by the pressure difference between the ionosphere and the deep space.

The argument as to whether the polar wind was subsonic or supersonic was settled when measurements clearly indicated that the flow was supersonic (Johnson, 1979; Shelley et al., 1982; Moore et al., 1986). Also, Chappell et al. (1987, 2000) showed that the polar wind outflow rates were sufficiently high to populate the plasma sheet.

12.2 POLAR WIND ENVIRONMENT

As the solar wind flows past Earth, a dynamo electric field (**E**) is induced that extends across the polar cap from dawn to dusk when the interplanetary magnetic field (IMF) is southward. This "convection" electric field maps down to ionospheric altitudes along the highly conducting **B**-field lines, and at F-region altitudes and above, the ionospheric drifts in an **E** × **B** (anti-sunward) direction across the polar cap (Figure 12.2). At lower latitudes, the **B**-field lines are closed, and in this region, there is a return (sunward) ionospheric flow. Field-aligned currents and particle precipitation occur in the region that separates the anti-sunward and sunward ionospheric flow. However, the ionosphere also has a tendency to corotate with Earth, and this motion is driven by a corotational electric field. When this corotational drift is added to magnetospheric convection, the streamlines of the ionospheric flow take the form as shown in Figure 12.2. Equatorward of the auroral oval on the dawn side, the

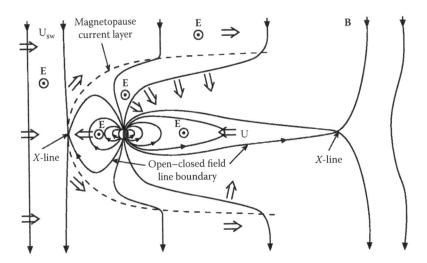

FIGURE 12.1 Diagram showing the plasma flow directions and the magnetic and electric fields in the terrestrial magnetosphere. The Sun is to the left, the north is at the top, and the south is at the bottom. The induced electric field points from dawn to dusk across the polar cap (out of the plane of the figure). (From Lyons, L.R., *Rev. Geophys.*, 30, 93, 1992. With Permission.)

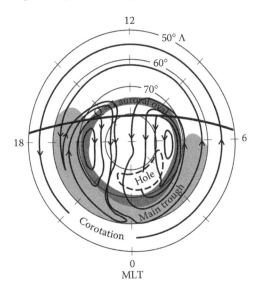

FIGURE 12.2 Diagram showing the drift of the high-latitude ionosphere and the resulting plasma density features in an MLT-invariant latitude coordinate system. The solid lines with arrow show plasma convection trajectories at 300 km and the solid curved line is the terminator. The features shown are the polar hole, the quiet-time oval, and the main electron density trough. (From Brinton, H. C. et al., *J. Geophys. Res.*, 83, 4767, 1978. With Permission.)

corotational and convection drifts are in the same direction, and they add to produce a stronger sunward drift, but on the dusk side, the drifts are in opposite directions and a stagnation region appears. At still lower latitudes, the ionosphere simply corotates with Earth.

The ionospheric features that form in the polar region therefore depend on the strength and direction of plasma convection, the intensity of auroral precipitation, and the location of the terminator. Figure 12.2 shows slow convection, weak precipitation, and equinox. In this case, the dayside plasma drifts slowly across the terminator and into darkness, where it decays due to $O^+ + N_2$ recombination, and this leads to a polar hole. As the plasma drifts into the nocturnal oval, plasma production due to particle precipitation produces enhanced ionization levels. In the dusk stagnation region, the plasma decays to low values and then begins a slow eastward drift, which leads to the "main" electron density trough. However, the situation is completely different if the IMF is southward and the plasma convection is strong. In this case, the anti-sunward drift of the plasma across the terminator acts to produce a tongue of ionization in the polar cap. Also, when the IMF turns northward, the plasma flow can be sunward in the polar cap. Adding in the fact that the convection and precipitation patterns are highly structured and dynamic means that the high-latitude ionosphere and polar wind are highly structured and dynamic, because the primary source of the polar wind results from the resonant charge exchange reaction $O^+ + H \Leftrightarrow O + H^+$.

12.3 CLASSICAL POLAR WIND

For the classical polar wind, the mathematical description is straightforward because it is based on the hydrodynamic continuity and momentum equations (Schunk and Nagy, 2009):

$$\frac{\partial n_s}{\partial t} + \nabla \cdot \left(n_s \mathbf{u}_s \right) = P_s - L_s \tag{12.1}$$

$$n_s m_s \left[\frac{\partial \mathbf{u}_s}{\partial t} + \left(\mathbf{u}_s \cdot \nabla \right) \mathbf{u}_s \right] + \nabla P_s + \nabla \cdot \boldsymbol{\tau}_s - n_s m_s \mathbf{G}$$
$$- n_s e_s \left(\mathbf{E} + \mathbf{u}_s \times \mathbf{B} \right) = \sum_t n_s m_s \nu_{st} \left(\mathbf{u}_t - \mathbf{u}_s \right) \qquad (12.2)$$

where:

n_s is the density for species s

$\mathbf{u}_s$ is the drift velocity

P_s is the production rate

L_s is the loss rate

m_s is the mass

$p_s = n_s k T_s$ is the partial pressure

T_s is the temperature

$\boldsymbol{\tau}_s$ is the stress tensor

e_s is the charge

ν_{st} is the momentum transfer collision frequency between species s and t

t is the time

∇ is the spatial gradient

$\mathbf{E}$ is the electric field

$\mathbf{B}$ is the magnetic field

$\mathbf{G}$ is the gravitational acceleration

k is the Boltzmann constant

The plasma outflow in the polar region is parallel to Earth's magnetic field. For H⁺, the upflow begins above the F-region peak, and the drift velocity increases with altitude, becoming supersonic above about 800–1200 km. The basic characteristics of the supersonic H⁺ outflow can be obtained by invoking several simple, but realistic, assumptions: specifically, (1) daytime steady-state conditions; (2) the ionosphere is isothermal and composed of H⁺, electrons, and neutral atomic oxygen (subscript n); (3) the stress tensor $\boldsymbol{\tau}_s$ is negligible; and (4) there are no field-aligned currents so that $n_e = n_i$ and $\mathbf{u}_e = \mathbf{u}_i$, where the subscript e is for electrons and i is for H⁺. Also, as it turns out, the production and loss of H⁺ is not important at the altitudes where H⁺ is supersonic.

With these assumptions, the H⁺ and electron momentum equations reduce appreciably, and if you add them, the result is

$$n_i m_i u_i \frac{du_i}{dr} + k\left(T_e + T_i \right) \frac{dn_i}{dr} + n_i m_i g = -n_i m_i \nu_{in} u_i \quad (12.3)$$

where r is the coordinate along $\mathbf{B}$ and $\mathbf{G} = -g\mathbf{b}$ ($\mathbf{b}$ is a unit vector along $\mathbf{B}$, positive in the upward direction). Equation 12.3 can be expressed in the following form:

$$u_i \frac{du_i}{dr} + \frac{V_s^2}{n_i} \frac{dn_i}{dr} + g = -\nu_{in} u_i \qquad (12.4)$$

where V_s is the ion-acoustic speed:

$$V_s = \left[\frac{k\left(T_e + T_i \right)}{m_i} \right]^{\frac{1}{2}} \qquad (12.5)$$

The density gradient in Equation 12.4 can be related to the velocity gradient by using the continuity equation (12.1), which in the steady-state case with negligible production and loss processes becomes

$$\nabla \cdot \left(n_i u_i \right) = \frac{1}{A} \frac{d}{dr} \left(n_i u_i A \right) = 0 \qquad (12.6)$$

where the divergence is taken in a curvilinear coordinate system and A is the cross-sectional area of the flux tube. Note that $A \approx r^3$ for radial ion outflow along dipolar field lines near Earth's magnetic poles. Using Equation 12.6, one obtains

$$\frac{1}{n_i} \frac{dn_i}{dr} = -\frac{1}{u_i} \frac{du_i}{dr} - \frac{1}{A} \frac{dA}{dr} \qquad (12.7)$$

and substituting this result into Equation 12.4 yields

$$\left(u_i^2 - V_s^2 \right) \frac{1}{u_i} \frac{du_i}{dr} - \frac{V_s^2}{A} \frac{dA}{dr} + g = -\nu_{in} u_i \quad (12.8)$$

In terms of the ion-acoustic Mach number,

$$M = \frac{u_i}{V_s} \qquad (12.9)$$

Equation 12.8 can be expressed as follows:

$$\frac{dM}{dr} = \frac{M}{M^2 - 1} \left(\frac{1}{A} \frac{dA}{dr} - \frac{g}{V_s^2} - \frac{\nu_{in}}{V_s} M \right) \quad (12.10)$$

This equation is a first-order, nonlinear, ordinary differential equation for the Mach number. Note that the equation contains singularities at $M = \pm 1$—at the points of transition from the subsonic to supersonic flow in the upward ($M = 1$) or downward ($M = -1$) directions. Equation 12.10 is a good representation of H⁺ outflow in the classical polar wind.

Sample solutions from this type of hydrodynamic formulation are given in Figure 12.3 (Raitt et al., 1975). The ion density and drift velocity profiles shown in this figure correspond to steady-state solutions of the

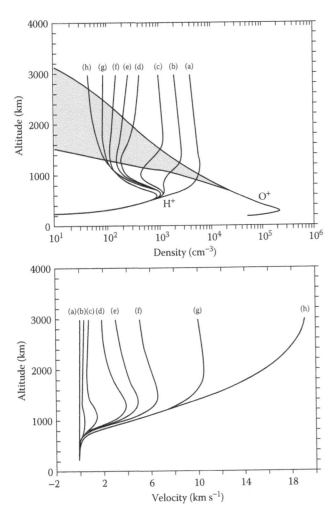

FIGURE 12.3 Calculated H+ density (top) and field-aligned outflow velocity (bottom) profiles for Earth's daytime ionosphere at high latitudes. The different profiles correspond to different assumed H+ outflow velocities at 3000 km: (a) 0.06, (b) 0.34, (c) 0.75, (d) 2.0, (e) 3.0, (f) 5.0, (g) 10.0, and (h) 20 km s⁻¹. The range of O+ density profiles is shown by the shaded region, with the upper O+ curve associated with high H+ outflow and vice versa, where 1 cm⁻³ = 10⁶ m⁻³. (From Raitt, W. J. et al., *Planet Space Sci.*, 23, 1103–1117, 1975. With Permission.)

H+ and O+ continuity, momentum, and energy equations assuming that O+ is gravitationally bound. The different sets of H+ profiles shown in the figure correspond to different "assumed" H+ escape velocities at the upper boundary (3000 km). The H+ ions are produced in and above the F-region by the resonant charge exchange reaction O+ + H ⇔ O + H+, and then the H+ ions drift to higher altitudes. As the assumed H+ escape velocity at 3000 km is increased, the upward H+ velocity increases and the H+ density decreases at altitudes above 600–700 km. Curve (Figure 3a) shows basically

a diffusive equilibrium profile (very small upward H+ velocity); curves (Figure 3b–e) correspond to subsonic outflows, curve (Figure 3f) shows the transonic outflow; and curves (Figure 3g–h) correspond to supersonic outflows. Simulations like these have shown that the H+ escape flux ($n_i u_i$ at 3000 km) increases to a saturation limit as the H+ escape velocity at the upper boundary increases. Basically, the maximum flux that can be drawn from the ionosphere is determined by how much H+ can be created from O+.

As noted previously, the ionosphere at high latitudes continually drifts across the polar region moving into and out of the different high-latitude regions (Figure 12.2). To account for this motion, Schunk and Sojka (1989, 1997) constructed a global ionosphere–polar wind model that extends from 90 to 9000 km for magnetic latitudes greater than 50°. In the E- and F-regions, the model calculates time-dependent, three-dimentional distributions for the electron and ion (NO+, O₂+, N₂+, N+, O+) densities, field-aligned drift velocities, and temperatures from diffusion and heat conduction equations. In the topside ionosphere, the time-dependent, nonlinear, hydrodynamic equations for H+ and O+ are solved self-consistently with the lower ionospheric equations. The transport equations are solved as a function of altitude for convecting plasma flux tubes. The three-dimensional ionosphere–polar wind distributions are obtained by following many plasma flux tubes (up to 1500). This global ionosphere–polar wind model takes account of particle hearing, ion-neutral frictional heating, the magnetic mirror force, anisotropic ion temperatures, supersonic ion outflow, shock formation, ion energy increase during plasma expansions, as well as a myriad of E- and F-region processes.

The initial global ionosphere–polar wind simulations were conducted for an idealized geomagnetic storm. In these simulations, empirical models were used for the plasma convection (Heppner and Maynard, 1987) and particle precipitation (Hardy et al., 1985) patterns, and these empirical models were varied in time to mimic a $K_p = 6$ geomagnetic storm. The idealized storm contained growth, main, and decay phases. During increasing magnetic activity, the particle precipitation and plasma convection patterns expanded, the particle precipitation became more intense, and the convection speeds increased. The reverse occurred during declining magnetic activity. First, for quiet conditions ($K_p = 1$), a diurnally reproducible, global, ionosphere–polar wind system was simulated. Then, from 0400 to 0500 UT, there was an exponential increase in magnetic activity (growth phase); from 0500 to 0600 UT, the magnetic activity was held constant (main phase); from 0600 to 1000 UT, there was an exponential decrease in magnetic activity back to quiet conditions (decay phase); and then the global simulation was continued for several more hours. The particle precipitation and plasma convection patterns varied continuously and smoothly during this idealized storm.

Idealized storm simulations were conducted for four geophysical cases (winter and summer solstices at solar maximum and minimum). Each simulation involved 1000 convecting plasma flux tubes, which provided a 130–200 km horizontal spatial resolution in the polar cap. Figure 12.4 shows a snapshot of the H⁺ density distribution versus altitude and latitude across the polar region at the end of the storm's main phase (0600 UT) for the winter, solar maximum case. Note the spatial structure in the H⁺ density distribution with altitude and latitude. Specifically, note the "bite-out" in the H⁺ density at altitudes from 1400 to 2000 km on the night side at latitudes between 80° and 65°.

The main results obtained from the Schunk and Sojka (1989, 1997) global ionosphere–polar wind "storm" simulations are as follows: (1) The polar wind outflow

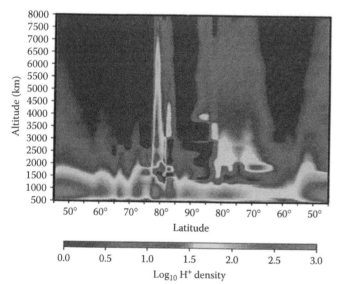

FIGURE 12.4 Snapshot of the H⁺ density distribution versus altitude and latitude across the polar region from noon to midnight. The snapshot is for the end of the storm's main phase (0600 UT) and the geophysical conditions are for winter and solar maximum, where $1 \text{ cm}^{-3} = 10^6 \text{ m}^{-3}$. (From Schunk, R. W., and J. J. Sojka, *J. Geophys. Res.*, 102, 11625–11651, 1997; Schunk, R. W., *Geophys. Monogr.*, 109, 195–206, 1999. With Permission.)

is similar to a campfire with flickering flames; (2) H⁺ bite-outs with altitude can occur at various times during the storm; (3) H⁺ "blowouts" can occur throughout the polar region after the storm commencement; (4) O⁺ becomes the dominant ion to altitudes as high as 9000 km in the polar region during the storm. However, in the classical polar wind, O⁺ does not have sufficient energy to escape, and as the storm subsides, O⁺ flows back down to the ionosphere; (5) the polar wind exhibits a day–night asymmetry (solar zenith effect) due to the variation of T_e across the terminator; (6) both propagating and stationary "vertical polar wind jets" typically occur; (7) the temporal variation of the polar wind at high and low altitudes can be opposite; and (8) ion counterstreaming vertical flows can occur in the polar cap.

12.4 GENERALIZED POLAR WIND

Enhanced upward flows of thermal ions have been measured in the cusp, polar cap, and nocturnal oval by incoherent scatter radars and satellites (Lockwood et al., 1985; Yau et al., 1985; Tsunoda et al., 1989; Yeh and Foster, 1990; Loranc et al., 1991; Wahlund et al., 1992), and these measurements verify that the H⁺ polar wind occurs throughout the entire high-latitude region. The measured plasma upflows are associated with elevated ion and electron temperatures, enhanced upward O⁺ drift velocities, and increased O⁺ densities in the topside ionosphere, and these measurements are in agreement with the Schunk and Sojka (1989, 1997) classical polar wind simulations. However, satellite measurements have also shown that O⁺ is an important ion in the magnetosphere, particularly during storms. They have been observed in several magnetospheric regions, including the plasma sheet, lobe, and distant tail, and they have been shown to significantly affect magnetospheric processes (Peterson et al., 2006, 2008; Yau et al., 2007). Therefore, because O⁺ ions need an energy increase of more than 10 eV (1 eV = 1.602×10^{-19} joule) to escape the ionosphere, nonclassical energy sources must be operating in the polar wind.

Figure 12.5 is a schematic diagram that shows the various nonclassical processes that affect ion outflow (generalized polar wind [GPW]). On the dayside, elevated electron temperatures and escaping photoelectrons (~30 eV; 1 eV = 1.602×10^{-19} joule) are important energy sources for escaping ions. The photoelectrons act to drag the thermal ions with them. In the cusp and aurora, unstable field-aligned currents can excite waves over a range of altitudes that can then accelerate ions

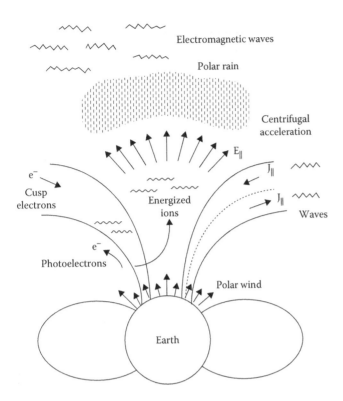

FIGURE 12.5 Schematic diagram showing the nonclassical processes that affect ion outflow from the high-latitude ionosphere. (From Schunk, R. W., and J. J. Sojka, *J. Geophys. Res.*, 102, 11625–11651, 1997. With Permission.)

both parallel and perpendicular to **B**. This can result in the formation of ion beams, conics, and pancake velocity distributions, all of which have sufficient energy to escape the ionosphere. The convection of the ion beams and conics from the cusp and into the polar cap can then lead to unstable plasma conditions as they pass through the slower moving background polar wind. Also, the interaction of the cold polar wind electrons with the hot magnetospheric (polar rain) electrons can result in field-aligned, upward-directed, double-layer electric fields in the polar cap (at ~4000 km), which can accelerate the escaping ionospheric ions. At higher altitudes in the polar cap (~6000 km), electromagnetic wave turbulence can significantly affect the ion outflow through the perpendicular wave heating that is associated with wave–particle interactions (WPIs). In addition, centrifugal acceleration acts to increase ion outflow velocities as the plasma flux tubes convect across the polar cap (cf. Gombosi and Nagy, 1989; Horwitz et al., 1994; Barakat et al., 2003; Banerjee and Gavrishchaka, 2007; Lemaire et al., 2007; Schunk, 2007; Tam et al., 2007; Schunk and Nagy, 2009; Khazanov et al., 2012, 2013; and references therein).

As an example of a nonclassical polar wind process, it is instructive to consider the interaction of the cold upflowing polar wind electrons with the hot magnetospheric electrons (polar rain, showers, and squall) that exist at high altitudes above the polar region. Lemaire and Scherer (1978) suggested that such an interaction could be important, and they modeled the interaction of the upflowing plasma in the cusp with warm magnetosheath particles. They found that because of the hot–cold plasma interaction, a parallel electric field (double layer) can form at about 20,000 km altitude, which can energize the thermal ions in the polar wind. Hot–cold plasma interactions also occur in the polar cap, and the resulting semikinetic simulations (Barakat and Schunk, 1984; Ho et al., 1992; Barakat et al., 1998) were consistent with the previous measurements (Winningham and Gurgiolo, 1982). The simulations indicated that when the cold upflowing polar wind electrons encounter the hot magnetospheric electrons, a parallel electric field forms that acts to separate the hot and cold electron populations. This electric field is directed away from Earth and acts to reflect the cold upflowing electrons and energize the escaping H⁺ and O⁺ ions. The H⁺ energy gain varies from a few electronvolts to about 2 keV (1 eV = 1.602×10^{-19} joule), depending on the hot electron density and temperature.

To show the effect of the hot–cold electron interaction, Barakat et al. (1998) selected a plasma flux tube trajectory from the "idealized" storm simulation of the classical polar wind conducted by Schunk and Sojka (1997). Figure 12.6 shows the trajectory. The plasma flux tube that followed this trajectory started at location "a" on the nightside (4:20 magnetic local time [MLT]; 65° magnetic) at 0300 UT, which was just before the start of the storm. The plasma then convected sunward, turned anti-sunward, convected across the polar cap (segment b–c), moved through the nightside oval, and then convected sunward again. Note that the auroral oval and plasma convection pattern changed continuously with time as the plasma flux tube followed this trajectory. The effect of the hot–cold electron interaction was simulated with a macroscopic particle-in-cell (PIC) model that contained two million particles and covered the altitude range from 2000 km to $8R_E$ (Earth radius). The PIC model included ionospheric H⁺ and O⁺ ions and both hot magnetospheric and cold ionospheric electrons. The ions were kinetic (PIC), whereas the two electron populations were described by simple fluid equations

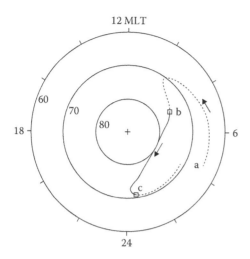

FIGURE 12.6 A plasma flux tube trajectory that starts in the dawn sector of the northern polar region. The segment b–c shows the region where the plasma crosses the polar cap. The geophysical conditions are for winter and solar maximum. (From Barakat, A. R. et al., *J. Geophys. Res.*, 103, 29289–29303, 1998. With Permission.)

(Boltzmann relation). As the plasma flux tube followed the trajectory in Figure 12.6, the time-dependent boundary conditions at 2000 km that were needed for the PIC simulation were adopted from the Schunk and Sojka (1997) storm simulation. Note that the convecting flux tube of plasma was only subjected to the hot magnetospheric electrons in the polar cap (segment b–c in Figure 12.6). In the polar cap, the hot electron density and temperature at 2000 km were held constant, whereas the values above 2000 km were calculated with the macroscopic PIC model. The boundary conditions for the hot electrons were imposed at 2000 km because measurements were available to guide the selection of appropriate density and temperature values.

The trajectory in Figure 12.6 was followed for three simulations; each simulation with a different hot electron temperature but the same value for the hot electron density. The hot electron density was 1% of the cold electron density at point a, and the three temperature values were 30, 100, and 300 times the cold electron temperature at point a (reference location). After the hot electron density and temperature values at 2000 km were determined, they were held constant and only applied in the polar cap, as noted previously. Figure 12.7 shows the PIC simulation results for the three cases. For each case, offset H$^+$ density profiles are plotted at 750 s intervals as the plasma moves along the trajectory shown in Figure 12.6. For the temperature ratio of 30, the cold

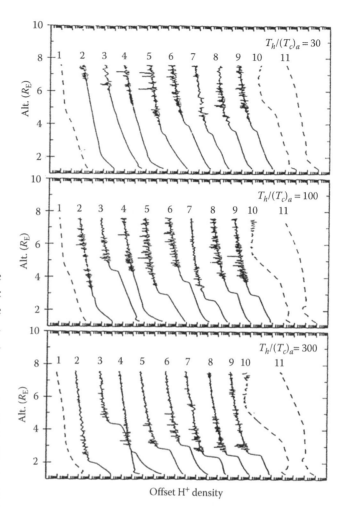

Offset H$^+$ density

FIGURE 12.7 Offset H$^+$ density profiles for hot/cold electron temperature ratios of 30 (top), 100 (middle), and 300 (bottom), and for a hot/cold electron density ratio of 0.01 at the lower boundary altitude of 2000 km. The ratios are relevant to the thermal electron parameters at the reference point a. The interval between successive profiles is 750 s. The solid H$^+$ density profiles correspond to the times when the plasma flux tube is in the polar cap. (From Barakat, A. R. et al., *J. Geophys. Res.*, 103, 29289–29303, 1998. With Permission.)

ionospheric electrons and hot magnetospheric electrons tend to penetrate each other, and a weak, distributed, upward, and parallel (to **B**) electric field is created. However, for hot or cold electron temperature ratios of 100 and 300, the interaction of the hot and cold electrons results in the formation of a "double-layer" electric field. This parallel electric field occurs at an altitude where the H$^+$ density displays a sharp decrease with altitude. Note that the altitude and magnitude of the parallel electric field vary as the plasma flux tube convects across the polar cap; this behavior is primarily determined by the balance of the magnetospheric and ionospheric electron pressures. Because the magnetospheric electron pressure at 2000 km is held constant, the variability shown in Figure 12.7 is due to the variability in the underlying ionosphere. As expected, the escaping H⁺ ions are accelerated as they cross the parallel electric field. For the three cases shown in Figure 12.7, the H⁺ drift velocity above the double layer varies from 10 to 110 km s⁻¹. These values agree with typical H⁺ velocities measured by the POLAR spacecraft (Moore et al., 1997).

The interaction of the cold ionospheric and hot magnetospheric electrons also has a direct effect on the underlying ionosphere. Specifically, the cold polar wind electrons gain energy from the hot "polar rain" electrons, and this energy is subsequently conducted down into the underlying ionosphere, thereby raising the ionospheric electron temperature (Schunk et al., 1986). The elevated electron temperature then increases the ion temperature and the enhanced temperatures modify the ionospheric density. This is shown in Figure 12.8, which provides a snapshot of recent ionospheric simulations

of the effect of downward electron heat flows into the high-latitude ionosphere (David et al., 2011). The results are shown for solar minimum, winter, and quiet magnetic conditions, and the snapshot corresponds to 0500 UT. Three topside (1400 km) electron heat flow values ($Q_T = 0$, 0.5, and 1.5×10^{10} eV cm⁻² s⁻¹, where 1 eV cm⁻² s⁻¹ = 1.602×10^{-15} joule m⁻² s⁻¹) were adopted in three separate simulations, and the largest value is consistent with values deduced from measurements (Bekerat et al., 2007). For the largest downward electron heat flow, the peak density in the F-region (N_mF_2) can change by up to a factor of 10 in some regions of the polar cap.

WPIs are another nonclassical process that is very effective in providing energy for escaping ionospheric ions. This process operates over a wide range of altitudes and is important in the cusp, nocturnal oval, and polar cap. WPIs act to heat the ions in a direction perpendicular to the geomagnetic field, and the ions are then expelled via the magnetic mirror force. The outflowing ions have energies that vary from about 10 eV to 17 keV

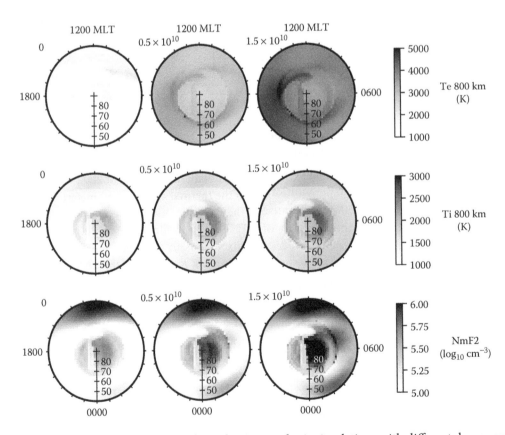

FIGURE 12.8 Comparison of plasma parameters from three ionospheric simulations with different downward electron heat flows into the ionosphere (Q_T). The simulations were for solar medium ($F_{10.7} = 160$), winter (day 357), and quiet ($K_p = 2$) conditions. The plasma parameters are noted on the right, and the downward heat flow values are 0 (left column), 0.5 (middle column), and 1.5×10^{10} (right column) eV cm⁻² s⁻¹; 1 eV cm⁻² s⁻¹ = 1.602×10^{-15} joule m⁻² s⁻¹. (From David, M. et al., *J. Atmos. Solar-Terr. Phys.* doi:10.1016/j.jastp.2011.08.009, 2011. With Permission.)

(Sharp et al., 1974; Shelley et al., 1982; Collin et al., 1987; Horwitz et al., 1992). WPIs are very effective in the cusp, where they energize both atomic and molecular ions to transverse energies in the 10–50 eV range (1 eV = 1.602×10^{-19} joule). The gradient-**B** force acts to drive the energized ions upward, and simultaneously, the magnetospheric electric field acts to drive the ions in an anti-sunward direction across the polar cap. The lower energy heavy ions fall back to Earth, whereas the more energetic ions escape to the plasma sheet. This process is known as the cleft ion fountain (Lockwood et al., 1985).

Numerous modeling studies have been conducted to determine the effects that WPI have on polar wind ions. A Monte Carlo approach was used to study the transverse heating of O^+ due to a cyclotron resonance with broadband electromagnetic turbulence (Chang et al., 1986; Retterer et al., 1987), and O^+ conics were formed that were in agreement with measurements. Ganguli et al. (1988) studied the effects of WPI on field-aligned transport in the auroral return current region, including both the electrostatic ion cyclotron instability (perpendicular ion heating) and the anomalous resistivity (electron heating). The feedback between the plasma conditions and WPI was included, and it resulted in a strong perpendicular heating and density cavity formation. Brown et al. (1991) included WPI effects in a macroscopic PIC model of an auroral magnetic flux tube and showed that the combined action of perpendicular ion heating and velocity filter effects can lead to a variety of ion velocity distributions (rings, conics, etc.). In the polar cap, the turbulence levels are much lower than those on auroral field lines. However, Barghouthi (1997) included an altitude variation for an "imposed" wave turbulence level and showed that although the effects of WPI are larger in the auroral region than in the polar cap, they are important in both domains.

The various nonclassical polar wind processes operate at different locations and altitudes, and they operate in conjunction with the classical polar wind processes. Because the high-latitude plasma convects horizontally through the auroral oval and across the polar cap, a given nonclassical polar wind process only has a finite length of time to operate. Also, a given nonclassical process, such as WPIs, must compete with other processes, such as hot magnetospheric electrons. Therefore, when different nonclassical processes are included in the same simulation, the conclusion as to the importance of a given process may change. To address this issue,

Barakat and Schunk (2006) constructed a "global" GPW model. The GPW model is a fluid–PIC hybrid model, with a standard ionospheric model at low altitudes (90–1200 km) and PIC ions/fluid electrons at high altitudes (1200 km—several Earth radii). The solutions are obtained as a function of altitude for convecting plasma flux tubes, including both classical and nonclassical processes. Typically, 1000 plasma flux tubes are followed as they drift across the high-latitude region, and each flux tube contains one million ions, so a global simulation contains one billion particles.

The global GPW model includes H^+ and O^+ ions, classical polar wind processes, self-collisions and H^+– O^+ collisions, photoelectrons, body forces (electrostatic, gravity, magnetic mirror), WPIs, low-altitude auroral energy sources, cold ionospheric and hot magnetospheric electrons, plasma instabilities, and centrifugal acceleration. The global GPW model calculates ion velocity distributions as well as the standard plasma parameters, such as ion and electron densities, drift velocities, and temperatures.

The first global GPW simulations were conducted for an idealized geomagnetic storm, and four geophysical cases were modeled (summer and winter at solar maximum and minimum). The storm contained growth (1 h), main (1 h), and decay (4 h) phases. During the growth phase, the plasma convection and auroral precipitation patterns expanded, and the electric field strengths and precipitation levels intensified. The reverse occurred during the decay phase. During the main phase, the convection and precipitation patterns remained constant. Figure 12.9 shows snapshots of the O^+/H^+ density ratio just before the commencement of the idealized storm (0400 UT; left panel) and in the middle of the storm's main phase (0530 UT; right panel), and at altitudes of 1500, 2500, and 18,000 km. The important results obtained from the set of global GPW simulations are as follows:

1. During geomagnetic storms, there are large upward O^+ fluxes throughout the polar region and O^+ becomes the dominant ion at almost all altitudes.

2. The nonclassical processes supply sufficient energy to the O^+ ions so that they eventually escape into the magnetosphere. This is in contrast to the results from the hydrodynamic polar wind simulations, where during the storm the O^+ ions flow upward and become the dominant ion at high altitudes, but then flow downward and return to the ionosphere

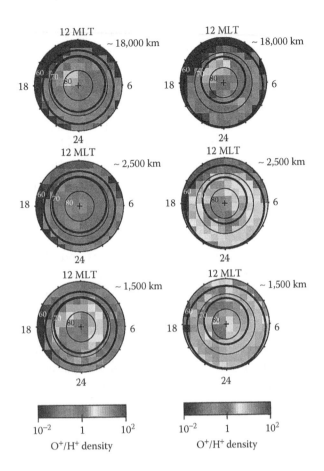

FIGURE 12.9 Snapshots of the O$^+$/H$^+$ density ratio from a global GPW simulation of an idealized geomagnetic storm. The snapshots are for the times just before the commencement of the idealized storm (0400 UT; left panel) and in the middle of the storm's main phase (0530 UT; right panel), and at altitudes of 1500, 2500, and 18,000 km. The coordinates of the dial plots are MLT and magnetic latitude (greater than 60°) in the Northern Hemisphere. The thick black lines show the boundaries of the auroral oval. (From Barakat, A. R., and R. W. Schunk, *J. Geophys. Res.*, 111, A12314, 2006. With permission.)

when the storm subsides. The classical polar wind ions (O$^+$) do not have sufficient energy to escape.

3. Basically, all of the O$^+$ ions that reach 4000–5000 km acquire sufficient energy from nonclassical processes to escape the ionosphere.

4. The escaping ionospheric ions have non-Maxwellian velocity distributions, with the shape and severity depending on the location, altitude, and storm phase. The non-Maxwellian ion velocity distributions that are obtained in the global GPW storm simulations include beam, pancake, conic, toroid, bi-Maxwellian, double-peaked, counterstreaming, and elongated tail distributions.

12.5 NEUTRAL POLAR WIND

Early simulations of neutral particle escape fluxes predicted small values (Tinsley et al., 1986; Hodges, 1994). By contrast, the NASA image for magnetopause-to-aurora global exploration (IMAGE) spacecraft measured large escape fluxes (~1–4 × 10^9 cm^{-2} s^{-1}, where 1 cm^{-2} s^{-1} = 10^4 m^{-2} s^{-1}) of neutral atoms from the high-latitude ionosphere, but the measurements were perplexing in that the neutrals appeared to be coming from all directions (Wilson et al., 2003, Wilson and Moore, 2005). However, global ionosphere–polar wind simulations had indicated that there is a significant increase in the H$^+$ and O$^+$ escape fluxes during geomagnetic storms, and that they are time dependent and spatially nonuniform. Therefore, during geomagnetic storms, large fluxes of superthermal neutrals (H and O) could be created via charge exchange between escaping polar wind ions and the background neutrals in the upper atmosphere. Subsequent simulations by Gardner and Schunk (2004, 2005) that modeled both the polar wind ions and the charge exchange neutrals were able to explain the surprising neutral particle measurements and established the existence of a neutral polar wind.

The escaping polar wind ions (H$^+$ and O$^+$) undergo three characteristic motions. They spiral around the geomagnetic field, flow up and out of the topside ionosphere, and drift across the polar region in response to magnetospheric electric fields. The electric fields cause the plasma to convect into and out of sunlight, the cusp, the polar cap, the nocturnal oval, and the main electron density trough. During their transit, the ions can undergo charge exchange reactions with the thermal and hot geo-coronal neutrals that exist in the upper atmosphere (Figures 12.10 and 12.11). On the one hand, the upflowing O$^+$ and H$^+$ ions can just pass through the upper atmosphere and escape, but they can also have charge exchange reactions with both H and O atmospheric neutrals. The charge exchange reactions would produce upflowing superthermal O and H atoms. These reactions would also produce nonflowing H$^+$ and O$^+$ ions, which would then be accelerated upward by the ambipolar electric field in the polar wind. The initial velocities of the neutral particles created by charge exchange are equal to the velocities of the H$^+$ and O$^+$ parent ions just before the charge exchange reactions. Consequently, at high altitudes, neutral streams of superthermal H and O atoms are created that predominantly flow in the vertical direction (the "neutral polar wind"), whereas at low

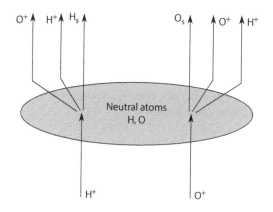

FIGURE 12.10 Diagram showing the production of neutral particles created by charge exchange as the polar wind ions traverse the neutral upper atmosphere. (From Gardner, L. C., and R. W. Schunk, *J. Geophys. Res.*, 109, A05301, 2004. With Permission.)

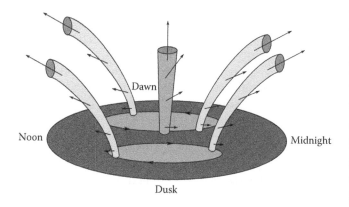

FIGURE 12.11 Diagram showing the three-dimensional flow of the neutral polar wind. Note that the superthermal neutrals move in all directions after creation. (From Gardner, L. C., and R. W. Schunk, *J. Geophys. Res.*, 109, A05301, 2004. With Permission.)

altitudes the superthermal neutrals tend to "move in all directions" owing to the motion of the parent ions, that is, ion gyration, ion upflow, and plasma convection (Gardner and Schunk, 2004, 2005).

12.6 SPACE WEATHER IMPACT

It has been clearly established that the O^+ outflow associated with the GPW has a significant effect on the magnetosphere, and energized O^+ ions have been observed in the plasma sheet, lobe, and distant tail. Consequently, recent modeling efforts have been focused on incorporating multiple ion species in global, magnetohydrodynamic, magnetospheric models (e.g., Winglee et al., 2002; Glocer et al., 2009; Wiltberger et al., 2010; Brambles et al., 2011). These and other studies indicate that O^+ can affect the dayside

reconnection, reduce the cross-tail potential, induce substorms, and modify the plasma sheet. Because the O^+ outflow is continuous time dependent and spatially structured, the magnetospheric electric fields, currents, and particle precipitation will also exhibit these features. This, in turn, will cause the ionosphere–thermosphere system to be highly structured and to vary markedly from hour to hour, particularly during geomagnetic storms and substorms. Strong variations can occur at all latitudes, longitudes, and altitudes.

Unfortunately, hourly variations that occur in the ionosphere–thermosphere system can have a detrimental effect on numerous high-tech operations and systems. They can affect high-frequency (HF) communications, over-the-horizon radars, global positioning satellite (GPS) navigation systems, surveillance, electric power grids, ocean drilling operations, deep space tracking, the Federal Aviation Administration (FAA's) wide area augmentation system, and satellite drag and lifetimes. For example, the structure in the polar wind will cause the structure in the magnetospheric electric fields, currents, and precipitation, which will then cause the structure in the ionosphere and thermosphere. Consequently, these changes will affect HF propagation and cause scintillation, absorption, and deflection of HF waves across a range of frequencies. Also, the structure in the thermosphere will affect the aerodynamic drag on both low-Earth orbit satellites and space debris, which, in turn, affects satellite lifetimes and the tracking of the vast number of objects that are currently in space.

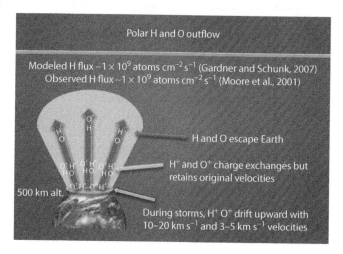

FIGURE 12.12 Schematic diagram showing the ion and neutral polar winds. (Courtesy of Stefan Thonnard, Physics Department, Utah State University, Logan, Utah, 2013). In the figure, $1\ cm^{-2}\ s^{-1} = 10^4\ m^{-2}\ s^{-1}$.

It is instructive to estimate the loss of H during Earth's lifetime, which relates to the loss of H_2O because O also escapes from Earth. Assuming that the loss starts at an altitude of 500 km (radius of ~6.9×10^8 cm), the loss occurs above 60° in both hemispheres (area of ~3×10^{18} cm²), the loss occurs over four billion years (~1.3×10^{17} s), and the outflow rate is 1×10^9 cm⁻² s⁻¹, the total loss of H over four billion years is ~3.8×10^{44} atoms. However, there are 7×10^{25} H atoms per liter of water, so the total water loss is about 5×10^{18} L. According to *Wikipedia*, the estimated total water on Earth now is 1.4×10^{21} L. Therefore, the loss of H due to the ion and neutral winds amounts to less than 0.5% of the water on Earth today (calculations by Stefan Thonnard Physics Department, Utah State University, Logan, Utah) (Figure 12.12).

REFERENCES

Axford, W. I. 1968. The polar wind and the terrestrial helium budget. *J. Geophys. Res.*, 73: 6855–6859.

Axford, W. I., and C. O. Hines. 1961. A unifying theory of high-latitude geophysical phenomena and geomagnetic storms. *Can. J. Phys.*, 39: 1433–1464.

Banerjee, S., and V. V. Gavrishchaka. 2007. Multimoment convecting flux tube model of the polar wind system with return current and microprocesses. *J. Atmos. Solar-Terr. Phys.*, 69: 2071.

Banks, P. M., and T. E. Holzer. 1968. The polar wind. *J. Geophys. Res.*, 73: 6846–6854.

Banks, P. M., and T. E. Holzer. 1969. Features of plasma transport in the upper atmosphere. *J. Geophys. Res.*, 74: 6304.

Barakat, A. R., H. G. Demars, and R. W. Schunk. 1998. Dynamic features of the polar wind in the presence of hot magnetospheric electrons. *J. Geophys. Res.*, 103: 29289–29303.

Barakat, A. R., and R. W. Schunk. 1984. Effect of hot electrons on the polar wind. *J. Geophys. Res.*, 88: 9771–9784.

Barakat, A. R., and R. W. Schunk. 2006. A 3-dimensional model of the generalized polar wind. *J. Geophys. Res.*, 111: A12314. doi:10.1029/2006JA011662.

Barakat, A. R., R. W. Schunk, and H. G. Demars. 2003. Seasonal and solar cycle dependence of the generalized polar wind with low-altitude auroral ion energization. *J. Geophys. Res.*, 108(A11): 1405.

Barghouthi, I. A. 1997. Effects of wave-particle interactions on H⁺ and O⁺ outflow at high latitude: A comparative study. *J. Geophys. Res.*, 102: 22065–22075.

Bauer, S. J. 1966. The structure of the topside ionosphere. In *Electron Density Profiles in the Ionosphere and Exosphere*, ed. J. Frihagen, 387–397. New York: North-Holland.

Bekerat, H. A., R. W. Schunk, and L. Scherliess. 2007. Estimation of the high-latitude electron heat flux using DMSP plasma density measurements. *J. Atmos. Solar-Terr. Phys.*, 69: 1029–1048.

Brambles, O. J., W. Lotko, B. Zhang, M. Wiltberger, J. Lyon, and R. J. Strangeway. 2011. Magnetosphere sawtooth oscillations induced by ionospheric outflow. *Science*, 332(6): 1183. doi:10.1126/science.1202869.

Brinton, H. C. et al. 1978. The high-latitude winter F-region at 300 km: Thermal plasma observations from AE-C. *J. Geophys. Res.*, 83: 4767.

Brown, D. G., G. R. Wilson, J. L. Horwitz, and D. L. Gallagher. 1991. Self-consistent production of ion conics on return current auroral field lines: A time-dependent, semi-kinetic model. *Geophys. Res. Lett.*, 18: 1841–1844.

Chang, T. et al. 1986. Transverse acceleration of oxygen ions by electromagnetic ion cyclotron resonance with broadband left-hand-polarized waves. *Geophys. Res. Lett.*, 13: 636–639.

Chappell, C. R., B. L. Giles, T. E. Moore, D. E. Delcourt, P. D. Craven, and M. O. Chandler. 2000. The adequacy of the ionospheric sources in supplying magnetospheric plasma. *J. Atmos. Solar-Terr. Phys.*, 62: 421–436.

Chappell, C. R., T. E. Moore, and J. H. Waite Jr. 1987. The ionosphere as a fully adequate source of plasma for the Earth's magnetosphere. *J. Geophys. Res.*, 92: 5896–5910.

Collin, H. L., W. K. Peterson, and E. G. Shelley. 1987. Solar cycle variation of some mass dependent characteristics of up flowing beams of terrestrial ions. *J. Geophys. Res.*, 92: 4757.

David, M., R. W. Schunk, and J. J. Sojka. 2011. The effect of downward electron heat flow and electron cooling processes in the high-latitude ionosphere. *J. Atmos. Solar-Terr. Phys.* 73: 2399–2409, doi:10.1016/j.jastp.2011.08.009.

Dessler, A. J., and F. C. Michel. 1966. Plasma in the geomagnetic tail. *J. Geophys. Res.*, 71: 1421–1426.

Dungey, J. W. 1961. Interplanetary magnetic field and the auroral zones. *Phys. Rev. Lett.*, 6: 47–48.

Ganguli, S. B., P. J. Palmadesso, and H. G. Mitchell. 1988. Effects of electron heating on the current driven electrostatic ion cyclotron instability and plasma transport processes along auroral field lines. *Geophys. Res. Lett.*, 15: 1291–1294.

Gardner, L. C., and R. W. Schunk. 2004. Neutral polar wind. *J. Geophys. Res.*, 109: A05301. doi:10.1029/2003JA010291.

Gardner, L. C., and R. W. Schunk. 2005. Global neutral polar wind model. *J. Geophys. Res.*, 110: A10302. doi:10.1029/2005JA011029.

Glocer, A., G. Tóth, T. Gombosi, and D. Welling. 2009. Modeling ionospheric outflows and their impact on the magnetosphere, initial results. *J. Geophys. Res.*, 114(A5): 5216.

Gombosi, T. I., and A. F. Nagy. 1989. Time-dependent modeling of field-aligned current-generated ion transients in the polar wind. *J. Geophys. Res.*, 94(A1): 359–369.

Hardy, D. A., M. S. Gussenhoven, and E. Holeman. 1985. A statistical model of auroral electron precipitation. *J. Geophys. Res.*, 90: 4229–4248.

Heppner, J.-P., and N. C. Maynard. 1987. Empirical high-latitude electric field models. *J. Geophys Res.*, 92: 4467–4489.

Ho, C. W., J. L. Horwitz, N. Singh, G. A. Wilson, and T. E. Moore. 1992. Effects of magnetospheric electrons on polar plasma outflow: A semikinetic model. *J. Geophys. Res.*, 97: 8425–8437.

Hodges, R. R., Jr. 1994. Monte Carlo simulation of the terrestrial hydrogen exosphere. *J. Geophys. Res.*, 99(A12): 23229–23248. doi:10.1029/94JA02183.

Horwitz, J. L. et al. 1992. The polar cap environment of outflowing O+. *J. Geophys. Res.*, 97: 8361.

Horwitz, J. L., C. W. Ho, H. D. Scarbro, G. R. Wilson, and T. E. Moore. 1994. Centrifugal acceleration of the polar wind. *J. Geophys. Res.*, 99(A8): 15051–15064.

Johnson, R. G. 1979. Energetic ion composition in the Earth's magnetosphere. *Rev. Geophys.*, 17: 696–705.

Khazanov, G. V., A. Glocer, M. W. Liemohn and E. W. Himwich. 2013. Superthermal electron energy interchange in the ionosphere-plasmasphere system. *J. Geophys. Res.*, 118: 110. doi:10.1002/jgra.50127.

Khazanov, G. V., I. Khabibrakhmanov, and A. Glocer. 2012. Kinetic description of ionospheric outflows based on the exact form of Fokker-Planck collision operator: Electrons. *J. Geophys. Res.*, 117(A11): A11203. doi:10.1029/2012JA018082.

Lemaire, J. F. et al. 2007. History of kinetic polar wind models and early observations. *J. Atmos. Solar-Terr. Phys.*, 69: 1901.

Lemaire, J., and M. Scherer. 1978. Field aligned distribution of plasma mantle and ionospheric plasmas. *J. Atmos. Solar-Terr. Phys.*, 40: 337–342.

Lockwood, M. et al. 1985. The cleft ion fountain. *J. Geophys. Res.*, 90: 9736.

Loranc, M., W. B. Hanson, R. A. Heelis, and J.-P. St.-Maurice. 1991. A morphological study of vertical ionospheric flows in the high-latitude F-region. *J. Geophys. Res.*, 96: 3627–3646.

Lyons, L.R. 1992. Formation of auroral arcs via magnetosphere-ionosphere coupling. *Rev. Geophys.*, 30: 93.

Moore, T.E. et al. 1997. High-altitude observations of the polar wind. *Science*, 277: 349–351.

Moore, T. E., C. J. Pollick, R. L. Arnoldy, and P. M. Kintner. 1986. Preferential O+ heating in the topside ionosphere. *Geophys. Res. Lett.*, 13: 901–904.

Peterson, W. K., L. Andersson, B. C. Callahan, H. L. Collin, J. D. Scudder, and A. W. Yau. 2008. Solar-minimum quiet time ion energization and outflow in dynamic boundary related coordinates. *J. Geophys. Res.*, 113: A07222. doi:10.1029/2008JA013059.

Peterson, W. K., H. L. Collin, O. W. Lennartsson, and A. W. Yau. 2006. Quiet time solar illumination effects on the fluxes and characteristic energies of ionospheric outflow. *J. Geophys. Res.*, 111: A11S05. doi:10.1029/2005JA011596.

Raitt, W. J., R. W. Schunk, and P. M. Banks. 1975. A comparison of the temperature and density structure in high and low speed thermal proton flows. *Planet Space Sci.*, 23: 1103–1117.

Retterer, J. M. et al. 1987. Monte Carlo modeling of oxygen ion conic acceleration by cyclotron resonance. *Phys. Rev. Lett.*, 59: 148–151.

Schunk, R. W. 1999. Ionospheric outflow, in Sun-Earth plasma connections. *Geophys. Monogr.*, 109: 195–206.

Schunk, R. W. 2007. Time-dependent simulations of the global polar wind. *J. Atmos. Solar-Terr. Phys.*, 69: 2028–2047.

Schunk, R. W., and A. F. Nagy. 2009. *Ionosphere*, 2nd edition. Cambridge: Cambridge University Press.

Schunk, R. W., and J. J. Sojka. 1989. A three-dimensional time-dependent model of the polar wind. *J. Geophys. Res.*, 94: 8973–8991.

Schunk, R. W., and J. J. Sojka. 1997. Global ionosphere-polar wind system during changing magnetic activity. *J. Geophys. Res.*, 102: 11625–11651.

Schunk, R. W., J. J. Sojka, and M. D. Bowline. 1986. Theoretical study of the electron temperature in the high-latitude ionosphere for solar maximum and winter conditions. *J. Geophys. Res.*, 91: 12041.

Sharp, R. D. et al. 1974. Energetic O+ ions in the magnetosphere. *J. Geophys. Res.*, 79: 1844.

Shelley, E. G., W. K. Peterson, A. G. Ghielmetti, and J. Geiss. 1982. The polar ionosphere as a source of energetic magnetospheric plasma. *Geophys. Res. Lett.*, 9: 941–944.

Tam, S. W. Y., T. Chang, and V. Pierrard. 2007. Kinetic modeling of the polar wind. *J. Atmos. Solar-Terr. Phys.*, 69: 1984.

Tinsley, B. A. et al. 1986. Monte Carlo models for the terrestrial exosphere over a solar cycle. *J. Geophys. Res.*, 91: 13631.

Tsunoda, R. T. et al. 1989. Dayside observations of thermal ion upwellings at 800 km altitude: An ionospheric signature of the cleft ion fountain. *J. Geophys. Res.*, 94: 15277–15290.

Wahlund, J.-E., H. J. Opgenoorth, I. HaÈggstroÈm, K. J. Winser, and G. O. L. Jones. 1992. EISCAT observations of topside ionospheric ion outflows during auroral activity: Re-visited. *J. Geophys. Res.*, 97: 3019–3037.

Wilson, G. R., and T. E. Moore. 2005. Origins and variation of terrestrial energetic neutral atoms outflow. *J. Geophys. Res.*, 110: A02207. doi:10.1029/2003JA010356.

Wilson, G. R., T. E. Moore, and M. R. Collier. 2003. Low-energy neutral atoms observed near the Earth. *J. Geophys. Res.*, 108(A4): 1142. doi:10.1029/2002JA009643.

Wiltberger, M., W. Lotko, J. G. Lyon, P. Damiano, and V. Merkin. 2010. Influence of cusp O+ outflow on magnetotail dynamics in a multifluid MHD model of the magnetosphere. *J. Geophys. Res.*, 115: A00J05. doi:10.1029/2010JA015579.

Winglee, R. M., D. Chua, B. Brittnacher, and G. K. Parks. 2002. Global impact of ionospheric outflows on the dynamics of the magnetosphere and cross-polar potential. *J. Geophys. Res.*, 107(A9): 1237.

Winningham, J. D., and C. Gurgiolo. 1982. DE-2 photoelectron measurements consistent with a large scale parallel electric field over the polar cap. *Geophs. Res. Lett.*, 9: 977.

Yau, A. W., P. H. Beckwith, W. K. Peterson, and E. G. Shelley. 1985. Long-term (solar cycle) and seasonal variations of upflowing ionospheric ion events at DE-1 altitudes. *J. Geophys. Res.*, 90: 6395.

Yau, A. W., A. Takumi, and W. K. Peterson. 2007. The polar wind: Recent observations. *J. Atmos. Solar-Terr. Phys.*, 69: 1936–1983.

Yeh, H.-C., and J. C. Foster. 1990. Storm time heavy ion out flow at mid-latitude. *J. Geophys. Res.*, 95: 7881.

CHAPTER 13

Far Ultraviolet Imaging of the Aurora

Larry J. Paxton and Yongliang Zhang

CONTENTS

WE WILL BRIEFLY REVIEW the FUV techniques (such as filter and spectrograph based) used in recent missions (such as Viking, Polar, imager for magnetopause-to-aurora global exploration [IMAGE], thermosphere ionosphere mesosphere energetics and dynamics [TIMED], and Defense Meteorological Satellite Program), key discoveries or major advances from these missions, applications to space weather nowcast/forecast, and future need of next-generation FUV imagers with high sensitivity, spatial resolution, and cadence. This chapter will also place the specification and observation of the aurora in context by introducing the current issues in our understanding of the response of the upper atmosphere during disturbed times, remote sensing of the magnetosphere using auroral observations, and how auroral inputs and the limitations of our understanding of those inputs affect our ability to nowcast and forecast the state of the ionosphere and upper atmosphere.

13.1 INTRODUCTION

13.1.1 Why FUV?

Why do we need to image the aurora? What are we trying to learn from this? What are the advantages to use FUV? These are the fundamental questions that we will look at here with a focus on understanding how we can characterize the ionosphere–thermosphere (IT) system drivers. In addition, the aurora is a visible manifestation of space weather and can be spectacular whether viewed from the ground and/or space (Figure 13.1). The motivation for imaging the aurora arises from two threads. The first is that of studying the morphology of auroral forms that provide the context for our science and a way to connect to the public. The second is concerned with understanding part of energy flow in the near-Earth space and its impact on the IT system.

Because of strong O_2 absorption of FUV emissions and high O_2 densities in the low thermosphere/

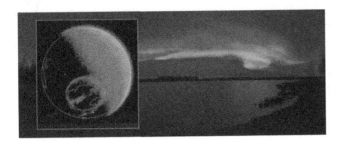

FIGURE 13.1 Auroral images obtained from space-based FUV and ground-based visible observations. From the ground, auroras have mystified humans since we began to question the world. The space age revealed more mystery—the Theta Auroral Oval (inset) and the challenge of understanding the phenomena. (Courtesy of NASA/APOD; from http://www.universetoday.com/117503/solved-the-mystery-of-earths-theta-aurora/ by Tim Reyes on December 24, 2014.)

atmosphere, FUV emissions from altitudes ~50 km or below (Rees, 1989) will be significantly reduced when they are observed from a spacecraft, which is typically at an altitude of a few hundred kilometers and above. The principal advantages of FUV observations are as follows:

- They image under any conditions (day or night, moon up or down, clear or cloudy sky) and the lower atmosphere is essentially black at these wavelengths.

- The algorithms and techniques for understanding the observations are mature.

- The technology is mature and instruments can be small with relatively low data rate.

- They provide global contextual information (i.e., inputs and response).

- They can be applied to a wide range of investigations.

However, FUV observations also have some limitations:

- The instruments require technologies (e.g., gratings, coatings, filters, optical designs) that, while well understood, are supported by a relatively small market.

- They must be calibrated in a vacuum in laboratory.

- Stellar calibration is usually considered to be the only practical means of maintaining calibration traceability on-orbit.

However, despite the excellent auroral pictures or movies that can be obtained in visible wavelengths (e.g., the ground auroral image in Figure 13.1), visible or infrared (IR) observations are subject to a number of issues: limited to nightside and clear day without moon, difficult to retrieve parameters of auroral particles, and so on. It is impractical to monitor the auroral activities continuously at all local times and over a long period of time without the above issues. Overall, FUV observations offer many advantages over visible and IR observations.

This subject (FUV observation of auroral emissions) is, potentially, incredibly broad, and a reasonable treatment requires that only a particular aspect of the subject be examined. In a recent book (Zhang and Paxton, 2015), auroral dynamics and its link to space weather are reviewed in 19 chapters from different perspectives—largely from the outside looking in. That is to say, the book reflects the view of the aurora as the means of understanding the coupling of Earth to the magnetosphere, solar wind, and, hence, Sun. Each one of those chapters and many other chapters, books, and articles, including the book that this chapter is part of, reflect a particular perspective on aurora. There are many other excellent references describing the aurora from various perspectives, including Brekke (2013), Keiling et al. (2012), and many American Geophysical Union (AGU) monographs on various aspects of auroral physics. The reader is urged to consult Akasofu (1974a) for a review of the early history of auroral studies. Other useful early resources include Chamberlain (1961), Eather (1967), Omholt (1971), and Vallance-Jones (1974). In this chapter, we will examine imaging from space with an emphasis on what we learn about the atmospheric response.

13.1.2 Observation Techniques in FUV

There are two ways to detect auroral emissions in FUV from a spacecraft (1) using a fixed band-pass filter or (2) spectrograph imaging. We can generally divide our consideration of the space-based auroral observations into those made with point- (0D), one- (1D), and two-dimensional (2D) detectors. The 0D or point measurement device is typically a photomultiplier tube with optics and a filter. A 1D detector is a line array. It may record a portion of the spectrum or a segment of an image. If an 1D detector is used in a spectrograph, it may be swept across the target region to build up an image in space. Similarly, a 2D detector can be used to

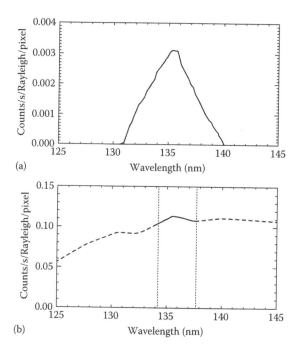

(a)

(b)

FIGURE 13.2 Instrument responsive curves between 125 and 145 nm for (a) IMAGE SI-13 (fixed filter) and (b) TIMED/ GUVI (spectrograph). The GUVI 135.6 channel data covers 134.2–137.7 nm. Emissions from 131 to 140 nm contribute to the SI-13 intensities. (Adapted from Zhang, Y. et al., *J. Geophys. Res.*, 109, A10308, 2004. With Permission.)

image a scene in two spatial dimensions, or it can be used at the output of an imaging spectrograph to image in one spatial dimension and in the spectral domain. By sweeping the instrument field of view, a 2D image in the spatial domain can be formed with spectral information at each point in space.

Figure 13.2a shows a typical example of fixed filter (band-pass) responsivity curve from IMAGE SI-13 instrument (Mende et al., 2000a, 2000c). The SI-13 peak responsivity is between 135 and 136 nm to capture the OI 135.6 emission. Emissions other than 135.6 nm (down to 131 nm and up to 140 nm) also contribute to the observed SI-13 intensities. The sharpness of the band-pass filter depends on its design. Complete rejection of out-of-band contributions is difficult to achieve. Figure 13.2b shows the responsivity curve of TIMED/ Global Ultraviolet Imager (GUVI) (Paxton and Meng, 1999; Paxton et al., 1998, 1999, 2002, 2003; Christensen et al., 2003). Despite that the GUVI responsivity curve is rather flatter than the SI-13 curve, GUVI is able to cleanly reject out-of-band contribution because its bandpass can be selected (e.g., 134.2–137.7 nm in Figure 13.2b; Zhang et al., 2004). Note that the bandpass can be dynamically

adjusted (e.g., 135.5–136.5 nm) and out-of-band contribution can be eliminated. Both fixed filter and spectrographic imaging techniques have their advantages and disadvantages: (1) Fixed filter allows 2D imaging at a high cadence rate, but it is difficult to reject out-of-band contribution and make simultaneous measurements over many different wavelengths. (2) Spectrograph imaging allows to simultaneously measure emissions at different wavelengths and offer excellent out-of-band rejection, but it usually needs a scanning mirror to make a 2D image and thus limits the time resolution of the images. A fixed filter imager is a good candidate for auroral morphology study, whereas a spectrographic imager allows to reliably retrieve quantitative parameters of auroral particles (precipitating electrons and/or ions, the energy source of aurora).

A spectrograph takes light incident at its entrance slit and breaks the light into its constituent wavelengths and disperses this light over a range of angles. Prism and grating spectrographs are available to perform this function at most wavelengths—but prism is not an option in the FUV because most common optical materials absorb all FUV light. Gratings are used in FUV instead. A simple point detector, such as a photomultiplier tube, is placed at the exit slit of the spectrograph. The spectrograph must then be "scanned" in wavelength, by moving either the grating or the detector, in order to produce a spectrum. The method of implementing the scan depends on the spectrograph mount chosen. Generally speaking, grating-based spectrographs are restricted to, at most, a factor of 2 in the difference between the longest wavelength and the shortest wavelength images. With a 2D detector at the focal plane of a spectrograph, an image with spatial and spectral information could then be, in principle, obtained. Here the limitations of spectrograph design became important. A spectrograph produces a focused image but not at all wavelengths or all spatial points. New technology has remedied this to some degree: Instead of using simple plane gratings as in the early Ebert spectrometers or the spherical gratings of the Rowland spectrograph (e.g., Samson, 1967), a grating with a toroidal shape could be used to improve the imaging properties of the system.

Accurate calibration of FUV instruments is necessary to insure the accuracy of the auroral products of FUV measurements. Some users are satisfied with imagery of the morphological behavior of the aurora. From the IT community's viewpoint, accurate measurements of the intensity and a thorough understanding of the instrument characterization are required in order to make use of the measurements. Preflight laboratory calibration provides a vital reference point for instrument performance. On-orbit stellar calibration is required in order for us to be able to continue to insure that the instrument calibration has not changed.

13.1.3 History of FUV Auroral Observations from Space

UV remote sensing has undergone a profound change in the past 20 years as the instruments that provide these measurements have become more capable of making high-quality, spectrally pure measurements. As more bandwidth becomes available for the downlink of the data, 2D detectors can be used. This key technology—a reliable, robust, solar-blind detector that can readily be produced via a commercial process rather than as a

"one-off" in the laboratory—has enabled the production of operational and research instruments that take the advantage of the wealth of spectral signatures of upper atmospheric processes that can be observed in the spectral band between about 115 and 180 nm (also called the "FUV"). The techniques and phenomena of UV remote sensing are described in the works of Samson (1967), Samson and Ederer (2000), Huffman (1992), Meier (1991), Paxton and Anderson (1992), Paxton and Meng (1999), Paxton et al. (1999), and Paxton et al. (2003), among others.

Table 13.1 summarizes the auroral imaging experiments that have flown or are expected to fly in the near future. FUV observations of the aurora were obtained more than 30 years ago. These early observations demonstrated that the aurora could be detected against Earth's FUV background emission. Joki and Evans (1969) and Hicks and Chubb (1970) obtained nadir FUV measurements with a clear local maxima in intensity recorded

TABLE 13.1 Summary of Space-Based Auroral Observations at Optical Wavelengths

Satellite	Type	Launch Date	Instruments	Reference
OGO series	Spin	mid-1960s	FUV photometers 6300Å photometer	Hicks and Chubb (1970)
Apollo 16	Fixed	1972	FUV imager	Carruthers and Page (1976)
AE series	Spin	1972	Two channel photometers, NUV–NIR	Hays (1973)
ISIS-2	Spin	1973	391.4, 557.7, and 630.0 nm photometers	Anger et al. (1973), Shepherd et al. (1973), Cogger and Anger (1973)
STP72-1	Spin	1972	EUV photometer	Weller and Meier (1974)
STP78-1	Spin	1978	EUV spectrometer	Bowyer et al. (1981)
S3-4	Fixed	1978	FUV spectrometer	Huffman et al. (1980), Ishimoto et al. (1989, 1992)
KYOKKO	Fixed	1978	FUV TV camera	Kaneda (1979)
DMSP	Fixed	1974–	Broadband visible cross-track scanner	Rogers et al. (1974)
DE-1	Spin	1981	FUV and visible scanned photometers	Frank et al. (1981)
HILAT	Fixed	1983	FUV and visible cross-track spectrometers	Meng and Huffman (1984)
Polar BEAR	Fixed	1986	FUV and visible cross-track spectrometers	Meng and Huffman (1987)
Viking	Spin	1986	FUV photometric imager	Anger et al. (1987b), Cogger et al. (1988)
Akebono				Kaneda and Yamamoto (1991), Oguti et al. (1990)
Freja				
VIS	Fixed		Visible and FUV imagers	Frank et al. (1995)
UVI	Fixed		FUV imager	Torr et al. (1995)
SSUSI	Fixed	2003, 2006, 2009, 2014	FUV cross-track SIS	Paxton et al. (1992a, 1992b)
MSX	Fixed	1996	FUV-NIR imaging spectrographs	Carbary et al. (1994)
GUVI	Fixed	2001	FUV cross-track SIS	Paxton et al. (1999), Christensen et al. (1994, 2003)
IMAGE FUV	Spin	2000	FUV spectrographs and imager	Mende et al. (2000a, 2000b, 2000c)
ICON FUV		2017?	FUV imager	Rider et al. (2015)
GOLD		2017?	FUV imager	Eastes et al. (2008), Eastes (2009)

? means that the expected launch date is unknown at this time.

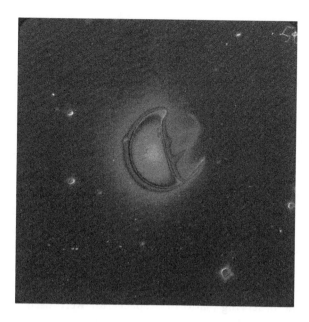

FIGURE 13.3   Picture of Earth in UV light, taken from the surface of the moon by the crew of Apollo 16. The dayside reflects a lot of UV light from Sun, but the nightside shows bands of UV emission from the aurora caused by charged particles. The UV camera was operated by astronaut John W. Young on the Apollo 16 lunar landing mission. This pseudo-color image was artificially reproduced from the original black and white UV camera photograph. (From http://spaceflight.nasa.gov/gallery/images/apollo/apollo16/hires/s72-40821.jpg.)

at auroral latitudes. Auroral UV spectra obtained by the spectrometer on OGO-4 were reported by Gerard and Barth (1976). Carruthers and Page (1972, 1976) and Carruthers et al. (1976) obtained the first global images of Earth at FUV wavelengths. They used an electron-bombarded film-based camera that was placed on the surface of the moon by the Apollo 16 astronauts. They obtained broadband FUV images longward of the Lyman α line and images that included H Lyman α. The system demonstrated the utility of global FUV imaging but clearly had a number of drawbacks not the least of which was the use of film. High Earth orbit in or near the equatorial plane leads to a degraded spatial resolution at the poles. The ISIS-2 mission was able to observe the aurora optically (Anger et al., 1973; Cogger and Anger, 1973; Shepherd et al., 1973) and correlate visible wavelength observations with energetic particle precipitation.

To put these advances in perspective, we should consider Figure 13.3. This image obtained by the Apollo 16 astronauts using an electronographic camera illustrates the features that can be observed in this spectral region. This one "broadband" image (i.e., one that responds to light over a relatively broad spectral region—in this case from shortward of the H Lyman α line at 121.6 nm to a long wavelength end at around 180 nm; details of auroral

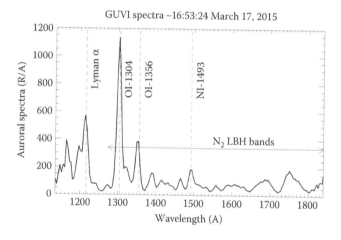

FIGURE 13.4   An example of FUV spectra of auroral emissions from TIMED/GUVI. The major spectral features are identified. The relative amplitude of the features depends on the atmospheric composition, the excitation process, and the viewing geometry.

spectra from TIMED/GUVI shown in Figure 13.4) captures a huge range of phenomena, including hot UV stars, Earth's geocorona, the aurora, equatorial arcs, the dayglow, and the limb profile of the emissions. This image also illustrates some of the challenges of remote sensing—the appearance of stars in the limb profiles, the contrast between the limb and the disk intensity (note that the limb is actually brighter than the disk), the contrast between the day and the night, and the appearance in the same part of the image of more than one phenomenon (e.g., the dayglow in the auroral region or geocoronal emissions in the nightside observations of the equatorial arcs). The ideal instrument would enable one to separate out each of these effects and, from that, add to our understanding of the physics and chemistry of Earth's upper atmosphere as well as its connection to the space environment.

Figure 13.3 also shows the potential of UV remote sensing: The broad range (fixed filter) of observed phenomena, the global view of the coupling of the aurora to the atmosphere, and the possibility of viewing both auroral regions. As we shall see, we can obtain "simultaneous" views of the aurora at both poles from LEO; the requirement is for more than one observing platform. These days, there are many options for flying instruments in space, especially in LEO. Auroral imagers, flying on LEO platforms, can enable imaging of both poles. Geosynchronous orbit cannot effectively image the aurora. Eccentric orbits such as Tundra or Molniya orbits can be used but, generally speaking, are much more expensive in terms of implementing a solution as they often require a dedicated launch. There are two factors that we also must bear in mind: First, we must decide which questions we are going to answer, and then what the mission design drivers (and associated costs) are for the questions we choose to address.

FUV observations of the aurora were started more than 30 years ago. These early observations demonstrated that the aurora could be detected against Earth's FUV background emission. Joki and Evans (1969) and Hicks and Chubb (1970) obtained nadir FUV measurements with a clear local maxima in intensity recorded at auroral latitudes. The ISIS-2 mission was able to observe the aurora optically (Anger et al., 1973; Cogger and Anger, 1973; Shepherd et al., 1973) and correlate visible wavelength observations with energetic particle precipitation. Correlation with ground site measurements (e.g., Lui et al., 1975a, 1975b, 1977) was able to delineate the morphology of proton precipitation in a limited sense. No H emission capability was available; however, the ISIS-2 mission was the first to demonstrate the value of coordinated space- and ground-based observations. The KYOKKO mission used an image memory tube with a KBr photocathode and a MgF_2 window to image the aurora (Kaneda, 1979). The choice of the photocathode determines the longest wavelength nominally thought

of as having a significant red response, whereas the window material determines the short wavelength limit. In principle, the bandpass should have peaked over the range of 120–140 nm. Any residual response in the photocathode, even if several orders of magnitude lower than the band-pass center response, would have precluded sunlit observations of the aurora.

In the 1980s, there were the first real, systematic imaging missions. There was a trend toward imaging in more than one bandpass and an effort to obtain quantitative information from the data. The Dynamics Explorer mission focused on understanding the coupling among the magnetosphere, the ionosphere, and the thermosphere (Hoffman and Schmerling, 1981). The instrument complement was comprehensive, and two dedicated spacecraft were used to obtain the observations from a high-altitude vantage point (DE-1) and low-altitude or *in situ* observations (DE-2). Three spin-scan auroral imaging (SAI) photometers (Frank et al., 1981) were used to image Earth from high altitude at visible and FUV wavelengths (see the global auroral image from DE-1 in Figure 13.1). The three imaging photometers could obtain data at the same time. They were mounted 120° apart on the spinning spacecraft. See Table I and II in Frank et al. (1981) for a description of the bandpasses. The vacuum UV photometer system filters were sufficiently broad that the strong atomic oxygen emissions and hydrogen emissions were mixed (see Fig. 8 in Frank et al. (1981) and the discussion in Rairden et al. (1986)). In principle, the entire spectrum could be modeled and then multiplied by the spectral response of the instrument. This approach requires the ability to model the spectrum quickly and efficiently, and remove the observational geometry (e.g., Craven et al., 1994; Gladstone, 1994; Strickland et al., 1994, 1999; Immel et al., 1997; Nicholas et al., 1997).

The Defense Meteorological Satellite Program (DMSP) Operational Line Scanner (OLS) had been obtaining auroral imagery at visible wavelengths since the mid-1970s (e.g., Akasofu, 1974a, 1974b, 1976; Rogers et al., 1974), but the bright sunlit disk of Earth prevents that sensor from imaging the aurora during daylight conditions. OLS imagery can be correlated with energetic particle data from the auroral electron and ion spectrometer (known as the SSJ/4 and the SSJ/5 sensor). SSJ/4 and SSJ/5 record the ion and electron energy spectra from 30 eV to 30 keV in 20 energy channels with two ranges for electrons and two for ions. The SSJ/4 data provided a data resource for auroral studies and climatological studies (Hardy et al., 1985, 1987, 1989) and identification of the manifestation of magnetospheric boundaries (e.g., Newell et al., 1996). A Special Sensor Ultraviolet Spectrographic Imager (SSUSI) was added to DMSP F16 through F20 (Paxton et al., 1992a, 1992b). More details will be discussed in Section 13.1.2 devoted to SSUSI later. Although Table 13.1 summarizes the major missions using FUV techniques, we discuss examples of auroral observations from some of the satellites that are included to illustrate the features and quality of the imagers in the rest of this section.

The Auroral Ionospheric Mapper (AIM) instrument (Schenkel et al., 1985) on the Defense Nuclear Agency's HILAT satellite (P83-1) was designed to demonstrate the value to the Department of Defense of imaging the sunlit aurora. HILAT was launched on June 27, 1983, into an 830 km altitude circular orbit having an 82° inclination. The AIM's instantaneous field of view (FOV) was quite small (0.4° × 1.7°: across track and along track, respectively). In order to provide contiguous coverage on the ground so that the FOV from one scan matched up with that on the following one, the instrument had to scan over its entire 134.8° field of regard in 3 s, which reduced its effective sensitivity. The scan mirror fed a simple spectrograph that allowed a single wavelength to be selected. AIM, as simple as it was, proved the power of monochromatic imaging by obtaining the first image of the aurora under sunlit conditions (Meng and Huffman, 1984). By using a movable grating, a single selectable 3 nm wide bandpass in the 110–190 nm range could be specified. As a part of the AIM sensor, two fixed wavelength photometers, one at 391.4 nm and the other at 630 nm, were also available. These photometers had a 1 nm bandpass. The HILAT satellite did provide some interesting and useful comparisons between space-based and ground-based observations (e.g., Robinson et al., 1987; Steele and McEwen, 1990). A principal limitation was that the data were collected episodically due to ground station constraints. Figure 13.5 shows two FUV auroral images in the Northern Hemisphere from HILAT satellites. The images were taken at OI-1356Å and NI-1493Å, respectively.

The Aurora and Ionospheric Remote Sensing (AIRS) instrument on the Polar BEAR satellite provided four color images of the aurora. This instrument was launched on November 13, 1986, into a 1000-km circular polar orbit. The images were obtained in two

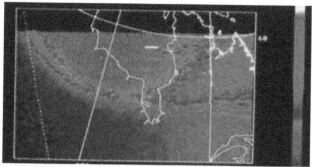

(a)

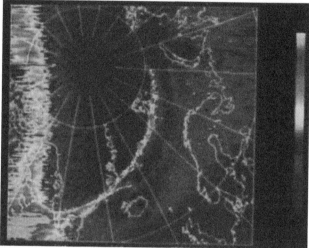

(b)

FIGURE 13.5 FUV auroral images from HILAT satellite (a) at OI-1356Å 10:47–10:51 UT, July 16, 1983, and (b) at NI-1493Å 19:17–19:28 UT, July 23, 1983. (Adapted from Huffman, R. E. et al., *Radio Sci.*, 20, 425–430, 1985. With Permission.)

visible wavelengths and two FUV wavelengths. A simple spectrometer and detector system was used with the elaboration that two photomultipliers were located at the spectrometer's focal plane. Placing two detectors at a fixed spatial separation meant that the two detectors simultaneously sampled wavelengths that were of a fixed spectral separation. AIRS could be operated in a spectral mode in which data were obtained as the grating's angular position was stepped through the full range of angles. The spectral image thus obtained provides a FUV spectrum along the orbital track. There are two important disadvantages to this operational mode: The spectra are not obtained simultaneously and there is no cross-track information on the spatial structure of the aurora. AIRS usually operated in a horizon-to-horizon cross-track scanning mode. A 2D image is built up by using the scan mirror motion in the cross-track direction and the motion of the spacecraft in the along-track

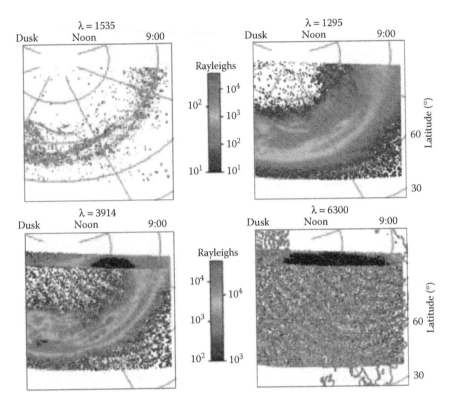

FIGURE 13.6 Auroral images from Polar BEAR satellite at four different wavelengths: 1535, 1295, 3914, and 6300Å (FUV to visible). The 630 nm system did not work as planned. The 391.4 nm system worked well but, as it was sensitive to visible light, could be saturated by sunlight or glint. The FUV channels performed as expected. The images were obtained at Sondrestromfjörd Station on December 26, 1996, from 01:43 to 01:51 UT. (Adapted from Paxton and Meng, *Johns Hopkins APL Tech. Dig.*, 20, 556, 1999. With Permission.)

direction. Examples of auroral images from Polar BEAR are shown in Figure 13.6. The images clearly demonstrate the value of UV imaging: high-quality auroral oval in UV channels and no auroral seen in the visible (6300Å) channel. Polar BEAR data were analyzed using the techniques outlined in Robinson et al. (1987) to establish auroral parameters (Robinson et al., 1992).

Polar BEAR demonstrated that spectral imagery of the aurora had utility that extended beyond morphological applications. Robinson et al. (1992) compared Polar BEAR UV data with coincident Sondrestrom radar data. In 1986 and 1987, the Sondrestrom incoherent scatter radar in Greenland was operated in coordination with selected overpasses of the Polar BEAR satellite. The FUV signatures of the N_2 Lyman–Birge–Hopfield bands were used to deduce the average energy and flux of an equivalent pure electron aurora. From these two quantities, an altitude profile of ionization could be calculated. From an assumption of stationarity, the equilibrium E-region electron density profile (EDP) was deduced, which could then be compared to profiles measured by the incoherent

scatter radar. Robinson et al. (1987) report that the derived profiles agree well with the measured profiles in both the peak electron density and the altitude of the peak. Differences between the measured and derived profiles below the peak are attributed to deviations between the actual spectral shape and the Maxwellian shape assumed in the modeling. Above the peak, the observed differences may be due to the longer times required for the electron density to reach its steady-state value. Basu et al. (1993) analyzed another sequence of Polar BEAR overflights and again reported good agreement between the inferred and measured E-region EDP. The mutual consistency of these different sets of measurements provides confidence in the ability of the different techniques to remote sense large- and small-scale plasma density structures in the E-region at least during sunspot minimum when the convection-dominated high-latitude F-region is fairly weak. In Section 13.3, we investigate the utility of the elaboration of this idea. Polar BEAR data availability was limited and compromised by a pointing stability problem. The satellite was a very low-cost demonstration flight and as such relied upon simple systems for telemetry and attitude control. The telemetry was broadcast in real time and could only be captured during overflight of a ground station. The satellite was gravity gradient stabilized and subject to significant excursions from a nadir orientation. Such excursions, when on the order of a few degrees, are difficult to detect.

The S3-4 satellite included two Ebert–Fastie spectrometers (Fastie, 1963) that spanned the 110–290 nm range (Huffman et al., 1980). The spacecraft was in a 22:30–10:30 local solar time orbit with an inclination of 96.5° and an altitude that varied from 260 to 160 km. The UV instruments shared a common grating drive and were consequently stepped together. The spectra were obtained in 22 s with a 0.1, 0.5, or 2.5 nm bandpass. Examples of the S3-4 observations are given by Ishimoto et al. (1989, 1992).

The Viking experiment was launched on February 22, 1986, into a 98.8° inclination orbit with an apogee of 13,530 km and a perigee of 817 km. The UV imager was described by Anger et al. (1987a, 1987b) and Murphree and Cogger (1988). Viking used two cameras with fixed wavelength ranges: one covered about 134–180 nm and the other 124–160 nm. The imaging instrument is very similar to that implemented on the IMAGE mission. See Murphree et al. (1993) for a discussion of the Viking data and the use of UV imagery. Figure 13.7 shows two examples of Viking UV auroral images with a global coverage.

The Freja mission employed the basic procedure used successfully in the Viking program and allowed UV image repetition rates of 6 s with an exposure time of 0.3 s. Simultaneously, exposing two broadband UV cameras,

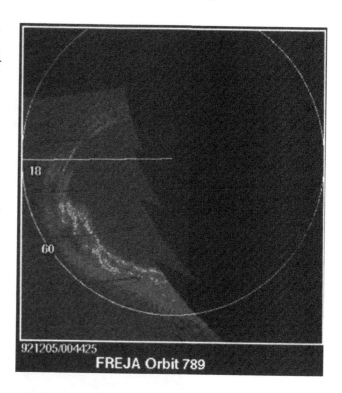

FIGURE 13.8 Five auroral images from Freja satellite (December 5, 1992) mapped in magnetic coordinates. (Adapted from Murphree, J. S. et al., *Space Sci. Rev.*, 70, 421–446, 1994. With Permission.)

the UV instrument generates its normal operational mode 264.6 kb per image pair. Results from the operations confirmed the the design approach and suggested that such UV imagers can be routinely included on challenging low-cost missions such as Freja (Murphree et al., 1994) (see example in Figure 13.8).

The Akebono or EXOS-D satellite was launched on February 21, 1989, into a 75° inclination orbit with an initial apogee of 10,500 km and a perigee of 274 km. The investigation focused on the magnetosphere and polar ionosphere. The instrument complement included a FUV imager along with electric and magnetic fields and *in situ* energetic particle measurements (Tsuruda and Oya, 1993). The FUV imager (Oguti et al., 1990) obtained broadband FUV images covering Lyman α through N_2 Lyman-Birge-Hopfield (LBH) band emission features (see Figure 13.9).

The Polar mission (Acuna, 1995) obtained UV images from two instruments: visible imaging system (VIS) (Frank et al., 1995) and ultraviolet imaging system (UVI) [Torr et al., 1995]. The VIS instrument contains a FUV camera called the Earth camera in addition to the low- and medium-resolution visible cameras.

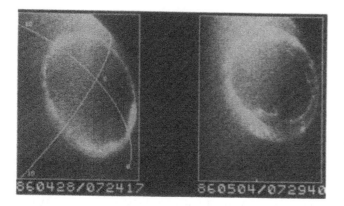

FIGURE 13.7 Viking UV aurora on April 28, 1986 (left) and May 4, 1986 (right). Below each image is the date and time (UT). (Adapted from Cogger, L. L. et al., *Physica Scripta*, 37, 432–436, 1988. With Permission.)

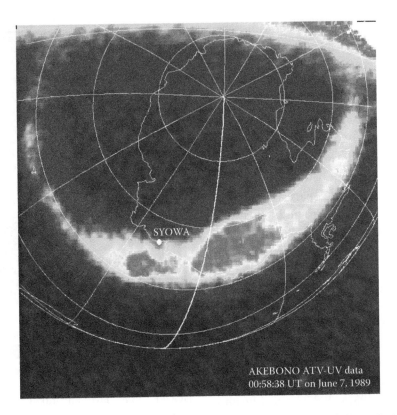

FIGURE 13.9 An UV auroral image obtained by Akebono satellite at 00:58:38 UT, June 7, 1989. (http://polaris.nipr.ac.jp/~dbase/e/100/e/100_10_Akebono_at_Syowa_e.htm.)

Frank et al. (1995) describes the filters (Figure 13.9). A review of the database seems to indicate that this filter is infrequently used. The UVI instrument operates in either of four filters: OI 130.4, OI 135.6, N_2 LBH short (~150 nm), or N_2 LBH long (~170 nm). Images similar to the pair shown in Figure 13.3 (except for the substitution of the Hα for the H Lyman α image) could then be obtained in coordination with Earth camera data or use of simultaneous imagery from UVI. Polar is a spinning spacecraft with the UVI and VIS instruments mounted on a despun platform. Due to implementation issues, the effective spatial resolution of the imagers was on the order of 100 km. Due to highly elliptical orbit of Polar satellite, Polar UVI is able to continuously view global aurora over many hours (see examples in Figure 13.10).

The Midcourse Space Experiment (MSX) (Paxton et al., 1996; Paxton and Meng, 1999) was the first flight of a hyperspectral sensor suite in space. Hyperspectral sensors provide simultaneous spectral and spatial information at hundreds or even thousands of wavelengths. The Ultraviolet and Visible Imagers and Spectrographic Imager (UVISI), with four cameras and five imaging spectrographs, provides spectral images from 110 to 890 nm in 1350 bands. MSX was launched on April

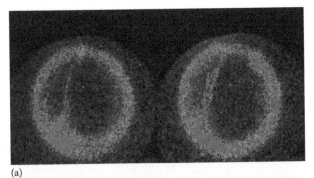

(a)

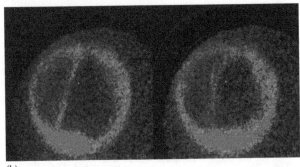

(b)

FIGURE 13.10 Auroral images (LBHS) from Polar UVI on January 11, 1997. The UTs for the four images are 07:35:30 and 07:53:54 (a), and 08:46:02 and 09:16:42 (b), respectively. (Adapted from Newell, P. T. et al., *J. Geophys. Res.*, 104, 95–104, 1999. With Permission.)

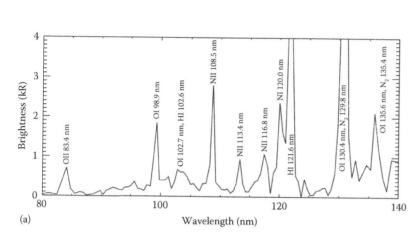

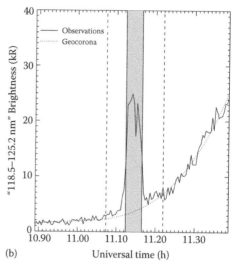

FIGURE 13.11 FUV spectrum (a) observed at a magnetic latitude of −63° and a total brightness (b, solid line) between 118.5 and 125.2 nm over magnetic latitude from −67° to −57.5° from MSX (STP78-1). The dotted line indicates inferred geocoronal intensities. (Adapted from Galand, M. et al., *J. Geophys. Res.*, 107, 2002. With Permission.)

24, 1996, into a 900 km circular polar orbit with a local time near the terminator. Only a rather limited number of data collection events suitable for scientific use were performed. The UVISI imagers could be pointed at a point on the ground or in space and consequently were able to obtain tomographic (3D) views of the aurora. MSX observations of proton aurora were published (Strickland et al., 2001). Galand et al. (2002) describe this chapter in the context of other FUV spectroscopic measurements (see many atomic emission lines in the left panel of Figure 13.11). The right panel of Figure 13.11 shows bright proton aurora (Lyman α ~131.6 nm) over the background geocorona.

IMAGE FUV (Mende et al., 2000a, 2000b, 2000c) inaugurated a new era in auroral imaging as part of the National Aeronautics and Space Administration (NASA) IMAGE mission. The IMAGE mission (Burch, 2000) was designed to image the magnetosphere. IMAGE contained a wide suite of imaging instruments (optical, energetic neutral atom [ENA], and radio) that were designed to explore magnetospheric processes. The IMAGE FUV experiment consisted of a Wideband Imaging Camera (WIC) and two spectrographic imagers SI-12 and SI-13. These instruments are mounted on a spinning spacecraft in an elliptical polar orbit with apogee at latitudes from 40° to 90° in the Northern Hemisphere. The spin period of the IMAGE spacecraft is 2 min, and this was chosen to enable the tracking of magnetospheric substorms. By using a design that provides these three channels at the same time, IMAGE FUV

is able to provide 2 min resolution imagery of energy deposition processes. Mende et al. (2003) review the results from the IMAGE FUV investigation. The ability to measure proton precipitation provided key insights into magnetospheric dynamics and the onset of substorms. Figure 13.12 shows an example of simultaneous measurement of proton and electron auroras by IMAGE FUV. Due to highly elliptical orbit of IMAGE, the FUV instrument is able to monitor global aurora most of time.

The next-generation imagers, SSUSI and GUVI, were built and awaiting launch when IMAGE flew. Fortunately, there was some significant overlap among IMAGE, GUVI, and SSUSI. The GUVI and SSUSI sensors incorporate a large instantaneous FOV (almost 12°), full simultaneous wavelength coverage, and imaging along the slit direction, and are accurately calibrated and fully characterized. They produce horizon-to-horizon images with coverage of one limb (the one looking away from Sun). Figure 13.13 shows typical simultaneous auroral images in five "colors" mapped in the magnetic latitude and local time coordinates from TIMED/GUVI. The simultaneous LBH short (LBHS) and LBH long (LBHL) images allow us to retrieve information of precipitating particles without worrying the errors caused by moving auroras. In the following paragraphs, we will provide more details on SSUSI and GUVI because they are still in operation and their data products are available to public. However, earlier missions (such as Polar and IMAGE) have stopped their operations already.

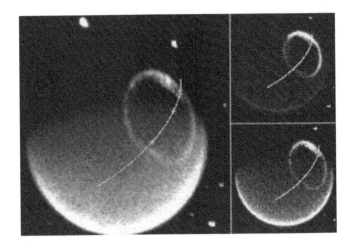

FIGURE 13.12 Images taken by WIC (left), SI-12 (upper right), and SI-13 (lower right) during the FAST orbit 15226 (June 24, 2000, 06:22:20 UT). The ability to make observations of the aurora during simultaneous observations from the ground or space is of great assistance in elucidating auroral processes. The footprint of the FAST track is given by the lines through each image; the position at the snapshot time of the image is shown by a diamond. Local midnight is at the top of the image and FAST first crosses the night part of the auroral oval and then moves to the dayside part. Plus signs mark the location of FAST at 06:20, 06:30, and 06:40, respectively. (Adapted from Frey, H. U. et al., *Geophys. Res. Lett.*, 28, 1135–1138, 2001. With Permission.)

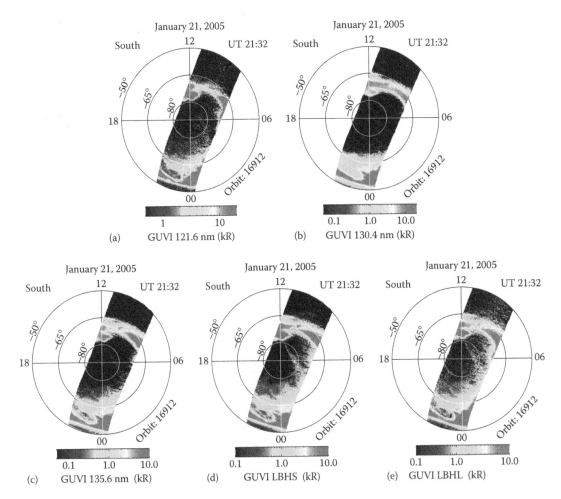

FIGURE 13.13 Auroral images from TIMED/GUVI (~21:32 UT, January 21, 2005) in five colors: (a) 121.6 nm, (b) 130.4 nm, (c) 135.6 nm, (d) LBHS (140–150 nm), and (e) LBHL (165–180 nm). This particular swath is about 2500 km across and has a "pixel" size of 25 km.

The data from TIMED/GUVI and DMSP/SSUSI are particularly helpful to understanding the auroral energy inputs and the response of the system. These instruments are nearly identical in design, measuring FUV airglow emissions such as H, O, and N_2, and absorption by O_2 in five spectral bands simultaneously (Paxton et al., 1992a, 1992b, 1998, 1999, 2002, 2003, 2004; Paxton and Meng, 1999; Humm et al., 1998, 1999; Morrison et al., 2002; Christensen et al., 2003). SSUSI also includes three photometers, in addition to the spectrograph it shares with GUVI. The SSUSI photometers measure nightside 630 (red line), 629 (background), and 428 (N_2 first negative band) signatures. This was originally intended as a test of the auroral LBHS/LBHL algorithms.

SSUSI and GUVI benefited from our institutional experience and the growth of first-principles physics-based auroral radiative transfer computer codes. Armed with theoretical predictions and practical experience, we were able to make decisions about implementation of these next-generation sensors. The instrument design space is constrained by the range phenomena expected to be encountered. Because SSUSI's customer, DMSP, was particularly interested in the measurement of ionospheric parameters, the system had to accommodate an interscene dynamic range of a factor of 10,000. Atmospheric physics models predicted that the dayside intensity at 130.4 nm is as great as 40,000 Rayleighs and the nightside intensities are a few Rayleighs. Using these predictions together with a model of the instrument's end-to-end response, we were able to identify the system design drivers. A slit mechanism, which gives a factor of 4 control over the input rate, and high-speed photon-counting electronics are used to meet the measurement requirements.

The GUVI and SSUSI scanning imaging spectrograph (SIS) have two modes of operation. The imaging mode produces horizon-to-horizon line scan images at all wavelengths in the instrument bandpass. Five colors or selected wavelength intervals are stored in the spacecraft data system for later downlink (see Paxton et al., 1992a, 1992b, 1999, 2004, for discussion). In the spectrograph mode, the entire FUV spectrum is obtained at one selected look angle. The spectrograph mode is intended for instrument characterization and used during "ground truth" observing sequences. SSUSI and/or GUVI environmental parameters produced from data collected during overflights of ground-based facilities will be compared to that facility's measurements of the

same parameters. This "ground truthing" of the data is a necessary step in assuring that the instrument and the data reduction algorithms are performing as advertised.

The SIS consists of a cross-track scanning mirror at the input to a telescope (a 75 mm focal length off-axis parabola system with a 25 mm × 50 mm clear aperture) and a Rowland circle spectrograph. The SIS is an $f/3$ system with a toroidal grating. Two 2D photon-counting detectors are located at the focal plane of the spectrograph. The operating detector is selected by a "pop-up" mirror that is moved into or out of the optical path to direct light from the grating onto one of the two detectors. The detectors employ a position-sensitive anode to determine the photon event location. For convenience, we refer to the quantization of the position determination on the detector as defining a "pixel." The imaging spectrograph builds multispectral images by scanning spatially across the satellite track. One dimension of the detector array is binned by the detector electronics processor into 16 spatial elements (these are the spatial extent of the slit and are oriented parallel to the spacecraft track). The other dimension consists of 160 spectral bins over the range of 115–180 nm. The scan mirror sweeps the 16-spatial element footprint from horizon to horizon perpendicular to the spacecraft motion, producing one frame of 16 cross-track lines in 22 s. The large field of view creates a significant overlap on the disk away from nadir. This, and the slower scan period, increases the effective sensitivity of SSUSI and GUVI by a factor of 10–100 over that of AIRS and AIM.

GUVI is an imaging spectrometer on the NASA TIMED spacecraft. GUVI is a modified version of the SSUSI instrument. These modifications are largely concerned with the details of the implementation on the TIMED spacecraft, for example, a scan range from 80° to −60°, 14 spatial pixels rather than 16, and so on. TIMED focuses on the mesosphere and lower thermosphere (60–180 km), the least explored and least understood region of the atmosphere, from its 74° inclination 600 km orbit. A comprehensive global picture is needed to understand this region, where *in situ* measurements are difficult. The basic physics that controls this region is understood, but the details are difficult to model: the inputs from above and below are not well known. In this region, the atmospheric temperature and temperature gradients reach their largest values, the composition changes from molecular to predominantly atomic, and complex chemical and electrodynamic processes

TABLE 13.2   Major FUV Emission Features

	HI (121.6 nm)	OI (130.4 nm)	OI (135.6 nm)	N₂ (LBHS)	Nitric Oxide	N₂ (LBHL)
AURORAL ZONE (Disk)	Region of proton precipitation H precipitation Doppler shift	Auroral boundary and amount of column O_2 present[a] outflow O Doppler shift	Region of electron and (possibly) proton precipitation outflow O Doppler shift	Used with LBHL to form E_0 and the ionization rate and conductance information HP Radar clutter Charging	NO production	Measure of the effective precipitating flux, used with LBHL to form E_0 and the ionization rate and conductance information
AURORAL ZONE (Limb)	H profile	O outflow	O outflow	N_2 outflow	NO outflow	N_2 outflow
DAYSIDE LIMB	H profiles and escape rate[a]	Amount of O_2 absorption[a]	O altitude profile	Amount of O_2 as seen in absorption	NO profile	N_2 Temperature
DAYSIDE DISK	Column H	Amount of O_2 absorption[a]	Used with LBHS to form O/N_2	N_2, solar EUV	NO column density	Solar EUV
NIGHTSIDE LIMB	H profile and escape rate	Ion/ENA precipitation	EDP HmF2 NmF2 T_{plasma}	Ion/ENA precipitation characteristic energy		Ion/ENA precipitation characteristic energy

Source: Paxton, L. J., and D. E. Anderson. Ultraviolet remote sensing of Venus and Mars. In *Venus and Mars: Atmospheres, Ionospheres, and Solar Wind Interactions. Geophysical Monograph 66*, eds. J. G. Luhmann, M. Tatrallyay, and R. O. Pepin, pp. 113–190, 1992.

[a] The difference in the intensity of the atomic oxygen lines and the ratio of N_2 LBH bands is indicative of the amount of O_2 absorption or, equivalently, the column O_2 present in the line of sight

become the major determinants of composition. The combination of these effects prevents an adequate global description of the upper atmospheric "weather." It is known that the global structure of this region can be perturbed during stratospheric warming events and solar-terrestrial storms, but the overall structure and dynamics of the response are not understood. An important component of this overall question is that of energy balance. To begin to understand that we must determine the auroral inputs and be able to specify the neutral atmosphere, GUVI will provide these key parameters. The fundamental difference between TIMED and the DMSP mission is the difference between exploration and monitoring: TIMED seeks an understanding, whereas DMSP focuses on the knowledge of current conditions. Table 13.2 summarizes the major FUV emissions and associated applications.

13.2 SCIENCE CONTRIBUTIONS OF FUV OBSERVATIONS

13.2.1 Morphology

Global FUV auroral imagers made significant contributions in improving our understanding of auroral dynamics and detected many new auroral features. Those auroral dynamics and new features have been reported in many papers and books. Here we just list a

small part of them, such as hemispheric conjugation, IP shocked aurora, polar rain aurora, cusp aurora, auroral undulation, and ring current auroras (RCAs).

Conjugate aurora: Global auroral imagers, such as IMAGE FUV and Polar UVI, sometimes observed the aurora in two hemispheres simultaneously. Those observations indicate that large-scale features in the aurora are often similar and conjugated between the hemispheres (see Figure 13.14). Figure 13.15 shows a nonconjugate aurora (theta aurora in the Northern Hemisphere and non-theta

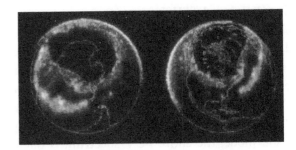

FIGURE 13.14   Polar/VIS Earth Camera (130.4 nm) images (left to right) at 16:38:20, 16:39:13, and 16:40:07 UT on October 24, 2002. The auroral ovals and substorms in both hemispheres are conjugated. (Adapted from "Polar/VIS Science Report" by John Sigwarth et al., January 28, 2005. With Permission.)

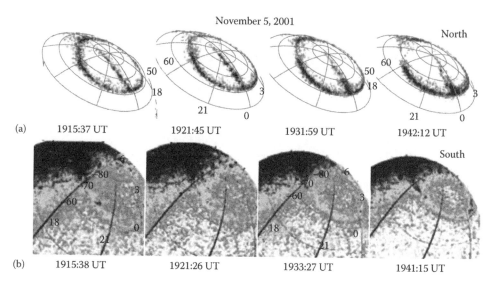

November 5, 2001

(a) 1915:37 UT   1921:45 UT   1931:59 UT   1942:12 UT   North

(b) 1915:38 UT   1921:26 UT   1933:27 UT   1941:15 UT   South

FIGURE 13.15 Nonconjugated theta aurora on November 5, 2001: (a) IMAGE SI-13 images from the Northern Hemisphere and (b) Polar VIS Earth images from the Southern Hemisphere. (Adapted from Østgaard, N. et al., *Geophys. Res. Lett.*, 30, 2125, 2003. With Permission.)

aurora in the Southern Hemisphere). Nonconjugate aurora often relates to high magnetic activity. More example of nonconjugate aurora will be discussed in Section 13.2.2.

IP shocked aurora: When an IP shock hits the magnetosphere, it compresses the magnetosphere resulting in enhanced aurora propagating from the dayside toward the nightside as revealed by Polar UVI auroral data (Zhou and Tsurutani, 1999). Also based on Polar UVI observations, Liou and Sibeck (2014) found that nightside auroral intensities varied periodically (~2.5 mHz) for about 2.5 hours after an IP slow shock under a southward IMF. Coincident Geotail measurements indicated ultralow frequency (ULF) waves with the same frequency in the magnetosphere, suggesting that a standing ULF poloidal mode may be able to change the loss cone, resulting in periodic auroral precipitation (see an example in Figure 13.16).

Polar rain aurora: Intense polar rain electron (especially kiloelectron volt electrons) could cause detectable FUV auroral emissions. Figure 13.17 shows two examples of structured polar rain aurora (dawn–dusk aligned bars) (Zhang et al., 2007). These structures were occurred under a high solar wind speed and likely caused by the Kelvin–Helmholtz (K–H) instability on the high-latitude magnetopause (Zhang et al., 2013).

Cusp aurora: The cusp aurora (Figure 13.18) became a spot and moved to higher latitudes during a northward IMF (see an example in Figure 13.19). This is due to the fact that the location of antiparallel reconnection on the dayside low-latitude magnetopause (under a southward IMF) moves to the high-latitude magnetopause and becomes lobe reconnection (under a northward IMF). The spatial size of the reconnection is also reduced. This leads to a small cusp or cusp aurora. When the IMF is strongly northward for a long time (many hours), double cusp reconnections in two hemispheres could create new closed field lines and the dayside auroras (especially the auroral arcs) extend into the polar cap, join the nightside auroral oval, and make a fully closed magnetosphere (no or few open field lines) (Zhang et al. 2009).

Auroral undulation: Wavy structure or undulation sometimes develops at the equatorward edge of the auroral oval (Zhang et al., 2005b and references therein). An example is shown in Figure 13.19. Note the undulation occurred in both proton (121.6 nm) and electron auroras (135.6 nm, LBHS and LBHL) with a same pattern, suggesting that the particle sources (precipitating electrons and protons) are modulated by the same processes, velocity shear or Rayleigh–Taylor instability. This underlines the importance for having simultaneous observations

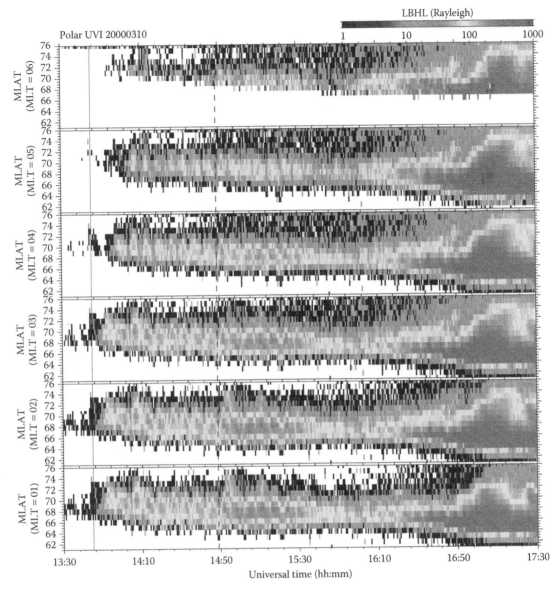

FIGURE 13.16   Auroral keograms at fixed magnetic local times (left: 01–06 MLT and right: 19–24 MLT) derived from the Polar UVI images. (Adapted from Liou, K., and D. G. Sibeck, *Geophys. Res. Lett.*, 41, 2014. With Permission.)

of the electron and proton auroras (i.e., for the N_2 LBH emissions and the H Lyman α line at 121.6 nm) (Figure 13.20).

RCAs: RCAs are the aurora emissions caused by precipitating ions from the ring current (Zhang et al., 2001, 2003b, 2004, 2005, 2006, 2008), and they are mostly at sub-auroral latitude. Figure 13.21 shows a sequence of IMAGE SI-12 (proton aurora) images on November 8, 2000, where a dayside detached aurora (DDA) between 60° and 70° of magnetic latitudes following an IP shock (Zhang et al., 2003b). The detached aurora may appear as a proton spot (PS) (Figure 13.22) (Frey et al., 2001). Different names

were used in literature, such as DDA (Zhang et al., 2001, 2003b), duskside detached aurora (DuDDA), nightside detached aurora (NDA) (Zhang et al., 2005), PS (Frey et al., 2001), and neutral particle aurora (NPA) (Zhang et al., 2006). Note that NPA is not controlled by Earth's magnetic field and can be seen at all latitudes between the northern and southern auroral oval (see Figure 13.23) and the equatorial NPA intensities peaked at the time with the largest disturbance storm time index (Dst) decreasing rate (see Figure 13.24). RCAs (including NPA) are due to hot-ion pitch angle scattering and/ or charge exchange. Figure 13.25 summarizes the features of RCA.

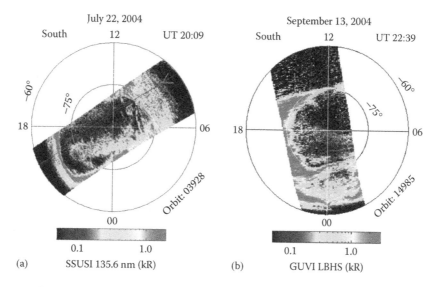

FIGURE 13.17   Two examples of polar rain aurora with a dawn–dusk aligned auroral bar(s): (a) ~20:09 UT, July 2004 from DMSP F16 SSUSI and (b) ~22:39 UT on September 13, 2004, from TIMED/GUVI. (Adapted from Zhang, Y. et al., *Geophys. Res. Lett.*, 34, L20114, 2007. With Permission.)

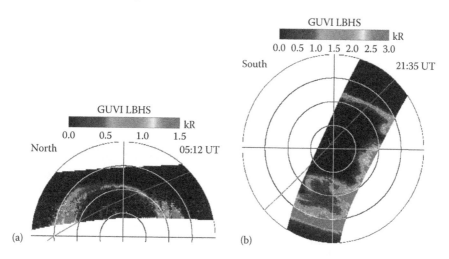

FIGURE 13.18   Thin cusp aurora with broad MLT coverage from TIMED/GUVI: (a) (~05:12 UT on June 17, 2003) and (b) (~21:35 UT on May 29, 2003) under a southward IMF. (Adapted from Zhang, Y. et al., *J. Geophys. Res.*, 110, A02206, 2005. With Permission.)

13.2.2 Remote Sensing of Magnetosphere

Because RCA (except for NPA) is due to hot-ion pitch-angle scattering that requires hot ions with temperature anisotropy ($T_{per}/T_{para} > 1$) and right cold-ion density to match the cyclotron resonance condition, it tends to occur at the plasmapause where cold-ion density changes significantly over a short distance, and it is easy to satisfy the resonance condition. Simulations also show that ring current temperature anisotropy occurs at the edge of ring current flux (Zhang et al., 2008). All of these features indicate that RCA is well correlated with plasmapause and/or ring current edge during storm times.

Observations of RCA can be used to estimate the location of the plasmapause and/or ring current flux edge.

Structures in polar rain aurora (Zhang et al., 2007, 2013) were likely to the K–H waves on the high-latitude magnetopause under high solar wind speed conditions. Observations of PRA can be used to monitor K–H waves and high solar wind speed. Polar rain electrons also show energy-latitude dispersion near the poleward boundary of the nightside auroral oval (Zhang et al., 2011). The dispersion has been reproduced in simulations (Wing and Zhang, 2015). The simulations also provided an opportunity to assess the existing algorithms of estimation tail

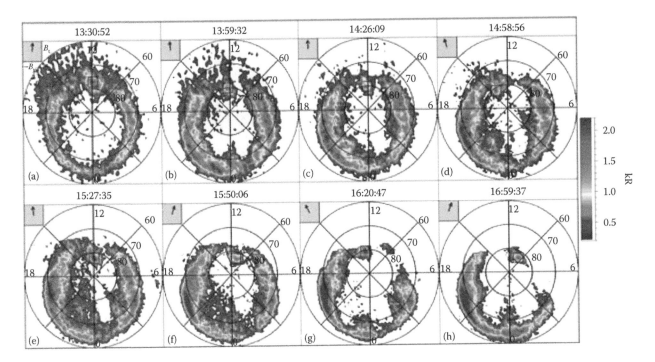

FIGURE 13.19 Auroral images from IMAGE FUV SI-12 under a northward IMF. (Adapted from Frey, H. U. et al., *Nature*, 426, 533–537, 2003. With Permission.)

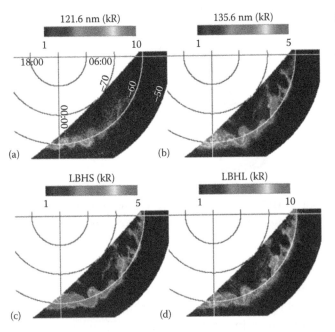

FIGURE 13.20 GUVI images in the Southern Hemisphere around 08:50 UT, October 4, 2002. Undulation was seen on the nightside edge of the auroral oval (both proton and electron auroras). The elements of the figure (a, b, c, and d) are obtained in different subregions of the FUV spectral range. They correspond to emissions from atomic hydrogen in the Lyman a line, atomic oxygen at 135.6nm and the molecular nitrogen LBH short (145-160nm) and LBH long 165-180nm) bands, respectively. (Adapted from Zhang, Y. et al., *J. Geophys. Res.*, 110, A08211, 2005b. With Permission.)

reconnection location using the dispersion (Shirai et al., 1997; Alexeev et al., 2006), which tend to underestimate the distance between Earth and the tail reconnection location. Zhang and Wing (2015) improved the algorithm. Results from the new algorithm on real events are consistent with the tail reconnection location dependence on IMF condition consistent with Geotail *in situ* measurements.

The size of the cusp (extent in magnetic local time) depends on the magnitude and sign of IMF B_z. B_z is the North-South component of the interplanetary magnetic field. For example, as Zhang et al. (2005a) report, the cusp extends over 6 h in magnetic local time for southward IMF. Frey et al. (2003) reported that during northward IMF, there was only a rather small PS. Thus, imaging the cusp aurora provides a measure of the dayside reconnection region.

13.2.3 Particle Precipitation

In addition to the contribution to auroral morphologic studies, FUV observations also allow us to estimate the precipitating particle's properties, such as total energy flux (Q) and mean energy (E_0) on a global scale. Based on Boltzmann three-constituent (B3C) and atmospheric ultraviolet radiance integrated code (AURIC) models (Strickland et al., 1999 and reference therein), the GUVI LBHS and LBHL intensities have been converted

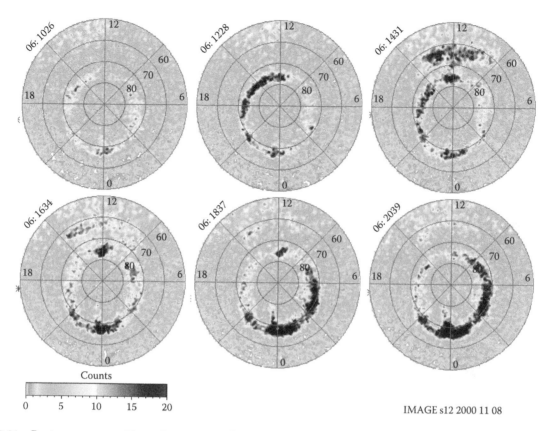

FIGURE 13.21   Proton aurora on November 8, 2000, from IMAGE FUV SI-12. (Adapted from Zhang et al., *J. Geophys. Res.*, 108, 8001, 2003b.)

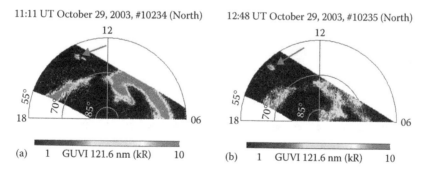

FIGURE 13.22   PSs from TIMED/GUVI on October 29, 2003. (a) The small-scale behavior observed at 11:11 UT. On the next pass (b), the spot had, apparently, moved.

to E_0 and Q maps (see the top panel in Figure 13.25) (Zhang and Paxton, 2008). Figure 13.25c and d shows the comparison with roughly coincident DMSP electron measurements and they appear to agree. The GUVI auroral data over 6 years (2002–2007) have been used to construct a K_p-dependent auroral model (Zhang and Paxton, 2008) covering K_p from 0 to 9. This extends the DMSP particle data-based auroral model (Hardy et al., 1987), which is valid for K_p from 0 to 6.

With the development of the GUVI auroral model, the GUVI or SSUSI auroral images can be used to estimate auroral hemispheric power (HP) and global equatorward boundary, despite the partial auroral oval coverage in GUVI and SSUSI images (see two examples in Figure 13.26). The GUVI model-based equatorial auroral boundary has been included in IRI-2012 (Zhang et al., 2010; Bilitza et al., 2014).

13.2.4 Auroral Ionosphere

Furthermore, auroral ionospheric characteristics (such as E-region peak density and height) an height-integrated conductance can be estimated using

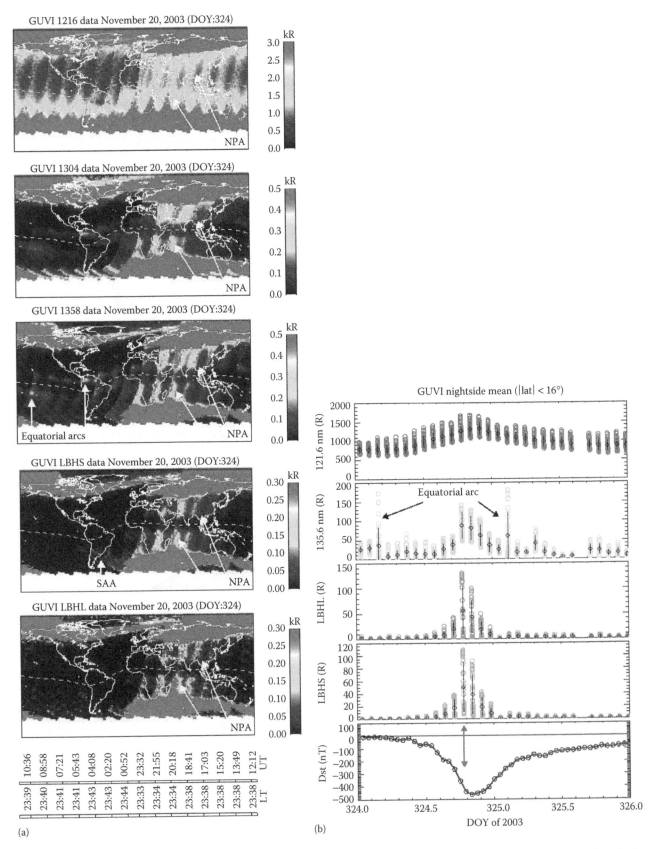

FIGURE 13.23 Nightside auroral emissions in five GUVI colors covering latitudes from southern to northern auroral ovals (a on the left) and equatorial mean intensity shown in b (|latitude| < 16°). The emissions between the southern to northern auroral ovals were caused by precipitation of energetic NPA. (Adapted from Zhang, Y. et al., *J. Geophys. Res.*, 111, A09307, 2006. With Permission.)

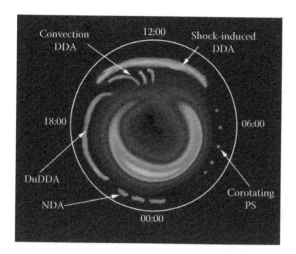

FIGURE 13.24 A composite sketch of RCA. Different names were used in literature, such as DDA, DuDDA, NDA, and PS. (From Zhang, Y., and L. J. Paxton, *Geophys. Res. Lett.* 33, L13107, 2006.)

FUV-based auroral measurements. Figure 13.27 shows comparison of SSUSI and radar HmE (height of the maximum electron density in the E-region ionosphere) and NmE (number density of the maximum in the E-region ionosphere) obtained on January 13, 2005. Note that these auroral ionospheric parameters provide important constraints for global models.

13.3 APPLICATIONS FOR SPACE WEATHER

Geomagnetic storms, especially in their manifestation as auroral displays, are the spectacular evidence of the coupling of Earth to space. These space weather events, and the much more common, lower intensity, high-latitude energy inputs, drive the thermospheric circulation pattern. This change in the usual upwelling from the sub-solar point that arises from the deposition of solar extreme UV (EUV) light causes important and poorly

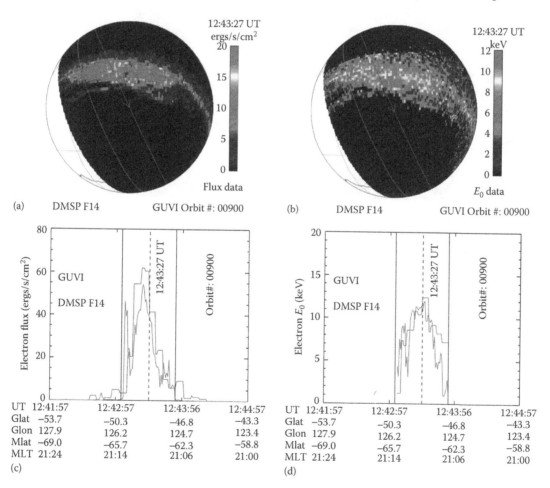

FIGURE 13.25 Maps of estimated electron energy flux (a) and mean energy (b) using GUVI data for orbit 00900 on February 6, 2002. The grid size is 30–30 km². The tracks of TIMED and DMSP F14 spacecraft are indicated from left to right. The tip of the arrow indicates the location of TIMED and DMSP F14 at 12:43:27 UT. (c and d) Comparison between results from GUVI and DMSP F14 along the DMSP F14 track. The two vertical lines indicate the region where the DMSP F14 electron energy flux is above 1.0 erg/(cm² s). (Adapted from Zhang, Y., and L. J. Paxton, *J. Atmos. Solar-Terr. Phys.*, 70, 1231–1242, 2008. With Permission.)

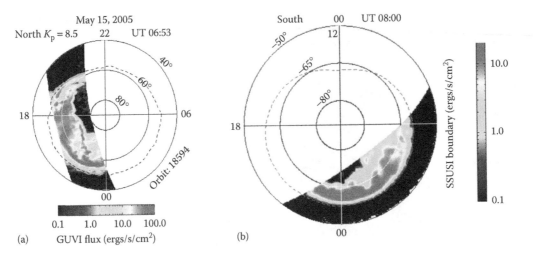

FIGURE 13.26   GUVI (a) and DMSP F18 SSUSI (b) electron energy flux maps with global equatorward boundary (~06:53 UT on May 15, 2005, and ~08:00 UT, October 25, 2011, respectively).

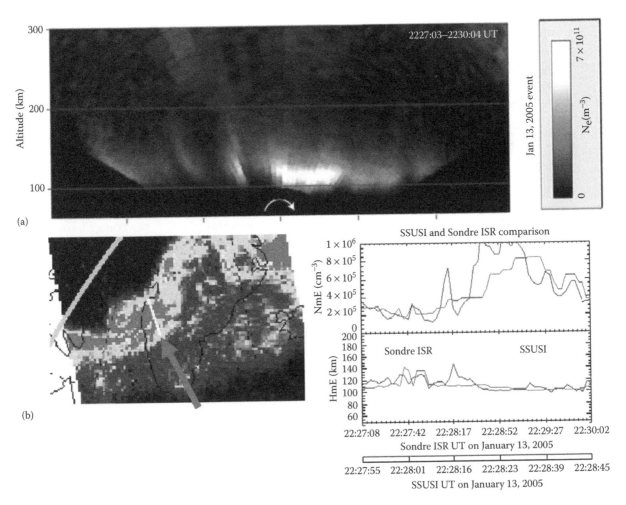

FIGURE 13.27   (a) Auroral ionospheric E-layer EDPs obtained from the Sondrestrom Incoherent Scattering Radar between 22:27 and 22:30 UT on January 13, 2005. (b) DMSP F16 SSUSI auroral image over Greenland. The white bar over intense aurora (indicated by a solid arrow) shows the scan track of the Sondrestrom Incoherent Scattering Radar. (b, right) NmE and HmE along the white bar from SSUSI (the line with more variation) and the radar. (Adapted from Zhang, Y. et al., *Adv. Space Res.*, 46, 1055–1063, 2010. With Permission.)

understood changes in the composition and dynamics of the upper atmosphere.

FUV remote sensing can also help us understand the distribution of conductivity in the polar region. In the most general sense, there are three sources of ionization in the polar region: solar EUV flux, broad precipitation throughout the polar cap, and that seen in the auroral oval. The high-latitude ionosphere plays a role in determining the global state of the magnetosphere (see, e.g., Ridley et al., 2004); magnetospheric currents close through the ionosphere, thereby enabling the occurrence of magnetospheric convection. Ionospheric conductivity determines the flow of these currents. Solar EUV flux can be either measured directly or the dayglow, the photons emitted on the dayside, can be used to determine the effective solar EUV flux at high latitudes. In addition, many of the diagnostics (the signatures of the response of the system) can be observed in the FUV (e.g., as illustrated in Figure 13.3).

The thermosphere is the sink for the energy that is transferred through the system into the high-latitude region. Aside from the local impacts of this energy deposition, there are significant changes in the density, composition, winds, and temperature that propagate outward from the high-latitude region. From a practical standpoint, these perturbations are important because they change the satellite drag environment and have an impact on LEO space mission operators.

Impulsive energy deposition also produces a wave that is observed as a traveling atmospheric disturbance. These waves in the neutrals couple to the ionosphere to create traveling ionospheric disturbances (TIDs). TIDs are important for systems that rely on predictable ionospheric structure and behavior. The practical aspect of this phenomenon manifests itself in problems with global navigation satellite system (GNSS) geolocation, prediction of HF radiowave propagation for communications, and over-the-horizon radar systems.

The impact of high-latitude heating events on thermospheric composition has been examined from many perspectives. We note that the details of the high-latitude convection pattern and the location of the energy deposition and its timing can have significant impacts on the global composition and circulation (e.g., Prölss and Craven, 1998; Zhang et al., 2003b, 2004). Early work (Roble and Ridley, 1987) showed that relatively soft particles (~1–2 keV) produce enough ionization in the summer lower ionosphere to be competitive with solar photoionization;

ion-neutral drag then alters the dynamics in the lower thermosphere. This illustrates the intimate connection between the neutral dynamics and the auroral ionization distribution. In the absence of sunlight, particle precipitation controls the dynamical structure of both the upper and the lower thermosphere. For instance, in that early study, both polar cap temperature and wind velocity increased due to auroral precipitation. We know that the dusk cell convection velocities are enhanced and that enhanced Joule heating produces a significant temperature increase in the polar cap.

During a geomagnetic storm, intense Joule and particle heating cause strong upwelling of the atmosphere around the auroral oval and lead to upward transport of oxygen-depleted or nitrogen-rich air from the lower thermosphere into the F-region (for earlier studies, see Mayr and Volland, 1972; Mayr et al., 1978; Prölss, 1980; Fuller-Rowell and Evans, 1987; Buonsanto, 1999; Strickland et al., 2001). Horizontal neutral winds then redistribute this nitrogen-rich/oxygen-depleted (N_2/O-depleted) air over much of the high-latitude region, and part of the middle-latitude and/or equatorial regions (Evans et al., 1995; Strickland et al., 1995; Strickland et al., 2001, 2004; Zhang et al., 2003a, 2004, 2010, 2011). However, energetic particle precipitation creates enhanced oxide (NO) densities that are transported to mid- and low latitudes following enhanced thermospheric convection (Gerard and Barth, 1977; Barth, 1992; Roble, 1992; Baker et al., 2001; Barth et al., 2003). At low latitudes, the predominant source of NO is solar soft X-rays (2–10 nm) (e.g., Swaminathan et al., 1998; Barth et al., 2003]. One of the primary loss mechanisms of NO is photodissociation by solar UV irradiance. The effective lifetime of NO is about 1 day under sunlit conditions, but the lifetime can be much longer in the winter under dark conditions. NO plays a strong role in the thermospheric energy balance (e.g., Mlynzcak et al., 2003; Lu et al., 2010) as it emits efficiently in the IR and has a long lifetime. It is the source of the terminal ion in the lower ionosphere and can catalytically destroy ozone in the stratosphere. The global thermospheric temperature and density are therefore significantly impacted by NO enhancements (Zhang et al., 2014).

From the perspective of coupling of the high-latitude inputs to the response of the upper atmosphere, understanding storm-time NO variations is an important and urgent problem: Unless we address this key parameter, it will be difficult to make progress toward a deeper understanding of the thermospheric dynamics. For instance, Knipp et al. (2013) reported the significant overestimation of the thermospheric neutral density by models during sheath-enhanced geomagnetic storms. The study suggested that enhanced NO density, due to this high-latitude source, is likely the causative mechanism for the much lower observed thermospheric neutral densities than those predicted by models.

When we consider thermospheric dynamics, we generally view the drivers as being confined to auroral processes, that is, particle precipitation, Joule heating, the Poynting flux, and ohmic dissipation in field-aligned currents. However, there is recent evidence that understanding the polar cap processes is important (Huang et al., 2014). Hajra et al. (2014) examined a high intensity long duration continuous AE activity (HILDCAA) (where AE is the auroral electrojet index) event and found that a large amount of the energy was deposited in the polar cap region. This is in accord with an earlier study from the same research group (Guarnieri, 2006).

Other instruments, ground and space based, also have their strengths and weaknesses. The instruments that we use to explore near-Earth geospace have shaped the way that we understand the system. Ground-based systems can be used to determine the flux and characteristic energy of precipitating particles (e.g., Strickland et al., 1989). Strickland et al. (1983) describe the application of multispectral imagery to the determination of the average or characteristic energy (E_0) and flux (Q) of auroral particles. The technique requires the simultaneous imagery of two wavelength bands: One must be where absorption by O_2 is important (usually called the N_2 LBH short or LBHS band) and the other must be where the O_2 absorption is not important (usually called the LBH long or LBHL band). The exact formulation of the bandpasses determines the mathematical relationship that translates the intensity into Q and E_0. The application of these relationships has been described elsewhere (see, e.g., Zhang and Paxton, 2008).

The application of the two-color technique was outlined by Paxton et al. (1992a, 1992b) for use in the SSUSI imaging spectrographs. Germany et al. (1994) used this technique as well. Because the emissions of the LBHS and LBHL originate from the same species, their ratio is nearly independent of compositional changes. Germany et al. (1997) applied this method to images from the Ultraviolet Imager (UVI) on board the Polar satellite. They found the inferred energy parameters (E_0 and Q)

generally agree in magnitude and morphology with selected DMSP flights. The difference between the estimated and *in situ* measured E_0 and Q were due to errors in the N_2 LBH cross sections, the modeling process, instrumental calibration, and pixel-dependent Poisson signal statistics (Germany et al., 1997). It is important to note that the Polar UVI LBHS and LBHS images used for the E_0 and Q estimation were not simultaneous. The dynamic nature of the aurora (whether we speak of the time-varying intensity and/or flux at a location or the magnetospheric view of the dynamical reconfiguration of the electric and magnetic fields) places constraints on the time between samples. One might well argue that it is better to have simultaneous multispectral imagery if one seeks energy inputs.

13.4 CHALLENGES AND FUTURE IMPROVEMENTS

The key challenges in FUV observations are to have continuous and global observations with high sensitivity and resolutions (spectral, space, and time). Meeting these challenges depends on advances in technologies. Other challenges are how to use the FUV measurements in different applications, such as inputs for global magnetosphere, ionosphere, and thermosphere models. Here we focus on the discussion of the challenges in applications.

One of the first things that models the thermospheric global circulation addressed was the need to incorporate auroral energy inputs into the model (e.g., Roble and Ridley, 1987) in a somewhat realistic manner. In the nearly three decades, since that time there has been a steady evolution in the sophistication of the approaches used. As our models have become more sophisticated, we have come to the realization that a "smooth" single parameter model of the particle inputs or the extent of the auroral oval (Holzworth and Meng, 1975; Wallis and Budzinski, 1981; Spiro et al., 1982; Whalen, 1983; Hardy et al., 1985; Hardy et al., 1987, 1989; Roble and Ridley, 1987; Zhang and Paxton, 2008). These models have been valuable tools for monitoring space weather and served to provide inputs to global simulations. However, there is still a room for improvement. Most of the previous global auroral models are based on precipitating charged particle measurements by satellites at a few fixed local times (e.g., DMSP satellites). These measurements are done along the satellite's tracks that cover a very limited magnetic local time (MLT) range, even when data from a number of satellites are used. The use of a basis function that depends on the latitude and MLT leads to distortions or inaccuracies due to the limited local time coverage of the data that constrains the functional fit. However, global FUV imagers, such as TIMED/GUVI, IMAGE/FUV, and Polar/UVI, can provide a wide MLT coverage. The technique has been developed to estimate the two key auroral parameters, mean electron energy (E_0) and electron energy flux (Q), from the FUV auroral data.

One of the biggest questions in the study of the high-latitude inputs is—what are the important spatial and temporal scales for energy deposition? Energy can be dissipated in the upper atmosphere by direct deposition and through the dissipation of energy from the upper atmospheric current system due to the highly variable and highly structured conductivity distribution. Spatial scales in the aurora range from meters to hundreds of kilometers. This energy input, aside from varying with location, is also highly variable with time. We know from ground-based and space-based observations that the timescales for the variation of the auroral input range from milliseconds to hours (or longer). In addition, there is clear evidence that past inputs (or preconditioning) are important in setting the stage for the evolution of the system. The energy inputs are only part of the problem. The response of the high-latitude system involves the conversion of particle and electric field energy into momentum—in particular the momentum of the neutrals. This response while slower to evolve persists longer and may well have a profound role in determining the evolution of the high-latitude system.

Correlative analysis of *in situ* and remote sensing measurements of the high-latitude inputs and response will enable us to understand how to capture the important drivers needed for accurate forecasts and nowcasts.

The energy inputs are specified from (1) *in situ* measurements of the particle energy and flux spectrum, such as by the DMSP satellite SSJ/4 and/or SSJ/5 instrument (e.g., Mitchell et al., 2013); (2) auroral observations from space; and (3) ground-based images (e.g., Donovan et al., 2006). For the first type of measurement, the distribution function of the electron and ion precipitation is being directly measured (within a certain energy range). Optical measurements of the aurora from satellites have the advantage that they have significantly better spatial extent in their measurements but are limited by the satellite's orbit and the inversion techniques that are used

to derive precipitation characteristics from radiances. Ground-based optical measurements are not limited by orbit but are limited by several factors: They can only be used during Sun- and Moon-down conditions, and the sky must be free of clouds. The spatial extent of the measurements is also often limited, even for large networks. In addition, visible wavelength measurements are subject to uncertainties that arise from the scattering of light by the atmosphere.

Assimilative models can provide a self-consistent to modeling these inputs. Because more energetic particles deposit their energy deeper into the thermosphere, and the different wavelengths of the auroral emissions also originate from different regions of the thermosphere, most techniques utilize ratioing images of the aurora in different wavelength bands. Lookup tables then relate the observed spectral ratio to the characteristic energy of the precipitating particles. There are many assumptions that go into these ratioing techniques (e.g., it depends on the energy dependence of the precipitating particles and the background atmosphere). The resulting ionization rate profile and the derived EDP then have inherent assumptions about the background atmosphere (e.g., Sharber et al., 1993, 1996). When these are then converted in conductivity profiles, this is further compounded because the conductivity depends on the electron-neutral and ion-neutral collision frequencies, which depend on the electron and ion density profiles and the temperatures of the electrons, ions, and neutrals (Rasmussen et al., 1988; Senior, 1991; Lilensten et al., 1996; Schunk and Nagy, 2000).

At high latitudes, the EDP plays a critical role in determining the energetics of the upper atmosphere. For example, low-energy particles deposit their energy higher and, because the local density is lower, may cause a greater temperature increase locally, than would be seen for the same total energy input but a higher mean energy. For higher mean energies, the ionization will occur lower in the thermosphere, where the ion loss rate is faster. The total electron column density may be lower for the same flux.

Auroral processes also change the energy balance of the high-latitude region through indirect means. Auroral particles cause the dissociative excitation of nitrogen, leading to the creation of nitric oxide. An increase in NO leads to a rapid cooling of the thermosphere, which drives departures from the typical, climatological mean distribution. This cooling effect has been termed the "NO thermostat" (Mlynczak et al., 2003).

Data assimilation (see, e.g., Bust et al., 2004; Schunk et al., 2004) provides the means for directly incorporating physical measurements into computational models of the physical processes in the upper atmosphere. These rapidly evolving techniques are similar in a broad sense to those used earlier to estimate ionospheric conductivity. The assimilative mapping of ionospheric electrodynamics (AMIE) model creates the best-fit electric potential distribution consistent with high-latitude ionospheric observations that constrain the high-latitude convection pattern and ionospheric conductivities. The AMIE procedure is an optimally constrained, weighted, least-squares fit of the electric potential distribution to diverse types of high-latitude ionospheric observations that give information about the high-latitude convection pattern and ionospheric conductivities (Richmond and Kamide, 1988; Richmond, 1992). In the future, datasets such as SuperDARN, SuperMAG, and AMPERE will be excellent additions to these global assimilative models as they provide direct measurements of the electrodynamic signature of high-latitude inputs.

13.5 SUMMARY

New approaches to using data from ground and space assets offer the potential for a better understanding of the coupled Sun–Earth system and, most importantly, the response of the coupled atmospheric system to the external drivers (particle precipitation, induced currents, penetration electric fields, etc). FUV instruments operating now provide the Heliophysics research community and the space weather operational user with the opportunity to extend our knowledge and capabilities. We have begun to understand the coupled space environment and can make predictions.

One of the fundamental questions in the study of the high latitude inputs is—what are the important spatial and temporal scales for energy deposition? Energy can be dissipated in the upper atmosphere by direct deposition and through the dissipation of energy from the upper atmospheric current system due to the highly variable and highly structured conductivity distribution. Spatial scales in the aurora range from meters to hundreds of kilometers. This energy input, aside from varying with location, is also highly variable with time. We know from ground-based and space-based observations that the timescales for the variation of the auroral input range from milliseconds to hours (or longer). In addition, there is clear evidence that past inputs (or preconditioning) are important in setting the stage for the evolution of the system. The energy inputs are only part of the problem. The response of the high latitude system involves the conversion of particle and electric field energy into momentum—in particular the momentum of the neutrals. This response while slower to evolve persists longer and may well have a profound role in determining the evolution of the high-latitude system.

Correlative analysis of *in situ* and remote sensing measurements of the high-latitude inputs and response will enable us to understand how to capture the important drivers in our models. To do this, we will determine the important spatial and temporal scales for each of these phenomena based on the empirical evidence. We will then determine whether these spatial and temporal scales are adequately represented in our first principle models. If not, then we must consider how to capture those terms. Doing that, within the programmatic constraints that we live under, will require new approaches that integrate data from a variety of sources in order to elucidate the processes at work.

FUV remote sensing can play an important role in developing our understanding of the fundamental processes that connect Earth's atmosphere to Sun. Auroral imaging is the means by which we can image magnetospheric processes through their interaction, along magnetic field lines, with Earth's atmosphere. FUV imaging also provides us with quantitative information on the response of the upper atmosphere (see Table 13.2). The scientific community needs to continue to fly FUV auroral imagers if we are to continue to have global insight into the coupling of the ionosphere and magnetosphere and the response of the atmosphere to external forcing.

ACKNOWLEDGMENTS

The authors thank Margaret Simon for editorial assistance and Misty Crawford for manuscript production assistance.

REFERENCES

Acuna, M. H. The global geospace science program and its investigations. *Space Sci. Rev.* 71 (1995): 5.
Akasofu, S. I. The aurora and the magnetosphere: The Chapman memorial lecture. *Planet. Space Sci.* 22 (1974a): 885.
Akasofu, S. I. Recent progress in studies of DMSP auroral photographs. *Space Sci. Rev.* 19 (1976): 169.

Akasofu, S. I. A study of auroral displays photographed from the DMSP-2 satellite and from the Alaska meridian chain of stations. *Space Sci. Rev.* 16 (1974b): 617.

Alexeev, I. V. et al. Remote sensing of a magnetotail reconnection X-line using polar rain electrons. *Geophys. Res. Lett.* 33 (2006): L19105. doi:10.1029/2006GL027243.

Anger, C. D. et al. ISIS-II scanning auroral photometer. *Appl. Opt.* 12 (1973): 1853.

Anger, C. D. et al. Scientific results from the Viking Ultraviolet imager: An introduction. *Geophys. Res. Lett.* 14 (1987a): 383–386.

Anger, C. D. et al. An ultraviolet auroral imager for the Viking spacecraft. *Geophys. Res. Lett.* 14 (1987b): 387–390.

Baker, D. N. et al. Relationships between precipitating auroral zone electrons and lower thermospheric nitric oxide densities: 1998–2000. *J. Geophys. Res.* 106 (2001): 24465.

Barth, C. A. Nitric oxide in the lower thermosphere. *Planet. Space Sci.* 40 (1992): 315.

Barth, C. A. et al. Global observations of nitric oxide in the thermosphere. *J. Geophys. Res.* 108 (2003): 1027. doi:10.1029/2002JA009458.

Basu, S., S. Basu, R. Eastes, R. E. Huffman, R. E. Daniell, P. K. Chaturvedi, C. E. Valladares, and R. C. Livingston. Remote sensing of auroral E region plasma structures by radio, radar, and UV techniques at solar minimum. *J. Geophys. Res.* 98(A2) (1993): 1589–1602. doi:10.1029/92JA01655.

Bilitza, D. et al. The international reference ionosphere 2012—a model of international collaboration. *J. Space Weather Space Clim* 4(A07) (2014). doi:10.1051/swsc/2014004.

Bowyer, S. et al. Continous-readout extreme ultraviolet airglow spectrometer. *Appl. Opt.* 20 (1981): 477–486.

Brekke, A. *The Polar Upper Atmosphere*, Springer (2013).

Buonsanto, M. J. Ionospheric storms—A review. *Space Sci. Rev.* 88 (1999): 563.

Burch, J., IMAGE mission overview. *Space Sci. Rev.* 91(1) (2000). doi:10.1023/A:1005245323115.

Bust, G. S., T. W. Garner, and T. L. Gaussiran II. Ionospheric Data Assimilation Three-Dimensional (IDA3D): A global, multisensor, electron density specification algorithm. *J. Geophys. Res.* 109 (2004): A11312. doi:10.1029/2003JA010234.

Carbary, J. F., E. H. Darlington, K. J. Heffernan, T. J. Harris, C.-I. Meng, M. J. Mayr, P. J. Mcevaddy, and K. Peacock. Ultraviolet and visible imaging and spectrographic imaging (UVISI) experiment. *Proceedings of the SPIE 2217, Aerial Surveillance Sensing Including Obscured and Underground Object Detection, (22 July 1994).* doi:10.1117/12.179940.

Carruthers, G. R., and T. Page. Apollo 16 far-ultraviolet camera/spectrograph: Earth observations. *Science* 177 (1972): 788–791.

Carruthers, G. R., and T. Page. Apollo 16 far ultraviolet imagery of the polar auroras, tropical airglow belts, and general airglow. *J. Geophys. Res.* 81 (1976): 483–496.

Carruthers, G. R., T. Page, and R. R. Meier. Apollo 16 Lyman alpha imagery of the hydrogen geocorona. *J. Geophys. Res.* 81(10) (1976): 1664–1672. doi:10.1029/JA081i010p01664.

Chamberlain, J. W. *Physics of the Aurora and Airglow.* ISBN: 9780875908571 (1961). Academic Press: London, UK.

Christensen, A. B. et al. Initial observations with the Global Ultraviolet Imager (GUVI) in the NASA TIMED satellite mission. *J. Geophys. Res.* 108(A12) (2003): 1451. doi:10.1029/2003JA009918.

Christensen, A. B., R. L. Walterschied, M. N. Ross, C. Meng, L. Paxton, D. Anderson, G. Crowley, S. Avery, R. Meier, and D. Strickland. Global ultraviolet imager for the NASA TIMED mission. In *SPIE Optical Spectroscopic Techniques and Instrumentation for Atmospheric and Space Research*, eds. Jinxue Wang and Paul B. Hays, vol. 2266, pp. 451–466 (1994). doi: 10.1117/12.187583.

Cogger, L. L., and C. D. Anger. The OI 5577Å airglow experiment on the ISIS 2 satellite. *J. Atmos. Solar-Terr. Phys.* 35 (1973): 2081, 2084.

Cogger, L. L., J. S. Murphree, and C. D. Anger. High space and time resolution ultraviolet auroral images from the Viking spacecraft. *Physica Scripta* 37 (1988): 432–436.

Craven, J. D. et al. Variations in FUV dayglow with intense auroral activity. *Geophys. Res. Lett.* 21 (1994): 2793–2796.

Donovan, E. et al. The THEMIS All-Sky Imager Array—System design and initial results from the prototype imager. *J. Atmos. Solar-Terr. Phys.*(2006), 1472–1487. doi:10.1016/j.jastp.2005.03.27.

Eastes, R. NASA mission to explore forcing of Earth's space environment. *Eos Trans. AGU* 90(18) (2009): 155–155. doi:10.1029/2009EO180002.

Eastes, R. W. et al. Global-scale Observations of the Limb and Disk (GOLD): New observing capabilities for the ionosphere-thermosphere. In *Midlatitude Ionospheric Dynamics and Disturbances*, AGU Geophysical Monograph Series, eds. Kintner, P. M., A. J. Coster, T. Fuller-Rowell, T. Mannucci, M. Mendillo, and R. Heelis., vol. 181, pp. 319–326 (2008). American Geophysical Union: Washington, DC. doi:10.1029/181GM29.

Eather, R. H. Auroral proton precipitation and hydrogen emissions. *Rev. Geophys.* 5(3) (1967): 207–285. doi:10.1029/RG005i003p00207.

Evans, J. S., D. J. Strickland, and R. E. Huffman. Satellite remote sensing of thermospheric O/N$_2$ and solar EUV: 2. Data analysis. *J. Geophys. Res.* 100 (1995): 12227.

Fastie, W. G. Instrumentation for ultraviolet rocket spectrometry. *J. Quant. Spectrosc. Radiat. Transf.* 3 (1963): 507.

Frank, L. A. et al. Global auroral imaging instrumentation for the Dynamics Explorer Mission. *Space Sci. Instr.* 5 (1981): 369.

Frank, L. A. et al. The visible imaging system (VIS) from the polar space craft. *Space Sci. Rev.* 71 (1995): 297–328.

Frey, H. U. et al. The electron and proton aurora as seen by IMAGE-FUV and FAST. *Geophys. Res. Lett.* 28 (2001): 1135–1138. doi:10.1029/2000GL012352.

Frey, H. U., T. D. Phan, S. A. Fuselier, and S. B. Mende. Continuous magnetic reconnection at Earth's magnetopause. *Nature* 426 (2003): 533–537. doi:10.1038/nature02084.

Fuller-Rowell, T., and D. Evans. Height-integrated Pedersen and Hall conductivity patterns inferred from TIROS–NOAA satellite data. *J. Geophys. Res.* 92 (1987): 7606.

Galand, M., D. Lummerzheim, A. W. Stephan, B. C. Bush, and S. Chakrabarti. Electron and proton aurora observed spectroscopically in the far ultraviolet. *J. Geophys. Res.* 107(A7) (2002): SIA 14-1–SIA 14-14. doi:10.1029/2001JA000235.

Gerard, J. C., and C. A. Barth. OGO-$ observations of the ultraviolet auroral spectrum. *Planet. Space Sci.*, 24 (1976): 1059–1063. doi:10.1016/0032-0633.

Gerard, J. C., and C. A. Barth. High-latitude nitric oxide in the lower thermosphere. *J. Geophys. Res.* 82 (1977): 674–680.

Germany, G. A. et al. Remote determination of auroral energy characteristics during substorm activity. *Geophys. Res. Lett.* 24 (1997): 995.

Germany, G. A. et al. Use of FUV auroral emissions as diagnostic indicators. *J. Geophys. Res.* 99 (1994): 383.

Gladstone, G. R. Simulations of DE-1 FUV airglow images. *J. Geophys. Res.* (1994): 11441–11448. doi:10.1029/93JA03525.

Guarnieri, F. L. The nature of auroras during high-intensity long-duration continuous AE activity (HILDCAA) events: 1998–2001. In *Recurrent Magnetic Storms: Corotating Solar Wind Streams*, Geophys. Monogr. Ser., eds. B. T. Tsurutani et al., vol. 167, 235 pp. American Geophysical Union: Washington, D.C. (2006). doi:10.1029/167GM19.

Hajra, R., E. Echer, B. T. Tsurutani, and W. D. Gonzalez. Solar wind-magnetosphere energy coupling efficiency and partitioning: HILDCAAs and preceding CIR storms during solar cycle 23. *J. Geophys. Res. Space Phys.* 119 (2014): 2675–2690. doi:10.1002/2013JA019646.

Hardy, D. A., M. S. Gussenhoven, and D. Brautigam. A statistical model of auroral ion precipitation. *J. Geophys. Res.* 94 (1989): 370.

Hardy, D. A., M. S. Gussenhoven, and E. Holeman. A statistical model of auroral electron precipitation. *J. Geophys. Res.* 90 (1985): 4229.

Hardy, D. A., M. S. Gussenhoven, R. Raistrick, and W. McNeil. Statistical and functional representation of the pattern of auroral energy flux, number flux, and conductivity. *J. Geophys. Res.* 92 (1987): 12275.

Hays, P. B. The visible airglow experiment on Atmosphere Explorer. *Radio Sci.* 8 (1973): 369.

Hicks, G. T., and T. A. Chubb. Equatorial aurora/airglow in the far ultraviolet. *J. Geophys. Res.* 75(31) (1970): 6233–6248. doi:10.1029/JA075i031p06233.

Hoffman, R. A., and E. R. Schmerling. Dynamics Explorer program: An overview. *Space Sci. Instr.* 5 (1981): 345–348.

Holzworth, R. H., and C.-I. Meng. Mathematical representation of the auroral oval. *Geophys. Res. Lett.* 2 (1975): 377–380.

Huang, C. Y., Y.-J. Su, E. K. Sutton, D. R. Weimer, and R. L. Davidson. Energy coupling during the August 2011 magnetic storm. *J. Geophys. Res. Space Phys.* 119 (2014): 1219–1232. doi:10.1002/2013JA019297.

Huffman, R., F. LeBlanc, J. Larrabee, and D. Paulsen. Satellite vacuum ultraviolet airglow and auroral observations. *J. Geophys. Res.* 85(A5) (1980): 2201–2215. doi:10.1029/JA085iA05p02201.

Huffman, R. E. *Atmospheric Ultraviolet Remote Sensing.* Academic Press: Boston, MA (1992).

Huffman, R. E., D. C. Larrabee, and F. d. LeBlanc, and C. I. Meng. Ultraviolet remote sensing of the aurora and ionosphere for C3I system use. *Radio Sci.* 20 (3) (1985): 425–430.

Humm, D. C. et al. Design and performance of the Global Ultraviolet Imager (GUVI). In *Proceedings of the SPIE X-Ray and Gamma-Ray Instrumentation for Astronomy IX*, vol. 3445, pp. 2–12 (1998). SPIE: Bellingham, WA.

Humm, D. C., B. S. Ogorzalek, M. J. Elko, D. Morrison, L. J. Paxton. Optical calibration of the Global Ultraviolet Imager (GUVI). In *SPIE Ultraviolet Atmospheric and Space Remote Sensing: Methods and Instrumentation II*, vol. 3818 (1999).

Immel, T. et al. Variations in Earth's FUV dayglow within the polar caps (abstract). *EOS Trans. AGU* 78(46) Fall meet. (1997): F250.

Ishimoto, M. et al. Anomalous UV auroral spectra During a large magnetic disturbance. *J. Geophys. Res.* 94 (1989): 6955–6960.

Ishimoto, M. et al. Night UV spectra (1100–2900A) at mid and low latitude during a magnetic storm. *Geophys. Res. Lett.* 19 (1992): 813–816.

Joki, E. G., and J. E. Evans Satellite measurements of auroral ultraviolet and 3914-A radiation. *J. Geophys. Res.* 74(19) (1969): 4677–4686. doi:10.1029/JA074i019p04677.

Kaneda, E. Auroral TV observation by KYOKKO. *Proc. Jpm. IMS Symp.* 12 (1979): 146–151.

Kaneda, E., and T. Yamamoto. Auroral substorms observed by UV-imager on Akebono. In *Magnetospheric Substorms*, eds. J. R. Kan, T. A. Potemra, S. Kokubun, and T. Iijima. American Geophysical Union: Washington, D.C. (1991). doi:10.1029/GM064p0235.

Keiling, A., E. Donovan, F. Bagenal, and T. Karlsson. *Auroral Phenomenology and Magnetospheric Processes: Earth and Other Planets.* An AGU Geophysical Monograph, vol. 197 (2012). Wiley: New York.

Knipp, D. et al. Thermospheric damping response to sheath-enhanced geospace storms. *Geophys. Res. Lett.* 40 (2013): 1263–1267. doi:10.1002/grl.50197.

Lilensten, L., P. L. Blelly, W. Kofman, and D. Alcayde. Auroral ionospheric conductivities: A comparison between experiment and modeling, and theoretical f10.7-dependent model. *Ann. Geophysicae* 14 (1996): 1297.

Liou, K., and D. G. Sibeck. Study of a global auroral Pc5 pulsation event with concurrent ULF waves. *Geophys. Res. Lett.* 41 (2014): 6547–6555. doi:10.1002/2014GL060755.

Lu, G., M. G. Mlynczak, L. A. Hunt, T. N. Woods, and R. G. Roble. On the relationship of Joule heating and nitric oxide radiative cooling in the thermosphere. *J. Geophys. Res.* 115 (2010): A05306. doi:10.1029/2009JA014662.

Mayr, H. G., I. Harris, and N. W. Spencer. Some properties of upper atmosphere dynamics. *Rev. Geophys.* 16 (1978): 539.

Lui, A. T. Y. et al. Simultaneous observations of particle precipitations and auroral emissions by Isis 2 satellite in the 19–24 MLT sector. *J. Geophys. Res.* 82 (1977): 2210.

Lui, A.T.Y. et al. The topology of the auroral oval as seen by the ISIS-2 scanning auroral photometer. *J. Geophys. Res.* 80 (1975a): 1795–1804.

Lui, A. T. Y., C. D. Anger, and S.-I. Akasofu. The equatorward boundary of the diffuse aurora and auroral substorms as seen by the Isis 2 auroral scanning photometer. *J. Geophys. Res.* 80(25) (1975b): 3603–3614. doi:10.1029/JA080i025p03603.

Mayr, H. G., and H. Volland. Magnetic storm effects in the neutral composition. *Planet. Space Sci.* 20 (1972): 379.

Meier, R. R. Ultraviolet spectroscopy and remote sensing of the upper atmosphere. *Space Sci. Rev.* (ISSN 0038-6308), vol. 58 (1991): 1–185.

Mende, S. B. et al. Far ultraviolet imaging from the IMAGE spacecraft, 1, System design. *Space Sci. Rev.* 91 (2000a): 243.

Mende, S. B. et al. Far ultraviolet imaging from the IMAGE spacecraft. 2. Wideband FUV imaging. *Space Sci. Rev.* 91 (2000b): 271–282. doi:10.1023/A:1005227915363.

Mende, S. B. et al. Far ultraviolet imaging from the IMAGE spacecraft, 3, Spectral imaging of Lyman alpha and OI 135.6 nm. *Space Sci. Rev.* 91 (2000c): 287.

Mende, S. B., C. W. Carlson, H. U. Frey, L. M. Peticolas, and N. Østgaard. FAST and IMAGE-FUV observations of a substorm onset. *J. Geophys. Res.* 108 (2003): 1344. doi:10.1029/2002JA009787, A9.

Meng, C.-I, and R. E. Huffman. Preliminary observations from the Auroral and Ionospheric Remote Sensing imager. *Johns Hopkins APL Tech. Dig.* 8 (1987): 303.

Meng, C.-I, and R. E. Huffman. Ultraviolet imaging from space of the aurora under full sunlight. *Geophys. Res. Lett.* 11 (1984): 315–318.

Mitchell, E. J., P. T. Newell, J. W. Gjerloev, and K. Liou. OVATION-SM: A model of auroral precipitation based on SuperMAG generalized auroral electrojet and substorm onset times. *J. Geophys. Res.* 118 (2013): 3747–3759. doi:10.1002/jgra.50343.

Mlynczak, M. et al. The natural thermostat of nitric oxide emission at 5.3 μm in the thermosphere observed during the solar storms of April 2002. *Geophys. Res. Lett.* 30(21) (2003): 2100. doi:10.1029/2003GL017693.

Morrison, D., L. J. Paxton, H. Kil, Y. Zhang, B. S. Ogorzalek, and C. Meng. On-orbit calibration of the Special Sensor Ultraviolet Scanning Imager (SSUSI): A far-UV imaging spectrograph on DMSP F-16. In *SPIE Optical Spectroscopic Techniques and Instrumentation for Atmospheric and Space Research IV*, eds. A. M. Larar, and M. Mlynczak, vol. 4485, pp. 328–337 (2002). SPIE: Bellingham, WA.

Murphree, J. S. et al. The Freja ultraviolet imager. *Space Sci. Rev.* 70 (1994): 421–446.

Murphree, J. S. et al. Interpolation of optical substorm onset observations. *J. Atmos. Social-Terr. Phys.* 55 (1993): 1159.

Murphree, J. S., and L. L. Cogger. The application of CCD detectors to UV imaging from a spinning satellite. *Ultraviolet Technology II, Proc. Soc. Opt. Eng.* 932 (1988): 42.

Newell, P., Y. Feldstein, Y. Galperin, and C. I. Meng. Morphology of nightside precipitation. *J. Geophs. Res.* 101(A5) (1996): 10737–10748.

Newell, P. T. et al. Dynamics of double-theta aurora: Polar UVI study of January 10–11, 1997. *J. Geophys. Res.* 104(A1) (1999): 95–104.

Nicholas, A. C. et al. A survey of large-scale variations in thermospheric oxygen column density with magnetic activity as inferred from observations of the FUV dayglow. *J. Geophys. Res.* 102 (1997): 4493.

Oguti, T., Kaneda, E., Ejiri, M., Kadokura, A., and Sasaki, S. Studies of aurora dynamics by aurora-TV on the Akebono (EXOS-D) satellite. *J. Geomag. Geoelectricity* (ISSN 0022-1392) 42(4) (1990): 555–564.

Omholt, A. *The Optical Aurora: Physics and Chemistry in Space 4* (1971). Springer Verlag: New York.

Østgaard, N., S. B. Mende, H. U. Frey, L. A. Frank, and J. B. Sigwarth. Observations of non-conjugate theta aurora. *Geophys. Res. Lett.* 30(21) (2003): 2125. doi:10.1029/2003GL017914.

Paxton, L. J., and D. E. Anderson. Ultraviolet remote sensing of Venus and Mars. In *Venus and Mars: Atmospheres, Ionospheres, and Solar Wind Interactions. Geophysical Monograph 66*, eds. J. G. Luhmann, M. Tatrallyay, and R. O. Pepin, pp. 113–190 (1992). American Geophysical Union: Washington, DC.

Paxton, L. J. et al. Global ultraviolet imager (GUVI): Measuring composition and energy inputs for the NASA Thermosphere Ionosphere Mesosphere Energetics and Dynamics (TIMED) mission. In *SPIE Optical Spectroscopic Techniques and Instrumentation for Atmospheric and Space Research III*, vol. 3756, pp. 265–276 (1999). SPIE: Bellingham, WA.

Paxton, L. J. et al. GUVI : A Hyperspectral imager for geospace. In *Instruments, Science, and Methods for Geospace and Planetary Remote Sensing. Proc. SPIE*, eds. C. A. Nardell, P. G. Lucey, J.-H. Yee, J. B. Garvin, vol. 5660, pp. 228–240 (2004). SPIE: Bellingham, WA.

Paxton, L. J., and C.-I. Meng. Auroral imaging and space-based optical remote sensing. *Johns Hopkins APL Tech. Dig.* 20 (1999): 556.

Paxton, L. J., C.-I. Meng, D. E. Anderson, and G. J. Romick. MSX—A multi-use space experiment. *APL Tech. Dig.* 17(1) (1996): 19–34.

Paxton, L. J., C.-I. Meng, G. H. Fountain, B. S. Ogorzalek, E. H. Darlington, J. Goldsten, S. Geary, D. Kusnierkiewicz, S. C. Lee, K. Peacock. Special Sensor UV Spectrographic Imager (SSUSI): An instrument description. In *Instrumentation for Planetary and Terrestrial Atmospheric Remote Sensing, SPIE*, vol. 1745, pp. 2–16 (1992a). SPIE: Bellingham, WA.

Paxton, L. J., C.-I. Meng, G. H. Fountain, B. S. Ogorzalek, E. H. Darlington, J. Goldsten, and K. Peacock. SSUSI: Horizon-to-horizon and limb-viewing spectrographic imager for remote sensing of environmental parameters. *Ultraviolet Technology IV, SPIE* 1764 (1992b): 161–176.

Paxton, L. J., D. Morrison, H. Kil, Y. Zhang, B. S. Ogorzalek, and C. Meng. Validation of remote sensing products produced by the Special Sensor Ultraviolet Scanning Imager (SSUSI): A far UV imaging spectrograph on DMSP F-16. In *SPIE Optical Spectroscopic Techniques and Instrumentation for Atmospheric and Space Research IV*, eds. Allen M. Larar, Martin G. Mlynczak, vol. 4485, pp. 338–348 (2002). SPIE: Bellingham, WA.

Paxton, L. J., D. Morrison, D. J. Strickland, M. G. McHarg, Y. Zhang, B. Wolven, H. Kill, G. Crowley, A. B. Christensen, and C.-I. Meng. The use of far ultraviolet remote sensing to monitor space weather. *Adv. Space Res.* 31(4) (2003): 813–818.

Paxton, L. J., D. J. Strickland, M. Weiss, and C.-I. Meng. Interactive data analysis and display of far ultraviolet data. *COSPAR Adv. Space Res.* 22(11) (1998): 1577–1582.

Prölss, G. W. Magnetic storm associated perturbations of the upper atmosphere: Recent results obtained by satellite-borne gas analyzers. *Rev. Geophys.* 18 (1980): 183.

Prölss, G. W., and J. J. Craven. Perturbations of the FUV day-glow and ionospheric storm effects. *Adv. Space Sci.* 22 (1998): 129.

Rairden, R. L., L. A. Frank, and J. D. Craven. Geocoronal imaging with dynamics explorer. *J. Geophys. Res.* 91 (1986): 13613.

Rasmussen, C. R. et al., A photo-chemical equilibrium model for ionospheric conductivity. *J. Geophys. Res.* 93 (1988): 9831–9840.

Rees, M. H. *Physics and Chemistry of the Upper Atmosphere.* Cambridge Atmospheric and Science Series, eds. J. T. Houghton, M. J. Rycroft, and A. J. Dessler, p. 101 (1989). Cambridge University Press: London, UK.

Richmond, A. Assimilative mapping of ionospheric electrodynamics. *Adv. Space Res.* 12 (1992): 59.

Richmond, A., and Y. Kamide. Mapping electrodynamic features of the high-latitude ionosphere from localized observations: Technique. *J. Geophys. Res.* 93 (1988): 5741–5759.

Rider, K. et al. Where the Earth's weather meets space weather. *IEE Aerospace* (2015). http://ieeexplore.ieee.org/xpl/articleDetails.jsp?arnumber=7119120.

Ridley, A. J., T. I. Gombosi, and D. L. DeZeeuw. Ionospheric control of the magnetosphere: Conductance. *Ann. Geophysicae* 22 (2004): 567–584.

Robinson, R., T. Dabbs, J. Vickrey, R. Eastes, F. Del Greco, R. Huffman, C. Meng, R. Daniell, D. Strickland, and R. Vondrak. Coordinated measurements made by the Sondrestrom radar and the Polar BEAR Ultraviolet Imager. *J. Geophys. Res.* 97(A3) (1992): 2863–2871. doi:10.1029/91JA02803.

Robinson, R. M., R. R. Vondrak, K. Miller, T. Dabbs, and D. Hardy. On calculating ionospheric conductances from the flux and energy of precipitating electrons. *J. Geophys. Res.* 92 (1987): 2565.

Roble, R. G. The polar lower thermosphere. *Planet. Space Sci.* 40 (1992): 271.

Roble, R. G., and E. C. Ridley. An auroral model for the NCAR thermospheric general circulation model (TGCM). *Ann. Geophys.* 5A (1987): 369.

Rogers, E. H. et al., Auroral photography from a satellite, *Science* 183 (1974) 951.

Samson, J. A. R. *Techniques of Vacuum Ultraviolet Spectroscopy.* Pied Publications: Lincoln, NE (1967).

Samson, J. A. R., and D. L. Ederer. *Techniques of Vacuum Ultraviolet Spectroscopy.* Academic Press (2000), New York.

Schenkel, F. W., B. S. Ogorzalek, J. C. Larrabee, F. J. LeBlanc, and R. E. Huffman. Ultraviolet daytime auroral and ionospheric imaging from space. *Applied Optics* 24(20) (1985): 3395.

Schunk, R. W. et al. Global Assimilation of Ionospheric Measurements (GAIM). *Radio Sci.* 39 (2004): RS1S02. doi:10.1029/2002RS002794.

Schunk, R. W., and A. F. Nagy. *Ionospheres, Physics, Plasma Physics, and Chemisty.* Cambridge Atmospheric and Space Science Series, Cambridge University Press: New York (2000).

Senior, C. Solar and particle contributions to auroral height-integrated conductivities from EISCAT data: a statistical study. *Ann. Geophysicae* 9 (1991): 449.

Sharber, J. R. et al. Observations of the UARS particle environment monitor and computation of ionization rates in the middle and upper atmosphere during a geomagnetic storm. *Geophys. Res. Lett.* 20 (1993): 1319.

Sharber, J. R. et al. Validation of UARS PEM electron energy deposition. *J. Geophys. Res.* 101 (1996): 9571–9582.

Shepherd, G. et al. ISIS-II atomic oxygen red line photometer, *Appl. Opt.* 12 (1973): 1767–1774. doi:10.1364/AO.12.001767.

Shirai, H. et al. Drop-off of the polar rain flux near the plasma sheet boundary. *J. Geophys. Res.* 102 (1997): 2271.

Spiro, R. W., P. H. Reiff, and L. J. Maher. Precipitating electron energy flux and auroral zone conductances—An empirical model. *J. Geophys. Res.* 87 (1982): 8215.

Steele, D. P., and D. J. McEwen. Electron auroral excitation efficiencies and intensity ratios. *J. Geophys. Res.* 95(A7) (1990): 10321–10336. doi:10.1029/JA095iA07p10321.

Strickland, D. J., J. Bishop, J. S. Evans, T. Majeed, R. J. Cox, D. Morrison, G. J. Romick, J. F. Carbary, L. J. Paxton, and C.-I. Meng. Midcourse space experiment/ultraviolet and visible imaging and spectrographic imaging limb observations of combined proton/hydrogen/electron aurora. *J. Geophys. Res.* 106 (2001): 65.

Strickland, D. J., J. Bishop, J. S. Evans, T. Majeed, P. M. Shen, R. J. Cox, R. Link, R. E. Huffman. Atmospheric Ultraviolet Radiance Integrated Code (AURIC): Theory, software architecture, inputs, and selected results. *J. Quant. Spectrosc. Radiat. Transf.* 62 (1999): 689.

Strickland, D. J., R. E. Daniell, and J. D. Craven. Negative ionospheric storm coincident with DE-1 observed thermospheric disturbance on October 14, 1981. *J. Geophys. Res.* 21(049) (2001): 106.

Strickland, D. J., J. S. Evans, and L. J. Paxton. Satellite remote sensing of thermospheric O/N_2 and solar EUV 1. Theory. *J. Geophys. Res.* 100 (1995): 12217.

Strickland, D. J. et al. A model for generating global images of emisions from the thermosphere. *Applied Optics* 33 (1994): 3578–3594.

Strickland, D. J., J. R. Jasperse, and J. A. Whalen. Dependence of auroral FUV emission on the incident electron spectrum and neutral atmosphere. *J. Geophys. Res.* 88 (1983): 8051.

Strickland, D. J., R. R. Meier, J. H. Hecht, and A. B. Christensen. Deducing composition and incident electron spectra from ground-based auroral optical measurements: Theory and model results. *J. Geophys. Res.* 94(A10) (1989): 13527–13539. doi:10.1029/JA094iA10p13527.

Strickland, D. J., R. R. Meier, R. L. Walterscheid, J. D. Craven, A. B. Christensen, L. J. Paxton, D. Morrison, and G. Crowley. Quiet time seasonal behavior of the thermosphere seen in the far ultraviolet dayglow. *J. Geophys. Res.* 109 (2004): A01302.

Swaminathan, P. K. et al. Nitric oxide abundance in the mesosphere/lower thermosphere region: Roles of solar soft X rays, superthermal N(^{4}S) atoms, and vertical transport. *J. Geophys. Res.* 103(A6) (1998): 11579–11594.

Torr, M. R. et al. A far ultraviolet imager for the international solar-terrestrial physics mission. *Space Sci. Rev.* 71 (1995): 329.

Tsuruda, K., and H. Oya. Introduction to the Akebono (EXOS D) Project. *J. Geophys. Res.* 98(A7) (1993): 11123–11125. doi:10.1029/92JA02231.

Vallance-Jones, A. *Aurora.* ISBN: 978-90-277-0273-9. *Geophysics and Astrophysics Monographs* (1974). Springer Verlag: New York.

Wallis, D. D., and E. E. Budzinski. Empirical model of height integrated conductivities. *J. Geophys. Res.* 86 (1981): 125.

Weller, C. S., and R. R. Meier. First satellite observaitons of the He+ 304A radiation and its interpolation. *J. Geophys. Res.* 79 (1974): 1572.

Whalen, J. A. A quantitative description of the spatial distribution and dynamics of the energy flux in the continuous aurora. *J. Geophys. Res.* 88 (1983): 7155.

Wing, S., and Y. Zhang. The nightside magnetic field line open-closed boundary and polar rain electron energy-latitude dispersion. *Ann. Geo Comm.* 33 (2015): 39–46.

Zhang, Y. et al. Double dayside detached auroras: TIMED/GUVI Observations. *Geophys. Res. Lett.* 31 (2004): L10801. doi:10.1029/2003GL018949.

Zhang, Y. et al. Nightside detached auroras due to precipitating protons/ions during intense magnetic storms. *J. Geophys. Res.* 110 (2005): A02206. doi:10.1029/2004JA010498.

Zhang, Y. et al. Sudden solar wind dynamic pressure enhancements and dayside detached auroras: IMAGE and DMSP observations. *J. Geophys. Res.* 108(A4) (2003b): 8001. doi:10.1029/2002JA009355.

Zhang, Y., C.-I. Meng, L. J. Paxton, D. Morrison, B. Wolven, H. Kil, P. Newell, S. Wing, and A. B. Christensen. Far-ultraviolet signature of polar cusp during southward IMF Bz observed by TIMED/Global Ultraviolet Imager and DMSP. *J. Geophys. Res.* 110 (2005a): A01218. doi:10.1029/2004JA010707.

Zhang, Y., and L. J. Paxton. *Auroral Dynamics and Space Weather.* ISBN: 978-1-118-97870-2. AGU Monograph (2015). American Geophysical Union: Washington, DC.

Zhang, Y., and L. J. Paxton. Dayside convection aligned auroral arcs. *Geophys. Res. Lett.* 33 (2006): L13107. doi:10.1029/2006GL026388.

Zhang, Y., and L. J. Paxton. An empirical Kp-dependent global auroral model based on TIMED/GUVI data. *J. Atmos. Solar-Terr. Phys.* 70 (2008): 1231–1242.

Zhang, Y., L. J. Paxton, D. Bilitza, and R. Doe. Near real-time assimilation in IRI of auroral peak E-region density and equatorward boundary. *Adv. Space Res.* 46 (2010): 1055–1063.

Zhang, Y., L. J. Paxton, and T. Immel. Sudden enhancement of solarwind dynamic pressure and dayside detached aurora. In *AGU Fall Meeting, Abstract* SM41B-0814 (2001).

Zhang, Y., L. J. Paxton, and H. Kil. Large-scale structures in the Polar Rain. *Geophys. Res. Lett.* 40 (2013): 5576–5580. doi:10.1002/2013GL058245.

Zhang, Y., L. J. Paxton, and H. Kil. Nightside polar rain aurora boundary gap and its applications for magnetotail reconnection. *J. Geophys. Res.* 116 (2011): A11214.

Zhang, Y., L. J. Paxton, H. Kil, C.-I. Meng, S. B. Mende, H. U. Frey, and T. J. Immel. Negative ionospheric storms seen by the IMAGE FUV instrument. *J. Geophys. Res.* 108 (A9) (2003a): 1343. doi:10.1029/2002JA009797.

Zhang, Y., L. J. Paxton, J. U. Kozyra, H. Kil, and P. C. Brandt. Nightside thermospheric FUV emissions due to energetic neutral atom precipitation during magnetic superstorms. *J. Geophys. Res.* 111 (2006): A09307. doi:10.1029/2005JA011152.

Zhang, Y., L. J. Paxton, and A. T. Y. Lui. Polar rain aurora. *Geophys. Res. Lett.* 34 (2007): L20114. doi:10.1029/2007GL031602.

Zhang, Y., L. J. Paxton, D. Morrison, A. T. Y. Lui, H. Kil, B. Wolven, C.-I. Meng, and A. B. Christensen. Undulations on the equatorward edge of the diffuse proton aurora: TIMED/GUVI observations. *J. Geophys. Res.* 110 (2005b): A08211. doi:10.1029/2004JA010668.

Zhang, Y., L. J. Paxton, D. Morrison, D. Marsh, and H. Kil. Storm-time behaviors of O/N_2 and NO variations. *J. Atmos. Solar-Terr. Phys.* 114 (2014): 42–49.

Zhang, Y., L. J. Paxton, D. Morrison, B. Wolven, H. Kil, C.-I. Meng, S. B. Mende, and T. J. Immel. O/N$_2$ changes during 1–4 October 2002 storms: IMAGE SI-13 and TIMED/GUVI observations. *J. Geophys. Res.* 109 (2004): A10308. doi:10.1029/2004JA010441.

Zhang, Y., L. J. Paxton, N. P. Newell, and C.-I. Meng. Does the polar cap disappear under an extended strong northward IMF? *J. Atmos. Solar-Terr. Phys.* 71 (2009): 2006–2012. doi:10.1016/j.jastp.2009.09.005.

Zhang, Y., L. J. Paxton, and Y. Zheng. Interplanetary shock induced ring current auroras. *J. Geophys. Res.* 113 (2008): A01212. doi:10.1029/2007JA012554.

Zhang, Y., and S. Wing. Determining magnetotail reconnection location from polar rain energy dispersion. *J. Atmos. Solar-Terr. Phys.* 130–131 (2015): 75–80.

Zhou, X.-Y., and B. T. Tsurutani. Rapid intensification and propagation of the dayside aurora: Large scale interplanetary pressure pulses (fast shocks). *Geophys. Res. Lett.* 26(8) (1999): 1097–1100. doi:10.1029/1999GL900173.

CHAPTER 14

Ionospheric Electrodynamics

Arthur D. Richmond

CONTENTS

14.1 INTRODUCTION

This chapter is revised from a previously published book chapter [1]. The free electrons and ions in Earth's ionosphere make it electrically conducting. Current flows are connected with the magnetosphere above and, to a much weaker extent, with the poorly conducting atmosphere below. One of the important generators of the ionospheric current is the ionospheric wind dynamo, which operates as upper atmospheric winds that move the electrically conducting medium through Earth's magnetic field, creating an electromotive force that drives currents and causes electric polarization fields to develop. Other current-generation mechanisms exist that are associated with the interaction of the solar wind with the magnetosphere [2] and, to a much less significant extent, with electrified clouds in the troposphere [3]. In the daytime ionosphere, the largest currents flow between 90 and 200 km; this general region is sometimes called the dynamo region. The currents and electric fields interact with the dynamics of the ionospheric plasma and neutral air in and above the dynamo region. We call the electrical phenomena and their interacting dynamical effects ionospheric electrodynamics. The historical development of the field has been described in [4–7].

The currents in and above the ionosphere produce magnetic perturbations that can be sensed at the ground and in space. The currents associated with the ionospheric dynamo have regular, smooth, daily variations. The main part of these currents is often referred to as S_q (for solar quiet) or S_R (for solar regular); there is also a smaller component related to lunar periods, which is sometimes referred to as L. By contrast, the currents associated with solar-wind/magnetosphere interactions are highly variable: When they are weak, one speaks of a magnetically quiet period, and when they are strong, one speaks of a magnetically disturbed period, the most dramatic manifestation of which is the magnetic storm. During disturbed periods, the magnetospherically produced electric fields and currents can dominate over those produced by the ionospheric wind dynamo. It is therefore important to distinguish between magnetically quiet and disturbed periods when analyzing ionospheric electrodynamics.

Ionospheric electrodynamics depends, among other things, on the conductivity of the ionosphere and on the strength of thermospheric winds. Both of these depend on the flux of solar ultraviolet radiation absorbed in the upper atmosphere, which varies considerably with solar activity. Major changes in the solar ultraviolet irradiance occur over the 11-year solar activity cycle; between solar minimum and solar maximum, the total ionizing extreme ultraviolet flux varies by more than a factor of 2. Thus, it is natural to expect significant changes in ionospheric electrodynamics with the solar cycle. A measure of solar activity often used is the index S_a, representing the flux of solar radio emissions at a wavelength of 10.7 cm, in units of 10^{-22} W m^{-2} Hz^{-1}.

The conductivity of the ionosphere and magnetosphere is highly anisotropic with respect to the geomagnetic field direction. Electrodynamic features are therefore strongly organized with respect to the geomagnetic field, and it is common to use magnetic coordinates to organize the observations and to do model simulations. Figure 14.1 shows a map of quasi-dipole latitude and longitude over Earth at a height of 110 km for the year 2015 [8,9]. (Secular change of the geomagnetic field causes slow changes of the map over decades.) Geomagnetic field lines thread through points with the same absolute values of magnetic latitude lying along the same magnetic longitude. Magnetic longitude is also used to define magnetic local time (MLT), for which 12 MLT lies on the magnetic longitude of the subsolar point at sufficiently high altitude that the geomagnetic field is essentially a dipole, with a 1-h increase of MLT for each 15° increase of magnetic longitude.

At middle and low magnetic latitudes, electric current flows freely along geomagnetic field lines between the northern and southern hemispheres, and closes across field lines in the lower ionosphere where the conductivity transverse to the field is the greatest. The current is associated with small perturbations of the geomagnetic field observable on the ground and in space. Along the magnetic equator is an enhanced band of current flowing in the lower ionosphere at day, called the equatorial electrojet (EEJ) [10–12]. The sensitivity of the EEJ to ionospheric electric fields of various sources makes its magnetic perturbations a valuable source of information about global electrodynamics.

14.2 IONOSPHERIC CONDUCTIVITY

The upper atmosphere is ionized primarily by solar ultraviolet and X-ray radiation and, at higher latitudes, by the precipitation of energetic charged particles from the magnetosphere [13]). Starlight and cosmic rays are minor ionization sources that have some influence in the nightside ionosphere and at low altitudes. The primary ions produced are N_2^+ and O^+, but these are reactive with the neutral gases, and the nitrogen ions are rapidly converted to O_2^+ and NO^+. The dominant ions present are therefore NO^+, O_2^+, and O^+, with the molecular ions predominant below 150 km and O^+ predominant

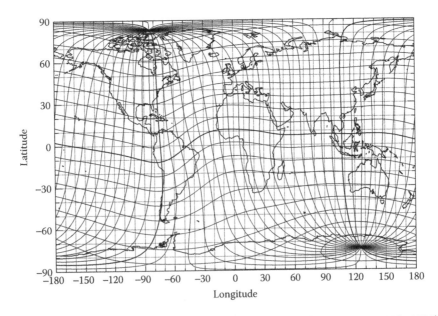

FIGURE 14.1 Quasi-dipole coordinates (Richmond, A.D., *J. Geomagn. Geoelectr.*, 47, 191–212, 1995) at 110 km altitude, derived from the International Geomagnetic Reference Field (Thebault, E. et al., *Earth Planets Space* 67, 2015) for epoch 2015.0. The dashed grid represents geographic coordinates, whereas lines of constant apex latitude and longitude are drawn at 10° intervals. The thick contours represent the magnetic equator and magnetic longitudes of 0° (Americas) and 180° (Asia).

above 200 km, and a mixture in between. Negative ions and more complex positive ions become important only below about 90 km, where conductivities are relatively small. Between about 90 and 150 km, the sum of the densities of NO^+ and O_2^+ is approximately equal to the electron number density N. These molecular ions NO^+ and O_2^+ recombine with electrons with a reaction coefficient represented by the symbol α. These ions are usually near photochemical equilibrium, meaning that their production rate is in balance with their chemical loss rate. If Q_e is the electron production rate, then between 90 and 150 km photochemical equilibrium implies that

$$Q_e = \alpha N^2 \qquad (14.1)$$

or

$$N = \sqrt{\frac{Q_e}{\alpha}} \qquad (14.2)$$

In reality, the recombination coefficient is somewhat different for NO^+ and O_2^+, so α in the above equations represents an effective mean value, weighted by the relative densities of the two ions. In addition, α is approximately inversely proportional to the electron temperature. A characteristic value of α is 3×10^{-13} m^3 s^{-1}. Equation 14.2 shows that the electron density in the lower ionosphere is proportional to the square root of the ionization production rate. When direct sunlight disappears, only weak starlight and solar radiation scattered from the geocorona remain, and the electron density is greatly reduced, unless there is production at high latitudes by energetic particles precipitating from the magnetosphere.

The O^+ ions at higher altitudes do not often recombine directly with electrons, but rather react first with molecular neutral constituents to form molecular ions that can then recombine with electrons. Equations 14.1 and 14.2 become invalid at these heights. As neutral densities become small at high altitudes, the loss of O^+ ions becomes slow, and ions can be transported significant distances before being lost, so that photochemical equilibrium breaks down.

The primary ionization source at day, solar extreme ultraviolet light, varies significantly with the level of solar activity. The electron density also shows important variations with solar activity. Figure 14.2 shows representative vertical profiles of electron density for day and night conditions, and for low and high levels of solar activity [14].

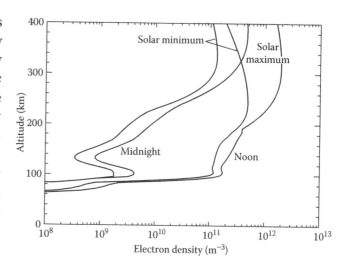

FIGURE 14.2 Electron densities at 44.6° N, 2.2° E from the 1990 International Reference Ionosphere at noon and midnight on March 21, for solar minimum and maximum conditions (annual average sunspot numbers 15 and 160, respectively). (Bilitza, D., ed., *International Reference Ionosphere 1990*, Greenbelt, MD, 1990.)

The ionospheric region around the minor peak in density near 105 km is known as the *E*-region, whereas the larger density region above 150 km is known as the *F*-region. The *E*-region and lower part of the *F*-region undergo relatively greater variations in electron density between day and night than does the upper *F* region. In the auroral regions, ionization by energetic magnetospheric electrons and ions is highly variable (e.g., [15]), and often exceeds ionization by solar radiation.

Conductivities are calculated by treating the ions and electrons as fluids in force balance, where the forces are the Lorentz force, the collisional forces with neutrals and with other charged particles, the gravity, and the pressure gradient. Above 90 km, nearly all ions have a single positive charge equal to the magnitude of the electron charge, e, and it is adequate to consider all ion species together as a single fluid of the number density equal to that of the electrons, N. The ion mass m_i, velocity $\mathbf{v}_i$, and temperature T_i, and the collision frequencies of ions with neutrals ν_{in} and with electrons ν_{ie} then represent mass density-weighted averages over all ion species. The current density $\mathbf{J}$ is related to N and to the ion and electron velocities $\mathbf{v}_i$ and $\mathbf{v}_e$ by

$$\mathbf{J} = Ne(\mathbf{v}_i - \mathbf{v}_e). \qquad (14.3)$$

For ions and electrons, the force balance conditions are, respectively, as follows:

$$Ne(\mathbf{E} + \mathbf{v}_i \times \mathbf{B}) - Nm_i\nu_{in}(\mathbf{v}_i - \mathbf{v}_n) - Nm_i\nu_{ie}(\mathbf{v}_i - \mathbf{v}_e)$$
$$+ Nm_i\mathbf{g} - \nabla(NkT_i) = 0 \tag{14.4}$$

$$-Ne(\mathbf{E} + \mathbf{v}_e \times \mathbf{B}) - Nm_e\nu_{en}(\mathbf{v}_e - \mathbf{v}_n)$$
$$- Nm_e\nu_{ei}(\mathbf{v}_e - \mathbf{v}_i) + Nm_e\mathbf{g} - \nabla(NkT_e) = 0 \tag{14.5}$$

where:

m_e and T_e are the electron mass and temperature, respectively

$\mathbf{v}_n$ is the velocity of the neutral gas

ν_{en} and ν_{ei} are the electron-neutral and electron-ion collision frequencies, respectively

$\mathbf{E}$ and $\mathbf{B}$ are the electric and magnetic fields, respectively

$\mathbf{g}$ is the acceleration of gravity

k is the Boltzmann constant

In reality, the collision frequencies are tensors rather than scalars in the presence of a magnetic field [16,17]. Above 80 km, the electron-neutral collision frequency is about 1.4 times as large for motions perpendicular to **B** as for motions along **B** [18]; below this height, off-diagonal terms in the collision-frequency tensor can also be significant. ν_{ei} is similarly anisotropic, but we shall be concerned only with its value for motions along **B**. For ions, the anisotropy of the collision frequency is unimportant. Table 14.1 lists the formulas for the collision frequencies as functions of density, composition,

and temperature of the colliding species. These expressions are based on a combination of laboratory measurements and theory. The electron gravitational term is always much smaller than the other terms in Equation 14.5 and is neglected.

An expression for the current density parallel to **B** is derived from the component of Equation 14.5 parallel to **B**, which in rearranged form is

$$Nm_e\nu_{en\parallel}(\upsilon_e - \upsilon_n)_\parallel + Nm_e\nu_{ei\parallel}(\upsilon_e - \upsilon_i)_\parallel$$
$$= -Ne(E_\parallel - E_a) \tag{14.6}$$

$$E_a = -\frac{1}{Ne}\frac{\partial(NkT_e)}{\partial s} \tag{14.7}$$

where:

the subscripts "∥" denote the direction along **B** and s is the distance along **B**

E_a is called the ambipolar electric field, which from Equation 14.6 is seen to be the value of $E_\parallel$ in the absence of collisional drag on the electrons

At low altitudes, where the term involving electron-neutral collisions in Equation 14.6 is significant, the ion motion along **B** is also strongly affected by collisions with neutrals, to the extent that the parallel ion and neutral velocities are essentially the same. Therefore, $\upsilon_{n\parallel}$ can be replaced by $\upsilon_{i\parallel}$, and the electron-ion velocity difference along **B** is found to be

TABLE 14.1   Formulas for Collision Frequencies

$\nu_{in}(NO^+)$	$=$	$\left[4.35(N_{N_2} + N_{O_2})R_i^{-0.11} + 1.9N_OR_i^{-0.19}\right] \times 10^{-16}\,\mathrm{m^3s^{-1}}$
$\nu_{in}(O_2^+)$	$=$	$\left[4.3N_{N_2} + 5.2N_{O_2} + 1.8N_OR_i^{-0.19}\right] \times 10^{-16}\,\mathrm{m^3s^{-1}}$
$\nu_{in}(O^+)$	$=$	$\left[5.4N_{N_2}R_i^{-0.20} + 7.0N_{O_2}R_i^{0.05} + 6.7N_OR_i^{0.5}(0.96 - 0.135\log_{10}R_i)^2\right] \times 10^{-16}\,\mathrm{m^3s^{-1}}$
$\nu_{en\perp}$	$=$	$\left[7.2N_{N_2}R_e^{0.95} + 5.2N_{O_2}R_e^{0.79} + 1.9N_OR_e^{0.85}\right] \times 10^{-15}\,\mathrm{m^3s^{-1}}$
$\nu_{en\parallel}$	$=$	$\left[4.6N_{N_2}R_e^{0.90} + 4.3N_{O_2}R_e^{0.55} + 1.5N_OR_e^{0.83}\right] \times 10^{-15}\,\mathrm{m^3s^{-1}}$
$\nu_{ei\parallel}$	$=$	$\left(27.6 \times 10^{-6}\,\mathrm{s^{-1}m^3K^{3/2}}\right)NT_e^{-3/2}$

Note: Formulas for collisions of NO^+ and O_2^+ with N_2 and O_2 are based on [19]. Formulas for collisions of NO^+ and O_2^+ with O and of O^+ with N_2 and O_2 are based on Table 3 of [20]. The formula for collisions of O^+ with O is based on [21]. The formula for $\nu_{en\perp}$ is based on Table 2 of [22], as parameterized by [18]. The formula for $\nu_{en\parallel}$ takes into account the factor $g_\sigma(\alpha)$ of [22], with α estimated from the temperature dependence of ν_{en}. The formula for $\nu_{ei\parallel}$ is the Spitzer conductivity, with the value of the Coulomb logarithm set to 15.

N_{N_2}, N_{O_2}, N_O, N are the number densities of N_2, O_2, O, and electrons, respectively.
$R_i = (T_i + T_n)/1000\,\mathrm{K}$.
$R_e = T_e/300\,\mathrm{K}$.
T_i, T_n, T_e are the temperatures of ions, neutrals, and electrons, respectively.

$$(\upsilon_e - \upsilon_i)_{\parallel} = -\frac{e(E_{\parallel} - E_a)}{m_e(\nu_{en\parallel} + \nu_{ei\parallel})} \qquad (14.8)$$

The electric current parallel to **B** is then

$$J_{\parallel} = \sigma_{\parallel}(E_{\parallel} - E_a) \qquad (14.9)$$

$$\sigma_{\parallel} = \frac{Ne^2}{m_e(\nu_{en\parallel} + \nu_{ei\parallel})} \qquad (14.10)$$

where $\sigma_{\parallel}$ is the parallel conductivity.

To derive expressions for motions and currents perpendicular to the magnetic field, we begin by neglecting the effects of collisions between ions and electrons in Equations 14.4 and 14.5. This turns out to be an excellent approximation (unlike what we found for motions parallel to **B**), because the magnetic component of Lorentz force is several orders of magnitude larger than the rate of momentum transfer between electrons and ions. Let us use primes to denote velocities and electric fields in the frame of reference of the neutral air, that is,

$$\mathbf{v}'_{i,e} = \mathbf{v}_{i,e} - \mathbf{v}_n \qquad (14.11)$$

$$\mathbf{E}' = \mathbf{E} + \mathbf{v}_n \times \mathbf{B} \qquad (14.12)$$

Then the component of Equation 14.4 perpendicular to **B**, upon rearrangement, gives

$$-Ne\mathbf{v}'_i \times \mathbf{B} + Nm_i\nu_{in}\mathbf{v}'_{i\perp} = Ne\mathbf{E}'_{\perp} + \mathbf{F}_{i\perp} \qquad (14.13)$$

$$\mathbf{F}_i = Nm_i\mathbf{g} - \nabla(NkT_i) \qquad (14.14)$$

where the subscript ⊥ signifies the component perpendicular to **B**, and $\mathbf{F}_i$ is the force per unit volume on the ions due to the ion pressure gradient and gravity. To solve Equation 14.13 for $\mathbf{v}'_i$, take its cross product with $e\mathbf{B}/(m_i\nu_{in})$, add the result to Equation 14.13, and rearrange, yielding

$$\mathbf{v}'_{i\perp} = \frac{\nu_{in}\Omega_i\mathbf{E}'_{\perp} - \Omega_i^2\mathbf{b}\times\mathbf{E}'_{\perp}}{B(\nu_{in}^2 + \Omega_i^2)} + \frac{\nu_{in}\mathbf{F}_{i\perp} - \Omega_i\mathbf{b}\times\mathbf{F}_{i\perp}}{Nm_i(\Omega_i^2 + \nu_{in}^2)} \qquad (14.15)$$

where **b** is a unit vector in the direction of **B** and

$$\Omega_i = \frac{eB}{m_i} \qquad (14.16)$$

is the angular gyrofrequency for the ions, describing their gyration in the geomagnetic field. Similar equations are obtained for electrons:

$$\mathbf{v}'_{e\perp} = \frac{-\nu_{en\perp}\Omega_e\mathbf{E}'_{\perp} - \Omega_e^2\mathbf{b}\times\mathbf{E}'_{\perp}}{B(\nu_{en\perp}^2 + \Omega_e^2)} + \frac{\nu_{en}\mathbf{F}_{e\perp} - \Omega_e\mathbf{b}\times\mathbf{F}_{e\perp}}{Nm_e(\Omega_e^2 + \nu_{en\perp}^2)} \qquad (14.17)$$

$$\mathbf{F}_e = -\nabla(NkT_e) \qquad (14.18)$$

$$\Omega_e = \frac{eB}{m_e} \qquad (14.19)$$

Figure 14.3 shows the height variations of the collision frequencies $\nu_{in}, \nu_{en\perp}$, and $\nu_{en\parallel} + \nu_{ei\parallel}$ for daytime solar minimum conditions, along with the angular gyrofrequencies of the ions and electrons [23,14]. The ion-neutral and electron-neutral collision frequencies are proportional to the neutral density, which falls off exponentially with increasing altitude in the upper atmosphere, changing by a factor of 10^5 between 80 and 300 km. The ion-electron and electron-ion collision frequencies, however, are approximately proportional to the electron number density N and peak near the height of maximum electron density. In the F–region, the gyrofrequencies are much larger than the collision frequencies for both ions and electrons, and the first terms on the right-hand sides of Equations 14.15 and 14.17 are usually much larger than the second terms related to $\mathbf{F}_i$ or $\mathbf{F}_e$. In this case, the approximate solutions for the ion and electron velocities are

$$\mathbf{v}_{i\perp} \approx \mathbf{v}_{e\perp} \approx \frac{\mathbf{E}\times\mathbf{b}}{B} = \mathbf{v}_E \qquad (14.20)$$

where $\mathbf{v}_E$ is the so-called $\mathbf{E}\times\mathbf{b}$ velocity.

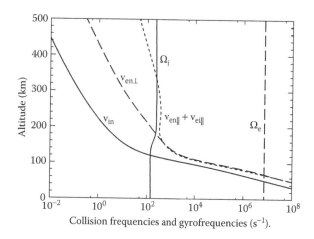

FIGURE 14.3 Collision frequencies $\nu_{in}, \nu_{en\perp}$ (for motion perpendicular to **B**) and $\nu_{en\parallel} + \nu_{ei\parallel}$ (for motion along **B**), and gyrofrequencies Ω_i and Ω_e at 44.6° N, 2.2° E, for medium solar activity ($S_a = 120$) on March 21. Collision frequencies are derived from the formulas in Table 14.1, using neutral densities from the MSISE-90 model (Hedin, A.E., *J. Geophys. Res.*, 96, 1159, 1991) and the electron densities and temperatures from the 1990 International Reference Ionosphere model (Bilitza, D., ed., *International Reference Ionosphere 1990*, Greenbelt, MD, 1990). The magnetic field is from the International Geomagnetic Reference Field epoch 1983.2.

By subtracting Equation 14.17 from 14.15, we can obtain the velocity difference $\mathbf{v}_{i\perp} - \mathbf{v}_{e\perp}$, which, when multiplied by Ne, gives the electric current density perpendicular to $\mathbf{B}$. An expression for Ohm's law can then be obtained that expresses the total current density $\mathbf{J}$ in terms of the electric field that exists in the frame of reference of the moving neutral gas, $\mathbf{E}'$, and the current density $\mathbf{J}_{pg}$ associated with gravity and plasma pressure gradients:

$$\mathbf{J} = \sigma_P \mathbf{E}'_\perp + \sigma_H \mathbf{b} \times \mathbf{E}'_\perp + \sigma_\parallel (E_\parallel - E_a)\mathbf{b} + \mathbf{J}_{pg} \quad (14.21)$$

$$\sigma_P = \frac{Ne}{B}\left(\frac{\nu_{in}\Omega_i}{\nu_{in}^2 + \Omega_i^2} + \frac{\nu_{en\perp}\Omega_e}{\nu_{en\perp}^2 + \Omega_e^2} \right) \quad (14.22)$$

$$\sigma_H = \frac{Ne}{B}\left(\frac{\Omega_e^2}{\nu_{en\perp}^2 + \Omega_e^2} + \frac{\Omega_i^2}{\nu_{in}^2 + \Omega_i^2} \right) \quad (14.23)$$

$$\mathbf{J}_{pg} = \frac{\nu_{in}\Omega_i \mathbf{F}_{i\perp} - \Omega_i^2 \mathbf{b} \times \mathbf{F}_{i\perp}}{B(\Omega_i^2 + \nu_{in}^2)} + \frac{-\nu_{en\perp}\Omega_e \mathbf{F}_{e\perp} - \Omega_e^2 \mathbf{b} \times \mathbf{F}_{e\perp}}{B(\Omega_e^2 + \nu_{en\perp}^2)} \quad (14.24)$$

σ_P and σ_H are called the Pedersen and Hall conductivities, respectively. For currents perpendicular to $\mathbf{B}$, the Pedersen conductivity gives the component in the direction of $\mathbf{E}'_\perp$, whereas the Hall conductivity gives the component perpendicular to both $\mathbf{E}'$ and $\mathbf{B}$. The component $\mathbf{v}_n \times \mathbf{B}$ of $\mathbf{E}'$, representing the difference between the electric field in the reference frame of the moving neutrals and the electric field in Earth reference frame, is often called the dynamo electric field.

It should be noted that the above derivation of ionospheric conductivities rests on the assumption that the collision frequencies are independent of the fluid velocities. This is a reasonable assumption so long as the relative velocities are small with respect to the thermal velocities, on the order of 300–1000 m s^{-1} for ions and neutrals, depending on composition and temperature of the gases. However, the differential velocities of the ion and neutral gases can sometimes be comparable to the thermal velocities, especially in the auroral regions, where strong electric fields associated with magnetospheric processes can exist. Under such circumstances, the above formulas can become inaccurate.

Figure 14.4 shows typical midlatitude vertical profiles of the daytime conductivity components for low solar activity. At all altitudes above 80 km, the parallel conductivity is much larger than the Pedersen and Hall conductivities, attaining a value on the order of 100 S m^{-1} in the upper ionosphere. This value depends on

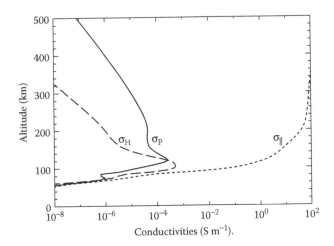

FIGURE 14.4 Noontime parallel ($\sigma_\parallel$), Pedersen (σ_P), and Hall (σ_H) conductivities at 44.6° N, 2.2° E, for medium solar activity on March 21. The electron density is the noon profile of Figure 14.2, and the collision frequencies are those of Figure 14.3.

the electron temperature, varying as $T_e^{3/2}$, but it is practically independent of the electron density. The Pedersen conductivity peaks at an altitude around 125 km during the day, whereas the Hall conductivity peaks around 105–110 km. At a given altitude, both σ_P and σ_H are essentially proportional to the electron density. The ratio between the Hall and Pedersen conductivities is greater than 1 between about 70 and 125 km, and maximizes a little below 100 km with a value of about 36, implying an angle between the current and the electric field of 88.4° in the plane perpendicular to $\mathbf{B}$.

At night, when the E-region electron density decays strongly away, the largest Pedersen conductivities can sometimes be in the ionospheric F-region, above 200 km. This is especially true during high levels of solar activity, when F-region Pedersen conductivities are particularly large, both because of large electron densities and large neutral densities that lead to large ion-neutral collision frequencies. Because the F-region electron density is highly variable, the F-region conductivity also shows a great deal of variability. The Hall conductivity also varies in proportion to the electron density but always peaks in the E-region.

The very large parallel conductivity prevents the establishment of any significant potential differences along magnetic field lines, so that the field lines are essentially equipotential. (The potential associated with the ambipolar electric field is generally much smaller than potential differences that develop perpendicular to $\mathbf{B}$.)

Consequently, the perpendicular electric field, which is determined by the electric potential difference between adjacent field lines, is nearly constant with height over most of Earth. (An exception to this near-height constancy of **E** occurs where geomagnetic field lines are almost horizontal, in the vicinity of the magnetic equator.) For this reason, it is conceptually useful to consider the height-integrated ionospheric Pedersen and Hall conductivities, or Pedersen and Hall conductances, when analyzing the current-carrying capacity of the ionosphere. In the sunlit ionosphere, for values of the solar zenith angle χ less than 80°, the following formulas are found to give representative values of the Pedersen and Hall conductances at midlatitudes:

$$\int \sigma_P dz = (11\,S)\left(\frac{S_a}{S_0}\right)^{1.1}\left(\frac{B}{B_0}\right)^{-1.6}(\cos\chi)^{0.5} \quad (14.25)$$

$$\int \sigma_H dz = (14\,S)\left(\frac{S_a}{S_0}\right)^{0.5}\left(\frac{B}{B_0}\right)^{-1.3}(\cos\chi)^{0.8} \quad (14.26)$$

The values of the normalizing constants are $S_0 = 100$ and $B_0 = 5\times10^{-5}$ T. The conductances have a fairly strong dependence on the geomagnetic field strength B, which for these formulas is assumed to be evaluated at an altitude of 125 km. There are additional seasonal modulations of the conductances: For example, the Pedersen conductance tends to be lower in summer and higher in winter than suggested by Equation 14.25, due to the so-called winter anomaly in F-region electron densities, according to which the winter electron densities are greater than the summer densities at middle latitudes.

The conductances on the nightside of Earth are considerably more complicated than on the dayside, because of the large variability of the electron density in response to movements of the plasma, to changes in thermospheric conditions, and to additional ionization sources. The high-latitude nightside ionosphere displays the greatest degree of variability, due to the highly variable nature of auroral precipitation. Particle precipitation can also influence the low-latitude nighttime ionosphere enough to affect the Pedersen conductance at solar minimum (e.g., [24]).

14.3 THERMOSPHERIC WINDS

Ohm's law, as expressed by Equation 14.21, relates the current density to the electric field as would be measured in a reference frame moving at the wind velocity $\mathbf{v}_n$. It is

the wind that drives the dynamo current, which, together with the relatively small current $\mathbf{J}_{pg}$, leads to the creation of the electric field **E**. The distribution of electric fields and currents therefore depends very much on the characteristics of global winds in the atmosphere above 90 km, called the thermosphere.

The dominant effect driving winds in the thermosphere is the diurnal variation in the absorption of solar ultraviolet radiation, which heats and expands the dayside thermosphere, creating day-to-night horizontal pressure gradients. Because of the atmosphere's much greater horizontal extent than vertical extent, the motions are constrained to be predominantly horizontal: Vertical winds are typically only on the order of 1% the horizontal wind magnitude. At high latitudes, especially during magnetospheric disturbances, Joule heating and momentum transfer by strong electric currents connected to the magnetosphere are also important forcing mechanisms. An additional important effect, especially in the lower thermosphere, is the upward propagation of global wave features from the atmosphere below, especially atmospheric tides [25]. The winds associated with both the diurnal heating of the thermosphere and the upward-propagating tides are periodic, with main periods of 1 day or harmonics thereof.

Figure 14.5 shows results from a model simulation of horizontal thermospheric winds at heights of approximately 300 km (F-region, top) and 125 km (E-region, bottom), at 0 UT for September equinox solar minimum conditions. The subsolar point is at the equator and at the edges of these diagrams, that is, 180° E and W longitude. Some influence of forcing by rapidly convection ions at high magnetic latitudes is seen, though for this simulation that forcing is relatively weak (polar cap electric potential drop of 30 kV). At middle and low latitudes, the 300 km winds are generally larger at night (central portion of Figure 14.5) than during day. At 125 km, the winds have longitude variations that are dominated by atmospheric tides propagating up from the lower atmosphere. There is a tendency for the entire wind pattern to migrate westward with the apparent position of Sun, although at 125 km there exist tidal wind components that do not migrate with the solar position, and at 300 km the high-latitude winds tend to migrate westward with respect to magnetic coordinates rather than geographic coordinates.

During magnetospheric disturbances, when high-latitude ion convection becomes much stronger, the

Neutral wind at approximately 300 km

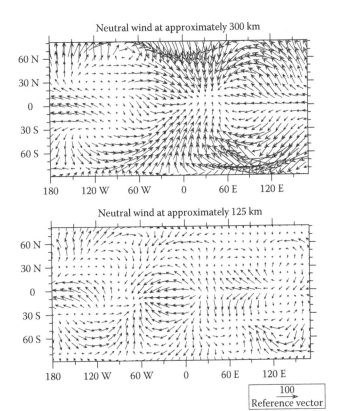

Neutral wind at approximately 125 km

FIGURE 14.5 Horizontal wind velocity at altitudes of approximately 300 km (top) and 125 km (bottom) as a function of geographic longitude and latitude, calculated by the National Center for Atmospheric Research Thermosphere–Ionosphere–Electrodynamics General Circulation Model for September equinox, 0 UT, with $S_a = 70$ and cross-polar cap potential = 30 kV. Midnight is at 0° longitude. Both migrating and nonmigrating tides at the lower boundary are included.

high-latitude winds also become stronger, and the Joule dissipation of electric energy affects not only the local but also the global thermospheric circulation. The heated air rises at high latitudes, and there is a general outflow of air above the region of peak heating, centered roughly at an altitude of 125 km. The equatorward flow at upper midlatitudes is acted upon by the Coriolis force, producing a westward motion. The altered global winds produce a disturbance dynamo effect [26].

14.4 THEORY OF IONOSPHERIC ELECTRIC FIELDS AND CURRENTS

If the distributions of thermospheric winds and ionospheric conductivities are known, the electric fields and currents generated by dynamo action can be calculated. In addition to Ohm's law (Equation 14.21), the

conditions of current continuity and of electrostatic fields must be satisfied:

$$\nabla \cdot \mathbf{J} = 0 \tag{14.27}$$

$$\mathbf{E} = -\nabla \Phi \tag{14.28}$$

where Φ is the electrostatic potential. Combining Equations 14.21, 14.27, and 14.28 gives a partial differential equation for Φ:

$$\nabla \cdot [\sigma_P (\nabla \Phi)_\perp + \sigma_H \mathbf{b} \times \nabla \Phi + \sigma_\parallel (\nabla \Phi)_\parallel]$$
$$= \nabla \cdot [\sigma_P \mathbf{v}_n \times \mathbf{B} + \sigma_H \mathbf{b} \times (\mathbf{v}_n \times \mathbf{B}) - \sigma_\parallel E_a \mathbf{b} + \mathbf{J}_{pg}] \tag{14.29}$$

Solving the above equation requires boundary conditions that represent the effects of electric coupling with the magnetosphere above and with the lower atmosphere below. Because the atmosphere below 80 km is a very poor conductor, electric current coupling between the ionosphere and the lower atmosphere is usually negligible. An adequate lower boundary condition for Equation 14.29 at the base of the ionosphere is thus obtained from the requirement that the vertical component of electric current density vanish there.

At altitudes well above the main ionosphere, the assumptions behind Ohm's law (Equation 14.21) become less valid, and suitable boundary conditions for Equation 14.29 are more difficult to determine. The plasma is not necessarily in force balance. Collisions become unimportant, and pressures become anisotropic. The terms involving Pedersen and Hall conductivities in Equation 14.21 become negligible. In the magnetosphere, the parallel electric field $E_\parallel$ still has a tendency to balance the ambipolar field E_a, although electrons of kiloelectron volt energies come to dominate the electron pressure gradient along the magnetic field that determines E_a. A partial imbalance of $E_\parallel$ and E_a can exist in regions where ion flows are accelerated parallel to the magnetic field. In any case, $E_\parallel$ is usually much smaller than $E_\perp$, although parallel potential drops of thousands of volts can sometimes develop along auroral geomagnetic field lines. $\mathbf{J}_{pg}$ becomes dominated by the pressure gradients of energetic magnetospheric ions and electrons, rather than the relatively cold ionospheric particles. Convergence of $\mathbf{J}_{pg}$ is balanced by divergence of $\mathbf{J}_\parallel$, leading to geomagnetic field-aligned currents into and out of the ionosphere. Because of the complexity of coupling with the magnetosphere, practical solutions of the partial differential equation represented by Equation 14.29 generally assume either a given distribution of field-aligned current at the top of the ionosphere or a boundary condition on the electric potential at high latitudes (e.g., [27]).

The fact that the conductivity is extremely anisotropic necessitates the utilization of a geomagnetic field-oriented coordinate system in order to carry out numerical solutions of Equation 14.29. One of the coordinates will vary along the geomagnetic field, whereas the other two coordinates must be defined so as to be constant along geomagnetic field lines. For practical purposes, a major simplification can be made to Equation 14.29 by additionally assuming that Φ is constant along field lines, even through the magnetosphere, and that $\sigma_\parallel$ is infinite. The component of $\mathbf{J}$ along $\mathbf{B}$ is then no longer determined by Ohm's law, but rather by the condition that it assume whatever value is necessary in order to ensure that the current density remains divergence free, so that Equation 14.27 is satisfied everywhere. The partial differential equation for Φ can then be reduced from three to two dimensions upon multiplying Equation 14.29 by $1/B$ and integrating along field lines. In doing so, note that

$$\nabla \cdot \mathbf{J}_\parallel = \nabla \cdot \left(\frac{J_\parallel \mathbf{B}}{B} \right) = \mathbf{B} \cdot \nabla \left(\frac{J_\parallel}{B} \right) = B \frac{\partial}{\partial s} \left(\frac{J_\parallel}{B} \right) \quad (14.30)$$

where s is the distance along a field line. The integral of $(\nabla \cdot \mathbf{J}_\parallel)/B$ thus equals the difference between $(J_\parallel / B)$ at the boundaries of the integration. For closed field lines that connect the southern and northern magnetic hemispheres, $J_\parallel$ goes to zero at both ends, owing to the vanishing conductivity at the base of the ionosphere, and so the integral also vanishes. For polar cap field lines that extend into interplanetary space, $J_\parallel$ needs to be specified at some point above the ionosphere. If we define the two coordinates that are constant on field lines as x_1 and x_2 such that $\mathbf{B}$ is in the direction of $\nabla x_2 \times \nabla x_1$, the following two-dimensional partial differential equation is obtained:

$$\frac{\partial}{\partial x_1} \left(S_{11} \frac{\partial \Phi}{\partial x_1} + (S_{12} + \Sigma_H) \frac{\partial \Phi}{\partial x_2} \right)$$
$$+ \frac{\partial}{\partial x_2} \left((S_{21} - \Sigma_H) \frac{\partial \Phi}{\partial x_1} + S_{22} \frac{\partial \Phi}{\partial x_2} \right) \quad (14.31)$$
$$= \frac{\partial D_1}{\partial x_1} + \frac{\partial D_2}{\partial x_2} \pm (J_\parallel A)^{top}$$

where:

$$S_{ij} = \int (\nabla x_i \cdot \nabla x_j) \sigma_P A ds \quad (14.32)$$

$$\Sigma_H = \int \sigma_H ds \quad (14.33)$$

$$D_i = \int \nabla x_i \cdot (\sigma_P \mathbf{v}_n \times \mathbf{B} + \sigma_H B \mathbf{v}_n + \mathbf{J}_{pg}) A ds \quad (14.34)$$

$$A = |\nabla x_1 \times \nabla x_2|^{-1} \quad (14.35)$$

where $(\pm J_\parallel A)^{top}$ is the upward field-aligned current passing through an area of unit increments of x_1 and x_2 at the top of the ionosphere (the "+" sign applying to the Southern Hemisphere and the "−" sign to the Northern Hemisphere). For closed field lines, the integrals in Equations 14.32 through 14.34 are taken along the entire field line between the base of the ionosphere in each magnetic hemisphere and $(\pm J_\parallel A)^{top}$ is zero, but for open polar field lines, the integrals are through the thickness of the ionosphere in only a single hemisphere and $(\pm J_\parallel A)^{top}$ is generally nonzero. The solutions of Equation 14.31 for Φ in the closed field region and in the two open field regions have to match on field lines lying at the closed open boundary, which couples the partial differential equations at that boundary. For simplified geometries like a purely dipolar geomagnetic field, x_1 and x_2 can be selected to be orthogonal, so that S_{12} and S_{21} vanish. Once Equation 14.31 has been solved for Φ, $\mathbf{E}$, and $\mathbf{J}_\perp$ (the component of $\mathbf{J}$ perpendicular to $\mathbf{B}$) can be determined using Equations 14.28 and 14.21, respectively. $\pm J_\parallel$ can then be calculated by integrating $-(\nabla \cdot \mathbf{J}_\perp)/B$ with respect to s from the base of a field line to the desired point, and multiplying the result by the local value of B.

Close to the magnetic equator, the combination of the nearly horizontal geomagnetic field and the large difference between σ_P and σ_H in the lower ionosphere gives rise to the EEJ [10–12]. During the daytime, when the ionospheric electric field in the equatorial region is usually eastward, a downward component of the Hall current is produced. In order to maintain current continuity, a counterbalancing upward component of the Pedersen current must flow, necessitating an upward component of electric field. Because the Pedersen conductivity is small in the lower part of the dynamo region compared with the Hall conductivity, the vertical field is much larger than the original eastward electric field. This strong upward electric field produces the strong eastward Hall current of the EEJ.

In regions where the EEJ is not changing rapidly with longitude, Equation 14.31 can be solved in an approximate way for the vertical polarization electric field in terms of the eastward field and the winds. Let x_1 be in the magnetic eastward direction and x_2 be in the direction generally upward (with an additional poleward tilt as one moves off the magnetic equator). Let us assume that east–west gradients of the electric field can be neglected. Because the component of electric field in the x_2 direction is proportional to $(\partial\Phi/\partial x_2)$, this means that the x_1-derivative of $(\partial\Phi/\partial x_2)$ is zero, and consequently

$$\frac{\partial}{\partial x_2}\left(\frac{\partial\Phi}{\partial x_1}\right)=\frac{\partial}{\partial x_1}\left(\frac{\partial\Phi}{\partial x_2}\right)=0, \quad (14.36)$$

signifying that $(\partial\Phi/\partial x_1)$ is independent of x_2, that is, constant in height and latitude.

Let us assume that the current density in the x_2 direction (upward/poleward), integrated along a geomagnetic field line, essentially vanishes, meaning that

$$\Sigma_H\frac{\partial\Phi}{\partial x_1}-S_{22}\frac{\partial\Phi}{\partial x_2}+D_2=0 \quad (14.37)$$

where S_{21} has been neglected in comparison with Σ_H. Within the lower equatorial ionosphere ($\pm 10°$ magnetic latitude, 80–200 km altitude), x_1 and x_2 can be approximated as actual spatial distances perpendicular to **B** in the eastward and upward/poleward directions, respectively (neglecting the fact that in reality the distance between two field lines varies slightly along the field), so that ∇x_1 and ∇x_2 are the unit vectors. In addition, B can be considered nearly constant over the region. Then the electric field components perpendicular to **B**, which we label E_1 in the eastward direction and E_2 in the upward/poleward direction, can also be considered to be constant along field lines. With these simplifications, and neglecting the very small term $\mathbf{J}_{pg}$, Equation 14.37 can be solved for E_2 as

$$E_2=\frac{\Sigma_H}{\Sigma_P}E_1-\frac{\int(\sigma_P\upsilon_1+\sigma_H\upsilon_2)Bds}{\Sigma_P} \quad (14.38)$$

where:

$$\Sigma_P=\int\sigma_P ds \quad (14.39)$$

The contribution to E_2 of the first term on the right of Equation 14.38 is very large for field lines peaking below 110 km, where $\Sigma_H\gg\Sigma_P$. This is the primary source of the strong polarization field responsible for the EEJ current. The contribution to E_2 of the wind terms in Equation 14.38 depends both on the wind velocity and on its variation along the magnetic field line. An eastward wind υ_1 contributes negatively to E_2. In fact, it can be seen that an eastward wind that is constant along the field line would produce an electric field equal to $-\upsilon_1 B$, which in turn would be associated with an eastward $\mathbf{E}\times\mathbf{B}$ velocity of exactly υ_1, and the ions and electrons would move at the same velocity as the neutral wind. The contribution of the vertical/meridional wind component υ_2 to the wind-related integral in Equation 14.38 tends to be much less important.

The eastward current density, obtained by using Equation 14.38 in Equation 14.21, is

$$J_1=\left(\sigma_P+\frac{\Sigma_H}{\Sigma_P}\sigma_H\right)E_1$$
$$+\sigma_H\left[\left(\upsilon_1-\frac{\int\sigma_P\upsilon_1 ds}{\int\sigma_P ds}\right)-\left(\frac{\sigma_P}{\sigma_H}\upsilon_2+\frac{\int\sigma_H\upsilon_2 ds}{\int\sigma_P ds}\right)\right]B \quad (14.40)$$

The multiplier of E_1 on the right-hand side of the above equation is sometimes called the Cowling conductivity, although the usual expression of the Cowling conductivity is derived for a one-dimensional geometry, for which Σ_H/Σ_P is replaced by σ_H/σ_P. (The two expressions give similar results below 105 km at the equator.) It is much larger than either the Pedersen or the Hall conductivity below 110 km at the magnetic equator. The terms dependent on υ_1 in Equation 14.40 would cancel if υ_1 were constant along **B**; thus, only spatially varying east–west winds affect the current.

Models of the EEJ using standard conductivities like those in Figure 14.4 predict a peak EEJ current density near 100 km altitude that is both lower and more intense than observations indicate (e.g. [18]). Reference [28] showed that plasma irregularities observed in the lower part of the EEJ are of sufficient magnitude to strongly reduce the vertical polarization field and horizontal current there, and to raise the heights of peak field and current, provided it is assumed that the irregularities are very highly aligned with the geomagnetic field. Reference [18] showed that the discrepancies could be largely removed by arbitrarily increasing the electron-neutral collision frequency $\nu_{en\perp}$ by a factor of around 4.

14.5 OBSERVATIONS OF IONOSPHERIC ELECTRIC FIELDS AND CURRENTS

Most of the available observations that give us electric field information are ion drift measurements in the ionosphere above 200 km, either by radar or by spacecraft. The component of the electric field perpendicular to the geomagnetic field is readily derived from these drift measurements by inverting Equation 14.20:

$$\mathbf{E}_\perp = -\mathbf{v}_i \times \mathbf{B} \qquad (14.41)$$

The observations show a high degree of variability, even on magnetically quiet days. The variability tends to be greater at night than during day. However, after the data are averaged over a number of days, a regular pattern emerges.

Figure 14.6 shows the MLT variations of the average quiet-day drifts for middle- and low-latitude incoherent scatter radars (Table 14.2) for different seasons and levels of solar activity. (For St. Santin, the solar cycle dependence of the drifts has not yet been determined.) In order to have sufficient data in each season, only three seasons are used for the year, centered around June, December, and the combined equinoxes: J months (May–August), E months (March, April, September, and October), and D months (November–February). Note that the seasons are ordered differently for the St. Santin and MU radars than for the American-sector radars in Figure 14.6, because geometrical considerations suggest that the solstitial variations might be opposite in the Eurasian and American longitude sectors, owing to the opposite relative placement of the magnetic and geographic equators in these two sectors (see Figure 14.1). Because

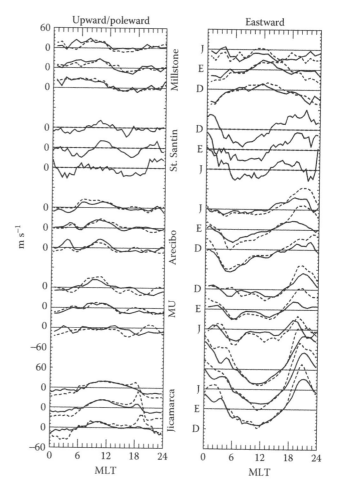

FIGURE 14.6 Average quiet-day $\mathbf{E} \times \mathbf{B}$ drift components perpendicular to the geomagnetic field over the incoherent scatter radars listed in Table 14.2. The mean altitude is 300 km. Tick marks are every 10 m s^{-1}. Solid lines are for low solar activity, and dotted lines for high solar activity (different criteria were used to define low and high activities for the different data sets). Averages for the months of November–February are denoted by D (December solstice); for March, April, September, and October by E (equinox); and for May–August by J (June solstice). Note that the seasons are ordered differently for the St. Santin and MU radars than for the other radars (see text).

TABLE 14.2 Middle- and Low-Latitude Incoherent Scatter Radars Observing Ionospheric Drifts

Radar	Geographic Latitude	Geographic Longitude	Magnetic Latitude	MLT−UT	Reference
Millstone Hill	42.6°	−71.5°	53.7°	−4.4	[34]
St. Santin	44.6°	2.2°	39.6°	0.5	[35]
Arecibo	18.3°	−66.8°	29.8°	−4.2	[36]
MU	34.9°	136.1°	27.3°	9.1	[37]
Jicamarca	−11.9°	−76.0°	1.0°	−5.1	[38]

Source: Richmond, A.D., Ionospheric electrodynamics. In *Handbook of Atmospheric Electrodynamics*, ed. H. Volland, vol. II, 249–290, Boca Raton, FL: CRC Press, 1995.

the electric fields equalize along the geomagnetic field between conjugate points in the Northern and Southern Hemispheres, seasonal variations of the electric fields and **E** × **B** drifts are not great, but differences between the J and D solstices can occur in association with the offset magnetic and geographic equators, among other factors. The upward/poleward drift component (perpendicular to the tilted geomagnetic field lines in the magnetic meridian) tends to be positive in the morning and negative in the afternoon at upper middle latitudes, with the phase shifting somewhat later at low latitudes. The general phase of the diurnal variation of the eastward drift reverses between the magnetic latitudes of Millstone Hill (54°) and St. Santin (40°). Different criteria were used to define low and high activity for the different data sets shown in Figure 14.6, so quantitative comparisons of the solar cycle variations among the various curves are not possible. However, some qualitative features are found to be consistent among the curves: for example, the nighttime east–west drifts are generally stronger for high solar activity. At Jicamarca, Peru, a notable phenomenon in the upward/poleward drift is the strong solar activity amplification of the peak that occurs just after sunset. This feature is generally attributed to the F-region dynamo [29–33].

Magnetic disturbances produce fluctuating electric fields over the entire globe. In addition to the fluctuations, the "average" **E** × **B** drifts are altered during disturbed periods. On average, the low-latitude upward/meridional **E** × **B** drift during the main phase of a magnetic storm tends to increase from around 10 MLT to around 21 MLT, with a peak after sunset, and to decrease from around 21 MLT to around 10 MLT, with a strong minimum shortly before sunrise [39]. At midlatitudes, other observations show a general average westward shift of the zonal drift at all times, most strongly during the night (e.g., [40,41]). Observations of equatorial **E** × **B** vertical drifts have sometimes shown a decrease in magnitude following magnetic storms [42–44], suggesting possible disturbance dynamo effects.

Ionospheric currents have been measured by rocket-borne magnetometers and by incoherent scatter radars [45]. Figure 14.7 shows a collection of current profiles obtained from rocket-borne magnetometers near the magnetic equator off the coast of South America, showing the EEJ [1, 46–48].

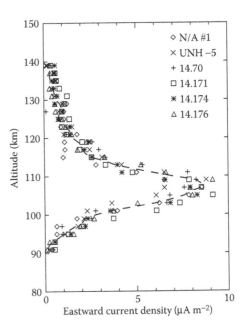

FIGURE 14.7 Electric current density around noon near the magnetic dip equator over Peru, measured by rocket-borne magnetometers. Values have been normalized to a ground-level magnetic perturbation of 100 nT at the nearby Huancayo observatory. Symbols are for the current density measured on each of six flights. Dashed line shows mean profile of the measured current density. The profile measured on flight N/A #1 is from (Shuman, B.M., *J. Geophys. Res.*, 75, 3889, 1970); that on flight UNH-5 is from (Maynard, N.C., *J. Geophys. Res.*, 72, 1863, 1967); and those on flights 14.70, 14.171, 14.174, and 14.176 are from (Davis, T.N. et al., *J. Geophys. Res.*, 72, 1845, 1967). (From Richmond, A.D., Ionospheric electrodynamics, in *Handbook of Atmospheric Electrodynamics*, ed. H. Volland, vol. II, pp. 249–290, CRC Press, Boca Raton, FL, 1995. With Permission.)

14.6 GEOMAGNETIC VARIATIONS

Perturbations in the geomagnetic field were one of the earliest phenomena of ionospheric electrodynamics to be observed, and magnetic perturbation data exist for a long period of time, from points widely spread over Earth. These data have provided us with some of our most detailed information about global ionospheric electrodynamics and its variability (e.g., [5,49–51]).

In the standard terminology of geomagnetism, the geomagnetic vector is given by three components labeled H (horizontal intensity), D (declination), and Z (vertical component, defined to be positive downward), or else X (geographically northward), Y (geographically eastward), and Z. Earth's main field has a strength at

Earth's surface of 25,000–60,000 nT. Typical quiet-day perturbations at midlatitudes are usually less than 100 nT, whereas disturbed perturbations in the auroral zone rarely exceed 2000 nT, much smaller than the main field. An increase in H represents a vector perturbation component in the magnetic northward direction (the direction of a compass needle). An algebraic increase in D represents a vector perturbation component in the magnetic eastward direction. An algebraic increase in Z represents a vector perturbation component in the downward direction.

Figure 14.8 shows the average S_q variations at equinox for a chain of magnetometers in the American longitude sector [1]. The solid and dashed curves represent average data for solar cycle minimum and maximum conditions, respectively. In general, the solar cycle and seasonal variations are what one would expect on the basis of ionospheric conductivity variations: larger perturbations at solar maximum and in summer. A strong

EEJ enhancement in the H component is evident at Huancayo, Peru. Spacecraft in low-Earth orbit can also measure the magnetic perturbations. Whereas the H perturbation at the magnetic equator is positive on the ground, it is negative above the EEJ current.

Variability in the geomagnetic perturbations is caused both by variable magnetospheric activity and by changes in the winds and conductivities in the dynamo region. The auroral electrojets are the most prominent manifestation of the variability in magnetospheric activity; these are connected by geomagnetic field-aligned currents with the outer magnetosphere. The field-aligned currents, along with magnetospheric ring currents, and currents at the magnetopause and in the magnetotail contribute to ground-level magnetic perturbations seen at middle and low latitudes, and are responsible for the fact that disturbance magnetic perturbations occur even at night, in the absence of significant ionospheric conductivity. Daytime geomagnetic disturbances are often

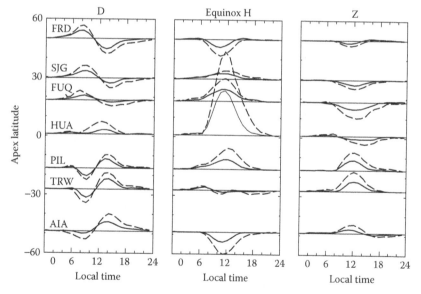

FIGURE 14.8 Average quiet-day magnetic perturbations at several stations in the American longitude sector for equinox months (March, April, September, October). D = magnetically eastward; H = magnetically northward; Z = downward. Station names and their geographical coordinates, from north to south, are as follows: Fredericksburg (38.2 N, 282.6 E), San Juan (18.1 N, 293.8 E), Fuquene (5.5 N, 286.3 E), Huancayo (12.0 S, 284.7 E), Pilar (31.7 S, 296.1 E), Trelew (43.2 S, 294.7 E), and Argentine Island (65.2 S, 295.7 E). Solid lines are for solar cycle minimum (1964–1965 or 1985–1986), and dashed lines are for solar cycle maximum (1957–1958, 1967–1968, or 1979–1980). Values are adjusted to a mean solar radio flux value of S_a =75 (solar minimum) or 200 (solar maximum). The perturbations are measured from a baseline defined to be the average value between 0 and 3 local time, after removal of a linear trend. The scale is 50 nT per 10°. Local time, in hours, is defined as universal time + longitude/(15°/h). (Adapted from Richmond, A.D., Ionospheric electrodynamics, in *Handbook of Atmospheric Electrodynamics*, ed. H. Volland, vol. II, pp. 249–290, CRC Press, Boca Raton, FL, 1995. With Permission.)

amplified in the equatorial EEJ with respect to nearby low-latitude locations (e.g., [52]), reflecting the penetration of disturbance electric fields to the equator.

Significant day-to-day changes in the geomagnetic perturbations are observed not only on magnetically disturbed days but also on quiet days. It is generally believed that these changes are due in large part to changes in the propagation conditions for tides entering the dynamo region from below. For example, strong effects during stratospheric sudden warmings are associated with modulation of both solar and lunar atmospheric tides [53]. The day-to-day variability is found to be only weakly correlated between stations widely separated in longitude (e.g., [54]).

ACKNOWLEDGMENTS

The National Center for Atmospheric Research is sponsored by the National Science Foundation. This work was supported in part by NASA grant NNX14AE08G and by NSF grant AGS-1135446. I am grateful to M. Buonsanto, B. Fejer, and W. Oliver for providing digital files of their data shown in Figure 14.6 and to A. Maute for providing Figure 14.5.

REFERENCES

1. Richmond, A.D. 1995. Ionospheric electrodynamics. In *Handbook of Atmospheric Electrodynamics*, ed. H. Volland, vol. II, 249–290. Boca Raton, FL: CRC Press.

2. Cowley, S.W.H. 2000. Magnetosphere-ionosphere interactions: A tutorial review. In *Magnetospheric Current Systems, Geophys. Monograph 118*, ed. S.-I. Ohtani, R. Fujii, M. Hesse, and R.L. Lysak, 91–106. Washington, DC: American Geophysical Union.

3. Rycroft, M.J., and R.G. Harrison. 2011. Electromagnetic atmosphere-plasma coupling: The global atmospheric electric circuit. *Space Sci. Rev.* doi:10.1007/s11214-011-9830-8.

4. Stewart, B. 1882. Terrestrial magnetism. In *Encyclopaedia Britannica*, 9th ed., vol. 16, 159–184.

5. Chapman, S., and J. Bartels. 1961. *Geomagnetism.* 2nd ed. London: Oxford University Press.

6. Matsushita, S. 1967. Solar quiet and lunar daily variation fields. In *Physics of Geomagnetic Phenomena*, ed. S. Matsushita and W.H. Campbell, 301. New York: Academic Press.

7. Brekke, A., and J. Moen. 1993. Observations of high latitude ionospheric conductances. *J. Atmos. Terr. Phys.* 55:1493.

8. Richmond, A.D. 1995. Ionospheric electrodynamics using Magnetic Apex Coordinates. *J. Geomagn. Geoelectr.* 47:191–212.

9. Thebault, E. et al. 2015. International geomagnetic reference field: The 12th generation. *Earth Planets Space* 67: 1–19. doi:10.1186/s40623-015-0228-9.

10. Forbes, J.M. 1981. The equatorial electrojet. *Rev. Geophys. Space Phys.* 19:469.

11. Reddy, C.A. 1989. The equatorial electrojet. *Pure Appl. Geophys.* 131:485.

12. Rastogi, R.G. 1989. The equatorial electrojet: Magnetic and ionospheric effects. In *Geomagnetism*, ed. J.A. Jacobs, vol. 3, 461. San Diego, CA: Academic Press.

13. Schunk, R.W., and A.F. Nagy. 2009. *Ionospheres: Physics, Plasma Physics and Chemistry.* 2nd ed. Cambridge: Cambridge University Press.

14. Bilitza, D., ed. 1990. *International Reference Ionosphere 1990.* NSSDC 90-22, Greenbelt, MD.

15. Newell, P.T., T. Sotirelis, and S. Wing. 2009. Diffuse, monoenergetic, and broadband aurora: The global precipitation budget. *J. Geophys. Res.* 114:A09207. doi:10.1029/2009JA014326.

16. Shkarofsky, I.P., T.W. Johnston, and M.P. Bachynski. 1966. *The Particle Kinetics of Plasmas.* Reading, MA: Addison-Wesley.

17. Hill, R.J., and S.A. Bowhill. 1977. Collision frequencies for use in the continuum momentum equations applied to the lower ionosphere. *J. Atmos. Terr. Phys.* 39:803.

18. Gagnepain, J., M. Crochet, and A.D. Richmond. 1977. Comparison of equatorial electrojet models. *J. Atmos. Terr. Phys.* 39:1119.

19. Viehland, L.A., and E.A. Mason. 1995. Transport properties of gaseous ions over a wide energy range, IV. *At. Data Nucl. Data Tables* 60:37–95.

20. Mason, E.A. 1970. Estimated ion mobilities for some air constituents. *Planet Space Sci.* 18:137.

21. Pesnell, W.D., K. Omidvar, and W.R. Hoegy. 1993. Momentum transfer collision frequency of O^+-O. *Geophys. Res. Lett.* 20:1343–1346.

22. Itikawa, Y. 1971. Effective collision frequency of electrons in atmospheric gases. *Planet Space Sci.* 19:993.

23. Hedin, A.E. 1991. Extension of the MSIS thermosphere model into the middle and lower atmosphere. *J. Geophys. Res.* 96:1159.

24. Rowe, J.F., Jr., and J.D. Mathews. 1973. Low-latitude nighttime *E* region conductivities. *J. Geophys. Res.* 78:7461.

25. Forbes, J.M. 1995. Tidal and planetary waves. In *The Upper Mesosphere and Lower Thermosphere*, ed. R.M. Johnson and T.L. Killeen, 67–87, Washington, DC: Amercian Geophysical Union.

26. Blanc, M., and A.D. Richmond. 1980. The ionospheric disturbance dynamo. *J. Geophys. Res.* 85:1669.

27. Richmond, A.D., and A. Maute. 2013. Ionospheric electrodynamics modeling. In *Modeling the Ionosphere-Thermosphere System*, ed. J. Huba, R. Schunk, and G. Khazanov, 57–71. Washington, DC: American Geophysical Union, Geophysical Monograph 201. doi:10.1029/2012GM001331. [published online as Richmond, A.D., and A. Maute. 2014. Ionospheric electrodynamics modeling. In *Modeling the*

Ionosphere-Thermosphere System, ed. J. Huba, R. Schunk and G. Khazanov. Chichester: John Wiley & Sons. doi:10.1002/9781118704417.ch6.]

28. Ronchi, C., R.N. Sudan, and P.L. Similon. 1990. Effect of short-scale turbulence on kilometer wavelength irregularities in the equatorial electrojet. *J. Geophys. Res.* 95:189.

29. Heelis, R.A., P.C. Kendall, R.J. Moffett, D.W. Windle, and H. Rishbeth. 1974. Electrical coupling of the *E*- and *F*-regions and its effect on *F*-region drifts and winds. *Planet Space Sci.* 22:743.

30. Rishbeth, H. 1981. The *F*-region dynamo. *J. Atmos. Terr. Phys.* 43:387.

31. Eccles, J.V. 1998. Modeling investigation of the evening prereversal enhancement of the zonal electric field in the equatorial ionosphere. *J. Geophys. Res.* 103:26709–26719.

32. Richmond, A.D., T.-W. Fang, and A. Maute. 2015. Electrodynamics of the equatorial evening ionosphere: 1. Importance of winds in different regions. *J. Geophys. Res. Space Phys.* 120:2118–2132. doi:10.1002/2014JA020934.

33. Richmond, A.D., and T.-W. Fang. 2015. Electrodynamics of the equatorial evening ionosphere: 2. Conductivity influences on convection, current, and electrodynamic energy flow. *J. Geophys. Res. Space Phys.* 120:2133–2147. doi:10.1002/2014JA020935.

34. Buonsanto, M.J., M.E. Hagan, J.E. Salah, and B.G. Fejer. 1993. Solar cycle and seasonal variations in *F* region electrodynamics at Millstone Hill. *J. Geophys. Res.* 98:15677.

35. Blanc, M., and P. Amayenc. 1979. Seasonal variations of the ionospheric $E \times B$ drifts above Saint-Santin on quiet days. *J. Geophys. Res.* 84:2691.

36. Fejer, B.G. 1993. F region plasma drifts over Arecibo: Solar cycle, seasonal, and magnetic activity effects. *J. Geophys. Res.* 98:13645.

37. Oliver, W.L., Y. Yamamoto, T. Takami, S. Fukao, M. Yamamoto, and T. Tsuda. 1993. Middle and Upper atmosphere radar observations of ionospheric electric fields. *J. Geophys. Res.* 98:11615.

38. Fejer, B.G., E.R. de Paula, S.A. Gonzalez, and R.F. Woodman. 1991. Average vertical and zonal *F* region plasma drifts over Jicamarca. *J. Geophys. Res.* 96:13901.

39. Huang, C.-S. 2015. Storm-to-storm main phase repeatability of the local time variation of disturbed low-latitude vertical ion drifts. *Geophys. Res. Lett.* 42:5694–5701. doi:10.1002/2015GL064674.

40. Heelis, R.A., and W.R. Coley. 1992. East-west ion drifts at mid-latitudes observed by Dynamics Explorer 2. *J. Geophys. Res.* 97:19461.

41. Pedatella, N.M., and J.M. Forbes. 2011. Electrodynamic response of the ionosphere to high-speed solar wind streams. *J. Geophys. Res.* 116:A12310. doi:10.1029/2011JA017050.

42. Fejer, B.G., M.F. Larsen, and D.T. Farley. 1983. Equatorial disturbance dynamo electric fields. *Geophys. Res. Lett.* 10:537.

43. Sastri, J.H. 1988. Equatorial electric fields of ionospheric disturbance dynamo origin. *Ann. Geophysicae* 6:635.

44. Scherliess, L., and B.G. Fejer. 1997. Storm time dependence of equatorial disturbance dynamo zonal electric fields. *J. Geophys. Res.* 102:24037–24046.

45. Harper, R.M. 1977. A comparison of ionospheric currents, magnetic variations, and electric fields at Arecibo. *J. Geophys. Res.* 82:3233.

46. Shuman, B.M. 1970. Rocket measurement of the equatorial electrojet. *J. Geophys. Res.* 75:3889.

47. Maynard, N.C. 1967. Measurements of ionospheric currents off the coast of Peru. *J. Geophys. Res.* 72:1863.

48. Davis, T.N., K. Burrows, and J.D. Stolarik. 1967. A latitude survey of the equatorial electrojet with rocket-borne magnetometers. *J. Geophys. Res.* 72:1845.

49. Matsushita, S., and W.H. Campbell, eds. 1967. *Physics of Geomagnetic Phenomena*. New York: Academic Press.

50. Kane, R.P. 1976. Geomagnetic field variations. *Space Sci. Rev.* 18:413.

51. Yamazaki, Y. et al. 2011. An empirical model of the quiet daily geomagnetic field variation. *J. Geophys. Res.* 116:A10312. doi:10.1029/2011JA016487.

52. Yamazaki, Y., and M.J. Kosch. 2015. The equatorial electrojet during geomagnetic storms and substorms. *J. Geophys. Res. Space Phys.* 120:2276–2287. doi:10.1002/2014JA020773.

53. Yamazaki, Y. 2014. Solar and lunar ionospheric electrodynamic effects during stratospheric sudden warmings. *J. Atmos. Solar-Terr. Phys.* 119:138–146. doi:10.1016/j.jastp.2014.08.001.

54. Kane, R.P. 1972. Longitudinal spread of equatorial S_q variability. *J. Atmos. Terr. Phys.* 34:1425.

CHAPTER 15

Simulating Space Weather

Tamas I. Gombosi

CONTENTS

THIS CHAPTER REVIEWS THE evolution of global-scale space weather simulations from the early days when the first numerical models were developed to the present when a new generation of physics-based models is able to outperform empirical models. We will review the various physics-based approaches and summarize their advantages and disadvantages. The chapter also summarizes validation studies and the present state of community use. Finally, we describe the first efforts to transition from physics-based space weather models to operational space weather forecasting.

15.1 INTRODUCTION

Numerical simulation and modeling have become increasingly essential to basic and applied space physics research for two primary reasons. First, the heliosphere and magnetosphere are vast regions of space with relatively few *in situ* measurements. To understand the global behavior of this complex system, numerical simulations provide the ability to "stitch together" observations from different regions to provide insight into the interpretation of data. The second reason for the increased reliance upon simulations is that the models themselves have evolved to a point where their physical content and numerical robustness, flexibility, and improving ease of use inspire researchers to apply them to intriguing scenarios with a new measure of confidence. To be sure, many shortcomings and questions remain for even the most advanced models in terms of inclusion of important physical mechanisms, spatial and temporal domains that can be addressed, and thorny technical numerical issues to be dispatched. Nonetheless, it can be safely stated that modeling has over the past several years crossed a threshold whereby they have made the transition from the arcane preserves of specialists to practical tools with widespread application.

Global computational models based on first principles mathematical descriptions of the physics represent a very important component of efforts to understand plasma phenomena associated with the solar system, including the large-scale solar corona, the solar wind, its interaction with planetary magnetospheres, comets, the interstellar medium, and the initiation, structure, and evolution of solar eruptive events. Presently, and in the foreseeable future, numerical models based on the equations of magnetohydrodynamics (MHD) are the only self-consistent mathematical descriptions that can span the enormous distances associated with large-scale phenomena in space. Although providing only a relatively low-order approximation to the actual behavior of plasmas, MHD models have been used successfully to simulate many important space plasma processes and provide a powerful means for significantly advancing the understanding of such processes.

Global MHD models for space science applications were first developed in the early 1980s [1–4]. In the late 1990s and early 2000s, the performance, robustness, as well as the level of sophistication of the global models dramatically improved to the point that simulations became the "third branch" of investigation methods complementing observations and theory. Most of the presently operational models are based on single-fluid ideal MHD with numerical dissipation (resistivity, diffusion, viscosity, etc.) "mimicking" physical processes missing from the governing equations. More advanced "research codes" incorporate physically more accurate descriptions of these processes by using resistive MHD, multispecies and multifluid descriptions (multispecies description uses separate continuity equations for the ion components but assumes that they all have the same bulk velocity and temperature, whereas multifluid approximation uses separate continuity, momentum,

and energy equations for all ion species.), the Hall term, anisotropic plasma pressure, and other improvements.

Most global space weather simulation codes were primarily designed for a single physics problem. These include the Lyon–Fedder–Mobarry (LFM) code [5,6], the OpenGGCM code [7,8], the Watanabe–Sato code [9,10], the grand unified magnetosphere–ionosphere coupled simulation (GUMICS) code [11], the Integrated Space Weather Prediction Model [12,13] of Earth's magnetosphere, the Magnetohydrodynamics around a sphere (MAS) code [14,15], Usmanov's heliosphere code [16,17], Hayashi's corona model [18], the SIP-CESE model [19], Odstrcil's ENLIL [20,21] model of the solar corona, and the inner heliosphere and Pogorelov's outer heliosphere model [22]. More general-use models include the widely used ZEUS code [23], Ogino's planetary magnetosphere code [24], Tanaka's 3D global MHD model [25], Winglee's multifluid Hall MHD code [26,27], Toth's general MHD Versatile Advection Code [28], and the University of Michigan's Block Adaptive-Tree Solar-Wind Roe-Type Upwind Scheme (BATS-R-US) [29,30] model.

About a decade ago, space weather models started to couple global MHD description of large-scale phenomena with regional models describing critical smaller scale regions or processes. This technology was used for space weather simulations by the University of Michigan's space weather modeling framework (SWMF) [30,31], the OpenGGCM [32], and the Center for Integrated Space Weather Modeling (CISM) [33,34].

The most recent development of space weather modeling is the embedding of kinetic simulation regions in global MHD simulations. Two groups initiated this approach: a group located at the UCLA used MHD solutions to provide initial and static boundary conditions for a particle-in-cell (PIC) simulation [35], whereas the University of Michigan group two-way coupled a global Hall MHD code to embedded PIC simulation boxes [36].

The first part of this chapter will briefly summarize the equations used by global space weather models and point out the various approximations used by the various codes. We use our BATS-R-US multiphysics extended MHD model and SWMF to illustrate the various techniques. The second part of the chapter will discuss the validation of global space weather codes, the transition to operational space weather forecasting, and the most recent efforts to incorporate improved physics in the next-generation models.

15.2 SPACE WEATHER: A MULTISCALE PROBLEM

The richness of spatial and temporal scales represents a fundamental challenge in modeling space weather. Figure 15.1 shows typical temporal and spatial scale sizes for the solar corona, the inner heliosphere, and the magnetosphere. One can see that both the spatial and temporal scales span a factor of $2^{28} \approx 3 \times 10^8$. If one tries to simulate the entire space weather system with a single simulation, the volume ratio of the smallest to largest cells is about 2×10^{25}, a prohibitively huge number. It is obvious that one needs a series of regional models coupled together by an efficient computational framework to simulate, and eventually forecast, space weather.

The CISM has developed a loosely coupled framework [33] in which each model runs as a separate executable. The executables are coupled either by flat files or by a general communication library Intercom [37], and the grid interpolation can be handled by the Overture library [38]. The use of these libraries has been demonstrated with a limited number of models and couplings so far [33]. The CISM framework minimizes changes to the original physics models. The SWMF [30,31] has followed a different strategy. Each physics domain in the SWMF corresponds to a component. Each component is represented by one or more component versions. A component version is a physics model as well as the appropriate wrappers and couplers. The components are compiled into libraries, and they are linked to the core of the framework and the shared libraries to form a single executable. The SWMF distributes the components over a parallel machine, and executes and couples them in an efficient manner [39] using the Message Passing Interface library for communication. More recently, frameworks have been playing a critical role in coupling sub-gridsize kinetic phenomena (such as magnetic reconnection) to large-scale simulations [see 36].

The conceptual structure of SWMF is shown in Figure 15.2. There are over a dozen components represented by the boxes. The thick arrows show how the domains are coupled together. In an actual simulation,

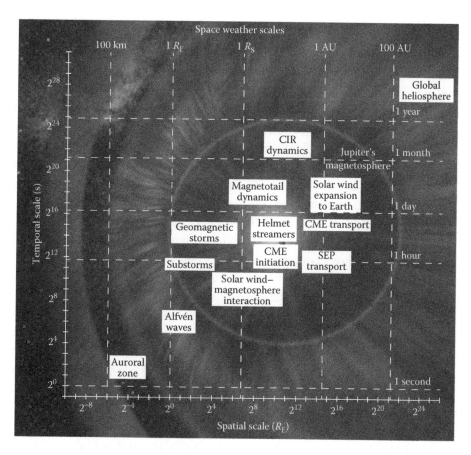

FIGURE 15.1 Spatial and temporal scales of various space weather processes.

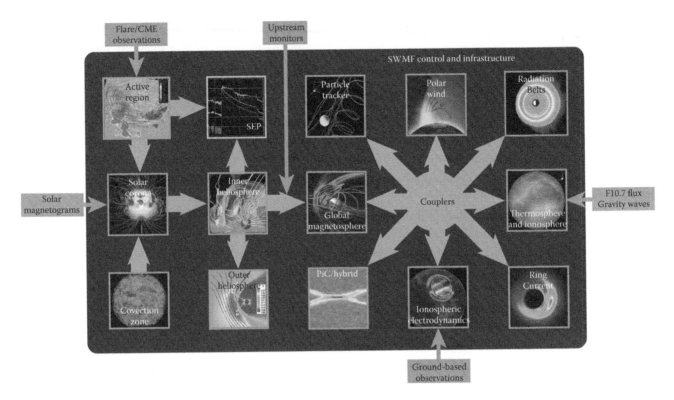

FIGURE 15.2 Components (boxes) and their couplings (thick arrows) in the SWMF. External input is indicated by the thin arrows.

one can use any meaningful subset of the components. If the simulation starts from the Sun, it is typically driven by magnetogram (synoptic or synchronic) data and flare/CME observations. Simulations restricted to magnetospheric components are usually driven by the solar wind data obtained by satellites upstream of Earth, for example, Advanced Composition Explorer (ACE), Wind, or Geotail. We also use the F10.7 solar flux for some of the empirical relationships in the ionosphere and thermosphere models. The SWMF has a layered architecture. The top layer is an optional graphical user interface. The second layer contains the control module, which is responsible for distributing the active components over the parallel machine, executing the models, and coupling them at the specified frequencies. The third layer contains the physics domain components. Each component can have multiple versions. Each component version consists of a physics model with a wrapper and one or more couplers. The wrapper is an interface with the control module, whereas each coupler is an interface with another component. The physics models can also be compiled into stand-alone executables. The fourth and

lowest layer contains the shared library and the utilities that can be used by the physics models as well as by the SWMF core.

Table 15.1 lists the currently available SWMF components. In practice, the SWMF is almost never run with all components coupled together, but we typically use a subset of the available components. For example, we can run the solar corona and the inner heliosphere models driven by solar synoptic magnetograms, or the global magnetosphere, inner magnetosphere, and ionosphere electrodynamics models are run together driven by satellite observations of the solar wind.

As Table 15.1 shows, several components can be represented by the BATS-R-US code. In each of these models, and in many other applications, BATS-R-US solves different sets of equations: radiative, ideal, Hall, two-fluid, anisotropic, semirelativistic, multispecies, or multifluid MHD. In addition to the basic equations, there are various source terms that also change from application to application: collisions, charge exchange, chemistry, photoionization, recombination, and so on. The boundary and initial conditions vary greatly as well. There are currently about 40 equation modules

TABLE 15.1   SWMF Components

Domain	ID	Model
Solar convection zone	CZ	Finite-difference Spherical Anelastic Model [40] coupled to the Stagger code [133]
Transition region	TR	Threaded Field Line Model [41]
Solar corona	SC	AWSoM [42]
Active region	AR	Titov–Dèmoulin [43], Gibson–Low [44], breakout [45,46], regional subsurface [47]
Inner heliosphere	IH	Block Adaptive-tree Solar-wind Roe-type Upwind Scheme (BATS-R-US) [48,29]
SEPs	SP	Kota [49], Field Line Advection Model for Particle Acceleration [50], Adaptive Mesh Particle Simulator (AMPS) [51]
Outer heliosphere	OH	BATS-R-US [29,52,53]
Global magnetosphere	GM	BATS-R-US [29,54]
Inner magnetosphere	IM	RCM [55,56], CRCM [57,58], Ring Current–Atmosphere Interactions Model [59] coupled to a 3D equilibrium magnetic field code (SCB) [60], Hot Electron Ion Drift Integrator [61]
Radiation belt	RB	Radiaion Belt Environment [62]
Plasmasphere	PS	Dynamic Global Core Plasma Model [63]
Polar wind	PW	Polar Wind Outflow Model [64]
Ionospheric electrodynamics	IE	Ridley Ionosphere Model [65], Weimer [66]
Upper atmosphere	UA	Global Ionosphere–Thermosphere Model [67], Mass Spectrometer and Incoherent Scatter Model [68], International Reference Ionosphere [69]
Particle tracker	PT	AMPS [51]
Particle in cell	PC	Implicit Particle-in-Cell method in 3D [70–72], Alternating-Order interpolation (ALTOR) [73]
Hybrid	HY	ALTOR [73]

and over 40 user modules (obviously not all combinations are possible), which means that BATS-R-US can be configured for quite a few different applications. An overview of the BATS-R-US code is shown in Figure 15.3.

In Section 15.3, we briefly outline the main physics assumptions and approximations used in the various continuum description of the space environment. These approximations are at the core of the BATS-R-US multiphysics code.

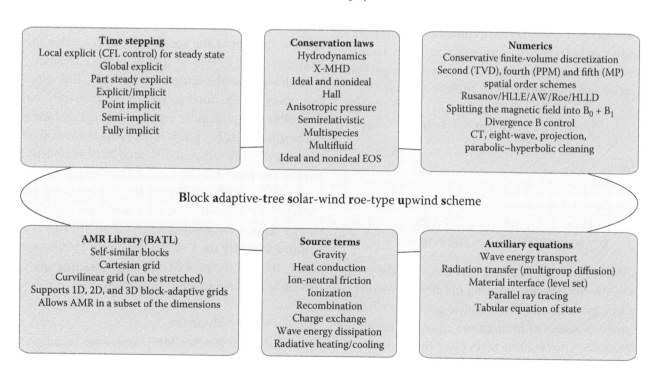

FIGURE 15.3   Overview of the BATS-R-US multiphysics code.

15.3 CONTINUUM APPROXIMATIONS

15.3.1 Kinetic Equation

All continuum descriptions of space plasmas can be derived as an approximation to the low-order velocity moments of the plasma kinetic equation and Maxwell's equations. In classical kinetic theory, the state of a dilute or rarefied gas (plasma) is prescribed by a distribution function for the particle velocities, F, whose time evolution in phase space is governed by the kinetic equation. In general, this equation is a nonlinear integro-differential equation in terms of seven independent variables. The kinetic equation for an ideal gas can be formulated in terms of the of the particle random velocity c, position coordinate x, and time t [see 74]:

$$\frac{\partial F_s}{\partial t} + (\mathbf{u}_s + c_s) \cdot \nabla F_s - \left[\frac{\partial \mathbf{u}_s}{\partial t} + (\mathbf{u}_s + c_s) \cdot \nabla \mathbf{u}_s - \mathbf{a}_s \right]$$
$$\cdot \nabla_{cs} F_s = \frac{\delta F_s}{\delta t} \qquad (15.1)$$

where:
   $\mathbf{u}_s$ is the bulk velocity vector of the species "s"
   $\mathbf{a}_s$ is the acceleration due to external forces
   $\delta F_s / \delta t$ is the collision operator representing the time rate of change of the particle distribution function produced by particle collisions

In Equation 15.1, we explicitly included the dependence of the random velocity on time and location, therefore t, x, and c_s can be treated as independent variables. In space physics, the external acceleration acting on the particles is gravity and the Lorentz force, therefore we use

$$\mathbf{a}_s = \mathbf{g} + \frac{q_s}{m_s} [\mathbf{E} + (\mathbf{u}_s + c_s) \times \mathbf{B}] \qquad (15.2)$$

where:
   m_s is the particle mass
   q_s is the particle charge
   $\mathbf{g}$, $\mathbf{E}$, and $\mathbf{B}$ are the gravitational acceleration and electric and magnetic fields, respectively

Finally, we note that Equation 15.1 describes the evolution of the distribution function for any given component of a gas mixture. The complete gas is described by the coupled system of Boltzmann's equations for all the components. The coupling takes place through the self-generated electromagnetic field and direct intermolecular interactions described by the collision terms.

Equation 15.1 is supplemented by Maxwell's equations for the electromagnetic fields:

$$\text{Coulomb's law: } \nabla \cdot \mathbf{E} = \frac{1}{\varepsilon_0} \rho_c \qquad (15.3)$$

$$\text{No magnetic monopoles: } \nabla \cdot \mathbf{B} = 0 \qquad (15.4)$$

$$\text{Faraday's law: } \frac{\partial \mathbf{B}}{\partial t} = -\nabla \times \mathbf{E} \qquad (15.5)$$

$$\text{Ampère's law: } \mu_0 \mathbf{j} = \nabla \times \mathbf{B} - \frac{1}{c^2} \frac{\partial \mathbf{E}}{\partial t} \qquad (15.6)$$

where:
   c is the speed of light
   ε_0 and μ_0 are the electric permittivity and magnetic permeability of vacuum, respectively

Equations 15.1 and 15.3 through 15.6 are coupled through the charge and electric current densities:

$$\rho_c = \sum_{s=all} q_s \iiint\limits_{-\infty}^{+\infty} d^3 c_s F_s \qquad (15.7)$$

$$j = \sum_{s=all} q_s \mathbf{u}_s \iiint\limits_{-\infty}^{+\infty} d^3 c_s F_s \qquad (15.8)$$

In most MHD simulations, the displacement current is neglected in Ampère's law (Equation 15.6). However, the LFM [5,6] and OpenGGCM [7,8] codes keep this term and use it to lower the speed of light to a value that is large compared to typical speeds in space plasmas, but much lower than the actual speed of light. This approximation, called the Boris correction [75], not only makes it possible to significantly increase the explicit time step in the simulation, but it also reduces the numerical resistivity [see 76]. However, it also modifies the underlying physics and one has to be careful when using it.

On scales much larger than the Debye length, space plasmas are quasi-neutral, and therefore, the electric charge density is approximately zero. In effect, this requires to satisfy the $\nabla \cdot \mathbf{E} = 0$ condition. This, however, is not enforced in space MHD codes and this approximation results in a small inconsistency.

15.3.2 Velocity Moments

All global space weather MHD codes are based on the low-order velocity moments of the distribution function, F_s, which are as follows:

Mass density: $\rho_s = m_s \langle F_s \rangle = \iiint\limits_{-\infty}^{+\infty} d^3 c_s F_s$ (15.9)

Pressure tensor: $P_{s_{ij}} = m_s \langle c_{s_i} c_{s_j} F_s \rangle$

$$= \iiint\limits_{-\infty}^{+\infty} d^3 c_s c_{s_i} c_{s_j} F_s$$ (15.10)

Heat flow tensor: $Q_{s_{ijk}} = m_s \langle c_{s_i} c_{s_j} c_{s_k} F_s \rangle$

$$= \iiint\limits_{-\infty}^{+\infty} d^3 c_s c_{s_i} c_{s_k} c_{s_j} F_s$$ (15.11)

We note that, by definition, $\langle c_s F_s \rangle = 0$.

The zeroth, first, and second moments of the plasma kinetic equation yield the following transport equations for the density, bulk velocity vector, and pressure tensor, respectively [see 77–81]:

$$\frac{\partial \rho_s}{\partial t} + \nabla \cdot (\rho_s \mathbf{u}_s) = \frac{\delta \rho_s}{\delta t}$$ (15.12)

$$\rho_s \frac{\partial \mathbf{u}_s}{\partial t} + \rho_s (\mathbf{u}_s \cdot \nabla) \mathbf{u}_s + \nabla \cdot P_s - \rho_s \mathbf{g} - q_s n_s (\mathbf{E} + \mathbf{u}_s \times \mathbf{B})$$
$$= \rho_s \frac{\delta \mathbf{u}_s}{\delta t}$$ (15.13)

$$\frac{\partial P_s}{\partial t} + (\mathbf{u}_s \cdot \nabla) P_s + P_s (\nabla \cdot \mathbf{u}_s) + (P_s \cdot \nabla) \mathbf{u}_s$$
$$+ [(P_s \cdot \nabla) \mathbf{u}_s]^T + \frac{q_s}{m_s} (\boldsymbol{\tau}_s \times \mathbf{B} + [\boldsymbol{\tau}_s \times \mathbf{B}]^T)$$ (15.14)
$$+ \nabla \cdot Q_s = \frac{\delta P_s}{\delta t}$$

where $n_s = \rho_s / m_s$, and we introduced the stress tensor, $\boldsymbol{\tau}_s = p_s \mathbf{I} - P_s$ (**I** is the identity tensor and p_s is the scalar pressure, $p_s = \mathrm{Tr}[P]/3$).

Equations 15.12 through 15.14 provides 10 equations for 20 unknown functions (the scalar ρ_s, vector $\mathbf{u}_s$, symmetric two-index tensor P_s, and symmetric three-index tensor Q_s) as well as the six unknown components of the electric and magnetic field vectors. Obviously, additional information is needed.

It is interesting to point out that the 10 lowest velocity moments (ρ_s, $\mathbf{u}_s$, and P_s) uniquely define the so-called 10-moment distribution function that ensures hyperbolicity and realizability (in plain English: all eigenvalues of the associated equations are real and the distribution function is positive semidefinite everywhere) [78,82]:

$$G_s = n_s \left(\frac{m_s}{2\pi k_B} \right)^{3/2} \sqrt{\frac{1}{\Delta_s}} \exp\left(-\frac{1}{2} \Theta_{s_{ij}}^{-1} c_{s_i} c_{s_j} \right)$$ (15.15)

where:

k_B is the Boltzmann constant
$\Theta_s = P_s / \rho$ is the acoustic speed tensor
$\Delta_s = \det \Theta_s$

This means that if we neglect the source terms, the electromagnetic effects and the heat flow term equations (15.12 through 15.14) are hyperbolic for "any" physically meaningful pressure tensor.

15.3.3 Heat Flow Tensor

The pressure tensor can be obtained from the third moment of the kinetic equation. This derivation, however, is cumbersome and it only leads to a local approximation of the heat flux. It has been pointed out by several authors that in plasmas the heat flow has a significant nonlocal component as well [see 81,83,84]. Fourier-type approximations were suggested for the local heat flux by several authors [78,81], whereas other considered collisionless heat conduction [85]. For ions and neutrals, we use the collisional local heat conduction obtained by Groth and McDonald [78] using the Levermore closure [82]:

$$Q_{s_{ijk}} = -\tau_{s_c} \left[P_{s_{ij}} \frac{\partial}{\partial x_l} \left(\frac{P_{s_{lk}}}{\rho_s} \right) + P_{s_{ik}} \frac{\partial}{\partial x_l} \left(\frac{P_{s_{lj}}}{\rho_s} \right) + P_{s_{jk}} \frac{\partial}{\partial x_l} \left(\frac{P_{s_{li}}}{\rho_s} \right) \right]$$ (15.16)

where τ_{s_c} is the collisional timescale of species s.

For electrons, we consider both collisional and collisionless heat conduction components and assume that the heat conduction is mainly along the magnetic field. As a first, simple approximation, we use a generalization of the empirical relation obtained by Bale et al. [86]:

$$Q_{e_{ijk}} = -\alpha_e \frac{3}{2} p_e \upsilon_e b_i b_j b_k$$ (15.17)

where:

p_e is the scalar electron pressure
$\upsilon_e = (8k_B T_e / \pi m_e)^{1/2}$ is the electron thermal speed
$\mathbf{b} = \mathbf{B}/B$ is the unit vector along the magnetic field line
the function α_e is given by

$$\alpha_e = 1.07 \mathrm{Kn}_e e^{-5\mathrm{Kn}_e} + 0.3(1 - e^{-5\mathrm{Kn}_e})$$ (15.18)

where Kn_e is the electron Knudsen number defined as

$$Kn_e = \frac{\lambda_e}{L_e} \qquad (15.19)$$

where:

λ_e is the electron mean free path ($\lambda_e = \tau_{ec}\,\upsilon_e$)

L_e is the characteristic length scale of the electron temperature variation ($L_e^{-1} = |\nabla_\| T_e|/T_e$). In the collisional regime ($Kn_e \ll 1$), $\alpha_e \propto Kn_e$, and we reproduce the Spitzer-Härm thermal conductivity [87]. In the transition and collisionless regime ($Kn_e \geq 1$,) α_e becomes constant and the electron heat flux is about 30% of the free-streaming (or saturation) heat flux. The variation of α_e as a function of electron Knudsen number is shown in Figure 15.4.

15.3.4 Electromagnetic Quantities

Equation 15.16 or 15.17 provides 10 additional equations for the heat flow components and thus fully defines the fluid quantities in Equations 15.12 through 15.14. To be able to solve the coupled system of equations, we still need equations describing the evolution of the electromagnetic variables, **E** and **B**.

The source terms in Maxwell's equations (15.3 through 15.6) can be obtained by assuming charge quasineutrality ($\rho_c \approx 0$) and by dropping the displacement current in Ampère's law (Equation 15.6). Charge neutrality yields an expression for the electron density, and the simplification of Equation 15.6 expresses the electric density with the help of the magnetic field vector:

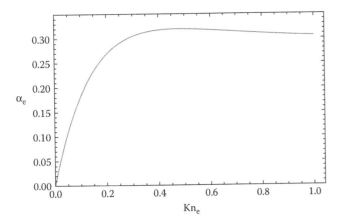

FIGURE 15.4 Variation of the quantity α_e as a function of the electron Knudsen number. (After C. Wu et al., *Geophys. Res. Lett.*, 8, 523–526, 1981. With Permission.)

$$n_e = \sum_{s=ions} Z_s n_s \qquad (15.20)$$

$$\mathbf{j} = \frac{1}{\mu_0} \nabla \times \mathbf{B} \qquad (15.21)$$

where Z_s is the charge state of species s.

The time evolution of the magnetic field vector, **B**, is described by Faraday's law (also called the induction equation) given by Equation 15.5, which connects the curl of the electric field to the time derivative of the magnetic field.

Finally, we need an equation for the electric field vector. This equation, called the "generalized Ohm's law," can be obtained by multiplying the momentum equations (Equation 15.13) for the individual species by a factor of q_s/m_s and adding them together. After some algebra, this yields the following equation:

$$(\mathbf{E} + \mathbf{u}_e \times \mathbf{B}) + \frac{1}{en_e}\nabla \cdot P_e - \frac{m_e}{en_e}\nabla$$
$$\cdot \left(\sum_{s=ions} Z_s n_s \mathbf{u}_s \mathbf{u}_s - n_e \mathbf{u}_e \mathbf{u}_e \right) = -\frac{m_e}{e^2 n_e}\frac{\delta \mathbf{j}}{\delta t} \qquad (15.22)$$

where we neglected the time derivative of the current density and used the fact that the electron mass is much smaller than the mass of any ion species. On the left-hand side of Equation 15.22, one can neglect the last term because it is multiplied by the very small electron mass-to-charge ratio. The term on the right-hand side is also multiplied by m_e, but this term describes the resistive (ohmic) dissipation of the electric current density and it is usually kept in the equation. These simplifications lead to the following form of generalized Ohm's law:

$$(\mathbf{E} + \mathbf{u}_e \times \mathbf{B}) + \frac{1}{en_e}\nabla \cdot P_e - \eta_e \mathbf{j} = 0 \qquad (15.23)$$

where η_e is the electric resistivity ($\eta_e \mathbf{j} \approx -(m_e/e^2 n_e)\delta \mathbf{j}/\delta t$).

Most MHD codes do not directly solve for the electron velocity but derives it from the ion velocities using the assumption of charge neutrality combined with Equation 15.8:

$$\mathbf{u}_e = \sum_{s=ions} \frac{Z_s n_s \mathbf{u}_s}{n_e} - \frac{\mathbf{j}}{en_e} = \mathbf{u}_+ - \frac{\mathbf{j}}{en_e} \qquad (15.24)$$

where $\mathbf{u}_+$ is the charge-averaged ion velocity. We note that in general $\mathbf{u}_+ \neq \mathbf{u}_i$, where $\mathbf{u}_i$ is the mass averaged ion velocity ($\mathbf{u}_i = \sum_{s=ions}\rho_s \mathbf{u}_s / \sum_{s=ions}\rho_s$). With the help of Equation 15.24, Ohm's law becomes the following:

$$\mathbf{E} = -\mathbf{u}_+ \times \mathbf{B} + \frac{1}{en_e}\mathbf{j} \times \mathbf{B} - \frac{1}{en_e}\nabla \cdot P_e + \eta_e \mathbf{j} \quad (15.25)$$

The physical interpretation of the four terms on the right-hand side are motional electric field, Hall term, electron pressure gradient (or ambipolar electric field) term, and resistive term, respectively.

15.3.5 Multifluid MHD

The generalized Ohm's law has some very interesting consequences in the case of multiple ion species. Let us substitute Equation 15.25 into the multifluid momentum equation (15.13) and consider an ionized species, s.

$$\rho_s \frac{\partial \mathbf{u}_s}{\partial t} + \rho_s(\mathbf{u}_s \cdot \nabla)\mathbf{u}_s + \nabla \cdot P_s + \frac{Z_s n_s}{n_e}\nabla \cdot P_e - \rho_s \mathbf{g}$$
$$\hspace{5cm} (15.26)$$
$$-\frac{Z_s n_s}{n_e}\mathbf{j} \times \mathbf{B} - \delta \mathbf{j}_s \times \mathbf{B} = \eta_e \mathbf{j} + \rho_s \frac{\delta \mathbf{u}_s}{\delta t}$$

where we introduced the drift current density, $\delta \mathbf{j}_s = Z_s e n_s (\mathbf{u}_s - \mathbf{u}_+)$. This describes the current density carried by an individual ion species in the coordinate system moving with the positive charge bulk velocity, $\mathbf{u}_+$.

Equation 15.26 reveals several interesting features of multifluid MHD. First, it shows that the impacts of the ambipolar electric field and the $\mathbf{j} \times \mathbf{B}$ force are distributed among the various ion species according to their fractional charge density ($Z_s e n_s$ divided by the total charge density en_e). Second, an interesting new term appears in the multifluid momentum equation, $\delta \mathbf{j}_s \times \mathbf{B}$. This term results in the gyration of each ionized species with respect to the positive charge bulk velocity, $\mathbf{u}_+$, thus capturing some aspects of the kinetic physics that is generally absent in single-fluid MHD.

The full set of multifluid MHD equations can be solved in SWMF/BATS-R-US [30] and in Winglee's multifluid code [26]. The multifluid LFM code neglects the ambipolar electric field and simplifies the drift current density term [88]. There are other multifluid and/or Hall codes under development, but they did not reach the level of maturity when they could be used for space weather simulations. Multifluid Hall MHD is clearly at the forefront of global space weather simulations.

Another important aspect of equations is the full pressure tensor, P_s. Most space weather MHD codes assume isotropic pressure tensors, $P_s = p_s\mathbf{I}$, where p_s is the scalar pressure and $\mathbf{I}$ is the unit matrix. Presently, no global space weather code incorporates the full pressure tensor and only initial work has been published

in this direction [81]. MHD with anisotropic pressure (where the pressure components parallel and perpendicular to the magnetic field are treated as independent quantities) has been implemented in BATS-R-US [79,89].

15.3.6 Single-Fluid MHD

The most commonly used approximation in global space weather codes is single-fluid MHD. The governing equations can be obtained by summing the multifluid equations (given by Equations 15.12 through 15.14) for all species (ions, electrons, neutrals):

$$\frac{\partial \rho}{\partial t} + \nabla \cdot (\rho\mathbf{u}) = \frac{\delta\rho}{\delta t} \quad (15.27)$$

$$\rho\frac{\partial \mathbf{u}}{\partial t} + \rho(\mathbf{u}\cdot\nabla)\mathbf{u} + \nabla\cdot P - \rho\mathbf{g} - \rho_c\mathbf{E} - \mathbf{j}\times\mathbf{B} = \rho\frac{\delta\mathbf{u}}{\delta t} \quad (15.28)$$

$$\frac{\partial P}{\partial t} + (\mathbf{u}\cdot\nabla)P + P(\nabla\cdot\mathbf{u}) + (P\cdot\nabla)\mathbf{u} + [(P\cdot\nabla)\mathbf{u}]^\mathrm{T}$$
$$-[(\mathbf{j}-\rho_c\mathbf{u})(\mathbf{E}+\mathbf{u}\times\mathbf{B}) + (\mathbf{E}+\mathbf{u}\times\mathbf{B})(\mathbf{j}-\rho_c\mathbf{u})]$$
$$+\left\{ \left(\sum_{s=\text{all}}\frac{q_s}{m_s}\tau_s^*\right)\times\mathbf{B} + \left[\left(\sum_{s=\text{all}}\frac{q_s}{m_s}\tau_s^*\right)\times\mathbf{B}\right]^\mathrm{T}\right\}$$
$$\hspace{5cm} (15.29)$$
$$+\nabla\cdot Q = \frac{\delta P}{\delta t}$$

where we introduced the following single-fluid quantities:

$$\rho = \sum_{s=\text{all}} \rho_s \quad (15.30)$$

$$\mathbf{u} = \frac{1}{\rho}\sum_{s=\text{all}} \rho_s\mathbf{u}_s \quad (15.31)$$

$$P = \sum_{s=\text{all}} P_s^* \sum_{s=\text{all}} [P_s + \rho_s(\mathbf{u}_s - \mathbf{u})(\mathbf{u}_s - \mathbf{u})] \quad (15.32)$$

$$Q_{ijk} = \sum_{s=\text{all}} Q_{sijk} + \sum_{s=\text{all}} P_{sij}(u_{s_k} - u_k)$$
$$+ \sum_{s=\text{all}} P_{sik}(u_{s_j} - u_j) + \sum_{s=\text{all}} P_{sjk}(u_{s_i} - u_i) \quad (15.33)$$
$$+ \sum_{s=\text{all}} \rho_s(u_{s_i} - u_i)(u_{s_j} - u_j)(u_{s_k} - u_k)$$

$$\frac{\delta\rho}{\delta t} = \sum_{s=\text{all}} \frac{\delta\rho_s}{\delta t} \quad (15.34)$$

$$\rho \frac{\delta \mathbf{u}}{\delta t} = \sum_{s=\text{all}} \frac{\delta(\rho_s \mathbf{u}_s)}{\delta t} - \mathbf{u}\frac{\delta \rho}{\delta t} \qquad (15.35)$$

$$\frac{\delta P}{\delta t} = \sum_{s=\text{all}} \frac{\delta P_s^*}{\delta t} \qquad (15.36)$$

There are some very interesting new terms appearing in Equations 15.28 and 15.29. In the single-fluid momentum equation (15.28), the well-known $\mathbf{j} \times \mathbf{B}$ term appears, introducing a new driving force for conducting fluids. The more interesting new terms are in the single-fluid pressure equation. In addition to the familiar hydrodynamic terms in the first line of (15.29), the term in the second line describes Joule heating. In effect, this term is the symmetrized ($\mathbf{jE} + \mathbf{Ej}$) diad measured in the single-fluid frame of reference (moving with velocity $\mathbf{u}$).

The term in the curly braces describes the interaction of the stress tensor with the magnetic field. In most cases, this term is negligible, but it becomes important in the case of anisotropic and nongyrotropic plasmas. In reality, this term greatly simplifies because the charge per mass ratio for electrons is some three orders of magnitude larger than the same ratio for ions.

Presently, no global space weather model uses the full pressure tensor, but several groups (including our group at the University of Michigan) are working on incorporating it in the simulation.

15.3.7 Anisotropic MHD

A simpler approximation of the MHD equations assumes that the random motion of the particles is tied to the magnetic field lines (e.g., [77,90,91]). In this approach, the pressure tensor of species s simplifies to the following:

$$P_s = p_{s\perp} I + \left(p_{s\parallel} - p_{s\perp} \right) \mathbf{bb} \qquad (15.37)$$

where $p_{s\parallel}$ and $p_{s\perp}$ are the pressure components parallel and perpendicular to the magnetic field unit vector, $\mathbf{b}$. It is also assumed that heat transport takes place only along magnetic field lines; therefore, the heat flux tensor can be written as

$$Q_{s\,ijk} = h_{s\perp}\left(\delta_{ij}b_k + \delta_{ik}b_j + \delta_{jk}b_i\right)$$
$$+ \left(h_{s\parallel} - 3h_{s\perp}\right)b_i b_j b_k \qquad (15.38)$$

where $h_{s\perp}$ and $h_{s\parallel}$ describe the field-aligned random transport of perpendicular and parallel pressure components, respectively.

Substituting Equations 15.37 and 15.38 into the multifluid equations (15.12 through 15.14) yields the governing equations of multifluid anisotropic MHD:

$$\frac{\partial \rho_s}{\partial t} + \nabla \cdot (\rho_s \mathbf{u}_s) = \frac{\delta \rho_s}{\delta t} \qquad (15.39)$$

$$\rho_s \frac{\partial \mathbf{u}_s}{\partial t} + \rho_s (\mathbf{u}_s \cdot \nabla)\mathbf{u}_s + \nabla p_{s\perp} - \rho_s \mathbf{g} - q_s n_s$$
$$[\mathbf{E} + \mathbf{u}_s \times \mathbf{B}] = -B\nabla_{\parallel}\left(\frac{p_{s\parallel} - p_{s\perp}}{B}\mathbf{b}\right) + \rho_s \frac{\delta \mathbf{u}_s}{\delta t} \qquad (15.40)$$

$$\frac{\partial p_{s\parallel}}{\partial t} + (\mathbf{u}_s \cdot \nabla)p_{s\parallel} + p_{s\parallel}(\nabla \cdot \mathbf{u}_s) + 2p_{s\parallel}\mathbf{b}\cdot(\nabla_{\parallel}\mathbf{u}_s)$$
$$+ B\nabla_{\parallel}\left(\frac{h_{s\parallel}}{B}\right) = -2\frac{h_{s\perp}}{B}\nabla_{\parallel}B + \frac{\delta p_{s\parallel}}{\delta t} \qquad (15.41)$$

$$\frac{\partial p_{s\perp}}{\partial t} + (\mathbf{u}_s \cdot \nabla)p_{s\perp} + 2p_{s\perp}(\nabla \cdot \mathbf{u}_s) + p_{s\perp}\mathbf{b}\cdot(\nabla_{\parallel}\mathbf{u}_s)$$
$$+ B\nabla_{\parallel}\left(\frac{h_{s\perp}}{B}\right) = \frac{h_{s\perp}}{B}\nabla_{\parallel}B + \frac{\delta p_{s\perp}}{\delta t} \qquad (15.42)$$

where $\nabla_{\parallel} = \mathbf{b}\cdot\nabla$.

15.3.8 Ideal MHD

In the ideal MHD approximation, the heat flux is neglected and the parallel and perpendicular pressure components are assumed to be equal, $p_s = p_{s\parallel} = p_{s\perp}$. The pressure equation can now be obtained by multiplying Equation 15.42 by two and adding it to Equation 15.41. This yields the following multifluid equations:

$$\frac{\partial \rho_s}{\partial t} + \nabla \cdot (\rho_s \mathbf{u}_s) = \frac{\delta \rho_s}{\delta t} \qquad (15.43)$$

$$\rho_s \frac{\partial \mathbf{u}_s}{\partial t} + \rho_s (\mathbf{u}_s \cdot \nabla)\mathbf{u}_s + \nabla p_s - \rho_s \mathbf{g} - q_s n_s$$
$$\left(\mathbf{E} + \mathbf{u}_s \times \mathbf{B}\right) = \rho_s \frac{\delta \mathbf{u}_s}{\delta t} \qquad (15.44)$$

$$\frac{\partial p_s}{\partial t} + (\mathbf{u}_s \cdot \nabla)p_s + \frac{5}{3}p_s(\nabla \cdot \mathbf{u}_s) = \frac{\delta p_s}{\delta t} \qquad (15.45)$$

One can obtain the governing equations of single-fluid ideal MHD by summing Equations 15.43 through 15.45 for all species:

$$\frac{\partial \rho}{\partial t} + \nabla \cdot (\rho \mathbf{u}) = \frac{\delta \rho}{\delta t} \qquad (15.46)$$

$$\rho \frac{\partial \mathbf{u}}{\partial t} + \rho(\mathbf{u} \cdot \nabla)\mathbf{u} + \nabla p - \rho \mathbf{g} - \mathbf{j} \times \mathbf{B} = \rho \frac{\delta \mathbf{u}}{\delta t} \quad (15.47)$$

$$\frac{\partial p}{\partial t} + (\mathbf{u}_s \cdot \nabla)p + \frac{5}{3} p(\nabla \cdot \mathbf{u}) = \frac{\delta p}{\delta t} \quad (15.48)$$

where we assumed charge neutrality. These equations are supplemented by the ideal Ohm's law (obtained from Equation 15.25 by neglecting the Hall, ambipolar, and resistive terms, and assuming that $\mathbf{u}_+ = \mathbf{u}$):

$$\mathbf{E} = -\mathbf{u} \times \mathbf{B} \quad (15.49)$$

Finally, a set of governing equations is completed by Ampère's law and Faraday's law:

$$\frac{\partial \mathbf{B}}{\partial t} = -\nabla \times \mathbf{E} \quad (15.50)$$

$$\mu_0 \mathbf{j} = \nabla \times \mathbf{B} \quad (15.51)$$

15.4 SUN–EARTH SIMULATIONS

Space weather originates at the Sun and it takes somewhere between 12 and 36 h for the space storm to reach Earth. At the present time, we do not have the ability to predict when a solar active region will erupt, the travel time through the interplanetary medium represents the time period available to forecast the terrestrial impact of any given solar eruption.

Presently, global space weather models fall into three categories: global solar corona models that typically extend from the photosphere to about 20–30 solar radii ($R_\odot$) [14–18,42,92,93], inner heliosphere models that start at 20–30 $R_\odot$ and extend to the orbits of Jupiter and Saturn [16,17,20,21,92,94], and magnetosphere–ionosphere models that describe the region starting from the L_1 point to hundreds of Earth radii (R_E) downstream from Earth [5–11,24–27,95–98]. These models can be coupled together by the SWMF [31].

15.4.1 Solar Corona

Presently, there are two global solar corona simulation codes available for the space weather community at the Community Coordinated Modeling Center (CCMC): the MAS model developed by the group at Predictive Science Inc., San Diego, California [99–101], and the AWSoM model [42,93,102] based on SWMF/BATS-R-US [30,103].

The MAS model is a 3D time-dependent resistive MHD model extending from the low corona to 30 $R_\odot$.

It first uses a potential field model and a Parker solar wind solution to determine the initial plasma and magnetic field parameters and then solves Maxwell's equations and the plasma continuity, momentum, and energy equations to get a steady-state MHD solution. Two versions of the MAS model are available at the CCMC. The older model uses a polytropic energy equation with an empirical correction at 30 $R_\odot$ for solar wind speed based on the distance of field line photospheric footpoints from the closest coronal hole boundary [101]. The new MAS model has the option to solve a thermodynamic energy equation [104]. The new model includes plasma acceleration due to Alfvén wave pressure gradient (but this term does not appear in the energy equation), radiative loss, collisional and collisionless heat conduction, and a coronal heating function that is adjusted to obtain reasonable solutions.

Figure 15.5 shows synthetic emission images obtained with the thermodynamic MAS model using three different coronal heating models (exponential decrease with a scale length of 0.7 $R_\odot$, an observation-based empirical model [105], and a combined model that also includes active region heating). It also displays the corresponding observations.

The AWSoM model [48,106,107,108,102,93,42,48,93, 102,106–108] simultaneously solves for the plasma density, velocity, ion and electron pressures, and parallel and antiparallel propagating Alfvénic turbulence that is mainly responsible for coronal heating and solar wind acceleration. The injection of Alfvén wave energy at the inner boundary is such that the Poynting flux is proportional to the magnetic field strength. The 3D magnetic field topology is simulated using data from photospheric magnetic field measurements. This model does not impose open–closed magnetic field boundaries; those develop self-consistently. The physics include the following:

1. The model employs three different temperatures, namely, the isotropic electron temperature and the parallel and perpendicular ion temperatures. The firehose, mirror, and ion-cyclotron instabilities due to the developing ion temperature anisotropy are accounted for.

2. The Alfvén waves are partially reflected by the Alfvén speed gradient and the vorticity along the field lines. The resulting counterpropagating waves

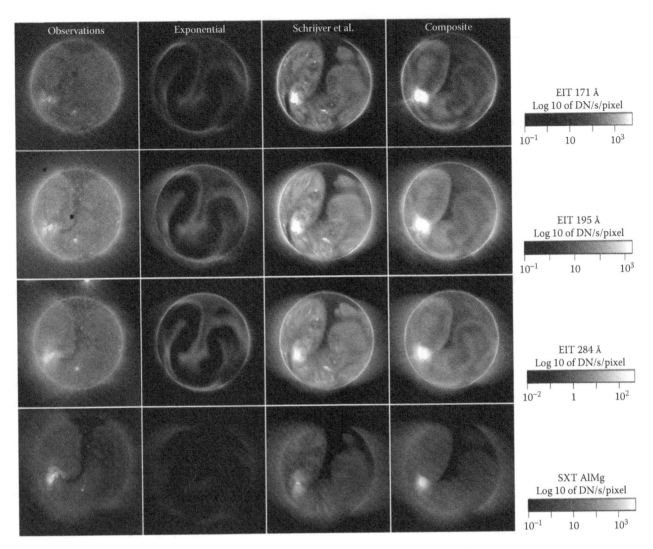

FIGURE 15.5 Comparison of observed and synthetic emission images obtained with the thermodynamic MAS model. The first column shows the observed emission, and each remaining column shows the computed emission for various coronal heating models. Rows show emission in the EIT 171, 195, and 284 Å bands and in the SXT AlMg configuration. (From R. Lionello et al., *Astrophys. J.*, 690, 902–912, 2009. With permission.)

are responsible for the nonlinear turbulent cascade. The balanced turbulence due to uncorrelated waves near the apex of the closed field lines and the resulting elevated temperatures are addressed.

3. To apportion the wave dissipation to the three temperatures, results of the theories of linear wave damping and nonlinear stochastic heating are incorporated.

4. Collisional and collisionless electron heat conduction is incorporated into the model.

The AWSoM model uses the numerical schemes of the BATS-R-US MHD solver and the overarching SWMF [30]. The SWMF is a software framework for modeling various space physics domains in a single coupled model. It has been used, besides space weather applications for the coupled Sun–Earth system, for many planetary, moon, and comet applications as well as the outer heliosphere. The AWSoM model uses a 3D spherical grid with the radial coordinate ranging from 1 to 24 $R_\odot$. The grid is highly stretched toward the Sun with the smallest radial cell size $\Delta r = 10^{-3}\ R_\odot$ to numerically resolve the steep density gradients in the upper chromosphere. The transition region is artificially broadened (similar to that as described in References [102] and [104]) to be able to resolve this region. The grid is block decomposed using the block-adaptive tree library [30]. This library is a tool to create, load balance, and message pass the adaptive refined mesh and solution data. The grid blocks consist of 6 × 4 × 4 mesh cells. Inside $r = 1.7 R_\odot$, the angular resolution is 256 cells in longitude and 128 cells in latitude corresponding to an angular cell size of 1.4°, whereas outside that radius the grid is one level less refined. The system of equations is solved in the heliographic rotating frame by including centrifugal and Coriolis forces in the momentum equation and adding the centrifugal contribution to the ion energy equations. A uniform solar rotation is assumed with a 25.38-day period. For steady-state simulations, local time stepping is used, which speeds up the convergence relative to time accurate simulations. During the steady-state convergence, an additional level of mesh refinement is used at the heliospheric current sheet.

Figure 15.6 shows the comparison between the synthetic Solar Dynamics Observatory/Atmospheric Imaging Assembly (SDO/AIA) images with the images observed by AIA [109] onboard SDO. The model results are obtained with the MP5 limiter. The wavelengths indicated at the top of each panel correspond to various characteristic temperatures. These line-of-sight (LOS) images were produced by assuming that the plasma is optically thin for all the considered wavelengths. We note that the steady-state simulation was performed for a synoptic magnetogram, while the observation is for the time March 7, 2011, 20:00 UT, and consequently, the model cannot reproduce the time-dependent activity during the rotation. Also the polynomial extrapolation toward the pole in the CR2107 magnetogram might distort the high latitudinal region somewhat unfavorably. The observed polar coronal holes are somewhat wider than the coronal holes of the new model. In spite of these limitations, the agreement is quite good between simulations and observations.

15.4.2 Coronal Mass Ejections

Presently, there are two inner heliosphere models available for community use at the CCMC: ENLIL [21] and SWMF [30]. These models can be used to simulate the interplanetary propagation of coronal mass ejections [110–112].

The ENLIL model is used in combination with the WSA empirical model that connects solar synoptic magnetograms to solar wind conditions at a heliocentric distance of $sim20 R_\odot$ [113].

ENLIL is a time-dependent, ideal fluid approximation to a 3D model for which equal temperatures and densities are assumed for electrons and protons, and microscopic processes are neglected [20,114]. The computational domain grid spans 30°–150° in heliospheric colatitude, 0°–360° in longitude, and 0.1 AU ($21.5R_\odot$) to 1.7 AU in radial distance. ENLIL specifies the solar wind velocity (V), plasma density (n), mean plasma temperature (T), and interplanetary magnetic field (IMF) (B) throughout the computational domain. The combined WSA–ENLIL modeling thus provides numerical predictions of the ambient solar wind in the inner heliosphere.

To model CMEs and other solar energetic particle (SEP) events in addition to the background solar wind, the Cone tool is used in conjunction with WSA–ENLIL. Cone is an iterative graphic tool that uses a cone shape to estimate initial CME parameters from white-light coronagraph images of the CME [115,116]. From observations, a CME expands radially outward with constant angular width and radial direction and speed to the first order, allowing the cone shape to provide a basis for estimating its initial radial speed, size, location, and direction of propagation. The cone-shaped CME is then integrated into WSA–ENLIL for its evolution with the background solar wind by adding a pressure pulse at the inner ENLIL boundary parameterized by the Cone-derived radial speed, angular width, source location, and direction of propagation, as well as the time at which the CME front crosses $21.5 R_\odot$. ENLIL uses hydrodynamic ejecta, so the simulated CMEs are not configured with the magnetic cloud structure. Likewise, the cone-shaped parameterization from the coronagraph images does not initialize other plasma parameters of the CME, namely, density and temperature. The initial CME density and temperature are free parameters and are by default equal to four times and one time the typical ambient fast-wind mean values, respectively [117]. These parameters can be modified so that a modeled CME matches 1 AU *in situ* observations more accurately. Because CMEs are integrated into the ENLIL numerical grid along with the background solar wind conditions, CMEs evolve along with the background solar wind, allowing for acceleration or deceleration of the CMEs.

A color representation of the radial component of the solar wind velocity (V_r) in the heliospheric plane on June 1, 2011, from the WEC model is shown in Figure 15.7. The model results depict fast-moving CMEs near Earth and a moderate solar wind stream region in the ecliptic plane during this time. According to the model, the ~550 km/s fast stream passed Earth several days prior to the

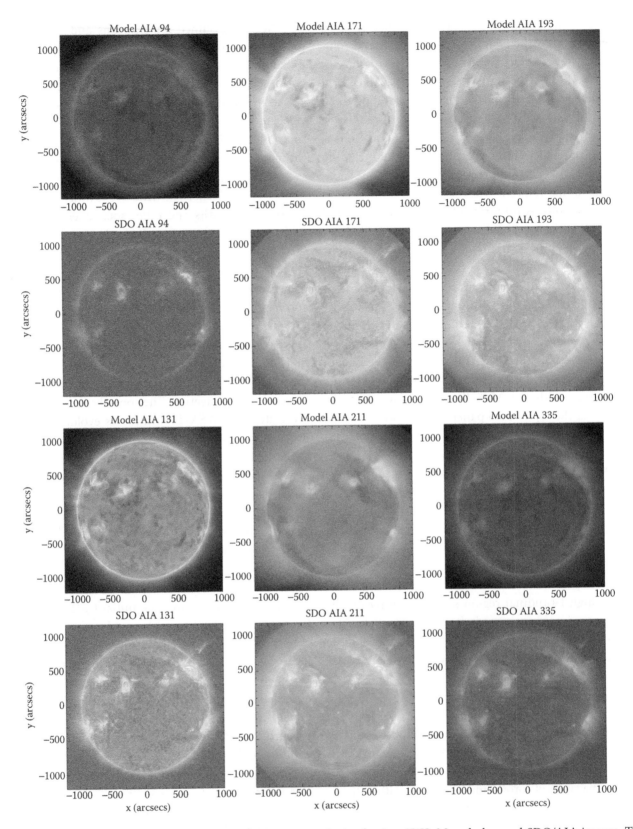

FIGURE 15.6 Comparison between synthesized AIA images obtained using AWSoM and observed SDO/AIA images. Top panels (from left to right) show AIA synthesized images for 94, 171, and 193 Å. Panels in the second row show observational SDO/AIA images for those wavelengths. Panels in the third row show AIA synthesized images for 131, 211, and 355 Å. Bottom panels show observational SDO/AIA images for those wavelengths. The observation time is March 7, 2011, 20:00 UT. (From B. van der Holst et al., *Astrophys. J.*, 782, 81, 2014. With permission.)

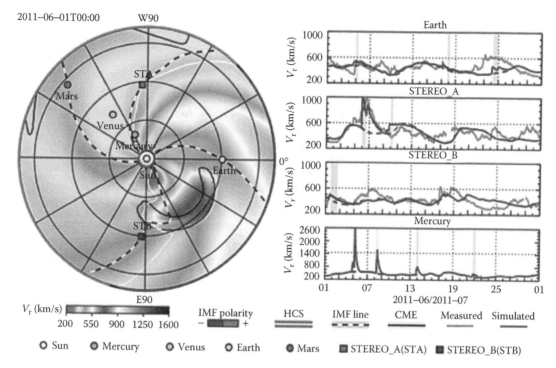

FIGURE 15.7 Modeled radial solar wind speed, viewed from the north ecliptic pole, in the ecliptic plane from the WSA-ENLIL-Cone (WEC) model for June 1, 2011. The locations of spacecraft, planets, and the Sun are indicated by small symbols. The inner boundary of the model (where WSA outputs are used) is denoted by the central circle filled in white. Other features are listed in the legend. The plots on the right give V_r as simulated (dark; solid line is WEC and dashed is WSA-ENLIL) and measured (light) at Mercury, Earth, and the STEREO spacecraft for the month of June 2011. The shaded white areas indicate the CMEs, as incorporated into the ENLIL results with the Cone extension. (From R. M. Dewey, *J. Geophys. Res.*, 120, 5667–5685, 2015. With permission.)

snapshot but had yet to envelop STEREO-A, and Earth had yet to interact with the energetic particles. The *in situ* observations from the ACE, STEREO-A, and STEREO-B spacecraft at Earth and at several longitudes in the figure match well with the prediction and capture much of the disturbed conditions. The Cone extension captures the disturbed conditions at STEREO-A on June 7 particularly well compared with the WSA–ENLIL model only (dashed line). Such confirmation at a variety of locations within the inner heliosphere indicates that the solar wind conditions are predicted well by the model, as illustrated in Figure 15.7 [118]. The WEC model does a satisfactory job of characterizing the solar wind near, behind, and ahead of Earth locations at 1 AU as seen by these three spacecraft over this period.

The SWMF is also capable to simulate CMEs in the corona and the inner heliosphere. It uses user-defined coronal magnetic flux ropes to initiate solar eruptions [110,119,120]. In contrast to the Cone model [115,116], SWMF uses magnetic free energy to drive the CMEs. This improvement allows SWMF to simulate the interplanetary magnetic field. Although this capability is still untested and it certainly needs improvements, it offers the potential of forecasting the critically important IMF B_z component that plays a very important role in controlling the geo-effectiveness of CMEs.

In a detailed study of the March 7, 2011 CME [110], the location of the flux rope was chosen to match the position of the large filament that existed before the eruption, along the polarity inversion line. The left panel of Figure 15.8 shows the H_α observation on March 7, 2011, 07:53:37 UT that is ~12 h before the CME. One can see the filament in AR 11164 clearly. In the right panel of Figure 15.8, the position and configuration of the flux rope in the simulation are shown. The current of the flux rope is set to 2.0×10^{12} A. The length is 60 Mm and the radius is 9 Mm. The total mass of plasma in the flux rope is set to 10^{16} g, which is within the typical range of observed CME mass [121] and a good estimate for large ones (e.g., [122]). Based on these parameters, the total free energy included in this flux rope is calculated to be

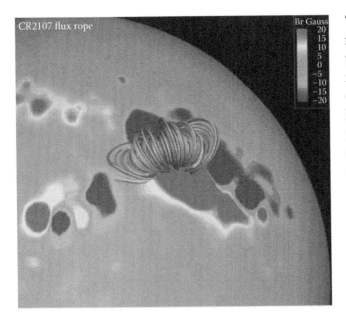

FIGURE 15.8 *Left*: The H$_\alpha$ observation on March 7, 2011, showing the filament location. *Right*: The Titov–Démoulin flux-rope setup in the CME simulation. The color scale on the Sun shows the radial magnetic field strength. (From M. Jin et al., *Astrophys. J.*, 773, 50, 2013. With permission.)

3.9×10^{33} erg. For this active region, the free energy of the flux rope is a bit more energetic than in reality.

Fifteen minutes after the CME is initiated, a heat precursor develops in both the one temperature (1T) and two temperatures (2T) cases. In the 1T case, the heat precursor is caused by electron heat conduction applied to the single shock-heated fluid along open field lines. In the 2T model, we can still see the heat precursor in the electron temperature due to the energy exchange between shock-heated protons and electrons near the Sun as well as the adiabatic compression at the shock. However, the strength of the heat precursor is much smaller than in the 1T case, with the highest temperature inside the heat precursor being less than 2 MK. For the proton temperature, the shock structure is well captured. Due to the shock heating, the proton temperature reaches ~85 MK at the shock region. Particle collisions are too infrequent to affect the large difference in the proton and electron temperatures found behind the shock. With the increasing of the shock speed, the difference between electron and proton temperatures becomes larger.

It is also interesting to compare some aspects of the 1T and 2T CME simulations. There is ~500 km/s difference between the 1T and 2T CME speeds. Because the CMEs were initiated with identical flux ropes, the energy input is the same in the 1T and 2T models.

The speed difference is mainly caused by the nonradial flow in the 2T CME. The shock Alfvén Mach number, the compression ratio, and the angle between the magnetic field and the shock normal are three key parameters for the diffusive shock acceleration of SEPs. The higher shock Alfvén Mach number and higher compression ratio can result in the higher energy of the accelerated SEPs. Comparing 1T and 2T models, there are two major differences. First, in the 1T model, the shock attains a maximum speed in about a minute after CME initiation. However, in the 2T model, the acceleration process continues to about 5 min. The maximum shock speeds (~3000 km/s) are similar in the 1T and 2T models. The other difference is the magnitude of the shock Alfvén Mach number. In the 2T CME, the shock achieves larger shock Alfvén Mach number than in the 1T CME. The shock Alfvén Mach number of 1T CME is ~4–5 during the whole evolution, whereas in the 2T case, the Alfvén Mach number is larger than ~5 with a maximum of ~7.

Figure 15.9 shows the 3D CME evolution for both 1T and 2T models at 5, 10, and 20 min after the initiation. The isosurfaces represent the radial velocity of 1000 km/s. The color scale on the isosurfaces shows the temperature (1T)/proton temperature (2T). The field lines are colored by the density so that we can roughly see the propagation of the CME material as well as the shock positions. The gray scale on the surface of the Sun shows the magnetic field strength. In the 1T model, the plasma is heated by the shock and cools by heat conduction to reach 15 MK at 5 min and gradually cools down due to the adiabatic expansion and heat conduction behind the CME-driven shock. In the 2T model, because the CME-driven shocks exceed the proton sound speed, the protons are dissipatively heated by the shock and reach ~90 MK at 5 min.

The morphology of the CME is quite different in the 1T and 2T models after ~10 min. In the 1T model, the shape of the velocity isosurface shows a radial expansion, whereas in the 2T model, it seems that the expansion has a nonradial component. This nonradial flow is caused by the thermal pressure gradient in the CME sheath. Becasue the proton population is not falsely attributed electron heat conduction, the energy of shock-heated protons cannot be effectively transferred from behind the shock. Therefore, the thermal pressure and pressure gradient in the 2T model are much higher than in the 1T model. This pressure gradient pushes

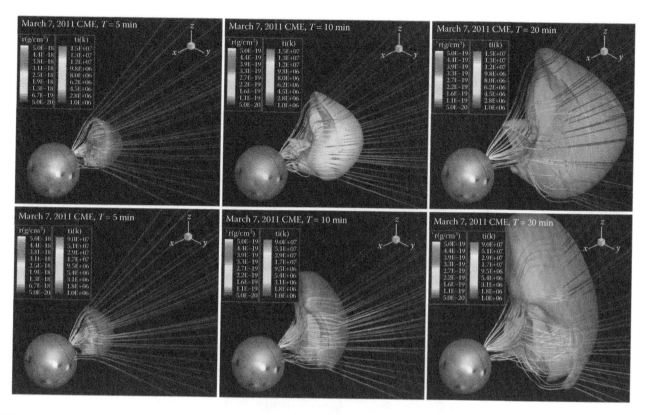

FIGURE 15.9   3D CME evolution of 1T and 2T models. The isosurface represents $V_r = 1000$ km/s. The gray scale on the isosurface shows the proton temperature. The field lines are shaded by plasma density. (From M. Jin et al., *Astrophys. J.*, 773, 50, 2013. With permission.)

the plasmas poleward and causes the nonradial flows. The nonradial flows are only evident near the pole due to the simpler magnetic structure and smaller density in this region. The observed nonradial flows have latitudinal dependence in that the poleward expansion is more preferred than the equatorward expansion. The 2T CME simulation result is consistent with these observations.

In Figure 15.10, the SOHO/LASCO C2 (1.5–6.0 $R_\odot$) CME white-light image is compared with both the 1T and 2T model synthesized white light images. The color scale shows the relative intensity changes. In this study, we use an unstable flux rope to initiate the CME; therefore, the three-component structure of the CME cannot be correctly reproduced. For the Titov–Dèmoulin [43] flux rope model, the flux rope structure and erupting filament material coincide with each other. However, in the three-component structure, they are represented by "dark cavity" and "bright core," respectively. For this event, the CME-driven shock has a typical "double-front" morphology, in which the faint front is caused by the shock and the bright front is the coronal plasma piled up at the top of the erupting flux rope. In both the

1T and 2T models, we can see the bright front. However, the faint front is not obvious in the 1T model but is evident in the 2T model. There are some fine structures behind the shock, which could be related to the disturbance after the shock passing.

Jin and his colleagues [110] conclude that the 1T model produces significant errors in CME-driven shocks. Because collisions are so infrequent, the electrons and protons thermally decouple on the timescale of the CME propagation. In order to produce the physically correct CME structures and CME-driven shocks, an explicit treatment of electron heat conduction in conjunction with proton shock heating is needed in the CME simulation. However, there is still room to improve the models: (1) because the electron heat conduction is treated with a diffusive formulation, the heat flux can transfer with speeds higher than the electron thermal speed, which is unrealistic; (2) the Joule heating of the electrons in the reconnection region also needs to be addressed by using explicit resistivity instead of energy conservation; and (3) the Spitzer heat condition is only applicable to the collision-dominated regions near the Sun. After ~10 $R_\odot$, the plasmas become free-streaming due to infrequent

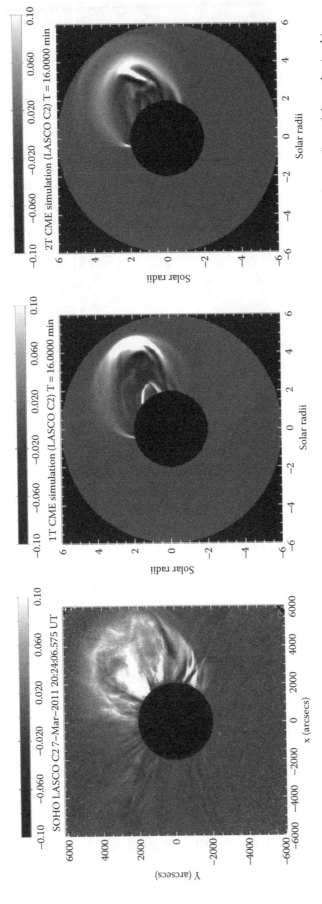

FIGURE 15.10 Comparison of the SOHO/LASCO C2 CME white-light image (left panel) with the 1T (middle panel) and 2T (right panel) model-synthesized images for the March 7, 2011 CME event. The color scale shows the relative intensity changes. (From M. Jin et al., *Astrophys. J.*, 773, 50, 2013. With permission.)

collisions. Nevertheless, CME simulations came a long way and they are becoming quite realistic.

15.4.3 Magnetosphere Simulations

Presently, there are four global magnetosphere simulation codes available for community use at the CCMC: SWMF/BATS-R-US [103], GUMICS [123], LFM [5], and OpenGGCM [124].

GUMICS is a global solar wind–magnetosphere–ionosphere coupling model. Its solar wind and magnetospheric part is based on solving the ideal MHD equations, and its ionosphere part is based on solving the electrostatic current continuity equation. Although the code is only first-order accurate, it uses automatically refined Cartesian octogrid and temporal subcycling to speed up the computation. The computational box is from –224 to +32 R_E in GSM X and from –64 to +64 R_E in Y and Z directions [123].

The LFM code [5] is an integrated simulation model for the global magnetosphere–ionosphere system. The heart of the model is a time-dependent, ideal MHD calculation of the state of the magnetosphere. This MHD part is tightly coupled to a model for the polar ionospheres and is driven by solar wind plasma and magnetic field data upwind of the calculation domain. While lacking important physical processes in both the magnetosphere and the ionosphere, the code is a self-consistent first principles-based model. The architecture of the LFM is based on a number of important design considerations: (1) total variation diminishing (TVD) transport with the highest possible resolving power; (2) div **B** = 0 maintained to roundoff accuracy on a staggered grid; (3) the grid is preadapted to the problem, packing more resolution where needed; and (4) integrated ionospheric model.

The outer boundary of the LFM model is handled as either inflow or outflow. The solar wind in the vicinity of Earth is generally highly supersonic, so inflow is appropriate for those directions where the solar wind passes into the computational domain. The actual inflow conditions are usually described from an appropriately time-shifted time series of solar wind data observed by satellite. Other regions of the outer boundary use simple outflow boundaries. The computational domain is large enough that flows have reaccelerated back to supersonic by the time they reach the edge of the grid.

The low-altitude boundary of the LFM is handled by incorporating a tightly coupled model for the ionospheric electric field. The MHD calculation is stopped at some altitude between 2 and 3 R_E. Field-aligned currents are calculated and mapped along dipole field lines to the ionosphere where they are used as the source term for the height-integrated potential equation. The calculated potential is then mapped back out to the MHD lower boundary where it is used to determine a boundary condition for the velocity and electric field. Details of the calculation of the ionospheric conductances can be found in Reference [125].

The LFM grid is a distorted spherical grid with azimuthal symmetry about the polar axis. The polar axis points in the x-direction of the solar-magnetospheric (SM) coordinate system (roughly toward the Sun). In the $(r - \Theta)$ planes, the grid is distorted to place resolution chosen by the user. The MHD grid usually covers the domain from about 30 R_E upwind to 300 R_E downwind of Earth and roughly 100 R_E out to the sides. The complementary ionospheric grid is a mapping of the inner surface of the MHD (magnetospheric) domain to the two polar ionospheres. This typically covers the region from the pole to 45°–60° latitude.

The SWMF is also used by CCMC to simulate the magnetosphere. The standard setup includes upstream solar wind driving of the global magnetosphere component (represented by BATS-R-US [30]) coupled to the Ridley height-integrated ionosphere model [126] and either the Rice Convection Model (RCM) [103] or the Comprehensive Ring Current Model (CRCM) [58].

As an example of magnetosphere simulations, we show results from CCMC runs modeling the impact of a solar filament on the magnetosphere [127]. When a significant IMF B_y component is present, the cusp is shifted away from noon in both hemispheres. For positive IMF B_y, the shift is toward the dusk side in the Northern Hemisphere and the dawn side in the Southern Hemisphere. For negative IMF B_y, the opposite occurs. Figure 15.11 shows the magnetospheric configuration at 20:05 UT on January 21, 2005, in the SWMF simulation with RCM under northward IMF conditions with a significant eastward (positive) B_y component that existed during this interval. The regions of minimum magnetic field strength mark the cusps in the upper panel, which is a magnetospheric cross section at $x = 2 R_E$. The cusps are shifted to the dusk side in the Northern Hemisphere and the dawn side in the Southern Hemisphere as expected. The lines in the figure are magnetic field lines projected into the (x, z) plane. Planes intersecting the two cusps are given in the figure—one intersecting the southern cusp at $y = -3 R_E$ and the other intersecting the northern cusp at $y = +2 R_E$.

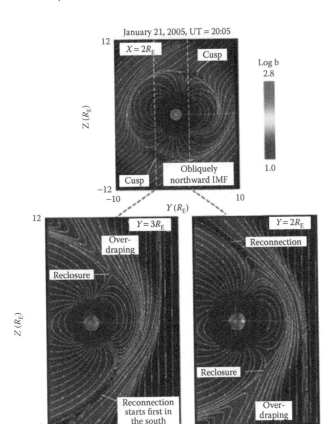

FIGURE 15.11 Cross sections of magnetic field strength from the SWMF simulation showing double lobe reconnection during the strong IMF B_y and northward IMF conditions associated with the initial formation of the cold dense plasma sheet at 20:05 UT on January 21, 2005. The top shows the weak.

Reconnection in the Southern Hemisphere between the IMF and a lobe field line creates an open field line that overdrapes into the Northern Hemisphere. The newly opened field line moves sunward as it straightens out and then moves toward noon, and also toward the Northern Hemisphere cusp region under the influence of the solar wind flow. The result is a region of overdraped open field lines just outside the magnetopause on the dusk side that are connected to the Southern Hemisphere and on the dawn side that are connected to the Northern Hemisphere. An X-line develops in the Northern Hemisphere as the overdraped field line from the Southern Hemisphere reaches the cusp null in the Northern Hemisphere. At this site, Southern Hemisphere open field lines are reclosed and subsequently sink into the magnetosphere. Farther tailward, the reconnection process returns IMF to the interplanetary space. As soon as a closed field line is created, solar wind plasma is trapped within the magnetosphere forming a boundary region of magnetosheath-like

plasma on closed field lines. A similar process occurs in the Northern Hemisphere on the dusk side, shown in Figure 15.11 in the $y = +2$ R_E plane.

15.4.4 Simulating Geomagnetic Activity

With the recent advances of geospace models, they are now capable to predict physical conditions that can create major disruptions of technological systems. One of the most vulnerable systems is the North American Power Grid. The current worst-case scenarios range from a wide-scale voltage and the system collapses to catastrophic loss of a large number of high-voltage power transformers. The space weather modeling and forecasting community is responding to this need by supporting the operational utilization of the latest advancements in science. A few years ago, the National Oceanic and Atmospheric Administration's Space Weather Prediction Center (NOAA/SWPC) requested the CCMC to evaluate geospace models available at the CCMC for possible transition to operations. This effort included the participation of model developers, as well as the CCMC, SWPC, and the broader scientific community. The results of this evaluation were published a few years ago by a CCMC-led group [54], and here we briefly summarize the main conclusions.

The validation effort focused on the models' capability to reproduce the observed rapid fluctuation of the ground magnetic field. The primary argumentation for this choice is that the time derivative of the ground magnetic field (referred to as "dB/dt") can be used as an indicator for the level of geomagnetically induced electric field or geoelectric field on the surface of Earth. The geoelectric field, in turn, is the primary physical quantity driving geomagnetically induced currents (GICs). Consequently, although numerous additional complexities such as ground conductivity, conductor system configuration, and other engineering details including high-voltage power transformer design are critical for more detailed assessment of the threat, dB/dt can be used as an indicator for a potential GIC hazard. Further, if data from an upstream monitor such as the National Aeronautics and Space Administration (NASA)'s Advanced Composition Explorer (ACE) spacecraft is used to produce dB/dt, one can generate short lead time (15–30 min) forecast estimates of the potential hazard.

Six geospace storm events were chosen for the study. Solar wind bulk plasma and the interplanetary magnetic field observations carried out by Solar Wind Electron, Proton, and Alpha Monitor (SWEPAM) and magnetometer (MAG) instruments onboard ACE for the events were used as inputs in the analyses. Solar wind observations were propagated to model inflow boundaries by ballistic propagation and the *x*-component (geocentric solar magnetospheric [GSM] coordinate system) of the interplanetary magnetic field was set to zero. Although solar wind propagation constitutes a source for modeling errors, identical uncertainties were introduced for all models in the specification of the inflow boundaries. For each event, the model performance was evaluated by comparing the observed versus predicted ground d*B*/d*t*. Sixty-second geomagnetic observatory recordings were used to provide the observed signal. Twelve geomagnetic observatories (magnetometer stations) were selected based on the global spatial and temporal coverage. The quiet-time baseline level was determined visually for each station and for each event, and the baseline was removed from the magnetic field data to obtain the disturbance field. Small data gaps with length of no more than few minutes were patched by means of linear interpolation. The modeled magnetic field data were resampled by means of spline interpolation to match the time stamps of the observations. For this comparison, five models were analyzed by the CCMC:

1. LFM: LFM coupled with ionospheric electrodynamics [5]

2. Weigel: empirical model [128]

3. OpenGGCM: global MHD coupled with a coupled thermosphere–ionosphere model [32]

4. Weimer: empirical model [129]

5. SWMF: BATS-R-US coupled with Ridley Ionosphere Model (RIM) and RCM [103]

The length of the individual analysis windows was selected to be 20 min, and the d*B*/d*t* thresholds were 0.3, 0.7, 1.1, and 1.5 nT/s. The selected thresholds represent values that both span lower and higher ranges of rates of change and are also in the "mid-range" in a sense that enough threshold crossing could be detected for good statistics. The analysis calculated the number of correctly predicted threshold crossings *H* (hits), the number of false alarms *F*, the number of missed crossings *M*, and the number of correctly predicted no crossings *N*. The set *H,F,M,N* can be used to compute a number of different metrics quantifying the performance of individual models. In this study, three metrics proposed

by the NOAA's SWPC were selected for use in the final analyses. The selected metrics are probability of detection (POD), probability of false detection (POFD), and Heidke Skill Score (HSS).

The final metrics-based analyses were carried out for each individual model, and the corresponding contingency tables with elements *H,F,M,N* were generated for each model for each event and station for d*B*/d*t* thresholds of 0.3, 0.7, 1.1, and 1.5 nT/s. The summary results are also integrated separately over high-latitude and mid-latitude stations. Figure 15.12 shows the corresponding POD and POFD for all participating models. It is quite clear that for a given set of stations, events, and metrics, the SWMF provides the highest POD and HSS for most of the thresholds. As an indication that large d*B*/d*t* events are still a challenge to capture accurately, for threshold 1.5 nT/s, none of the models is capable of providing POD or HSS greater than 0.5.

15.5 TRANSITION TO OPERATIONS

With the advance of physics-based space weather models, the transition to operational forecasting started in the early 2010s [130]. The WSA–ENLIL heliosphere model was the first model selected by the NOAA's SWPC for operational use.

The modeling system consists of two main parts:

1. A semiempirical near-Sun module (WSA) that approximates the outflow at the base of the solar wind

2. A 3D-dimensional MHD numerical model (ENLIL) that simulates the resulting flow evolution out to Earth

The WSA module is driven by observations of the solar surface magnetic field accumulated over a solar rotation and composited into a synoptic map; this input is used to drive a parameterized model of the near-Sun expansion of the solar corona, which provides input for the interplanetary module to compute the quasi-steady (ambient) solar wind outflow. Finally, when an Earth-directed CME is detected in coronagraph images from the NASA spacecraft, these images are used to characterize the basic properties of the CME, including the speed, direction, and size. This input "cone" representation is injected into the preexisting ambient flow, and the subsequent transient evolution forms the basis for

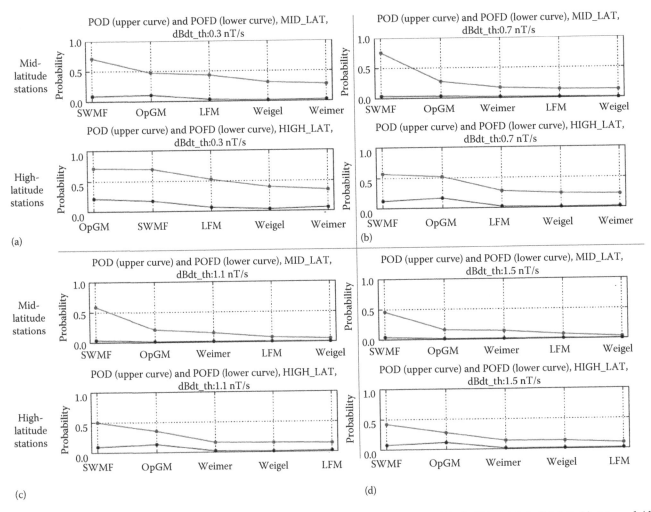

FIGURE 15.12 POD (upper curve) and POFD (lower curve) for the d*B*/d*t* thresholds (nT/s): (a) 0.3; (b) 0.7; (c) 1.1; and (d) 1.5. The models are ordered according to their POD. The model with the largest POD is the leftmost in all panels. (From A. Pulkkinen et al., *Space Weather*, 11, 369–385, 2013. With permission.)

the prediction of the CME's arrival time at Earth, its intensity, and its duration.

Key aspects of the system include the following:

- Automated, robust input data ingest from redundant sources, including synoptic magnetic maps for the ambient flow

- Automated updating of the ambient solar wind flow several times per day

- A flexible database system to facilitate operations, support publicly accessible archival storage at the NOAA's National Geophysical Data Center, and enable ongoing verification and validation efforts

- The WSA–ENLIL model has been in operational use since 2012, and it is one of the most frequently used space weather products of the NOAA's SWPC

As an example of the operational space weather product, Figure 15.13 shows the velocity structure for three CMEs in the outburst at the time the largest CME was predicted to reach Earth, about 15:00 UTC on February 17. Image at left depicts the velocity structure in the ecliptic plane, looking down from above the solar North Pole. Earth is the to the right, and the positions of the two STEREO spacecraft (A, top circle; B, bottom circle) are also shown; velocity is gauged by the color scale at top. Image at right shows a north–south cut along the Sun–Earth line. This model prediction proved to be a little early, with the main CME actually arriving at around 01:00 UTC on February 18.

The second physics-based space weather model that was selected for transition to operation is the geospace component of SWMF (BATS-R-US coupled with RIM and RCM [103]). This selection was based on the

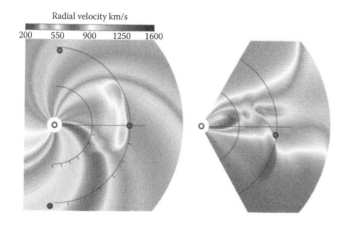

Radial velocity km/s

200   550   900   1250   1600

FIGURE 15.13   Sample output for WSA–ENLIL cone run for the February 15, 2011 multiple coronal mass ejection event. (From A. Parsons et al., *Space Weather*, 9, S03004, 2011. With permission.)

CCMC's evaluation of geospace models [54] that showed that SWMF/Geospace was capable to predict weak-to-moderate dB/dt events with probabilities exceeding 0.5 (see Figure 15.12). Although there is plenty of room for improvements, this performance justified the effort to transition SWMF/Geospace to operational use at the NOAA's SWPC.

The most exciting new feature of the SWMF/Geospace operational model is its ability to forecast regional space weather conditions. Currently available space weather forecasting tools are pretty much limited to predicting planetary indices, such as the K_p index [131]. The new tool, however, will be able to forecast the local time regional K index that will enable a more nuanced response from the operators of sensitive technological systems. Figure 15.14 shows one of the operational displays of the new SWMF/Geospace operational tool. It is driven by observed solar wind conditions at the L_1 point (about a million kilometers upstream of Earth) and will give a 30–60 min warning time for space weather events.

15.6 SUMMARY AND FUTURE DIRECTIONS

Physics-based global space weather models came a long way from the early days when the first 3D global MHD models were developed. They reached the point when numerical space weather prediction starts to beat empirical models. Forecasting space weather has many similarities to hurricane forecasting. 24^h hurricane forecasts can predict the track and landfall location of major hurricanes with ~85 km accuracy, whereas local intensity prediction is the most uncertain quantity. This accuracy is achieved by using multiple models to develop ensemble forecasts. In many respects, large space weather events are similar to hurricanes. Geomagnetic superstorms have a similar frequency to category five hurricanes. The terrestrial impacts of space weather events depend on their

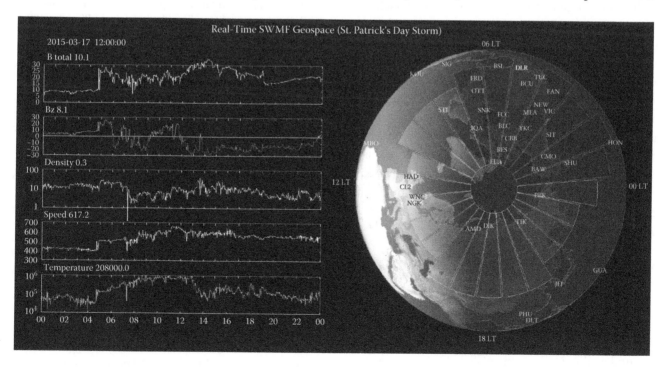

FIGURE 15.14   Local time regional K-index prediction display for SWMF/Geospace developed at the NOAA's SWPC. (Courtesy of George Millward, CU/CIRES and NOAA/SWPC.)

track through the interplanetary medium and on their intensity. Just like in the case of hurricanes, it is important to forecast the track and intensity of space weather events to minimize their impact on the economy. The National Space Weather Program is an interagency initiative to prepare the country to deal with technological vulnerabilities associated with the space environment. The NOAA's National Hurricane Center is responsible for accurate hurricane forecasting, and the NOAA's SWPC is responsible for civilian space weather forecasting for the country. The national goal is to achieve 120^h regional forecasts of large space weather events [132]. We need to accomplish global forecasts for the space environment, forecast the path and strength of solar "hurricanes," and predict the intensity in different geographical regions. The US government is increasingly recognizing the national need for improved space weather modeling, and we can expect a major push in this direction in the near future.

The next natural step in physics-based space weather modeling is improving the fidelity of magnetic reconnection in the global simulations. This is an extremely challenging problem, because magnetic reconnection usually takes place in very small regions compared to the entire computational domain. In these regions, however, a realistic simulation must resolve both the electron and ion spatial and temporal scales.

The current-generation space weather models use either numerical dissipation or some empirical resistivity models to establish reconnection. The next major advance will come from models that couple a kinetic code with an MHD or an extended MHD (XMHD) code and restrict the kinetic code to a small part of the computational domain. Even though the first attempts for such models have already been published [35,36], these efforts are still in their infancy. The next decade or so will witness major advances in space weather modeling when the coupled XMHD-kinetic codes become more robust, realistic, and, most importantly, computationally affordable.

ACKNOWLEDGMENTS

This work was supported by the NASA's Magnetospheric Multiscale Mission (MMS) under grant NNX14AE75G and by the NASA–National Science Foundation (NSF) partnership for collaborative space weather modeling under the NSF grant AGS-1322543. I acknowledge high-performance computing support from the Yellowstone system (ark:/85065/d7wd3xhc) provided by the National Center for Atmospheric Research's (NCAR) Computational and Information Systems Laboratory, sponsored by the NSF. I am also thankful for the use of the NASA Pleiades supercomputer system at Ames Research Center, Moffett Field, CA.

REFERENCES

1. S. H. Brecht, J. Lyon, J. A. Fedder, and K. Hain. A simulation study of east-west IMF effects on the magnetosphere. *Geophys. Res. Lett.*, 8:397–400, 1981.

2. S. H. Brecht, J. Lyon, J. A. Fedder, and K. Hain. A time-dependent three-dimensional simulation of the Earth's magnetosphere: Reconnection events. *J. Geophys. Res.*, 87:6098–6108, 1982.

3. J. N. LeBoeuf, T. Tajima, C. F. Kennel, and J. M. Dawson. Global simulations of the three-dimensional magnetosphere. *Geophys. Res. Lett.*, 8:257–260, 1981.

4. C. C. Wu, R. J. Walker, and J. M. Dawson. A three-dimensional MHD model of the Earth's magnetosphere. *Geophys. Res. Lett.*, 8:523–526, 1981.

5. J. G. Lyon, J. A. Fedder, and C. M. Mobarry. The Lyon–Fedder–Mobarry (LFM) global MHD magnetospheric simulation code. *J. Atmos. Solar-Terr. Phys.*, 66:1333–1350, 2004.

6. J. G. Lyon, J. A. Fedder, and J. G. Huba. The effect of different resistivity models on magnetotail dynamics. *J. Geophys. Res.*, 91:8057–8064, 1986.

7. J. Raeder, J. Berchem, and M. Ashour-Abdallah. The importance of small scale processes in global simulations: Some numerical experiments. In T. Chang and J. R. Jasperse, editors, *The Physics of Space Plasmas* (1998), vol. 14, pp. 403–414. MIT Press, Cambridge, MA, 1995.

8. J. Raeder, R. J. Walker, and M. Ashour-Abdalla. The structure of the distant geomagnetic tail during long periods of northward IMF. *Geophys. Res. Lett.*, 22:349–352, 1995.

9. A. Usadi, A. Kageyama, K. Watanabe, and T. Sato. A global simulation of the magnetosphere with a long tail: Southward and northward interplanetary magnetic field. *J. Geophys. Res.*, 98:7503–7517, 1993.

10. K. Watanabe and T. Sato. Global simulation of the solar wind-magnetosphere interaction: The importance of its numerical validity. *J. Geophys. Res.*, 95:75–88, 1990.

11. P. Janhunen. GUMICS-3: A global ionosphere-magnetosphere coupling simulation with high ionospheric resolution. In W. Burke and T.-D. Guyenne, editors, *Proceedings of the ESA 1996 Symposium on Environment Modelling for Space-Based Applications*, pp. 233–239. European Space Agency, Noordwijk, The Netherlands, 1996.

12. G. L. Siscoe, N. U. Crooker, G. M. Erickson, B. U. Sonnerup, K. D. Siebert, D. R. Weimer, W. W. White, and N. C. Maynard. *Global Geometry of Magnetospheric Currents Inferred from MHD Simulations*, vol. 118 of *AGU Geophysical Monograph*, pp. 41–52. American Geophysical Union, Washington, D.C., 2000.

13. W. W. White, G. L. Siscoe, G. M. Erickson, Z. Kaymaz, N. C. Maynard, K. D. Siebert, B. U. Ö. Sonnerup, and D. R. Weimer. The magnetospheric sash and the cross-tail S. *Geophys. Res. Lett.*, 25(10):1605–1608, 1998.

14. J. A. Linker, Z. Mikić, D. A. Biesecker, R. J. Forsyth, S. E. Gibson, A. J. Lazarus, A. Lecinski, P. Riley, A. Szabo, and B. J. Thompson. Magnetohydrodynamic modeling of the solar corona during Whole Sun Month. *J. Geophys. Res.*, 104(A5):9809–9830, 1999.

15. J. A. Linker, Z. Mikić, and D. D. Schnack. Modeling coronal evolution. In J. J. Hunt, editor, *Proceedings of the Third SOHO Workshop*, Estes Park, CO, pp. 249–252. European Space Agency, Noordwijk, The Netherlands, 1994.

16. A. V. Usmanov. A global numerical 3-D MHD model of the solar wind. *Sol. Phys.*, 146:377–396, 1993.

17. A. V. Usmanov, M. L. Goldstein, B. P. Besser, and J. M. Fritzer. A global MHD solar wind model with WKB Alfvén waves: Comparison with Ulysses data. *J. Geophys. Res.*, 105:12675–12695, 2000.

18. K. Hayashi. An MHD simulation model of time-dependent global solar corona with temporally varying solar-surface magnetic field maps. *J. Geophys. Res.*, 118:6889–6906, 2013.

19. X. Feng, L. Yang, C. Xiang, S. T. Wu, Y. Zhou, and D. Zhong. Three-dimensional solar wind modeling from the Sun to Earth by a SIP-CESE MHD model with a six-component grid. *Astrophys. J.*, 723:300, 2010.

20. D. Odstrčil. Modeling 3-D solar wind structures. *Adv. Space Res.*, 32(4):497–506, 2003.

21. D. Odstrčil and V. J. Pizzo. Numerical heliospheric simulations as assisting tool for interpretation of observations by STEREO heliospheric imagers. *Sol. Phys.*, 259:297–309, 2009.

22. N. V. Pogorelov, G. P. Zank, and T. Ogino. Three-dimensional features of the outer heliosphere due to coupling between the interstellar and interplanetary magnetic fields. I. Magnetohydrodynamic model: Interstellar perspective. *Astrophys. J.*, 614(2):1007, 2004.

23. J. M. Stone and M. L. Norman. ZEUS-2D: A radiation magnetohydrodynamics code for astrophysical flows in two space dimensions. II. The magnetohydrodynamic algorithms and tests. *Astrophys. J. Suppl.*, 80:791–818, 1992.

24. T. Ogino. A three-dimensional MHD simulation of the interaction of the solar wind with the Earth's magnetosphere: The generation of field-aligned currents. *J. Geophys. Res.*, 91:6791–6806, 1986.

25. H. Washimi and T. Tanaka. 3-D magnetic field and current system in the heliosphere. *Space Sci. Rev.*, 78:85–94, 1996.

26. R. M. Winglee. Multi-fluid simulations of the magnetosphere: The identification of the geopause and its variation with IMF. *Geophys. Res. Lett.*, 25:4441–4444, 1998.

27. R. M. Winglee, W. Lewis, and G. Lu. Mapping of the heavy ion outflows as seen by IMAGE and multifluid global modeling for the 17 April 2002 storm. *J. Geophys. Res.*, 110:A12S24, 2005. doi:10.1029/2004JA010909.

28. G. Tóth. A general code for modeling MHD flows on parallel computers: Versatile advection code. *Astrophys. Lett. Commun.*, 34:245–250, 1996.

29. K. G. Powell, P. L. Roe, T. J. Linde, T. I. Gombosi, and D. L. De Zeeuw. A solution-adaptive upwind scheme for ideal magnetohydrodynamics. *J. Comput. Phys.*, 154(2):284–309, 1999.

30. G. Tóth, B. van der Holst, I. V. Sokolov, D. L. De Zeeuw, T. I. Gombosi, F. Fang, W. B. Manchester, X. Meng, D. Najib, K. G. Powell, Q. F. Stout, A. Glocer, Y.-J. Ma, and M. Opher. Adaptive numerical algorithms in space weather modeling. *J. Comput. Phys.*, 231:870–903, 2012.

31. G. Tóth, I. V. Sokolov, T. I. Gombosi, R. Chesney D, C. R. Clauer, D. L. De Zeeuw, K. C. Hansen, K. J. Kane, W. B. Manchester, R. C. Oehmke, K. G. Powell, A. J. Ridley, I. I. Roussev, Q. F. Stout, O. Volberg, R. A. Wolf, S. Sazykin, A. Chan, and Bin Yu. Space weather modeling framework: A new tool for the space science community. *J. Geophys. Res.*, 110(A12):A12226, 2005.

32. J. Raeder, Y. Wang, and T. J. Fuller-Rowell. *Geomagnetic Storm Simulation with a Coupled Magnetosphere-Ionosphere-Thermosphere Model*, vol. 125 of *AGU Geophysical Monograph Series*, pp. 377–384. American Geophysical Union, Washington, D.C., 2001.

33. C. C. Goodrich, A. L. Sussman, J. G. Lyon, M. A. Shay, and P. A. Cassak. The CISM code coupling strategy. *J. Atmos. Solar-Terr. Phys.*, 66:1469–1479, 2004.

34. J. G. Luhmann, S. C. Solomon, J. A. Linker, J. G. Lyon, Z. Mikic, D. Odstrcil, W. Wang, and M. Wiltberger. Coupled model simulation of a Sun-to-Earth space weather event. *J. Atmos. Solar-Terr. Phys.*, 66:1243, 2004.

35. M. Ashour-Abdalla, G. Lapenta, R. J. Walker, M. El-Alaoui, and H. Liang. Multiscale study of electron energization during unsteady reconnection events. *J. Geophys. Res.*, 120: 4784–4799, 2015.

36. L. K. S. Daldorff, G. Tóth, T. I. Gombosi, G. Lapenta, J. Amaya, S. Markidis, and J. U. Brackbill. Two-way coupling of a global hall magnetohydrodynamics model with a local implicit particle-in-cell model. *J. Comput. Phys.*, 268:236–254, 2014.

37. Y. Lee and A. Sussman. Efficient communication between parallel programs with Intercomm. Technical Report Tech. Rep. 2004-04, UMIACS, 2004.

38. D. L. Brown, W. D. Henshaw, and D. J. Quinlan. Overture: An object-oriented framework for serving partial differential equations on overlapping grids. In M. E. Henderson, C. R. Anderson, and S. L. Lyons, Editors, *Proceedings of the Object Oriented Methods for Interoperable Scientific and Engineering Computing*, pp. 245–255. SIAM: Philadelphia, PA, 1999.

39. G. Tóth. Flexible, efficient and robust algorithm for parallel execution and coupling of components in a framework. *Comput. Phys. Commun.*, 174:793, 2006.

40. Y. Fan. The three-dimensional evolution of buoyant magnetic flux tubes in a model solar convective envelope. *Astrophys. J.*, 676(1):680, 2008.

41. I. V. Sokolov, B. van der Holst, and T. I. Gombosi. Field-lines-threaded model for the low solar corona powered by the Alfven wave turbulence. *Astrophys. J.*, submitted, 2016.

42. B. van der Holst, I. V. Sokolov, X. Meng, M. Jin, W. B. Manchester, G. Tóth, and T. I. Gombosi. Alfvén wave solar model (AWSoM): Coronal heating. *Astrophys. J.*, 782:81, 2014.

43. V. S. Titov and P. Dèmoulin. Basic topology of twisted magnetic configurations in solar flares. *Astron. Astrophys.*, 351:701–720, 1999.

44. S. Gibson and B. C. Low. A time-dependent three-dimensional magnetohydrodynamic model of the coronal mass ejection. *Astrophys. J.*, 493:460–473, 1998.

45. S. K. Antiochos, C. R. DeVore, and J. A. Klimchuk. A model for solar coronal mass ejections. *Astrophys. J.*, 510:485–493, 1999.

46. B. van der Holst, W. Manchester, IV, I. V. Sokolov, G. Toth, T. I. Gombosi, D. DeZeeuw, and O. Cohen. Breakout coronal mass ejection or streamer blowout: The bugle effect. *Astrophys. J.*, 693(2):1178–1187, 2009.

47. F. Fang, W. Manchester, W. P. Abbett, and B. van der Holst. Simulation of flux emergence from the convection zone to the corona. *Astrophys. J.*, 714:1649–1657, 2010.

48. M. Jin, W. B. Manchester, B. van der Holst, J. R. Gruesbeck, R. A. Frazin, E. Landi, A. M. Vasquez, P. L. Lamy, A. Llebaria, A. Fedorov, G. Toth, and T. I. Gombosi. A global two-temperature corona and inner heliosphere model: A comprehensive validation study. *Astrophys. J.*, 7745:6, 2012.

49. J. Kóta, W. B. Manchester, J. R. Jokipii, D. L. De Zeeuw, and T. I. Gombosi. Simulation of SEP acceleration and transport at CME driven shocks. In G. Li, G. Zank, and C. T. Russell, editors, *The Physics of Collisionless Shocks*, pp. 201–206. American Institute of Physics, Melville, NY, 2005.

50. I. V. Sokolov, I. I. Roussev, T. I. Gombosi, M. A. Lee, J. Kóta, T. G. Forbes, W. B. Manchester, and J. I. Sakai. A new field line advection model for solar particle acceleration. *Astrophys. J.*, 616:L171–L174, 2004.

51. V. Tenishev, M. Combi, and B. Davidsson. A global kinetic model for cometary comae: The evolution of the coma of the Rosetta target comet Churyumov-Gerasimenko throughout the mission. *Astrophys. J.*, 685(1):659–677, 2008.

52. M. Opher, P. C. Liewer, T. I. Gombosi, W. B. Manchester, D. L. De Zeeuw, I. V. Sokolov, and G. Tóth. Probing the edge of the solar system: Formation of an unstable jet-sheet. *Astrophys. J.*, 591:L61–L65, 2003.

53. M. Opher, E. C. Stone, and T. I. Gombosi. The orientation of the local interstellar magnetic field. *Science*, 316(5826):875–878, 2007.

54. A. Pulkkinen, L. Rastätter, M. Kuznetsova, H. Singer, C. Balch, D. Weimer, G. Toth, A. Ridley, T. Gombosi, M. Wiltberger, J. Raeder, and R. Weigel. Community-wide validation of geospace model ground magnetic field perturbation predictions to support model transition to operations. *Space Weather*, 11(6):369–385, 2013.

55. F. Toffoletto, S. Sazykin, R. Spiro, and R. Wolf. Inner magnetospheric modeling with the Rice Convection Model. *Space Sci. Rev.*, 107:175–196, 2003.

56. R. A. Wolf, M. Harel, R. W. Spiro, G.-H. Voigt, P. H. Reiff, and C. K. Chen. Computer simulation of inner magnetospheric dynamics for the magnetic storm of July 29, 1977. *J. Geophys. Res.*, 87:5949–5962, 1982.

57. M.-C. Fok, T. E. Moore, and D. C. Delcourt. Modeling of inner plasma sheet and ring current during substorms. *J. Geophys. Res.*, 104(A7):14557–14569, 1999.

58. A. Glocer, M. Fok, X. Meng, G. Toth, N. Buzulukova, S. Chen, and K. Lin. CRCM + BATS-R-US two-way coupling. *J. Geophys. Res.*, 118(4):1635–1650, 2013.

59. V. K. Jordanova, M. Spasojevic, and M. F. Thomsen. Modeling the electromagnetic ion cyclotron wave-induced formation of detached subauroral proton arcs. *J. Geophys. Res.*, 112(A8):A08209, 2007.

60. S. Zaharia, V. K. Jordanova, M. F. Thomsen, and G. D. Reeves. Self-consistent modeling of magnetic fields and plasmas in the inner magnetosphere: Application to a geomagnetic storm. *J. Geophys. Res.*, 111(A11):A11S14, 2006.

61. R. Ilie, M. W. Liemohn, G. Toth, and R. M. Skoug. Kinetic model of the inner magnetosphere with arbitrary magnetic field. *J. Geophys. Res.*, 117(A4):A04208, 2012.

62. M.-C. Fok, R. A. Wolf, R. W. Spiro, and T. E. Moore. Comprehensive computational model of Earth's ring current. *J. Geophys. Res.*, 106(A5):8417–8424, 2001.

63. D. M. Ober, J. L. Horwitz, and D. L. Gallagher. Formation of density troughs embedded in the outer plasmasphere by subauroral ion drift events. *J. Geophys. Res.*, 102(A7):14595–14602, 1997.

64. A. Glocer, G. Toth, T. Gombosi, and D. Welling. Modeling ionospheric outflows and their impact on the magnetosphere, initial results. *J. Geophys. Res.*, 114:A05216, 2009.

65. J. A. Ridley and M. W. Liemohn. A model-derived storm time asymmetric ring current driven electric field description. *J. Geophys. Res.*, 107(A8), 2002. doi:10.1029/2001JA000051.

66. D. R. Weimer. Improved ionospheric electrodynamic models and application to calculating joule heating rates. *J. Geophys. Res.*, 110(A5):A05306, 2005.

67. A. J. Ridley, Y. Deng, and G. Tóth. The global ionosphere-thermosphere model. *J. Atmos. Solar-Terr. Phys.*, 68:839, 2006.

68. A. E. Hedin. Extension of the msis thermosphere model into the middle and lower atmosphere. *J. Geophys. Res.*, 96(A2):1159–1172, 1991.

69. D. Bilitza. The international reference ionosphere status 2013. *Adv. Space Res.*, 55(8):1914–1927, 2015.

70. J. U. Brackbill and D. W. Forslund. An implicit method for electromagnetic plasma simulation in two dimensions. *J. Comput. Phys.*, 46:271–308, 1982.

71. G. Lapenta, J. U. Brackbill, and P. Ricci. Kinetic approach to microscopic-macroscopic coupling in space and laboratory plasmas. *Phys. Plasmas*, 13:055904, 2006.

72. S. Markidis, G. Lapenta, and Rizwan-Uddin. Multi-scale simulations of plasma with iPIC3D. *Math. Comput. Simul.*, 80(7):1509–1519, 2010.

73. I. V. Sokolov. Alternating-order interpolation in a charge-conserving scheme for particle-in-cell simulations. *Comput. Phys. Commun.*, 184(2):320–328, 2013.

74. T. I. Gombosi. *Gaskinetic Theory*. Cambridge University Press: Cambridge, 1994.

75. J. P. Boris. A physically motivated solution of the Alfvén problem. Technical Report NRL Memorandum Report 2167. Naval Research Laboratory: Washington, D.C., 1970.

76. T. I. Gombosi, G. Tóth, D. L. De Zeeuw, K. C. Hansen, K. Kabin, and K. G. Powell. Semi-relativistic magneto-hydrodynamics and physics-based convergence acceleration. *J. Comput. Phys.*, 177:176–205, 2002.

77. T. I. Gombosi and C. E. Rasmussen. Transport of gyration dominated space plasmas of thermal origin I.: Generalized transport equations. *J. Geophys. Res.*, 96:7759, 1991.

78. C. P. T. Groth and J. G. McDonald. Towards physically realizable and hyperbolic moment closures for kinetic theory. *Continuum Mech. Therm.*, 21(6):467–493, 2009.

79. X. Meng, G. Toth, and T. I. Gombosi. Classical and semirelativistic magnetohydrody-namics with anisotropic ion pressure. *J. Comput. Phys.*, 231, 2012.

80. C. Paty and R. Winglee. The role of ion cyclotron motion at ganymede: Magnetic field morphology and magnetospheric dynamics. *Geophys. Res. Lett.*, 33(10):L10106, 2006.

81. L. Wang, A. H. Hakim, A. Bhattacharjee, and K. Germaschewski. Comparison of multi-fluid moment models with particle-in-cell simulations of collisionless magnetic reconnection. *Phys. Plasmas*, 22(1):012108, 2015.

82. C. David Levermore. Moment closure hierarchies for kinetic theories. *J. Stat. Phys.*, 83:1021–1065, 1996.

83. P. Goswami, T. Passot, and P. L. Sulem. A landau fluid model for warm collisionless plasmas. *Phys. Plasmas*, 12(10):102109, 2005.

84. G. W. Hammett and F. W. Perkins. Fluid moment models for landau damping with application to the ion-temperature-gradient instability. *Phys. Rev. Lett.*, 64:3019–3022, 1990.

85. J. V. Hollweg. Some physical processes in the solar wind. *Rev. Geophys.*, 16(4):689–720, 1978.

86. S. D. Bale, M. Pulupa, C. Salem, C. H. K. Chen, and E. Quataert. Electron heat conduction in the solar wind: Transition from spitzer-härm to the collisionless limit. *Astrophys. J. Lett.*, 769(2):L22, 2013.

87. L. Spitzer and R. Härm. Transport phenomena in a completely ionized gas. *Phys. Rev.*, 89:977–981, 1953.

88. M. Wiltberger, W. Lotko, J. G. Lyon, P. Damiano, and V. Merkin. Influence of cusp O+ outflow on magnetotail dynamics in a multifluid MHD model of the magnetosphere. *J. Geophys. Res.*, 115:A00J05, 2010.

89. X. Meng, G. Toth, M. W. Liemohn, T. I. Gombosi, and A. Runov. Pressure anisotropy in global magnetospheric simulations: A magnetohydrodynamics model. *J. Geophys. Res.*, 117, 2012.

90. J. M. Burgers. *Flow Equations for Composite Gases*. Academic Press: New York, 1969.

91. R. W. Schunk. Mathematical structure of transport equations for multispecies flows. *Rev. Geophys.*, 15:429–445, 1977.

92. O. Cohen, I. V. Sokolov, I. I. Roussev, C. N. Arge, W. B. Manchester, T. I. Gombosi, R. A. Frazin, H. Park, M. D. Butala, F. Kamalabadi, and M. Velli. A semiempirical magnetohydrodynamical model of the solar wind. *Astrophys. J.*, 654:L163–L166, 2007.

93. B. van der Holst, W. B. Manchester, R. A. Frazin, A. M. Vásquez, G. Toth, and T. I. Gombosi. A data-driven, two-temperature solar wind model with Alfvén waves. *Astrophys. J.*, 725(1):1373–1383, 2010.

94. G. Tóth, D. L. De Zeeuw, T. I. Gombosi, W. B. Manchester, A. J. Ridley, I. V. Sokolov, and I. I. Roussev. Sun to thermosphere simulation of the October 28–30, 2003 storm with the Space Weather Modeling Framework. *Space Weather J.*, 5:S06003, 2007.

95. A. J. Ridley, D. L. De Zeeuw, T. I. Gombosi, and K. G. Powell. Using steady-state MHD results to predict the global state of the magnetosphere-ionosphere system. *J. Geophys. Res.*, 106:30067–30076, 2001.

96. A. J. Ridley, K. C. Hansen, G. Tóth, D. L. De Zueew, T. I. Gombosi, and K. G. Powell. University of Michigan MHD results of the GGCM metrics challenge. *J. Geophys. Res.*, 107(A10):1290, 2002. doi:10.1029/2001JA000253.

97. P. Song, T. I. Gombosi, D. L. De Zeeuw, K. G. Powell, and C. P. T. Groth. A model of solar wind–magnetosphere–ionosphere coupling for due northward IMF. *Planet Space Sci.*, 48:29–39, 2000.

98. D. T. Welling and A. J. Ridley. Validation of SWMF magnetic field and plasma. *Space Weather*, 8:S03002, 2010. doi:10.1029/2009SW000494.

99. J. A. Linker, Z. Mikić, R. Lionello, P. Riley, T. Amari, and D. Odstrcil. Flux cancellation and coronal mass ejections. *Phys. Plasmas*, 10(5):1971–1978, 2003.

100. Z. Mikić and J. A. Linker. Disruption of coronal magnetic field arcades. *Astrophys. J.*, 430:898–912, 1994.

101. P. Riley, J. A. Linker, and Z. Mikić. An empirically-driven global MHD model of the solar corona and inner heliosphere. *J. Geophys. Res.*, 106:15889–15902, 2001.

102. I. V. Sokolov, B. van der Holst, R. Oran, C. Downs, I. I. Roussev, M. Jin, W. B. Manchester, R. M. Evans, and T. I. Gombosi. Magnetohydro-dynamic waves and coronal heating: Unifying empirical and MHD turbulence models. *Astrophys. J.*, 764:23, 2013.

103. G. Tóth, O. Volberg, A. J. Ridley, T. I. Gombosi, D. L. De Zeeuw, K. C. Hansen, D. R. Chesney, Q. F. Stout, K. G. Powell, K. J. Kane, and R.C. Oehmke. A physics-based software framework for Sun-Earth connection modeling. In A. T. Y. Lui, Y. Kamide, and G. Consolini,

editors, *Multiscale Coupling of Sun-Earth Processes*, pp. 383–397. Elsevier, Amsterdam, The Netherlands, 2005.

104. R. Lionello, J. A. Linker, and Z. Mikić. Multispectral emission of the Sun during the first whole Sun month: Magnetohydrodynamic simulations. *Astrophys. J.*, 690(1):902–912, 2009.

105. C. J. Schrijver, A. W. Sandman, M. J. Aschwanden, and M. L. DeRosa. The coronal heating mechanism as identified by full-Sun visualizations. *Astrophys. J.*, 615(1):512–525, 2004.

106. X. Meng, B. van der Holst, G. Tóth, and T. I. Gombosi. Alfvén wave solar model (AW-SoM): Proton temperature anisotropy and solar wind acceleration. *Mon. Not. R. Astron. Soc.*, 454(4):3697–3709, 2015.

107. R. Oran, E. Landi, B. van der Holst, S. T. Lepri, A. M. Vásquez, F. A. Nuevo, R. Frazin, W. Manchester, I. Sokolov, and T. I. Gombosi. A steady-state picture of solar wind acceleration and charge state composition derived from a global wave-driven MHD model. *Astrophys. J.*, 806(1):55, 2015.

108. R. Oran, B. van der Holst, E. Landi, M. Jin, I. V. Sokolov, and T. I. Gombosi. A global wave-driven magnetohydrodynamic solar model with a unified treatment of open and closed magnetic field topologies. *Astrophys. J.*, 778:176–195, 2013.

109. J. R. Lemen, A. M. Title, D. J. Akin, P. F. Boerner, C. Chou, J. F. Drake, D. W. Duncan, C. G. Edwards, F. M. Friedlaender, G. F. Heyman, N. E. Hurlburt, N. L. Katz, G. D. Kushner, M. Levay, R. W. Lindgren, D. P. Mathur, E. L. McFeaters, S. Mitchell, R. A. Rehse, C. J. Schrijver, L. A. Springer, R. A. Stern, T. D. Tarbell, J.-P. Wuelser, C. J. Wolfson, C. Yanari, J. A. Bookbinder, P. N. Cheimets, D. Caldwell, E. E. Deluca, R. Gates, L. Golub, S. Park, W. A. Podgorski, R. I. Bush, P. H. Scherrer, M. A. Gummin, P. Smith, G. Auker, P. Jerram, P. Pool, R. Soufli, D. L. Windt, S. Beardsley, M. Clapp, J. Lang, and N. Waltham. The Atmospheric Imaging Assembly (AIA) on the Solar Dynamics Observatory (SDO). *Sol. Phys.*, 275(1–2):17–40, 2012.

110. M. Jin, W. B. Manchester, B. van der Holst, R. Oran, I. Sokolov, G. Toth, Y. Liu, X. D. Sun, and T. I. Gombosi. Numerical simulations of coronal mass ejection on 2011 March 7: One-temperature and two-temperature model comparison. *Astrophys. J.*, 773(1):50, 2013.

111. W. B. Manchester, A. Vourlidas, G. Toth, N. Lugaz, I. I. Roussev, I. V. Sokolov, T. I. Gombosi, D. L. De Zeeuw, and M. Opher. Three-dimensional MHD simulations of the 2003 October 28 coronal mass ejection: Comparison with LASCO coronagraph observations. *Astrophys. J.*, 684:1448–1460, 2008.

112. V. Pizzo, G. Millward, A. Parsons, D. Biesecker, S. Hill, and D. Odstrcil. Wang-Sheeley-Arge-Enlil cone model transitions to operations. *Space Weather*, 9, 2011.

113. C. N. Arge and V. J. Pizzo. Improvement in the prediction of solar wind conditions using near-real time solar magnetic field updates. *J. Geophys. Res.*, 105:10465, 2000.

114. D. Odstrčil, P. Riley, and X. P. Zhao. Numerical simulation of the 12 May 1997 interplanetary CME event. *J. Geophys. Res.*, 109(A2):A02116, 2004.

115. H. Xie, L. Ofman, and G. Lawrence. Cone model for halo CMEs: Application to space weather forecasting. *J. Geophys. Res.*, 109(A3):A03109, 2004.

116. X. P. Zhao, S. P. Plunkett, and W. Liu. Determination of geometrical and kinematical properties of halo coronal mass ejections using the cone model. *J. Geophys. Res.*, 107(A8): SSH 13–1–SSH 13–9, 2002.

117. M. L. Mays, A. Taktakishvili, A. Pulkkinen, P. J. MacNeice, L. Rastätter, D. Odstrčil, L. K. Jian, I. G. Richardson, J. A. LaSota, Y. Zheng, and M. M. Kuznetsova. Ensemble modeling of CMEs using the WSAENLIL+cone model. *Sol. Phys.*, 290(6):1775–1814, 2015.

118. R. M. Dewey, D. N. Baker, B. J. Anderson, M. Benna, C. L. Johnson, H. Korth, D. J. Gershman, G. C. Ho, W. E. McClintock, D. Odstrcil, L. C. Philpott, J. M. Raines, D. Schriver, J. A. Slavin, S. C. Solomon, R. M. Winslow, and T. H. Zurbuchen. Improving solar wind modeling at Mercury: Incorporating transient solar phenomena into the WSA-ENLIL model with the cone extension. *J. Geophys. Res.*, 120(7):5667–5685, 2015.

119. W. B. Manchester, T. I. Gombosi, I. I. Roussev, D. L. De Zeeuw, I. V. Sokolov, K. G. Powell, G. Tóth, and M. Opher. Three-dimensional MHD simulation of a flux-rope driven CME. *J. Geophys. Res.*, 109(A1):A01102, 2004. doi:10.1029/2002JA009672.

120. I. I. Roussev, T. G. Forbes, T. I. Gombosi, I. V. Sokolov, D. L. De Zeeuw, and J. Birn. A three-dimensional flux rope model for coronal mass ejections based on a loss of equilibrium. *Astrophys. J.*, 588:L45–L48, 2003.

121. R. A. Howard, N. R. Sheeley, M. J. Koomen, and D. J. Michels. Coronal mass ejections: 1979–1981. *J. Geophys. Res.*, 90(A9):8173–8191, 1985.

122. M. Jin, M. D. Ding, P. F. Chen, C. Fang, and S. Imada. Coronal mass ejection induced outflows observed with Hinode/EIS. *Astrophys. J.*, 702(1):27–38, 2009.

123. P. Janhunen, M. Palmroth, T. Laitinen, I. Honkonen, L. Juusola, G. Facskó, and T. I. Pulkkinen. The GUMICS-4 global MHD magnetosphereionosphere coupling simulation. *J. Atmos. Solar-Terr. Phys.*, 80:48–59, 2012.

124. J. Raeder, J. Berchem, and M. Ashour-Abdalla. The geospace environment modeling grand challenge: Results from a global geospace circulation model. *J. Geophys. Res.*, 103(A7):14787–14797, 1998.

125. J. A. Fedder, S. P. Slinker, J. G. Lyon, and R. D. Elphinstone. Global numerical simulation of the growth phase and the expansion onset for a substorm observed by Viking. *J. Geophys. Res.*, 100:19083–19093, 1995.

126. A. J. Ridley, T. I. Gombosi, and D. L. DeZeeuw. Ionospheric control of the magnetosphere: Conductance. *Ann. Geophys.*, 22:567–584, 2004.

127. J. U. Kozyra, M. W. Liemohn, C. Cattell, D. De Zeeuw, C. P. Escoubet, D. S. Evans, X. Fang, M.-C. Fok, H. U. Frey, W. D. Gonzalez, M. Hairston, R. Heelis, G. Lu,

W. B. Manchester, S. Mende, L. J. Paxton, L. Rastaetter, A. Ridley, M. Sandanger, F. Soraas, T. Sotirelis, M. W. Thomsen, B. T. Tsurutani, and O. Verkhoglyadova. Solar filament impact on 21 January 2005: Geospace consequences. *J. Geophys. Res.*, 119(7):5401–5448, 2014.

128. R. S. Weigel, A. J. Klimas, and D. Vassiliadis. Solar wind coupling to and predictability of ground magnetic fields and their time derivatives. *J. Geophys. Res.*, 108(A7):1298, 2003.

129. D. R. Weimer. An empirical model of ground-level geomagnetic perturbations. *Space Weather*, 11(3):107–120, 2013.

130. A. Parsons, D. Biesecker, D. Odstrcil, G. Millward, S. Hill, and V. Pizzo. Wang-sheeley-argeenlil cone model transitions to operations. *Space Weather*, 9(3):S03004, 2011.

131. J. Bartels, N. H. Heck, and H. F. Johnston. The three-hour-range index measuring geomagnetic activity. *Terr. Magnet. Atmos. Elec.*, 44(4):411–454, 1939.

132. NSWP. National Space Weather Program implementation plan. Technical Report FCM-P30-2010, Office of the Federal Coordinator for Meteorological Services and Supporting Research, Washington, D.C., 2010.

133. R. F. Stein, A. Lagerfjärd, Å. Nordlund, and D. Georgobiani. Solar flux emergence simulations. *Sol. Phys.*, 268(2):271–282, 2011.

Space Weather and the Extraterrestrial Planets

Henry B. Garrett

CONTENTS

16.1 INTRODUCTION

Space weather for the extraterrestrial planets, just as at Earth, takes many different forms. These range from radiation (solar particle events [SPEs], trapped radiation around the planets, or galactic cosmic rays [GCRs]), extreme temperatures and pressures at the surfaces of the planets, to, in a broader sense, dust storms and atmospheric weather. The interaction of space weather with each planet has unique characteristics that must be considered in designing reliable space systems. Although landers and rovers have been successfully deployed on the moon and Mars as well as on Titan and Venus, these environments all pose potential space weather threats to future spacecraft systems and astronauts. With the desire to fly ever-more complex spacecraft, robotic landers, and "aerobots," understanding the space weather threats is critical for designers. This chapter will cover a wide range of these interplanetary and planetary space weather environments with particular stress on the characteristics of concern in the design of a reliable space system.

16.2 SOLAR WIND

The Sun is the driving source behind most space weather in the solar system. Although descriptions of the Sun's immediate environment (the solar spectrum, Sun spots, solar flares, and coronal mass ejections) and its impact on Earth are covered in Chapters 1 and 2, here the extended "atmosphere" of the Sun, the solar wind, will be briefly addressed in the context of its impact on the other extraterrestrial bodies. Although the solar wind is the "ocean" which space missions must traverse, it has a relatively benign effect on spacecraft compared to the other sources of space weather in the form of SPEs, GCRs, and trapped radiation. However, as the solar wind is a fully ionized, electrically neutral plasma that flows outward from the Sun carrying the solar magnetic field with it, spacecraft in this interplanetary environment will experience charging and plasma interactions that can impact spacecraft instruments (Chapter 17). Table 16.1 summarizes some of the basic physical properties of the solar wind in the ecliptic plane at 1 AU.

TABLE 16.1   Characteristics of the Solar Wind at 1 AU in the Ecliptic Plane

Plasma Environment	Minimum	Maximum	Average
Flux (#/cm² s)	10^8	10^{10}	$2–3 \times 10^8$
Velocity (km s⁻¹)	200	2500	400–500
Density (#/cm³)	0.4	80	5 to >10
T (eV)	0.5	100	20
T_{max}/T_{avg}	1.0[a]	2.5	1.4
Helium ratio (N_{He}/N_H)	0	0.25	0.05
Flow direction	±15° from radial		~2° East
Alfvén speed (km/s)	30	150	60
Magnetic field			
$\|B(G)\|$	0.25	40	6
B vector	Polar component		Average in ecliptic plane
	Planar component		Average spiral angle ~45°

Source: Garrett, H. B., *Encyclopedia of Aerospace Engineering*, John Wiley & Sons, Hoboken, NJ, 2010. Copyright 2009 John Wiley & Sons Ltd.

Note: B stands for magnetic field. G is the units of the magnetic field, for example, "Gauss."

[a] Isotropic.

To estimate the solar wind environment at the other planets, the plasma density and the magnetic field properties scale roughly as r^{-2} and r^{-3} with distance, respectively (Garrett and Minow, 2007; Minow et al., 2007; and references therein for discussions of the solar wind and its interactions with spacecraft). Based on Table 16.1, nominal plasma parameters of the solar wind useful in estimating the flow field around a spacecraft and determining the surface potentials are listed in Table 16.2. For reference, typical values of spacecraft potential in the solar wind range from −20 V (in shadow) to +10 V (in sunlight) near Earth, with −100 V being the highest reported in the literature.

Of special interest are the Debye length (λ_D) and the electron (v_{te}) and ion (v_{ti}) thermal velocities as these help scale the dimensions of the plasma sheath around a spacecraft. As a special example, for a large 100 m × 100 m solar sail, the Debye length in the solar wind of

TABLE 16.2   Nominal Solar Wind Properties at 1 AU

Debye length	λ_D ~25 m
Electron thermal velocity	v_{te} ~2.65 × 10³ km s⁻¹
Ion thermal velocity	v_{ti} ~30 km s⁻¹
Acoustic speed	$c_s = (T_e/m_p)^{1/2}$ = ~62 km s⁻¹
Ion cyclotron frequency	Ω_i ~1 rad s⁻¹
Electron cyclotron frequency	Ω_e ~1.8 × 10³ rad s⁻¹

Source: Garrett, H. B., *Encyclopedia of Aerospace Engineering*, John Wiley & Sons, Hoboken, NJ, 2010. Copyright 2009 John Wiley & Sons Ltd.

~25 m is on the order of one-quarter the size of the solar sail. As this is the distance over which the sail potential is shielded by the plasma, it shows that the sail potential and its effects would extend well out from the sail's surface in the solar wind requiring scientific instruments to be mounted on an extended boom if pristine measurements are desired. As the thermal velocity of the ions is much lower than the solar wind velocity (which is hypersonic), a deep ion wake will be created behind a solar sail or any body for that matter in the plasma. By contrast, the electron thermal velocity is much higher than the solar wind velocity, and they would normally rapidly fill in this wake. They are prevented from doing so by the electric field produced by the slower ions. These features thus characterize the basic structure of the plasma sheath and wake around a spacecraft and provide insights into how the solar wind interacts with it.

To conclude this section, the solar wind plasma and its magnetic field interact on a much larger scale with the planets and their plasma and magnetic fields creating the wake-like structures called magnetospheres. As an example of a planetary magnetosphere, Chapters 3 and 4 discuss Earth and its interactions with the solar wind in detail. In Section 16.4, the solar wind's interaction with each of the other planets is discussed. These interactions, as discussed in Chapter 2, are also coupled to the 11-year solar cycle of activity leading to "space weather" variations throughout the solar system just as at Earth. Ultimately, the solar wind itself interacts with the interstellar wind and stands it off much like Earth's magnetic field stands off the solar wind—this stand-off shock is now believed to have been detected by the Voyager 1 spacecraft at ~125 AU. In considering this interstellar space weather environment, a primary effect of the solar wind magnetic field is to cut off GCR particles with energies less than ~10 GeV nuc^{-1} (GCRs below this energy are primarily of solar origin).

16.3 PHYSICAL AND DYNAMICAL PROPERTIES OF THE PLANETS

The planets and their moons, the comets, and the asteroids that make up our solar system exhibit a wide range of physical properties. The most important of these are presented in Table 16.3 for the Sun, the eight planets, and three of the largest "dwarf planets." Based on the data from the JPL HORIZONS System (Giorgini et al., 1996), the table lists key properties such as equatorial radius, planetary flattening, g (the body's standard gravity or the acceleration duty to gravity), GM (the gravitational constant × the body's mass), total mass, mass ratio (GM Sun/GM planet), density, mean surface temperature, geometric albedo, and the solar constant at the body. The HORIZONS database also includes data for over 170,000 asteroids and comets, and 128 natural satellites.

In addition to the physical properties of the planets, Table 16.3 provides an overview of their orbital characteristics. These include the semimajor axis, eccentricity, inclination, escape velocity, mean orbit velocity, sidereal day, mean solar day, sidereal period, obliquity to orbit, and the right ascension and declination of the body's spin axis. Of particular importance to extraterrestrial space weather interactions are the planets' magnetic fields. The magnetic field characteristics include dipole tilt, dipole offset, and magnetic moment of the magnetic field. Mars and Venus have basically no intrinsic magnetic field (Mars does have a small localized remnant field in the form of "magnetic stripes" on the surface). This absence at Mars is believed to be the result of cooling of its metallic core, whereas the absence of a field at Venus results from its extremely slow rotation rate. Similarly, Earth's moon has no primary magnetic field, though it has regions of local magnetic enhancements (~1–100 nT) apparently associated with mascons (lunar gravitational anomalies). The latest maps of these fields, based on data from the Kaguya and Lunar Prospector missions, have been recently published (Tsunakawa et al., 2015).

16.4 MAGNETOSPHERES

A unique space weather feature of all bodies in the solar system is the way they interact with the solar wind. This interaction takes many different forms depending primarily on the strength of the body's magnetic field, its local plasma environment, and, in the absence of a strong magnetic field, its ionosphere. The first two of these, the magnetic field and plasma, interact by diverting the solar wind plasma and its entrapped magnetic field—charged particles, in the absence of time-varying electric fields and particle scattering, do not easily cross or diffuse across planetary magnetic field lines, thereby causing a magnetohydrodynamic interaction that generates a magnetosphere (the region around a body controlled by its magnetic field). In two dimensions, the interaction resembles the flow of water around a boat that creates a wake structure. As illustrated for Earth in Chapter 3, the interaction produces a complex, three-dimensional region bounded by the bow shock (the outer boundary of the magnetospheric shock), the magnetosheath (the region of turbulence behind the shock), and, interior to the magnetosheath, the magnetopause (the boundary marking the region controlled by the body's magnetic field where the magnetic field pressure of the planet balances the solar wind pressure).

Although all of the planets interact differently, of particular interest are the magnetospheres of Uranus and Neptune. In the case of Uranus, its spin axis is almost in the ecliptic plane, whereas its dipole is ~59° to the spin axis. The result is a complex, time-varying (on the timescale of the planetary rotation rate) magnetosphere. Neptune, in many other ways very similar to Uranus, has its rotation axis nearly normal to the ecliptic plane like the other major planets. It was thought before the Voyager flyby that its magnetic dipole would, like the other planets, be parallel to the spin axis. Instead, like Uranus, it is also highly tilted at ~47° to its spin axis. This creates an even more complex magnetospheric interaction—indeed the Neptunian magnetosphere exhibits a split magnetic "tail" (Bagenal, 1992)! See a recent movie of it at https://www.ras.org.uk/images/stories/press/NAM_2015/Wednesday8July/Neptune_magnetic_field_animation.mp4 (Masters et al., 2015).

The magnetospheres of Mars and Venus, in contrast to the "magnetized" planets, are the result of interactions between the solar wind and the ionospheres of the planets—this causes their bow shocks to be very close to the planets' surfaces. Illustrating this effect, Figure 16.1

TABLE 16.3 Physical Properties of the Planets and the Sun

	Sun	Mercury	Venus	Earth	Mars	Ceres	Jupiter	Saturn	Uranus	Neptune	Pluto	2003 UB313
						Physical Properties						
Equatorial radius (km)	6.96E+05	2439.7	6051.8	6378.136	3397	474	71492	60268	25559	24766	1151	~1400
Flattening	0	0	0	0.003352813	0.006476306	0	0.06487	0.09796	0.02293	0.0171	0	
g polar (m s^{-2})	274	3.701	8.87	9.832186369	3.69	0.28	27.01	12.14	9.19	11.41	0.655	
g equatorial (m s^{-2})	274	3.701	8.87	9.780326772	3.69	0.28	23.12	8.96	8.69	11	0.655	
go (m s^{-2})	274	3.701	8.87	9.82022	3.69	0.28	24.79	10.44	8.87	11.15	0.655	
GM (km^3 s^{-2})	132712439940	22032.09	324858.63	398600.44	42828.3	63.2	126686537	37931284.5	7793947	6835107	870	
Mass (kg)	1.9891E+30	3.302E+23	4.8685E+24	5.9736E+24	6.4185E+23	9.4715E+20	1.8986E+27	5.68461E+26	8.6832E+25	1.0243E+26	1.314E+22	
Mass ratio (GM Sun/GM planet)		6.0236E+06	4.0852E+05	3.3295E+05	3.0987E+06	2.0999E+09	1.0473E+03	3.4979E+03	2.2903E+04	1.9412E+04	1.3500E+08	
Core mass (kg)				1.84E+24			4.96E+25	5.84E+25	1.04E+23			
Core rad (km)		~1600	~3200	3480	~1700							
Density (gm cm^{-3})	1.408	5.427	5.204	5.515	3.933		1.326	0.687	1.318	1.638	1.75	
Moment of inertia	0.059	0.33	0.33	0.3308	3.66E-01		0.254	0.21	0.225			
Atmospheric pressure (bar)		1.00E-15	92	1	0.0056	0						
Mean temperature (K)	5778[a]	442.5	735	270	210		165 (1 bar)	134 (1 bar)	76 (1 bar)	72 (1 bar)	37.5	
Geometric albedo		0.106	0.65	0.367	0.15	0.1132	0.52	0.47	0.51	0.41	0.3	
Visual magnitude V(1,0)		-0.42	-4.4	-3.86	-1.52	3.34	-9.4	-8.88	-7.19	-6.87	-1	-1.1143
Solar constant (W m^{-2})	6.3180E+07	9936.9	2613.9	1367.6	589.2	178.7776775	50.5	15.04	3.71	1.47	0.89	0.295962975

(Continued)

TABLE 16.3 (*Continued*)   Physical Properties of the Planets and the Sun

	Sun	Mercury	Venus	Earth	Mars	Ceres	Jupiter	Saturn	Uranus	Neptune	Pluto	2003 UB313
					Dynamical Characteristics							
Semimajor axis (AU)		0.38709893	0.72333199	1.00000011	1.52366231	2.765813554	5.20336301	9.53707032	19.19126393	30.06896348	39.48168677	67.97682158
Eccentricity		0.20563069	0.00677323	0.01671022	0.09341233	0.078561009	0.04839266	0.0541506	0.04716771	0.00858587	0.24480766	0.43581589
Inclination (°)		7.00487	3.39471	0.00005	1.85061	10.6069597	1.3053	2.48446	0.76986	1.76917	17.14175	43.83410933
Escape velocity (km s^{-2})	617.7	4.435	10.361	11.186	5.027	0.45061632	59.5	35.5	21.3	23.5	1.3	3.6125
Mean orbit velocity (km s^{-1})		47.8725	35.0214	29.7859	24.1309	17.9091	13.0697	9.6624	5.4778 (6.81)	5.43	4.947	
Sidereal day (h)	609.12	1407.5088	−5832.5	23.93419	24.62	9.08	9.894	10.61	17.14	16.7	6.38675	
Mean solar day (day)		175.9421	116.749	1.002738	1.03E+00							
Sidereal period (years)		0.2408445	0.6151826	1.0000174	1.88081578	4.59984	11.856523	29.423519	83.747407	163.72321	248.0208	560.46627
Sidereal period (days)		87.968435	224.695434	365.25636	686.98	1680.12	4330.60	10746.94	30588.74	59799.9	90589.596	204710.30
Obliquity to orbit (°)	25 (ecliptic)	~0.1	177.3	23.45	25.19		3.12	26.73	97.86	29.56	122.53	
Hill's sphere radius (R_p)		94.4	167.1	234.9	319.8		740	1100	2700	4700		
Pole (RA in degrees J2000)	286.13	281.01	272.76	0	317.681		268.05	40.589	357.311	299.36	313.02	
Pole (DEC in degrees J2000)	63.87	61.45	67.16	90	52.881		64.49	83.537	−15.175	43.46	9.09	
Dipole tilt (°)		169		11.3			9.6	0	58.6	47		
Dipole offset (rp)				0.0725			0.131	0.04	0.3	0.55		
Magnetic moment (gauss Rp³)		0.0033		0.305	<1E−4		4.28	0.21	0.228	0.133		

Source: Giorgini, J. D. et al., *Bulletin of the American Astronomical Society, 28,* 1158, 1996. Copyright 2009 John Wiley & Sons Ltd. See text for an explanation of the different parameters. Values are based on the HORIZON system.
[a] Photosphere.

compares the relative sizes of the planets versus their magnetospheres as represented by the bow shocks normalized to the same distance. Jupiter, having the strongest magnetic field relative to the other planets, is not observable on this scale, whereas the bow shock is almost on the surface of Venus as it has no intrinsic magnetic field.

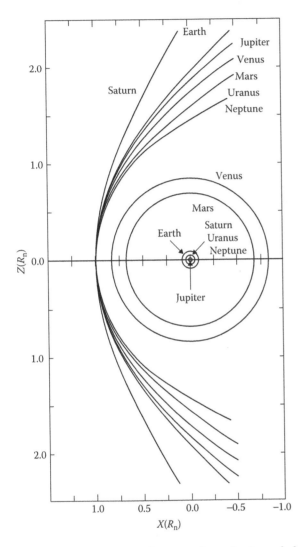

FIGURE 16.1 Comparisons between the scale sizes of planetary magnetospheres. The bow shock standoff distances of the magnetospheres are normalized to 1 with the planetary radii scaled to the same units. On this scale, Jupiter is not visible because of its intense magnetic field, whereas Earth and Saturn have similar sized magnetic fields when scaled by their radii. Venus and Mars in contrast are almost as large as their bow shocks. $Z(R_n)$, distance along planetary polar axis in units of the bow shock standoff distance; $X(R_n)$, distance along axis in the equatorial plane in units of the bow shock standoff distance. Direction is positive away from the Sun. (Adapted from Slavin, J. A. et al., *J. Geophys. Res., 89*, 2708–2714, 1984; Slavin, J. A. et al., *J. Geophys. Res., 90*, 6275, 1985; Copyright 2009 John Wiley & Sons Ltd. With Permission.)

Asteroids and comets also exhibit miniature versions of the planetary magnetospheres. Comets, because of their plasma and neutral atmosphere emissions, interact with the solar wind in a similar manner to Venus and Mars. In addition to the beautiful dust tail associated with a comet, this plasma interaction shows up as a visible plasma tail. This tail is created when the solar wind as it flows away from the Sun interacts and "picks up" the cometary plasma ions. Examples of cometary magnetospheres can be seen in Flammer (1991). An asteroid will likely have at best a very weak magnetic field, so it will interact much like a spacecraft creating a plasma wake structure like that for the large solar sail discussed earlier.

Interior to their magnetopauses, Earth, Jupiter, Saturn, Uranus, and Neptune have regions of closed magnetic field (i.e., the magnetic field lines are simply connected from one hemisphere to the next in a dipole-like configuration) and sufficient field strength to trap plasma and energetic charged particles. The latter particles form radiation belts, the regions of magnetically trapped energetic electron and proton fluxes that can seriously damage microelectronic components and damage and alter surface coatings, optical lenses, and polymer materials. In fact, the surface of Europa has been characteristically darkened by the Jovian radiation belts. The Jovian radiation belts are so intense that the synchrotron radiation they produce can easily be imaged at Earth by radio telescopes. Engineering models of the radiation belts for shielding design have been developed for Earth (the AE/AP models [Johnston et al., 2014; Sawyer and Vette, 1976; Vette, 1991; Vette et al., 1979]), Jupiter (GIRE [Divine and Garrett, 1983; Garrett et al., 2003, 2005a, 2012]), and Saturn (SATRAD [Garrett et al., 2005b]) and are available online. As a shielding design point, an ~1 MeV electron or an ~10 MeV proton will penetrate approximately 2.54 mm (100 mils) of aluminum, a standard spacecraft shield thickness assumed for shielding design purposes. Meridian contour plots of these three planetary belts for 1 MeV electron and 10 MeV proton fluxes are presented in Figure 16.2. Also shown is the author's estimate for the Uranian radiation belts for 1 MeV electrons and 5 MeV protons. For Earth, note the dual belt structure with protons dominating the inner belt, whereas electrons dominate the outer belt. Jupiter, because its magnetic field is proportionally ~20 times more intense than Earth's and Saturn's, can trap more particles. This results in an intense belt dominated by electrons. Saturn, because its famous

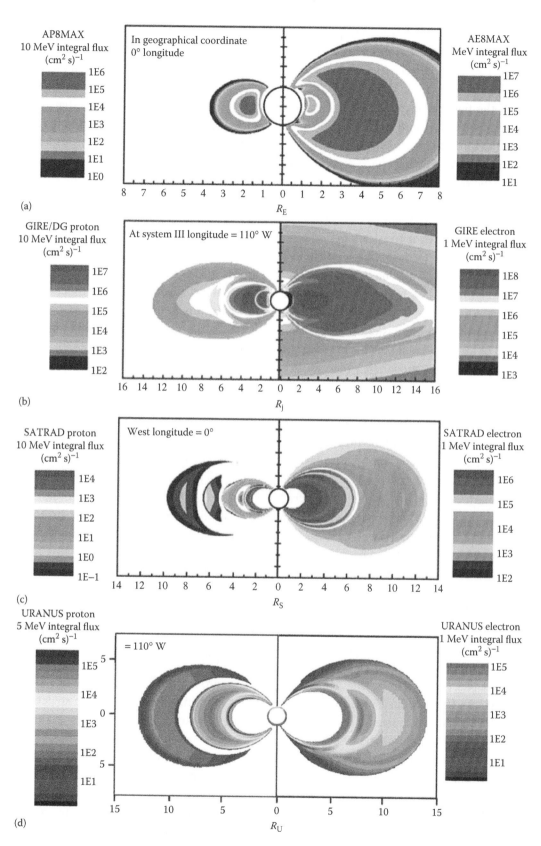

FIGURE 16.2 Meridional cross sections for the radiation belts of (a) Earth, (b) Jupiter, (c) Saturn, and (d) Uranus. Proton fluxes are for 10 MeV (5 MeV for Uranus) and electron fluxes for 1 MeV electrons, which have roughly the same "penetration" depths through a typical spacecraft shield. (From Garrett, H. B., *Encyclopedia of Aerospace Engineering*, John Wiley & Sons, Hoboken, NJ, 2010; Copyright 2009 John Wiley & Sons Ltd. With Permission.)

rings are located near the expected peak in trapped radiation, does not have a particularly pronounced radiation environment—the ring dust particles absorb and scatter the energetic radiation particles in the inner magnetosphere. Although the figure does not show it, Uranus's radiation belts, because they are tied to its magnetic field, are tilted ~59° to the spin axis.

At Jupiter and Earth, overlaying the radiation belts in their inner magnetospheres, there is a cold plasma environment with a mean energy of ~1–10 eV and with densities varying from ~10^6 cm^{-3} in their ionosphere at low altitudes (see later) down to ~100 cm^{-3} at high altitudes. This environment, termed the plasmasphere (see Chapter 11), dominates plasma interactions at low latitudes and altitudes. At higher latitudes near the polar caps, the auroras play an additional role for both planets. The auroras are the visual manifestation of the interaction of intense plasma beams of electrons and protons with the upper atmosphere. These beams (actually sheets) are an extension of the hot equatorial (10–100 keV) particles (for Jupiter, this corresponds to about 15–20 R_j (Jovian radii) and near-geosynchronous orbit at Earth) and extends up along the magnetic field lines to atmospheric altitudes at high latitudes (or about 70°). This region, called the plasma sheet, and the auroras that mark its base can produce spacecraft potentials approaching −20 kV on spacecraft surfaces at Earth and up to −4 kV at Jupiter (Garrett et al., 2008) and represent a significant threat to poorly designed spacecraft. See Chapter 17 for more discussion on surface charging.

Models of the plasmasphere and plasma sheet regions and auroras at Jupiter and Saturn (Bagenal and Delamere, 2011; Garrett et al., 2008) are currently under development. These models are of concern to designers as the plasma characteristics determine whether the cold plasmasphere currents that produce potentials of few tens of volts or the hot auroral currents dominate the spacecraft charging process. A meridian cross section of the cold electron component is presented in Figure 16.3 for Jupiter. Up to about 5 R_j, the Jovian plasmasphere approximates Earth's—a simple extension of the ionosphere out to near 5–6 R_j. Three issues, however, alter the environment beyond that distance: First, Io orbits at 5.9 R_j and emits a dense cloud of cold plasma made up of sulfur and sodium. Second, Jupiter's rotation period is 10 h producing a significant centrifugal force on the cold,

trapped iogenic plasma. The plasma is forced outward by this centrifugal force (Figure 16.3). Third, the mass of the plasma and the resulting centrifugal force are sufficient to severely distort the Jovian magnetic field—the energy in the magnetic field and that in the plasma are equal near 20 R_j. At this point, the magnetic field necessary for trapped radiation can no longer adequately confine the radiation and plasma, and the Jovian plasma is forced out into a broad plasma disk (its plasma sheet). It should be noted that similar cold plasma rings have been observed in conjunction with Europa at Jupiter and Titan at Saturn.

Jupiter, Saturn, Uranus, and Neptune all exhibit auroras and have the equivalent of the auroral zones and a plasma sheet like Earth. Indeed variations in the auroras at Jupiter and Uranus have been observed from Earth. In the case of Jupiter and Saturn, Garrett and Hoffman (2000b) estimated potentials associated with the hot plasma on the order of −4 kV at Jupiter and a few hundred volts negative at Saturn.

At Jupiter, another type of unusual plasma interaction is observed. Due to the rapid rotation of Jupiter, the $V \times B$ force induced by its magnetic field at Io produces a radially directed electric field on the order of ~1 MV across the moon (see Chapter 17 for a discussion on the electric fields generated by $V \times B$ forces). This field maps down to the planet's atmospheric surface and generates an aurora at the foot of the magnetic field line in both hemispheres. These auroral footprints can be seen from Earth by Hubble space telescope in association with the magnetic field lines that intersect Europa, Ganymede, and Callisto, and allow accurate mapping of the Jovian magnetic field lines.

16.5 ATMOSPHERES

With aero-capture (particularly at Mars) becoming a useful option, knowledge of the structure of a planet's atmosphere to determine drag is an important consideration for many missions. An atmosphere is described in terms of its composition, density, and temperature (or pressure) as functions of altitude above a reference altitude—taken to be the equivalent of one atmosphere of pressure for the gas giants or from the planet's surface for the terrestrial planets. ANSI/AIAA Standard G-003C-2008 (American National Standard, 2008) provides extensive references to models of Earth's neutral atmosphere. It also provides reference atmospheric

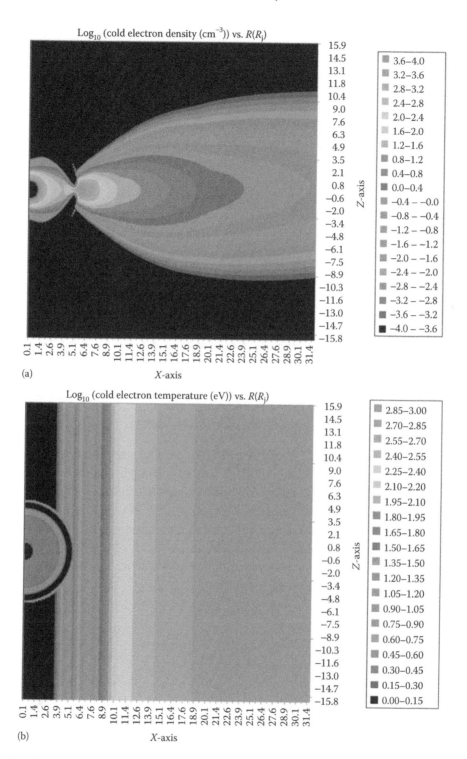

FIGURE 16.3 Meridional density (a) and temperature (b) contour plots for the cold electrons (1–1000 eV) at Jupiter. The cold ion contours are very similar. (Garrett, H. B. et al., *AIAA SPACE 2015 Conference and Exposition*, AIAA, Pasadena, CA, 2015.)

models for Mars (Mars-GRAM), Neptune (Neptune-GRAM), Titan (Titan-GRAM), Venus (Venus-GRAM), and the Venus International Reference Atmosphere (VIRA). Sample temperature height profiles for Venus, Earth, Mars, and Titan are presented in Figure 16.4a.

The profiles for the gas giants (Jupiter, Saturn, Uranus, and Neptune) are plotted in Figure 16.4b. Of primary concern to a mission designer, the density and pressure versus the altitude and temperature for each of the planets are plotted in Figure 16.5.

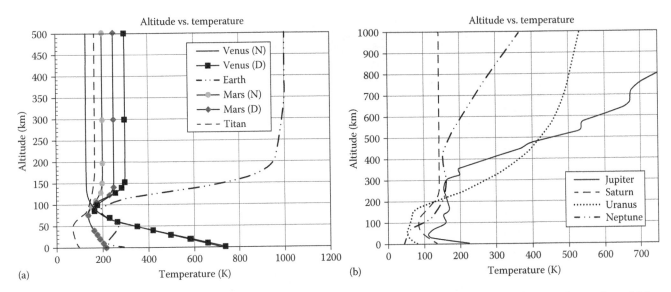

FIGURE 16.4   Temperature versus height profiles for (a) the terrestrial planets Venus (day [D] and night [N]), Earth, and Mars (day [D] and night [N]), and the moon Titan and (b) the gas giants Jupiter, Saturn, Uranus, and Neptune. (Data are adapted from American National Standard, Guide to reference and standard atmosphere models, August 27, 2008. ANSI/AIAA G-003C-2008 [Revision of G-003B-2004], 2008; Garrett, H. B., and R. W. Evans, *The Extra-Terrestrial Space Environment, a Reference Chart*, Jet Propulsion Laboratory, 2008; Lodders, K., and B. Fegley Jr., *The Planetary Scientist's Companion*, Oxford University Press, New York, NY, 371 pp., 1998; and references therein; Copyright 2009 John Wiley & Sons Ltd. With Permission.)

Table 16.4 provides atmospheric surface composition, temperature, and pressures for the planets and Titan (note that the surface conditions at Mercury are currently being updated by the MESSENGER spacecraft). The gas giant atmospheres are all referenced to ~1 ATM/10^5 Pa. Representing the most severe of the terrestrial environments, Venus' surface, with a pressure ~90× that of Earth and at a temperature of over 700°K, can have serious effects on landers. Although made up primarily of carbon dioxide (96.5%) which has unique effects at these temperatures and pressures, it also has caustic clouds containing sulfuric acid and compounds of chlorine and fluorine that can precipitate as an acid rain ("virga"). In addition, in the upper atmosphere, the Venusian winds can reach over 300 km h^{-1}.

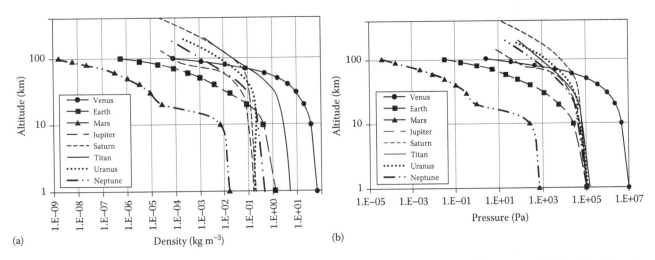

FIGURE 16.5   Atmospheric profiles for various planets. Plots for (a) altitude versus density and (b) altitude versus pressure.

(Continued)

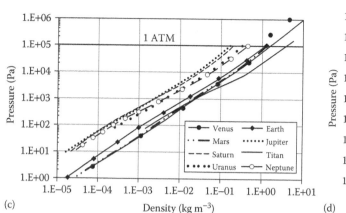

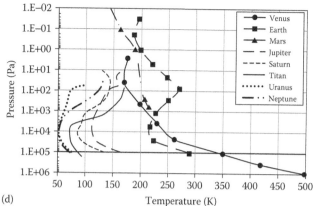

FIGURE 16.5 (Continued)   Atmospheric profiles for various planets. Plots for (c) pressure versus density and (d) pressure versus temperature. (Data are adapted from American National Standard, Guide to reference and standard atmosphere models, August 27, 2008. ANSI/AIAA G-003C-2008 [Revision of G-003B-2004], 2008; Garrett, H. B., and R. W. Evans, *The Extra-Terrestrial Space Environment, a Reference Chart*, Jet Propulsion Laboratory, 2008; Lodders, K., and B. Fegley Jr., *The Planetary Scientist's Companion*, Oxford University Press, New York, NY, 371 pp., 1998; and references therein; Copyright 2009 John Wiley & Sons Ltd. With Permission.)

TABLE 16.4   Surface Composition, Temperature, and Pressure for the Planets and Titan

Planet	Surface Composition	Temperature (K)	Pressure (Pa)
Mercury	O (42%), Na (29%), H_2 (22%), He (6%), K (0.5%)	90–700	Trace
Venus	CO_2 (96.5%), N_2 (3.5%), CO	735	9.2e6
Earth	N_2 (87%), O_2 (21%), H_2O (<4% varies)	288	1e5
Mars	CO_2 (95%), N_2 (2.7%), Ar (1.6%)	214	636
Jupiter	H_2 (86%), He (13%), CH_4 (1%–2%)	165	1e5
Saturn	H_2 (96%), He (3.25%), CH_4 (0.45%)	135	1e5
Titan	N_2 (65%–98%), Ar (<25%), CH_4 (2%–10%), H_2 (0.2%)	94	1.5e5
Uranus	H_2 (82.5%), He (15.2%), CH_4 (2.3%)	76.4	1e5
Neptune	H_2 (80%), He (19%), CH_4 (1%–2%)	71.5	1e5

Sources: American National Standard, Guide to reference and standard atmosphere models, August 27, 2008. ANSI/AIAA G-003C-2008 (Revision of G-003B-2004), 2008; Garrett, H. B., and R. W. Evans, *The Extra-Terrestrial Space Environment, a Reference Chart*, Jet Propulsion Laboratory, 2008; Lodders, K., and B. Fegley Jr., *The Planetary Scientist's Companion*, Oxford University Press, New York, NY, 371 pp., 1998; and references therein). Copyright 2009 John Wiley & Sons Ltd.

16.6 IONOSPHERES

Incoming ultraviolet (UV), extreme UV (EUV), and, to a lesser extent, particulate radiation will ionize a planet's ambient neutral atmosphere forming an ionosphere (Chapter 14). The ionospheric plasma affects radio waves propagating through it by absorbing and refracting them depending on frequency, and it is this distortion that allows measurements of the ionospheric density. As the atmospheric density increases with depth, the ion density increases proportionally. The UV/EUV radiation, however, is increasingly absorbed by the atmosphere so that the ionization rate falls off rapidly. As a result, the cold ionospheric plasma typically has energies representative of the background

neutrals of a fraction of an electronvolt or less (1 eV is equivalent to 11,604°K). The ionosphere, because of the interplay between incoming UV/EUV and the fall-off with altitude of the atmosphere and local chemistry, forms layered ionized structures called Chapman layers in the atmosphere. This layering can be seen in the sample ionospheric profiles presented in Figure 16.6. These profiles correspond to the total charge density as represented by electrons as a function of altitude. The relative compositions of the planetary ionospheres at low altitudes are tabulated in Table 16.5. *In situ* chemistry greatly complicates the ionospheric composition with altitude, and the reader should consult articles on each planet for details.

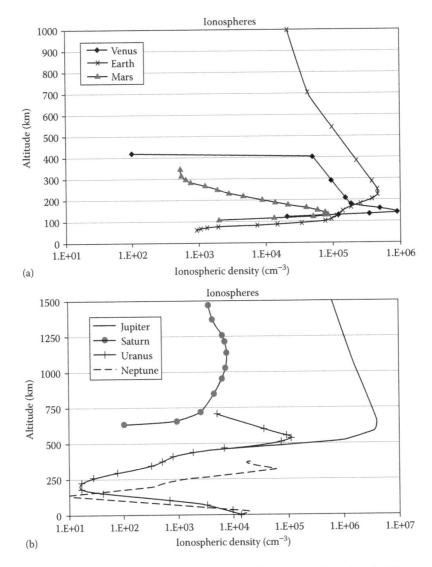

FIGURE 16.6 Representative vertical ionospheric density profiles for the planets. Note that the Venus profile is for the dayside and is truncated as it is compressed by the solar wind. The gas giants are referenced to an altitude at 1 atm/10^{-5} Pa. Plots of the ionospheres for the terrestrial planets (a) and for the gas giants (b). (Data are adapted from Garrett, H. B., and R. W. Evans, *The Extra-Terrestrial Space Environment, a Reference Chart*, Jet Propulsion Laboratory, 2008; and references therein; Copyright 2009 John Wiley & Sons Ltd. With Permission.)

TABLE 16.5   Representative Compositions at Various Altitude Ranges for Planetary Ionospheres

Planet	Species	Range (km)	Species	Range (km)
Venus	O_2^+	<200	O^+	200–400
Earth	NO^+,O_2^+	<200	O^+	200–400
Mars	O^+,O_2^+	<400	H^+	>400
Jupiter	$C_2H_5^+$	<200	H^+	>200
Saturn	$C_2H_5^+$	<1000	H_3^+	1000–6000
Uranus	$C_2H_9^+$	<400	H^+	>400
Neptune	$C_2H_9^+$	<250	CH_5^+	250–300

Sources: Adapted from Garrett, H. B., and R. W. Evans, *The Extra-Terrestrial Space Environment, a Reference Chart*, Jet Propulsion Laboratory, 2008; and references therein; Garrett, H. B., *Encyclopedia of Aerospace Engineering*, John Wiley & Sons, Hoboken, NJ, 2010. Copyright 2009 John Wiley & Sons Ltd. With Permission.

16.7 DUST ENVIRONMENTS

After radiation and thermal effects, dust is a major space weather hazard on the surfaces of Mars, moon, comets, and some asteroids of space systems. Mars has global dust storms that obscure the Sun for long periods, whereas the moon has fine dust that can become charged in the space environment and cling to everything. The primary sources of the dust are the particles that make up the surface regolith. These fine particulates, especially on the moon, are highly abrasive. They both pose physical and chemical biohazards to astronauts—the dust regolith (which resembles crushed glass) can damage the lungs, although the composition of the dust can be a chemical threat. On Earth, the atmosphere has largely oxidized dangerous metals and other chemicals, whereas on the moon and Mars, they still exist and pose a serious health risk. Perchlorates in particular are reactive chemicals that were detected in the martian soil by the National Aeronautics and Space Administration (NASA)'s Phoenix Lander and are highly oxidizing.

Although the force of the wind on Mars is only one-tenth that of Earth, Mars experiences the largest dust storms in the solar system. Some of these storms can cover the entire planet. They introduce micrometer-sized particles into the atmosphere that can remain suspended for many months and seriously limit solar array power generation. It is believed that the bright areas of Mars are areas of long-term dust precipitation, whereas the dark areas are regions where the lighter colored dust has been blown away leaving darker bedrock material. The dust particles in the air are estimated at <2–3 μm. Their total column mass during a dust storm is estimated at 10^{-3} g cm^{-2}. The dust composition is not well known but is probably similar to that of the martian soil with the iron content making it appear reddish.

The dust storm "season" on Mars coincides approximately with its southern spring–summer and occurs near the perihelion in Mars' orbit. The solar heating is ~45% higher than normal at perihelion and creates the strong atmospheric dynamics and winds required to raise dust. The dust storms can vary greatly in size and last up to several months. There are typically three martian dust storms per day globally in late spring and two per day in the early northern and mid-southern summers. Statistics for the occurrence frequency and size of the dust storms are presented in Table 16.6. Local dust storms are shorter in duration than the great dust storms and may not have an "explosive" expansion phase. Great dust storms form a continuum that can be subdivided into type A (large), type AB (globe-circling), and type B (globe-covering) (Evans, 1988). Dust devils are also a frequent occurrence (Evans, 1988) and are currently credited with keeping the Mars exploration rover (MER) solar arrays clean, thereby greatly extending the mission's lifetime.

Unlike the wind-driven processes on Mars, lunar dust is formed primarily by meteor impact on the lunar surface and is composed of silicate minerals, glass, and unoxidized metals (primarily iron). Lunar dust, because it is much more penetrating than terrestrial dust and, in the vacuum of the moon, can become charged and attracted to other materials, is potentially a major concern for space systems. Specifically, it is the component of the lunar soil particles smaller than 50 μm. The particle size distribution follows a log-normal distribution (Cauffman, 1973) that implies that half of the soil by mass is dust.

Whereas the dust at Mars is charged by the wind through triboelectric charging, the space plasma environment itself (e.g., the solar wind) can charge lunar dust and levitate it above the moon's surface. Surveyor images showed clouds of dust hovering up to a meter off the ground (Criswell, 1972), whereas astronauts reported seeing dust being lofted to high altitudes in jets (Abbas et al., 2007; Criswell, 1972; Stubbs et al., 2006). Dust winds were measured at the lunar terminator by the Apollo surface

TABLE 16.6   Classification of Martian Dust Storms

Type of Storm	Latitude of Origin	Season of Origin, L_s	Duration (Earth Days)	Nominal Extent (km)	Comments
Local	+70° to −75°	All	4–10	100	No rapid expansion
Type A	+50° to −60°	192° < L_s < 310° (110°, 163° rare)	70	2,500	Expansion East–West
Type AB	−20° to −60°	205°	90	20,000	Globe-circling
Type B	−20° to −60°	250° < L_s < 310°	100	Global	Globe-covering

Source: Evans, R. W., *Mars Atmosphere Dust-Storm Occurrence Estimate for Mars Observer*, Jet Propulsion Laboratory, 1988; Copyright 2009 John Wiley & Sons Ltd.

Type A are large dust storms, Type AB are globe-circling storms, and Type B are globe-covering storms.

L_s is the heliocentric longitude of Mars.

instruments (Grund and Colwell, 2007). The winds were stronger at sunrise and weaker at sunset. A theory of lunar dust charging that attempts to explain these observations is presented in the works of Freeman and Ibrahim (1975) and Stubbs et al. (2006). Briefly, the daytime environment is dominated by solar UV/EUV that causes photoelectron emission yielding potentials around +10 V, though extreme conditions can produce +25 to +200 V. Charging at the terminator is much more complex and depends on the location of the moon relative to the magnetosphere and the solar wind. Moving away from the terminator toward lunar midnight, the surface potential drops slowly making a sudden drop a few degrees sunward of the "optical terminator." The environment in the lunar shadow is marked by a steady drop in potential as local lunar midnight is approached. As charged dust particles are accelerated by a potential gradient (e.g., the local electric field), not the potential, the maximum acceleration of the dust is at the terminator. Kilovolt negative potentials in the lunar shadow or at the terminator (Freeman and Ibrahim, 1975) have also been reported with global charging patterns described in the work of Stubbs et al. (2006).

16.8 ASTEROIDS AND COMETS

As Johannes Kepler said in 1596, "Inter Jovem et Martem interposui planetam". 200 years later, on January 1, 1801, Father Giuseppe Piazzi discovered what he at first believed to be a new comet. Karl Friedrich Gauss, using a revolutionary new method for determining orbits, found that the orbit of the body was circular and at 2.8 AU—Kepler's missing planet. Within little over a year another "planet" was found, also at 2.8 AU. Thereafter many more discoveries followed so that by the late 1800's the newly found asteroids were becoming a distraction to astronomers—several hundred thousand asteroids have been catalogued to date.

Although the orbital characteristics of the asteroids are quite varied, a number of patterns have emerged. Of the three largest groups in the inner solar system, the primary population is referred to as the Main Belt and is made up of asteroids with orbits between Mars and Jupiter. The next group of asteroids is made up of those that cross the Earth's orbit, the Near-Earth Asteroids (NEAs). The third principal group is the Trojans, asteroids trapped at the Jovian Lagrange points 60° ahead and behind Jupiter in the same orbit.

Another means of classifying asteroids is by their apparent composition as estimated by their observed spectra. These are the C-type (which stands for dark carbonaceous objects), S-type (which stands for stony objects—metallic nickel–iron mixed with iron and magnesium silicates), and U (or M) for those that do not fit into either C or S (typically bright and metallic). Roughly, 75% of asteroids are believed to be of the C-type and 17% of the S-type.

At 974 km in diameter, Ceres (actually "1 Ceres") is the largest asteroid and contains about 25% of the mass in the asteroid belt. The NASA's Dawn spacecraft is currently in orbit around Ceres and in the process of mapping it. As the controversy over Pluto has demonstrated, the distinction between planets and asteroids have become blurred over the past few years, and Ceres is now classified not only as the largest asteroid but as one of the dwarf planets (or plutoids) along with Pluto, Haumea, Makemake, and Eris (many more dwarf planets are believed to exist in the Kuiper Belt, the trans-Neptunian disk surrounding the Sun). At present, the name "asteroid" has been limited to minor planets in the inner solar system. The names and sizes of the 24 largest asteroids are listed in Table 16.7.

Unlike the asteroids, comets, because of their much higher visibility, have been observed since antiquity. In reality, however, it is proving harder to clearly distinguish between the two—Comets appear to be asteroids far from the Sun, whereas some asteroids, particularly in eccentric orbits, may be "burned-out" comets. Comets are believed to be a mixture of ice, frozen gases (e.g., carbon monoxide, carbon dioxide, methane, ammonia, and organic compounds), dirt, and rocks—material left over from the formation of the solar system. As comets approach the Sun, they become active emitting gases, dust, and plasma. Currently, the European Space Agency's (ESA) Rosetta is in orbit around comet 67P/Churyumov–Gerasimenko and monitoring this activity in real time. Active comets such as 67P/Churyumov–Gerasimenko all have several visible features in common. First, there is the nucleus, the solid body at the center of the head of a comet. The nucleus is believed to be an icy body covered with a dark layer of dry dust or rock. This is surrounded by a bright region called the coma, which is a cloud of dust, water, and gases forming the atmosphere above the nucleus. As evidenced by the Rosetta mission, jets of material are also observed being emitted from the surface. A cloud of neutral hydrogen extends out millions of kilometers from the coma. Of course, the most visible feature of a comet is the tail(s). This is made up of

TABLE 16.7 Listing of the Largest Asteroids

Number	Name	Radius (km)	Semimajor Axis (AU)	Period (years)	Albedo	Absolute Magnitude (H)	Rotation Period (days)
1	Ceres[a]	474.00	2.76581	4.59984	0.11320	3.3400	9.0750
2	Pallas	266.00	2.77335	4.61866	0.15870	4.1300	7.8132
4	Vesta	265.00	2.36181	3.62975	0.42280	3.2000	5.3420
10	Hygiea	203.56	3.13659	5.55515	0.07170	5.4300	27.6230
511	Davida	163.03	3.16956	5.64295	0.05400	6.2200	5.1294
704	Interamnia	158.31	3.06260	5.35974	0.07428	6.2368	8.7270
52	Europa	151.25	3.10076	5.46022	0.05780	6.3100	5.6330
87	Sylvia	130.47	3.48689	6.51128	0.04350	6.7783	5.1840
31	Euphrosyne	127.95	3.14631	5.58099	0.05430	6.7400	5.5310
15	Eunomia	127.67	2.64335	4.29774	0.20940	5.1803	6.0830
16	Psyche	126.58	2.91966	4.98892	0.12030	5.8881	4.1960
65	Cybele	118.63	3.43267	6.35999	0.07060	6.6343	4.0410
3	Juno	116.96	2.66848	4.35918	0.23830	5.3300	7.2100
324	Bamberga	114.72	2.68516	4.40012	0.06280	6.8200	29.4300
624	Hektor	112.50	5.16799	11.74871	0.02500	7.1140	6.9210
451	Patientia	112.48	3.06144	5.35669	0.07640	6.5470	9.7270
107	Camilla	111.31	3.48732	6.51248	0.05250	6.8133	4.8439
532	Herculina	111.20	2.77324	4.61837	0.16940	5.8100	9.4050
48	Doris	110.90	3.11067	5.48643	0.06240	6.9094	11.8900
7	Isis	99.92	2.38618	3.68607	0.27660	5.5100	7.1390
24	Themis	99.00	3.12718	5.53015	0.06700	7.0800	8.3740
9	Metis	95.00	2.38636	3.68647	0.11800	6.2800	5.0790
6	Hebe	92.59	2.42468	3.77563	0.26790	5.7100	7.2745
8	Flora	67.95	2.20117	3.26579	0.24260	6.3436	12.7990

Source: Giorgini, J. D. et al., Bulletin of the American Astronomical Society, 28, 1158, 1996; Copyright 2009 John Wiley & Sons Ltd. Based on JPL HORIZONS System.

[a] Also a "dwarf planet."

two components: a dust tail composed of tiny dust particles driven off the nucleus by the escaping gases extending up to 10 million kilometers away from the Sun and an ion plasma tail trapped by the solar wind extending hundreds of million kilometers downwind.

As in the case of the asteroids, comets are classified in terms of their orbits. Comets with short and intermediate periods spend most of their orbits within that of Pluto. They are thought to originate in the Kuiper Belt and have periods of 200 years or less. Short-period comets are divided into families: for example, the Jupiter family (periods <20 years) or the Halley family (20–200 year periods). Figure 16.7 shows a histogram of the aphelia (maximum distance from the Sun) showing the associations of comet orbit families with the gas giants. Comets with periods of 200 years to thousands or more are termed long-period comets and have highly eccentric orbits with a variety of orbital inclinations. The long-period comets are believed to originate in the distant Oort cloud—a

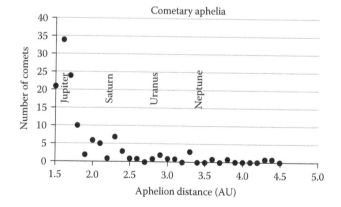

FIGURE 16.7 Histogram for short-period comet aphelia (the maximum distance of the comet from the Sun) as listed by the Minor Planet Center (Cambridge, MA) and *Wikipedia*. The abscissa is the natural logarithm of the aphelia expressed in astronomical units. The family grouping at Jupiter's orbit dominates, but the Saturn, Uranus, and Neptune family groups can also be seen. (Copyright 2009 John Wiley & Sons Ltd. With Permission.)

spherical cloud of frozen comets surrounding the solar system out to about 3 light-years. Gravitational perturbations by the nearby stars can put the comets onto elliptical orbits into the inner solar system.

With a number of recent flybys, intercepts, and even impacts (the NASA's Deep Impact mission) and landings on asteroids (the NASA's NEAR spacecraft "landed" on Eros) and comets (e.g., the ESA's Rosetta Philae lander), interest has been growing in these bodies. Given their great variety (some appear to be heavily cratered, whereas others are very smooth or look like piles of dust and ruble), they can interact with spacecraft in a number of possible ways. Asteroids have even been found to have small companions (Ida and its subsatellite Dactyl), which could pose collision threats during close encounters or during an asteroid recovery mission.

The dust and debris cloud associated with a comet, because of the possible hypervelocity impacts with a spacecraft, is another threat. Indeed, the dust and debris from a comet typically spread out along the comet's trajectory giving rise to meteor streams. These streams pose a direct threat to spacecraft passing through them (Mariner 4 is suspected of having encountered one and being damaged in the process), and spacecraft trajectories have been altered to avoid them. Although the atmospheres of comets are relatively low density (10^{12} cm^{-3}, roughly equivalent to 150 km altitude at Earth), there are visible jets that might be of concern to a lander. Indeed, Rosetta, at closest approach to its comet, had problems with the dust particles impacting its solar arrays and confusing its star tracker. Finally, the charged dust and gas near the surface of a comet or asteroid could contaminate the surfaces of a lander or affect an asteroid recovery mission much like lunar dust.

16.9 SUMMARY

It is challenging to provide sufficient information in a review of this nature on all aspects of the effects of space weather on extraterrestrial bodies. Instead, this chapter has sought to provide information on the most relevant space environments that a designer would need in evaluating a mission to the principal bodies in the solar system and at least point the way to the main environmental issues of concern. The references are intended to provide linkage to the details necessary for the more demanding reader and direct them to the means for mitigating the worst of the environmental concerns. To conclude this chapter, for general reference, the reader is referred to the many fine online sources on space environments such as Spaceweather.com (http://www.spaceweather.com/), the Jet Propulsion Laboratory home page (http://www.jpl.nasa.gov/), and SPENVIS—the Space Environment, Effects, and Education System (http://www.spenvis.oma.be/) to obtain further information on the critical environments and their effects. Several books devoted to the space environment and its effects that cover in detail many of the issues discussed in this section are as follows:

1. Garrett, H. B., and Pike, C. P., *Space Systems and Their Interactions with Earth's Space Environment*, AIAA Press, New York, 1980.

2. Hastings, D., and Garrett, H. B., *Spacecraft-Environment Interactions*, Cambridge University Press, Cambridge, 1996.

3. Tribble, A. C., *The Space Environment: Implications for Spacecraft Design*, Princeton University Press, Princeton, NJ, 2003.

4. Pisacane, V. L., *The Space Environment and Its Effects on Space Systems*, AIAA Press, Reston, VA, 2008.

5. Blockley, R., and Shyy, W., *Encyclopedia of Aerospace Engineering*, John Wiley & Sons, Hoboken, NJ, 2010.

ACKNOWLEDGMENTS

The portion of this work provided by H. B. Garrett was carried out at the Jet Propulsion Laboratory, California Institute of Technology, Pasadena, California, under a contract with the National Aeronautics and Space Administration (Washington, DC).

REFERENCES

Abbas, M., J. Spann, A. LeClair, J. Brasunas, and D. Jennings (2007). Lunar dust distributions from solar infrared absorption measurements with a fourier transform spectrometer. Paper presented at Workshop on Science Associated with the Lunar Exploration Architecture, NASA, Tempe, AZ, February 27–March 2, 2007.

American National Standard (2008). Guide to reference and standard atmosphere models. August 27, 2008. ANSI/AIAA G-003C-2008 (Revision of G-003B-2004).

Bagenal, F. (1992). Giant Planet Magnetospheres. *Annu. Rev. Earth Planet. Sci.*, 20, 289–328.

Bagenal, F., and P. A. Delamere (2011). Flow of mass and energy in the magnetospheres of Jupiter and Saturn. *J. Geophys. Res.*, 116, A05209(1–17). doi:10.1029/2010JA016294.

Cauffman, D. P. (1973). Recommendations concerning spacecraft charging in the magnetosphere. *SAMSO-TR-73-348; Aerospace TR-0074(9260-09)-5*, 33pp.

Criswell, D. R. (1972). Lunar dust motion. Paper presented at Proceedings of the Third Lunar Science Conference (Supplement 3, Geochimica et Cosmochimica Acta), Vol. 3, The MIT Press, Cambridge, MA.

Divine, T. N., and H. B. Garrett (1983). Charged particle distributions in Jupiter's magnetosphere. *J. Geophys. Res.*, *88* (September 9), 6889–6903.

Evans, R. W. (1988). *Mars Atmosphere Dust-Storm Occurrence Estimate for Mars Observer*. IOM 5137-87-124, Jet Propulsion Laboratory, Pasadena, CA.

Flammer, K. R. (1991). The global interaction of comets with the solar wind. In *Comets in the Post-Halley Era*, edited by M. N. R. L. Newburn, and J. Rahe, Vol. 2, pp. 1125–1141, Kluwer Academic Publishers, Amsterdam, The Netherlands.

Freeman, J. W., and M. Ibrahim (1975). Lunar electric fields, surface potential and associated plasma sheaths. *Moon*, *14*(September), 103–114. doi:10.1007/BF00562976.

Garrett, H. B. (2010). The interplanetary and planetary environments. In *Encyclopedia of Aerospace Engineering*, edited by R. Blockley and W. Shyy. John Wiley & Sons, Hoboken, NJ.

Garrett, H. B., and R. W. Evans (2008). *The Extra-Terrestrial Space Environment, a Reference Chart*. Jet Propulsion Laboratory, Pasadena, CA.

Garrett, H. B., R. W. Evans, A. C. Whittlesey, I. Katz, and I. Jun (2008). Modeling of the Jovian Auroral Environment and its effects on spacecraft charging. *IEEE Trans. Plasma Sci.*, *36*(5), 2440–2449. doi:TPS.2008.2004260.

Garrett, H. B., and A. Hoffman (2000). Comparison of spacecraft charging environments at the Earth, Jupiter, and Saturn. *IEEE Trans. Plasma Phys.*, *28*(6), 2048–2057.

Garrett, H. B., I. Jun, J. M. Ratliff, R. W. Evans, G. A. Clough, and R. W. McEntire (2003). *Galileo Interim Radiation Electron Model*. 72 pp. JPL Publication 03-006, The Jet Propulsion Laboratory, California Institute of Technology, Pasadena, CA.

Garrett, H. B., W. Kim, and R. W. Evans (2015). Updating the Jovian plasma and radiation environments—The latest results for 2015, AIAA 2015-4556. In *AIAA SPACE 2015 Conference and Exposition*. AIAA, Pasadena, CA.

Garrett, H. B., M. Kokorowski, I. Jun, and R. W. Evans (2012). *Galileo Interim Radiation Electron Model Update—2012*. 57 pp. JPL Publication 12-9, Jet Propulsion Laboratory, California Institute of Technology, Pasadena, CA.

Garrett, H. B., S. M. Levin, S. J. Bolton, R. W. Evans, and B. Bhattacharya (2005a). A revised model of Jupiter's inner electron belts: Updating the divine radiation model. *Geophys. Res. Lett.*, *32*(4), L04104(1–4). doi:10.1029/2004GL021986.

Garrett, H. B., and J. I. Minow (2007). *Charged Particle Effects on Solar Sails—Final Report*. NASA Report ISPT-SS-06-101. NASA Marshall Space Flight Center, Huntsville, AL.

Garrett, H. B., J. M. Ratliff, and R. W. Evans (2005b). *Saturn Radiation (SATRAD) Model*. 103 pp. Jet Propulsion Laboratory, Pasadena, CA.

Giorgini, J. D., D. K. Yeomans, A. B. Chamberlin, P. W. Chodas, R. A. Jacobson, M. S. Keesey, J. H. Lieske, S. J. Ostro, E. M. Standish, and R. N. Wimberly (1996). JPL's on-line solar system data service. *Bulletin of the American Astronomical Society*, *28*(3), 1158.

Grund, C. J., and J. E. Colwell (2007). Autonomous lunar dust observer for the systematic study of natural and anthropogenic dust phenomena on airless bodies. Paper presented at Workshop on Science Associated with the Lunar Exploration Architecture, NASA, Tempe, AZ.

Johnston, W. R., T. P. O'Brien, G. P. Ginet, S. L. Huston, T. B. Guild, and J. A. Fennelly (2014). AE9/AP9/SPM: New models for radiation belt and space plasma specification. In *Sensors and Systems for Space Applications VII*, edited by K. D. Pham and J. L. Cox. SPIE, Bellingham, WA. doi:10.1117/12.2049836.

Lodders, K., and B. Fegley Jr. (1998). *The Planetary Scientist's Companion*. 371 pp. Oxford University Press, New York.

Masters, A., J. Chittenden, and J. Eastwood (2015). *Neptune's Badly Behaved Magnetic Field*. Royal Astronomical Society, Piccadilly, London.

Minow, J. I., A. M. Diekmann, and W. C. Blackwell Jr. (2007). Status of the L2 and Lunar charged particle environment models. AIAA paper 2007-0910 in *45th AIAA Aerospace Sciences Meeting and Exhibit*. AIAA, Reno, NV.

Sawyer, D. M., and J. I. Vette (1976). *AP-8 Trapped Proton Environment for Solar Maximum and Solar Minimum*, 76-06, National Space Science Data Center/World Data Center-A for Rockets & Satellites, NASA Goddard Spaceflight Center, Greenbelt, MD.

Slavin, J. A., R. E. Holzer, J. R. Spreiter, and S. S. Stahara (1984). Planetary Mach cones: Theory and observation. *J. Geophys. Res.*, *89*(A5), 2708–2714. doi:10.1029/JA089iA05p02708.

Slavin, J. A., E. J. Smith, J. R. Spreiter, and S. S. Stahara (1985). Gas dynamic modeling of the Jovian and saturnian bow shocks: Solar wind flow about the outer planets. *J. Geophys. Res.*, *90*, 6275.

Stubbs, T. J., R. R. Vondrak, and W. M. Farrell (2006). A dynamic fountain model for lunar dust. *Adv. Space Res.*, *37*, 59–66.

Tsunakawa, H., F. Takahashi, H. Shimizu, H. Shibuya, and M. Matsushima (2015). Surface vector mapping of magnetic anomalies over the Moon using Kaguya and Lunar Prospector observations. *J. Geophys. Res. Planets*, *120*, 1160–1185. doi:10.1002/2014JE004785.

Vette, J. I. (1991). *The AE-8 Trapped Electron Model Environment*. 91-24, National Space Science Data Center/World Data Center-A for Rockets & Satellites, NASA Goddard Spaceflight Center, Greenbelt, MD.

Vette, J. I., M. J. Teague, D. M. Sawyer, and K. W. Chan (1979). Modeling the Earth's radiation belts. Paper presented at Solar-Terrestrial Prediction Proceedings, NOAA, Boulder, CO.

II

Some Space Weather Applications

CHAPTER 17

Spacecraft Charging

Henry B. Garrett

CONTENTS

17.1 INTRODUCTION

Spacecraft charging is a major environmental concern for the spacecraft design community and can take many forms. Charging typically results from the buildup of charge on surfaces (surface charging) or in dielectric surfaces (buried charge or, as used here, internal electrostatic discharge [IESD]). It also results from induced currents from asymmetric plasma flows or planetary magnetic fields. Strong evidence exists that space weather in the form of geomagnetic substorms and aurora and the trapped electron environments on Earth and other planets can cause spacecraft charging. In another form, as demonstrated by the International Space Station (ISS), the interaction of a spacecraft and a planetary ionosphere can generate a plasma wake that can distort the potentials around the vehicle. Finally, electric fields caused by the movement of a conducting body across a planetary magnetic field can induce currents in the structure.

The Earth appears to have the worst environment for spacecraft surface charging and can lead to surface potentials well in excess of −20 KV. While such potentials in and of themselves are not necessarily of concern, it is ultimately the buildup of differential potentials between surfaces that is dangerous, as it can lead to arc discharges between the surfaces, which can seriously damage electronic systems and cause material damage to solar arrays and optical surfaces. However,

of particular concern is IESD, as electron charge can accumulate on isolated surfaces or in isolated dielectrics (cabling or circuit boards) inside a spacecraft's protective Faraday cage and can arc directly to sensitive devices. Indeed, spacecraft charging is currently rated as the most serious space-weather-related source of damage to spacecraft (Koons et al., 2000). Surges in Earth's trapped electron belts or even passage through Jupiter's intense electron environment are well known to cause IESD and thus induce dangerous electrical upsets in electrical systems. This chapter reviews the basic characteristics of all these forms of spacecraft charging and discusses their effects, addresses the role of the space weather in affecting charging, and briefly indicates the methods for limiting the effects of charging on spacecraft systems. However, it is important to remember that it is ultimately the extremes of the space weather and the plasma and the energetic particles it energizes that are the sources of spacecraft charging effects.

The growing sophistication of spacecraft has led to increasing concerns over all aspects of interactions between spacecraft charging and space weather. Over the last three decades, numerous conferences have been held on the subject of spacecraft charging. These culminated with the 13th International Spacecraft Charging Technology Conference held in Pasadena, California, in 2014 (Garrett and Whittlesey, 2015). In addition, numerous books

(Garrett and Pike, 1980; Garrett and Whittlesey, 2011; Lai, 2011) have been written on the generic issues associated with plasma interactions or on specific areas such as surface charging. Still, as many researchers have demonstrated (Duck, 1994; Koons et al., 1988; Leung et al., 1981; Mullen and Gussenhoven, 1983, 1989; Mullen et al., 1986), charging (or, more correctly, differential charging followed by discharging) effects are a major source of spacecraft anomalies. Whether it is surface charging, internal charging, plasma interactions at low altitudes, or induced fields on tethers, the buildup of charge on or in spacecraft poses a continuing design problem for spacecraft builders.

Spacecraft charging is not a new issue for spacecraft designers (Garrett, 1981; Garrett and Whittlesey, 2000) and has been an area of concern to spacecraft users and operators since the early days of the Space Age. Initially, spacecraft charging was primarily concerned with the theory of simple probe charging and with rocket measurements of charging in the ionosphere. By the time Sputnik was launched in 1957, the formal foundations of charging theory (at least for the ionosphere) and the first tentative measurements by rockets had been completed. The early 1960s were characterized by the accurate measurement of charging on spacecraft and rockets. Self-consistent charging models were developed and factors such as secondary emission and photoelectron currents were included in these models. The publishing of E. C. Whipple's seminal thesis in 1965 (Whipple, 1965) on spacecraft surface charging marked the beginning of the modern phase of spacecraft charging studies. This thesis and reviews by Brundin (Brundin, 1963), Bourdeau (Bourdeau, 1963), and others established the basic components of charging theory and the range of observations as we know them today. The period from 1965 to 1980 was characterized by increasingly more sophisticated models of spacecraft surface charging, in situ measurements, and better definition of the space plasma environment. Giving impetus to the study of spacecraft charging, the first in situ observations of kilovolt potentials at geosynchronous orbit were reported by DeForest and McIlwain in 1971 (DeForest and McIlwain, 1971). This period ended with the flight of the Spacecraft Charging at High Altitudes (SCATHA) P78-2 spacecraft, which was designed to specifically study all aspects of in situ charging and map the environments that caused it. Reviews by Garrett (Garrett, 1981) and Whipple (Whipple, 1981) summarized these early theoretical and observational spacecraft charging results.

The engineering implications of those findings were addressed in the first handbooks on surface charging, *the NASA Spacecraft Charging Design Guidelines* (Purvis et al., 1984) and *MIL-STD-1541A* (Anon., 1987).

The progress since 1980 has in general been more in emphasis as there has been a major shift in attitude vis-a-vis the role of surface charging versus that of IESD caused by penetrating electrons. The former continues to be an important process. However, as surface charging and the elimination of differential potentials are now routinely addressed in spacecraft design, a growing proportion of spacecraft anomalies is now believed to be caused by "internal" charging (defined as charging that causes discharges inside the Faraday cage and near the internal electronics as opposed to discharges on the external visible surface of the spacecraft). To address this issue, the new NASA handbook *NASA-HDBK-4002A* (Whittlesey and Garrett, 2011) was recently issued, which combined the older surface and IESD design guidelines (Purvis et al., 1984; Whittlesey, 1999). With the importance of the ISS to the national space program, charging effects such as plasma wakes, which are unique to the low Earth orbit (LEO), have become of increasing concern. Finally, the continuing desire to use high voltages in space (especially for solar arrays) and to utilize electrodynamic tethers has led to increased research in these areas.

17.2 SURFACE CHARGING

Surface charging refers to charging effects and electrostatic discharge (ESD) effects on the outside of the spacecraft (generally the visible surface materials). Surface charging is defined by the current balance equation:

$$I_T(V) = -I_E(V) + \left(\begin{array}{c} I_I(V) + I_{SE}(V) + I_{SI}(V) \\ + I_{BSE}(V) + I_{PH}(V) \end{array} \right) \quad (17.1)$$

where:

V is the surface potential relative to space

I_T is the total current to the spacecraft surface at V = 0 at equilibrium when all the current sources balance

I_E is the incident negative electron current

I_I is the incident positive ion current

I_{SE} is the secondary emitted electron current due to I_E

I_{SI} is the secondary emitted electron current due to I_I

I_{BSE} is the backscattered electron current due to I_E

I_{PH} is the photoelectron current

The solution of Equation 17.1 can be quite complicated (Garrett, 1981; Whipple, 1965, 1981). Subject to various constraints (e.g., Poisson's equation and the time-independent collisionless Boltzmann or Vlasov equation), it is the fundamental relationship for determining surface potentials. Briefly, each of the current terms on the right-hand side of Equation 17.1 is determined as a function of potential to give so-called I-V curves. The equation is then solved (subject to the aforementioned constraints), so that $I_T(V) = 0$. A common procedure for geosynchronous orbit is to approximate the ambient environment in terms of Maxwellian or two Maxwellian plasma distributions. Then, dependent on the geometry, I-V curves for the electrons and ions can be readily estimated by simple analytic expressions. As material secondary emission properties have been shown to have a strong influence on surface charging (Garrett, 1981; Katz et al., 1986), the secondary, backscatter, and photoelectron current terms typically have to be included if quantitative estimates of the spacecraft potential are required. (Other current terms such as for artificial plasma beams may also be included in Equation 17.1, but these terms have not been discussed in this chapter.) One complication is the so-called "triple root." First recognized by Whipple (Whipple, 1965) and expanded on by subsequent authors, Equation 17.1 can have multiple roots, and in principle, the solution can jump between these "triple roots" (Laframboise, 1982; Prokopenko and Laframboise, 1980; Rothwell et al., 1976), causing sudden high-voltage jumps in the surface potential and arcing. In any event, Equation 17.1 is solved to give the spacecraft potential under a variety of conditions.

An example of a first-order solution of Equation 17.1 for the Earth's magnetosphere is presented in Figure 17.1 (Whittlesey and Garrett, 2011). Figure 17.1 is an approximation of the expected range of the charging threat in terms of surface potential as a function of altitude and latitude in the absence of photoemission and other secondary currents. Although representing an artificial, worst-case estimate of charging for a conducting satellite at that altitude and latitude, this figure provides a simple tool for mission planning; if a spacecraft's orbit passes through one of the high potential regions, a project should either take steps to mitigate surface charging or do an analysis to assess the risk from differential charging of the spacecraft.

As Figure 17.1 implies, the primary region of surface charging is in and near geosynchronous orbit and along

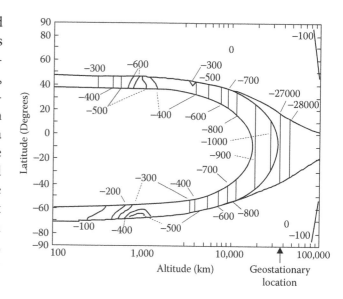

FIGURE 17.1 Surface charging potential contours (in the absence of sunlight) as a function of altitude and latitude. (Data from Evans, R. et al., Paper presented at the *Spacecraft Charging Technology Conference*, Naval Postgraduate School, Monterey, CA, 1989; Whittlesey, A. C., and H. B. Garrett, *Mitigating In-Space Charging Effects—A Guideline*, National Aeronautics and Space Administration, Washington, DC, 2011.)

the field lines extending down into the auroral zones, as it is characterized by hot (1–50 keV) electron and proton plasmas. The near-geosynchronous region was initially mapped by the ATS-5, ATS-6, and SCATHA satellites, and the characteristics of the environment were presented in a series of descriptive "atlases" (Garrett et al., 1978, 1979, 1981a, 1981b; Mullen et al., 1981) and worst case studies (Gussenhoven, 1983, 1985; Gussenhoven and Mullen, 1982; Purvis et al., 1984). Extensive statistical studies of the relationship between the characteristics of the geosynchronous plasma and the space weather environment have also been published (Baker, 2000; Choi et al., 2011; Fennel et al., 1983; Thomsen et al., 2007, 2013).

Of additional interest is the portion of the charging environment less than 1000 km in the polar regions. Although not as dramatic as geosynchronous charging, "low altitude" surface charging in this region is more common than originally thought (see review by Hastings [1995]), with potentials of ~1 kV having been observed (Anderson, 2012; Kawakita et al., 2004; Parker and Minow, 2013).

The spacecraft surface-charging environment has been mapped out for Jupiter and Saturn (Divine and Garrett, 1983; Garrett and Hoffman, 2000). The Voyager spacecraft may have observed large surface charging throughout the solar system — possibly tens of kV at Jupiter (Khurana et al., 1987) and −400 V at Uranus (Bridge et al., 1986). Many interplanetary spacecraft (e.g., Juno [Garrett et al., 2008]) are now being designed to minimize surface charging as a matter of course. These mitigation techniques are based on design guidelines and standards defined in NASA handbook *NASA-HDBK-4002A* (Whittlesey and Garrett, 2011) (see also [Purvis et al., 1984; Whittlesey, 1999]) and *MIL-STD-1541A* (Anon., 1987). The methods for controlling and mitigating surface charging were the direct outgrowth of the SCATHA mission (Fennell, 1982; Koons et al., 1981; Li and Whipple, 1988). Actual flight experience has repeatedly demonstrated the value of these methods. Indeed, they have consistently proven to be successful in limiting the effects of surface charging.

Although it is challenging to predict geomagnetic "weather" in terms of substorms with anything more than a half to one-hour lead time (Arnoldy, 1971; Feynman and Gabriel, 2000; Foster et al., 1971; Tsurutani and Gonzalez, 1987), it has been proven possible to estimate absolute surface charging levels at a given satellite location with some accuracy from the space weather indices (Garrett et al., 1979; Lam and Hruska, 1991) or, better still, in situ measurements of the plasma. In Garrett et al. (Garrett et al., 1980), data from plasma sensors on one geosynchronous spacecraft were successfully used to estimate charging levels at another spacecraft. These measurements, obtainable in near-real time, can be used to estimate charging levels at other spacecraft within several hours of local time around the observing spacecraft. This capability is demonstrated in Figure 17.2 (Garrett et al., 1980), where data from the US Air Force Defense Support Program (DSP) satellite were used to estimate the charging environment on the nearby ATS-6. The results also demonstrated that surface charging is primarily a function of the electron current at energies of a few 10s of keV or higher and that plasma injections at geosynchronous orbit associated with these current enhancements are related to space weather.

In contrast to predicting the overall spacecraft potential, estimating differential charging is a different matter altogether and requires intimate knowledge of spacecraft design. Sophisticated codes such as the NASCAP family of charging models (Mandell et al., 1984), the Spacecraft Plasma Interaction System (SPIS) (Roussel et al., 2008, 2012), and the Multi-Utility Spacecraft

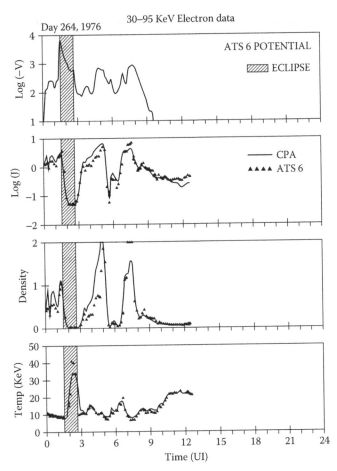

FIGURE 17.2 Measurements of the 30- to 80-keV electron channels from the Defense Support Program charged particle analyzer (CPA) instrument compared with the plasma and charging environment at ATS-6. (Reproduced from Garrett, H. B. et al., *J. Geophys. Res., 85*, 1155–1162, 1980. © American Geophysical Union.)

Charging Analysis Tool (MUSCAT) (Hosoda et al., 2007) are required to provide accurate estimates of the time evolution of the differential potentials. As a specific example of these codes, Nascap-2k (Davis et al., 2003; Mandell et al., 2006) is a widely used interactive toolkit to study plasma interactions with realistic spacecraft in three dimensions. It can model interactions that occur in tenuous (e.g., geosynchronous equatorial orbit or interplanetary missions) and in dense (e.g., LEO and the aurora) plasma environments. Capabilities include surface charging in geosynchronous and interplanetary orbits, sheath and wake structure and current collection in LEO, and auroral charging. External potential structure and particle trajectories are computed using a finite element method on a nested grid structure. The space charge surrounding a spacecraft can be treated either analytically, self-consistently with particle trajectories, or consistent with imported plume densities. Particle-in-cell capabilities are available to study dynamic plasma effects.

The Nascap-2k, MUSCAT, and SPIS codes require detailed descriptions of the material properties of the surfaces that are included in the surface charging computations. By locating large surface voltage gradients in a particular design, these codes can show where discharges may occur. The effect of changes in the surface materials or coatings in those areas on minimizing voltage gradients can then be evaluated.

Despite these advances, surface charging at geosynchronous orbit can pose a threat to spacecraft survivability (Koons et al., 1999; Lanzerotti et al., 1998; Leach and Alexander, 1995; Wilkinson, 1994). Indeed, evidence for the complexities associated with the surface charging and/or arcing process has emerged in the form of catastrophic, continuous arcs between adjacent solar cells on two high-powered spacecraft operating in geosynchronous orbit (Hoeber et al., 1998). Ground experiments and theories have shown that the most probable location for electrical discharges to occur on the surfaces of high-voltage solar arrays is at the so-called triple junction: the interface between a metallic interconnect, cover glass, and plasma (Cho and Hastings, 1991; Ferguson, 1989; Snyder et al., 1998). Although this type of arc is not believed to be able to cause substantial damage to a solar array, it has been hypothesized and demonstrated in the laboratory that such an arc can generate sufficient local heating to initiate outgassing and polymer pyrolysis (Katz et al., 1999; Snyder et al., 1998). This in turn can generate enough gas and plasma between biased solar cells to trigger long-duration arcs that can be maintained by the solar array and cause serious damage to an array. Fortunately, mitigation techniques (Frederickson, 1983; Katz et al., 1999; Snyder et al., 1998) (e.g., limiting the potential between adjacent solar cells and insulating the region between likely breakdown sites) have proven in testing to be very effective in reducing this problem. As in the case of surface charging, standards and handbooks that summarize these techniques and provide standard methods for hardening solar arrays and protecting other high-voltage systems in LEO have been developed by Ferguson and colleagues (Anon., 2007a, 2007b).

17.3 INTERNAL CHARGING

Internal charging, as used in this chapter, refers to the accumulation of electrical charge on interior, ungrounded metals or on or in the dielectrics inside a spacecraft. The key difference between "internal" and external or surface charging is that the surface ESDs are often loosely coupled to the victim circuits, whereas internal discharges may occur directly adjacent to the victim circuits. If a Faraday cage construction is employed, ESD events outside the Faraday cage — even if under thermal blankets — can be called "external" in this context. Figure 17.3 shows electron and proton ranges in aluminum versus energy. Since most satellites have an outer shell with aluminum equivalent thickness of at least 0.76 mm (or 30 mils) or more, internally deposited electrons require energies greater than 100–500 keV. Thus, electrons with energy of 100–500 keV or more are considered to be the primary space-weather-driven environment responsible for IESD (protons must have energies of 30 MeV or more to penetrate to similar depths). Although the fluxes are lower at these higher energies, any IESD spark they may cause is closer to the victim electronics than to the external ESDs and can therefore cause substantial upset or damage to satellite electronics.

During the passage of Voyager 1 by Jupiter on September 5, 1977 (Leung et al., 1986; Whittlesey and Leung, 1987), 42 identical electrical anomalies were observed. These anomalies were subsequently attributed to IESD. In particular, it was postulated that ~1 MeV electrons had penetrated the surface of a cable and built up a charge in it that was sufficient to cause arcing. The analysis of the data of SCATHA, the Combined Release and Radiation Effects Satellite (CRRES) (Violet and Frederickson, 1993), and DSP (Vampola, 1987) showed similar effects. Laboratory studies by Leung (Leung, 1983), Frederickson (Frederickson, 1980, 1983; Frederickson et al., 1992), and others demonstrated that internal charging (also called buried or deep dielectric charging) was a potential source of discharges. As a result, a series of IESD experiments were conducted on the CRRES spacecraft in 1990–1991 (Frederickson et al., 1992; Violet and Frederickson, 1993). These experiments, which exposed a variety of configurations of isolated conducting surfaces and dielectrics to the Earth's radiation environment, clearly demonstrated the reality of this effect. More than 4000 pulses were detected

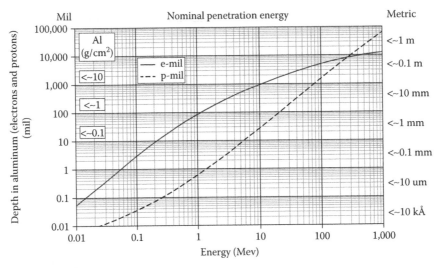

Notes:
1. Protons stop close to the mean depth shown, whereas electrons are deposited in a larger range around the given depth.
2. Surface charging: ~0–50 keV electrons.
3. Transition between surface and internal charging: ~50–100 keV.
4. Internal charging ~greater than 100 keV.
5. For GEO orbits, the practical range of interest for internal charging is 0.1–3 MeV (~3–110 mil of aluminum thickness).
6. Data for chart from ESTAR and PSTAR, at http://physics.nist.gov./Star.

FIGURE 17.3   Electron and ion penetration ranges in aluminum. (Data from Whittlesey, A. C., and H. B. Garrett, *Mitigating In-Space Charging Effects—A Guideline*, National Aeronautics and Space Administration, Washington, DC, 2011.)

during the 13-month lifetime of the CRRES spacecraft. As in the case of SCATHA for surface charging, CRRES marked a watershed in the study of internal charging. Since that time, the presence of internal charging continues to be investigated and reported (Fennell et al., 1999; Frederickson, 1996, 1999; Koons and Chen, 1999; Lai, 1998; Sorensen, 1996; Stassinopoulos and Brucker, 1996; Vampola, 1996a, 1996b; Wilkinson, 1994; Wrenn, 1995; Wrenn and Sims, 1996; Wrenn and Smith, 1996).

The IESD hazard to Earth-orbiting spacecraft arises primarily from the high-energy electron environments between ~1.5 Re and 4 Re (the inner belt) and near-geosynchronous orbit (the outer belt) in the equatorial plane. These regions are illustrated in Figure 17.4 (as a companion to Figure 17.1). Figure 17.4 can be used as a screening tool for internal charging problems for circular orbits. The orbits in the inner belt carry the greatest risk because of the enhanced high-energy electron fluxes, but geosynchronous equatorial orbit has exhibited more problems because this is a more populated region and contains high-value resources.

Charged insulators that are the greatest threat to internal electronics are those that are closest to the

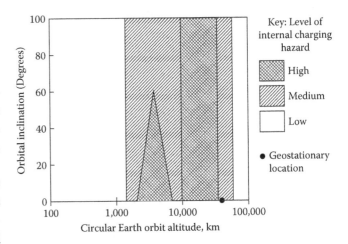

FIGURE 17.4   Internal charging threat for circular Earth orbits. (Data from Whittlesey, A. C., and H. B. Garrett, *Mitigating In-Space Charging Effects—A Guideline*, National Aeronautics and Space Administration, Washington, DC, 2011.)

victim electronics. Circuit board dielectrics with accumulated charges can be immediately adjacent to a victim integrated circuit and can couple nearly all their stored energy into the circuit if an ESD is unfavorably directed. *NASA-HDBK-4002A* (Whittlesey and Garrett, 2011) recommends no more than 3 cm² of open-circuit board area. The circuit board threat model assumes a typical FR4 board material about 80 mils thick, with no traces or metallization in that region. If traces or ground planes are located in that region, even if not visible on the surface of the board, the leakage paths increase and the threat level diminishes.

Cable insulation is another dielectric that poses an internal charging threat (Frederickson, 1996). The threat starts with charging in the dielectric (so that thinner-wire insulation is better) and enhances by small plasma plumes drifting onto the victim areas (such as exposed high-voltage terminals). Therefore, keeping the wire bundle tightly bound is better. Detailed guidelines are provided in *NASA-HDBK-4002A*.

Radiation-induced conductivity causes temporary or permanently enhanced conductivity due to ionizing radiation. Although it tends to mitigate internal charging, threat assessments and spacecraft design requirements are best done without assuming the benefits of this radiation-induced conductivity. In contrast, an effect that worsens IESD is that some dielectrics in space may become more resistive with time. Arcing or pulsing can begin some time after launch rather than beginning immediately on arriving on the station. This can be attributed in part to radiation damage to the material or to drying in the vacuum, which creates a greater surface and/or bulk resistivity in that dielectric.

The problem of internal charging will get worse in the future rather than getting better. Although the environment is not changing, economics are forcing less shielding, lighter-weight structural materials, and even the elimination of Faraday cage construction. In particular, Cubesats are becoming more common, and the desire to fly them in the inner radiation belt at medium Earth orbit, where IESD is most severe, is increasing. At the same time, integrated circuits are becoming smaller and of lower power and thus are more susceptible to damage by ESD. Designers will have greater challenges in the future, as design rules that might have worked in the past will need re-evaluation as the parts and structural design change.

Except for bulk conducting materials, electron charge will be deposited over a finite depth. Indeed, any electron with an energy more than a few electron volts will penetrate the surface. The depth of penetration and charge deposition is a function of stopping power, the energy of the impinging particles, and any electric fields normal to the surface (see Figure 17.3 for the penetration depth of energetic electrons and protons in aluminum). A common spacecraft surface configuration that will exhibit this behavior consists of an exposed dielectric material with a conductive backing connected to the spacecraft ground. Electron charge will accumulate (or diffuse away) in the dielectric volume over time as a function of the conductivity of the material and the imposed electric fields. If the charge accumulating in the dielectric induces a field greater than the breakdown strength of the material (typically of the order of 10^5–10^6 V/cm), a discharge can occur within the material or from the interior of the dielectric to one of its surfaces. This has been estimated to occur when the accumulated electron fluence exceeds a critical threshold of the order of 2×10^{10} electrons/cm². The latter number (10^{10}) has become a canonical rule of thumb; that is, if a surface or volume accumulates more than 10^{10} electrons/cm² in 10 h or less in Earth's orbit, IESD may be a problem (Whittlesey and Garrett, 2011). The 10 h interval arises from inflight data in the Earth's environment and represents the time in which charge can bleed off from a common dielectric. However, Bodeau (Bodeau, 2005, 2010) has suggested that the critical fluence level can be 10 times lower, which means that IESD may be more common than originally thought.

The computation of internal charging resembles surface charging calculations with the inclusion of space charge in the material volume. The basic problem is the calculation of the electric field and charge density in a self-consistent fashion over the three-dimensional space of interest. The primary difference between the two is the role that the conductivity of the material plays in the process. Poisson's equation must be solved, subject to the continuity equation in the dielectric. As a very simple example, consider a one-dimensional, planar approximation at a depth of X in the dielectric. The equation at X is then:

$$\varepsilon\left(\frac{dE}{d\tau}\right) + \sigma E = J \tag{17.2}$$

where:

E is the electric field at X

τ is time

σ is the conductivity in (ohm-m)$^{-1}$ ($= \sigma_o + \sigma_r$). Here, σ_o is the dark conductivity and σ_r is the radiation-induced conductivity (Frederickson et al., 1986)

ε is the dielectric constant

J is the incident particle flux (current density) at X, including the primary and secondary particles

A solution of this equation for σ and *J*, independent of time, is:

$$E = E_o \exp(-\sigma\tau/\varepsilon) + (J/\sigma)(1 - \exp(-\sigma\tau/\varepsilon)) \quad (17.3)$$

where E_o is the imposed electric field at $\tau = 0$.

Although these equations are only a crude approximation, they demonstrate the basic features of the radiation-induced charging. In particular, these equations demonstrate the importance of the charging time constant $\tau(= \varepsilon/\sigma)$. For many materials, τ ranges from 10 to 10^3 s. Some common dielectric materials used in satellites have even longer time constants of 3×10^5 s; low temperatures can further increase the time constants. In regions where the dose rate is high (enhancing the radiation conductivity), the *E* field comes to equilibrium rapidly. In lightly irradiated regions, where the time constant is long (the dark conductivity dominates), the field takes a long time to reach the equilibrium. Charging times for energies between 100–500 KeV and for flux levels typical of geosynchronous orbit are about 3–10 h. At lower charging rates, the material conductivity can leak off the charge, so that internal charging would not be a problem. In Figure 17.5, measurements of $E > 1.2$ MeV electrons at geosynchronous orbit by the GOES-2 satellite between July 1980 and May 1982 and star-sensor anomalies (assumed to be arc-related) on the DSP satellite are seen to be well-correlated, confirming this proposition (e.g., see Vampola, 1987).

Up to the publication of *NASA-HDBK-4002A* and the ISO standard 11221:2011 (Space systems—Space solar panels—Spacecraft charging induced electrostatic

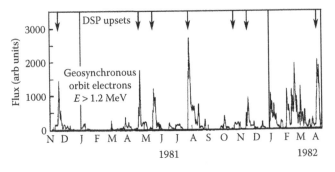

FIGURE 17.5 GOES-2 $E > 1.2$ MeV electron flux at geosynchronous orbit between November 1980 and May 1982 compared with star-sensor anomalies (indicated by vertical arrows) on the Defense Support Program satellite. (Data from Vampola, A. L., *J. Electrostatics, 20*, 21–30, 1987.)

discharge test methods [Anon., 2011]) for solar arrays in 2011, a consensus in the spacecraft engineering community as to what design features (and to what degree) are necessary to limit the charging effects was lacking. This lack of consensus resulted in several spacecraft suffering upsets that might have been avoided if proper guidelines were in place. As an example, on January 20 and 21, 1994, the Anik E1 and E2 spacecraft suffered serious upsets within hours of each other during severe space weather, which resulted in a brief loss of the one and a six-month outage of the other spacecraft (Lam et al., 2012). Subsequent analysis implicated internal charging as the cause (Baker et al., 1994, 1998).

Computer codes for modeling IESD are described in *NASA-HDBK-4002A* (Whittlesey and Garrett, 2011). This handbook also contains an overview of the overall subject of internal charging and provides a number of rules and techniques to mitigate both IESD and surface charging. It defines the primary Earth orbits and conditions where internal charging may be of concern, and provides a tutorial on the internal charging process and a checklist for designers to use in preventing internal charging (e.g., a basic set of design rules). The readers interested in additional methods of limiting IESD and testing for it may refer to *NASA-HDBK-4002A*.

17.4 LOW-ALTITUDE CHARGING

Spacecraft orbiting at low altitudes must also be concerned with charging. Owing to the complex effects of the structure, size, and shape on the magnetohydrodynamic flow fields in the LEO high-density plasma, hypersonic plasma interactions at low altitudes have always presented an analytic challenge. The desire to operate at increasingly higher solar array voltages has greatly added to the computational difficulties associated with this problem. Fortunately, with the continuing growth in computing speed, a number of spacecraft charging problems at low altitudes are, for the first time, yielding to numerical analysis. Intricate geometries, magnetic fields, changing composition, and high, imposed potentials can now all be effectively modeled—see review by Hastings (1995).

The low-altitude charging problem is best represented by the movement of a body through dense, cool ionospheric plasma. For a typical spacecraft, its characteristic dimensions are, in contrast to geosynchronous orbit, quite large compared with the plasma Debye length (typically of the order of ~10 cm). This factor makes current flow computations for complex geometries and field configurations difficult. However, the basic variations are related to first order to the neutral gas flow around a body. In Figure 17.6, predictions for a simple cylindrical geometry (Gurevich et al., 1970), as a function of altitude (and hence the composition and the space weather effects), are compared with the actual data from measurements on a small spacecraft (Samir and Wrenn, 1969). This figure demonstrates how current flow varies dramatically with angle relative to the spacecraft velocity vector and how the depth of the wake varies with altitude as the composition varies from mainly O^+ (~23% H^+) dominated at lower altitudes to ~94% H^+ dominated at higher altitudes, in going from ~600 to ~1200 km. Similar measurements have also been made for the Space Shuttle (Murphy et al., 1989; Samir et al., 1983, 1986; Stone et al., 1988) and the ISS.

When variations in the background magnetic field and ionosphere due to space weather, artificially imposed potentials, and complex geometries are included, the problem departs dramatically from simple neutral gas flow and can seldom be addressed analytically. As a sample of the difficulties that a typical problem can introduce, consider a large biased plate in LEO. Figure 17.7 is a plot of the plasma flow field for a large flat plate (representing a large solar array panel) at low altitudes. This

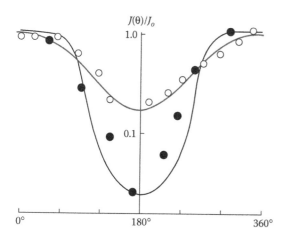

FIGURE 17.6 Normalized electron current ($j(\theta)/j_o$) versus angular position of the plasma probe on Explorer 31. 180° corresponds to the center of the wake (i.e., opposite the direction of movement). The closed circles correspond to a satellite altitude between 620 km and 720 km and a proton density of 23% of the charge density. The open circles correspond to a satellite altitude between 1175 km and 1275 km, where the proton density is 94% of the charge density. (Data from Samir, U., and G. L. Wrenn, *Planet Space Sci.*, *17*, 693, 1969, for other examples.)

figure (Wang et al., 1994) illustrates several possible variations. The first frame (Figure 17.7a) is for an unbiased plate in low-altitude ionospheric plasma at ~Mach 8, as would be typical of the ISS. Next, a small, isolated body is inserted in the flow field behind the plate and allowed to float to an equilibrium potential (e.g., an astronaut on extravehicular activity). This alters the flow field slightly (Figure 17.7b). Next, an externally imposed current source, such as an aurora, is applied. The main plate does not alter its potential substantially, but the smaller body begins to charge because it is shielded from the ionospheric plasma. In doing so, it substantially alters the wake flow (Figure 17.7c). In Figure 17.7d, a potential difference is applied between the plate and the small body (this might correspond to a crew module biased relative to the main arrays for the ISS). As the potential difference is increased, the flow field becomes even more altered.

Experimental work related to these wake phenomena (i.e., current collection by high voltages in a wake) have been carried out in the laboratory (Enloe et al., 1993) and in situ by the Shuttle Charging Hazards and Wake Studies (Enloe et al., 1995, 1997). The Shuttle Charging Hazards and Wake Studies consisted of plasma monitors and a bistable probe mounted on the Shuttle Wake Shield Facility (WSF). The experiment measured the plasma current in the wake of the WSF as a function of the negative potential of the probe (up to −5000 V relative to the WSF). The experiment was modeled using the programs NASCAP Potentials of Large Objects in the Auroral Region (Cooke et al., 1985; Lilley et al., 1989) and Dynamic Plasma Analysis Code (Mandell and Davis, 1990). The flight data and simulations indicated that the collected current had a power-law dependence on the potential but a less-than-linear dependence on the plasma density. However, the measurements at low voltages differed from the models, as the latter predicted a threshold for current collection at −100 V, which was not observed in the data (Davis et al., 1999).

Flow calculations introduce the rich variety of low-altitude plasma interactions being studied. For example, consider electrodynamic tethers (Parks and Katz, 1987; Hastings and Samanta Roy, 1993; Samanta Roy and Hastings, 1993; Samanta Roy et al., 1992). Multikilometer-long, thin conducting cables are now possible and have been demonstrated (e.g., the TSS-1 Shuttle experiment [Aguero et al., 2000; Dobrownoly and Melchioni, 1993; Oberhardt et al., 1993]). These electrodynamic tethers can be used to generate electricity.

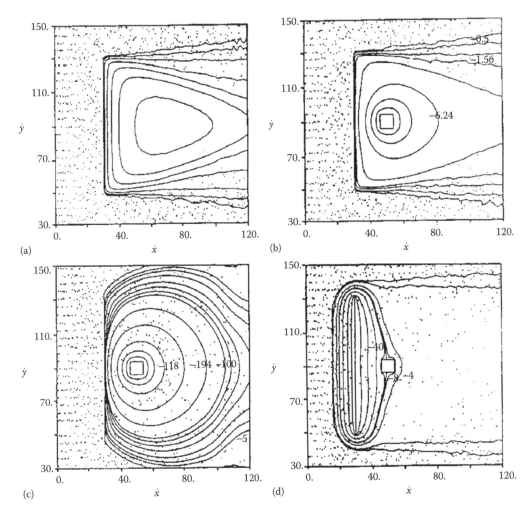

FIGURE 17.7 Low-altitude potential and ion flow contours for four different conditions (see text for an explanation of each frame). (Data from Wang, J. et al., *J. Spacecr. Rockets, 31,* 889–894, 1994.)

The basic principle is well-known and is contained in the Lorentz relationship:

$$V = \vec{v}_s \times \vec{B} \bullet \vec{L} \tag{17.4}$$

where:

  B is the magnetic field (vector)
  L is the tether length (vector)
  v_s is spacecraft velocity relative to the plasma
  V is potential across the tether

For a conducting object in LEO, the *vxB* electric field varies from a low of about 0.1 V/m at the equator to a maximum of 0.3 V/m over the polar caps. As Equation 17.4 states, the potential depends on the orientation of the tether relative to the *vxB* electric field vector. For a 10-km tether (easily possible with present technology),

a potential difference of up to 3000 V is possible—a spacecraft can, in principle, draw power from this voltage drop but at the price of a loss in the orbital altitude (Nobles, 1986; Vignoli et al., 1986). By biasing a tether, it can be used for spacecraft propulsion and orbit reboost (Estes et al., 2000; Gilchrist and Bilen, 2000; Johnson et al., 1998; Vas, 2000), as in the case of the plasma motor generator (Lilley et al., 1994) or the proposed propulsive small expendable deployer system experiment (Johnson et al., 1998). Tether use at Jupiter both as a power source and for propulsion (at Jupiter, the tether gains altitude as it draws power when dragged up to the co-rotation velocity of the ambient plasma torus) has also been studied by a number of authors (Sanmartin et al., 2008). Indeed, the high magnetic field over the jovian polar caps will induce potentials of up to 600 V across the Juno's rotating solar array at the closest approach (Garrett et al., 2008).

Observations of surface charging on low Earth orbit and polar orbit spacecraft have mainly come from the Defense Meteorological Satellite Program (DMSP) satellites. Papers (Besse et al., 1985; Gussenhoven, 1985) have reported potentials ranging from a few hundreds of volts to more than a kilovolt. Far from being a very rare event, moderate charging events (i.e., more than the ~100 V differential potentials normally believed to be the minimum necessary to cause surface arcing) are not uncommon for polar orbiting spacecraft in conjunction with polar aurora. However, discharge-induced anomalies are believed to be rare. Anderson and Koons in 1996 reported observing an operational anomaly on the DMSP F13 satellite (Anderson and Koons, 1996). On May 5, 1995, the microprocessor of the microwave imager experiment experienced a lock-up. At the time of the event, the spacecraft frame potential was estimated at −460 V and the surface potentials as high as −3 kV may have occurred in a ~6 s period. Cooke (1998) used the potentials of large objects in the auroral region code to simulate the charging of the DMSP satellite at the time of the event. His results indicated that the highest potential was only achieved by a few surfaces that had ion collection limited by their locations, perhaps explaining, in part, the rarity of such events (though surface material choices may be a more likely cause). The Advanced Earth-Observing Satellite II (Kawakita et al., 2004) is also believed to have failed after passage through an intense aurora. Both ground testing and modeling have substantiated this as a space-weather-induced event; apparently, an arc discharge, after passage through an intense aurora, coupled into the power system along a damaged power cable linking the solar array to the spacecraft. Given adequate measurements of ionospheric and geomagnetic activity, it should be possible to determine the probability of such events in real time (as in the case of geosynchronous charging) and take preventive measures.

Another area of low altitude charging interest is that associated with induced potentials due to biased surfaces, such as solar arrays. In addition to arcing, biased solar arrays at low altitude have been observed to drive plasma effects (e.g., broadband fluctuations extending beyond 1 MHz) (Sasaki, 1999). In a series of rocket and satellite experiments, the Department of Defense and NASA have completed several interesting studies on the effects of induced high potentials on solar arrays and of plasma beams on spacecraft potentials. Intended primarily to parameterize the ranges over which exposed

high-potential surfaces can be biased before arcing sets in and to demonstrate the control of the discharge process, two series of experiments stand out. The first are the solar array experiments associated with the advanced photovoltaic experiment (APEX) on the photovoltaic array space power (PASP) plus satellite (Ferguson et al., 1998; Guidice and Ray, 1996; Guidice et al., 1995). Launched into a 363 × 2550 km elliptical orbit on August 3, 1994, by a Pegasus rocket, this experiment consisted of a collection of several types of solar array cells. Ranging from solar concentrators to representative samples of the ISS arrays, the cells were biased over a range of voltages (+500 V) and their current collection and arcing characteristics were measured. In particular, the electron current collected by the so-called snap-over phenomena for positively biased solar arrays (Leung, 1983; Stevens, 1980) was studied. Likewise, arcing at large negative potentials was also monitored (Guidice and Ray, 1996; Guidice et al., 1995). Preflight and postflight simulations showed good agreement with the observations. In particular, earlier ground experiments had found that the arc rate increased with voltage and plasma density and decreased with temperature (Hastings et al., 1992a; Soldi et al., 1996). Other experiments (Ferguson, 1986; Grier, 1985; Miller, 1985) also found an increase in the arc rates with increasing voltage and plasma density. The PASP Plus results (Soldi et al., 1996) demonstrated that the arcing levels were indeed strongly dependent on the bias voltage. Cell temperature was also verified to be critical with higher arcing rates at low temperatures — there being a critical temperature above which no arcing occurred. Finally, although radiation flux was found to have no affect on the arcing, the ion flux was seen to have an effect, as expected (Hastings et al., 1992b; Soldi et al., 1996). The environmental and design issues associated with LEO charging and methods for mitigating them are summarized in a NASA design handbook (Anon., 2007a) and design standard (Anon., 2007b) prepared by Ferguson, and provide an excellent starting point for understanding them.

A final issue to be considered is the application of these results to the ISS. The major issues associated with the ISS are its huge size and high-voltage solar arrays. In addition to the former, various modeling efforts have attempted to address each of the high-voltage solar array features explicitly. For example, the floating potential and wake structure of the ISS and the likelihood of arcing have been extensively addressed by Hastings, Wang, and others (Hastings and Wang, 1989; Wang et al., 1994).

Plasma contactors (Hastings, 1987), in particular, play an important role in collecting current and controlling floating potentials on the ISS's solar arrays and in limiting sputtering and arcing (Hastings et al., 1992b). Other modeling programs specifically focusing on the ISS are summarized by Katz et al. (Katz et al., 1991). The practical aspects of the effects of these plasma interactions (i.e., the electromagnetic interference expected) are addressed by Murphy and Garrett (Murphy and Garrett, 1989). Recently, a NASA Engineering and Safety Center's Technical Assessment Report (Iannello, 2014) detailed the interactions of an astronaut with the ISS high-voltage solar arrays and the importance of the Plasma Contactor Unit in limiting the possibilities of arcing. However, these papers represent only a small portion of the plasma and charging studies, which will ultimately come from our utilization of the ISS in the years ahead.

17.5 CONCLUSION

To summarize, the study and analysis of spacecraft charging and its relationship to space weather have demonstrated a growing maturity. Surface charging continues to be recognized as a serious operational threat to spacecraft, but useful design guidelines are in place for its mitigation. These were made possible in large part by the success of the joint United States Air Force-NASA SCATHA program. The mitigation of IESD has become increasingly important, as its role as a major source of anomalies due to charging and/or arcing has become obvious. With the flight of CRRES and its IESD experiment, flight confirmation of this phenomenon now exists over the entire radiation belts. A formal update, *NASA-HDBK-4002A* (Whittlesey and Garrett, 2011), to the surface and internal charging design guidelines was recently completed. It details the testing and mitigation methods for limiting their effects. Just as modeling of low-altitude charging effects are yielding to detailed computer analysis and experiment, modern theory and evidence are converging on consistent models and techniques for controlling charging effects on solar arrays (in particular [Anon., 2007a] and [Anon., 2007b] by Ferguson for solar arrays). Successful conclusion of this process promises major advances in the utilization of high-voltage systems in the low-altitude space environment or for high-power solar propulsion systems. In particular, the use of tethers and high-voltage systems now appears possible if proper consideration is given to the details of the processes involved and the effects of the space weather. The ISS itself is proving to be a fertile laboratory setting for studying this and many other unusual plasma/charging interactions. Indeed, charging at Jupiter and other extraterrestrial locations are growing areas of interest. To conclude, the last few decades have seen substantial and meaningful progress in an important scientific and engineering area of research — spacecraft charging and its relation to space weather.

ACKNOWLEDGMENTS

This work was carried out by the Jet Propulsion Laboratory, California Institute of Technology, under a contract with the National Aeronautics and Space Administration. The handbook *NASA-HDBK-4002A* can be obtained online from: https://standards.nasa.gov/documents/detail/3314877 and a book version (Garrett and Whittlesey, 2011) can be downloaded from: http://descanso.jpl.nasa.gov/SciTechBook/series3/ChgingBook--110629-RibbonC.pdf.

REFERENCES

Aguero, V. M., B. E. Gilchrist, S. D. Williams, W. J. Burke, L. Krause, and L. C. Gentile (2000). Current Collection Model Characterizing Shuttle Charging During the Tethered Satellite System Missions. *J. Spacecr. Rockets*, 37(2), 212–217.

Anderson, P. C. (2012). Characteristics of Spacecraft Charging in Low Earth Orbit. *J. Geophys. Res.*, 117(A07308). doi:10.1029/2011JA016875.

Anderson, P. C., and H. C. Koons (1996). Spacecraft Charging Anomaly on a Low-Altitude Satellite in an Aurora. *J. Spacecr. Rockets*, 33(5), 734–738.

Anon. (1987). Electromagnetic Compatibility Requirements for Space Systems. *MIL-STD 1541A*. United States Air Force, US Government Printing Office.

Anon. (2007a). *Low Earth Orbit Spacecraft Charging Design Handbook*. NASA-HDBK-4006. NASA, Washington, DC.

Anon. (2007b). *Low Earth Orbit Spacecraft Charging Design Standard*. NASA-STD-4005. NASA, Washington, DC.

Anon. (2011). ISO 11221:2011, Space Systems—Space Solar Panels—Spacecraft Charging Induced Electrostatic Discharge Test Methods. International Organization for Standardization, Geneva, Switzerland.

Arnoldy, R. L. (1971). Signature in the Interplanetary Medium for Substorms. *J. Geophys. Res.*, 76(22), 5189–5201.

Baker, D. N. (2000). The Occurrence of Operational Anomalies in Spacecraft and Their Relationship to Space Weather. *IEEE Trans. Plasma Sci.*, 28, 2007–2016.

Baker, D. N., J. H. Allen, S. G. Kanekal, and G. D. Reeves (1998). Disturbed Space Environment May Have Been Related to Pager Satellite Failure. *EOS, Trans. Amer. Geophys. Union*, 79(40), 477–483.

Baker, D. N., S. G. Kanekal, J. B. Blake, B. Klecker, and G. Rostoker (1994). Satellite Anomalies Linked to Electron Increase in the Magnetosphere. *EOS, Trans. Amer. Geophys. Union, 75*, 401.

Besse, A. L., A. G. Rubin, and D. A. Hardy (1985). Charging of DMSP/F6 in Aurora on 10 January 1983. Paper presented at *Proceedings of the Spacecraft Environmental Interactions Technology Conference*-1983, U.S. Air Force Academy, Colorado Springs, CO.

Bodeau, M. (2005). Going Beyond Anomalies to Engineering Corrective Action, New IESD Guidelines Derived From A Root-Cause Investigation. In *2005 Space Environmental Effects Working Group*. Aerospace Corporation, El Segundo, CA.

Bodeau, M. (2010). High Energy Electron Climatology that Supports Deep Charging Risk Assessment in GEO, AIAA 2010-1608. Paper presented at *48th AIAA Aerospace Sciences Meeting*, Orlando, FL.

Bourdeau, R. E. (1963). On the Interaction between a Spacecraft and an Ionized Medium. *Space Sci. Rev., 1*, 719–728.

Bridge, H. S. et al. (1986). Plasma Observations Near Uranus: Initial Results from Voyager 2. *Science, 233*(July), 89–93.

Brundin, C. L. (1963). Effects of Charged Particles on the Motion of an Earth Satellite. *AIAA Journal, 1*, 2529–2538.

Cho, M., and D. E. Hastings (1991). Dielectric Charging Processes and Arcing Rates of High-Voltage Solar Arrays. *J. Spacecr. Rockets, 28*, 698.

Choi, H.-S., J. Lee, K.-S. Cho, Y.-S. Kwak, I.-H. Cho, Y.-D. Park, Y.-H. Kim, D. N. Baker, G. D. Reeves, and D.-K. Lee (2011). Analysis of GEO Spacecraft Anomalies: Space Weather Relationships. *Space Weather, S06001*(9). doi:10.1029/2010SW000597.

Cooke, D. L. (1998). Simulation of an Auroral Charging Anomaly on the DMSP Satellite. Paper presented at *36th Aerospace Sciences Meeting and Exhibit*, American Institute of Aeronautics and Astronautics, Reno, NV.

Cooke, D. L., I. Katz, M. J. Mandell, J. R. Lilley Jr., and A. J. Rubin (1985). Three Dimensional Calculation of Shuttle Orbiter Charging in Polar Orbit. Paper presented at *Spacecraft Environmental Interactions Technology Conference* 1983, U.S. Air Force Academy, Colorado Springs, CO.

Davis, V. A., M. J. Mandell, D. L. Cooke, and C. L. Enloe (1999). High-Voltage Interactions in Plasma Wakes: Simulation and Flight Measurements from the Charge Hazards and Wake Studies (CHAWS) Experiment. *J. Geophys. Res., 104*(A6), 12445–12459.

Davis, V. A., M. J. Mandell, B. M. Gardner, I. G. Mikellides, L. F. Neergaard, D. L. Cooke, and J. I. Minow (2003). Validation of Nascap-2k Spacecraft-Environment Interactions Calculations. In *8th Spacecraft Charging Technology Conference*. NASA, Huntsville, AL.

DeForest, S. E., and C. E. McIlwain (1971). Plasma Clouds in the Magnetosphere. *J. Geophys. Res., 76*, 3587.

Divine, T. N., and H. B. Garrett (1983). Charged Particle Distributions in Jupiter's Magnetosphere. *J. Geophys. Res., 88*(September 9), 6889–6903.

Dobrownoly, M., and E. Melchioni (1993). Electrodynamic Aspects of the First Tethered Satellite Mission. *J. Geophys. Res., 98*, 13761–13778.

Duck, M. J. (1994). Surface Charging and its Prevention. In *The Behavior of Systems in the Space Environment*, edited by R. N. DeWitt, D. P. Duston and A. K. Hyder, pp. 867–872. Kluwer Academic Publishers, Doredecht, The Netherlands.

Enloe, C. I., J. T. Bell, D. L. Cooke, D. A. Hardy, R. K. Kirkwood, J. W. R. Lloyd, and M. D. Violet (1995). The Charging Hazards and Wake Studies (CHAWS) Experiment. Paper presented at *33rd Aerospace Sciences Meeting and Exhibit*, January 9–12, American Institute of Aeronautics and Astronautics, Reno, NV.

Enloe, C. I., D. L. Cooke, S. Meassick, C. Chan, and M. F. Tautz (1993). Ion Collection in a Spacecraft Wake: Laboratory Simulations. *J. Geophys. Res., 98*, 13635–13644.

Enloe, C. I. et al. (1997). High-Voltage Interactions in Plasma Wakes: Results from the Charging Hazards and Wake Studies (CHAWS) Flight Experiments. *J. Geophys. Res., 102*(A1), 425–433.

Estes, R. D., J. Sanmartin, and M. Martinez-Sanchez (2000). Performance of Bare-Tether Systems Under Varying Magnetic and Plasma Conditions. *J. Spacecr. Rockets, 37*(2), 197–204.

Evans, R., H. B. Garrett, S. Gabriel, and A. C. Whittlesey (1989). A Preliminary Spacecraft Charging Map for the Near Earth Environment. Paper presented at the *Spacecraft Charging Technology Conference*, Naval Postgraduate School, Monterey, CA.

Fennell, J. F. (1982). Description of P78-2/SCATHA Satellite and Experiments. In *The IMS Source Book: Guide to the International Magnetospheric Study Data Analysis*, edited by C.T. Russell and David J. Southwood, pp. 65–81. American Geophysical Union, Washington, DC.

Fennell, J. F., H. C. Koons, and J. B. Blake (1999). A Deep Dielectric Charging Environmental Specification for Geosynchronous and HEO/Molniya Satellites. Paper presented at *Government Microcircuit Applications Conference*, Monterey, CA.

Fennel, J. F. et al. (1983). *A Review of SCATHA Satellite Results: Charging and Discharging. Spacecraft/Plasma Interactions and Their Influence on Field and Particle Measurements*, pp. 3–11. European Space Agency, Noordwijk, NL.

Ferguson, D. (1986). *The Voltage Threshold for Arcing for Solar Cells in LEO-Flight and Ground Test Results*. NASA TM-87259, NASA Glenn Research Center, Cleveland, OH.

Ferguson, D. (1989). Solar Array Arcing in Plasmas. Paper presented at 3rd *Annual Workshop on Space Operations Automation and Robotics*, Houston, TX.

Ferguson, D., G. Hillard, D. Snyder, and N. Grier (1998). The Inception of Snapover on Solar Arrays: A Visualization Technique. Paper presented at *36th Aerospace Sciences Meeting and Exhibit*, American Institute of Aeronautics and Astronautics, Reno, NV.

Feynman, J., and S. B. Gabriel (2000). On Space Weather Consequences and Predictions. *J. Geophys. Res.*, *105*(A5), 10543–10564.

Foster, J. C., D. H. Fairfield, K. W. Ogilvie, and T. J. Rosenberg (1971). Relationship of Interplanetary Parameters and Occurrence of Magnetospheric Substorms. *J. Geophys. Res.*, *76*(28), 6971–6975.

Frederickson, A. R. (1983). Electrostatic Charging and Discharging in Space Environments. Paper presented at Proceedings of the *10th International Symposium on Discharges and Electrical Insulation in Vacuum*, Columbia, SC.

Frederickson, A. R. (1999). Quantitative Guidelines for Charged Spacecraft Derived from the Physics of Discharges. Paper presented at *Government Microcircuit Applications Conference*, Monterey, CA.

Frederickson, A. R. (1980). Radiation Induced Dielectric Charging. In *Space Systems and Their Interactions with the Earth's Space Environment*, edited by H. B. Garrett and C. P. Pike, pp. 386–412. AIAA Press, New York.

Frederickson, A. R. (1996). Upsets Related to Spacecraft Charging. *IEEE Trans. Nucl. Sci.*, *43*(2), 426–441.

Frederickson, A. R., D. B. Cotts, J. A. Wall, and F. L. Bouquet (1986). *Spacecraft Dielectric Material Properties and Spacecraft Charging*. AIAA, Washington, DC.

Frederickson, A. R., E. G. Holeman, and E. G. Mullen (1992). Characteristics of Spontaneous Electrical Discharges of Various Insulators in Space Radiation. *IEEE Trans. Nucl. Sci.*, *V-39*(6), 1773–1782.

Garrett, H. B. (1981). The Charging of Spacecraft Surfaces. *Rev. Geophys.*, *19*, 577–616.

Garrett, H. B., R. W. Evans, A. C. Whittlesey, I. Katz, and I. Jun (2008). Modeling of the Jovian Auroral Environment and Its Effects on Spacecraft Charging. *IEEE Trans. Plasma Sci.*, *36*(5), 2440–2449. doi: TPS.2008.2004260.

Garrett, H. B., and A. Hoffman (2000). Comparison of Spacecraft Charging Environments at the Earth, Jupiter, and Saturn. *IEEE Trans. Plasma Phys.*, *28*(6), 2048–2057.

Garrett, H. B., R. E. McInerney, S. E. DeForest, and B. Johnson (1979). *Modeling of the Geosynchronous Orbit Plasma Environment-Part 3, ATS-5 and ATS-6 Pictorial Data Atlas*. AFGL-TR-79-0015. AFGL, Bedford, MA.

Garrett, H. B., E. G. Mullen, E. Ziemba, and S. E. DeForest (1978). *Modeling of the Geosynchronous Plasma Environment, 2, ATS-5 and ATS-6 Statistical Atlas*. AFGL-TR-78-0304. AFGL, Bedford, MA.

Garrett, H. B., and C. P. Pike (1980). *Space Systems and Their Interactions with Earth's Space Environment*. AIAA, New York, NY.

Garrett, H. B., A. G. Rubin, and C. P. Pike (1979). Prediction of Spacecraft Potentials at Geosynchronous Orbit. Paper presented at *Solar-Terrestrial Prediction Proceedings*, Vol. II, NOAA Environmental Research Laboratories, Boulder, CO.

Garrett, H. B., D. C. Schwank, and S. E. DeForest (1981a). A Statistical Analysis of the Low-energy Geosynchronous Plasma Environment—I. Electrons. *Planet Space Sci.*, *29*, 1021–1044.

Garrett, H. B., D. C. Schwank, and S. E. DeForest (1981b). A Statistical Analysis of the Low-energy Geosynchronous Plasma Environment—II. Protons. *Planet Space Sci.*, *29*, 1045–1060.

Garrett, H. B., D. C. Schwank, P. R. Higbie, and D. N. Baker (1980). Comparison between the 30–80 keV Electron Channels on ATS-6 and 1976-059A During Conjunction and Application to Spacecraft Charging Prediction. *J. Geophys. Res.*, *85*, 1155–1162.

Garrett, H. B., and A. C. Whittlesey (2015). Guest Editorial: Spacecraft Charging Technology. *IEEE Trans. Plasma Sci.*, *43*(9), p. 2775. doi:10.1109/TPS.2015.2466837.

Garrett, H. B., and A. C. Whittlesey (2011). *Guide to Mitigating Spacecraft Charging Effects*. John Wiley & Sons, Hoboken, NJ.

Garrett, H. B., and A. C. Whittlesey (2000). Spacecraft Charging, an Update. *IEEE Trans. Plasma Phys.*, *28*(6), 2017–2028.

Gilchrist, B. E., and S. G. Bilen (2000). Simulated Bare Electrodynamic Tethers in a Dense, Flowing, High-Speed Plasma. Paper presented at *11th Annual NASA-JPL-MSFC Advanced Space Propulsion Research Workshop*, Jet Propulsion Laboratory, Pasadena, CA.

Grier, N. (1985). Plasma Interaction Experiment II (PIX-II): Laboratory and Flight Results. Paper presented at *Spacecraft Environmental Interactions Technology*-1983, U.S. Air Force Academy, Colorado Springs, CO.

Guidice, D., and K. Ray (1996). PASP Plus Measurements of Space Plasma and Radiation Interactions on Solar Arrays. Paper presented at *34th Aerospace Sciences Meeting and Exhibit*, American Institute of Aeronautics and Astronautics, Reno, NV.

Guidice, D., P. Severance, H. Curtis, and M. Piszczor (1995). Investigation of Space-Environment Effects on Photo-Voltaic Technologies by the PASP Plus Experiment1. Paper presented at *33rd Aerospace Sciences Meeting and Exhibit*, January 9–12, American Institute of Aeronautics and Astronautics, Reno, NV.

Gurevich, A. V., L. P. Pitaevskii, and V. V. Smirnov (1970). Ionospheric Aerodynamics. *Sov. Phys. Usp. (Eng. Trans.)*, *99*(1–2), 595.

Gussenhoven, M. S. (1983). Geosynchronous Environment for Severe Spacecraft Charging. *J. Spacecr. Rockets*, *20*, 26–34.

Gussenhoven, M. S. (1985). High Level Spacecraft Charging in the Low Altitude Polar Environment. *J. Geophys. Res.*, *90*, 11009.

Gussenhoven, M. S., and E. G. Mullen (1982). A Worst Case Spacecraft Charging Environment as Observed by SCATHA. Paper presented at *AIAA 20th Aerospace Sciences Meeting*, April 24, 1979, Orlando, FL.

Hastings, D. E. (1987). Theory of Plasma Contactors Used in the Ionosphere. *J. Spacecr. Rockets*, *24*, 250–256.

Hastings, D. E. (1995). A Review of Plasma Interactions with Spacecraft in Low Earth Orbit. *J. Geophys. Res.*, *100*(A8), 14457–14483.

Hastings, D. E., M. Cho, and H. Kuninaka (1992a). The Arcing Rate for a High Voltage Solar Array: Theory, Experiments and Predictions. *J. Spacecr. Rockets*, *29*(4), 538–554.

Hastings, D. E., M. Cho, and J. Wang (1992b). The Space Station Freedom Structure Floating Potential and the Probability of Arcing. *J. Spacecr. Rockets*, *29*, 830–834.

Hastings, D. E., and R. Samanta Roy (1993). A Brief Overview of Electrodynamic Tethers. In *The Behavior of Systems in the Space Environment*, edited by R. N. DeWitt, D. P. Duston and A. K. Hyder, pp. 825–835. Kluwer Academic Publishers, Dordrecht, the Netherlands.

Hastings, D. E., and J. Wang (1989). Induced Emission of Radiation from a Large Space-Station-Like Structure in the Ionosphere. *J. Spacecr. Rockets*, *27*(4), 438–445.

Hoeber, C. F., E. A. Robertson, I. Katz, V. A. Davis, and D. B. Snyder (1998). Solar Array Augmented Electrostatic Discharge in GEO. Paper presented at the *17th International Communications Satellite Systems Conference and Exhibit*, February 23–27, American Institute of Aeronautics and Astronautics, Yokohama, Japan.

Hosoda, S., S. Hatta, T. Muranaka, J. Kim, N. Kurahara, M. Cho, H. Ueda, K. Koga, and T. Goka (2007). Verification of Multi-Utility Spacecraft Charging Analysis Tool (MUSCAT) Via Laboratory Test. In *45th AIAA Aerospace Sciences Meeting and Exhibit*, Reno, NV.

Iannello, C. (2014). *International Space Station (ISS) Plasma Contactor Unit (PCU) Utilization Plan Assessment Update.* NESC-RP-13-00869, 224 pp. NASA, Washington, DC.

Johnson, L., R. D. Estes, E. Lorenzini, M. Martinez-Sanchez, J. Sanmartin, and I. Vas (1998). Electrodynamic Tethers for Spacecraft Propulsion. Paper presented at the *36th Aerospace Sciences Meeting and Exhibit, American Institute of Aeronautics and Astronautics*, Reno, NV.

Katz, I., V. A. Davis, and D. B. Snyder (1999). Mitigation Techniques for Spacecraft Charging Induced Arcing on Solar Arrays. Paper presented at the *36th Aerospace Sciences Meeting and Exhibit*, January 11–14, American Institute of Aeronautics and Astronautics, Reno, NV.

Katz, I., G. Jongeward, and R. Rantanen (1991). Characterization of the Space Station Freedom External Environment. Paper presented at the *29th Aerospace Sciences Meeting*, January 7–10, Reno, NV.

Katz, I., M. Mandell, G. Jungeward, and M. S. Gussenhoven (1986). The Importance of Accurate Secondary Electron Yields in Modeling Spacecraft Charging. *J. Geophys. Res.*, *91*(A13), 13739–13744.

Kawakita, S., H. Kusawake, M. Takahashi, H. Maejima, J.-H. Kim, S. Hosoda, M. Cho, K. Toyoda, and Y. Nozaki (2004). Sustained Arc Between Primary Power Cables of a Satellite. Paper presented at the *2nd International Energy Conversion Engineering Conference*, August 16–19, 2004, American Institute of Aeronautics and Astronautics, Providence, R I.

Khurana, K. K., M. G. Kivelson, T. P. Armstrong, and R. J. Walker (1987). Voids in Jovian Magnetosphere Revisited: Evidence of Spacecraft Charging. *J. Geophys. Res.*, *92*(A12), 13399–13408.

Koons, H. C. (1981). *Spacecraft Charging Results from the SCATHA Satellite.* AFSC—Space Division, El Segundo, CA.

Koons, H. C., and M. W. Chen (1999). An Update on the Statistical Analysis of MILSTAR Processor Upsets. Paper presented at *Government Microcircuit Applications Conference*, Monterey, CA.

Koons, H. C., J. E. Mazur, R. S. Selesnick, J. B. Blake, J. F. Fennell, J. L. Roeder, and P. C. Anderson (1999). *The Impact of the Space Environment on Space Systems.* Rpt. No.TR-99(1670)-1. The Aerospace Corporation, El Segundo, CA.

Koons, H. C., J. E. Mazur, R. S. Selesnick, J. B. Blake, J. F. Fennell, J. L. Roeder, and P. C. Anderson (2000). The Impact of the Space Environment on Space Systems. In *6th Spacecraft Charging Technology Conference.* AFRL/USAF, Bedford, MA.

Koons, H. C., J. F. Mizera, P. F. Fennell, and D. F. Hall (1981). Spacecraft Charging—Results from the SCATHA Satellite. *Astronaut. Aeronaut.* pp. 44–47.

Koons, H. C., P. F. Mizera, J. L. Roeder, and J. F. Fennell (1988). A Severe Spacecraft-Charging Event on SCATHA in September 1982. *J. Spacecr. Rockets*, *25*(3), 239–243.

Laframboise, J. G. (1982). Is There a Good Way to Model Spacecraft Charging in the Presence of Space Charge Coupling, Flow, and Magnetic Fields?. Paper presented at *Proceedings of the Workshop on Natural Charging of Space Structures in Near Earth Orbits*, AFGL, Bedford, MA.

Lai, S. T. (1998). A Survey of Spacecraft Charging Events. Paper presented at *AIAA 36th Aerospace Sciences Meeting and Exhibit*, American Institute of Aeronautics and Astronautics, Reno, NV.

Lai, S. T. (2011). *Fundamentals of Spacecraft Charging: Spacecraft Interactions with Space Plasma*, 272 pp. Princeton University Press, Princeton, NJ.

Lam, H.-L., D. H. Boteler, B. Burlton, and J. Evans (2012). Anik-E1 and E2 Satellite Failures of January 1994 Revisited. *Space Weather*, *10*(S10003), 13. doi:10.1029/2012SW000811.

Lam, H.-L., and J. Hruska (1991). Magnetic Signatures for Satellite Anomalies. *J. Spacecr. Rockets*, *28*(1), 93–99.

Lanzerotti, L. J., K. LaFleur, and C. G. Maclennan (1998). Geosynchronous Spacecraft Charging in January 1997. *Geophys. Res. Lett.*, *25*(15), 2967–2970.

Leach, R. D., and M. B. Alexander (Eds.) (1995). *Failures and Anomalies Attributed to Spacecraft Charging.* National Aeronautics and Space Administration, Washington, DC.

Leung, M. S., M. B. Tueling, and E. R. Schanuss (1981). *Effects of Secondary Electron Emission on Charging.* TOR-0081(6470-02)-2. The Aerospace Corporation, El Segundo, CA.

Leung, P. (1983). Discharge Characteristics of a Simulated Solar Array. *IEEE Trans. Nucl. Sci., NS-30*, 4311.

Leung, P., A. C. Whittlesey, H. B. Garrett, P. A. Robinson Jr., and T. N. Divine (1986). Environment-Induced Electrostatic Discharges as the Cause of Voyager 1 Power-On Resets. *J. Spacecr. Rockets, 23*(3), 323–330.

Li, W.-W., and E. C. Whipple (1988). A Study of SCATHA Eclipse Charging. *J. Geophys. Res., 93*(A9), 10041–10046.

Lilley, J. R., Jr., D. L. Cooke, G. A. Jongeward, and I. Katz (1989). POLAR User's Manual. *GL-TR-89-0307*, AF Geophysics Laboratory, Hanscom AFB, Bedford, MA.

Lilley, J. R., Jr., A. Greb, I. Katz, V. A. Davis, J. E. McCoy, J. Galofaro, and D. C. Ferguson (1994). Comparison of the Theoretical Calculations with Plasma Motor Generator (pmg) Experimental Data. AIAA 94-0328, American Institute of Aeronautics and Astronautics, Reston, VA.

Mandell, M. J., and V. A. Davis (1990). User's Guide to NASCAP/LEO. *SSS-R-8507300-R2*, NASA Lewis Research Center, Cleveland, OH.

Mandell, M. J., V. A. Davis, B. M. Gardner, I. G. Mikellides, D. L. Cooke, and J. Minor (2006). Nascap-2k—An Overview. *IEEE Trans. Plasma Science, 34*, 2084.

Mandell, M. J., P. R. Stannard, and I. Katz (1984). NASCAP Programmer's Reference Manual. *Rep. SSS-84-6638*, S-Cubed, La Jolla, CA.

Miller, W. L. (1985). An Investigation of Arc Discharges on Negatively Biased Dielectric-Conductor Samples in a Plasma. Paper presented at *Spacecraft Environmental Interactions Technology*-1983, U.S. Air Force Academy, Colorado Springs, CO.

Mullen, E. G., and M. S. Gussenhoven (1983). SCATHA Environmental Atlas. *AFGL-TR-83-00002/ADA131456*, AFGL, Bedford, MA.

Mullen, E. G., and M. S. Gussenhoven (1989). SCATHA Environmental Atlas. Vol. II, *AFGL-TR-89-0249(II)*. AFGL, Bedford, MA.

Mullen, E. G., M. S. Gussenhoven, D. A. Hardy, T. A. Aggson, B. G. Ledly, and E. C. Whipple (1986). SCATHA Survey of High-Level Spacecraft Charging in Sunlight. *J. Geophys. Res., 91*(A2), 1474–1490.

Mullen, E. G., D. A. Hardy, H. B. Garrett, and E. C. Whipple (1981). P78-2 SCATHA Environmental Data Atlas. Paper presented at *Spacecraft Charging Technology* 1980. U.S. Air Force Academy, Colorado Springs, CO.

Murphy, G. B., and H. B. Garrett (1989). Interactions between the Space Station and the Environment: A Preliminary Assessment of EMI. Paper presented at *Proceedings of the 3rd Annual SOAR Workshop*, July 25–27, Houston, TX.

Murphy, G. B., D. L. Reasoner, A. Tribble, N. D'Angelo, J. S. Pickett, and W. S. Kurth (1989). The Plasma Wake of the Shuttle Orbiter. *J. Geophys. Res., 94*, 6866.

Nobles, W. (1986). Electrodynamic Tethers for Energy Conversion. Paper presented at *NASA/AIAA/PSN International Conference on Tethers in Space*, September 17–19, Arlington, VA.

Oberhardt, M. D., D. A. Hardy, I. Katz, M. P. Gough, and D. C. Thompson (1993). Vehicle Charging as Measured by the Shuttle Potential and Return Electron Experiment Aboard TSS-1. *Rep. AIAA 93-0699*, American Institute of Aeronautics and Astronautics, New York, NY.

Parker, L. N., and J. I. Minow (2013). Survey of DMSP Charging during the Period Preceding Cycle 24 Solar Maximum. Paper presented at *Space Weather Workshop*, April 16–19, Boulder, CO.

Parks, D., and I. Katz (1987). Theory of Plasma Contactors for Electrodynamic Tethered Satellite Systems. *J. Spacecr. Rockets, 24*, 245–249.

Prokopenko, S. M. L., and J. G. Laframboise (1980). High-Voltage Differential Charging of Geostationary Spacecraft. *J. Geophys. Res., 85*, 4125–4131.

Purvis, C. K., H. B. Garrett, A. C. Whittlesey, and N. J. Stevens (1984). Design Guidelines for Assessing and Controlling Spacecraft Charging Effects. *NASA Technical Paper 2361*, NASA Glenn Research Center, Cleveland, OH.

Rothwell, P. L. et al. (1976). Simulation of the Plasma Sheath Surrounding a Charged Spacecraft. In *Spacecraft Charging by Magnetospheric Plasmas*, edited by A. Rosen, pp. 121–133. MIT Press, Cambridge, MA.

Roussel, J., G. Dufour, J. C. Mateo-Velez, B. Thiebault, B. Andersson, D. Rodgers, A. Hilgers, and D. Payan (2012). SPIS Multi-Timescale and Multi-Physics Capabilities: Development and Application to GEO Charging and Flashover Modeling. *IEEE Trans. Plasma Sci., 40*, 183.

Roussel, J., F. Rogier, G. Dufour, J. C. Mateo-Velez, J. Forest, A. Hilgers, D. Rodgers, L. Girard, and D. Payan (2008). SPIS Opensource Code: Methods, Capabilities, Achievements, and Prospects. *IEEE Trans. Plasma Sci., 36*, 2360.

Samanta Roy, R., and D. E. Hastings (1993). A Brief Overview of Electrodynamic Tethers. In *The Behavior of Systems in the Space Environment*, edited by R. N. DeWitt, D. Dwight and A. K. Hyder, pp. 825–836. Kluwer Academic, Norwell, MA.

Samanta Roy, R., D. E. Hastings, and E. Ahedo (1992). A Systems Analysis of Electrodynamic Tethers. *J. Spacecr. Rockets, 29*(3), 415–424.

Samir, U., N. H. Stone, and K. H. Wright Jr. (1986). On Plasma Disturbances Caused by the Motion of the Space Shuttle and Small Satellites. *J. Geophys. Res., 91*(A1), 277–285.

Samir, U., and G. L. Wrenn (1969). The Dependence of Charge and Potential Distribution around a Spacecraft on Ionic Composition. *Planet Space Sci., 17*, 693.

Samir, U., K. H. Wright Jr., and N. H. Stone (1983). The Expansion of a Plasma into a Vacuum: Basic Phenomena and Processes and Applications to Space Plasma Physics. *Rev. Geophys., 21*(7), 1631–1646.

Sanmartin, J. R., M. Charro, E. C. Lorenzini, H. B. Garrett, C. Bombardelli, and C. Bramanti (2008). Electrodynamic Tether at Jupiter—I: Capture Operation and Constraints. *IEEE Trans. Plasma Sci., 36*(5), 2450–2458. doi:TPS.2008.2002580.

Sasaki, S. (1999). Plasma Effects Driven by Electromotive Force of Spacecraft Solar Array. *Geophys. Res. Lett.*, *26*(3), 1809–1812.

Snyder, D. B., D. C. Ferguson, B. V. Vayner, and J. T. Galofaro (1998). New Spacecraft-Charging Solar Array Failure Mechanism. Paper presented at the *6th Spacecraft Charging Technology Conference*, AF Research Laboratory, Hanscom AFB, Bedford, MA.

Soldi, J. D., D. E. Hastings, D. Hardy, D. Guidice, and K. Ray (1996). Arc Rate Predictions and Flight Data Analysis for the PASP Plus Experiment. Paper presented at the *34th Aerospace Sciences Meeting and Exhibit, American Institute of Aeronautics and Astronautics*, Reno, NV.

Sorensen, J. (1996). An Engineering Specification of Internal Charging. Paper presented at *ESA Symposium Proceedings on Environment Modeling for Space-Based Applications*, European Space Agency, Noordwijk, NL.

Stassinopoulos, E. G., and G. J. Brucker (1996). Radiation Induced Anomalies in Satellites. Paper presented at *AIAA 34th Aerospace Sciences Meeting and Exhibit*, American Institute of Aeronautics and Astronautics, Reno NV.

Stevens, N. J. (1980). Review of Interactions of Large Space Structures with the Environment. In *Space Systems and Their Interactions with Earth's Space Environment*, edited by H. B. Garrett and C. P. Pike, pp. 437–454. AIAA Press, New York.

Stone, N. H., K. H. Wright Jr., U. Samir, and K. S. Hwang (1988). On the Expansion of Ionospheric Plasma into the Near-Wake of the Space Shuttle Orbiter. *Geophys. Res. Lett.*, *15*(1169), 1169–1172.

Thomsen, M. F., M. H. Denton, B. Lavraud, and M. Bodeau (2007). Statistics of plasma fluxes at geosynchronous orbit over more than a full solar cycle. *Space Weather*, S03004(5). doi:10.1029/2006SW000257.

Thomsen, M. F., M. G. Henderson, and V. K. Jordanova (2013). Statistical properties of the surface-charging environment at geosynchronous orbit. *Space Weather*, *11*, 237–244. doi:10.1002/swe.20049.

Tsurutani, B. T., and W. D. Gonzalez (1987). The Cause of High-Intensity Long-Duration Continuous AE Activity (HILDCAAS): Interplanetary Alfven Wave Trains. *Planet Space Sci.*, *35*(4), 405–412.

Vampola, A. L. (1987). Thick Dielectric Charging on High-Altitude Spacecraft. *J. Electrostatics*, *20*, 21–30.

Vampola, A. L. (1996a). The Nature of Bulk Charging and its Mitigation in Spacecraft Design. Paper presented at WESCON, IEEE, Anaheim, CA.

Vampola, A. L. (1996b). The ESA Outer Zone Electron Model Update. Paper presented at *ESA Symposium Proceedings on Environment Modeling for Space-Based Applications*, European Space Agency, Noordwijk, NL.

Vas, I. E. (2000). Space Station Reboost with Electrodynamic Tethers. *J. Spacecr. Rockets*, *37*(2), 153.

Vignoli, M., M. Miller, and M. Matteoni (1986). Power Generation with Electrodynamic Tethers. Paper presented at *NASA/AIAA/PSN International Conference on Tethers in Space*, September 17–19, Arlington, VA.

Violet, M. D., and A. R. Fredrickson (1993). Spacecraft Anomalies on the CRRES Satellite Correlated With the Environment and Insulator Samples. *IEEE Trans. Nucl. Sci.*, *40*(6), 1512–1520.

Wang, J., P. Leung, H. Garrett, and P. Murphy (1994). Multibody-Plasma Interactions: Charging in the Wake. *J. Spacecr. Rockets*, *31*(5), 889–894.

Whipple, E. C., Jr. (1965). The Equilibrium Electric Potential of a Body in the Upper Atmosphere and in Interplanetary Space. *NASA X-615-65-296*, NASA.

Whipple, E. C., Jr. (1981). Potentials of Surfaces in Space. *Rep. Prog. Phys.*, *44*, 1197–1250.

Whittlesey, A. C. (1999). Avoiding Problems Caused by Spacecraft On-Orbit Internal Charging Effects. *NASA-HDBK-4002*, NASA.

Whittlesey, A. C., and H. B. Garrett (2011). *Mitigating In-Space Charging Effects—A Guideline*. NASA-HDBK-4002A, National Aeronautics and Space Administration, Washington, DC.

Whittlesey, A. C., and P. Leung (1987). Space Plasma Charging—Lessons from Voyager. Paper presented at *AIAA 25th Aerospace Sciences Meeting*, January 12–15, Reno, NV.

Wilkinson, D. C. (1994). National Oceanic and Atmospheric Administration's Spacecraft Anomaly Data Base and Examples of Solar Activity Affecting Spacecraft. *J. Spacecr. Rockets*, *31*(2), 160–165.

Wrenn, G. L. (1995). Conclusive Evidence for Internal Dielectric Charging Anomalies on Geosynchronous Communications Spacecraft. *J. Spacecr. Rockets*, *32*(3), 514–520.

Wrenn, G. L., and A. J. Sims (1996). *Internal Charging in the Outer Zone and Operational Anomalies*. Geophysical Monograph 97, American Geophysical Union, Washington, DC.

Wrenn, G. L., and R. J. K. Smith (1996). The ESD Threat to GEO Satellites: Empirical Models for Observed Effects Due to Both Surface and Internal Charging. Paper presented at *ESA Symposium Proceedings on Environment Modeling for Space-Based Applications*, European Space Agency, Noordwijk, NL.

CHAPTER 18

Satellite Orbital Drag

Eftyhia Zesta and Cheryl Y. Huang

CONTENTS

18.1 INTRODUCTION

Satellites orbiting the low Earth orbit (LEO) are within the Earth's upper atmosphere and suffer continuous collisions with the gas molecules of the atmosphere. As a result of the friction caused by these collisions, these satellites decelerate and their orbits decay, that is, the orbits' altitude is lowered until reentry. This orbital decay is typically referred to as "orbital drag."

The effects of orbital drag are typically significant at LEOs with altitudes less than 500–600 km, where the atmospheric density is still significant. However, during geomagnetic storms, when the Earth's upper atmosphere heats and expands, satellites at altitudes up to 800 km and greater can suffer decelerations and orbit disturbances due to orbital drag. In 1979, uncontrolled orbital decay degraded the orbit of the Skylab space station and it reentered the Earth's atmosphere and disintegrated. The International Space Station has to continuously correct its orbit because of the effects of orbital drag. Earth observing satellites, such as Aqua and Terra, at altitude ~700 km, typically experience orbit disturbances during magnetic storms, primarily due to the thermosphere heating and upwelling during such disturbed periods.

The Earth's upper atmosphere, at altitude greater than 80 km, is a mixture of neutral and ionized atoms and molecules. The ionized portion of the atmosphere called the ionosphere stretches from altitude ~80 km to more than 1000 km and is created by ionization resulting from solar ultraviolet (UV) illumination of the Earth's dayside atmosphere. The ionosphere overlaps with, and mechanically couples to, the neutral atmosphere, the thermosphere. The ion density is much less than the neutral density; therefore, the neutrals directly contribute to and control the orbital drag that satellites at LEO experience. However, the ionosphere is strongly coupled to the neutrals through ion-neutral interactions, and, at the same time, the ionosphere is electromagnetically coupled to the magnetosphere. Magnetospheric dynamic changes driven by the solar wind quickly couple to the ionosphere electrodynamically and to the thermosphere through ion-neutral interactions.

Thermospheric density is not constant; in fact, it can be very dynamic during geomagnetically active times. The source of dynamic changes of the thermosphere is the changes in the solar extreme UV (EUV) radiation and magnetospheric energy input to the ionosphere–thermosphere (IT) system. Solar radiation typically varies in 10s of days and up to months and years, whereas magnetospheric energy input varies in timescales as short as minutes and seconds. The most dramatic changes in magnetospheric energy input occur during the times of storms, when the whole magnetosphere is strongly driven by the solar wind.

When additional energy in inserted into the thermosphere, either by the Sun or by the magnetosphere, thermospheric layers heat and expand. The LEO satellites suddenly find themselves inside denser gas and experience higher drag force. The longer variation timescales of the solar EUV radiation elicit smoother and more global changes in the thermosphere. In contrast, magnetospheric energy input can vary abruptly in response to sudden changes in the solar wind; it flows into the IT system through the high latitudes initially and subsequently spreads globally. Understanding and predicting the localized and time-varying patterns by which magnetospheric energy is absorbed by the IT system and the neutral density response are currently a major challenge in the IT community.

It is a critical need for the maintenance of space object catalog and for collision avoidance to be able to specify and predict the 3D neutral density at LEO altitudes. Catalog maintenance refers to the routine tracking of objects by the US Space Surveillance Network (SSN). Collision avoidance requires accurate orbit propagation models in order to assess the risk of catastrophic impacts between two or more space objects or the impact of a man-made asset by small debris.

The US SSN is tasked with the cataloging and tracking of all space objects in LEO with size greater than ~10 cm. Presently, more than 21,000 objects are a part of this catalog (Achieving and Maintaining Orbit; National Research Council [NRC], 1995; Fedrizzi et al., 2012). A major part of the Space Surveillance project is to provide collision warnings for high-value satellites, such as the International Space Station and other critical assets. Debris as small as 1 cm can damage operational satellites (Crowther, 2003). The problem of debris accumulation became acute in 2007, when an intentional collision of an antisatellite weapon with the Chinese Fengyun 1C satellite produced more than 2500 trackable objects and increased the size of the SSN catalogue by 25%. In 2009, the accidental collision of Iridium 33, a US operational communications satellite, and Cosmos 2251, a Russian decommissioned communications satellite, at altitude of 790 km created clouds of debris that spread to altitudes down to 200 km within months (NASA, 2009).

Orbit propagation models are used to determine the location of space objects for purposes of tracking them or for reentry predictions. Gravity and atmospheric drag are the forces that determine the object's orbit, and neutral density is the major contributor to orbital drag changes. This chapter is mainly devoted to the specification and prediction of the neutral density and the issues that face our community in doing so. In Sections 18.4 and 18.7, we will discuss all the factors that affect orbital drag in more specific terms.

18.2 FUNDAMENTALS OF THE THERMOSPHERE

The thermosphere is commonly regarded as the part of the atmosphere between 90 km and 500 km and can be characterized by temperature, $T(z)$, which increases with altitude z, as shown in Figure 18.1. The lower boundary coincides with the mesopause, where the lowest temperatures in the atmosphere are found. Temperatures increase asymptotically to approximately 1000 K at altitudes greater than 500–700 km. The region above this altitude is regarded as the exosphere, where mean free paths of atoms allow for rapid transfer of heat. In empirical models, the thermosphere can be characterized by an exospheric temperature, T_∞, which specifies the asymptotic temperature at the upper boundary of the thermosphere. Once the temperature at any altitude, latitude, and longitude is given, the global distribution of temperatures is specified (Jacchia, 1970, 1977).

Figure 18.2 shows a schematic of the distribution of the major atomic and molecular species as a function of altitude. Over the region from 100 km to 500 km, which is

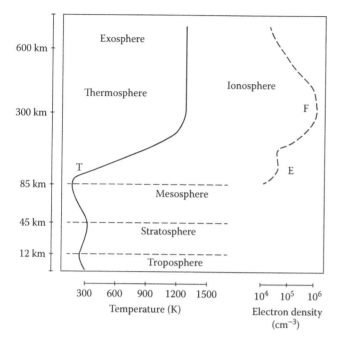

FIGURE 18.1 Schematic of atmosphere. (Courtesy of Bhamer, Wikipedia.)

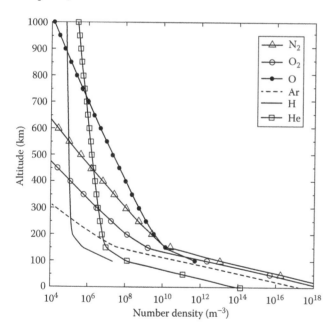

FIGURE 18.2 MSIS vertical profiles of major atmospheric constituents. (Courtesy of Y. Huang.)

of interest to this chapter, the major constituent is atomic oxygen. However, the minor constituents such as carbon dioxide and nitric oxide, which cool the thermosphere by radiation at low altitudes, are not included in Figure 18.2 (Roble et al., 1987; Maeda et al., 1989; Mlynczak et al., 2003).

The temperature of the thermosphere increases as a result of (1) solar radiation in the UV, EUV, and soft X-ray frequency range; (2) dynamic heating due

to magnetospheric and solar wind sources at high latitudes; and (3) heating and conduction generated by atmospheric waves generated at low altitudes. The primary effects of solar irradiance are direct heating, ionization, dissociation, and excitation of atmospheric atoms and molecules O, O_2, and N_2. Magnetospheric and solar wind energy inputs cause heating and generation of neutral winds. Atmospheric waves propagate upward and dissipate in the thermosphere. The major mechanisms for thermospheric cooling are via infrared radiation of CO_2 below 120 km and NO above 120 km.

It is generally assumed that the thermosphere has higher vertical than horizontal structure. In the vertical direction, hydrostatic equilibrium is dominant, so that gravity is balanced by gas pressure. In general, the gas pressure, p, varies with altitude, z, as

$$p(z) = \frac{p_0 \exp\left[-(z - z_0)\right]}{H}$$

where:

H is defined as the pressure scale height

$$H = \frac{kT}{mg}$$

where:

k is the Boltzmann constant
T is the temperature
m is the atmospheric mass
g is the gravitational acceleration

A second assumption commonly made in empirical models of orbital drag is that of diffusive equilibrium, in which the atmospheric constituents are assumed to have reached a steady state. This assumption is violated during solar wind forcing, when vertical energy input and winds create regions of upwelling and downwelling pressures (Rishbeth, 1987). Physics-based models typically derive the atomic and mass densities under the assumption of diffusive equilibrium at the upper boundaries.

Given these assumptions, it is also assumed that as the thermosphere expands due to increased Joule heat, the density at constant altitude increases. Thus, a local measure of increased or decreased neutral density can be an indirect measure of heating or cooling of the atmosphere. This is the basis of interpretation of neutral densities derived from spaceborne accelerometers flown in recent years.

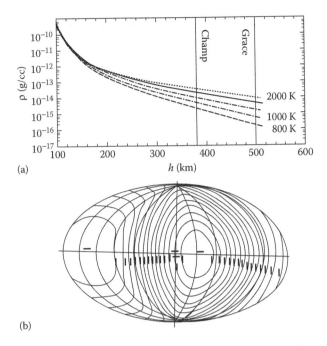

(a)

(b)

FIGURE 18.3 (a) Mass density profiles as functions of altitude for five selected values of T_∞. Vertical lines bound the range of altitudes of CHAMP and GRACE orbits in 2004. (b) Example of J77 T_∞ distributions plotted on a latitude and local time grid. (From Burke, W. J. et al., *Ann. Geophys., 27*, 2035–2044, 2009a. With permission.)

Figure 18.3 illustrates the distribution of exospheric temperatures from the Jacchia (1977) model at geographic latitude and longitude coordinates (Figure 18.3a, top). In Figure 18.3b (bottom) are shown the mass densities from the model for five selected values of T_∞. The orbital altitudes of the Gravity Recovery and Climate Experiment (GRACE) and Challenging Minisatellite Payload (CHAMP) spacecraft during 2004 are indicated (Burke et al., 2009a).

The dominant impact of solar irradiance on the thermosphere is the absorption of solar UV, EUV, and X-ray radiation. Solar irradiance is highly variable, peaking during solar flares, when the intensity of X-ray burst can increase by orders of magnitude. The net effect of radiation is to heat the main constituents, N_2, O_2, and O. About 50% of the solar energy deposited on the dayside of the thermosphere goes into local heat (Torr et al., 1980). The remainder is taken up by the dissociation of O_2 into O atoms and subsequent radiation to space.

The second source of energy to the thermosphere is the solar wind, which couples to the magnetosphere–IT system. The variability due to solar wind forcing is highest during geomagnetic storms, when electric currents increase to high levels at high latitudes. In the auroral

zones, where conductivity maximizes at low altitudes (100–120 km), there is an increase in Joule heating of ions, which transfer energy to neutrals via collisions. In the polar cap, conductivity peaks at higher altitudes, greater than 200 km, and is generally lower than in the auroral zones. However, as the neutral gas is substantially less dense at higher altitudes, the heat transfer per unit mass is high. Note that at *F*-region altitudes, the collision rate falls and energy transfer is via ion drag (Richmond, 1982).

The effect of forcing from below is the generation of upward-propagating waves, which permeate the thermosphere. Although this impact creates substantial structure in the form of migrating and nonmigrating tides, the effect on satellite drag, which depends on cumulative variability in atmospheric density, is usually taken to be small relative to the energy input from solar and solar wind sources. Waves are typically ignored in empirical models of drag. In physics-based models, a lower boundary term ascribed to eddy diffusion can be added to simulate the effect of upward wave propagation (Qian et al., 2009).

The effect of neutral winds in the thermosphere is relatively unknown because of the scarcity of observations. However, there is substantial evidence that during magnetic storms, the assumptions of hydrostatic and diffusive equilibrium are violated, as vertical winds in excess of 100 m s^{-1} have been recorded at high latitudes (Spencer et al., 1982; Rees et al., 1984; Conde and Dyson, 1995; Innis and Conde, 2001).

18.3 THE ORBITAL DRAG EQUATION

The largest uncertainty in determining orbits for the satellites operating in LEO is the atmospheric drag.

The equation that determines drag is:

$$a_D = -\frac{1}{2} C_D \frac{A_{\text{ref}}}{M} \rho V^2 \qquad (18.1)$$

where:

ρ is the atmospheric total mass density

A_{ref}, M, and C_D are the satellite's projected reference area, mass, and drag coefficient, respectively

V is the total atmospheric velocity relative to the satellite velocity ($V_{\text{S/C}} + V_{\text{W}}$)

Drag is the most difficult force to model, mainly because of the multiple unknown parameters in Equation 18.1, necessitating multiple assumptions that can be largely inaccurate. One can safely assume that the area (A_{ref}) that the satellite presents to the ambient gas is known with good attitude information. The mass of the satellite (M) is also known. This leaves the remaining three parameters, C_D, ρ, and V, to be unknown. It is generally accepted that of these three parameters, the density contributes most to the drag, whereas the drag coefficient and the relative velocity of the gas are considered "known."

Most of the drag observations to date are measurements of the satellite deceleration, either from ground tracking radar or from accelerometers on board satellites. The typical analysis process assumes a neutral wind pattern in the evaluation of V and then calculates the ρC_D parameter (the product of the density and the drag coefficient) from the measured acceleration. However, the calculation of C_D has been fraught with unrealistic assumptions and is a constant error parameter in the derivation of ρ (Marcos et al., 2010; Vallado and Finkelman, 2014). In most calculations and models, values between two and four are considered good estimates for C_D. The drag coefficient is a macroscopic parameter that defines the process of momentum transfer from the ambient gas to the satellite, and it depends on the physical properties of the ambient gas, the satellite geometry, the physical properties of the surface (or surfaces) of contact, the satellite orientation within the gas, and the relative velocity of the satellite with respect to the ambient gas (Marcos et al., 2010 and references therein). The drag coefficient equation can be complex for simple shapes of satellites and practically unknown for more complex shapes. Vallado and Finkelman (2014) gave a good review on how the boundary values of two and four for the drag coefficient are determined and how insufficient and potentially inaccurate they are (Figure 2 of Vallado and Finkelman, 2014). Sentman (1961) derived the values of drag coefficients for basic shapes of satellites, for example, spherical, rectangular, and cylindrical), in altitudes from 150 km to 300 km. Vallado and Finkelman (2014) pointed out that much higher values are operationally possible for the drag coefficient and that the range is potentially much wider at altitudes greater than 300 km. This means the drag coefficient can introduce errors in the calculation of neutral density of more than a factor of two.

Errors in the relative velocity, $V = V_{S/C} + V_W$, have been routinely ignored but can be potentially large, especially because V is squared in Equation 18.1. $V_{S/C}$ is the satellite inertial velocity, whereas V_W is the atmospheric rotational velocity plus any residual winds at the altitude of the satellite (Marcos et al., 2010). The atmospheric rotational velocity as a function of latitude is well-known, but the residual winds are poorly observed, poorly modeled, poorly understood, and can be quite large at the altitudes of most operational satellites, particularly during storms. The Dynamic Explorer 2, the last mission that observed in situ winds with accuracy, measured winds with velocity up to 1000 m/s or more during storms (Killeen and Roble, 1988; Bruinsma et al., 2004; Sutton et al., 2007). Such winds can produce an error in Equation 18.1 up to 20%–25% at high latitudes.

All of these errors translate into the calculation of neutral density from deceleration measurements. All the empirical models that exist today, even for operational purposes, are built with these inherent errors. All comparisons of physics-based models are made with densities calculated with these inherent errors. In short, the most substantial limitations in the successful specification and prediction of neutral density and orbital drag are the inherent errors in all the empirical models and measurement-derived densities and the complete lack of absolute validation of such measurements.

18.4 SPACE WEATHER EFFECTS

18.4.1 Solar Radiation

The variations in the UV solar radiation, which heat the thermosphere, are due to two components. The first is related to the solar rotational modulation of active regions and the second is ascribed to the long-term evolution of the solar magnetic field. Passage of active regions during a solar rotation produces a periodicity of approximately 27 days, whereas the main solar magnetic field variation produces irradiance variations more than 11 years, commonly referred to as the solar cycle. The solar index F_{10} is the historic index used to represent both effects, and this continues to be the primary index used to quantify solar input to the thermosphere and ionosphere.

However, the F_{10} index is measured at 10.7 cm wavelength, which is not a direct measure of the UV radiation and is not absorbed by the atmosphere. A number of other indices have been developed to represent EUV and far UV radiations, based on the measurements from the Solar and Heliospheric Observatory and the National Oceanic and Atmospheric Administration's operational satellites, which carry Solar Backscatter Ultraviolet spectrometers and X-ray spectrometers (Bowman et al., 2008). Other data, which are not fully exploited, are from the thermosphere ionosphere mesosphere energetics and dynamics/solar extreme ultraviolet (EUV) experiment (TIMED/SEE), Solar Dynamics

Observatory/EUV Variability Experiment (SDO/EVE), and Geostationary Operational Environmental Satellite-R Series/Extreme Ultraviolet and X-ray Irradiance Sensors (GOES-R/EXIS) instruments, which can also provide alternatives to the F_{10} index.

Solar flares can cause substantial variation in the thermospheric density, with the extent of the perturbation dependent on the irradiance as a function of the wavelength. The EUV flares create disturbances at high altitudes (Qian et al., 2011), whereas X-ray ultraviolet (XUV) dominates at lower altitudes, less than 150 km. A modeling study of the direct impact of EUV radiation on the neutral density showed an enhancement of 45 K in temperature at 400 km, which was ascribed to the 25–105 nm waveband (Huang et al., 2013). In addition to direct heating, flares can be associated with coronal mass ejections (CMEs), which are associated with magnetic storms.

18.4.2 Solar Wind Forcing of the Coupled Ionosphere–Thermosphere System

The coupled IT system is a completely driven system. This means that in order to fully specify the system, one must understand the different sources of forcing and the manner by which the system responds to such energy input. Solar EUV heating is the major driver of thermospheric dynamics, accounting for 75%–80% of the energy

input. The major solar EUV variation is with 11-year solar cycle, which can cause density variation of a factor of 10 at altitude of 400 km. Geomagnetic activity from auroral and polar cap heating processes provides most of the remaining 20%–25% of the thermospheric energy input. During times of strong geomagnetic activity, that is, storms, the magnetospheric forcing can well exceed the solar EUV input, as shown in Figure 18.4, adapted from Knipp et al. (2005), who used three empirical models to simulate the solar EUV, and the magnetospheric energy as Joule heating and particle power inputs for three solar cycles from 1975 to 2003. Knipp et al. (2005) showed that only the Joule heating part of the magnetospheric energy input is substantial for thermospheric heating; however, in Section 18.4.3 below, we discuss the observations that demonstrate that this is not the case, particularly during storms. Storms are abrupt, raising the energy input on short timescales, providing localized input at high latitudes, and mobilizing and strengthening the winds, which ultimately circulate the localized heating to global scales.

The solar wind and interplanetary magnetic field (IMF) vary continuously, driving the magnetosphere dynamically and providing varying energy to the IT system. All of the changes, large and small, are reflected in analogous thermospheric changes. The average solar wind

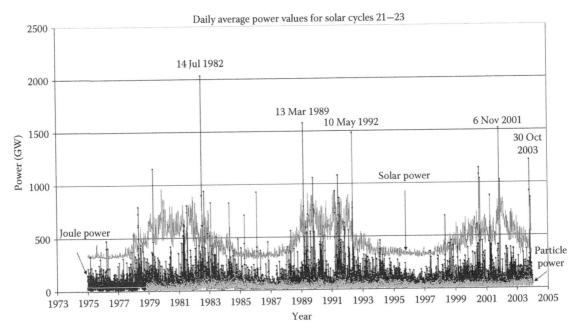

FIGURE 18.4 Relative power of solar EUV (top gray line) and geomagnetic energy input to the thermosphere during two solar cycles. The dark-black line is the Joule heating power and the bottom light-gray line is the precipitating particle power. All the black spikes represent geomagnetic energy input during disturbances and it is the dominant input during those times. (From Knipp, D. J. et al., *Sol. Phys.*, 224(1–2), 495–505, 2004. With permission.)

speed is 300–400 km/s but can be as high as 1000 km/s or more. The IMF substantially changes in amplitude, but its direction also varies on timescales as short as seconds and minutes, affecting how the solar wind energy is coupled to the magnetosphere. The most geoeffective solar wind disturbances, in terms of energy transfer, are shocks, sharp solar wind dynamic pressure enhancements, and storms.

Shocks and sudden dynamic pressure enhancements abruptly compress the magnetosphere, enhancing all large-scale magnetospheric currents in order to establish and maintain new boundaries in the more compressed space (e.g., Zesta et al., 2000; Boudouridis et al., 2003). They also enhance auroral precipitation (Zesta et al., 2000; Zhou et al., 2003, 2009; Liou, 2006; Laundal and Østgaard, 2008). As a result, more energy enters the IT system and ionospheric currents are enhanced (Zesta et al., 2000), resulting in more Joule heating of the ions and momentum transfer to the neutrals.

Geomagnetic storms are even more dynamic, global, and long lasting in their geoeffectiveness. Geomagnetic storms are triggered either by interplanetary CME (ICMEs) (e.g., Cane and Richardson, 2003) or by co-rotating interaction regions (CIRs) (e.g., Tsurutani et al., 2006). The ICMEs are preceded by a shock and then by a slowly rotating, high-amplitude southward magnetic field with a large B_y component. The ICMEs drive strong magnetospheric convection with strong energy input at high latitudes and a strong build-up of the ring current and particle energization in the inner magnetosphere. The CIRs are caused by fast solar wind catching up slower solar wind streams and are characterized by a region of slowly increasing speed of the solar wind (typically up to 800 km/s or more), accompanied by large-amplitude, turbulent magnetic field and bound by regions of high dynamic pressure (e.g., Tsurutani et al., 2006). The CIRs trigger storms of significantly less strength than the ICMEs (Richardson et al., 2002), resulting in disturbance storm time (D_{ST}) values that typically peak at −50 to −70 nT. However, the CIR storms can last much longer (many days) than the ICME storms, continuously driving the magnetosphere at a lower rate but for a more extended period and can therefore produce overall greater thermospheric density and orbital disturbances than the ICME storms (Chen et al., 2012). Even more important, for prediction purposes, the CIRs have typical 9- or 27-day periodicity (Lei et al., 2008), as the region of the corona from where the fast solar wind is emerging rotates with the solar rotation. The effects of the different types of

storms on the thermosphere have begun to be assembled only now, after ~14 years of continuous data collection from CHAMP, GRACE, and Gravity Field and Steady-State Ocean Circulation Explorer (GOCE) satellite missions and currently from the Swarm mission. This has started to be and should continue to be an active research area if the thermospheric density response is to be better specified and predicted for better orbital predictions.

18.4.3 Magnetospheric Energy Input to the Ionosphere–Thermosphere System: Partitioning and Dissipation

18.4.3.1 Poynting Flux

The magnetosphere is electromagnetically coupled to the ionosphere through the Earth's magnetic field lines, so that fields (**E** and **B**) couple from the outer magnetosphere down to ionospheric fields. Field-aligned currents communicate information from the magnetosphere and its dynamic coupling with the solar wind. The solar UV radiation provides the energy to maintain the ionosphere through ionization, and the solar wind coupling with the magnetosphere provides additional energy in the form of perturbations.

Energy enters the IT system in two forms: electromagnetic (EM), in the form of Poynting flux, and kinetic, in the form of precipitating particles. The majority of the magnetospheric volume maps down to the high latitudes, so both types of energy input occur at high latitudes: auroral, subauroral, and polar regions. The way the upper atmosphere responds to the magnetospheric energy input depends on the prior state of the upper atmosphere, the type of energy input, and the spatial and temporal structure of the energy input.

The Poynting flux vector, **S**, represents the directional energy density of an EM field:

$$S = \frac{E \times B}{\mu_0} \qquad (18.2)$$

where:

E and **B** are the electric and magnetic fields
μ_0 is the permeability of free space

In the case of energy directed toward the ionosphere, both quasi-static and EM wave fields are of importance to the ionosphere. In the above equation, **B** includes the background geomagnetic field, the quasi-static component from the large-scale field-aligned currents, and EM wave fields. The background field, B_0, produces strong, steady horizontal Poynting flux that does not dissipate energy within the ionosphere. Therefore, as many works have shown, it is only useful to consider the perturbation field δB of quasistatic and wave perturbations that create downward pointing flux (Kelley et al., 1991; Thayer and Semeter, 2004; Richmond, 2010).

Poynting's theorem, derived from Maxwell's equations, is the EM energy conservation equation for a particular volume:

$$\frac{\partial W}{\partial t} + \nabla \cdot S + j \cdot E = 0 \qquad (18.3)$$

where:

$W = (\varepsilon_0 E^2 / 2) + (B^2 / 2\mu_0)$ is the EM energy density
S is the Poynting flux
j is the current density within the volume
E is the electric field applied in the volume

Equation 18.3 shows that the time rate of change of the EM energy is balanced by the gradient of the Poynting flux entering the volume and the energy dissipated as Joule heating within the volume. For the ionosphere, prior works have carefully demonstrated that for a well-defined volume, no loss of EM energy occurs and the conservation equation becomes:

$$\nabla \cdot S = -j \cdot E \qquad (18.4)$$

Equation 18.4 describes the dissipation of Poynting flux within the ionosphere and is the basis of the common assumption that all EM energy input is equivalent to Joule heating. The volume typically considered is bound at the top by the top of the ionosphere, say 1000 km, and at the bottom by the layer just below the ionosphere, where no Poynting flux goes through, say 80 km (Kelley et al., 1991; Gary et al., 1994, 1995; Huang and Burke, 2004; Thayer and Semeter, 2004). Richmond (2010) showed that the sides of the volume should be equipotentials, and Gary et al. (1994) argued that if the volume over which the Poynting theorem is applied is not selected carefully, then the assumption of no EM energy loss through the volume is not accurate and Joule heating is overestimated. The upper and lower boundaries of the volume are easier to identify as the top and bottom of the ionospheric layer. The vertical boundaries have been the subject of the two papers that have clarified the more accurate application of Poynting's theorem within the ionosphere. For example, vertical boundaries of the volume cannot be within regions of auroral acceleration regions because large parallel potentials exist there and the field lines are not equipotentials.

Equation 18.4 demonstrates the energy conversion in the inertial frame of reference. However, in order to uncover more information on how the energy input is partitioned inside the IT system, the term is best written in the center-of-mass velocity frame, which is identical to the neutral-wind frame of reference, since the ions contribute minimally to the total mass. In that frame of reference, the magnetic field and current density remain unchanged, but the electric field transforms to $E' = E + V_n \times B$, where V_n is the neutral wind and E is the inertial frame field. The total EM energy dissipation becomes:

$$j \cdot E = j \cdot E' + V_n \cdot j \times B \qquad (18.5)$$

The first term on the right-hand side of Equation 18.5 is the typical Joule heating and the second term is the mechanical transfer of energy from EM forces directly to the neutral gas (Lu et al., 1995; Thayer et al., 1995; Thayer and Semeter, 2004). The Joule heating is also referred to as "frictional heating" because it enhances the ion drift, creating a differential velocity between ions and neutrals and thus enhancing the ion-neutral collisions resulting in the heating of the ions and eventually of the neutrals. The mechanical transfer term is also referred to as "momentum transfer" directly to the neutrals, increasing the winds. In the total energy budget, winds are often ignored and all energy is assumed to go into Joule heat, as in the global convection models (GCMs). It is important to understand and know how the incoming Poynting flux is partitioned into heating and momentum transfer, because the changes in temperature, composition, and winds are dramatically different when driven by a heat source as compared to a momentum source (Thayer and Semeter, 2004).

18.4.3.2 Particle Precipitation

High latitudes are the focus of magnetic field lines threading through 90% of the magnetospheric volume and are continuously subjected to particle precipitation with sources in the magnetosphere and instigated by dynamic interactions with the solar wind. Precipitating particles constitute the kinetic energy input from the magnetosphere to the IT system. As shown in Figure 18.4, the study by Knipp et al. (2005) demonstrated that, as determined by empirical models, particles contribute a small proportion of the total energy input during magnetic activity. However, this is not in agreement with prior and more recent observational studies. Vickrey et al.

(1982) and Doyle et al. (1986), using ground and satellite observations, found that daily averages of energy flux from particles and Joule heating are comparable within a factor of two in the auroral oval and polar latitudes, but there is a tendency of the two to be anticorrelated, so either of the two energy sources could dominate in a particular locality. Furthermore, the empirical models of particle precipitation are based on satellite observations that do not observe the thermal and suprathermal parts of the electron spectrum and that contribute to the particle fluxes precipitated to altitudes greater than 200 km.

While proton fluxes are more intense in the equatorward part of the aurora, fluxes from precipitating electrons are the most intense throughout the auroral oval and in the polar cap. Electrons fluxes directly change the plasma density and the chemistry of the upper atmosphere by ionizing neutrals and energizing ions and neutrals, creating bright aurorae. The ionizing collisions alter the ionospheric conductivity, which in turn can affect how and where currents flow and Poynting flux dissipates. The interdependence between Poynting flux and precipitation is strong and depends on both the prior state of the upper atmosphere and the characteristics and timescales of the energy input.

It has been quantitatively shown (e.g., Rees et al., 1983; Thayer and Semeter, 2004; Richards, 2013) that the energy spectrum of the precipitating electrons provides gas heating at different altitudes, depending on the characteristic energy of the precipitating profile. The gas heating energy comes from exothermic chemical reactions between the increased ionized population and the neutrals. Figure 18.5, adapted from Thayer and Semeter (2004), shows calculations of ionization rates and energy input for three different precipitating electron distributions. In the left panel, three curves of energy distributions are shown, which were measured by the Fast Auroral Snapshot Explorer satellite (FAST), as it approached the poleward boundary of an active auroral oval. The dashed curve has more suprathermal electrons, whereas the other two curves include energetic electrons. The middle panel gives the corresponding ion production (bottom axis) and energy deposition (upper axis) profiles using the Mass Spectrometer and Incoherent Scatter (MSIS) neutral atmosphere as the background. The right panel gives the ionospheric density profile for different times up to 60 s after the precipitating flux. Both the energy deposition and electron density profiles demonstrate how differently the energy is deposited in altitude, depending on the energy profile of the precipitating population. Richards (2013) took this work further and directly calculated the direct neutral heating that would result from different precipitating profiles, and similar to what is shown in Figure 18.5, he found that soft electron distributions typically heat gases at the F-layer and above (~400 km),

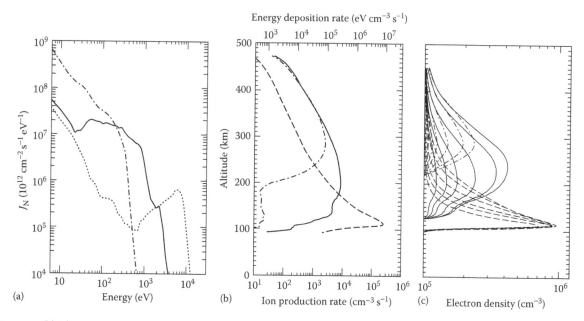

FIGURE 18.5 (a) Three examples of precipitating electron energy distribution functions, as observed by FAST, (b) simulated ionization rates (bottom axis) and energy deposition rates (upper axis) for the three distribution functions, and (c) plasma density profiles at different times after precipitation and up to 60 s. (From Thayer, J. P., and J. Semeter, *J. Atmos. Sol. Terr. Phys.*, 66, 807–824, 2004. With permission.)

increasing Pedersen conductivity in that region. More energetic precipitating electron profiles heat the gases near the *E*-layer, at similar altitudes where Joule heating takes place. It is, therefore, critical to understand such complex interactions in order to accurately specify and predict the state of the neutral atmosphere.

18.4.3.3 The Prior State of the Upper Atmosphere

Traveling atmospheric disturbances are generated at the locus of Joule heating of the thermosphere (Richmond, 1978; Balthazor and Moffett, 1997). They have been observed in accelerometer data (Bruinsma and Forbes, 2007) as a series of neutral density maxima, which were observed on successive orbits in the same local time. The maxima appear at different latitudes, from which a direction of propagation can be inferred. In most reported cases, TADs propagate from high to low latitudes (Bruinsma and Forbes, 2007; Huang et al., 2016). From the location and propagation direction, the region where Joule heating initiates can be deduced.

Figure 18.6 demonstrates how important is the prior state of the upper atmosphere, in the manner by which energy is being dissipated. The figure is reproduced from Thayer and Semeter (2004) with permission from the authors and demonstrates how winds affect the amount of dissipated energy and the amount of neutral heating. On the top panel are the observations from the Sondrestrom Incoherent Scatter Radar, showing the altitude distribution rate for the *j.E* term, which includes both frictional heating and mechanical transfer (solid line), and just the joule heating (dashed line). No difference between the two exists in the bottom-side *E*-layer (less than 120 km), but in the topside *E*-layer, the energy deposition that includes the mechanical transfer to winds is a factor of 3 smaller than the energy deposition that would exist if all energy was going into Joule heating. The bottom panel shows calculations of the neutral gas heating rates for the two curves in the top panel. Precipitating electron distributions with differing energy profiles can affect the altitude of heating and interact with how winds allow energy absorption at different altitudes.

Recent works, with simultaneous observations by CHAMP and GRACE, have demonstrated that the thermospheric response to magnetic storms is highly variable in altitude (Thayer et al., 2012; Liu et al., 2014). The altitudinal distribution of the

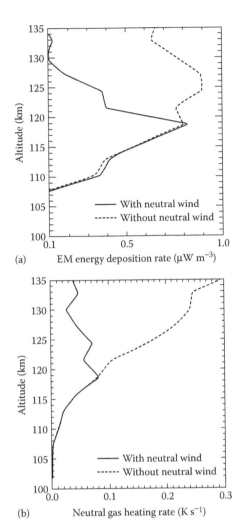

FIGURE 18.6 Ground radar observations of (a) *E*-region EM energy deposition rate, when measuring the quantity that is strictly Joule heating (dashed curve) and when measuring the term that includes both Joule heating and the momentum transfer to winds (solid curve) and (b) neutral gas heating rates based on the curves in (a). (From Thayer, J. P. and J. Semeter, *J. Atmos. Sol. Terr. Phys.*, 66, 807–824, 2004. With permission.)

thermospheric response is strongly dependent on the composition changes associated with geomagnetic activity. Particle precipitation and vertical winds, along with the prior state of the upper atmosphere, determine the composition before and during the geomagnetic activity.

Given the conventional emphasis in the IT studies on energy input to the auroral zone, it is surprising to note that thermospheric heating appears to be strongly focused at polar latitudes. From the earliest detection of neutral density, based on an analysis of the Sputnik 3 orbit, Jacchia (1959) showed a clear enhancement of the density over the northern pole during a magnetic storm. Liu et al. (2010) and Huang et al. (2016) confirmed this result in the studies of polar cap neutral densities, which showed clear maxima at polar latitudes in both hemispheres. This aspect of IT coupling and energy dissipation is relatively unexplored in current research and may prove useful, as higher accuracy in orbital drag predictions is sought.

18.4.4 Duration of Space Weather Impact

On the basis of the empirical studies of the change in exospheric temperature, as prescribed by the Jacchia models (1970, 1977), Burke et al. (2009b) derived a set of equations to describe the decay of temperature with time for a magnetic storm. They postulated that on a global scale, the energy in the thermosphere relaxes to a prestorm state as a driven-dissipative system, following the general form:

$$\frac{\mathrm{d}E_{\mathrm{thSW}}}{\mathrm{d}t} = \alpha_E \varepsilon_{\mathrm{VS}} - \frac{E_{\mathrm{thSW}}}{\tau_E}$$

where:

E_{thSW} is the thermal energy of the thermosphere due to solar wind forcing

α_E is a constant defining the coupling to the solar wind

$\varepsilon_{\mathrm{VS}}$ represents the interplanetary electric field, simplified as the Volland–Stern model field (Ejiri, 1978; Burke, 2007)

τ_E represents the decay time

This equation follows the same form as the driven-dissipative equation for changes in the D_{ST} index (Burton et al., 1975). The similarity between the two equations is based on the similarity of the temporal variations of thermospheric mass density and the D_{ST} index (Burke et al., 2009b).

The two constants, α_E and τ_E were determined empirically to be 5.5×10^{15} [(J/h)/(mV/m)] and 6.5 h, respectively. While these numbers are only approximate, they provided the basis for the Jacchia–Bowman 2008 (JB08) model (Bowman et al., 2008), as well as the extension of the Weimer (2005a) model to include cooling due to nitric oxide emissions (Weimer et al., 2011).

18.4.5 Thermospheric Cooling

The primary cooling of the thermosphere takes place by radiation by CO_2 and NO in the infrared portion of the spectrum (Kockarts, 1980; Sharma and Wintersteiner, 1990). The CO_2 emissions dominate below 120 km, whereas NO cooling occurs at higher altitudes, between 100 km and 200 km. While both molecular species respond to magnetic storms, NO responds more promptly to energy input and may be more important in regulating thermospheric energy balance. As described by Mlynczak et al. (2003), NO serves as a natural "thermostat" for the thermosphere.

NO is produced by the following exothermic reactions (Sharma et al., 1998):

$$N(^4S) + O_2 \rightarrow NO + O$$

$$N(^2D) + O_2 \rightarrow NO + O$$

Detection of emissions from the excited NO molecules provides a direct measure of the heat emitted by the thermosphere. The Sounding of the Atmosphere using Broadband Emission Radiometry (SABER) experiment on the TIMED satellite has provided extensive measurements of NO emissions for the past 13 years since the launch of TIMED in December 2001.

The TIMED satellite orbits at 74° inclination at an altitude of 625 km. SABER is a limb-scanning radiometer, which measures radiance altitude profiles along the orbit track from 400 km tangent height to the Earth's surface. Detailed descriptions of the derivation of the radiative cooling rates are given by Mlynczak et al. (2010).

Figure 18.7 shows the measured daily global power emitted in the NO and CO_2 band for the period 2002–2015, in which the solar cycle effect can be seen in the overall downward trend. In addition, large spikes occur, which coincide with the magnetic storms when the heat content of the thermosphere increases. A modeling study by Lu et al. (2010) showed that the global Joule heating predicted by physics-based modeling of a magnetic storm correlates well (correlation coefficient of 0.89) with the NO power measured by SABER, with a time lag of one day between the model predictions of Joule heat and the SABER observations of NO emissions.

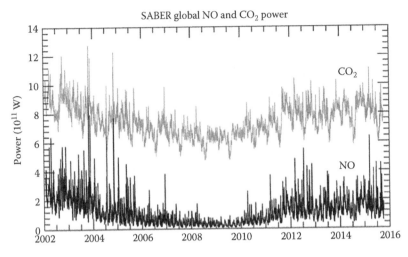

FIGURE 18.7   Daily global NO and CO_2 power at 100–200 km from 2002 to 2016. (Courtesy of L. Hunt.)

18.5 SPECIFICATION AND FORECAST OF THERMOSPHERIC DENSITY: EMPIRICAL MODELS AND ERROR ANALYSIS

Operational empirical density models used today are based on the Jacchia model series (Jacchia, 1970 and all prior versions), typically named the J70 model. Empirical models constructed by limited observations lack accuracy in detail and dynamics, whereas physics-based models can complement with the understanding of the different processes and their timescales. However, the coupled IT system and its response to external and internal drivers are so complex that physics-based models require a number of simplified assumptions, which in turn limit the accuracy and applicability of these models.

The IT community is clearly at a critical stage, as the demand for better orbit prediction is increasing, but the progress of predictive models is limited by a number of factors and does not match the need. This will be discussed in Section 18.7. Here, we briefly describe five empirical models: the J70, the MSIS group of models, the Drag Temperature Model, the JB08, and the operational High Accuracy Satellite Drag Model (HASDM).

The Jacchia model (Jacchia [1970] and references therein) used all the available low-resolution satellite drag data to create analytical representations of thermospheric temperature profiles as a function of latitude, local time, season, solar activity, and geomagnetic activity. The temperature profiles are linked to a single exospheric temperature, $T\infty$, which depends on solar activity, represented by the $F_{10.7}$ index, and geomagnetic activity, represented by the A_p index. Changes in $T\infty$ are the result of solar and geomagnetic activity. Height profiles of the major constituents (N_2, O, Ar, and He) are calculated as a function of the new $T\infty$, and coupled with the theoretical temperature profiles through the hydrostatic equation, they directly produce altitude profiles of density.

The MSIS series of models incorporates new composition (measured by satellites) and temperature (measured by ground radars) observations that became available in the 1970s (Hedin et al., 1974). The most updated version of this model is the NRL-MSISe-00 (Picone et al., 2002), and it includes more updated observations and more refined responses to geomagnetic activity, with dependence on the extended history of the present geomagnetic activity.

The drag temperature model is a variant of the MSIS and J70 models and is based on the orbital drag observations and measured temperatures from ground radar (Bruinsma et al., 2003). The limitation of this model in comparison to other models is the relatively simple dependence on geomagnetic activity (Prölss, 2011).

All of these models have similar errors, in the 15%–20% one-sigma range (Marcos et al., 2007, 2010). These errors were reduced significantly by the Air Force operations' HASDM (Storz et al., 2005), which is based on the J70 empirical density model. To improve the tracking of the density errors, observations from the SSN for ~80

18.6 PHYSICS-BASED MODELS

18.6.1 Ionosphere–Thermosphere Models (CTIPe, TIE-GCM, and GITM)

calibration satellites were processed directly to fit the J70 density profiles to produce corrected density profiles. Thermospheric density correction parameters were computed along with the trajectories of the 80 calibration satellites in a single estimation process named the Dynamic Calibration Atmosphere (Storz et al., 2005). The HASDM improved density specification and reduced errors down to the 4%–8% (Marcos et al., 2009). It has been the Air Force Space Command operational model and is updated with new density profiles every 3 h.

The JB08 model (Bowman et al., 2008) replaced the J70 model inside HASDM. The JB08 upgrades include better solar indices, semiannual effect formulation, and local time variation of density. Most important, JB08 includes an improved geomagnetic storm index, providing density responses during storms that were not captured by the A_p index. The HASDM is currently running with the JB08 as its basic empirical model and it uses approximately 150 calibration satellites. While HASDM has been a game changer for density specification, based on the corrections from the calibration satellites every 3 h, its core model, the JB08, provides the routine up to 72-h forecasts.

Validation of these models by using CHAMP observations was presented by Marcos et al. (2010). They found that the JB08 standard deviations in model-data comparisons were smaller than all other models. They also found that the most accurate density specifications were produced by the HASDM data assimilative model capability, as shown in Figure 18.8.

The GCMs were a major step forward in physics-based modeling of the IT system, providing physical descriptions of processes previously reduced to empirical data. A series of coupled IT three-dimensional general circulation models have been developed over recent decades (Rees et al., 1980; Dickinson et al., 1981, 1984; Roble et al., 1987; Fuller-Rowell et al., 1988, 2000; Ridley and Deng, 2006). The models solve the fundamental momentum, energy, and continuity equations for neutral and ion species in global self-consistent physics-based dynamic descriptions of the IT system. They have been widely used to simulate a wide range of electrodynamic processes, including energy deposition and dissipation, at high latitudes under active conditions. The three models discussed here are the Coupled Thermosphere Ionosphere Plasmasphere Electrodynamics (CTIPe), the National Center for Atmospheric Research (NCAR's) Thermosphere Ionosphere Electrodynamics Global Circulation Model (TIE-GCM), and the Global Ionosphere–Thermosphere Model (GITM), all of which have been submitted to the Community Coordinated Modeling Center (CCMC) (http://ccmc.gsfc.nasa.gov/) for runs on demand and model assessment.

The CTIPe model is a nonlinear, coupled thermosphere–ionosphere–plasmasphere physics-based code that includes a self-consistent electrodynamics scheme for the computation of dynamoelectric fields. There

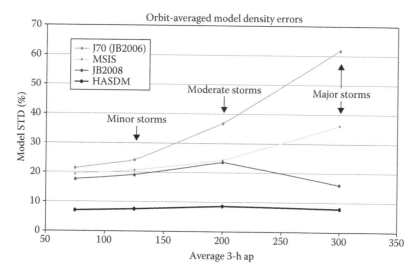

FIGURE 18.8 Empirical model orbit-averaged density errors for J70, MSIS, JB08, and HASDM. (From Marcos, F. A. et al., Toward next level of satellite drag modeling, Toronto, Ontario, Canada, 2010. With permission.)

are four distinct components that run simultaneously and are coupled: a global thermosphere, a high-latitude ionosphere, a mid- and low-latitude ionosphere and/or plasmasphere, and an electrodynamical calculation of the global dynamoelectric field (Fuller-Rowell and Rees, 1980; Codrescu et al., 2012). The thermospheric component is divided into a geographic latitude × longitude grid with resolution 2° × 18°. The vertical resolution is defined in terms of logarithm (pressure) from a lower boundary at altitude 80 km to altitudes above 500 km. The primary inputs to the model are the high-latitude electric field, typically provided by the Weimer (2005a) empirical model, auroral electron precipitation from the empirical model by Fuller-Rowell and Evans (1987), and EUV radiation (Solomon and Qian, 2005). Full descriptions of the model with examples of output can be found in Codrescu et al. (2012). The CTIPe is run in near-real time at the Space Weather Prediction Center, and model output is available for comparison, with observations, at http://helios.swpc.noaa.gov/ctipe/index.html.

The TIE-GCM provides a three-dimensional, nonlinear representation of the coupled IT system. A self-consistent calculation of ionospheric wind dynamic effects (Richmond et al., 1992) is included. The primary external drivers of the model are solar irradiance, magnetospheric energy, and tidal perturbations at the lower boundary of the model. Magnetospheric energy inputs include particle precipitation and high-latitude electric fields, which drive convection. The current version of TIE-GCM (v1.95) uses a fixed grid with spatial resolution of 2.5°× 2.5°× half-scale height (longitude × latitude × altitude) with 10-min temporal resolution. As with CTIPe, the vertical extent of the model is defined by pressure, with typical range of 97–500 km. The model can be downloaded directly from NCAR's website: http://www.hao.ucar.edu/modeling/tgcm/.

The GITM is the most recent of the widely used GCMs. It differs from CTIPe and TIE-GCM as follows: (1) spatial resolution is flexible; (2) the assumption of hydrostatic equilibrium is dropped, allowing for vertical acceleration; (3) the adoption of an altitude grid instead of scaling by pressure; (4) advection is solved explicitly; and (5) chemistry is solved explicitly without the assumption of chemical equilibrium. As a result, the time step in GITM is 2–4 s. The model uses the same magnetospheric drivers as CTIPe and TIEGCM,

viz., high-latitude electric fields and energetic particle precipitation.

The most widely used drivers to simulate the high-latitude electric field on the IT system are empirical models by Weimer (2005a), hereafter referenced as W05, and Heelis et al. (1982). A more recent empirical model was proposed by Cosgrove et al. (2014). An alternative is the assimilative mapping of ionospheric electrodynamics (AMIE) (Richmond and Kamide, 1988), which relies on the assimilation of magnetometer output and an inversion technique to derive electric potentials. More rarely, the output from magnetohydrodynamic (MHD) models such as the Block Adaptive-Tree Solar wind Roe-Type Upwind Scheme (BATS-R-US) (Powell et al., 1999), can be used as input, or MHD models (BATS-R-US, open geospace general circulation model [OpenGGCM] [Raeder et al., 1998]) can be coupled directly to IT models, which obviates the need for separate high-latitude electric field specification.

The empirical models by Weimer and Heelis are the most widely used drivers in IT modeling. One criticism of this approach is that the empirical models, by necessity, average over large amounts of data, which leads to the loss of the dynamic features common to magnetic storms. Thus, the usefulness of this approach is questionable. Data assimilation (DA) schemes coupled to the GCMs have been mentioned (Codrescu et al., 2012), but results have not been published at the time of writing. The AMIE, which can assimilate real-time data, was originally based on the assimilation of ground magnetometer data. Later upgrades have expanded the data input to include satellite-borne magnetometers. However, at high latitudes in the Northern hemisphere, there remains a lack of sufficient coverage. Availability of data from the Southern hemisphere is even more limited, so both hemispheres cannot be equally simulated.

The conductivity in physics-based models is derived from separate empirical models (Fuller-Rowell and Evans,1987; Roble and Ridley, 1987), which specifies particle precipitation at high latitudes. Note that the electric field and particle precipitation drivers are typically derived independently and are not consistent with each other.

We show the results from an assessment study that compares the outcome of model runs carried out by the CCMC for a magnetic storm that occurred in December 2006 (Shim et al., 2015). Figure 18.9 shows observed and

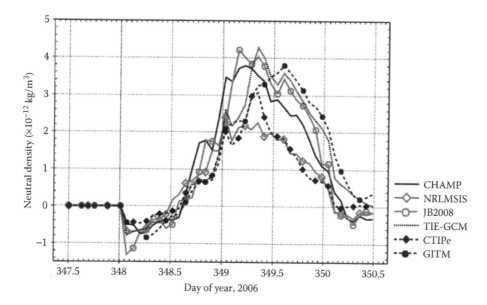

FIGURE 18.9　Observed and modeled neutral mass density data for a magnetic storm in December 2006 with baseline subtraction. (Courtesy of J. Shim.)

modeled data for the event, where all data have been shifted to zero by a point-by-point subtraction of the quiet-time neutral density obtained during the previous quiet day (day 347). Orbit-averaged neutral mass density from CHAMP satellite is shown as a sold line, together with model output from empirical models, the US Naval Research Laboratory MSIS (2000) and JB08 (Bowman et al., 2008), and GCMs TIE-GCM, CTIPe, and GITM. All three GCMs use W05 for the high-latitude electric field. The particle input from Roble and Ridley (1987)

has been used in the TIE-GCM run, and the Fuller-Rowell and Evans (1987) model is used with CTIPe and GITM.

In Figure 18.10, a comparison of the effects of imposing different high-latitude electric field drivers on CTIPe is shown for modeled orbit-averaged neutral mass densities, together with the CHAMP neutral mass density data for the same storm as in Figure 18.9. The same point-by-point subtraction of the quiet-time density on day 347 has been done. The drivers include inputs from empirical

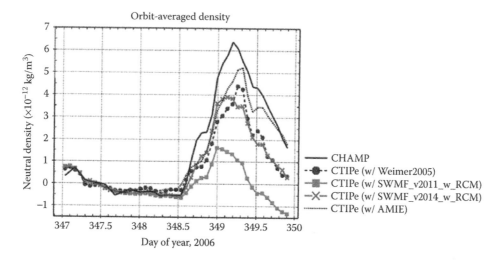

FIGURE 18.10　Orbit-averaged mass density from CHAMP observations and CTIPe model with different high latitude potential for the December 2006 storm. The SWMF electric fields are derived from the BATS-R-US and RCM-coupled models. (Courtesy of J. Shim.)

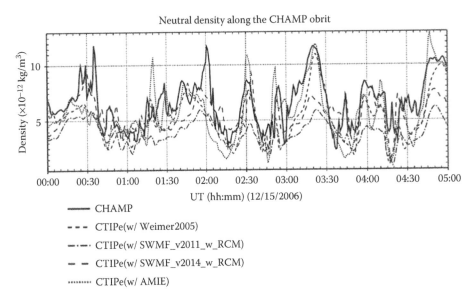

FIGURE 18.11 Mass density along CHAMP orbit, with CTIPe model mass densities produced using empirical, MHD, and assimilative drivers. (Courtesy of J. Shim.)

(W05), MHD (space weather modeling framework_rice convection model [SWMF_RCM]), and assimilative (AMIE) models. The W05 and AMIE models have been described above. The SWMF_RCM model couples BATS-R-US to the Rice Convection Model. This coupled model computes the electric field, which is passed to CTIPe in this study. Two versions were tested, labeled v2011 and v2014. More information about SWMF models can be found at http://ccmc.gsfc.nasa.gov/models/modelinfo.php?model=SWMF/BATS-R-US%20with%20RCM.

Figure 18.11 shows the CHAMP data along the satellite orbit for the December 2006 storm, with CTIPe as the physics-based model, using the same drivers shown in Figure 18.10. It can be seen that at times, there is general agreement between the model output and data, but there are times when the results are highly inconsistent.

The conclusions of the CCMC model assessment are that the performances of the empirical and physics-based models with different drivers depend on the metrics applied. Results vary with location, timing, and magnitude of neutral density variations. Furthermore, the skill scores obtained for other events not reported in the study are not consistent with those for the December 2006 storm. It should be noted that the GCMs are not clearly superior to the empirical models in these assessments.

18.7 FUTURE DIRECTIONS

The goals in the field of space weather impact on orbital drag are to (1) specify and (2) forecast drag with high accuracy. The operational requirements of the Department of Defense are to forecast thermospheric density to an accuracy of 5% 72 h in advance. Currently, empirical models are still the mainstay of operational application, despite the widespread belief that improved high-accuracy forecasting can only be achieved by using physics-based models. As discussed in Section 18.6, the orbit-averaged data from empirical and physics-based models show some agreement after baseline adjustments, but in point-by-point comparisons, the GCMs do not perform well. In this section, we consider what the possible impediments to improved performance by physics-based models may be and how these can be remediated.

Recent observational studies have reported deviations from the conventional view of high-latitude IT energetic processes, in which the auroral zones are the locus of all substantial energy inputs (Huang et al., 2016). On the basis of the estimates of the energy required to heat the thermosphere to observed levels, Burke et al. (2010) and Huang et al. (2014a) suggested that energy must enter the polar cap and not be included in empirical models of energy deposition (Weimer, 2005a; Cosgrove et al., 2014). Direct observations of Poynting flux from the Defense Meteorological Satellite System spacecraft confirm this suggestion. During the main phases of magnetic storms, substantial levels of Poynting flux were recorded sporadically at all local times and latitudes poleward of the convection reversal boundaries used to denote the polar cap boundary (Huang et al., 2015). In addition, polar rain, often set to zero in many models, can be significant in creating Pedersen conductivity at polar latitudes (Huang et al., 2014b).

Evidence that energy enters the polar cap is provided by many years of observations of thermospheric density and composition at high latitudes. The earliest recording of a storm-related enhancement in neutral density was shown by Jacchia (1959). A clear enhancement in neutral density was found over the Northern polar cap during a storm. Further evidence of changes in composition due to density enhancements in the polar caps was provided by Prölss (1997) and Prölss et al. (1988). More recently, polar cap enhancements in neutral density during moderate storms from 2002 to 2005 were described by Liu et al. (2010). It should be noted that there is additional energy superimposed on general polar cap energy input, which enters the cusp, heating neutrals in this region (Lühr et al., 2004). However, the results shown by Liu et al. (2010) show energization of neutrals at all local times at polar latitudes.

In the standard empirical models widely used as input to the GCMs, the polar cap is generally devoid of energy input either from Poynting flux or from particle precipitation. The reason for this discrepancy between the empirical models (Heelis et al., 1982; Fuller-Rowell and Evans, 1987; Roble and Ridley, 1987; Weimer, 2005a; Cosgrove et al., 2014) is fundamentally because of methodology. Empirical models are created by averaging over large volumes of data, typically sorted by solar and solar wind input: K_p, IMF, solar wind density, and speed. In the process of averaging, followed

by functional fits to the data, dynamic features that are the essence of magnetic storms are smoothed to the point of disappearance. Thus, empirical drivers will not and cannot capture the dynamic nature of storms. The discrepancy between model and observations is shown in Figure 18.12. The Weimer and Cosgrove model predictions of Poynting flux are compared with the observed values from the Defense Meteorological Satellite System. A second problem is that the models that describe the high-latitude electric fields and particle precipitation are derived separately and are not consistent with each other.

The AMIE model offers an alternative to the empirical drivers. This model assimilates magnetometer data in order to specify the electric field. The problem with AMIE is the lack of sufficient coverage, a problem that is particularly marked in the Southern hemisphere, and in both hemispheres at polar latitudes. An additional problem with this approach is the lack of response of ground magnetometers to Pedersen currents, which are the primary currents at polar latitudes (Huang and Burke, 2004). The best AMIE results are obtained when more than ground magnetometer data, such as auroral precipitation and radar convection, are assimilated.

There is no simple remedy to the driver problem. In the three GCMs discussed in this chapter, all dynamics are provided by the drivers. If the drivers are empirical models of energy input, the GCMs cannot provide faithful predictions of magnetic storm dynamics, which include accurate forecasts of thermospheric densities.

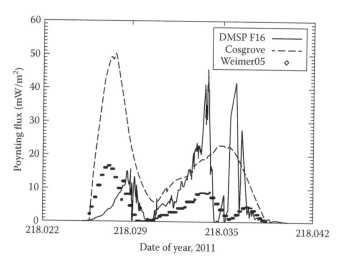

FIGURE 18.12 Comparison of model and observed Poynting flux along the DMSP F16 orbit during a magnetic storm that occurred in August 2011. (Courtesy of Y. Huang.)

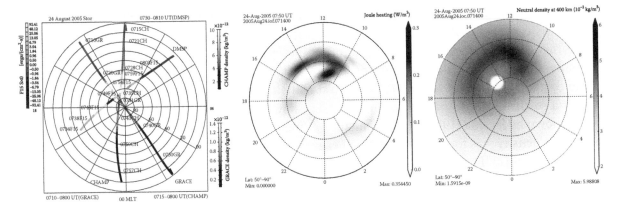

FIGURE 18.13 (Left) Observations of Poynting flux from DMSP and neutral density responses from CHAMP and GRACE after the shock that preceded the storm of August 24, 2005. (Center) Simulated neutral density response by the coupled OpenGGCM-CTIM model. (Right) Simulated Joule heating by the coupled OpenGGCM-CTIM model. (Courtesy of Hyunju Connor and Eftyhia Zesta.)

There has been also some emphasis on coupled models between global magnetospheric MHD models and IT physics-based models, which could in theory provide an end-to-end solution, starting from the Sun and ending at the Earth's surface. Such an example is shown in Figure 18.13, which demonstrates how the open global geospace circulation model - coupled thermosphere ionosphere model (OpenGGCM-CTIM) coupled model can more accurately reproduce the high-latitude heating during a storm's main phase. Despite some limited success, such coupled models have worked better for magnetospheric studies than for actual thermospheric studies (Connor et al., 2014). The MHD modes may produce dynamic EM energy input with solar wind coupling but can still not produce realistic precipitation patterns. Furthermore, the coupling with IT models is far from the state of the art for the merging of two physical systems that function with inherently incompatible time and length scales. Such couplings have produced complicated problems, with incompatible boundary conditions, numerical diffusion, and temporal/spatial resolution. As a result, the current trend is away from large-scale coupled models and toward the GCMs driven by empirical models of the electric field and particle precipitation. While better progress of the coupled models may still be possible in the future, this does not seem to be having much traction now. In addition, the GCMs driven by empirical inputs do not solve the biggest problem of accurately specifying and forecasting thermospheric responses.

One possible solution to the forecast problem is to couple DA schemes, which assimilate widely available data to the GCMs. A suite of DA codes is available at the Data Assimilation Research Testbed (Anderson et al., 2009), a community facility for DA using ensemble Kalman filter techniques developed at NCAR. The routines have been used in limited applications in assimilating neutral density and total electron content (Lee et al., 2012; Matsuo et al., 2013), but the algorithms cannot currently be applied to assimilation of GCM drivers (Matsuo, private communication, 2014). A number of other ensemble Kalman filters or Gauss–Markov algorithms have been used in DA to assimilate ionospheric measurements as state variables in models such as Global Assimilation of Ionospheric Measurements (USU-GAIM), the Global Assimilative Ionospheric Model (JPL-GAIM), the Ionospheric Data Assimilation Four-Dimensional (IDA4D) model, and others (Schunk, 2002; Bust et al., 2004; Hajj et al., 2004; Pi et al., 2004; Scherliess et al., 2004). This is a new area of IT model development that needs wider application and validation but have the potential to change the current state of the art. The basic principle in DA is to use observations to adjust model values in the way HASDM uses the neutral densities obtained from the calibration satellites to shift the Jacchia values to specify the actual thermosphere. This approach reduced the error in HASDM specification from an average of 15% to the 4%–8% level.

In principle, any observation can be the variable which is assimilated and modified, and any model can be the basis of assimilation. In practice, the density of observations is limited both in space and in time, and complex models are more difficult to couple to assimilative schemes. The most common state variable is the electron density, which can be detected from Global Positioning System satellites by ground receivers as total electron content, by satellites, and from an extensive array of ground-based radars and ionosondes. When assimilated into a basic empirical model, this yields a conductivity profile that should improve on empirical model input. Unfortunately, the highest data density is at low and mid latitudes, where Global Positioning System coverage is high, not over the polar regions, where energy enters the IT system. Assimilation of other variables such as neutral density, electric fields, and winds, is more difficult because of the sparseness of the available data. Thus, the question of how useful DA will be may depend on coherence scales in time and space, on the geoeffectiveness of small-scale (1–10 km), short-lived (seconds to a few minutes) structures, on the persistence of the state of the system, and on how well DA can be merged with empirical models where data are not available. These are the active areas of research, which may be productive in the future. Further details on the status of DA as an ionospheric modeling tool can be found in reviews by Schunk et al. (2011), Bust and Mitchell (2008), and Mandrake et al (2005).

Finally, the DA techniques need to be coupled with wide community efforts for better observations, particularly at the high latitudes. The Heliophysics decadal survey (Committee on a Decadal Strategy for Solar and Space Physics [Heliophysics], 2012) has directed the procurement of large strategic missions such as Dynamic and geospace dynamics constellation (GDC) that will focus on the IT responses to energy inputs. The advances in small satellite and Cubesat technology and launch opportunities are the primary fertile ground for our community to observationally study the temporal and spatial scales of magnetospheric forcing and the responses of the thermosphere in ways that directly apply to orbital drag specification and forecasting. Such observations in combination with DA techniques can provide the breakthrough that is much needed.

REFERENCES

Achieving and maintaining orbit. http://earthobservatory. nasa.gov/Features/OrbitsCatalog/ (link is external).

Anderson, J., T. Hoar, K. Raeder, H. Liu, N. Collins, R. Torn, A. Avellano (2009). The data assimilation research testbed: A community facility. *Bull. Amer. Meteor. Soc.*, 90, 1283–1296. http://dx.doi.org/10.1175/2009BAMS2618.1.

Balthazor, R. L. and R. J. Moffett (1997). A study of atmospheric gravity waves and travelling ionospheric disturbances at equatorial latitudes. *Ann. Geophys.*, 15, 1048–1056.

Boudouridis, A., E. Zesta, R. Lyons, P. C. Anderson, and D. Lummerzheim (2003). Effect of solar wind pressure pulses on the size and strength of the auroral oval. *J. Geophys. Res. Space Phys. (1978–2012)*, 108, 8012. doi:10.1029/2002JA009373, A4.

Bowman, B. R., W. K. Tobiska, F. A. Marcos, C. Y. Huang, C. S. Lin, and W. J. Burke (2008). A new empirical thermospheric density model JB2008 using new solar and geomagnetic indices. Paper presented at the AIAA/AAS Astrodynamics Specialist Conference, American Institute of Aeronautics and Astronautics, Honolulu, HI, 18–21 August.

Bruinsma, S., D. Tamagnan, and R. Biancale (2004). Atmospheric densities derived from CHAMP/STAR accelerometer observations. *Planet Space Sci.*, 52, 297–312. doi:10.1016/j.pss.2003.11.004.

Bruinsma, S., G. Thullier, and F. Barlier (2003). The DTM-2000 empirical thermosphere model with new data assimilation and constraints at lower boundary: Accuracy and properties. *J. Atmos. Sol. Terr. Phys.*, 65, 1053–1070. doi:10.1016/S1364-6826(03)00137-8.

Bruinsma, S. L., and J. M. Forbes (2007). Global observation of traveling atmospheric disturbances (TADs) in the thermosphere. *Geophys. Res. Lett.*, 34(14), L14103. doi:10.1029/2007GL030243.

Burke, W. J. (2007). Penetration electric fields: A Volland-Stern approach. *J. Atmos. Sol. Terr. Phys.*, 69, 1114–1126.

Burke, W. J., C. Y. Huang, and R. D. Sharma (2009a). Stormtime dynamics of the global thermosphere and equatorial ionosphere. *Ann. Geophys.*, 27, 2035–2044.

Burke, W. J., C. Y. Huang, D. R. Weimer, J. O. Wise, G. R. Wilson, C. S. Lin, and F. A. Marcos (2010). Energy and power requirements of the global thermosphere during the magnetic storm of November 10, 2004. *J. Atmos. Sol. Terr. Phys.*, 72, 309–318. doi:10.1016/j.jastp.2009.06.005.

Burke, W. J., C. S. Lin, M. P. Hagan, C. Y. Huang, D. R. Weimer, J. O. Wise, L. C. Gentile, and F. A. Marcos (2009b). Storm time global thermosphere: A driven-dissipative thermodynamic system. *J. Geophys. Res.*, 114, A06306.doi:10.1029/2008JA013848.

Burton, R. K., R. L. McPherron, and C. T. Russell (1975). An empirical relationship between interplanetary conditions and Dst. *J. Geophys. Res.*, 80, 4204.

Bust, G. S., T. W. Garner, and T. L. Gaussiran II (2004). Ionospheric Data Assimilation Three Dimensional (IDA3D): A global, multi-sensor, electron density specification algorithm. *J. Geophys. Res.*, 109, A11312. doi:10.1029/2003JA010234.

Bust, G. S., and C. N. Mitchell (2008). History current state, and future directions of ionospheric imaging. *Rev. Geophys.*, 46, RG1003. doi:10.1029/2006RG000212.

Cane, H. V., and Richardson, I. G. (2003). Interplanetary coronal mass ejections in the near-Earth solar wind during 1996–2002. *J. Geophys. Res.*, 108, 1156. doi:10.1029/2002JA009817, A4.

Chen, G. M., Xu, J., Wang, W., Lei, J., and Burns, A. G. (2012). A comparison of the effects of CIR-and CME-induced geomagnetic activity on thermospheric densities and spacecraft orbits: Case studies. *J. Geophys. Res.*, *117*, A08315. doi:10.1029/2012JA017782.

Codrescu, M. V., C. Negrea, and M. Fedrizzi, T. J. FullerRowell, A. Dobin, N. Jakowsky, H. Khalsa, T. Matsuo, and N. Maruyama (2012). A real-time run of the coupled thermosphere ionosphere plasmasphere electrodynamics (CTIPe) model. *Space Weather*, *10*, S02001. doi:10.1029/2011SW000736.

Committee on a Decadal Strategy for Solar and Space Physics (Heliophysics), SS Board, Aeronautics and Space Engineering Board, Division on Engineering and Physics Sciences and National Research Council (2012). doi:10.17226/13060.

Conde, M., and P. L. Dyson (1995). Thermospheric vertical winds above Mauson, Antarctica. *J. Atmos. Terr. Phys.*, *57*, 589–596.

Connor, H. K., Zesta, E., Ober, D. M., and Raeder, J. (2014). The relation between transpolar potential and reconnection rates during sudden enhancement of solar wind dynamic pressure: OpenGGCM-CTIM results. *J. Geophys. Res. Space Phys.*, *119*(5), 3411–3429.

Cosgrove, R. B., H. Bahcivan, S. Chen, R. J. Strangeway, J. Ortega, M. Alhassan, Y. Xu, M. Van Welie, J. Rehberger, S. Musielak, and N. Cahill (2014). Empirical model of Poynting flux derived from FAST data and a cusp signature. *J. Geophys. Res.*, *119*, 411–430. doi:10.1002/2013JA019105.

Crowther, R. (2003). Orbital debris: A growing threat to space operations. *Philos. Trans. R. Soc. Lond. Ser. A*, *361*, 157–168. doi:10.1098/rsta.2002.1118.

Dickinson, R. E., E. C. Ridley, and R. G Roble (1984). Thermospheric general circulation with coupled dynamics and composition. *J. Atmos. Sci.*, *41*, 205–219.

Dickinson, R. E., E. C. Ridley, and R. G. Roble (1981). A three dimensional general circulation model of the thermosphere. *J. Geophys. Res.*, *86*, 1499–1512.

Doyle, M. A., W. J.Burke, D. A. Hardy, P. F. Bythrow, F. J. Rich, and T. A. Potemra (1986). A simple model of auroral electrodynamics compared with HILAT measurements. *J. Geophys. Res. Space Phys.*, *91*(A6), 6979–6985.

Ejiri, M. (1978). Trajectory traces of charged particles in the magnetosphere. *J. Geophys. Res.*, *83*, 4798.

Fedrizzi, M., T. J. Fuller-Rowell, and M. V. Codrescu (2012). Global Joule heating index derived from thermospheric density physics-based modeling and observations. *Space Weather*, *10*, S03001. doi:10.1029/2011SW000724.

Fuller-Rowell, T. J., M. V. Codrescu, and P. Wilkinson (2000). Quantitative modeling of the ionospheric response to geomagnetic activity. *Ann. Geophys.*, *18*(7), 766–781. doi:10.1007/s00585-000-0766-7.

Fuller-Rowell, T. J., and D. S. Evans (1987). Height-integrated Pedersen and Hall conductivity patterns inferred from the TIROS-NOAA satellite data. *J. Geophys. Res.*, *92*, A7, 7606–7618.

Fuller-Rowell, T. J., and D. Rees (1980). A three-dimensional, time dependent global model of the thermosphere. *J. Atmos. Sci.*, *37*, 2454.

Fuller-Rowell, T. J., D. Rees, S. Quegan, R. J. Moffett, G. J. Bailey (1988). Simulations of the seasonal and universal time variations of the high-latitude thermosphere and ionosphere using a coupled, three-dimensional model. *Pure Appl. Geophys.*, 127, 189–217. doi:10.1007/BF00879811.

Gary, J. B., R. A. Heelis, W. B. Hanson, and J. A. Slavin (1994). Field-aligned Poynting flux observations in the high-latitude ionosphere. *J. Geophys. Res.*, *99*(A6), 11417–11427. doi:10.1029/93JA03167.

Gary, J. B., R. A. Heelis, and J. P. Thayer (1995). Summary of field-aligned Poynting flux observations from DE2. *Geophys. Res. Lett.*, *22*(14), 1861–1864.doi:10.1029/95GL00570.

Hajj, G. A., B. D. Wilson, C. Wang, X. Pi, and I. G. Rosen (2004). Data assimilation of ground GPS total electron content into a physics-based ionospheric model by use of the Kalman filter. *Radio Sci.*, *39*, RS1S05. doi:10.1029/2002RS002859.

Hedin, A. E., H. G. Mayr, C. A. Reber, N. W. Spencer, and G. R. Carignan (1974). Empirical model of global thermospheric temperature and composition based on data from the Ogo 6 quadrupole mass spectrometer. *J. Geophys. Res.*, *79*(1), 215–225.

Heelis, R. A., J. K. Lowell, and R. W. Spiro (1982). A model of the high-latitude ionospheric convection pattern. *J. Geophys. Res.*, *87*(A8), 6339–6345.

Huang, C. Y., and W. J. Burke (2004). Transient sheets of field-aligned current observed by DMSP during the main phase of a magnetic superstorm. *J. Geophys. Res.*, *109*, A06303. doi:10.1029/2003JA010067.

Huang, C. Y., Y. Huang, Y.-J.Su, E. K. Sutton, M. Hairston, and W. R. Coley (2016). Ionosphere-thermosphere (IT) response to solar wind forcing during magnetic storms. *J. Space Weather Space Clim.*, 6, A4. doi: 10.1051/swsc/2015041.

Huang, C. Y., Y.-J. Su, E. K. Sutton, D. R.Weimer, and R. L. Davidson (2014a). Energy coupling during the August 2011 magnetic storm. *J. Geophys. Res. Space Phys.*, *119*. doi:10.1002/2013JA019297.

Huang, Y., C. Y. Huang, Y.-J.Su, Y. Deng, and X. Fang (2014b). Ionization due to electron and proton precipitation during the August 2011 storm. *J. Geophys. Res. Space Phys.*, *119*, 3106–3116. doi:10.1002/2013JA019671.

Huang, Y., A. D. Richmond, Y. Deng, P. C. Chamberlin, L. Qian, S. C. Solomon, R. G. Roble, and Z. Ziao (2013). Wavelength dependence of solar irradiance enhancement during X_class flares and its influence on the upper atmosphere. *J. Atmos. Sol. Terr. Phys.*, *115*, 87. doi:10.1016/j.jastp.2013.10.011.

Innis, J. L., and M. Conde (2001). Thermospheric vertical wind activity maps derived from Dynamics Explorer-2 WATS observations. *Geophys. Res. Lett.*, *28*, 3747–3850.

Jacchia, L. G. (1959). Corpuscular radiation and the acceleration of artificial satellites. *Nature*, *183*, 1662–1663.

Jacchia, L. G. (1970). New static models of the thermosphere and exosphere with empirical temperature profiles. SAO Special Report 313.

Jacchia, L. G. (1977). Thermospheric temperature, density and composition: A new model. SAO Special Report 375.

Kelley, M. C., D. J. Knudsen, and J. F. Vickrey (1991). Poynting flux measurements on a satellite: A diagnostic tool for space research. *J. Geophys. Res., 96*(A1), 201–207. doi:10.1029/90JA01837.

Killeen, T. L., and R. G. Roble (1988). Thermosphere dynamics: Contributions from the first 5 years of the Dynamics Explorer Program. *Rev. Geophys., 26*(2), 329–367. doi:10.1029/RG026i002p00329.

Knipp, D. J., W. K. Tobiska, and B. A. Emery (2004). Direct and indirect thermospheric heating sources for solar cycles 21–23. *Sol. Phys., 224*(1–2), 495–505.

Kockarts, G. (1980). Nitric oxide cooling in the terrestrial thermosphere. *Geophys. Res. Lett., 7,* 137–140. doi:10.1029/GL007i002p00137.

Laundal, K. M., and N. Østgaard (2008). Persistent global proton aurora caused by high solar wind dynamic pressure. *J. Geophys. Res., 113, A08231.* doi:10.1029/2008JA013147.

Lee, I.-T., T. Matsuo, A. D. Richmond, J. Y. Liu, W. Wang, C. H. Lin, J. L. Anderson, and M. Q. Chen(2012). Assimilation of FORMOSAT-3/COSMIC electron density profiles into a coupled thermosphere/ionosphere model using ensemble Kalman filtering. *J. Geophys. Res., 117,* A10318. doi:10.1029/2012JA017700.

Lei, J., J. P. Thayer, J. M. Forbes, E. K. Sutton, and R. S. Nerem (2008). Rotating solar coronal holes and periodic modulation of the upper atmosphere. *Geophys. Res. Lett., 35,* L10109. doi:10.1029/2008GL033875.

Liou, K., P. T. Newell, T. Sotirelis, and C.-I.Meng (2006). Global auroral response to negative pressure impulses. *Geophys. Res. Lett., 33,* L11103. doi:10.1029/2006GL025933.

Liu, R., H. Lühr, S.-Y. Ma (2010). Storm-time related mass density anomalies in the polar caps observed by CHAMP. *Ann.Geophys., 28*(1), 165–180.

Liu, X., J. P. Thayer, A. Burns, W. Wang, and E. Sutton (2014). Altitude variations in the thermosphere mass density response to geomagnetic activity during the recent solar minimum. *J. Geophys. Res. Space Phys., 119*(3), 2160–2177.

Lu, G., M. G. Mlynczak, L. A. Hunt, T. N. Woods, and R. G. Roble (2010). On the relationship of Joule heating and nitric oxide radiative cooling in the thermosphere. *J. Geophys. Res., 115,* A05306. doi:10.1029/2009JA014662.

Lu, G., A. D. Richmond, B. A. Emery, and R. G. Roble (1995). Magnetosphere-ionosphere-thermosphere coupling: Effect of neutral winds on energy transfer and field-aligned current. *J. Geophys.Res. Space Phys., 100*(A10), 19643–19659.

Lühr, H., M. Rother, W. Köhler, P. Ritter, and L. Grunwaldt (2004). Thermospheric up-welling in the cusp region: Evidence from CHAMP observations. *Geophys. Res. Lett., 31,* L06805. doi:10.1029/2003GL019314.

Maeda, S., T. J. Fuller-Rowell, and D. S. Evans (1989). Zonally averaged dynamical and compositional response of the thermosphere to auroral activity during September 18–24, 1984. *J. Geophys. Res., 94*(A12), 16869–16883.

Mandrake, L., B.Wilson, C.Wang, G. Hajj, A. Mannucci, and X. Pi (2005). A performance evaluation of the operational Jet Propulsion Laboratory/University of Southern California Global Assimilation Ionospheric Model (JPL/USC GAIM). *J. Geophys. Res., 110,* A12306. doi:10.1029/2005JA011170.

Marcos, F. A., W. J. Burke, and S. T. Lai (2007). *Thermospheric Space Weather Modeling,* Air Force Research Laboratory Report, ADA471447, Defense Technical Information Center.

Marcos, F. A., C. Y. Huang, C. S. Lin, J. M. Retterer, and S. H. Delay (2009). Evaluation of recent thermospheric neutral density models.Executive Summary, For Dr. Lance Menthe, RAND Corp.

Marcos, F. A., S. T. Lai, C. Y. Huang, C. S. Lin, J. M. Retterer, S. H. Delay, and E. Sutton. (2010). Toward next level of satellite drag modeling.AIAA Atmospheric and Space Environments Conference 2–5 August, Toronto, Ontario, Canada.

Matsuo, T., I.-T. Lee, and J. L. Anderson (2013). Thermospheric mass density specification using an ensemble Kalman filter. *J. Geophys. Res. Space Phys., 118,* 1339–1350. doi:10.1002/jgra.50162.

Mlynczak, M., F. J. Martin-Torres, J. Russell, K. Beaumont, S. Jacobson, J. Kozyra, M. Lopez-Puertas, B. Funke, C. Mertens, L. Gordley, R. Picard, J. Winick, P. Wintersteiner, and L. Paxton (2003). The natural thermostat of nitric oxide emission at 5.3 mm in the thermosphere observed during the solar storms of April 2002. *Geophys. Res. Lett., 30*(21), 2100. doi:10.1029/2003GL017693.

Mlynczak, M. G., L. A. Hunt, B. T. Marshall, F. J. Martin-Torres, C. J. Mertens, J. M. Russell III, E. Remsberg, M. Lopex-Puertas, R. Picard, J. Winick, P. Wintersteiner, R. E. Thompson, and L. L. Gordley (2010). Observations of infrared radiative cooling in the thermosphere on daily to multiyear timescales from the TIMED/SABER instrument. *J. Geophys. Res., 115,* A03309.doi:10.1029/2009JA014713.

NASA (2009). Satellite collision leaves significant debris clouds *(PDF). Orbital Debris Q. News, 13*(2), 1–2.

National Research Council (NRC) (1995). *Orbital Debris: A Technical Assessment.* National Academy of Sciences, Washington, D.C.

Pi, X., C. Wang, G. A. Hajj, G. Rosen, B. D. Wilson, and A. J. Mannucci (2004). Assimilative modeling of low-latitude ionosphere. *Proceedings of the IEEE Plans,* pp. 543–550.

Picone, J. M., A. E. Hedin, D. P. Drob, and A. C. Aikin (2002). NRLMSISE-00 empirical model of the atmosphere: Statistical comparisons and scientific issues. *J. Geophys. Res., 107*(A12), 1468. doi:10.1029/2002JA009430.

Powell, K., P. Roe, T. Linde, T. Gombosi, and D. L. De Zeeuw (1999). A solution-adaptive upwind scheme for ideal magnetohydrodynamics. *J. Comput. Phys., 154,* 284–309.

Prölss, G. W. (2011). Density perturbations in the upper atmosphere caused by the dissipation of solar wind energy. *Surv. Geophys*, *32*(2), 101–195.

Prölss, G. W. (1997). Magnetic storm associated perturbations of the upper atmosphere. In *Magnetic Storms*. Geophysics Monograph 98, edited by B. T. Tsurutani, W. D. Gonzalez, Y. Kamide and J. K. Arballo, pp. 227–241. AGU, Washington, D.C.

Prölss, G. W., M. Roemer, J. W. Slowey (1988). Dissipation of solar wind energy in the earth's upper atmosphere: The geomagnetic activity effect. *Adv. Space Res.*, *8*(5–6), 215–261.

Qian, L., A. G. Burns, P. C. Chamberlin, and S. C. Solomon (2011). Variability of thermosphere and ionosphere responses to solar flares. *J. Geophys. Res.*, *116*, A10309. doi:10.1029/2011JA016777.

Qian, L., S. C. Solomon, and T. J. Kane (2009). Seasonal variation of thermospheric density and composition. *J. Geophys. Res.*, *114*, A01312. doi:10.1029/2008JA013643.

Raeder, J., J. Berchem, and M. Ashour-Abdalla (1998). The geospace environment grand challenge: Results from a global geospace circulation model. *J. Geophys. Res.*, *103*, 14787.

Rees, D., T. J. Fuller-Rowell, and R. W. Smith (1980). Measurements of high latitude thermospheric winds by rocket and ground-based techniques and their interpretation using a three-dimensional time-dependent dynamical model. *Planet Space Sci.*, *28*, 919–932.

Rees, D., R. W. Smith, P. J. Charles, F. G. McCormac, N. Lloyd, and A. Steen (1984). The generation of vertical thermospheric winds and gravity waves at auroral latitudes I: Observations of vertical winds. *Planet Space Sci.*, *32*, 667–684.

Rees, M. H., B. A. Emery, R. G. Roble, and K. Stamnes (1983). Neutral and ion gas heating by auroral electron precipitation. *J. Geophys. Res. Space Phys.*, *88*(A8), 6289–6300.

Richards, P. G. (2013). Reevaluation of thermosphere heating by auroral electrons. *Adv. Space Res.*, *51*(4), 610–619.

Richardson, I. G., H. V. Cane, and E. W. Cliver. (2002). Sources of geomagnetic activity during nearly three solar cycles (1972–2000). *J. Geophys. Res.*, *107*(A8), 1187. doi:10.1029/2001JA000504.

Richmond, A. D. (1978). The nature of gravity wave ducting in the thermosphere. *J. Geophys. Res.*, *83*(A4), 1385–1389. doi:10.1029/JA083iA04p01385.

Richmond, A. D. (2010). On the ionospheric application of Poynting's theorem. *J. Geophys. Res.*, *115*, A10311. doi:10.1029/2010JA015768.

Richmond, A. D. (1982). Thermospheric dynamics and electrodynamics. In *Solar-Terrestrial Physics, Principles and Theoretical Foundations*, edited by R. L. Carovillano and J. M. Forbes. D. Reidel, Dordrecht, The Netherlands.

Richmond, A., and Y. Kamide (1988). Mapping electrodynamic features of the high-latitude ionosphere from localized observations: technique. *J. Geophys. Res.*, *93*, 5741.

Richmond, A., E. Ridley, and R. Roble (1992). A thermosphere/ionosphere general circulation model with coupled electrodynamics. *Geophys. Res. Lett.*, *19*, 369.

Ridley, A. J., and Y. Deng (2006). The global ionosphere-thermosphere model. *J. Atmos. Sol. Terr. Phys.*, *68*. doi:10.1016/j.jastp.2006.01.008.

Rishbeth, H., T. J. Fuller-Rowell, and D. Rees (1987). Diffusive equilibrium and vertical motion in the thermosphere during a severe magnetic storm: A computational study. *Planet Space Sci.*, *35*(9), 1157–1165.

Roble, R., and E. Ridley (1987). An auroral model for the NCAR thermospheric general circulation model (TGCM). *Ann. Geophys.*, *5A*, 369.

Roble, R. G., E. C. Ridley, and R. E. Dickinson (1987). On the global mean structure of the thermosphere. *J. Geophys. Res.*, *92*(A8), 8745–8758.

Scherliess, L., R. W. Schunk, J. J. Sojka, and D. C. Thompson (2004). Development of a physics-based reduced state Kalman filter for the ionosphere. *Radio Sci.*, *39*, RS1S04. doi:10.1029/2002RS002797.

Schunk, R. W. (2002). Global Assimilation of Ionospheric Measurements (GAIM). Paper presented at Ionospheric Effects Symposium, Office of Naval Research, Alexandria, VA.

Schunk, R. W., L. Scherliess, and D. C. Thompson (2011). Ionosphere data assimilation: Problems associated with missing physics. In *Aeronomy of the Earth's Atmosphere and Ionosphere, edited by M. A. Abdu and D. Pancheva, Springer, the Netherlands*. doi:10.1007/978-94-007-0326-133.

Sentman, L.H. (1961). *Free Molecule Flow Theory and Its Application to the Determination of Aerodynamic Forces*. Lockheed Missile and Space Co., LMSC-448514, AD 265-409 (available from National Technical Information Service, Springfield, VA).

Sharma, R. D., H. Dothe, and J. W. Duff (1998). Model of the 5.3 mm radiance from NO during the sunlit thermosphere. *J. Geophys. Res.*, *103*, 14753–14758.

Sharma, R. D., and P. P. Wintersteiner (1990). Role of carbon dioxide in cooling planetary thermospheres. *Geophys. Res. Lett.*, *17*, 2201–2204. doi:10.1029/GL017i012p02201.

Shim, J. S., L. Rastaetter, K. M. Kuznetsova, E. C. Kalafatoglu, and Y. Zheng (2015). Assessment of the predictive capability of IT models at the Community Coordinated Modeling Center. Presented at Ionospheric Effect Symposium, Alexandria VA.

Solomon, S. C., and L. Qian (2005). Solar extreme-ultraviolet irradiance for general circulation models. *J. Geophys. Res.*, *110*, A10306. doi:10.1029/2005JA011160.

Spencer, N. W., L. E. Wharton, G. R. Carignan, J. C. Maurer (1982). Thermosphere zonal winds, vertical motions and temperature as measured from Dynamics Explorer. *Geophys. Res. Lett.*, *9*, 983–956.

Storz, M. F., B. R. Bowman, M. J. I. Branson, S. J. Casali, and W. K. Tobiska (2005). High accuracy satellite drag model (HASDM). *Adv. Space Res.*, *36*(12), 2497–2505.

Sutton, E. K., R. S. Nerem, and J. M. Forbes (2007). Density and winds in the thermosphere deduced from accelerometer data. *J. Spacecr. Rockets, 44,* 6.

Thayer, J. P., X. Liu, J. Lei, M. Pilinski, and A. G. Burns (2012). The impact of helium on thermosphere mass density response to geomagnetic activity during the recent solar minimum. *J. Geophys. Res., 117,* A07315, doi:10.1029/2012JA017832.

Thayer, J. P., and J. Semeter (2004). The convergence of magnetospheric energy flux in the polar atmosphere. *J. Atmos. Sol. Terr. Phys., 66,* 807–824.

Thayer, J. P., J. F.Vickrey, R. A. Heelis, and J. B. Gary (1995). Interpretation and modeling of the high-latitude electromagnetic energy flux. *J. Geophys. Res.Space Phys., 100*(A10), 19715–19728.

Torr, M. R., D. G. Torr, and P. G. Richards (1980). The solar ultraviolet heating efficiency of the midlatitude thermosphere. *Geophys. Res. Lett., 7.* doi:10.1029/GL007i005p00373.

Tsurutani, B. T., W. D.Gonzalez, A. L. Gonzalez et al. (2006). Corotating solar wind streams and recurrent geomagnetic activity: A review. *J. Geophys. Res., 111,* A07S01. doi:10.1029/2005JA011273.

Vallado, D.A., and D. Finkleman (2014). A critical assessment of satellite drag and atmospheric density modeling. *Acta Astronaut., 95,* 141–165. http://dx.doi.org/10.1016/j.actaastro.2013.10.005.

Vickrey, J. F., R. R. Vondrak, and S. J. Matthews (1982). Energy deposition by precipitating particles and Joule dissipation in the auroral ionosphere. *J. Geophys. Res. Space Phys., 87*(A7), 5184–5196.

Weimer, D. R. (2005a). Improved ionospheric electrodynamic models and application to calculating Joule heating rates. *J. Geophys. Res., 110,* A05306. doi:10.1029/2005JA010884.

Weimer, D. R. (2005b). Predicting surface geomagnetic variations using ionospheric electrodynamic models. *J. Geophys. Res., 110,* A12307. doi:10.1029/2005JA011270.

Weimer, D. R., B. R. Bowman, E. K. Sutton, and W. K. Tobiska (2011). Predicting global average thermospheric temperature changes resulting from auroral heating. *J. Geophys. Res., 116,* A01312. doi:10.1029/2010JA015685.

Zesta, E., H. J. Singer, D. Lummerzheim, C. T. Russell, L. R. Lyons, and M. J. Brittnacher (2000). The effect of the January 10, 1997, pressure pulse on the magnetosphere-ionosphere current system. *Geophys. Monogr. -Am. Geophys. Union, 118,* 217–226.

Zhou, X.-Y., K. Fukui, H. C. Carlson, J. I. Moen, and R. J. Strangeway (2009). Shock aurora: Ground-based imager observations. *J. Geophys. Res., 114,* A12216. doi:10.1029/2009JA014186.

Zhou, X.-Y., R. J. Strangeway, P. C. Anderson, D. G. Sibeck, B. T. Tsurutani, G. Haerendel, H. U. Frey, and J. K. Arballo (2003). Shock aurora: FAST and DMSP observations. *J. Geophys. Res., 108*(A4), 8019. doi:10.1029/2002JA009701.

CHAPTER 19

Space Weather Effects on Communication and Navigation

Keith M. Groves and Charles S. Carrano

CONTENTS

NUMEROUS COMMUNICATIONS AND NAVIGATION services for both commercial and military users are currently provided by space-based platforms. Common examples include automobile navigation systems that use the global positioning system (GPS) signals for positioning data and credit card transactions that require the exchange of information via satellite communications; the number of technologies relying on satellites continues to expand rapidly. Such services may be vulnerable to space weather effects as the radio waves used to transmit information pass through the Earth's ionosphere. The ionosphere is a partially ionized region of the upper atmosphere. The ionized gas substantially modifies the atmosphere's refractive index, inducing a variety of propagation effects that may distort the phase and/or amplitude of radio signals, thereby degrading the performance of a given radio frequency (RF) technology. Radio waves transiting the ionosphere experience the phase advance and group delay phenomenon, which manifests itself as a ranging error in single-frequency GPS and radar applications. Because the delay is dispersive, with an inverse-square frequency dependence, systems that employ two or more appropriately separated frequencies can successfully correct the ionospheric delay. This technique was demonstrated very successfully in dual-frequency GPS receivers and has been adopted by every other major global navigation satellite system (GNSS), including Global Navigation Satellite System (GLONASS), Galileo, and BeiDou. When ionization exhibits substantial structure or irregularities on scale sizes from tens to hundreds of meters, variable phase perturbations across the wave front can result in diffraction, essentially constructive and destructive self-interferences of the wave. This phenomenon, known as scintillation, is characterized by fluctuations in phase and amplitude, which affect both single- and dual-frequency systems. In this chapter, we examine the conditions under which scintillations occur in the natural ionosphere and describe the extent to which scintillations can disrupt the performance of the space-based communications and navigation systems, as well as both ground- and space-based radar applications. Furthermore, we consider potential strategies to mitigate ionospheric impacts, briefly explore the use of these technologies as sensors to monitor the state of the ionosphere, and detect the conditions in which system performance may be compromised by space weather. Let us begin with a brief overview of the fundamental propagation theory.

19.1 WAVE PROPAGATION IN THE IONOSPHERE

The propagation of radio waves through the ionosphere is described by Maxwell's equations coupled with the equations of motion for the particles comprising the ionospheric medium. For simplicity, we treat the ionosphere as a cold, collisionless, unmagnetized, quasi-neutral plasma, a model that adequately describes the propagation of radio waves through the ionosphere at frequencies above about 50 MHz. Such a medium is frequency-dispersive, but in the absence of magnetic field effects, it is isotropic.

19.1.1 The Vector Wave Equation

A suitable starting point for the analysis is the time-harmonic form of Maxwell's equations in a source-free region (Chew, 1995; Rino, 2011).

$$\nabla \times \mathbf{E} = i\omega\mu_0\mathbf{H} \tag{19.1}$$

$$\nabla \times \mathbf{H} = -i\omega\varepsilon\mathbf{E} \tag{19.2}$$

where:

    $\mathbf{E}$ is the electric field (volt/m)
    $\mathbf{H}$ is the magnetic field (ampere/m), and a time variation
        for these fields of the form $\exp(-i\omega t)$ is implied
    ω is the angular frequency

The medium is characterized by two material properties, which we assume to be spatially inhomogeneous but unchanging in time. The magnetic permeability in the ionosphere is essentially unchanged from its value in free space, μ_0 (henry/m). The electric permittivity ε (farad/m) incorporates all the spatial variation of the ionospheric medium and differs from the free-space value ε_0.

Taking the curl of Equation 19.1 and using Equation 19.2, Maxwell's equations may be expressed in terms of the electric field alone.

$$\nabla \times \nabla \times \mathbf{E} - \omega^2\varepsilon\mu_0\mathbf{E} = 0 \tag{19.3}$$

Using the vector identity $\nabla \times \nabla \times \mathbf{E} = -\nabla^2\mathbf{E} + \nabla\nabla \cdot \mathbf{E}$, the above equation can be rewritten as:

$$\begin{aligned}\nabla^2\mathbf{E} + \omega^2\varepsilon\mu_0\mathbf{E} &= \nabla\nabla \cdot \mathbf{E} \\ &= -\nabla(\mathbf{E} \cdot \nabla\log\varepsilon)\end{aligned} \tag{19.4}$$

The equivalence $\nabla\nabla \cdot \mathbf{E} = -\nabla(\mathbf{E} \cdot \nabla\log\varepsilon)$ is obtained by taking the divergence of (19.2). The result of Equation 19.4 is referred to as the vector wave equation for an inhomogeneous isotropic medium. The term on the right-hand side of Equation 19.4 is the only one through which the components of the electric field are coupled. Therefore, it is referred to as the polarization coupling term. Provided the variations in the ionosphere occur on scale sizes substantially larger than the wavelength, the polarization coupling term may be neglected (Nickisch and Franke, 1996) and the result takes the form of the vector Helmholtz equation.

$$\nabla^2\mathbf{E} + \omega^2\varepsilon\mu_0\mathbf{E} = 0 \tag{19.5}$$

19.1.2 Constitutive Relation for the Plasma

To determine the electric permittivity, the equation of motion for the charged particles comprising the plasma are used to express the electric displacement $\mathbf{D}$ and polarization $\mathbf{P}$ fields in terms of the applied electric field $\mathbf{E}$. The resulting expression is called the constitutive relation for the plasma. The displacement, polarization, and electric fields are related according to the following:

$$\mathbf{D} = \varepsilon\mathbf{E} = \varepsilon_0\mathbf{E} + \mathbf{P} \tag{19.6}$$

Before determining how the plasma responds to an electromagnetic wave, we consider how the plasma responds in the absence of an externally imposed electric field if the electrons are displaced a small distance from the ions. In this case, a polarization electric field develops and acts to restore the electrons to their equilibrium positions. Because the restorative force is proportional to the displacement distance, the charged particles undergo simple harmonic motion when free to move, with angular frequency:

$$\omega_p^2 = \frac{Ne^2}{m\varepsilon_0} \qquad (19.7)$$

where:
  N is the number of particles per unit volume
  m is the particle mass (kg)
  e is the (signed) particle charge in Coulombs

This is called the plasma frequency for each charged particle species, and it represents the frequency of oscillations in the plasma due to a locally imposed perturbation. For high-frequency (HF) perturbations, only the electrons have the time to respond because of their small inertia, whereas the ions remain comparatively stationary because of their much larger inertia. Hence, we ignore ion motion and consider only the oscillations of the electrons.

Now, suppose an electromagnetic wave travels through the plasma. The oscillating electric field imparts a Lorenz force on the electrons, setting them into simple harmonic motion.

$$m\ddot{\mathbf{r}} = e\mathbf{E} \qquad (19.8)$$

$$\mathbf{r} = -\frac{e}{m\omega^2}\mathbf{E} \qquad (19.9)$$

The displacement $\mathbf{r}$ of the electrons from their equilibrium position creates a dipole moment per unit volume, which is the electric polarization.

$$\mathbf{P} = Ne\mathbf{r} = -\frac{Ne^2}{m\omega^2}\mathbf{E} = -\frac{\omega_p^2}{\omega^2}\varepsilon_0\mathbf{E} \qquad (19.10)$$

The electric permittivity of the plasma follows from Equations 19.6 and 19.10.

$$\varepsilon = \varepsilon_0\left(1 - \frac{\omega_p^2}{\omega^2}\right) \qquad (19.11)$$

This result expresses the permittivity of the ionosphere as a function of the radio wave frequency and plasma frequency.

19.1.3 The Dispersion Relation

By substituting Equation 19.11 into the wave equation (19.5), the latter can also be expressed in terms of radio wave frequency and plasma frequency,

$$\nabla^2\mathbf{E} + \frac{\omega^2}{c^2}\left(1 - \frac{\omega_p^2}{\omega^2}\right)\mathbf{E} = 0 \qquad (19.12)$$

where:
  $c = 1/(\mu_0\varepsilon_0)^{1/2}$ is the speed of light in vacuum

Furthermore, if we assume that locally the solution behaves as a plane wave and substitute

$$\mathbf{E}(\mathbf{r}) = \hat{\mathbf{E}}(\mathbf{k})\exp(i\mathbf{k}\cdot\mathbf{r}) \qquad (19.13)$$

in Equation 19.12, we find that the equation is satisfied, and therefore, waves can propagate, provided that

$$\omega^2 = c^2k^2 + \omega_p^2 \qquad (19.14)$$

In the above equation, $\mathbf{k}$ is the vector wavenumber in the medium (which differs from its free-space value) and $k = |\mathbf{k}|$. The condition in Equation 19.14 is referred to as the dispersion relation for the plasma. It determines the rate at which different Fourier components of a wave pulse will disperse because of the variation of their phase velocity with frequency. The phase velocity is the rate at which the phase of a carrier wave advances:

$$v_p = \frac{\omega}{k} = c\left(1 - \frac{\omega_p^2}{\omega^2}\right)^{-1/2} \qquad (19.15)$$

The group velocity is the rate at which the envelope of a wave pulse travels:

$$v_g \equiv \frac{\partial\omega}{\partial k} = c\left(1 - \frac{\omega_p^2}{\omega^2}\right)^{1/2} \qquad (19.16)$$

Note that these velocities are real valued only when $\omega \geq \omega_p$, but otherwise, they are complex. Therefore, the radio wave frequency must exceed the plasma frequency in order for the wave to propagate; otherwise, the wave is attenuated in the form of a standing (evanescent) wave. It can be shown that no energy is transferred to the plasma in the latter case (given our assumptions about the medium), and the wave is simply reflected without loss. Assuming that the wave propagates, the phase velocity is faster than the speed of light, whereas the group velocity is slower than the speed of light.

The former situation does not violate special relativity because the wave carrier does not transmit information. Only the wave packet carries information, and this travels slower than the speed of light.

19.1.4 Refractive Index

The refractive index is a property of the medium defined as the speed of light in a vacuum divided by the speed at which waves propagate in the medium. For the ionosphere, we can define a phase refractive index and a group refractive index:

$$n_p = c / v_p = \left(1 - \frac{\omega_p^2}{\omega^2}\right)^{1/2}$$

$$n_g = c / v_g = \left(1 - \frac{\omega_p^2}{\omega^2}\right)^{-1/2} \tag{19.17}$$

Note that the phase and group refractive indices are related reciprocally; that is, they deviate from unity below and above by the same relative amount. If the wave frequency is much larger than the plasma frequency, as is the case at very HF (VHF) and higher frequencies, then the phase and group refractive indicies are well approximated by their first-order Taylor series expansions about unity:

$$n_p \approx 1 - \frac{1}{2}\frac{\omega_p^2}{\omega^2}$$

$$n_g \approx 1 + \frac{1}{2}\frac{\omega_p^2}{\omega^2} \tag{19.18}$$

These forms are commonly quoted in the literature.

19.1.5 Phase Advance and Group Delay

Next, we consider the propagation path of a radio wave from a space-borne transmitter through the ionosphere to a receiver on the ground. We are often interested in the phase advance and group delay of the measured radio signal relative to the true range between the transmitter and the receiver. Fermat's principle states that the propagation path taken is the one for which the optical path length is stationary with respect to variations in the path. The optical path length between the transmitter (T) and the receiver (R) is:

$$L = \int_T^R n\, ds \tag{19.19}$$

where:

> ds is a differential length along the path taken by the radio wave from T to R

The geometric path length or true range from T to R is:

$$\rho = \int_T^R ds_0 \tag{19.20}$$

where:

> ds_0 is a differential length along straight line path from T to R

The phase advance I_f^Φ and group delay I_f^P are defined as the difference $L - \rho$, where n is taken to be either the phase refractive index or the group refractive index. If the bending of the propagation path is neglected, $ds = ds_0$, then these are readily computed to be (in units of meters):

$$I_f^\Phi \approx -\frac{e^2}{2m\varepsilon_0(2\pi)^2}\frac{1}{f^2}\int_T^R N\,ds = -40.30\,\mathrm{TEC}/f^2$$

$$I_f^P \approx \frac{e^2}{2m\varepsilon_0(2\pi)^2}\frac{1}{f^2}\int_T^R N\,ds = +40.30\,\mathrm{TEC}/f^2 \tag{19.21}$$

The assumption of straight line propagation is appropriate for satellite signals transmitted at VHF and higher frequencies. The error incurred in making this assumption at the GPS L1 frequency is of the order of a few centimeters (Kashcheyev et al., 2012). Both phase advance and group delay are proportional to the total electron content (TEC), which is the number of electrons per unit volume integrated along the propagation path, with units of el⁻/m². This quantity is most often reported in terms of TEC units or TECU, where $1\,\mathrm{TECU} = 10^{-16}\,\mathrm{el^-/m^2}$.

19.1.6 Radio Wave Scintillation

Next, we consider the strategies for modeling the scintillation of electromagnetic waves due to propagation through irregularities in the distribution of the ionospheric plasma. Combining Equations 19.12 and 19.17, the Helmholtz equation governing the electric field can be written as:

$$\nabla^2 \mathbf{E} + k_0^2 n^2 \mathbf{E} = 0 \tag{19.22}$$

where $k_0 = \omega/c$ is the wavenumber of the wave when propagating in vacuum.

Now, let us assume the phase refractive index $n = n_p$ consists of small random fluctuations about its free-space value (unity), that is, $n = 1 + \Delta n$. If we neglect a term that is quadratic in Δn, Equation 19.22 can be rewritten as:

$$\nabla^2 \mathbf{E} + k_0^2 (1 + 2\Delta n)\mathbf{E} = 0 \qquad (19.23)$$

This equation is difficult to solve for two reasons. First, the random refractive index appears as multiplicative factor for the unknown electric field. Second, the Helmholtz equation supports the waves that interact with each other and with the medium and propagate in all directions. One fruitful strategy to solve the Helmholtz equation mitigates the second difficulty. The idea is to include in the approximate solution only those waves that, because of scattering from ionospheric irregularities, propagate within a narrow angular cone around the principal direction of propagation. This is called the narrow-angle scatter approximation. The benefit is that the character of the equation is changed from elliptic to parabolic, thereby enabling a solution by forward-marching techniques. The least restrictive of this type of approximation is called the forward propagation equation, discussed in the book by Rino (2011) and also in the paper by Rino and Carrano (2011). Here, we follow a simpler approach that gives essentially equivalent results for scintillation at VHF and higher frequencies. Assume that the solution can be expressed in terms of a complex amplitude $\mathbf{U}$, which varies slowly compared to the fast oscillation of the wave along the principal propagation direction, z.

$$\mathbf{E}(\mathbf{r}) = \mathbf{U}(\mathbf{r})\exp(ik_0 z) \qquad (19.24)$$

Substituting this into the Helmholtz equation and imposing the narrow-angle condition $|\partial^2 U/\partial z^2| \ll |k_0 \partial U/\partial z|$ leads to the so-called parabolic wave equation:

$$\nabla_\perp^2 U + 2ik_0 \frac{\partial U}{\partial z} + 2k_0^2 \Delta n U = 0 \qquad (19.25)$$

where the operator $\nabla_\perp^2 = \partial^2/\partial x^2 + \partial^2/\partial y^2$ is called the transverse Laplacian (Yeh and Liu, 1982). Note that since the components of the electric field vector are not coupled in the Helmholtz equation, Equation 19.22 may be applied to each of the field components separately. Hence, we dispense with the vector notation for convenience.

Now, suppose the refractive effects of the ionosphere on the wave can be replaced by a thin phase-changing screen centered at $z = 0$. Because the screen is thin,

diffractive effects do not have time to accumulate, so that Equation 19.25 becomes

$$\nabla_\perp^2 U + 2k_0^2 \Delta n U = 0 \qquad (19.26)$$

This has the solution:

$$U(x, y, 0^+, \omega) = U(x, y, 0^-, \omega)\exp[i\Delta\phi(x, y, \omega)] \qquad (19.27)$$

where:

$U(x, y, 0^-, \omega)$ is the complex amplitude of the wave just before crossing the screen

$U(x, y, 0^+, \omega)$ is the complex amplitude just after the screen

The exponential term imparts a phase change at the screen equal to the refractive index variations integrated through the ionosphere along the principal propagation direction.

$$\Delta\phi(x, y, \omega) = k_0 \int \Delta n(x, y, \xi, \omega)\, d\xi \qquad (19.28)$$

Below the ionosphere ($z > 0$), we retain the diffraction term but assume that refractive index fluctuations are negligible:

$$\nabla_\perp^2 U + 2ik_0 \frac{\partial U}{\partial z} = 0 \qquad (19.29)$$

This equation is readily solved by the Fourier transform methods. Given the Fourier transform of the complex amplitude just after the wave traverses the screen,

$$\hat{U}(\kappa_x, \kappa_y, 0^+, \omega) = \iint U(x, y, 0^+, \omega)\exp[-i(\kappa_x x + \kappa_y y)]\, dx\, dy \qquad (19.30)$$

Equation 19.29 can be used to obtain the complex amplitude at any location past the screen.

$$U(x, y, z, \omega) = \frac{1}{(2\pi)^2} \iint \hat{U}(\kappa_x, \kappa_y, 0^+, \omega)\exp\left[-i\frac{\kappa_x^2 + \kappa_y^2}{2k_0} z\right]$$
$$\exp[i(\kappa_x x + \kappa_y y)]\, d\kappa_x\, d\kappa_y. \qquad (19.31)$$

The result of Equation 19.31 is equivalent to the well-known Fresnel–Kirchhoff diffraction formula. It is important to note that because $\Delta n \propto \Delta N$, the screen phase, and hence the resulting scintillation on the ground, is determined by the integrated electron density variations along the path. The scintillations also depend on the distance past the screen and the frequency of the transmitted wave. From the complex amplitude on the ground, one can compute the intensity $I = |U|^2$ and phase $\varphi = \arg(U)$ of the received signal, as well as the statistical moments (Carrano et al., 2012b). Please note the important distinction between $\Delta\phi$ and ϕ in the notation used here. The former ($\Delta\phi$) represents the path-integrated phase caused by refractive effects and is used as the phase screen, whereas the latter (φ) is the phase scintillation measurement, which includes the effects of diffraction through Equation 19.31.

It is straightforward to evaluate Equation 19.31 for multiple frequencies to compute the correlation of scintillations at different frequencies and coherence bandwidth (Rino, 2011; Carrano et al., 2012b). For a transmitted broadband pulse, one can synthesize the time-dependent pulse on the ground by integrating Equation 19.31 over all frequencies:

$$\mathbf{E}(\mathbf{r}, t) = \frac{1}{2\pi} \int_{-\infty}^{\infty} \mathbf{U}(\mathbf{r}, \omega) \exp\left[i(k_0 z - \omega t)\right] d\omega \quad (19.32)$$

In practice, it is more efficient to reconstruct the fluctuating envelope of the pulse by using the techniques described by Knepp (1983). This technique has been used, for example, to evaluate spread spectrum correlation peak broadening due to scintillation (Bogusch et al., 1981).

19.1.6.1 Statistical Theory of Scintillation

In the statistical theory of scintillation (Rino, 2011; Rino and Carrano, 2011), a statistical model is assigned to the refractive index fluctuations in the ionosphere and the preceding analysis is repeated to obtain the statistics of signal fluctuations on the ground. If the irregularities are homogeneously distributed in a layer of thickness L centrally located about $z = 0$, the phase change imparted by the ionosphere is

$$\Delta\phi(x, y, \omega) = k_0 \int_{-L/2}^{L/2} \Delta n(x, y, \xi, \omega) d\xi \quad (19.33)$$

Furthermore, if it is assumed that the refractive index variations decorrelate over a vertical distance $L/2$, it can be demonstrated that

$$\Phi_{\Delta\phi}(\kappa_x, \kappa_y) = k_0^2 L \Phi_{\Delta n}(\kappa_x, \kappa_y, 0) \quad (19.34)$$

where:

$\Phi_{\Delta\phi}(\kappa_x, \kappa_y)$ is the 2D spectral density function (SDF) of path-integrated phase

$\Phi_{\Delta n}(\kappa_x, \kappa_y, \kappa_z)$ is the 3D SDF of the refractive index fluctuations in the ionosphere

Under disturbed ionospheric conditions, the SDF of electron density variations (and therefore also of refractive index variations) is most commonly observed to take the form of an inverse-power law extending over several decades. Given an assumed SDF of the refractive index variations, the Fresnel–Kirchhoff diffraction formula (represented in Equation 19.31) may be used to determine the statistical moments of the scintillation on the ground. The analysis is difficult in the general case but is straightforward when the scintillation is weak, in which case, the SDF of intensity Φ_I and phase Φ_φ are the filtered versions of the SDF of path-integrated phase variations:

$$\Phi_I(\kappa_x, \kappa_y) = 4\sin\left[\frac{\kappa_x^2 + \kappa_y^2}{2k_0} z\right] \Phi_{\Delta\phi}(\kappa_x, \kappa_y)$$

$$\Phi_\varphi(\kappa_x, \kappa_y) = \cos\left[\frac{\kappa_x^2 + \kappa_y^2}{2k_0} z\right] \Phi_{\Delta\phi}(\kappa_x, \kappa_y) \quad (19.35)$$

A derivation of these results may be found in the review paper by Yeh and Liu (1982). The factors multiplying $\Phi_{\Delta\phi}$ in Equation 19.35 are referred to as the Fresnel filter functions. Note that the intensity scintillations are strongly suppressed by the sine factor for wavenumbers smaller than the Fresnel frequency $\omega_F = \sqrt{2k_0/z}$, whereas the phase scintillations are not suppressed. As a consequence, weak-intensity scintillations are insensitive to the large-scale ionospheric structure, whereas weak-phase scintillations are dominated by it in an inverse-power law environment. This scenario changes when the scintillation is strong. Strong-intensity scintillation is influenced (and may even be dominated) by the large-scale ionospheric structure via focusing effects (Carrano and Rino, 2015a). Strong-phase scintillation also does not follow the weak scatter prediction (19.35); instead, it may be dominated by rapid phase transitions coincident with deep signal fades, which are a consequence of strong scatter effects (Carrano et al., 2013).

The statistics of intensity and phase scintillations are most commonly characterized in term of statistical moments. The scintillation index S_4 is defined as the standard deviation of intensity scintillations normalized by the mean

$$S_4^2 = \frac{\langle I^2 \rangle - \langle I \rangle^2}{\langle I \rangle^2} \quad (19.36)$$

where $\langle \cdot \rangle$ indicates an ensemble average. Formally, S_4^2 is given by the integration of Φ_I over wavenumber. In practice, trends in the intensity scintillation are removed before calculating S_4 by, for example, dividing it by its smoothed version obtained from a sliding boxcar average. Another commonly used metric of scintillation is the intensity correlation length. It is defined as the spatial separation at which the normalized correlation function of intensity equals 1/2 (some authors alternatively use 1/e). The correlation length is independent of the irregularity strength in weak scatter and increases with the Fresnel scale (and hence with distance from the screen). Because phase is not high-pass Fresnel filtered, like the intensity, the phase measurements are detrended to remove the contribution from large-scale ionospheric structure and also the effects that are not caused by the ionosphere. The most important of the latter is the contribution from the changing geometric range to the transmitter (integrated Doppler), which tends to be much larger than the contribution due to scintillation. The phase scintillation index σ_φ is defined as the standard deviation of the detrended phase scintillations.

$$\sigma_\varphi^2 = \left\langle \varphi^2 \right\rangle - \left\langle \varphi \right\rangle^2 \tag{19.37}$$

Formally, σ_φ^2 is given by the integration of Φ_φ, where the latter is understood to be the SDF of detrended phase. We note that the transmitter and irregularities are generally in motion, so that the diffraction pattern described in Equation 19.31 sweeps past a receiver on the ground, leading to temporal variations in the measured signal. In this case, it is common practice to use temporal averages, rather than ensemble averages, when computing S_4 and σ_φ. Owing to this motion, the decorrelation time of intensity is the direct measurement, rather than the correlation length.

To get familiar with the ranges of the scintillation indices for a given event, the plot in Figure 19.1 shows an example of a scintillated complex GPS signal and the corresponding values of S_4 and σ_φ. The data were recorded with NovAtel's GSV4004b receiver at Ascension Island on March 13, 2002. Both intensity and phase were detrended before statistical quantities were calculated. The structures of the amplitude and phase indices are characteristically similar, as expected for the activity that is not in the strong scatter regime.

19.1.6.2 Rino's Weak Scatter Scintillation Model

The model for weak ionospheric scintillation proposed by Rino (1979) accounts for several important aspects of the problem that have not been considered in the previous analysis: (1) oblique propagation through the irregularities, (2) irregularity drift and transmitter motion, and

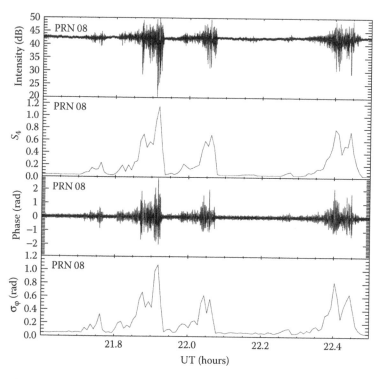

FIGURE 19.1 Raw GPS signal intensity and phase and corresponding scintillation indices, S_4 and σ_φ, from a Novatel GSV4004b receiver at Ascension Island, March 13, 2002.

(3) anisotropy of the irregularities due to their elongation in the direction of the magnetic field. Rino assumed the following inverse-power law form for the SDF of path-integrated phase, after one-way passage through a thin layer of field-aligned anisotropic ionospheric irregularities:

$$\Phi_{\Delta\phi}(\kappa) = \frac{abC_p}{(\kappa_0^2 + A\kappa_x^2 + B\kappa_x\kappa_y + C\kappa_y^2)^{(p+1)/2}} \quad (19.38)$$

In the above equation, p represents the spectral index or the slope of the phase SDF, which is one less than the spectral index of the 3D SDF of electron density fluctuations. The irregularities are assumed to have outer scale L_0, and $\kappa_0 = 2\pi/L_0$ is the outer scale wavenumber. The parameters a and b are the scaling factors that elongate the contours of constant correlation along and transverse to the magnetic field, respectively. The transverse wavenumbers in the geomagnetic east and geomagnetic north directions are symbolized by κ_x and κ_y, respectively. The coefficients A, B, and C depend on the directions of propagation and the magnetic field and are obtained by relating the propagation line of sight and the irregularity axes. Equations for A, B, and C are given in (Rino, 1979). The parameter C_p is the strength of the 3D spatial spectrum of the path-integrated phase evaluated at wavenumber 1 rad m^{-1}:

$$C_p = r_e^2 \lambda^2 C_s L \sec\theta \quad (19.39)$$

In Equation 19.39, r_e is the classical electron radius (2.8179×10^{-15} m), $\lambda = 2\pi/k_0$ is the radio wavelength in free space, C_s is the strength of the 3D spatial spectrum of electron density fluctuations, L is the thickness of the scattering layer, and θ is the propagation (nadir) angle at the ionospheric penetration point.

Rino (1979) showed that the S_4 index for this model can be expressed as

$$S_4^2 = C_p \rho_F^{p-1} F_S(p)\wp(p) \quad (19.40)$$

The Fresnel scale that appears in Equation 19.40 is $\rho_F^2 = z\sec\theta/k_0$, where z is the height of the scattering layer and k_0 is the free-space signal wavenumber. The quantity $F_S(p)$ depends only on p and is

$$F_S(p) = \frac{\Gamma[(5-p)/4]}{2^{(p-1)/2}\sqrt{\pi}(p-1)\Gamma[(p+1)/4]} \quad (19.41)$$

The parameter $\wp(p)$ is a combined geometry and propagation factor, which is given in Equation 19.34 of Rino (1979). We note that Equation 19.40 is valid when the

scintillations are weak and the phase spectral index lies in the range $1 < p < 5$. Under these conditions, S_4 increases with the total integrated electron density fluctuation, distance past the phase screen, and the wavelength.

Consistent with Rino's power-law irregularity model and Taylor's assumption of frozen-in (nonevolving) flow, a 1D time series of phase fluctuations measured by a receiver on the ground is characterized by 1D temporal spectrum (from Equations 16–18 in Rino, 1979).

$$\varphi(f) = \frac{T}{(f_0^2 + f^2)^{p/2}} \quad (19.42)$$

where:

f is the frequency
f_0 is the outer scale frequency

The 1D phase spectral strength sampled at the frequency 1 Hz is

$$T = C_p F_T(p) G V_{eff}^{p-1} \quad (19.43)$$

In the above equation, G is the geometry enhancement factor for phase, defined in (Rino, 1979). The effective scan velocity, V_{eff}, relates the spatial sizes of the structures encountered along the satellite scan to the temporal frequencies at which they appear in the measured 1D time series of phase. For example, the frequency at which the outer scale occurs in the spectrum of measured phase is $f_0 = V_{eff}/L_0$. The quantity $F_T(p)$ depends only on the spectral index and is

$$F_T(p) = \frac{\sqrt{\pi}\Gamma[p/2]}{(2\pi)^{p+1}\Gamma[(p+1)/2]} \quad (19.44)$$

Next, we present equivalent results in terms of the phase scintillation index σ_φ. Assuming that the time constant of the high-pass filter used to detrend the phase measurements satisfies $\tau_c \ll L_0/V_{eff}$, the phase spectrum is well approximated by an unmodified power law over the range of interest (i.e., the outer scale can be ignored). In this case, the defining relationship gives

$$\sigma_\varphi^2 = \int_{-\infty}^{\infty} \frac{T}{(f_0^2 + f^2)^{p/2}} df \approx 2T \int_{\tau_c^{-1}}^{\infty} f^{-p} df = \frac{2T}{p-1}\tau_c^{p-1} \quad (19.45)$$

Using Equation 19.43, the above equation can be written:

$$\sigma_\varphi^2 = \frac{2}{p-1} C_p F_T(p) G[\tau_c V_{eff}]^{p-1} \quad (19.46)$$

Hence, σ_φ increases with the total integrated electron density fluctuation, wavelength, and the effective length $L_{eff} = \tau_c V_{eff}$, which is a measure of the largest irregularity scale size to be admitted through the phase detrend filter.

Some of the implications of this theory are as follows. The strength of both intensity and phase scintillations is controlled by the total integrated electron density fluctuations (ΔN), which enters the theory through the quantity $C_s L$. This product is often reported in the literature referenced to a 1 km length scale and is referred to as $C_k L$ (Secan et al., 1995). The wavelength dependence enters through the phase spectral strength C_p and the Fresnel scale ρ_F. Equations 19.40 and 19.46 show that S_4 and σ_φ scale with wavelength as $S_4 \propto \lambda^{(p+3)/4}$ and $\sigma_\varphi \propto \lambda$, respectively. Both S_4 and σ_φ depend on the irregularity strength, but only σ_φ depends on satellite and irregularity motion through the effective scan velocity. The ratio of σ_φ/S_4 is independent of the scattering strength. In fact, for the typical case of highly elongated rod-like irregularities ($a \gg 1, b = 1$), phase spectral index $p = 3$ and detrend filter time constant $\tau_c = 10$ s, the ratio σ_φ/S_4 is nearly equal to the effective scan velocity divided by the Fresnel scale (Carrano et al., 2015b). This observation has been used to infer the effective scan velocity from the measurements of S_4 and σ_φ in the equatorial zone. Once the effective scan velocity is known, the zonal irregularity drift may be readily estimated (Carrano et al., 2015b). Given the substantially faster irregularity drift commonly encountered at high latitudes, the ratio σ_φ/S_4 tends to be large, which for a given irregularity strength is conducive to producing the so called "phase without amplitude scintillations." As discussed later in this chapter, these cases should not be considered scintillations at all, because the phase scintillations observed are not due to diffraction effects but are instead caused by large swaths of ionospheric TEC being swept past the receiver very quickly.

The characteristics of strong scintillation may differ markedly from the behavior described above. The intensity scintillations may develop substantial departures from power-law behavior because of the focusing effects and nonlinear interactions. As the strength of the scatter increases, the S_4 index ceases to follow the weak scatter behavior and ultimately saturates at unity (possibly with substantial overshoot) and thus ceases to depend on the irregularity strength. As this saturation condition develops, phase mixing effects cause the correlation length

of intensity scintillations to decrease with increasing perturbation strength, and as such, it becomes a more useful diagnostic tool for inferring irregularity strength than the S_4 index. The theory of strong scintillation is still an area of active research. The interested reader is referred to Carrano and Rino (2015a) for details.

19.2 SCINTILLATION MORPHOLOGY

Having established the propagation theory, we now consider the occurrence of the requisite electron density irregularities associated with space weather. For this purpose, it is convenient to characterize the scintillation environment in three distinct regions, namely high-, mid-, and low latitudes. Before exploring the morphologies of these regions, it is important to clarify the definition of "space weather" and consider the role of the Sun with respect to ionospheric irregularities. The traditional understanding of space weather phenomena includes anomalous magnetosphere–ionosphere–thermosphere responses driven by impulsive solar events (ISEs). Yet, with the exception of mid-latitudes where they are the sole driver of irregularity formation, ISEs account for a small minority of global scintillation activity and can even act to stabilize equatorial regions. Although high- and low-latitude processes are also influenced by the Sun, they can routinely generate irregularities even in the absence of solar perturbations. This relatively frequent yet variable activity, controlled primarily by magnetic field geometry and internal processes in the ionosphere-thermosphere system, also constitutes space weather. More details concerning global scintillation morphology can be found in Aarons (1982). In this chapter, we provide an overview and some updates that are consistent with current understanding. Figure 19.2 provides a simplified picture that shows nominal L-band scintillation strength as a function of location and local time during solar maximum and minimum periods. At solar minimum, L-band scintillations are weak or absent, as electron densities decrease substantially and often fall below the levels required to strongly perturb L-band signals ($\sim 10^6$ cm^{-1}). From the figure, it is clear that the most severe L-band scintillation occurs in the high-density postsunset low-latitude ionosphere. This is the region where so-called "bubbles" or depletions occur, causing strong perturbations in electron density (ΔN).

The high-latitude environment also has regular scintillation and can be divided into two subregions, the polar cap and the auroral zone. Of these, the polar cap

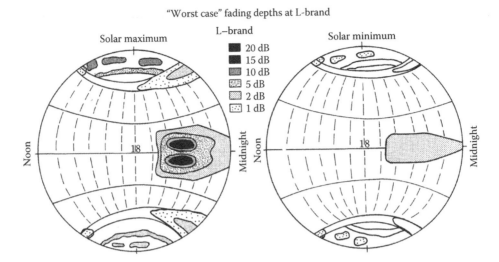

FIGURE 19.2 Nominal global distribution of L-band scintillation for solar maximum and minimum. (Reproduced from Basu, S. and Groves, K.M. *Space Weather, Geophysical Monograph*, 125, 424–430, 2001. With permission.)

experiences higher levels of scintillation due to the presence of patches and arcs. Patches are localized structures that form in local winter during B_z south conditions, when density from the sunlit auroral zone is advected into the dark polar cap (Weber et al., 1984; Valladares et al., 1994). The variable advection process results in discrete injections of plasma ("patches") into the polar cap, typically 100–1000 km in size. Once inside the polar cap, electron density irregularities form because of the gradient drift instability, resulting in relatively strong scintillations from patch structures. The patches drift antisunward in the polar cap convection pattern and may persist for hours before decaying through diffusion and recombination or reaching the auroral zone, where they dissipate and transform into "blobs." B_z north conditions are favorable for the formation of polar cap arcs created by soft particle fluxes that subsequently structure via shear instability. Arcs have lower density than patches, and the associated scintillations are thus less severe. Both arcs and patches may coexist as the interplanetary magnetic field varies over the course of time, altering the convection pattern and modulating the relevant physical processes. In all cases, magnetic storms would increase both scintillation severity and the areas affected, as the auroral oval expands in response to magnetospheric forcing (Aarons and Allen, 1971).

Irregularities formed in the nightside polar cap from both arcs and patches may eventually drift into the auroral zone, where zonal drifts typically stretch the structured plasma and exposure to sunlight relaxes the spatial gradients, with a resulting decrease in scintillation.

Direct particle precipitation in the auroral zone can also contribute to enhanced electron density and irregularities, but the largest auroral enhancements occur in the E-region and even in the D-region, where it is more difficult to create and maintain irregularities because of relatively high electron/ion-neutral collision rates in these lower-altitude regions of the ionosphere. Unlike scintillations in other regions, high-latitude L-band scintillations generally exhibit high phase scintillation and relatively weak-amplitude scintillation. The primary reason for this is large plasma drifts at high latitudes. From scintillation theory, we know that phase scintillations are not subject to Fresnel cut-off in terms of scale sizes that contribute to the scintillation. At high-latitudes, horizontal drifts can easily reach 1000 m/s; thus, during a typical 30-s observing period, approximately 30 km of plasma sweeps across the RF link between the satellite and the ground. That is, scale sizes of up to 30 km contribute to the phase scintillation, whereas contributions to the amplitude scintillation remain limited by the Fresnel scale cut-off of about 1 km, regardless of the plasma drift speed. In fact, high-latitude phase scintillations are often reported in the absence of the amplitude scintillations (e.g., Ngwira et al., 2010; Prikryl et al., 2010). In these situations, scintillation may not even be occurring, and the phase "scintillations" are in fact just the phase variations due to TEC structures convecting at high speeds across the link. The resulting phase fluctuations may have the same impact on a receiver as true phase scintillations, but it is technically not correct to characterize this phenomenon as scintillation because no diffractive processes occur.

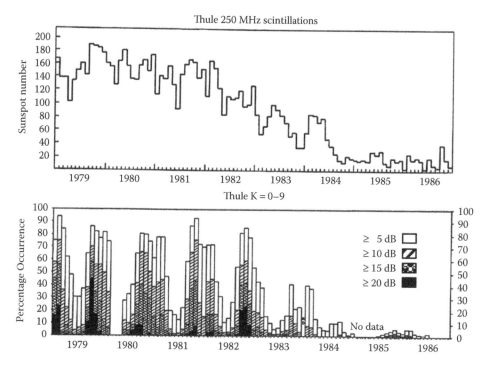

FIGURE 19.3 Sunspot number and VHF scintillation occurrence from 1979 to 1986 at Thule, Greenland. (Adapted from Basu, S. et al., *Radio Science*, 23, 363–378, 1988.)

The vast majority of high-latitude scintillations occur during polar winter, when the polar cap is primarily dark. Seasonal and solar cycle variations of 250 MHz scintillations from Thule, Greenland, are shown in Figure 19.3. The figure includes continuous observations for half a solar cycle beginning in 1979. The smoothed sunspot number is shown in the top panel and scintillation activity in the bottom panel. Two strong dependences are quite obvious. First, solar flux determines the extent and severity of scintillation in the polar cap. Second, the monthly activity clearly shows the near absence of scintillations during the summer in the Northern Hemisphere, when abundant sunshine creates plasma nearly continuously and stabilizes gradient drift and other instabilities. For more information, see Basu et al. (1988).

The overall activity level is quite high and scintillation can occur at any time of day, but there is a tendency for less activity during mid-late morning hours UT, particularly during the time period from September to December. Figure 19.3 includes all days. Although magnetically active periods account for just a handful of days each year, irregularities were present nearly every day. Note that observations at VHF are used to reveal this persistent amplitude scintillation at high-latitudes. It would be virtually impossible to see similar activity with GPS L-band measurements. This is not because irregularities are not present, but because the electron densities are simply not high enough to provide sufficient ΔN for substantial L-band scintillations in most of the solar cycle.

The ionosphere over mid-latitudes is normally quiescent and free from substantial irregularities. However, the response of the ionosphere to magnetic storms results in enhanced densities and irregularities, leading to mid-latitude scintillations (Basu et al., 2001b; Ledvina et al., 2002). The generation of mid-latitude irregularities during storms is most difficult to forecast and characterize, because the dynamics are driven by external electric and magnetic fields imposed by the magnetosphere. An example of mid-latitude GPS scintillation is shown in Figure 19.4. The duration and intensity vary, as would be expected, because of the dynamic storm-induced conditions. Strong mid-latitude irregularities are believed to form along the boundaries of storm-enhanced density (SED) features observed to form equatorward of the subauroral trough in the main phase of magnetic storms (Foster et al., 2002; Basu et al., 2005). However, our understanding of mid-latitude scintillations associated with magnetic storms lacks a firm connection to physical processes, and the morphology remains somewhat anecdotal; more research on this topic is needed in the future. In some sense, these events are more problematic

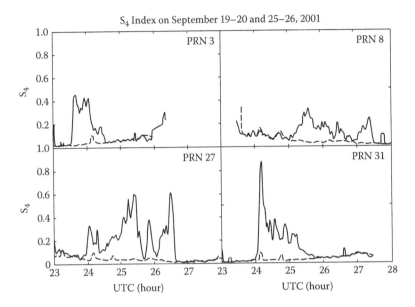

FIGURE 19.4 Mid-latitude GPS scintillations observed on September 19–20 (dashed) and September 25–26 (solid), showing storm-time enhancements on the latter days. Data were collected at Ithaca, New York. (Reproduced from Ledvina, B. M. et al., *Radio Science*, 39, 2004. With permission.)

from a societal impact perspective than the routine occurrence of scintillation at low- and high latitudes, simply because they are not expected and no mitigating actions are planned. The SED structures and irregularities associated with a strong magnetic storm can certainly rival the intensity of equatorial scintillations. Furthermore, mid-latitudes encompass major population centers, and scintillation activity has the potential to affect millions of people.

A large number of people are also potentially affected by low-latitude scintillation, which occurs routinely in a region comprising nearly one-third of the Earth's surface. The most severe natural scintillation occurs in low-latitude regions within about 15°–20° of the Earth's magnetic equator. A map showing the magnetic equator and the nominal scintillation "belt" region is provided in Figure 19.5. Scintillation in these regions is caused by gravitational Rayleigh–Taylor instability

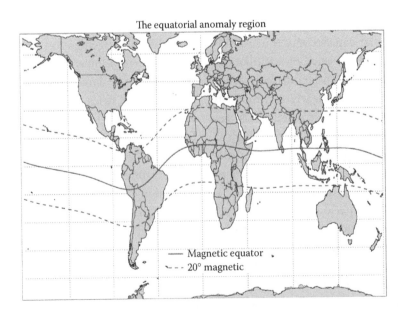

FIGURE 19.5 Global map showing the magnetic equator and the region within ±20° (dashed line), where the most intense space weather effects on RF systems occur. The area encompasses one-third of the Earth's surface.

(GRTI), driven by a sharp density gradient on the nighttime bottomside of the ionosphere. The physics of this instability has been studied extensively (Ott, 1978; Keskinen et al., 2003; Kudeki et al., 2007). We will not go into the details here, except to say that the process leads to the upwelling of low-density bottomside plasma into the high-density topside F region and strong local gradients subsequently become unstable to the gradient drift, and perhaps other instabilities, such that a broad spectrum of intense irregularities is generated. However, scintillation activity across the low-latitude region is not uniform with respect to latitude. The anomaly regions ($\pm$ ~15° latitude from the magnetic equator) boast the highest background densities, and, consequently, the most severe scintillations will be found there (Whalen, 2004).

The GRTI occurs only at low-latitudes, where the magnetic field is nearly parallel to the Earth's surface. Conceptually, the horizontal magnetic field effectively prevents electrons from moving vertically under the action of pressure and gravity to relax the sharp postsunset bottomside density gradient. This produces strong polarization electric fields, which cannot be shorted out in the evening, because E-region conductivity vanishes rapidly after sunset. This configuration is unstable to the GRTI. To gain more quantitative insight, we consider the linear growth rate for the GRTI and find that it reaches a maximum shortly after sunset.

$$\gamma = \frac{\Sigma_F}{\Sigma_F + \Sigma_E} \left[\frac{|E \times B|}{B^2} + U_n + \frac{g}{\nu_{eff}} \right] \frac{1}{N} \frac{\partial N}{\partial h} \quad (19.47)$$

Here, Σ_E and Σ_F correspond to the field-line-integrated E- and F-region conductivities, E and B refer to the electric and magnetic fields, respectively. In addition, U_n is the perpendicular neutral wind term, g is the acceleration due to gravity, and ν_{eff} is the effective collision frequency. N represents electron density and h represents height, so that the right-most factor corresponds to the vertical log density gradient.

After sunset, the inhibiting conductivity term approaches unity, as E-region conductivity vanishes. The enhanced postsunset electric field contributes both directly as a positive term in the growth rate and indirectly as it lifts the ionosphere higher to a less collisional environment (reduced ν_{eff}). In addition, the density gradient increases after sunset. The instability is thus constrained to occur only in the evenings, except for special cases in which the onset of a magnetic storm imposes penetration electric fields, which can lift the ionosphere abruptly and lead to fresh irregularity generation (Fejer et al., 1999). This instability has received a lot of attention in the literature. For a more in-depth analysis, see, for example, Sultan (1996) and previous references. Once created, irregularities typically exist for several hours, decaying slowly because of recombination and ambipolar diffusion, until sunrise regenerates the background density and the depletions disappear altogether.

The most severe time period for scintillations is 20:00–24:00 LT. After local midnight, the ionosphere typically drifts downward where recombination increases and the density falls rapidly. Irregularities are still present, but the overall decreased density results in a smaller total density perturbation (ΔN) and the scintillation weakens. Observationally, we find that L-band scintillations typically decay within an hour after midnight, whereas signals at lower frequencies (e.g., low ultra-HF [UHF] and VHF) can experience scintillation throughout the night and occasionally past sunrise. From the time they are created, the irregularities drift predominantly eastward with the background plasma. In the early evening, these drifts typically range from 100 to 150 m/s, decreasing in a roughly linear fashion to 25–50 m/s in the early morning hours. In rare solar maximum cases, where scintillations have persisted past sunrise, the eastward drift has been observed to reverse and become westward. Westward drifts can also be observed in an unusual postsunset period, when a strong disturbance dynamo electric field associated with an ISE prevails. The diurnal and seasonal dependencies of 250 MHz scintillations at one location are illustrated in Figure 19.6. The figure shows a year of VHF scintillation observations from a geostationary satellite received at Cuiabá, Brazil (15.1° S, 56.1° W), in 2011. The x-axis corresponds to day of year, whereas the y-axis shows time as hours past F-region (300 km) sunset; the gray scale represents the S_4 index.

We see that the onset time varies somewhat, depending on day of year within a range of 60–90 min past local F-region sunset. Strong scintillation then persists for 4–6 h, often followed by a period of moderate scintillation for up to two more hours. During this time, the irregularities may drift eastward from their origin, that is, a distance of ~2000 km or more. One can also see the direct effect of solar flux on scintillation intensity, and, conversely, the lack of importance for scintillation occurrence. Note that the first six weeks of 2011

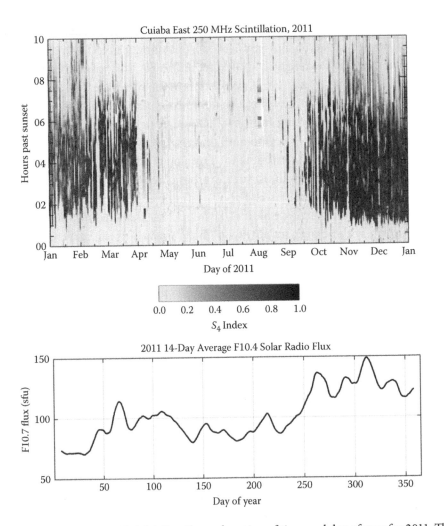

FIGURE 19.6 VHF scintillation activity at Cuiabá, Brazil, as a function of time and day of year for 2011. The reduced intensity at the beginning of the year is largely due to lower solar flux, as shown in the lower plot.

are characterized by less severe scintillation, typically reaching S_4 ~0.8, as opposed to the saturated $S_4 \geq 1.0$ that occurs during the last quarter. Comparing with the lower panel that shows the daily smoothed F10.7 flux, one can see that the increase in the intensity of scintillation correlates directly with the increase in F10.7, corresponding to a denser ionosphere. The occurrence frequency is not strongly modulated by the F10.7 flux, but lower densities result in weaker scintillation. Ultimately, there are fewer bubbles during solar minimum years, so the two parameters are not completely uncoupled, but the dominant effect of solar flux on equatorial scintillation is controlling the severity through the background densities.

This is convincingly demonstrated in Figure 19.7. Here, we plotted the upper tenth percentile of vertical total electron count and upper quartile of the largest S_4 for all GPS satellites above 30° elevation angle (chosen to minimize the effects of multipath and gradients), as well

as the monthly averaged F10.7 flux for each November from 2004 through 2014. We selected the most active time period from 2 to 4 h past local sunset, approximately 20:00–22:00 LT. The figure shows the strong correlation between the solar flux, TEC, and S_4 at low-latitudes over more than half a solar cycle. Simply put, solar flux controls electron density, and electron density controls scintillation strength. Conceptually, the result is not surprising, but the degree to which direct observations support the hypothesis is quite remarkable. From theoretical and experimental perspectives, we might expect even better agreement if peak electron density was used in the correlation, as this, along with slab thickness, is the dominant factor that controls scintillation severity. Peak density data were not available at this location, and the good agreement shown in the figure suggests that TEC can serve as a reasonable proxy for density at low-latitudes. It should be noted that Cuiabá was selected for

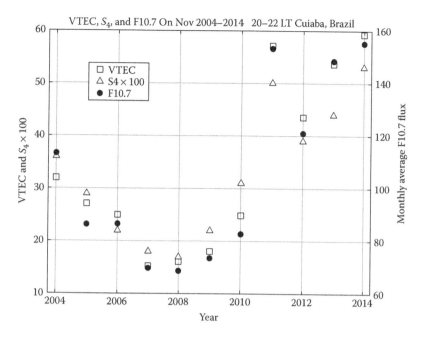

FIGURE 19.7 GPS S4, vertical TEC, and solar radio flux F10.7 in November for 11 years at Cuiabá, Brazil. The data show the strong correlation between the three parameters throughout the solar cycle. (Courtesy of C. Bridgwood.)

this exercise because of the high-quality long-term data set available from the site and because of its relatively modest magnetic latitude (~7° S MLat). A higher latitude site closer to the Appleton anomaly region would be susceptible to S_4 saturation and strong scintillation effects as solar flux increased and might therefore not show such a linear relationship between the parameters.

Another striking aspect of Cuiabá's annual activity, shown in Figure 19.6, is the abrupt seasonal behavior. First, notice that the scintillation "wet season" in Cuiabá ends early in April and resumes in mid-September. There is actually little true daily variability, except, perhaps, for the transition periods between ON and OFF seasons. Clearly, some dominant physical factor is responsible for the marked seasonal behavior that does not depend on variations in tides and neutral winds.

Recall the brief discussion on the physical mechanisms that lead to the onset of GRTI in the ionosphere and the strong polarization electric fields formed across the density gradient at the bottomside boundary. We pointed out that the horizontal magnetic field prevents the electrons from moving across **B** in response to electric fields. However, electrons can readily flow along **B** to the conductive E-region, which provides good cross-field electron mobility and the fields across a given magnetic flux tube can be shorted. At nighttime, the E-region disappears, but there are actually "two" E-regions, located at each end of a given magnetic flux tube, that control ionospheric

dynamics at the equator. If the E-region at one end of the flux tube remains sunlit for a period of time after the Sun has set on the other end, the sunlit region will still be able to supply the conductivity needed to effectively short out the instability. This hypothesis was first articulated by Tsunoda (1985), and it has since been supported by a combination of ground- and space-based observations (Gentile et al., 2006). Figure 19.8 shows a good example of the evidence in support of the hypothesis stated above.

The figure depicts the geographic and seasonal frequencies of occurrence of bubbles observed by the Defense Meteorological Satellite Program (DMSP) from 1999 to 2002, a solar maximum period. The data are not direct observations of scintillations but rather the observations of bubbles that reached the DMSP altitude of 800 km. The altitude bias is particularly worrisome during solar minimum, when the ionosphere is lower and few bubbles reach the DMSP orbit. Thus, although the absolute occurrence rates observed by DMSP are not equivalent to scintillation occurrence, the relative occurrence statistics should be valid for global comparisons. From the figure, one can see that active scintillation periods vary substantially with longitude. The two solid black lines in the plot mark the days of the year when the solar terminator and the magnetic field are in alignment. On these days, the E-regions on both ends of the local magnetic field lines would experience sunset at the same time and, according to the Tsunoda

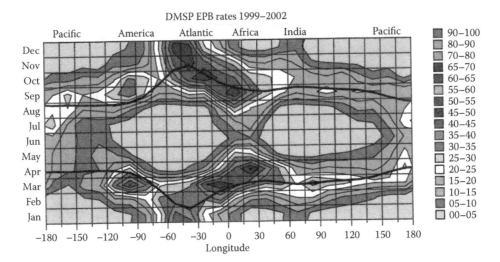

FIGURE 19.8 Percentage rates of equatorial plasma bubbles detected by DMSP from 1998 to 2002, showing the longitudinal variations as a function of day of year. The solid black lines indicate the two days of the year when the solar terminator and the magnetic field are aligned as a function of longitude, a period believed most favorable for scintillation occurrence. (Reproduced from Gentile, L. C. et al. *Radio Science*, 41, 2006. With permission.)

hypothesis, should be the most favorable for instability and subsequent irregularity formation.

From the plot, it is clear that this is usually true, but it is also clear that a substantial amount of scintillation occurs during periods when the condition is not satisfied, particularly over Africa and the Pacific. The role of gravity waves, wind shears, tides, and other thermospheric processes have long been debated to explain the daily variability of equatorial spread F (see, e.g., Haerendel et al., 1992; Fejer, 1997; Abdu et al., 2006). Indeed, ground-based scintillation observations in the African and Pacific sectors show a great deal of variability in scintillation occurrence. However, the analysis of stations across South America, extending to Cape Verde off the west coast of Africa, demonstrate low variability and strong seasonal control across the whole American and Atlantic longitude sectors, just as was shown for Cuiabá. In these regions, which are associated with westward declination of the Earth's magnetic field, thermospheric processes would appear to play a secondary role. Longitudinal differences in scintillation occurrence remains a topic of active research.

Finally, we should point out that ISEs often have a strong effect on low-latitude dynamics. Penetration electric fields present during the onset of magnetic storms reinforce quiescent postsunset eastward electric fields and thus enhance the vertical drift and scintillation occurrence in the appropriate local time sectors. Similarly, disturbance dynamo fields can lead to uplift and bubble formation

after local midnight. However, in other longitude sectors and during the later phases of storms, low-latitude scintillation is typically suppressed. More detailed information regarding storm-time effects on low-latitudes may be found in other studies (Fejer et al., 1999; Basu et al., 2001b, 2007, 2010; Abdu, 2012).

Table 19.1 contains a top-level summary of the factors that influence scintillation activity in different latitude regimes. As mentioned previously, solar events influence activity in all regions, but they are a requirement only for mid-latitude scintillation occurrence. The bulk of low-latitude scintillation activity, in particular, is largely controlled by processes internal to the coupled ionosphere–thermosphere system. Therein lies the reason to hope that scientists will be able to forecast these events in the foreseeable future.

19.3 SCINTILLATION IMPACTS ON RF SYSTEMS

The general effects and environments described in the previous two sections are essentially common to all RF propagation in the ionosphere, which depends only on fundamental parameters such as frequency, geometry,

TABLE 19.1   Factors Influencing Scintillation

Scintillation Region	Solar Events	Seasonal	Diurnal
High-latitude	Moderate	Strong	Weak
Mid-latitude	Strong (required)	Weak	Weak
Low-latitude	Moderate	Strong	Strong

scan velocity, and the irregularity spectrum. Having established where and when irregularities occur, let us now consider how RF systems are affected. Impacts to specific applications are primarily determined by the details of the system architecture, signal processing, and other engineering and physical attributes such as power, antennas, and waveforms. In the following section, we attempt to describe the degradation experienced by different types of RF applications in somewhat generic terms; specific examples are referenced and presented, but the objective is to establish the vulnerabilities of classes of technologies to space weather effects. The vulnerabilities identified refer to the information and services the systems provide and not to effects on the satellites themselves (see Chapters 19 and 20 for impacts on satellites).

19.3.1 Satellite Communications

We begin with narrow-band satellite communications, because the effects are readily understood and relatively predictable. The primary space weather concern for such systems is amplitude scintillation. When fades due to scintillation cause the received signal strength to drop below the noise floor of a given system, the transmitted

information is lost. An example of precisely that behavior is shown in Figure 19.9. In the example, a VHF signal transmitted by a geostationary satellite was received at Ascension Island (7.9° S, 14.4° W), situated in the equatorial scintillation belt. The figure shows how characters in the broadcast test message were lost as a function of scintillation fade depth. Because system performance is largely determined by signal-to-noise ratio (SNR), the S_4 index served as a useful metric. Research-grade hardware was used to collect the data in Figure 19.9. The actual fade depths at which such errors occur depends on the characteristics of a particular system configuration. Generally, telecommunication engineers establish nominal bit error rates as a function of SNR for their specific systems and can thus apply published and modeled scintillation statistics, such as S_4, to estimate performance impacts.

Increasing demand for global information distribution requires wide band communications systems that employ sophisticated spread spectrum waveforms to maximize usage of the available RF spectrum. Such systems may be vulnerable to both phase and amplitude scintillations, and the detailed performance impacts are less well-understood, partly because the signal-processing

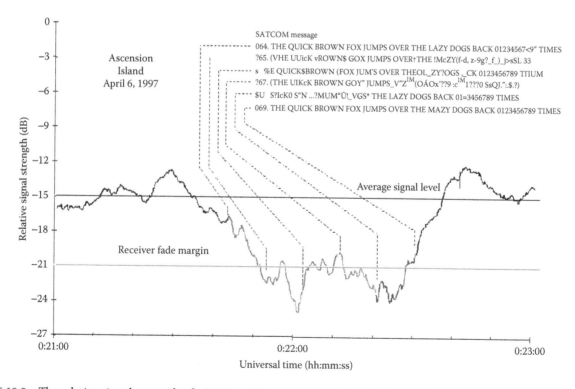

FIGURE 19.9 The relative signal strength of a VHF satellite communications signal and the corresponding lost characters in the data message, as the signal level falls below the fade margin for the system. The data were acquired at Ascension Island on April 6, 1997.

algorithms are complex and partly because the coherence bandwidth of the ionosphere has not been well-established. Phase modulation, also known as coding, is typically used to generate these signals. The baud rate of coded waveforms is $1/\tau_B$, where τ_B is the duration of the modulation period, known as the baud length. The shorter the baud length, the greater the bandwidth, which is directly related to the capacity to transfer information.

Consider as an example one recently deployed system, the Mobile User Objective System, a geostationary UHF communications system. The Mobile User Objective System is capable of transmitting information with multiple coding schemes. MUOS is capable of transmitting information with multiple coding schemes including binary phase-shift keying (BPSK), quadrature phase-shift keying (QPSK) and a wideband code-division multiple-access (WCDMA; mobile phone technology) waveform that employs a baud rate of about 4 MHz within a carrier bandwidth of approximately 5 MHz. Impacts on the relatively narrowband binary phase-shift keying and quadrature phase-shift keying waveforms can be estimated from amplitude scintillation considerations. The response of the wideband code-division multiple-access waveform may prove to be more subtle to quantify because amplitude fades due to scintillation are frequency dependent and may vary across the bandwidth of the signal, as discussed briefly in Section 19.1. Portions of the frequency spectrum of the signal may experience a fade, whereas other portions may not. Thus, some information can be transmitted, albeit at a reduced rate, even in strong scintillation. At other times, the whole 5-MHz band may fade in unison, so-called flat fading, and the signal may temporarily be lost altogether. In any case, the S_4 index alone is not sufficient to quantify performance impacts on wideband systems. A better analysis of such systems requires better knowledge of the temporal and frequency characteristics of scintillation-induced amplitude and phase fluctuations. These fluctuations are specified by parameters such as decorrelation time and coherence bandwidth, which are measures of the time and frequency separations, respectively, at which propagation effects on RF signals are no longer correlated. Decorrelation time is defined as the time lag at which the temporal autocorrelation function of the complex signal falls to a value of 0.5. Similarly, coherence bandwidth is the frequency offset at which the frequency-domain autocorrelation function dips to 0.5. Unfortunately, scintillation data collection systems have not traditionally recorded these parameters and the historical database is very limited. A combination of modeling, simulation, and observations is needed to make progress in assessing impacts on modern wide band communications systems.

Nevertheless, we conclude that most communications systems operating at or below a frequency of 400 MHz will experience performance disruptions and/or degradations at high-latitudes during solar maximum, during major magnetic storms at mid-latitudes, and throughout the solar cycle at low-latitudes. Systems operating at or below L-band frequencies (1–2 GHz) will be affected at low-latitudes during solar maximum. Next, we examine the impacts on space-based navigation systems and again find that S_4, while useful, provides an inadequate description of the relevant propagation characteristics.

19.3.2 Space-Based Navigation

The most widely used and studied GNSS is the GPS. GPS is a constellation of 31 satellites in six orbital planes, each transmitting a unique pseudorandom (PRN) phase code to provide accurate positioning and timing for civil and military users worldwide. The system became fully operational in 1990, and the original satellites broadcast signals on two frequencies, denoted L1 (1575 MHz) and L2 (1227 MHz). The most recent satellites now include a third frequency, known as L5 (1175 MHz), for civilian users; these satellites broadcast the civil coarse acquisition code on all frequencies (previously not present on L2). For more details on the technical specifications and operations of the GPS, see, for example, Parkinson et al. (1996).

Essentially, the system receives signals from all satellites in view simultaneously and correlates the various PRN codes with the observed signal to obtain the delay, or pseudorange, to each satellite. Trilateration calculations then determine the location of the receiver. As shown in Section 19.1, a portion of the measured delay is contributed by the ionosphere, and one must remove it to obtain accurate estimates of the desired geometric delay. Inexpensive single-frequency receivers utilize L1 only and are, by far, the most commonly used in automotive, aviation, recreation, and other applications. Using a single frequency it is not possible to estimate the delay due to the ionosphere, and a correction is required to improve positioning accuracy. The correction for standalone receivers is derived from a simple model that uses eight coefficients broadcast in an almanac message from the

GPS satellites (Klobuchar, 1987). The resulting correction is designed to account statistically for approximately one-half the actual ionospheric error. Even at mid-latitudes, the results are inadequate for many applications, and for users in the continental United States (CONUS), the Federal Aviation Administration has developed a complex space-based augmentation system that assimilates data from some 38 dedicated dual-frequency reference receivers to generate and broadcast improved model coefficients based on observed errors. The collective system is known as the Wide Area Augmentation System (WAAS). Europe has implemented a similar service known as the European Geostationary Navigation Overlay Service.

Performance of WAAS requires relatively uniform density gradients resolvable within the spatial separation of the reference stations. Magnetic storms commonly generate mid-latitude density gradients that exceed the accuracy bounds manageable by WAAS. However, the system performs continuous quality checks to insure integrity. Rather than broadcasting erroneous information to users, the system declares itself unusable and alternate navigation aids are employed for aviation. For more information on WAAS, see Loh et al. (1995). An example of typical WAAS coverage for the standard LPV200 service (localizer performance with vertical guidance, 200-ft minimum altitude; similar to CAT-1 accuracy) is shown in the top panel (a) of Figure 19.10. From the plot, the coverage area for 95% availability is 100% over CONUS and Canada and slightly less over Alaska. The high availability corresponds to a quiet space weather period, with minimal gradients present. The second plot in the lower panel (b), however, shows the effect of large-density gradients and variations during the main phase of a magnetic storm on February 27, 2014. In these circumstances, the WAAS system is unable to resolve the gradients and provide appropriate corrections on the spatial scales necessary to benefit users in the affected regions. At the 95% availability level, there is no LPV200 coverage in Alaska and Canada, and even CONUS coverage falls below 50% for a time. These and other related plots can be found on the Federal Aviation Administration's website: http://www.nstb.tc.faa.gov/. The February 2014 event presented here is hardly a notable storm relative to historical standards (min DST = −94; max K_p = 6). Since becoming operational in 2003, WAAS coverage has been reduced on numerous occasions because of

large spatial gradients associated with space weather events. Essentially, every magnetic storm results in degraded performance to some extent, starting at high-latitudes (Alaska) and extending southward, depending on the magnitude of the disturbance. The economic cost of these impacts is not inconsequential, as they are partially responsible for the addition of the new civil frequency on GPS satellites (L5) and the development of a new architecture for GPS-only flight navigation. The new system requires substantial investments in the ground, space, and flight segments and promises to improve safety, save time, and, eventually, reduce recurring operational costs.

The advantage of two or more frequencies for GNSS is immediately obvious when one considers the dispersive nature of ionospheric group delay. We start with the expressions for the pseudorange and carrier phase observations, measured by the receiver at two frequencies denoted by 1 and 2, respectively. For pseudorange, we have,

$$P_1 = \rho + c(\Delta t_r - \Delta t_s) + I_1^P + T + b_{1r}^P + b_{1s}^P + m_1^P + \varepsilon_1^P \quad (19.48)$$

$$P_2 = \rho + c(\Delta t_r - \Delta t_s) + I_2^P + T + b_{2r}^P + b_{2s}^P + m_2^P + \varepsilon_2^P \quad (19.49)$$

where:

P is the observed pseudorange
ρ is the geometric range
I^P is the ionospheric delay
T is the tropospheric delay
b is the instrumental bias for receiver and satellite
Δt_r is the receiver clock error (s)
Δt_s is the satellite clock error (s)
m^P is the multipath delay
ε^P is the thermal noise

Unless otherwise noted, all quantities have dimensions of meters. The equations for carrier phase delay in meters are similar, except that the observed phase is accurate to within an unknown number of cycles.

$$\Phi_1 = \rho + c(\Delta t_r - \Delta t_s) + I_1^\Phi + b_{1r}^\Phi + b_{1s}^\Phi \\ + \lambda_1 N_1 + m_1^\Phi + \varepsilon_1^\Phi \quad (19.50)$$

$$\Phi_2 = \rho + c(\Delta t_r - \Delta t_s) + I_2^\Phi + b_{2r}^\Phi + b_{2s}^\Phi \\ + \lambda_2 N_2 + m_2^\Phi + \varepsilon_2^\Phi \quad (19.51)$$

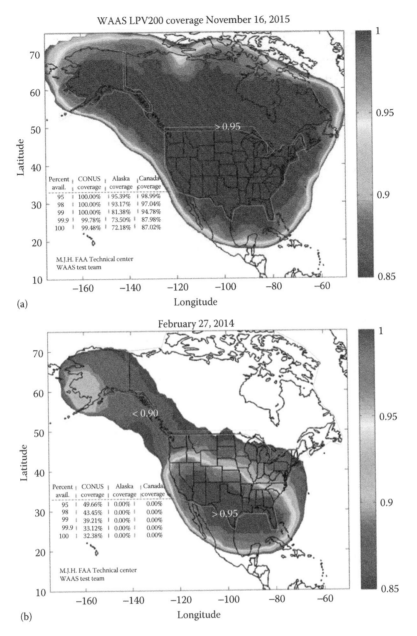

FIGURE 19.10 FAA plots displaying coverage for 95% LPV200 availability on (a) a typical quiet day on November 16, 2015, and (b) a moderate storm day, February 27, 2014. The coverage area for Alaska and Canada falls to zero (at 95% availability), whereas CONUS coverage falls below 50%.

Here, the terms represent the same quantities as those in the pseudorange equations, with the addition of I^Φ, ionospheric phase advance, and the absolute phase ambiguity term, λN, where λ is the wavelength and N is an unknown integer. Our objective here is to use the two frequency equations to cancel out the ionospheric delay, so that a reliable position fix can be achieved. Substituting the expression for ionospheric delay from Equation 19.21 and appropriately scaling and differencing Equations 19.50 and 19.51, we obtain the following ionosphere-free expression for phase:

$$\Phi = \frac{f_1^2}{f_1^2 - f_2^2}\Phi_1 - \frac{f_2^2}{f_2^2 - f_1^2}\Phi_2 \qquad (19.52)$$

One must still estimate the ambiguities and biases in order to determine accurate position and time, and this is usually done by using the pseudoranges. The key result here is that a two or three frequency receiver has the ability to cancel out the refractive ionosphere delay independently on each satellite link without external corrections and can obtain accurate positioning solutions, regardless of electron density gradients. We will now examine the extent to which a multifrequency receiver can cancel out the effects of scintillation.

To see the effect of scintillation on a multifrequency receiver, we apply the scaling equation, Equation 19.52, to the actual data obtained in a scintillation environment, as illustrated in Figure 19.11. The top panel in the figure shows the observed intensity of a moderately scintillated GPS L1 signal recorded at São José Dos Campos, Brazil (23.2° S, 45.9° W), in April 2012. The 50-Hz data show clear signatures of scintillation. Moving downward, the next two panels show 50-Hz phase measurements from the coarse acquisition code on both L1 and L2, respectively. The data have been detrended to remove the large-scale geometric phase component, and both frequencies show large rapid phase variations associated with scintillation. Next, we apply scaling and differencing to the respective phase measurements to obtain the ionosphere-free phase, which is expected to be zero for the detrended phase analysis performed here. However, the phase residual, plotted in the bottom panel of Figure 19.11, shows spiky signatures due to the diffractive contributions from

scintillation. This example demonstrates that multifrequency receivers are not able to filter or correct phase fluctuations due to scintillation. For more details, see Carrano et al. (2013). The next question is, do these diffractive effects have a material impact on the positioning capabilities of the receiver?

The ability to robustly track phase is a key element of modern GNSS technologies; all existing and planned systems employ phase-coded waveforms for absolute range determination. (Note that legacy GLONASS technology uses frequency diversity rather than code diversity to discriminate satellite signals.) One may then suspect that uncompensated phase distortions can adversely affect positioning and/or Doppler. Although experimental confirmation of this suspicion will be presented shortly, it is imperative to be aware that the complexity, diversity, and proprietary nature of GNSS technologies hinder efforts to develop meaningful error prediction algorithms applicable to a class of receivers or even to a specific receiver. Many field measurements with a variety of dual-frequency GPS receivers reveal unambiguously that scintillation causes substantial

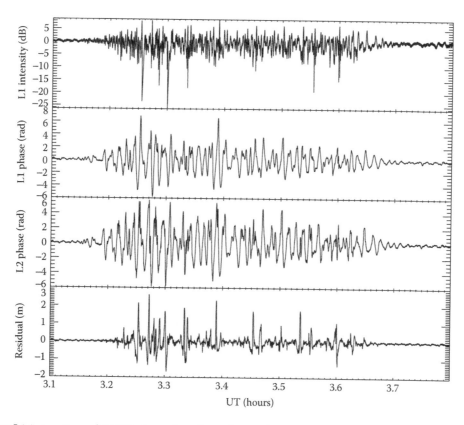

FIGURE 19.11 Raw L1 intensity and L1/L2 phase data from São José Dos Campos, Brazil, in April 2012. The bottom panel shows the phase residual after scaling and differencing L1 and L2 to remove ionospheric refractive effects. The finite residual comes from diffractive effects (scintillation), which multifrequency GNSS systems cannot correct.

positioning and Doppler errors (Groves et al., 2000; Skone et al., 2001; Thomas et al., 2004; Kintner et al., 2007).

McNeil et al. (2015) recently conducted a more systematic analysis of scintillation-induced GPS errors to examine the dependence on magnetic latitude during the most recent solar maximum period. Daytime observations were used to establish nominal truth baselines at a number of stations. These were then compared with positions reported during the most active scintillation period of the evening to assess scintillation-induced errors. An example of the horizontal and vertical errors observed at Ilheus, Brazil (14.8° S, 320.8° E), is shown in Figure 19.12. The distribution of large errors for the nighttime period is a result of scintillation. This was done for seven other stations in the American sector spanning a range of magnetic latitudes from the equator to more than 20°; the stations and their coordinates are listed in Table 19.2. In Figure 19.13, the histograms of errors for all sites are shown in order of increasing magnetic latitude. Noncoincidentally, the meridional structure of the errors is strikingly similar to the structure of the Appleton density anomaly, the region of most intense L-band scintillations. We see that GPS experiences little to no error at the magnetic equator, and performance is most affected from about 10°–20° magnetic latitude, a region centered about the anomaly. The results are consistent with our findings that high scintillation correlates with high electron density, but because GPS errors depend on other factors, such as the geometry and the number of links simultaneously affected, it is satisfying to verify where positioning accuracy may be most degraded by space weather. Figure 19.14 provides an idea of how the solar cycle affects GPS errors. Years of data from Ascension Island were analyzed for horizontal and vertical positioning errors as a function of solar flux. The errors remain relatively flat, until the monthly average F10.7 flux exceeds a value of about 100 to 120 sfu. As we saw in Figure 19.7 in Section 19.2, solar flux determines the density, and this appears to be the key driver for GPS navigation errors as well. Understanding how scintillated signals affect GNSS performance remains a challenge.

Previously, we demonstrated substantial observed phase errors due to diffraction effects on L1/L2 signals. Gherm et al. (2011) analyzed these effects in detail and estimated that the maximum diffraction-induced positioning errors for an individual link would not exceed 1 m,

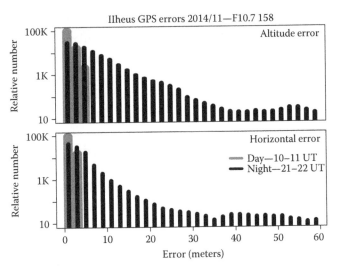

FIGURE 19.12 Horizontal and vertical GPS errors measured in November 2014 at Ilheus, Brazil. The daytime "truth" data is in gray and the nighttime scintillation-driven errors, extending to tens of meters, are plotted in black. (From McNeil, W.J., private communication, 2015. With permission.)

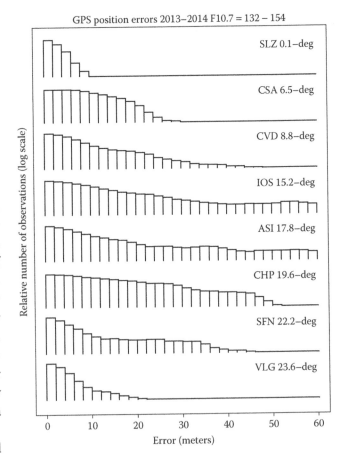

FIGURE 19.13 GPS position errors for eight sites plotted in order of increasing absolute magnetic latitude. Errors peak in and around either side of the Appleton anomaly region nominally located at approximately ±15° MLat.

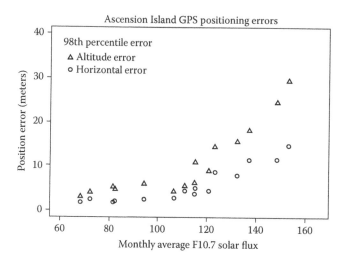

FIGURE 19.14 98th percentile horizontal and vertical errors reported by a stationary Ashtech Z-12 GPS receiver at Ascension Island as a function of solar radio flux, F10.7. Errors increase markedly when F10.7 exceeds 100–120 sfu.

TABLE 19.2 Stations Used in the Magnetic Latitude Study

Station	Code	Glat	Glon	Mlat
Sao Luis	SLZ	2.6 S	315.8 E	0.1
Cuiaba	CBA	15.1 S	303.9 E	6.5
Cape Verde	CVD	16.7 N	337.1 E	8.8
Ilheus	IOS	14.8 S	320.8 E	15.2
Ascension Island	ASI	8.0 S	345.6 E	17.8
Cachoeira Paulista	CHP	22.7 S	315.0 E	19.6
Santa Fe	SFN	33.0 S	299.4 E	22.2
Villegas	VLG	35.0 S	297.0 E	23.6

even for S_4 ~1. However, observed positioning errors frequently exceed tens of meters in both horizontal and vertical domains. Additional factors must then be contributing to the scintillation error budget. Figure 19.15 may offer some clues to help explain the disparity. The top panel shows the number of GPS satellites tracked by the receiver as a function of time. During the scintillation period, the receiver repeatedly drops and reacquires multiple satellites, at times tracking less than the minimum of four satellites required to determine the position and time. This results in a complete outage of navigation data. The abrupt loss of a satellite generates a sudden increase in the position dilution of precision and a corresponding spike in positioning error. The direction and magnitude of the error is determined by the location of the satellite, directly related to its effect on the position dilution of precision. As the number of tracked satellites decreases, the loss of a single additional link can result in large changes in accuracy instantaneously,

as evidenced by the spiky errors in the second panel of Figure 19.15. However, not all enhanced errors are attributable to sudden jumps, as the spikes appear to be supported by a slowly varying envelope of enhanced errors during the affected period. Therefore, it appears that there may be two distinct types of scintillation-induced errors on GNSS: (1) Abrupt errors associated with the loss of lock, and (2) secular ranging errors from processing scintillated signals. We will briefly examine both contributions in more detail to obtain a better understanding of how space weather impacts GNSS, at least on a qualitative level.

Most GNSS receivers use so-called phase-locked loops (PLL) to continuously track the code and carrier phase of incoming satellite signals. "Loss of lock" refers to the temporary failure of the tracking loop to maintain coherence with the PRN code used for absolute ranging. Failure to track the carrier phase is generally a

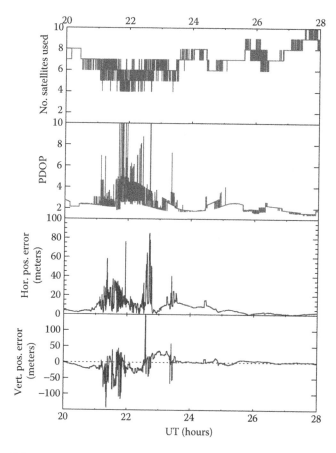

FIGURE 19.15 Four-panel plot showing vertical and horizontal positioning errors (lower panels) and the corresponding number of satellites tracked and position dilution of precision values as a function of time at Ascension Island on March 16, 2002.

less disruptive phenomenon, often referred to as "cycle slips." Both processes have been discussed extensively in the literature (Knight and Finn, 1998; Skone et al., 2001; Conker et al., 2003; Humphreys et al., 2005; Prikryl et al., 2010). The design parameters of the PLL are determined based on the needs of the application. A narrow-band PLL suitable for stationary and low-dynamics environments provides improved SNR. High-dynamics applications involving rapid geometric phase changes require wider bandwidth-tracking loops that subsequently admit more noise and reduce SNR. As we have seen, scintillation causes fluctuations in the observed phase. If the fluctuation rate exceeds the bandwidth of the PLL, the receiver may not be able to maintain lock and the signal will be lost; the signal will be reacquired when the rate of phase change decreases and again falls within the design parameters of the PLL. One might conclude that wider PLLs would therefore be less vulnerable to scintillation effects. However, during strong scintillation, the amplitude fades in conjunction with phase fluctuations. This may further reduce the lower SNR inherent in wide band PLLs and render the signal too noisy to track, with resulting loss of lock. Thus, depending on the tracking loop implementation, the receiver may be more sensitive to amplitude (wide band) or more sensitive to phase (narrow band) scintillations. In both cases, both the magnitude and rate of the fluctuations are important, and the combined effects of amplitude fades and phase changes exacerbate the difficulty of maintaining lock.

The secular enhancement of ranging errors during scintillations requires a different explanation and may be caused by multiple sources. The first stems from distortions to the complex received waveform, such that the peak in the code cross-correlation function is shifted from the true value or multiple peaks are formed. We have already discussed how the ionosphere can act to modify finite bandwidth waveforms, and the PRN codes employed by the GPS are simply L-band examples of such signals. Some receivers may perform internal checks on signal integrity and/or health that limit the use of degraded signals. Figure 19.16 shows an example of data from an Ashtech Z-12 at Ascension Island on March 16, 2002. The carrier-to-noise ratio is plotted as a function of time for L1 and L2, as well as the status of the signal with respect to use in navigation. The receiver attempts to track both signals continuously, as expected, but makes an internal decision not to use the information for navigation

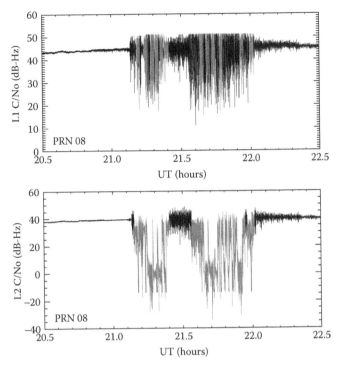

FIGURE 19.16   Raw L1 (top) and L2 (bottom) signals from PRN 08 on March 16, 2002, at Ascension Island. Gray lines denote the periods when the receiver did not use the signal to determine the position data.

(denoted by a gray line in the Figure 19.16) during periods when scintillation is occurring. Note especially the extended periods during which the system rejects information from the L2 signal. In itself this is not surprising, given that relative to L1, L2 is lower in frequency and thus more heavily scintillated, and the transmitted power level is 3 dB weaker. Moreover, the civil receiver cannot track the L2 Y-code signal directly and must resort to semi-codeless techniques, which incur a penalty of up to 13 dB in effective signal strength (Parkinson et al., 1996). Because the L2 signal is rejected, the unit essentially operates as a single-frequency receiver without measured ionosphere corrections. During scintillation events the ionosphere is typically in an extreme state with regard to density gradients, and single-frequency model corrections prove grossly inadequate. How the receiver actually functions during these periods is not known to the authors, and every receiver seems to perform a little differently in these environments, presumably determined by proprietary internal algorithms unique to each manufacturer. Nevertheless, whatever response the technology employs, it is certain to be less accurate than dual-frequency tracking, and this can explain enhanced ranging errors not associated with complete loss of lock in GNSS receivers.

A troubling aspect of predicting the GNSS response to scintillation is the lack of an appropriate propagation metric or index to characterize errors. The S_4 index has been shown to be a useful index for narrowband applications such as satellite communications. However, when applied to GNSS, it becomes obvious that S_4 has limited utility as an indicator of performance. Consider the histogram of S_4 samples presented in the upper panel of Figure 19.17. The data show all the scintillated samples from the Ashtech Z-12 receiver in use at Ascension Island, from March 5, 2002, to March 19, 2002. The samples indicated by the hatched marking correspond to a loss of lock (LOL) event. The distribution of loss of lock with S_4 is surprisingly broad, extending to relatively weak levels of scintillation. The lower panel of the figure shows the ratios of the two populations (all samples and those associated with loss of lock) to compute a probability of losing lock as a function of scintillation index; the line

shows the best-fit curve to the data. Although there is a clear correlation between increased S_4 and probability for losing lock, the data hardly establish a definite threshold for losing lock. The median probability corresponds to an S_4 of 0.7, whereas the 90% probability for losing lock is 0.97. We conclude that S_4 alone provides a relatively coarse indicator of the GPS link status. Figures of merit based on the phase index, σ_φ, are similarly indefinite. Neither S_4 nor σ_φ convey information regarding the temporal rates of change in amplitude or phase, which certainly influence the effect of scintillations on GNSS performance. Carrano and Groves (2010) explored these issues in considerably more detail and showed that combinations of known propagation parameters may serve as better discriminants for predicting the GPS behavior in scintillation. For example, decorrelation time was shown to correlate well with LOL events. Even still, no single metric appears to adequately explain receiver response, and a more comprehensive description of scintillated waveforms is required to better characterize the impact on GNSS. Even then, at best, one can hope to predict tracking behavior on an individual link. Additional parameters, such as link geometry, satellites affected, and internal signal processing algorithms, are all parts of the complex equation that must be solved to fully specify scintillation-induced GNSS navigation errors.

Regardless of the details, these results, combined with the analysis presented in Section 19.2, suggest that substantial effects on positioning performance are expected only when F10.7 reaches or exceeds a value of ~100–120 sfu, expected during 4–5 years of a typical 11-year solar cycle. For GNSS users, this may be a case of the glass half empty or half full, depending on one's outlook. However, solar minimum poses little threat to GNSS services and we continue to improve our assessment of where and when space-based navigation may be at risk because of space weather.

19.3.3 Other RF Systems

Although communication and navigation systems represent the largest communities of users impacted by space weather effects on radio wave propagation, other RF technologies are also subject to performance degradation. Examples include both ground- and space-based radar applications and HF propagation channels.

Ground-based radars that track objects in space operate through the ionosphere and are thus subject to the same group delay and scintillation effects as GNSS

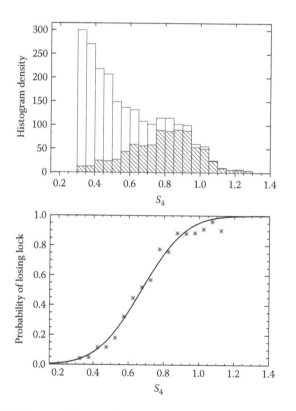

FIGURE 19.17 Top panel: Histogram of GPS S_4 samples from March 5–19, 2002 Ascension Island campaign, indicating the number of samples associated with a "loss of lock" event (cross-hatched fill). The probability for losing lock as a function of S_4 is shown in the lower panel (Ashtech Z-12 receiver).

and other RF signals. Systems operating at L-band and lower frequencies are the most susceptible. Some systems exploit dual frequencies to correct for ionospheric group delay in precisely the same manner as the GPS. The ALTAIR deep-space tracking radar, located in Kwajalein Atoll, the Marshall Island, transmits simultaneously at UHF (422 MHz) and VHF (158 MHz) to accurately remove ionospheric group delay for improved range estimates. However, as with GNSS, multiple frequencies do not correct the phase and amplitude variations associated with scintillations. Indeed, scintillation effects are substantially enhanced, as the signal passes through the irregularities twice. Because RF energy travels at the speed of light, the ray path is effectively stationary relative to a Fresnel scale during the time it takes the wave to bounce off the target and pass back through the ionosphere (2–3 ms). Therefore, the two-way phase perturbations will essentially be doubled in radians and the corresponding amplitude variations will be doubled in decibels. Figure 19.18 shows the measured radar cross section (RCS) recorded by ALTAIR while tracking a smooth calibration sphere 1.1 m in diameter on October 5, 2006. Under clear conditions, the RCS would essentially remain constant as the system applies accurate calibration constants and corrects for the changing range of the target. However, in this postsunset collection, both the UHF and VHF signatures show large amplitude variations during the beginning and end of the track, corresponding to the presence of irregularities between the target at 1000-km altitude and the radar.

Near the middle portion of the track, between approximately 160 s and 240 s, the radar beam intercepts weaker irregularities and the RCS becomes smoothly varying at UHF. As expected, the UHF signal experiences less perturbation than the VHF signal, which shows some evidence of irregularities throughout the track, as can be seen by the two-way S_4 data plotted in the lower panels. The S_4 values seem especially high, given that $S_4 > 1$ represents strong scatter and is not ordinary for one-way propagation. However, owing to the aforementioned "doubling effects", such values are not unusual for two-way propagation through the phase screen (Caton et al., 2009). Accurate cross-sectional measurements are not possible when irregularities are present. Moreover, the contribution of phase scintillations in this application manifests itself as Doppler errors. For a more extensive examination of these effects, see Cannon et al. (2006). The example presented here represents weak scatter during solar minimum conditions; solar maximum poses far more stressing environments.

Radars do represent a special case for space weather impacts. As they are relatively high-power active transmitters with sensitive receivers, they are subject to backscatter effects that have not yet been discussed. Although scintillations are associated with Fresnel-scale structures that are a few hundred meters to a kilometer, irregularities one-half the RF wavelength generate coherent backscatter through a process known as Bragg scattering when the radar beam is directed perpendicular to the field-aligned structures. ALTAIR frequently observes

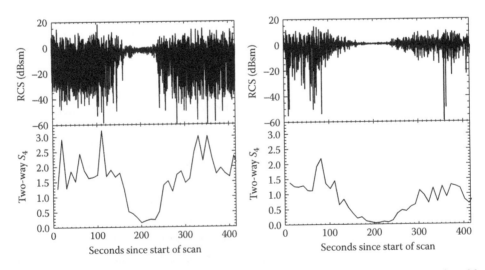

FIGURE 19.18 ALTAIR VHF (left) and UHF (right) measured RCSs for a 1.1-m diameter smooth calibration sphere on October 5, 2006. Two-way scintillation effects cause substantial fluctuations during all but the central portion of the object track. Two-way S_4 is plotted below each signal; values can exceed two at both frequencies even during solar minimum.

enhanced coherent scatter due to equatorial plasma bubbles, and experimenters often scan the beam perpendicular to B as a sensitive means of detecting the presence of small-scale irregularities. UHF and VHF coherent scatter volumetric cross sections of 30 dBsm (1000 m²) are common, and intense events may be even higher. Such large cross sections can represent a source of undesirable clutter for sensitive radars designed to detect targets as small as −30 to −40 dBsm (10⁻³–10⁻⁴ m²); if an antenna sidelobe one million times weaker (−60 dB) than the main beam illuminates a volume perpendicular to B, it can interfere with the desired target.

This is a real issue for mid-latitude radars operating at UHF. In this case, the culpable region of perpendicularity is in the E-region rather than in the F-region of the ionosphere. During periods of magnetic activity, a two-stream instability at high-latitudes creates small-scale irregularities in an altitude band from 100 to 110 km; (see, e.g., Hysell [2015] and references therein). A powerful, sensitive radar, such as the Millstone Hill UHF steerable incoherent scatter system, located in Westford, Massachusetts, can illuminate this region perpendicular to **B** with a sidelobe at 0° elevation angle. The coherent scatter returns from the sidelobe interfere or even mask the desired incoherent returns from the ionospheric F region (Foster and Erickson, 2000). During the famous March 1989 storm, a mid-latitude UHF surveillance radar actually observed more than 3000 "false" radar echoes, attributable to E-region clutter in less than 3 h. The physics of these processes are understood and their occurrence is predictable, but the effects of these processes range from being an annoying nuisance to seriously impacting desired operational outcomes.

Radars in space will experience the same types of impacts. As with other systems, the signatures of the anomalies are determined by the propagation effects and the details of the technical applications. In the case of space-based synthetic aperture radar (SAR), complex signal processing associated with image formation leads to an interesting manifestation of space weather phenomena.

Radars in space typically operate as SAR, so called because it is impractical to put large-scale antennas in orbit and the radars achieve high spatial resolution by "synthesizing" a large aperture by collecting and processing coherent samples over a region, as they move along their orbit. Collection for just 3–4 s above an area of interest provides an effective aperture size of 25–30 km for a space

radar in low Earth orbit. The technique is well established and effective when used from aircraft and space vehicles alike, under the assumption that the target scenes remain coherent over the data collection time period (Soumekh, 1999). A typical space-based application is known as coherent change detection. This technique utilizes repeat passes over a given scene and through coherent differencing of radar imagery, effectively functions as a sensitive interferometer, detecting minute changes in topography or scene content. When performing repeat-pass imagery over the Amazon Basin in Brazil, the Japanese PALSAR L-band radar aboard the Advanced Land Observing Satellite spacecraft detected strange signatures, namely blurring or smearing of the images, apparently along the direction of motion of the satellite. An example of a sharp image collected on December 25, 2007, is shown in Figure 19.19a, whereas a smeared image of the same scene observed just three months later on March 26, 2008, is presented in Figure 19.19b. The scene is still recognizable, but the extended streaks across the image create problems for coherent processing. Further investigation showed that the streaks were associated with ionospheric irregularities and the smearing axis was determined to be the direction

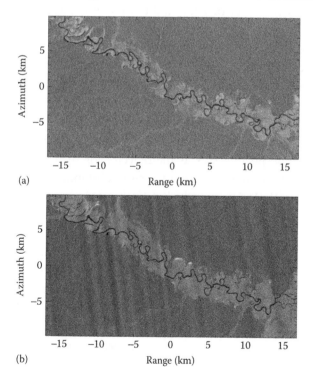

FIGURE 19.19 Advanced Land Observing Satellite PALSAR L-band SAR image of terrain in the Amazon River basin. Showing (a) crisp imagery and (b) imagery compromised by streaks due to ionospheric irregularities. Much stronger distortions are expected at solar maximum.

of the Earth's magnetic field rather than the satellite velocity. The effect arises from a combination of the long correlation distance of irregularities along the magnetic field direction, the observing geometry, and the signal processing used for SAR image formation. Several studies have investigated this phenomenon in more detail, and it appears possible to exploit L-band SAR data as a remote sensing tool for ionospheric irregularities and structure (Belcher and Rogers, 2009; Meyer, 2011; Carrano et al., 2012a; Rogers et al., 2014). It should also be noted that the observations shown here occurred during deep solar minimum, when the ionosphere was at or near the lowest-density state observed since the beginning of the modern space age. The effects of bubbles during solar maximum may appear less benign to future L-band SAR missions.

Finally, a chapter on the impacts of space weather to RF systems would not be complete without mentioning ionospheric effects on HF systems. Of course, the name familiarly applied to low latitude plasma bubbles, "Equatorial Spread F", originated with HF observations due to the extensive HF range spread the structures caused to vertical and near-vertical incidence ionosondes. Although the irregularities responsible for scintillations are of the order of 1 km and those that cause coherent radar backscatter are of the order of 1 m, effects on HF result from structures of the order of 10 km or greater in size. HF systems are special because the frequency is sufficiently close to the plasma frequency in the upper atmosphere and it can be refracted by the ionosphere back to the Earth's surface. Such systems are thus suitable for "over the horizon" communications and surveillance. Because the ionosphere serves as the reflector for such systems, even minor irregularities can cause substantial propagation path deviations, which drive common concerns for HF systems: Where does transmitted HF energy go, and from where do received HF signals originate? Ionospheric perturbations associated with both terrestrial and space weather can prevent a communication signal from reaching its desired recipient and cause substantial errors in estimating from where an HF radar return originated.

Spread F is an example of a severe and easily recognized modification of the propagation environment. More subtle and deleterious effects are caused by less auspicious space weather phenomena in the form of ionospheric gradients, tilts, and traveling ionospheric disturbances (TIDs). Examples showing the results of these phenomena for a specific link configuration are shown in Figure 19.20. The top panel shows the

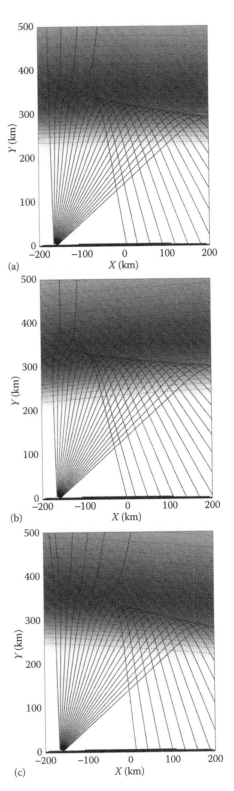

FIGURE 19.20 Ray tracing results for a 10 MHz Chapman-based ionosphere, with a peak plasma frequency of 10 MHz and a peak height of 350 km for (a) a uniform layer, (b) a layer with an upward tilt with height (2%/100 km) to the right, and (c) a layer tilted upward to the left of the transmitter. The fan of rays is the same for all cases, with 2.5° elevation step between rays. The signal frequency is 10.1 MHz.

propagation path for a 10.1 MHz signal in a uniform ionosphere defined by a Chapman layer with peak plasma frequency of 10 MHz at peak altitude of 350 km. The plots (b) and (c) illustrate changes caused by a 2% increase in height per 100 km to the right and left of the transmitter, respectively. Even this modest tilt, which could result from a gradient in density, height, or a TID, changes the ray paths substantially. In all the plots, the rays are launched overhead to lower angles, with a 2.5° step between rays. For the case of no tilt in the top panel, the fifth ray launched arrives near the arbitrary location of $x = 0$. In plot (b) where height increase to the right of the transmitter, the fifth ray reaches the ground some 40 km to the right, whereas the third ray, launched some 5° higher in elevation angle, meets the surface at $x = 0$. In addition, none of these rays reach the ground even when the tilt increases to the left of the transmitter.

The TIDs are wavelike perturbations usually associated with traveling atmospheric disturbances, which may be generated by tidal interactions, tropospheric perturbations, natural hazards such as earthquakes and tsunamis, and high-latitude heating associated with space weather activity (Hunsucker, 1982; Hocke and Schlegel, 1996; Komjathy et al., 2012). TID activity can be difficult to detect, and its effects on HF propagation are definitely difficult to model and predict. Unlike a gradient or tilt, the effects of TIDs are typically dynamic and cause time-varying and sometimes quasi-periodic propagation effects. A substantial body of scientific literature on HF propagation and TIDs has been developed over the past 50 years; far more detail on this topic can be found in McNamara (1991) and references therein. The field has recently enjoyed a resurgence in activity spurred by renewed interest in HF technologies. The impact of space weather on these systems is certain to remain a relevant and challenging topic in the foreseeable future.

19.4 MITIGATION STRATEGIES

So far, this chapter has dealt primarily with presenting impacts of the space environment on RF systems. Now, we turn from describing problems to considering potential solutions and mitigation strategies. The strategies presented here are not intended to be comprehensive in scope or detailed engineering roadmaps for defeating space weather impacts on specific systems. However, they do represent a sampling of generic concepts that the reader may find useful in developing an approach for applications of interest.

19.4.1 Frequency Diversity

Probably, the most obvious approach to defeating scintillation effects is moving to a higher frequency, as this always decreases the scattering strength that affects the radio wave and can improve the performance of a given communication or navigation service (note, however, that terrestrial weather propagation effects become more severe at higher frequencies). Diversity also refers to utilizing different portions of the RF spectrum, with the expectation that amplitude and phase variations in respective frequency bands are not highly correlated. Frequency-dependent fading means that some portions of the RF spectrum may be heavily attenuated, whereas the adjacent portions may not be, such that some rate of information transfer can be maintained. This type of fading is most common in a strong scatter environment, where the fading varies dynamically with frequency; the coherence bandwidth is a measure of the spectral separation needed to exploit frequency diversity in this way. A simple example of frequency diversity is illustrated in Figure 19.21. Plotted in gray lines are the L1 and L2 intensity for 120 s of a scintillated GPS signal. Both signals experience substantial fading, but the fades are uncorrelated in time. The dark-black line in the figure shows the combined signal, and it exhibits substantially less fading, except for a period between about

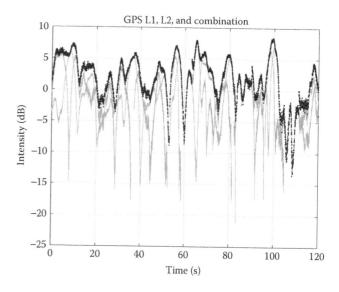

FIGURE 19.21 Scintillated L1 and L2 GPS signals shown in gray and the combined signal overplotted in black to demonstrate frequency diversity. Except for the period just after 100 s, when L2 is out completely, the combined signal stays several dB above the fade floor experienced by the individual channels.

100 s and 110 s, when L2 is experiencing a complete outage and L1 is the only available signal. Frequency diversity is part of the rationale for GPS expanding to a third civil frequency, L5.

Carrano et al. (2012b) explored the conditions under which the three GPS frequencies, L1, L2, and L5, would begin to decorrelate; they modeled the decorrelation as a function of scintillation strength and demonstrated the validity of the results with actual observations from the GPS satellites transmitting all three signals. They found that as scintillation strength increases, L1 decorrelates from L2 and L5 first, because the frequency difference is the greatest. Eventually, L2 and L5 become decorrelated, nominally when the coherence bandwidth near 1 GHz falls below the separation frequency of about 50 MHz. The point at which signals decorrelate is when frequency diversity becomes an effective countermeasure to scintillation. Information transmitted on multiple frequencies may be received continuously, as the amplitudes of the multiple signals vary independently and the "combined" signal maintains the nominal power margin for the link. However, this approach to performance enhancement does carry a cost, as information is transmitted redundantly on multiple channels, which is an unnecessary strategy in the absence of scintillation. Counterintuitively, applications designed to benefit from frequency diversity in this context may actually suffer more performance degradation in less intense frequency-independent "flat" fade environments and perform better when scintillation severity increases.

19.4.2 Spatial Diversity

Spatial diversity may be an effective means to reduce the impact of irregularities on RF systems by utilizing multiple separated data sources in space or on the Earth. Irregularities rarely, if ever, fill the entire sky and are usually localized in patches, auroral structures, mid-latitude storm-induced features, or equatorial plasma bubbles. The more signal sources available in space, the better the odds of having one or more clear channels to receive. In the case of the GPS, for example, a typical location may have 8–11 satellites in view at any given time, providing some level of spatial diversity. However, as we have seen from the previous section on GPS errors, the distribution of these satellites is frequently inadequate to provide enough clear channels and/or good geometry. Consider the improvement

obtained by utilizing all available GNSS, including GLONASS, Galileo, and BeiDou (Compass). An example of the increased coverage thus obtained with current satellites can be seen in Figure 19.22, where the sky tracks from the GPS and all other existing GNSS satellites for a location in the southwestern US are shown. There is both good news and bad news in this example. The additional GNSS satellites contribute almost twice as many measurement sources as the GPS during the period, bringing the total number to nearly 30. The bad news is that the actual spatial diversity provided by the additional sources is less compelling because of the similarities in orbital parameters adopted by the various systems; satellites track across similar regions of the sky, and other regions remain unsampled. If the measurements are clustered together, then the benefits may be reduced, as large structures, such as bubbles, can affect a large number of sources. It is too early to determine the impact of scintillation on combined GNSS positioning services, but the additional sources

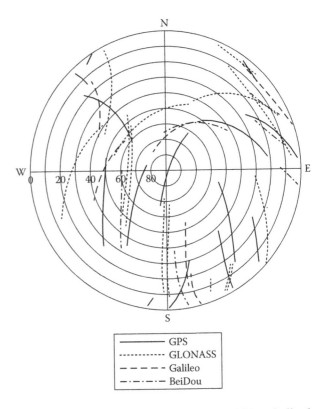

FIGURE 19.22 Sky coverage plot showing GPS and all other known GNSS satellites observed from a mid-latitude site. The tracks represent the azimuth/elevation trajectories of the satellites over a 2-h period. The inclusion of GNSS sources nearly triples the number of available beacons.

are certain to provide some improvement relative to GPS-only navigation. As the number of GNSS satellites on orbit increases in the future, we expect the severity of space weather impacts on these systems to decrease.

However, in most cases, the strategy of fielding additional satellites to minimize space weather impacts is not feasible, simply because of the prohibitive expense of space vehicles. A more practical solution for most systems, then, is to exploit spatial diversity on the ground. The diffraction pattern from scintillated sources in space has variations in signal strength on spatial scales of tens to hundreds of meters in the direction transverse to the magnetic field. Signals from antennas spaced beyond the so-called correlation distance of the fades can be combined to effectively mitigate scintillation effects. One might think that the minimum separation distance for effective spatial diversity would be the order of a Fresnel zone, given by $\sqrt{2\lambda z}$, where λ is the wavelength of the given RF system and z is the distance from the observer to the phase screen. For a system operating at 250 MHz, this is approximately 1 km in weak scatter, and the observed decorrelation distances are consistent with this estimate (van de Kamp et al., 2010). In stronger scatter, the distance decreases to a region smaller than a Fresnel zone, and a separation distance of ~100 m (transverse to **B**) can provide substantial improvement. An example is shown in Figure 19.23. Here, we have taken a signal at 250 MHz received on one antenna and combined it with the signal from a second antenna spaced 216 m from the first. The combined signal (thick black line) still shows evidence of scintillation, but the average amplitude is greatly improved. Referring to frequency diversity example presented in Figure 19.21, we see comparable results. In some ways, frequency diversity and spatial diversity are manifestations of the same diffraction phenomenon. Essentially, diffraction produces a frequency-dependent spatial pattern on the ground, such that a single antenna observes uncorrelated frequency-dependent fading for sufficiently diverse frequencies. Conversely, observing a single frequency diffraction pattern from one or more locations separated by the correlation distance on the ground (spatial diversity) yields similar uncorrelated time-varying signals.

19.4.3 Awareness of the Space Environment

A third mitigation strategy focuses on awareness of the space weather environment. Unlike terrestrial weather,

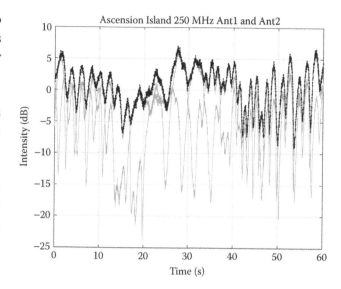

FIGURE 19.23 VHF scintillations observed on March 16, 2001, at Ascension Island, using two separate antennas spaced 216 m apart in the magnetic E-W direction. The two signals are plotted in gray, whereas the incoherent sum of the signals is plotted in black. The combined signal is considerably more uniform, exhibiting substantially fewer and shallower fades.

where a pilot can observe that a thunderstorm is nearby simply by glancing out the window or pausing to listen, with the exception of the aurora, space weather does not produce visible or audible signatures that can be sensed by people. Impacts caused by space weather may thus be attributed to other factors, such as equipment malfunctions or electromagnetic interference. Just the knowledge that space weather is affecting a particular system at a particular time can be valuable in terms of formulating a response to performance impacts. Knowledge of a future impact (forecast) can be even more valuable. Users may be able to schedule activities to avoid scintillation, reduce bandwidth, choose alternative communications systems, and generally adjust expectations for system performance. However, present capabilities to forecast electron density irregularities are primitive and limited. The dynamic high-latitude environment, controlled by complex solar interactions with the magnetosphere, admits generic activity forecasts, which are of little benefit to specific RF systems. Disturbances at mid-latitudes are similarly linked to the electric and magnetic fields, which drive high-latitudes, and more work is needed to understand the detailed coupling between these regions and develop meaningful forecasts. At low-latitudes, the situation is a bit more hopeful in the near-term, as most instabilities occur in the absence of impulsive solar activity, controlled instead by interactions internal to the ionosphere–thermosphere system. Even though the physics is fairly well understood in this region, the lack of data needed to sufficiently specify initial conditions has rendered true physics-based equatorial forecasts unreliable.

However, in all regions, after instabilities occur and irregularities form, it may be possible to detect and characterize the disturbances to at least provide space weather awareness to system users. This approach has been successfully demonstrated at low-latitudes by the Scintillation Network Decision Aid. Developed by the US Air Force to support satellite communications, the Scintillation Network Decision Aid consists of a network of automated scintillation-monitoring sites reporting data to a central computer server in the United States. The data drive an empirical model that produces simplified graphical products indicating the location and severity of scintillation impacts on UHF SATCOM (Groves et al., 1997). Because the behavior of equatorial structures is relatively predictable after they have formed, it is possible to generate an accurate nowcast up to 2 h in advance, accounting for the motion and evolution of the irregularities. Similar approaches may be successful in other latitudes aided by more research on the physics and dynamics of irregularities in the respective regions. Other systems, such as GNSS, can readily be supported in the same manner. The key is to characterize the irregularity strength accurately through a combination of sensors and model(s). Successful system support then requires one to understand the effects of scintillation on system performance. It is hoped that in longer term, increase in the availability of data and improved high-resolution models will yield full physics-based approaches that can address the true forecast needs of the space-based RF services community.

Other mitigation strategies being explored focus on engineering solutions such as adaptive tracking loop and communications bandwidths and protocols enabled by modern digital systems, signal processing, and software-defined radios (see, e.g., Humphreys et al., 2010; Carroll et al., 2014). The success of these approaches will be determined as the technologies evolve and mature. These approaches are mentioned here as they offer the potential to improve the performance of RF systems without substantial compromises in the hardware configuration or information bandwidth. For example, it already

appears that the current generation of GNSS receivers, which exploit newly available signals (GPS L2C and L5), performs better than the systems described in this chapter. However, one must recognize that the scintillation environment during the maximum of the most recent solar cycle (2011–2014) was substantially less severe than that of the last solar maximum period in 2000–2002. Therefore, direct comparisons are difficult to establish.

REFERENCES

Aarons, J., & Allen, R. S. (1971). Scintillation boundary during quiet and disturbed magnetic conditions. *Journal of Geophysical Research*, 76(1), 170–177.

Aarons, J. L. E. S. (1982). Global morphology of ionospheric scintillations. *Proceedings of the IEEE*, 70(4), 360–378.

Abdu, M. A. (2012). Equatorial spread F/plasma bubble irregularities under storm time disturbance electric fields. *Journal of Atmospheric and Solar-Terrestrial Physics*, 75, 44–56.

Abdu, M. A., Batista, P. P., Batista, I. S., Brum, C. G. M., Carrasco, A. J., & Reinisch, B. W. (2006). Planetary wave oscillations in mesospheric winds, equatorial evening prereversal electric field and spread F. *Geophysical Research Letters*, 33(7), L07107. doi:10.1029/2005GL024837.

Basu, S., Basu, S., Groves, K. M., MacKenzie, E., Keskinen, M. J., & Rich, F. J. (2005). Near-simultaneous plasma structuring in the midlatitude and equatorial ionosphere during magnetic superstorms. *Geophysical Research Letters*, 32(12), L12S05. doi:10.1029/2004GL021678.

Basu, S., Basu, S., Groves, K. M., Yeh, H. C., Su, S. Y., Rich, F. J., ... & Keskinen, M. J. (2001a). Response of the equatorial ionosphere in the South Atlantic region to the great magnetic storm of July 15, 2000. *Geophysical Research Letters*, 28(18), 3577–3580.

Basu, S., Basu, S., MacKenzie, E., Bridgwood, C., Valladares, C. E., Groves, K. M., & Carrano, C. (2010). Specification of the occurrence of equatorial ionospheric scintillations during the main phase of large magnetic storms within solar cycle 23. *Radio Science*, 45(5), RS5009. doi:10.1029/2009RS004343.

Basu, S., Basu, S., Rich, F. J., Groves, K. M., MacKenzie, E., Coker, C., ... & Becker-Guedes, F. (2007). Response of the equatorial ionosphere at dusk to penetration electric fields during intense magnetic storms. *Journal of Geophysical Research: Space Physics*, 112(A8), A08308. doi:10.1029/2006JA012192.

Basu, S., Basu, S., Valladares, C. E., Yeh, H. C., Su, S. Y., MacKenzie, E., ... & Bullett, T. W. (2001b). Ionospheric effects of major magnetic storms during the International Space Weather Period of September and October 1999: GPS observations, VHF/UHF scintillations, and in situ density structures at middle and equatorial latitudes. *Journal of Geophysical Research: Space Physics*, 106(A12), 30389–30413.

Basu, S., & Groves, K.M. (2001). Specification and forecasting of outages on satellite communication and navigation sysems. *Space Weather, Geophysical Monograph*, 125, 424–430.

Basu, S., MacKenzie, E., & Basu, S. (1988). Ionospheric constraints on VHF/UHF communications links during solar maximum and minimum periods. *Radio Science*, 23(3), 363–378.

Belcher, D. P., & Rogers, N. C. (2009). Theory and simulation of ionospheric effects on synthetic aperture radar. *IET Radar, Sonar & Navigation*, 3(5), 541–551.

Bogusch, R. L., F. W. Guigiano, D. L. Knepp, & A. H. Michelet (1981). Frequency selective propagation effects on spread-spectrum receiver tracking. *Proceedings of the IEEE*, 69, 787–796.

Cannon, P. S., Groves, K., Fraser, D. J., Donnelly, W. J., & Perrier, K. (2006). Signal distortion on VHF/UHF transionospheric paths: First results from the Wideband Ionospheric Distortion Experiment. *Radio Science*, 41(5), RS5S40. doi:10.1029/2005RS003369.

Carrano, C. S., & Groves, K.M. (2010). Temporal decorrelation of GPS satellite signals due to multiple scattering from ionosphericirregularities. In *Proceedings of the 23rd International Technical Meeting of The Satellite Division of the Institute of Navigation* (pp. 361–374). Portland, OR, September 2010.

Carrano, C. S., Groves, K. M., & Caton, R. G. (2012a). Simulating the impacts of ionospheric scintillation on L band SAR image formation. *Radio Science*, 47(4), RS0A12. doi:10.1029/2008RS004047.

Carrano, C. S., Groves, K. M., McNeil, W. J., & Doherty, P. H. (2012b). Scintillation characteristics across the GPS frequency band. In *Proceedings of the 25th International Technical Meeting of The Satellite Division of the Institute of Navigation* (pp. 1972–1989). Nashville, TN, September 2012.

Carrano, C. S., Groves, K. M., McNeil, W. J., & Doherty, P. H. (2013). Direct measurement of the residual in the ionosphere-free linear combination during scintillation. *Proceedings of the 2013 International Technical Meeting of The Institute of Navigation*. San Diego, CA, January 2013.

Carrano, C. S., & Rino, C. (2015a). A strong-scatter theory of ionospheric scintillations for two-component power law irregularity spectra. Presented at the *14th International Ionospheric Effects Symposium*, Alexandria, VA, May 12–14. http://ies2015.bc.edu/wp-content/uploads/2015/08/IES2015-Proceedings.pdf.

Carrano, C. S., Rino, C., Groves, K., & Doherty, P. (2015b). Inferring zonal irregularity drift from single-station measurements of amplitude (S_4) and phase (Sigmaphi) scintillations. Presented at the *14th International Ionospheric Effects Symposium*, Alexandria, VA, May 12–14. http://ies2015.bc.edu/wp-content/uploads/2015/08/IES2015-Proceedings.pdf.

Carroll, M., Morton, Y. J., & Vinande, E. (2014). Triple frequency GPS signal tracking during strong ionospheric scintillations over Ascension Island. In *Position,*

Location and Navigation Symposium-PLANS 2014, 2014 IEEE/ION (pp. 43–49). IEEE, Monterey, CA.

Caton, R. G., Carrano, C. S., Alcala, C. M., Groves, K. M., Beach, T., & Sponseller, D. (2009). Simulating the effects of scintillation on transionospheric signals with a two-way phase screen constructed from ALTAIR phase-derived TEC. *Radio Science*, 44(1), RS0L20. doi:10.1029/2011RS004956.

Chew, W. C. (1995). *Waves and Fields in Inhomogeneous Media* (Vol. 522). IEEE Press, New York.

Conker, R. S., El-Arini, M. B., Hegarty, C. J., & Hsiao, T. (2003). Modeling the effects of ionospheric scintillation on GPS/Satellite-Based Augmentation System availability. *Radio Science*, 38(1), 1–23.

Fejer, B. G. (1997). The electrodynamics of the low latitude ionosphere: Recent results and future challenges. *Journal of Atmospheric and Terrestrial Physics*, 59, 1465–1482.

Fejer, B. G., Scherliess, L., & Paula, E. D. (1999). Effects of the vertical plasma drift velocity on the generation and evolution of equatorial spread F. *Journal of Geophysical Research: Space Physics*, 104(A9), 19859–19869.

Foster, J. C., & Erickson, P. J. (2000). Simultaneous observations of E-region coherent backscatter and electric field amplitude at F-region heights with the Millstone Hill UHF radar. *Geophysical Research Letters*, 27(19), 3177–3180.

Foster, J. C., Erickson, P. J., Coster, A. J., Goldstein, J., & Rich, F. J. (2002). Ionospheric signatures of plasmaspheric tails. *Geophysical Research Letters*, 29(13), 1623. doi:10.1029/2002GL015067.

Gentile, L. C., Burke, W. J., & Rich, F. J. (2006). A climatology of equatorial plasma bubbles from DMSP 1989–2004. *Radio Science*, 41(5), RS5S21. doi:10.1029/2005RS003340.

Gherm, V. E., Zernov, N. N., & Strangeways, H. J. (2011). Effects of diffraction by ionospheric electron density irregularities on the range error in GNSS dual-frequency positioning and phase decorrelation. *Radio Science*, 46(3), RS3002. doi:10.1029/2010RS004624.

Groves, K. M., Basu, S., Quinn, J. M., Pedersen, T. R., Falinski, K., Brown, A., ... & Ning, P. (2000). A comparison of GPS performance in a scintillation environment at Ascension Island. *Proceedings of the 13th International Technical Meeting of the Satellite Division of The Institute of Navigation* (pp. 672–679). Salt Lake City, UT, September 2000.

Groves, K. M., Basu, S., Weber, E. J., Smitham, M., Kuenzler, H., Valladares, C. E., ... & Kendra, M. J. (1997). Equatorial scintillation and systems support. *Radio Science*, 32(5), 2047–2064.

Haerendel, G., Eccles, J. V., & Cakir, S. (1992). Theory for modeling the equatorial evening ionosphere and the origin of the shear in the horizontal plasma flow. *Journal of Geophysical Research: Space Physics*, 97(A2), 1209–1223.

Hocke, K., & Schlegel, K. (1996). A review of atmospheric gravity waves and travelling ionospheric disturbances: 1982–1995. *Annales Geophysicae*, 14(9), 917.

Humphreys, T. E., Psiaki, M. L., Ledvina, B. M., Cerruti, A. P., & Kintner, P. M., Jr. (2010). Data-driven testbed for evaluating GPS carrier tracking loops in ionospheric scintillation. *IEEE Transactions on Aerospace and Electronic Systems*, 46(4), 1609–1623.

Humphreys, T. E., Psiaki, M. L., Ledvina, B. M., & Kintner, P. M., Jr. (2005). GPS carrier tracking loop performance in the presence of ionospheric scintillations. In *Proceedings of the 18th International Technical Meeting of the Satellite Division of The Institute of Navigation*, September 13–16, Long Beach Convention Center, Long Beach, CA (pp. 156–167).

Hunsucker, R. D. (1982). Atmospheric gravity waves generated in the high-latitude ionosphere: A review. *Reviews of Geophysics*, 20(2), 293–315.

Hysell, D. L. (2015). The radar aurora. *Auroral Dynamics and Space Weather*, 215, 193.

Kashcheyev, A., Nava, B., & Radicella, S. M. (2012). Estimation of higher-order ionospheric errors in GNSS positioning using a realistic 3-D electron density model. *Radio Science*, 47, RS4008. doi:10.1029/2011RS004976.

Keskinen, M. J., Ossakow, S. L., & Fejer, B. G. (2003). Three-dimensional nonlinear evolution of equatorial ionospheric spread-F bubbles. *Geophysical Research Letters*, 30(16), 1855. doi:10.1029/2003GL017418.

Kintner, P. M., Ledvina, B. M., & De Paula, E. R. (2007). GPS and ionospheric scintillations. *Space Weather*, 5(9), S09003. doi:10.1029/2006SW000260.

Klobuchar, J. (1987). Ionospheric time-delay algorithm for single-frequency GPS users. *IEEE Transactions on Aerospace and Electronic Systems*, 3, 325–331.

Knepp, D. L. (1983). Multiple phase screen calculations of the temporal behavior of stochastic waves. *Proceedings of the IEEE*, 71(6), 722–737. doi:10.1109/PROC.1983.12660.

Knight, M., & Finn, A. (1998). The effects of ionospheric scintillations on GPS. In *Proceedings of the 11th International Technical Meeting of the Satellite Division of The Institute of Navigation*, September 15–18, Nashville, TN (pp. 673–685).

Komjathy, A., Galvan, D. A., Stephens, P., Butala, M. D., Akopian, V., Wilson, B., … & Hickey, M. (2012). Detecting ionospheric TEC perturbations caused by natural hazards using a global network of GPS receivers: The Tohoku case study. *Earth, Planets and Space*, 64(12), 1287–1294.

Kudeki, E., Akgiray, A., Milla, M., Chau, J. L., & Hysell, D. L. (2007). Equatorial spread-F initiation: Post-sunset vortex, thermospheric winds, gravity waves. *Journal of Atmospheric and Solar-Terrestrial Physics*, 69(17), 2416–2427.

Ledvina, B. M., Kintner, P. M., & Makela, J. J. (2004). Temporal properties of intense GPS L1 amplitude scintillations at midlatitudes. *Radio Science*, 39(1)., RS1S18. doi:10.1029/2002RS002832

Ledvina, B. M., Makela, J. J., & Kintner, P. M. (2002). First observations of intense GPS L1 amplitude scintillations at midlatitude. *Geophysical Research Letters*, 29(14), 1659. doi 10.1029/2002GL014770.

Loh, R., Wullschleger, V., Elrod, B., Lage, M., & Haas, F. (1995). The US Wide-Area Augmentation System (WAAS). *Navigation*, 42(3), 435–465.

McNamara, L. F. (1991). *The Ionosphere: Communications, Surveillance, and Direction Finding*. Krieger publishing company, Malabar, FL.

McNeil, W. J., Groves, K. M., & Carrano, C. S. (2015). A look at GPS positioning in solar cycle 24. Presented at the *14th International Ionospheric Effects Symposium*, Alexandria, VA, May 12–14. http://ies2015.bc.edu/wp-content/uploads/2015/08/IES2015-Proceedings.pdf.

Meyer, F. J. (2011). Performance requirements for ionospheric correction of low-frequency SAR data. *IEEE Transactions on Geoscience and Remote Sensing*, 49(10), 3694–3702.

Ngwira, C. M., McKinnell, L. A., &Cilliers, P. J. (2010). GPS phase scintillation observed over a high-latitude Antarctic station during solar minimum. *Journal of Atmospheric and Solar-Terrestrial Physics*, 72(9), 718–725.

Nickisch, L. J., & Franke, P. M. (1996). Finite difference time domain tests of random media propagation theory. *Radio Science*, 31(4), 955–963. doi:10.1029/96RS00874.

Ott, E. (1978). Theory of Rayleigh-Taylor bubbles in the equatorial ionosphere. *Journal of Geophysical Research: Space Physics*, 83(A5), 2066–2070.

Parkinson, B., Spilker, J. J., Axelrad, P., & Enge, P. (1996). *GPS: Theory and Applications* (vols 1 and 2). AIAA, Washington, DC.

Prikryl, P., Jayachandran, P. T., Mushini, S. C., Pokhotelov, D., MacDougall, J. W., Donovan, … & St.-Maurice, J.-P. (2010). GPS TEC, scintillation and cycle slips observed at high latitudes during solar minimum. *Annales Geophysicae*, 28, 1307–1316. doi:10.5194/angeo-28-1307-2010.

Rino, C. (1979). A power law phase screen model for ionospheric scintillation, 1 weak scatter. *Radio Science*, 14, 1135–1145.

Rino, C. L. (2011). *The Theory of Scintillation With Applications in Remote Sensing*. John Wiley & Sons, New York.

Rino, C. L., & Carrano, C. S. (2011). The application of numerical simulations in Beacon scintillation analysis and modeling. *Radio Science*, 46, RS0D02. doi:10.1029/2010RS004563.

Rogers, N. C., & Quegan, S. (2014). The accuracy of Faraday rotation estimation in satellite synthetic aperture radar images. *Geoscience and Remote Sensing, IEEE Transactions on*, 52(8), 4799–4807.

Secan, J. A., Bussey, R. M., Fremouw, E. J., & Basu, S. (1995). An improved model of equatorial scintillation. *Radio Science*, 30, 607–617.

Skone, S., Knudsen, K., & De Jong, M. (2001). Limitations in GPS receiver tracking performance under ionospheric scintillation conditions. *Physics and Chemistry of the Earth, Part A: Solid Earth and Geodesy*, 26(6), 613–621.

Soumekh, M. (1999). *Synthetic Aperture Radar Signal Processing* (pp. 5–16). Wiley, New York.

Sultan, P. J. (1996). Linear theory and modeling of the Rayleigh-Taylor instability leading to the occurrence of equatorial spread F. *Journal of Geophysical Research: Space Physics*, 101(A12), 26875–26891.

Thomas, R. M., Cervera, M. A., Ramli, A. G., Totarong, P., Groves, K. M., & Wilkinson, P. J. (2004). Seasonal modulation of GPS performance due to equatorial scintillation. *Geophysical Research Letters*, 31(18), L18806. doi:10.1029/2004GL020581.

Tsunoda, R. T. (1985). Control of the seasonal and longitudinal occurrence of equatorial scintillations by the longitudinal gradient in the integrated E-region Pedersen conductivity. *Journal of Geophysical Research*, 90, 447.

Valladares, C. E., Basu, S., Buchau, J., & Friis-Christensen, E. (1994). Experimental evidence for the formation and entry of patches into the polar cap. *Radio Science*, 29(1), 167–194. doi:10.1029/93RS01579.

van de Kamp, M. M., Cannon, P. S., & Watson, R. J. (2010). V/UHF space radars: Spatial phase decorrelation of transionospheric signals in the equatorial region. *Radio Science*, 45(4), RS4012. doi:10.1029/2009RS004226.

Weber, E. J., Buchau, J., Moore, J. G., Sharber, J. R., Livingston, R. C., Winningham, J. D., & Reinisch, B. W. (1984). F layer ionization patches in the polar cap. *Journal of Geophysical Research*, 89(A3), 1683–1694.

Whalen, J. A. (2004). Linear dependence of the postsunset equatorial anomaly electron density on solar flux and its relation to the maximum prereversal E × B drift velocity through its dependence on solar flux. *Journal of Geophysical Research*, 109, A07309. doi:10.1029/2004JA010528.

Yeh, K. C., & Liu, C. H. (1982). Radio wave scintillations in the ionosphere. *Proceedings of the IEEE*, 70, 324–360.

CHAPTER 20

Concluding Remarks

George V. Khazanov

SPACE WEATHER IS A cross-disciplinary scientific and engineering discipline and requires the international collaboration of physicists, engineers, aeronomers, and educators from different areas of space science research. Although space weather forecasts are not normally featured on the evening news, space weather does impact life on Earth in many ways. Our modern, technologically complex systems, including communications, transportation, and electrical power systems, can be disrupted and damaged by space weather storms. Exposure to radiation can be life-threatening to astronauts and commercial air travelers alike and has affected the evolution of life on Earth.

The study of space weather is a relatively young science. As such, it has many unanswered questions and unsolved mysteries. Although some of our data relevant to space weather, such as sunspot counts, go back many years, most of knowledge of the field stems from recent times. Supercomputers, satellite-borne instruments, and telescopes capable of imaging the Sun in many different parts of the electromagnetic spectrum are the recent developments, and each plays a large role in our developing understanding of solar science and space weather.

The space plasma, which is the "kitchen" of the space weather, is a very unique composition of different kinds of plasmas and electromagnetic fields. It covers a huge plasma energy range, with spatial and time variations of many orders of magnitude. Each of these populations has distinct features and contributes in a different way to the dynamic and energetic processes of the space plasma. Each has a unique coupling mechanism with surrounding plasma and electromagnetic fields, providing a very noticeable contribution to the physics of the space weather. Such a situation requires a very different kind of mathematical approach as the starting point for a proper theoretical description of the space weather phenomena. In this book, we have only attempted to outline some of the investigations in this area of research and provide to the reader the fundamentals of space weather formation.

Space weather was identified as one of the nine natural hazards that pose the greatest threat to the nation's security in the US Presidential Policy Directive. There is a 12% chance in the next 10 years that a solar storm of the magnitude of the Solar Storm of 1859 will hit the Earth, which will cause a damage of more than $2 trillion. Our forecasting capabilities of space weather are currently quite inadequate, and the nations are ill-prepared to deal with this impending danger. The authors of this book hope that this book will bring attention to this important area and provide an updated review of the status of this field and will be useful for PhD researchers and students who are learning about space physics and, in particular, about space weather.

Index

Note: Page numbers followed by f and t refer to figures and tables, respectively.

Printed and bound by CPI Group (UK) Ltd, Croydon, CR0 4YY

01/11/2024

01782600-0017